SOCIETÀ ITALIANA DI FISICA

RENDICONTI

DELLA

SCUOLA INTERNAZIONALE DI FISICA

"ENRICO FERMI"

CLV Corso

a cura di F. Mallamace and H. E. Stanley

Direttori del Corso

VARENNA SUL LAGO DI COMO

VILLA MONASTERO

1 – 11 Luglio 2003

La fisica dei sistemi complessi (Nuovi sviluppi e prospettive)

2004

SOCIETÀ ITALIANA DI FISICA

BOLOGNA-ITALY

ITALIAN PHYSICAL SOCIETY

PROCEEDINGS

OF THE

INTERNATIONAL SCHOOL OF PHYSICS

"ENRICO FERMI"

COURSE CLV

edited by F. MALLAMACE and H. E. STANLEY

Directors of the Course

VARENNA ON LAKE COMO

VILLA MONASTERO

1 – 11 July 2003

The Physics of Complex Systems (New Advances and Perspectives)

2004

IOS Press

AMSTERDAM, OXFORD, TOKIO, WASHINGTON DC

ISBN 1-58603-445-6 (IOS)
ISBN 88-7438-016-X (SIF)
Library of Congress Control Number: 2004108559

Production Manager
A. Oleandri

Copy Editor
M. Missiroli

jointly published and distributed by:

IOS PRESS
Nieuwe Hemweg 6B
1013 BG Amsterdam
The Netherlands
fax: +31 20 620 34 19
e-mail: order@iopress.nl

SOCIETÀ ITALIANA DI FISICA
Via Saragozza 12
40123 Bologna
Italy
fax: +39 051 581340
e-mail: order@sif.it

Distributor in the UK and Ireland
IOS Press/Lavis Marketing
73 Lime Walk
Headington
Oxford OX3 7AD
England
fax: +44 1865 75 0079

Distributor in the USA and Canada
IOS Press, Inc.
4502 Rachael Manor Drive
Fairfax, VA 22032
USA
fax: +1 703 323 3668
e-mail: iosbooks@iospress.com

Printed in Italy

Supported by Fondazione CARIPLO

Supported by the European Commission, Research DG, Human Potential Programme, High-Level Scientific Conferences, Contract No.: HPCF-CT-2002-00362

Supported by UNESCO-ROSTE, Contract No.: 875.599.3

Supported by Consiglio Nazionale delle Ricerche (CNR)

Supported by Istituto Nazionale di Fisica Nucleare (INFN)

Supported by Istituto Nazionale per la Fisica della Materia (INFM)

Supported by Dipartimento di Fisica, Università di Messina

Supported by Office of Naval Research (ONR)

Supported by Centro Sviluppo Materiali, Roma

To Edelweiss and Idahlia

In Memoriam

INDICE

Preface

It is widely known that complex systems and complex materials comprise a major interdisciplinary scientific field that draws on mathematics, physics, chemistry, biology, and medicine as well as such social sciences as economics. The role of statistical physics in this new field has been expanding. Statistical physics has shown how phenomena and processes in different research areas that have long been assumed to be unrelated can have a common description. Through the application of statistical physics, methods developed for studying order phenomena in simple systems and processes have been generalized to more complex systems (*e.g.*, polymers, biological macromolecules, and glasses) and processes (*e.g.*, chaos, turbulence, economics, jamming, and biological functions). The two conceptual pillars in this approach are scaling and universality.

This volume contains all the lectures presented at the CLV Course of the International School of Physics "Enrico Fermi" held on 1–11 July 2003 in Varenna, Italy. They focus on recent advances and perspectives in the physics of complex systems and provide both an overview of the field and a more detailed examination of the new ideas and unsolved problems that are currently attracting the attention of researchers. Although all of the lecturers have made important contributions within their individual areas, and most have made significant contributions to the field as a whole, this collection is pedagogically focused so that the material will be accessible to a wider audience. We hope that this published form of the lectures will augment the pedagogical utility of the presentations. Bear in mind that the field of complex systems is very broad and that no single volume can be truly comprehensive.

This book should be a useful reference work for anyone interested in this area, whether beginning graduate student or advanced research professional. It provides up-to-date reviews of cutting-edge topics compiled by leading authorities and is designed to both broaden the reader's competence within their own field and encourage the exploration of new problems in related fields.

The present subfield configuration of the physics of complex systems is suggested by the topic groupings used in this volume. These groupings provide a kind of "backbone" for the course and include i) scaling behavior, ii) supramolecular systems, polymers, associating polymers, polyelectrolytes and gels, iii) granular matter, iv) biological materials and systems, v) high-temperature superconductivity, vi) phase separation and out-of-equilibrium dynamics, vii) glass transition, supercooled fluids, and viii) geometrically constrained dynamics.

To facilitate the free interchange of ideas, the lecturers and seminar leaders also made themselves available as course participants. Their active presence made possible a constant updating of the topics under discussion, and kept the focus clearly on a "state-of-the-art" level. The Course schedule was such that extra time was allowed for informal discussion. Student participants were given special encouragement to participate in the poster sessions. Both the seminar contributions and the posters are included in this volume.

The book opens with an M. H. Cohen tribute to Enrico Fermi on the occasion of the 50th anniversary of the Varenna school (July 1953–July 2003). M. H. Cohen also reviews in detail the interplay between simplicity and complexity.

The opening lectures of B. Widom concern the hydrophobic effect, interfaces, and contact lines in three-phase coexistence. D. R. Nelson reports on recent developments in the Thomson problem in viruses, vesicles, and colloidosomes, and on the unzipping experiments carried out on single DNA molecules. S. Havlin and A. Bunde contribute a detailed examination of fractal correlations in medicine, biology, climate models, and global warming. E. I. Shakhnovic contributes a more specialized review of the field of protein folding, evolution, and design, including the most recent experimental work.

G. Parisi and C. Tsallis review important topics of general interest in statistical physics. Parisi also provides both an overview and a detailed examination of combinatorial optimization and Tsallis does the same for non-extensive statistical mechanics. K. A. Mueller considers intrinsic heterogeneous high-temperature superconductors from the point of view of complexity.

P.-G. de Gennes reports on the development of physical models and techniques used in the study of cellular adhesion. L. Blum discusses phase behavior in polyelectrolites and T. Schneider phase behavior in superconductivity. J. K. Percus reviews dynamical properties of fluids under tight confinement. Electrostatics in biological systems is discussed by M. Barbosa and F. Brochard considers properties of artificial cells.

The theory and applications associated with granular materials provide very precise tools for many of the topics discussed in the lectures, *e.g.*, structural arrest at the glass transition, gelification, and non-equilibrium pattern formation. A. Coniglio proposes a statistical mechanics approach, based on clustering and frustration, that highlights similarities between the jamming of granular materials and glassy systems. H. L. Swinney reports on original studies of spatial patterns and shocks in granulary media. Related to this area is the contribution of H. J. Herrmann.

More detailed aspects of the glassy state are treated in complementary lectures. K. Binder discusses the statistical mechanics of glass and the glass transition. S. H. Chen, using mode coupling theory, discusses the critical slowdown in attractive glassy systems characterized by the existence of two types of structurally arrested states.

The Course faculty included 13 lecturers and 10 invited seminar speakers. Most of the students and young researchers attending the Course presented original contributions in the poster session.

The aim of the Course was not only to present reviews of all the various subareas, but also to stimulate the search for a unified approach in complex systems and complex materials by gathering together participants from a variety of specialized backgrounds. We all worked very hard, attending nearly 60 hours of lectures, seminars, and discussions during the two-week period. We hope this effort was rewarding for everyone attending.

Three "ground rules" were adopted: i) each session was to consider coherent arguments, ii) lecturers could be interrupted at any time during the talks for questions, and iii) speakers, young researchers, and students were encouraged to mix during meals, coffee breaks, and the extended free period following lunch. The outcome was as we had hoped: numerous spontaneous discussions and lively debates took place in an honest, friendly atmosphere of work. Many of these discussions occurred in the beautiful gardens of Villa Monastero and Villa Cipressi. Judging from the positive and often enthusiastic reactions of the lecturers and participants during and after the Course, the event was extraordinarily successful in achieving its intended objectives.

It is fortunate that we are able to present in this volume the final versions of nearly all the lectures and seminars presented at the Course. So much of importance and of high quality was presented that the material defies condensation or hierarchical ordering: the participants—experienced scientists, postdocs, and graduate students—have contributed with skill and enthusiasm.

We are grateful for the invaluable contributions and suggestions of G. Maino and M. F. Shlesinger, scientific secretaries for the School. We give special thanks to Professor F. Bassani for his encouragement and advice during the preparation for the School. We also thank the General Committee of the Italian Physical Society (SIF), the European Commission, the Unesco-Roste, the National Organization for the New Technologies the Energy and the Environment (ENEA), the Istituto Nazionale di Fisica della Materia (INFM), the MIUR-PRIN02 project Unifying Concepts in Jamming, the US and European divisions of the Office of Naval Research (ONR), the Fondazione Bonino-Pulejo for their generous financial support. Last, but certainly not least, we offer our appreciation to B. Alzani, C. Vasini and M. Missiroli of the SIF for their valuable and excellent cooperation before, during, and after the School period, and for their heroic work in producing the present volume.

Before concluding, some personal remarks. First, we sincerely apologize to all the students whom, due to the finite size of the auditorium, we were not able to admit to the Enrico Fermi School. Secondly, we wish to thank Ente Villa Monastero for making their beautiful buildings and gardens available to us "invaders." Lastly, we thank the citizens of Varenna for their warmth and hospitality.

We offer this volume to the reader in the hope that it will provide something of the enjoyment and rewards that we and the other participants in the School experienced.

F. MALLAMACE and H. E. STANLEY

Società Italiana di Fisica
SCUOLA INTERNAZIONALE DI FISICA «E. FERMI»
CLV CORSO - VARENNA SUL LAGO DI COMO
VILLA MONASTERO 1 - 11 Luglio 2003

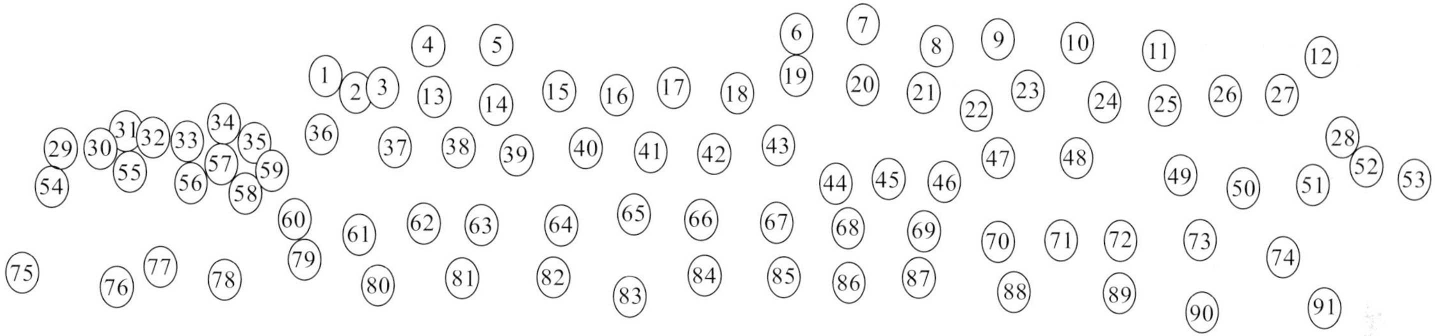

1) M. Broccio
2) A. Cavallo
3) J.K. Dreyer
4) A. Faraone
5) E. Pontecorvo
6) Y. Srebro
7) G. Baldi
8) P. Buenzli
9) O. Campas
10) E.J. Riberio Parteli
11) A. Nicolin
12) A. Gray-Weale
13) F. Zamponi
14) T. Hanney
15) N. Fusco
16) P. Pierobon
17) R. Cerbino
18) S. Moshel
19) N. Fujita
20) A. Gozolchiani
21) S.K. Das
22) M. Pica Ciamarra
23) P. Calabrese
24) T. Aste
25) A. Chatterji
26) P. Parruccini
27) M. Tarzia
28) A. Trovato
29) H. Katsuragi
30) C.J. Tessone
31) A. Furukawa
32) G. Baxter
33) A. Pluchino
34) D. Rybski
35) M. Wackenhut
36) S. Zschiegner
37) A. Astillero
38) H. Lavicka
39) L. Lanzi
40) R. Cohen
41) A. Alexander-Katz
42) J. Dzubiella
43) S. Karanikas
44) V. Livina
45) S. Imberti
46) V. De Grandis
47) A. Gambassi
48) G. Bormetti
49) P. Faisca
50) A. Lamura
51) A. Kudlay
52) A. Mora
53) W. Hu
54) F. Fusina
55) R. Brigatti
56) B. Alzani
57) N. Denesyuk
58) J. Lal
59) A. Ochab-Marcinek
60) A. Coniglio
61) B. Widom
62) T. Schneider
63) M.H. Cohen
64) S.H. Chen
65) D.R. Nelson
66) H.E. Stanley
67) F. Mallamace
68) M.E. Fisher
69) C. Tsallis
70) H.L. Swinney
71) G. Maino
72) P. Tartaglia
73) F. Mohamed
74) K. Shundyak
75) L. Corengia
76) J.N. Aqua
77) H. Wada
78) A. Hilfinger
79) V. Vitelli
80) T. Voigtmann
81) J. Eichner
82) F. Penna
83) M. Barbosa
84) A. Gholami
85) T. Schilling
86) F. Trudu
87) M. Micheelsen
88) M. Sega
89) A. Kimmel
90) B. Nadrowski
91) A. Chernova

Remembering Fermi

Morrel H. Cohen

Department of Physics and Astronomy, Rutgers University
136 Frelinghuysen Road, Piscataway, NJ 08854-8019 USA
and
Department of Chemistry, Princeton University
Washington Road, Princeton, NJ 08544-1009, USA

Since this School carries the name of Enrico Fermi, a few personal remarks about him seem appropriate. I did my graduate work at the University of California (Berkeley) during 1948-52. Emilio Segré, a student of Fermi at Rome in the 1920's and a colleague in the 20's and 30's, was a faculty member in the Berkeley Physics Department. At meetings of the Department journal club, he would read a handwritten note from Fermi every two weeks or so summarizing new work of Fermi and others at Chicago. Segré was to get the Nobel Prize later for his discovery of the antiproton with Chamberlain, also a student of Fermi's but from Chicago, yet he read those notes with awe in his voice as though they formed a direct communication from a god.

When I arrived at Chicago in 1952 as the most junior member of the Physics Department, I discovered that Fermi was indeed regarded as a god there. Goldberger, one of the brilliant young theorists in the Department, would however say of Fermi to counter that unhealthy tendency "He puts his pants on one leg at a time like the rest of us."

Fermi's intellectual powers were awe inspiring. He was particularly famous for the penetrating simplicity of his arguments. Despite his fame and his great accomplishments, he remained, in essence, a simple man. He slept relatively little, rose early, and already had new results before arriving at work each morning. He tried to spend 1/3 of his time learning from the work of others, 1/3 of his time teaching, and 1/3 of his time creating new knowledge.

In those days, there was one Physical Review, not 5, no separate Physical Review Letters, and the Physical Review came out monthly. Fermi read it cover to cover, idiosyncratically. If the title and abstract of a paper interested him, he'd read the introduction and the conclusions, skimming the body for the main points. If he remained interested, he'd derive the paper's results his own way, working on a huge unlined pad, ignoring the author's presentation. He would pass on to the next paper unless stimulated to go deeper, then recording his results in his notebooks and generating new physics.

The audience in the Chicago Theory Seminar was most imposing: Fermi, Wentzel, Maria and Joe Mayer, Gell-Mann, Goldberger, and others very bright but less well known. A speaker from elsewhere might last ten minutes before someone would raise a deep question the speaker could not answer. The audience would debate the answer at great speed, slowing down as they realized the answer was not forthcoming. Fermi, meanwhile, would have remained silent. Then, just as the speaker would be about to resume, Fermi would say "Well, I think that – – ––" and answer the question. We youngsters always wondered whether his mind went slowly but directly to the right answer or whether it rapidly explored many other possibilities on the way.

This school was started in 1953, precisely 50 years ago. Fermi lectured here in 1954, one year later. At that time he experienced stomach pains. Upon his return to Chicago in September, an advanced case of stomach cancer was diagnosed. He continued to work until two weeks before his death in November, exactly two months past his 53rd birthday, a very great loss.

After his death, the International School of Physics of the Italian Physical Society was most appropriately renamed the International School of Physics "Enrico Fermi".

I am grateful to Andrea Gambassi for his assistance.

Adhesion of soft objects

P.-G. de Gennes

Collège de France & Institut Curie - F-77231, Paris Cedex 05, France

1. – Introduction

Living cells are often able to recognize some partners, and to stick to them. Over the past 30 years, various families of proteins specific of these processes have been identified. Understanding the adhesion process is important in various fields, ranging from embryology to biomedical studies. From a physical point of view, these problems are novel:

a) the objects involved are very soft and require new methods of measurement,

b) the "stickers" can move and aggregate in an adhesion patch.

Our discussion concentrates on adhesion patches, and ignores the (many) measurements performed on a *single* sticker.

The primary model system allowing for the build-up of a patch is a *vesicle* (shown in fig. 1a) formed by a closed lipid bilayer —roughly similar to a cellular membrane [1]. Ideally, a vesicle formed by unsoluble lipids has 0 surface tension, because it has minimized its energy *versus* area. In actual fact, the vesicle often has a small, but nonvanishing, surface tension γ (10^{-4} times smaller than the surface tension of water): during the duration of an experiment, γ cannot relax to 0 because permeation of water through the bilayer is slow —resulting in a constraint of fixed volume.

We can measure γ "on the spot" if we use "heavy vesicles". The internal fluid of these vesicles has a relative density excess, $\Delta\rho/\rho \sim 10^{-2}$ (induced for instance by the addition of suitable sugars [2]). When facing a horizontal surface (fig. 1b), a heavy vesicle develops a "cushion": a thin water film separating the vesicle from the wall, with a thickness (of order 10 nm). This film is a consequence of the *Helfrich repulsion* [3] between a flexible, fluctuating membrane, and a wall (the wall inhibits some fluctuations, the entropy is decreased and the free energy is increased). However, the range of the Helfrich repulsion

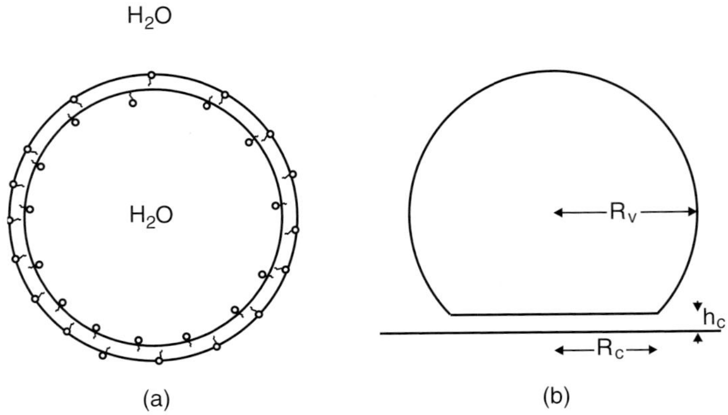

Fig. 1. – (a) A vesicle (or liposome) is based on a closed lipid bilayer of thickness ∼ 3 nanometers. The giant vesicles of interest here have a radius ∼ 10–20 micrometers. (b) A giant vesicle (with a heavy additive inside) becomes squeezed on its support. It is still separated from the support by a water film ("cushion") of radius R_c (micrometers) and thickness h_c (nanometers).

is truncated sharply by the surface tension γ, beyond a film thickness h_c given by

(1) $$h_c^2 = \frac{kT}{\gamma},$$

where kT is the thermal energy [4]. Thus the cushion thickness is essentially equal to h_c (and is significant because γ is so small).

The horizontal size $2R_c$ of the cushion results from a macroscopic balance between capillarity and gravity. It is given by [5]

(2) $$R_c = \kappa R_v^2,$$

where R_v is the vesicle radius, and κ^{-1} is the "Laplace length" defined by

(3) $$\kappa^2 = \frac{\Delta\rho g}{\gamma}$$

(g = gravitational acceleration).

If we measure R_c (by an interference technique) we can obtain γ using eqs. (2), (3).

If we wish to monitor γ, we can use a powerful set-up invented by E. Evans [6], based on aspiration in a micropipette (fig. 2). Using the Laplace formulas for capillarity pressures, one finds that the surface tension is related to the aspiration pressure Δ by the formula

(4) $$\Delta = 2\gamma\left(\frac{1}{R_p} - \frac{1}{R_v}\right),$$

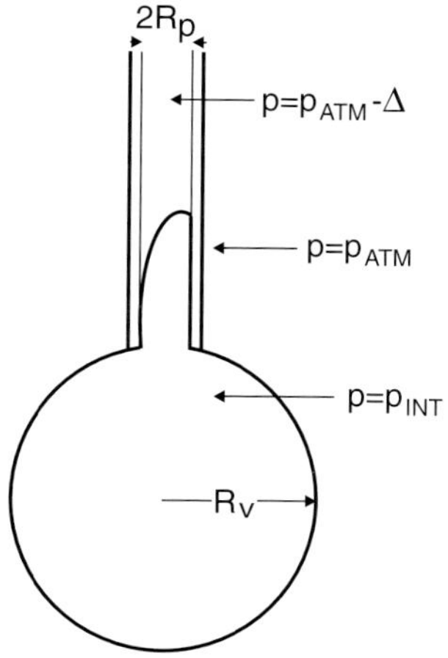

Fig. 2. – A vesicle under aspiration from a micropipette, imposing a pressure drop Δ.

where R_p is the pipette radius. Fixing Δ imposes a certain value of γ.

Knowing these basic instrumental features we can now start and discuss the adhesion properties. Section **2** is restricted to the statics. Section **3** presents one (single) regime of growth of a patch. But our reader should know that there are many other regimes which can occur in certain physical conditions (see sect. **6**). They are just summarized briefly here. Section **4** discusses the tear-out. Section **5** is a brief *résumé* of ideas concerning a vesicle rolling under flow.

2. – Static adhesion

2˙1. *Relation with osmotic pressures.* – Let us now focus our attention on the stickers. Their description requires two parameters:

$\Gamma =$ the number of stickers/unit area of membrane,

$U =$ the binding energy of one sticker to its receptor on the complementary surface.

We assume here that the stickers are *mobile*: both on the unbound bilayer and inside the patch. This is not unrealistic:

a) In an unbound belayer, proteins move with diffusion coefficients D in the range 10^{-9}–$10^{-8}\,\mathrm{cm}^2/\mathrm{s}$.

b) Inside the patch, we still expect mobility whenever the receptor molecule is itself floating inside another bilayer. Then the sticker/receptor pair can move. This occurs naturally for vesicle/vesicle contacts. It can also be obtained when one partner is a solid wall, provided that the wall is purposely covered with an extra bilayer.

What is the relation between the equilibrium adhesion energy W and the local sticker concentration Γ? At first sight most of us would write

$$W = \Gamma_i U, \tag{5}$$

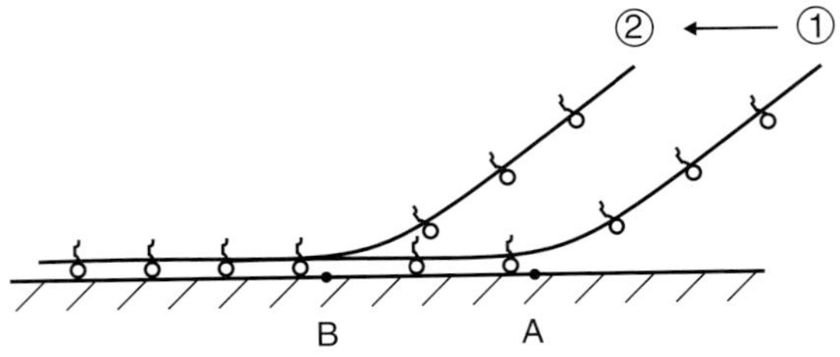

Fig. 3. – Role of the osmotic pressure of a 2-dimensional fluid of stickers. We peel out a piece of the adhesion patch, going from position 1 to position 2. One way of doing this is by tearing out the bonds between points A and B. But, if the stickers are mobile, it is better to shift them to the left of point B, fighting only the 2d osmotic pressure.

where Γ_i is the value inside the patch. But Bell, Dembo and Bongrand have shown long ago [7] that eq. (5) is wrong for mobile stickers! The correct formula is

$$W = \Pi(\Gamma_{\rm i}) - \Pi(\Gamma_{\rm out}), \tag{6}$$

where $\Pi(\Gamma)$ is the (2-dimensional) *osmotic pressure* of the fluid of stickers. $\Gamma_{\rm i}$ is the concentration inside the patch and $\Gamma_{\rm out}$ is the concentration immediately outside. Equation (6) can be understood from fig. 3, where we shrink the patch from position (1) to position (2). To do this, we could stupidly tear out the stickers which lie between point A and B. But it is less expensive to push them to the left of B, without any tear-out. Doing this, we need only to compress them slightly, *i.e.* to fight against the 2d osmotic pressure.

Equation (6) has two interesting limits:

a) If the concentration Γ is small, the osmotic pressure is that of an ideal gas, $\Pi = kT\Gamma$. Notice that this is significantly smaller than the naive prediction of eq. (5). (In most practical systems U/kT ranges from 4 to 10.)

b) If we go to the opposite limit, of highly packed stickers ($\Gamma_{\rm out}$ close to $\Gamma_{\rm i}$) one can show, using basic thermodynamics of chemical potentials, that eq. (6) returns to eq. (5). This is natural from the argument of fig. 3. If we cannot push more stickers to the left of B, we must tear out.

The Bell theorem (eq. (6)) is in some sense disappointing, at least in case (a): a measurement of the equilibrium adhesion energy W does not give us any direct information on the strength U of the adhesion!

2·2. *Measurements of W in static conditions.* – A first, classical, measurement is based on *contact angles.* If, near the contact line, the membrane makes an (equilibrium) angle $\theta_{\rm e}$ with respect to the support, we may obtain a relation between W and $\theta_{\rm e}$ through the Young relation [5], obtaining

$$W = \gamma(1 - \cos\theta_{\rm e}) \cong \frac{1}{2}\gamma\theta_{\rm e}^2. \tag{7}$$

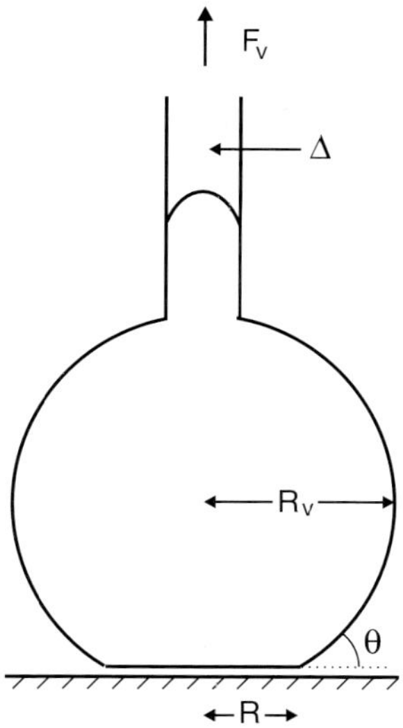

Fig. 4. – Principle of a "pull-out" experiment on an adhesion patch. The vesicle is bound to a pipette (aspiration Δ), and also a vertical force F_v is applied. In static (or slow) conditions the contact angle θ is always given by Young's law. But, for $F_\mathrm{v} > 0$, the contact radius R is smaller than $R_\mathrm{v} \sin\theta$.

A second type of experiment is based on "vertical pull-out" with a pipette (fig. 4). We first fix the surface tension, through a certain value of the aspiration pressure Δ. Then we pull mechanically on the pipette, applying a vertical force F_v. This force is transmitted to the solid at the bottom. The solid sees *two* forces:

- one along the contact line, of magnitude $\gamma 2\pi R \sin\theta_\mathrm{e}$;
- a pressure force (due to pressure excess, $2\gamma/R_\mathrm{v}$, inside the vesicle) integrated over the patch surface πR^2.

Thus

(8) $$F_\mathrm{v} = 2\pi\gamma\left[R\sin\theta_\mathrm{e} - \frac{R^2}{R_\mathrm{v}}\right] = 2\pi\gamma R\left(\theta_\mathrm{e} - \frac{R}{R_\mathrm{v}}\right).$$

If we plot this relation we see that for 0 force we return to the standard core $R = \theta_\mathrm{e} R_\mathrm{v}$. But, if we increase the force, R decreases. At a certain critical value, $R = \theta_\mathrm{e} R_\mathrm{v}/2$, the force has reached a maximum value

(9) $$F_{\mathrm{v\,max}} = \frac{\pi}{2}\gamma\theta_\mathrm{e} R_\mathrm{v} = \pi W R_\mathrm{v}.$$

At forces $F_\mathrm{v} > F_{\mathrm{v\,max}}$, the vesicle separates.... Thus eq. (9) allows for a measurement of W.

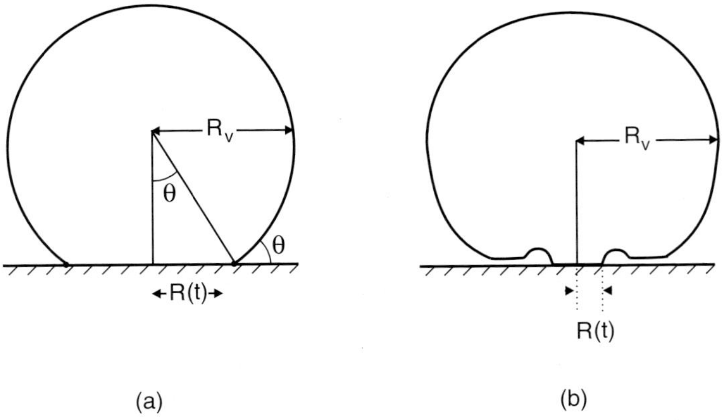

Fig. 5. – Two ideal types of growth for an adhesion patch. (a) Starting from a very small contact due to nonspecific adhesion, and waiting for sticker diffusion to enlarge the patch up to a much larger size $R(t)$. (b) Starting from a heavy vesicle, lying above its "cushion". Here growth requires only the elimination of the very thin water film defining the cushion.

2˙3. *Starting the patch.* – Think of a large vesicle floating above a wall. How do we convince it to start an adhesion process? I shall quote two methods.

a) If (apart from the stickers) there exists a small, nonspecific, adhesion between vesicle and wall (*e.g.*, Van der Waals) this will initiate a very small patch (within seconds). Then, later, the stickers will slowly migrate towards this region, and we can monitor this "mobilisation process". This, for a physicist, would be the best approach. But for a biophysicist, anxious to mimic a living cell, there is a strong dislike for nonspecific adhesion. Most cells avoid nonspecific adhesion by having on their surface a protective brush of hydrophilic polymers: the "glycolalix". This is also achieved with vesicles by grafting flexible polymers (PEG = polyethyleneglycol) on the outer surface.

b) When this is done, the vesicle does not spontaneously stick to the wall, and we must squeeze it against its partner. One elegant way to do this is based on the "heavy vesicles" (where the inner medium contains a sugar additive) which we presented in sect. **1**. We still have the problem with the "cushion", which creates a barrier against the nucleation of the patch. As usual in nature, nucleation will be usually associated with a defect, or a dust particle, on the solid surface.

The two preparation methods (a) and (b) lead to very different hydrodynamic scenarios for the patch growth (fig. 5). In case (a) the growth implies mainly the motion of water in a wedge of angle θ near the contact line. In case (b) we must dewet a thin water film (of thickness h_c —see sect. **1**). There is much less water to be moved, and the process is expected to be faster. In the present, short, presentation we focus on the simpler case (a), following [8]. For a discussion of case (b) our reader is referred to [9].

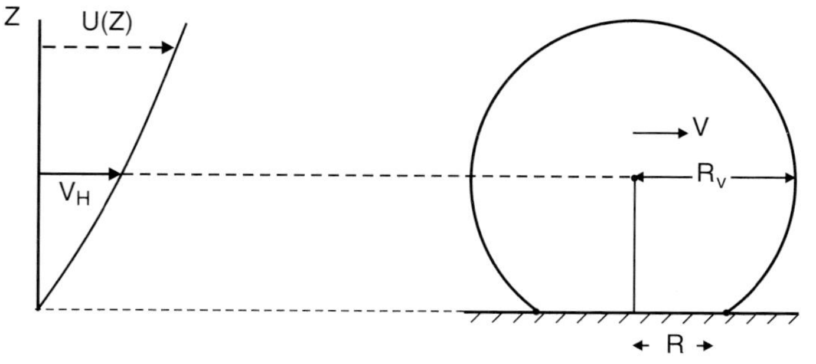

Fig. 6. – Principle of a rolling experiment. A vesicle, with a strong surface concentration of stickers (no migration) is sticking to a solid surface, but is pushed tangentially by a hydrodynamic flow (velocity V_{H}). In favorable conditions the vesicle rolls slowly on the surface (with a velocity $V \ll V_{\mathrm{H}}$) without ever leaving it. This is very different from the so-called "flux chamber" measurements, when the cells (or vesicles) hop around and sometimes stick because of a *single* bond.

3. – Patch growth: a simple example

We return here to the case shown in fig. 5a, and again assume that the stickers are mobile on all the membrane (including the patch region, of radius $R(t)$). The concentration profiles are shown in fig. 3. Very far from the patch, we still have the (small) initial concentration of stickers Γ_0. Around the patch, there is a depleted region, of radius $(Dt)^{1/2}$, where D is the diffusion coefficient. (We shall check later that $R \ll (Dt)^{1/2}$.) Inside the patch, we have a concentration Γ_{out} which is small when compared to Γ_0, and is negligible. Instead of solving the full diffusion equation, we shall use a simple sealing argument, which gives the same results (except for log factors).

The amount of stickers which have been pulled in is $\pi R^2\ (\Gamma_i - \Gamma_0) \cong \pi R^2 \Gamma_i$. This corresponds to all the stickers which have been reaped outside, in an area Dt. Thus the diffusion balance is (ignoring all coefficients)

$$\Gamma_{\mathrm{i}} R^2 = \Gamma_0 Dt. \tag{10}$$

Here we have two unknowns: Γ_{i} and $R = R_{\mathrm{v}}\theta$. To obtain an equation for the contact angle θ, we assume that a local equilibrium is reached at the contact line. This corresponds to the Young equation (eq. (7)) in which we insert the Bell equation (eq. (6)) for Γ_{i} (namely $W = \Pi(\Gamma_{\mathrm{i}})$ since Γ_{out} is negligible).

We must then choose a form for $\Pi(\Gamma)$. Two examples will be discussed.

a) Ideal gas behavior, $\Pi = kT\Gamma$. Combining this with eqs. (7) and (8), we arrive at the scaling law [8]

$$\theta^4(t) = \varepsilon \frac{Dt}{R_{\mathrm{v}}^2}\,, \tag{11}$$

where ε is a (small) dimensionless parameter

$$\varepsilon = \frac{2\Gamma_0 kT}{\gamma} . \tag{12}$$

Equation (9) describes a slow growth, combining capillary features (eq. (8)) with diffusion features (eq. (7)). Our assumption $R < (Dt)^{1/2}$ does hold, provided that

$$\varepsilon < \frac{Dt}{R_{\mathrm{v}}^2} . \tag{13}$$

There are no clear-cut experimental data to be compared with eq. (10). Most observations on growth, due principally to the group of E. Sackmann [10, 11]) correspond to more complex situations. But it will be important, in the future, to obtain data related to the present scenario.

4. – Tear-out processes

There are two main methods for the suppression of a patch.

a) We may attach the vesicle to a system transmitting a vertical force (for instance the pipette of fig. 4). The dynamical laws of separation for this case are discussed in refs. [8] and [12].

b) We can roll the vesicle along the support surface, using a tangential flow to push the vesicle (fig. 6). Then, on the rear side of the patch, we are peeling it out.

In the present (short) review we shall concentrate on the second case. We apply a dynamic flow with a certain shear rate γ_0. The horizontal flow velocity at the level of the vesicle is $V_{\mathrm{H}} \cong \gamma_0 R_{\mathrm{v}}$ and (very crudely) the resulting horizontal force is a Stokes force

$$F_{\mathrm{H}} = 6\pi\eta V_{\mathrm{H}} = 6\pi\eta\gamma_0 R_{\mathrm{v}}, \tag{14}$$

where η is the viscosity of water. On the moving contact line the adhesion energy W differs deeply from its static value. On the front side (sticking) W has a relatively small value W_+. On the rear side (peeling) it has a much larger value W_-. The length of action of W_- is the projected length of the contact line normal to the flow direction, *i.e.* $2R$. The force balance is (roughly)

$$2RW_-(V) = 6\pi\eta\gamma_0 R_{\mathrm{v}}^2 . \tag{15}$$

Here V is the advancing velocity of the rolling vesicles, and is much smaller than V_{H}. (We shall see that the tear-out adhesion energy W_- is dependent on V). Measuring R, γ_0, and V one can in principle arrive at a determination of $W_-(V)$. The whole scheme works (without extreme distortion of the vesicle) provided that the viscous stress $\eta\gamma$ is smaller than the Laplace pressure γ/R_{v}. This holds only in the limit $W \ll \gamma$.

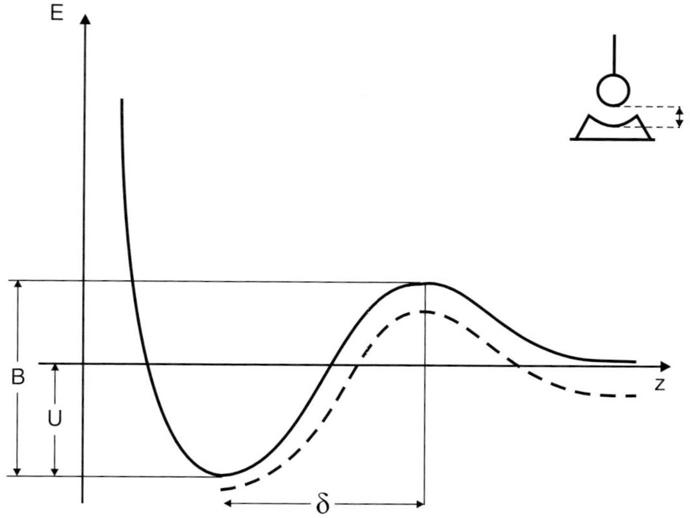

Fig. 7. – Principle of the Bell formula (eq. (16)) relating (vertical) exit velocity to (vertical) force for a single bond. $E(z)$ is the energy as a function of vertical displacement, displaying a bond energy U and a barrier energy $B \geq U$. Under a finite pull-out force f_z, the diagram is modified (dotted line) and the barrier is reduced, $\boldsymbol{B} \longrightarrow \boldsymbol{B} - f_z\delta$.

What is the theoretical structure of $W_-(V)$? This is the basic problem of transforming a *vertical* tear-out force (where stickers leave the floor) into a *horizontal* force on the contact line. The starting point is related to fundamental experiments of E. Evans [13] measuring tear-out processes under vertical forces on a *single* sticker. The principle is shown in fig. 7. When bound, the sticker is in a potential well with a minimum-U and a barrier height B ($\geq U$). The velocity of escape, under a vertical force f_z, is of the form

$$V_z = V_0 \exp\left[-\frac{B - f_z a}{kT}\right] \equiv V_1 \exp\left[\frac{f_z a}{kT}\right]. \tag{16}$$

Equation (16) is an early idea of Bell [14]. We can invert it relating f_z to the average velocity of extraction

$$f_z = \frac{kT}{a} \ln\left(\frac{V_z}{V_1}\right). \tag{17}$$

We shall now derive the tear-out energy $W_-(V)$ through a calculation of the dissipation near a contact line moving at the velocity v. The shape of the vesicle is characterized by a certain profile $z(x)$: along the profile we have stickers (of concentration Γ_i) moving upwards with a velocity $V_z = V\frac{\mathrm{d}z}{\mathrm{d}x}$. The dissipation is (per unit length of line) $T \equiv$

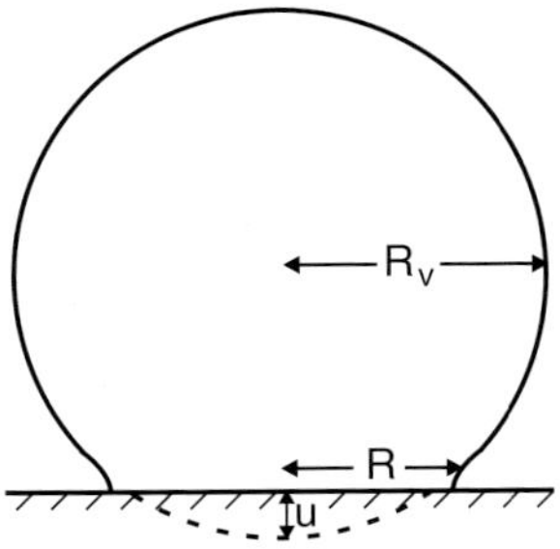

Fig. 8. – An elastic sphere (radius R_v) lying on an adhesive surface. The sphere is squeezed towards the surface by an amount u, and establishes contact over a radius R, related to u by the scaling law $\boldsymbol{R}^2 \simeq u\boldsymbol{R}_v$. All elastic distorsions are confined in a region of volume $\sim R^3$ above the patch. This "JKR situation" is crudely discussed in sect. **5**.

$$VW_-(V) = \int \mathrm{d}x f_z V_z \Gamma_{\mathrm{i}}$$

$$= \frac{kT}{a}\Gamma_{\mathrm{i}} V \int \frac{\mathrm{d}z}{\mathrm{d}x}\left[\ln\left(\frac{V}{V_1}\right) + \ln\left(\frac{\mathrm{d}z}{\mathrm{d}x}\right)\right] \tag{18}$$

$$= \frac{kT}{a}\Gamma_{\mathrm{i}} V z_{\mathrm{max}}(1+l), \tag{19}$$

where z_{max} is the separation distance, and l is a logarithmic correction,

$$l = \frac{1}{z_{\mathrm{max}}}\int \mathrm{d}z \ln\left(\frac{\mathrm{d}z}{\mathrm{d}x}\right), \tag{20}$$

which we shall drop in the following. (It is fortunate that the shape of the profile enters only in this corrective term!). Ultimately, we arrive at

$$W_-(V) = kT\Gamma_i \frac{z_{\mathrm{max}}}{a} \ln\left(\frac{V}{V_1}\right). \tag{21}$$

Equation (21) can be compared to some rolling simulations performed with a much more detailed molecular model by D. Hammer [15]. It gives a reasonable summary of Hammer's data. If we return to the force balance of eq. (15), we find that V becomes observable (in a typical example) for $\gamma_0 > 50\,\mathrm{s}^{-1}$. What is nice in eq. (21) is the fact that V_1 (eq. (16)) depends on the barrier heigth. Thus a measurement of $W_-(V)$ gives us information of the barrier height. This may, in principle, be compared with data from chemical kinetics on the unbinding of sticker/receptor pairs in solution.

5. – Extension towards less soft objects

A living cell has many differences from the above vesicle models of adhesion. One of them is purely mechanical. Cells contain both a cytoskeleton under the membrane, and a transient actin network in most of the internal volume.

For our purposes the cytoskeleton represents a mild complication: it increases the rigidity of the membrane, but does not necessarily upset our description. On the other hand, if we have a 3d gel, the build-up of an adhesive patch is seriously modified. We know from simple observations that something like a gel is present in many cells —at least in transient states. For instance, when performing aspiration with a micropipette, if a cell hops out the pipette, it often retains a certain memory of shape under aspiration: looking like the top end of a lemon!

We are thus led to think about the mobilisation of stickers at the surface of a viscoelastic gel, and the resulting adhesion patches. A complete analysis of a deformed elastic sphere facing a wall can, in principle, be constructed from the classic paper of Johnson, Kendall, and Roberts [16]. Here, we shall present a very crude, simplified version of their argument, which omits certain hysteretic features of the contact.

The basic energy for a deformed elastic sphere (shown in fig. 8) has the following structure (ignoring again all numerical coefficients):

$$E \sim R^3 \mu \left(\frac{u}{R}\right)^2 - WR^2 \,, \tag{22}$$

where μ is an elastic modulus, and u is the amplitude of squeezing. The first term is an elastic energy corresponding to a deformation u/R extending on a volume R^3 (where R is always the contact radius). We have roughly $u \sim R^2/R_{\mathrm{v}}^2$, where R_{v} is the radius of the original sphere). Equilibrium corresponds to the condition $\mathrm{d}E/\mathrm{d}R = 0$, giving

$$W = \frac{\mu R^3}{R_{\mathrm{v}}^2} \,. \tag{23}$$

We shall now discuss patch growth, using eq. (23) instead of the Young equation (7), and combining it with the diffusion balance of stickers (eq. (10)). Assuming for instance an ideal gas law for the osmotic pressures $\Pi(\Gamma)$ we arrive at

$$\frac{R^5}{R_{\mathrm{v}}^5} \cong \varepsilon_{\mathrm{e}} \frac{Dt}{R_{\mathrm{v}}^2} \,, \tag{24}$$

where ε_{e} is a dimensionless constant

$$\varepsilon_{\mathrm{e}} = \frac{2\Gamma_0 kT}{\mu R_{\mathrm{v}}} \,. \tag{25}$$

Equation (25) should hold if our cell is elastic at the time scale t of interest —*i.e.* when the Maxwell relaxation time τ of our viscoelastic fluid is longer than t. In the opposite limit ($t > \tau$) we return to a viscous fluid, of viscosity $\eta = \mu\tau$. Then we should make the substitution $\mu \rightarrow \eta/t$, and this leads to the growth law

$$\frac{R^5}{R_{\mathrm{v}}^5} = \frac{\Gamma_0 kTD}{\eta R_{\mathrm{v}}^3} \,. \tag{26}$$

On both eqs. (25) and (26) we see that viscoelastic features alter the exponent n in the growth rate ($R \sim t^n$). But this does not mean that the growth should be very slow. Consider for instance the elastic regime (eq. (25)) with $D = 10^{-8}\,\mathrm{cm}^2/\mathrm{s}$ and $\mu = 0.1\,\mathrm{atm}$. At a time t of 1 minute we expect a contact radius $R \sim 0.1 R_\mathrm{v}$.

6. – Conclusions

For ultrasoft objects (vesicles, ghosts of red blood cells, etc.) it seems feasible to describe adhesion in relatively simple macroscopic terms. However, we should insist here on the diversity of possible scenarios.

a) Whenever we reach diffusion lengths $(Dt)^{1/2}$ comparable to the *vesicle* size (R_v) the dynamics is expected to slow down.

b) We have assumed that the surface tension γ stayed constant. This is not true when the contact angle θ becomes large (typically $\theta > 0.2$). For a discussion of this effect see [9].

c) It may happen that the stickers do not move any more after bonding. This is also discussed in [9].

d) The osmotic pressure $\Pi(\Gamma)$ of the 2d fluid of stickers may have very different shapes, and may show a plateau due to a phase transition.

e) It may possibly happen that the reaction between sticker and receptor is slow: this deeply alters the scaling laws as explained in [9].

We do have some experimental data on growth: the most significant ones are described in ref. [10]. Here the vesicles are protected by a PEG brush. The stickers are peptides carrying a specific recognition sequence (RGD). The corresponding receptor on the wall is a protein (integrin) attached to a glass wall by physical adsorption. The stickers are stuck as soon as they bind. We are dealing here with "heavy" vesicles, which should build up a cushion. It is not clear, however, whether the "cushion" has really had time to form before the start-up of the adhesion process. The observed growth law is $R(t) \sim t^{1/2}$, and this is very far from what we should expect for dewetting of a cushion.

On the whole, all the ideas presented here are still highly conjectural. But it should become possible to prepare systems which correspond to all the scenarios which we have in mind.

* * *

These views have been constructed in cooperation with F. BROCHARD and P.-H. PUECH. We also benefited from very fruitful discussions with F. AMBLARD, E. SACKMANN, and P. NASSOY.

REFERENCES

[1] See for instance LIPOWSKY R., *Encyclop. Appl. Phys.*, **23** (1998) 199.
[2] PUECH P.-H. and BROCHARD F., to be published.
[3] HELFRICH W. and SERVUSS R., *Nuovo Cimento D*, **3** (1984) 137.

[4] Seifert U., *Adv. Phys.*, **46** (1997) 13.

[5] Brochard F., Quere D. and de Gennes P.-G., *Gouttes, Bulles, Perles et Ondes* (Belin, Paris) 2001.

[6] For a review, see Curtis A. and Lackie J. M., *Measuring Cell Adhesion* (Wiley, New York) 1990.

[7] Bell G., Dembo M. and Bongrand P., *Biophys. J.*, **45** (1985) 1051.

[8] Brochard F. and de Gennes P.-G., *Proc. Natl. Acad. Sci. USA*, **99** (2001) 7854.

[9] de Gennes P. G., Puech P.-H. and Brochard F., *Langmuir*, **19** (2003) 7112.

[10] Boulbitch A., Guttenberg Z. and Sackmann E., *Biophys. J.*, **91** (2001) 2743.

[11] Guttenberg Z., Bausch A., Hu B., Bruinsma R., Moroder L. and Sackmann E., *Langmuir*, **16** (2000) 8984.

[12] Brochard F. and de Gennes P.-G., *C. R. Acad. Sci. (Paris) Phys.*, **4** (2003) 281.

[13] Merkel R., Nassoy P., Leung A., Ritchie K. and Evans E., *Nature*, **397** (1999) 50.

[14] Bell G., *Science*, **200** (1978) 618.

[15] Chang K. C., Tees D. and Hammer D., *Proc. Natl. Acad. Sci. USA*, **97** (2000) 11262.

[16] Johnson K. L., *Contact Mechanics* (Cambridge University Press) 1985.

Statistical mechanics of undercooled fluids and the glass transition: Testing theoretical concepts by computer simulation

K. BINDER

Institut für Physik, Johannes Gutenberg-Universität - Mainz, Germany

1. – Introduction

1.1. *Main facts and some ideas about the glass transition.* – Amorphous materials such as window glass, plastic containers, etc. are familiar to everybody, and nevertheless the formation and resulting properties of such "ill-condensed matter" are very poorly understood. In particular, the theoretical understanding of the character of the glass transition from the undercooled fluid to the amorphous solid is still one of the "grand challenge problems of science", despite enormous efforts in the last decades [1-12]. Already understanding the geometric structure of how atoms are arranged in a glass is puzzling (figs. 1, 2): of course, it is clear that the glass lacks the regular periodic long-range order of the lattice structure that characterizes a crystal, at best there is short-range order. However, the problem is to say what is the difference between a glass and a snapshot picture of the corresponding fluid? From the point of view of an elastic scattering experiment, there is hardly any distinction (fig. 2) [13]: $S(Q)$ is very small and structureless at wave numbers $Q \leq 1\,\text{Å}^{-1}$, reflecting the fact that the compressibility (which is related to $S(Q \to 0)$) of the dense fluid is very small, and there are no strong long-range density fluctuations. At about $Q = 1.5\,\text{Å}^{-1}$ the intensity rises to the "first sharp diffraction peak", also called "amorphous halo"; its position is related (inversely) to the mean distance at which the atoms in the fluid are "packed". Consequently, due to thermal expansion there is a slight shift of the position with temperature. The second peak there reflects intramolecular correlations along the polymer chain, which are essentially tem-

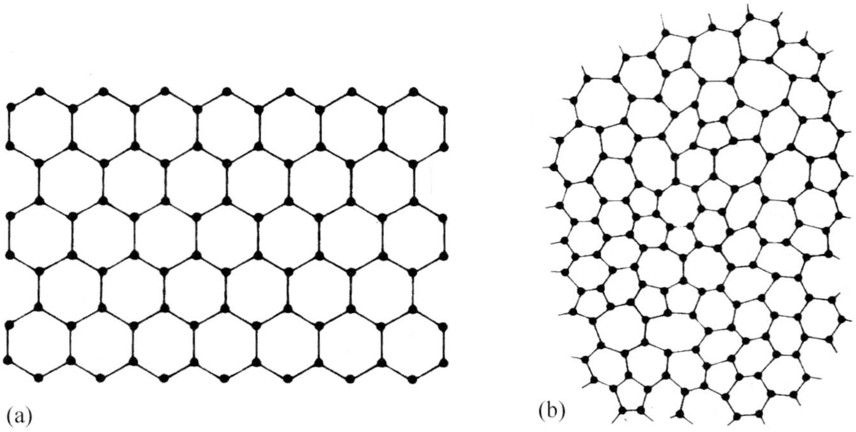

Fig. 1. – (a) The honeycomb lattice and (b) a two-dimensional continuous random network.

perature independent, since the potentials for lengths of covalent bonds and the angles between them are very stiff. Note that this structure seen in the second and further peak is strongly dependent on the chemical nature of the material that is studied; only the appearance of the first peak, the "amorphous halo", is similar for all glasses.

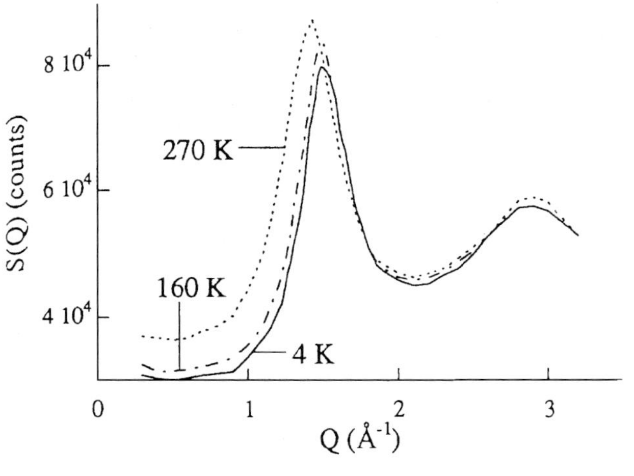

Fig. 2. – Static collective structure factor $S(Q)$, as observed from neutron scattering intensities, of liquid and amorphous polybutadiene, plotted against the scattering wave number Q. For this material, the glass transition temperature is $T_g = 180\,K$, and the critical temperature T_c of mode-coupling theory is estimated as $T_c = 220\,K$. Three temperatures are shown, $T = 270\,K$ (far above both T_g and T_c), $T = 160\,K$ and $T = 4\,K$. Note that the scattering background is not subtracted here, thus the zero of the ordinate axis for $S(Q)$ is not precisely known. From Arbe *et al.* [13].

This behavior of the structure factor suggests that (apart from thermal expansion, which could be eliminated anyhow by studying the glass transition at constant volume rather than constant pressure) the "snapshot" of atomic positions in the glass looks just the same as a "snapshot" taken in the liquid. This suggestion that a glass can be imagined as a "frozen liquid" is very strongly corroborated by the inspection of configurations generated by Molecular Dynamics simulations of suitable models for undercooled liquids and the corresponding glasses. But the glass is a solid, where atoms vibrate around their positions only, unlike the liquid where atoms diffuse: what is then the static order parameter, from the point of view of structure, that distinguishes fluid and glass, and is there such an order parameter at all?

Similarly, already in the supercooled fluid when the temperature is lowered towards the glass transition, one observes a dramatic increase of the structural relaxation time by up to 15 decades, often in a rather narrow temperature regime (fig. 3) [14]. The question arises: what changes in the structure (if any) occur that cause this dramatic slowing-down?

It should be emphasized, however, that not all glasses behave in the same way: in this so-called "Angell plot", the viscosity on a logarithmic scale is plotted *vs.* inverse temperature, in units of the glass transition temperature, so that one can compare many different systems on the same plot (note that the glass transition temperature is arbitrarily defined by requesting $\eta(T_{\rm g}) = 10^3$ poise in fig. 3, therefore all curves cross at $T_{\rm g}/T = 1$ simply due to this normalization). If the viscosity follows a simple Arrhenius law, $\eta(T) = \eta(\infty)\exp[A/T]$, we expect a straight line on the Angell plot, and in fact this is a good approximation for glass formers such as molten SiO_2. This type of behavior is classified as a "strong glass former", while systems that show a pronounced curvature on the Angell plot, indicative of a divergence of $\eta(T)$ at a nonzero temperature, are called "fragile glass formers". This divergence is often described by the Vogel [15]-Fulcher [16] relation

$$\ln \eta(T) \propto \frac{E_{\rm VF}}{(T - T_{\rm VF})}, \tag{1}$$

where $E_{\rm VF}$ is a characteristic energy, and $T_{\rm VF}$ (well below $T_{\rm g}$) is the (hypothetical) temperature where $\eta(T)$ would diverge, if the system did not fall out of the equilibrium at $T_{\rm g}$. However, the precise value of $T_{\rm VF}$ does depend on the temperature range over which $\eta(T)$ is fitted to eq. (1); also sometimes there is evidence that eq. (1) provides a good fit at temperatures distinctly higher than $T_{\rm g}$, while closer to $T_{\rm g}$ a simple Arrhenius relation works better. Some authors (*e.g.*, [17]) claim that near $T_{\rm g}$ actually a power law divergence

$$\eta(T) \propto \tau(T) \propto (T - T_{\rm f})^{-\Delta} \tag{2}$$

describes the data on viscosities or other measures of the structural relaxation time τ. Here Δ is a "critical exponent" that typically is rather large (of the order of 10), and

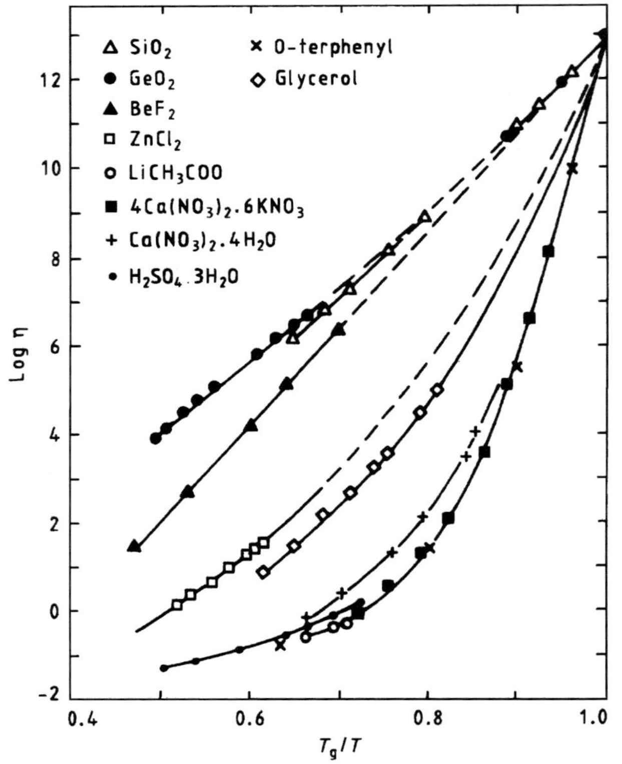

Fig. 3. – Viscosity $\eta(T)$ of various glass-forming liquids as a function of T_g/T, where the glass transition temperature T_g is arbitrarily defined by the condition that $\eta(T_g) = 10^{13}$ poise. From Angell and Sichina [14].

T_f is the critical temperature of the "ideal glass transition", with T_f a few degrees K below T_g. Equation (2) is motivated from the experience with spin glasses [3, 9]: spin glasses are produced by random dilution of a ferromagnet that has competing ferro- and antiferromagnetic interactions with nonmagnetic impurities (fig. 4). Disorder plus "frustration" (not all magnetic interactions can be satisfied in the ground state) cause the spins to freeze in random directions [3, 9]. Now for spin glasses it is rather well established [3,9] that this transition is an unconventional type of second-order transition, caused by a divergence of a correlation length ξ_{SG} at the freezing temperature T_f [3, 9]:

$$\xi_{SG} \propto (T - T_f)^{-\nu}, \tag{3}$$

ν being the associated critical exponent. Note that also for spin glasses the existence of this length does not imply that one would see any interesting behavior in a neutron magnetic scattering experiment: the correlation function $[\langle S_i S_j \rangle_T]_{av}$, where $\langle \cdots \rangle_T$ is the standard canonical ensemble thermal average and $[\cdots]_{av}$ stands for an average over the

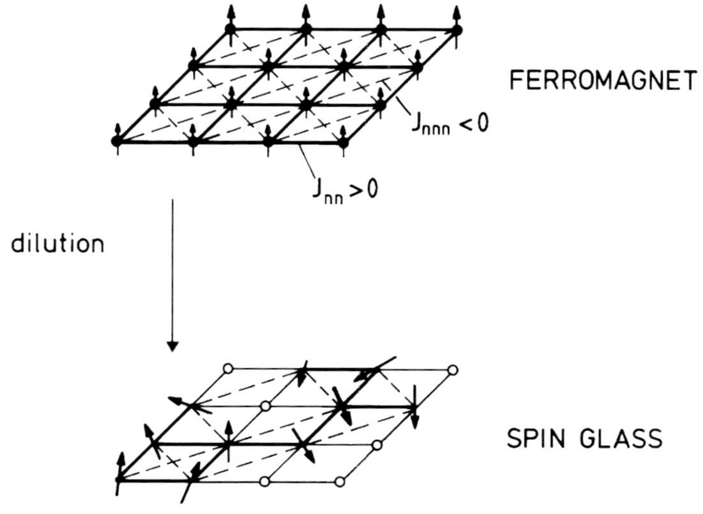

Fig. 4. – Schematic explanation on how quenched disorder by random dilution leads from a ferromagnet to a spin glass. It is assumed that nearest-neighbor interactions $J_{\rm nn}$ are ferromagnetic, while next-nearest-neighbor interactions $J_{\rm nnn}$ are weaker but antiferromagnetic. (An experimental example for this case would be the face-centered cubic ferromagnet EuS, where $J_{\rm nnn} \approx J_{\rm nn}/2$.) Dilution with nonmagnetic ions (*e.g.*, Sr) leads to a spin glass (*e.g.*, $Eu_xSr_{1-x}S$ for $0.13 \leq x \leq 0.5$) [3].

quenched disorder, does not contain any information on $\xi_{\rm SG}$, rather one has to consider a correlation function $g_{\rm SG}(\mathbf{r}_i - \mathbf{r}_j) = [\langle S_i S_j \rangle_T^2]_{\rm av}$ which is not experimentally accessible by scattering experiments. The experimental evidence in favor of eq. (3) is indirect, from the observation of a divergence of the static nonlinear magnetic susceptibility (which is related to $\sum_{i,j} g_{\rm SG}(\mathbf{r}_i - \mathbf{r}_j)$ [3]). The divergence of $\xi_{\rm SG}$ then leads to a divergent relaxation time via the standard mechanism of critical slowing-down

$$\tau(T) \propto [\xi_{\rm SG}(T)]^z \,, \tag{4}$$

which leads to eq. (2) with $\Delta = z\nu$. One finds also for spin glasses that z is unusually large, and hence data on spin glass relaxation times plotted *vs.* $1/T$ look rather similar to the Angell plot for fragile glass formers (fig. 3), and can even be fitted rather well to a Vogel-Fulcher relation.

In view of the success of eqs. (3), (4) for spin glasses, there has been an extensive search for a characteristic length $\xi_{\rm G}$ for structural glass transitions, that grows in the supercooled fluid as $T_{\rm g}$ is approached and hence can be made responsible for the glass transition, see, *e.g.*, [10, 18-21]. So far, these attempts have had limited success. The recent attempts to extract information on this issue from confinement effects on the glass transition will be discussed in sect. **3**.

Although the glass transition occurs in a great variety of systems (molten oxides, fluids from small organic molecules, polymer melts, molten metallic alloys, etc.) there is a surprising universality of many aspects of the behavior of physical properties near T_g. *E.g.*, time-dependent correlation functions or relaxation functions $\varphi(t)$ are found to decay with time t according to the Kohlrausch [22]-Williams-Watts (KWW) [23] law,

$$\varphi(t) \propto \exp\left[-\left(\frac{t}{\tau}\right)^{\beta_K}\right]; \tag{5}$$

here the exponent β_K is not universal, it differs from material to material (and even may be different for different properties), but often it does not depend on temperature. Therefore a scaling property named "time-temperature-superposition principle" [2] holds.

The physical mechanism responsible for the dramatic slowing-down of the relaxation as T_g is approached is still very controversial, and many ideas have been suggested, which will not be discussed further (*e.g.*, the concept of an "avoided critical point" [24] or the "free volume" approach [25]). Rather we focus on two theories only: the idea [26, 27] that the undercooled fluid must undergo a static phase transition to an "ideal glass" (in metastable equilibrium) to avoid an "entropy catastrophe" [28], and the mode-coupling theory of the glass transition [4, 29, 30] (fig. 5). The latter theory attempts to model the "cage effect": in a dense supercooled liquid, every particle is moving inside a confining cage formed by its nearest neighbors and only rarely a local density fluctuation occurs allowing a particle to "escape" from each cage. Of course, in the framework of a systematic microscopic theory for the dynamics of liquids these local density fluctuations have to be dealt with self-consistently, and in the framework of an approximation where unknown higher-order correlations are factorized this mechanism yields a nonlinear feedback effect, leading to the divergence of the "residence time" of a particle in a cage at the critical temperature T_c. This divergent relaxation time not only shows up in a power law singularity of the viscosity (fig. 5), but also the decay of the Fourier-transformed correlator of local density fluctuations then stops at some value f, rather than decaying towards zero. This means that density fluctuations can no longer completely relax, but rather remain frozen-in, and the system is no longer fully ergodic: therefore f is called the "nonergodicity parameter"; it is somehow reminiscent of the Edwards-Anderson order parameter [31] of spin glasses $q_{\mathrm{EA}} = \lim_{t\to\infty}[\langle S_i(0)S_i(t)\rangle_T]_{\mathrm{av}}$. When one approaches this dynamical transition from ergodic to nonergodic behavior from higher temperature, $\phi_q(t)$ gradually develops a plateau around f, whose "lifetime" gets longer and longer as T_c is approached, just as in eq. (2), with an exponent Δ typically close to $\Delta = 2$. However, this divergence comes about from the nonlinearity of the equation of motion for $\phi_q(t)$, and no diverging static length $\xi_{\mathrm{G}}(T)$ is involved in this singularity [4, 29, 30], unlike the mechanism of standard critical slowing-down (eqs. (3), (4)).

Now a comparison to experiment shows [29] that this "idealized MCT" describes real atomic and molecular fluids often rather well over a range of relaxation times from $10^{-12}\,\mathrm{s} \le \tau \le 10^{-8}\,\mathrm{s}$, while at lower temperatures drastic deviations occur: rather than diverging to infinity at T_c, τ and $\eta(T)$ stay finite at T_c, but they increase very rapidly with decreasing temperature. The interpretation of this discrepancy is that

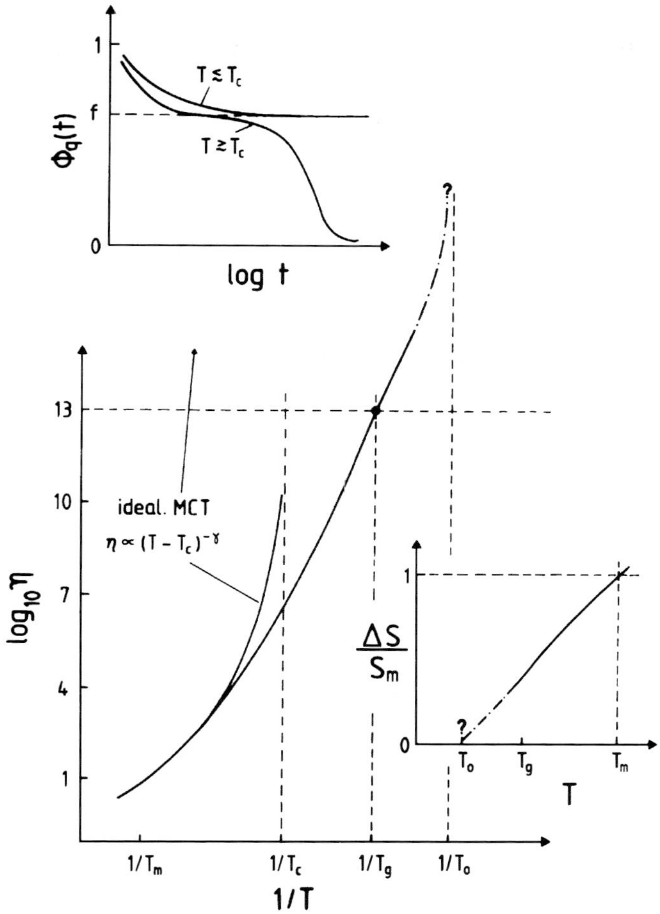

Fig. 5. – Schematic sketch of the logarithm of the viscosity η (or the structural relaxation time τ, respectively, assuming that $\eta = 1$ poise corresponds to $\tau = 10^{-12}$ s) of a supercooled glass-forming fluid of small molecules plotted *vs.* inverse temperature. From the melting temperature $T_{\rm m}$ to the glass transition temperature $T_{\rm g}$ the viscosity $\eta(T)$ rises by many decades. Here $T_{\rm g}$ is defined empirically by choosing $\eta(T = T_{\rm g}) = 10^{13}$ poise, as in fig. 3. The upper inset qualitatively illustrates the main prediction of idealized mode-coupling theory (MCT) [4] for the relaxation function $\phi_q(t)$ of density fluctuations at wave number q, corresponding to the peak of the static structure factor $S(q)$. While $\phi_q(t)$ for temperatures T larger than the critical temperature $T_{\rm c}$ decays with time t to zero in two steps, for $T \leq T_{\rm c}$ it decays only towards the "nonergodicity parameter" f. This theory would then also imply a power law divergence of the viscosity $\eta \propto (T - T_{\rm c})^{-\gamma}$, as indicated in the main part of the figure. The lower inset shows the excess configurational entropy ΔS (relative to the entropy of the corresponding crystal that forms in equilibrium at $T_{\rm m}$, normalized by $\Delta S(T_{\rm m}) = S_{\rm m}$) plotted *vs.* temperature. A linear extrapolation of ΔS *vs.* T yields the Kauzmann temperature $T_{\rm K}$ from the condition $\Delta S(T_{\rm K}) = 0$, and often $T_{\rm K}$ is compatible with the equality $T_{\rm K} = T_{\rm VF}$, $T_{\rm VF}$ being the temperature where $\eta(T)$ diverges if the Vogel-Fulcher law (eq. (1)) holds. Question marks emphasize that the validity of this linear extrapolation, as well as the validity of an extrapolation based on eq. (1), are very uncertain.

ideal MCT ignores thermally activated escape of particles from their cages, the so-called "hopping processes". In fact, extended versions of MCT have been proposed [32] where the divergence of τ at T_c is replaced by a smooth crossover from the power law to a simple Arrhenius law, $\eta \propto \exp[A/T]$, with a large value of the activation barrier A. However, the validity of this extended form of MCT is rather uncertain.

While in this MCT theory the claim is that the most interesting behavior occurs near T_c, at temperatures where $\eta(T)$ is still about 10 orders of magnitude smaller than at T_g, the entropy theory [26, 28] asserts that the most interesting behavior occurs at temperatures below T_g (fig. 5, lower insert): if one estimates experimentally [28] the "configurational entropy" of a supercooled fluid or glass by subtracting from the total (calorimetrically determined) entropy of the system the entropy of the corresponding crystal, one finds a rapid decrease. (This subtraction assumes that in the crystal and the glass or deeply supercooled fluid the vibrational entropy is essentially the same.) At T_g one roughly has $\Delta S/S_m \approx 1/3$ [7], and a linear extrapolation would yield $\Delta S(T) = 0$ at a "Kauzmann temperature" T_K, which is typically 20–50 K lower than T_g [28]. The consequence that for $T < T_K$ the entropy of the disordered supercooled fluid is less than the entropy of the more ordered crystal is called "Kauzmann's paradox", and the suggestion is that this paradoxical consequence is avoided by a static phase transition to an "ideal" glass phase at a temperature $T_0 \geq T_K$.

A theory that implies a transition according to this scenario was proposed by Gibbs and diMarzio [26] for the glassy freezing of polymer melts. They describe the polymers as self-avoiding walks on the (simple cubic) lattice, including also a fraction of vacant sites. In order to introduce temperature into the problem, they associate an energy cost ε whenever the walk makes a 90° turn on the lattice, while no energy cost occurs when the walk continues straight on. This model by construction has no entropy due to phonons but configurational entropy only. The task of statistical mechanics would be to calculate the free energy via the partition function, summing over all configurations that describe how one can put these chains on the lattice with their proper statistical weight, and hence obtain the (configurational) entropy ΔS. Of course, this task can only be carried out by making several very crude approximations (*e.g.*, the condition that chains must not intersect themselves or other chains is largely ignored, etc.). However, if one accepts these approximations one indeed finds a temperature T_K, where $\Delta S(T = T_K) = 0$, and for $T < T_K$ the configurational entropy of the high-temperature phase would be negative. Gibbs and diMarzio [26] suggest that at T_K the transition to the ideal glass takes place. They also found that T_K depends on the chain length N of the polymers as $T_K(N) = T_K(\infty) - \text{const}/N$. Since an analogous dependence of the glass transition temperature $T_g(N)$ is found experimentally (*e.g.*, [33]), $T_g(N) = T_g(\infty) - \text{const}/N$, the entropy theory enjoys great popularity in polymer science. (Note that for polymers T_g is defined by requiring that the structural relaxation time $\tau(T = T_g) = 100\,\text{s}$, since the viscosity of polymer melts does depend on chain length very strongly, $\eta(T) \propto N^3$ for entangled polymer melts, and can no longer be measured near T_g.) Furthermore, Adam and Gibbs [27] presented an extension of the theory to dynamics, suggesting that the decrease of $\Delta S(T)$ reflects the growth of large regions in which cooperative rearrangements must take place in order to effect the structural relaxation. They suggest that the free-energy barrier for these cooperative motions scales like $1/\Delta S(T)$, implying $\ln \tau \propto 1/\Delta S(T) \propto 1/(T - T_K)$, which hence yields the Vogel-Fulcher relation, eq. (1), if the identification $T_{VF} = T_K$ is made.

Now it is clear that this approach is also subject to various criticisms: i) Stillinger [34] has given arguments to show that in systems with short-range interactions the (configurational) entropy can be strictly zero only at $T = 0$. Thus, if a static transition to an ideal glass really occurs at some temperature T_0, the configurational entropy $\Delta S(T)$ of this glass ought to be nonzero for $0 < T < T_0$, and hence T_K can only be a lower bound to the actual transition temperature. ii) Since the energetics of the model at low temperatures favors configurations where the chains are stretched out linearly like a hard rod, it is natural to expect that the system undergoes a phase transition to an ordered low-temperature state (with crystalline or liquid crystalline order) and not a glass. However, this criticism can be avoided by considering a slightly more complicated model, namely the bond fluctuation model [35, 36] on the square or simple cubic lattice, where one gives an energetic preference to long bonds. In this model, there is no tendency towards rodlike configurations, chains at low temperature stay random-walk–like, but many signatures of a real glass transition are in fact found [37-39]. For this model the Gibbs-diMarzio theory can be worked out and yields an "entropy catastrophe" at $T_K > 0$ as the original model [40]. In contrast, if one uses the same model but with a bond-angle potential favoring chain stretching, one can show that an isotropic-nematic transition indeed pre-empts the glass transition [41].

At this point we would like to emphasize that the two concepts emphasized in fig. 5, namely the dynamical transition from ergodic to nonergodic behavior at T_c and a static transition to a kind of "ideal glass phase" at a lower temperature, need not be mutually exclusive. In fact, a model system for which one can show that in fact both transitions do occur is the infinite-range Potts glass with $p > 4$ states, see fig. 6 [42-47]. The Hamiltonian of this model is given by

$$\mathcal{H} = -\sum_{i<j} J_{ij}\big(p\,\delta_{\sigma_i\sigma_j} - 1\big), \qquad \sigma_i \in (1, 2, \ldots, p), \tag{6}$$

where the N Potts spins σ_i interact with bonds J_{ij} which are random variables, distributed according to a Gaussian distribution $P(J_{ij})$,

$$P(J_{ij}) = \big[(2\pi)^{1/2}\Delta J\big]^{-1} \exp\left[\frac{-(J_{ij} - J_0)^2}{[2(\Delta J)^2]}\right]. \tag{7}$$

To ensure a sensible thermodynamic limit, the first two moments of $P(J_{ij})$ are normalized as follows:

$$J_0 = [J_{ij}]_{\rm av} = \frac{\widetilde{J}_0}{(N-1)}, \qquad (\Delta J)^2 = \big[J_{ij}^2\big]_{\rm av} - \big[J_{ij}\big]_{\rm av}^2 = \frac{(\Delta\widetilde{J})^2}{(N-1)}. \tag{8}$$

(a)

(b)

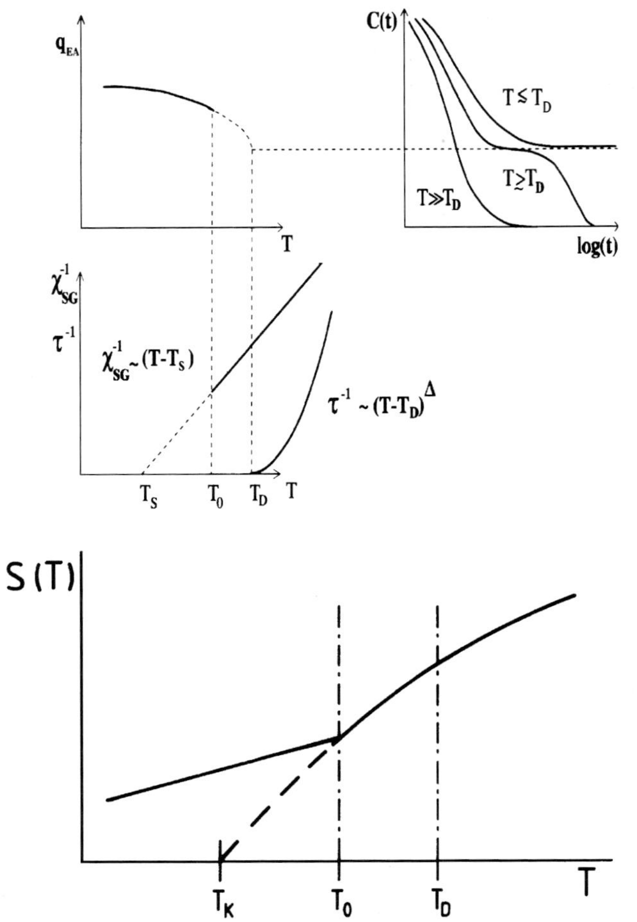

Fig. 6. – Qualitative sketch of the mean-field predictions for the p-state Potts glass model with $p > 4$. Part (a) shows the static glass order parameter q_{EA} *vs.* temperature; q_{EA} is nonzero for $T < T_0$ and jumps discontinuously to zero at $T = T_0$. The glass susceptibility χ_{SG} is finite at T_0; its (metastable) continuation would diverge at $T_{\mathrm{s}} < T_0$ in a Curie-Weiss–like fashion (eq. (10)). The relaxation time τ diverges already at the dynamical transition temperature $T_{\mathrm{D}} > T_0$ (eq. (11)). This divergence is due to the occurrence of a long-lived plateau of height q_{EA} in the time-dependent spin autocorrelation function $C(t)$ of the Potts spins. Part (b) shows the entropy $s(T)$ per spin *vs.* T; the entropy in the high-temperature phase has no singularity whatsoever at T_{D}, and when it is analytically continued to $T < T_0$ (broken curve) it would vanish at T_{K}. The true entropy of the glass phase (full curve for $T < T_0$) stays positive throughout. Note that the first-order glass transition in this model neither exhibits a latent heat nor a jump singularity of $s(T)$ at $T = T_0$, unlike conventional first-order transitions. From Brangian *et al.* [47].

Choosing $\Delta\widetilde{J} = 1$ to set the temperature scale and $\widetilde{J}_0 < (4-p)/2$ to avoid ferromagnetic order of the Potts spins, the model exhibits a transition from the disordered phase at a temperature T_0, which is second order for $p \leq 4$ and first order for $p > 4$ (note that for $p = 2$ the model reduces to the Sherrington-Kirkpartrick [48] infinite-range version of the Ising spin glass [3]). The entropy density of the disordered phase in this model can be shown to be [42]

$$s(T) = \ln p - \left[\frac{(p-1)}{4}\right]\left(\frac{k_{\mathrm{B}}T}{\Delta\widetilde{J}}\right)^{-2}, \tag{9}$$

i.e. a Kauzmann temperature occurs at $k_{\mathrm{B}}T_{\mathrm{K}}/\Delta\widetilde{J} = (1/2)\sqrt{(p-1)/\ln p}$. However, one always finds $T_0 > T_{\mathrm{K}}$, so T_{K} is only a lower bound to the static glass transition temperature, as expected. For $p > 4$ (where the glass transition is first order) two additional characteristic temperatures occur (fig. 6), namely a "spinodal temperature" T_{s} where the static spin glass type susceptibility diverges,

$$\chi_{\mathrm{SG}}^{-1} = \left[1 - \left(\frac{T_{\mathrm{s}}}{T}\right)^2\right], \quad \frac{k_{\mathrm{B}}T_{\mathrm{s}}}{\Delta\widetilde{J}} = 1, \tag{10}$$

and a temperature T_{D} of a dynamical transition (from ergodic to nonergodic behavior) where the relaxation time τ of the spin autocorrelation function of dynamical versions of this model diverges,

$$\tau \propto \left(1 - \frac{T_{\mathrm{D}}}{T}\right)^{-\Delta}, \tag{11}$$

Δ being the dynamical exponent, and $T_{\mathrm{D}} > T_0 > T_{\mathrm{s}}$ (while for $p \leq 4$ these three temperatures coincide, of course [3]). This situation is described qualitatively in fig. 6 [47]. Both T_0 and T_{D} depend on p; their precise values can be computed within the one-step replica symmetry-breaking scheme [43] only numerically [46].

Now the qualitative similarity between the spin autocorrelation function $C(t)$ in fig. 6(a) and the density fluctuation autocorrelation function in fig. 5 is not at all accidental: also the dynamics of the mean-field Potts glass is described by idealized mode-coupling theory [44, 45]. However, there is no proof that also the static transition of the Potts glass has a counterpart in the problem of the structural glass transition, fig. 5. After all, there is already frozen-in disorder present in the high-temperature phase of the Potts glass model by construction (eq. (8)), while in the supercooled fluid above T_{g} there is no quenched disorder. Furthermore real glasses are systems with short-range interactions among the particles, and hence should be compared to the Potts glass with short-range interactions (which has rather different properties [49]) rather than the mean-field model. Nevertheless, we shall return in the last part of this lecture series to this curious case of the Potts glass to study its properties more closely.

At this point, however, we want to discuss the general description of systems where a dynamical transition to nonergodic behavior has occurred (the region $T_0 < T < T_D$ in the Potts glass, for instance). Physically, this situation can be interpreted by stating that the phase space has split into many valleys separated by infinitely high barriers from each other, which are mutually inaccessible: if the system is stuck in one valley, it remains there for an infinitely long time before it could cross a barrier to move into another valley. Therefore a time average samples properties from a single valley (also called "ergodic component") only, while a full statistical-mechanics ensemble average would be an average over all valleys [3, 50, 51]. This latter average can be "sorted" according to the contributions from the different valleys which we label by ℓ, so that the free energy F_ℓ for the ℓ-th valley is given by

$$\exp\left[\frac{-F_\ell}{k_\mathrm{B}T}\right] = \sum_{\lambda\in\ell} \exp\left[\frac{-E_\lambda}{k_\mathrm{B}T}\right] = Z_\ell\,, \tag{12}$$

λ being a microstate (*e.g.*, a spin configuration of the Potts glass) associated with state ℓ. The partition function Z and the statistical weight of state ℓ are then given by

$$Z = \sum_\ell Z_\ell = \sum_\lambda \exp\left[\frac{-E_\lambda}{k_\mathrm{B}T}\right], \quad P_\ell = \frac{Z_\ell}{Z} \tag{13}$$

and canonical averages $\langle A\rangle_T$ of an observable A can similarly be decomposed:

$$\langle A\rangle_T = \frac{1}{Z}\sum_\lambda A_\lambda \exp\left[\frac{-E_\lambda}{k_\mathrm{B}T}\right] = \frac{1}{Z}\sum_\ell \left\{\frac{1}{Z_\ell}\sum_{\lambda\in\ell} A_\lambda \exp\left[\frac{-E_\lambda}{k_\mathrm{B}T}\right]\right\} Z_\ell\,. \tag{14}$$

Obviously, this can be written as

$$\langle A\rangle_T = \sum_\ell P_\ell \langle A\rangle_T^{(\ell)}, \qquad \langle A\rangle_T^{(\ell)} = \frac{1}{Z_\ell}\sum_{\lambda\in\ell} A_\ell \exp\left[\frac{-E_\lambda}{k_\mathrm{B}T}\right]. \tag{15}$$

This decomposition into contributions from individual "states" ℓ (or "valleys" or "ergodic components" ℓ) has an interesting consequence when we consider the free energy F, since F is not simply the average $\bar F$ of the free energies F_ℓ of the various components:

$$F = -k_\mathrm{B}T\ln Z, \qquad F_\ell = -k_\mathrm{B}T\ln Z_\ell = F - k_\mathrm{B}T\ln P_\ell \tag{16}$$

and hence

$$\bar F = \sum_\ell P_\ell F_\ell = F - k_\mathrm{B}T\sum_\ell P_\ell \ln P_\ell = F + TI\,, \tag{17}$$

where we have introduced the so-called "complexity" I,

$$I = -k_{\mathrm{B}} \sum_{\ell} P_\ell \ln P_\ell = k_{\mathrm{B}} \ln N^* \,, \tag{18}$$

where N^* is the "effective" number of states describing this ergodicity breaking. Correspondingly, considering that $F_\ell = U_\ell - TS_\ell$, $F = U - TS$, the entropy $\bar{S}$ averaged over all the valleys becomes ($U = \sum_\ell P_\ell U_\ell$ is the average interval energy)

$$\bar{S} = \sum_{\ell} P_\ell S_\ell = S - I \,. \tag{19}$$

Thus, we see that at an ergodic-to-nonergodic transition part of the entropy of the system disappears in order to become the complexity (only a single ergodic component $\ell = 1$ exists if the system is ergodic, of course). Therefore sometimes I is also called "configurational entropy" but this is a misnomer, because in the framework of the statistical mechanics of lattice models (*e.g.*, self-avoiding walks on lattices) the configurational entropy has a different and well-defined meaning. The interesting point about the complexity in the present context, however, is that for the mean-field Potts glass with $p > 4$ (and related models, such as the p-spin interaction spin glass [52,53]) I becomes nonzero at T_{D} discontinuously and vanishes continuously at the static transition T_0: the appearance of a nonzero static glass-order parameter means that the effective number of states N^* is no longer extensive [54]. For the spherical model version of the spin glass with p-spin interactions

$$\mathcal{H} = - \sum_{i_1 < i_2 < \cdots < i_p} J_{i_1 i_2} \cdots i_p \, S_{i_1} S_{i_2} \cdots S_{i_p} \,, \tag{20}$$

where the J's are drawn from a Gaussian distribution centered at $J_0 = 0$, and the spins S_i's are real valued but subject to the "spherical constraint" $\sum_i S_i^2 = N$, one can show that [53]

$$I(T \longrightarrow T_{\mathrm{D}}^-) = N\left(\frac{1}{2}\ln(p-1) + \frac{2}{p} - 1\right) \tag{21}$$

is the value of the complexity just below the dynamical transition.

The fact that a static glass transition can be characterized by the vanishing of a contribution of the entropy, namely the complexity, has led to the widespread speculation (*e.g.*, [54-56]) that the complexity can be linked to the entropy difference ΔS considered by Kauzmann [28]. However, while the complexity is zero in the high-temperature phase and suddenly becomes nonzero at the dynamical transition temperature T_{D}, the entropy difference ΔS of undercooled glass-forming fluids is clearly nonzero already at $T > T_{\mathrm{c}}$ and perfectly smooth when T_{c} is crossed (fig. 5).

1·2. *Testing mode-coupling theory with simulations of a Bead-Spring model for polymers.* – A simulation of chemically realistic models for polymer melts would be very demanding, since a flexible macromolecule exhibits nontrivial structure from the length scale of covalent bonds (≈ 1 Å) to the gyration radius (≈ 10^2 Å), and even far above the glass transition temperature the spectrum of relaxation times is spread out over many orders of magnitude [57]. Therefore it is advantageous to study a coarse-grained model of polymers only, where several covalent bonds along the backbone of the chain are integrated into an effective segment. This consideration justifies the use of a bead-spring model, where neighboring effective monomers along the chain interact with the finitely extensible nonlinear elastic (FENE) potential [58]

$$U_{\mathrm{F}} = \frac{1}{2} k R^2 \ln\left[1 - \left(\frac{r}{R_0}\right)^2\right], \quad R_0 = 1.5, \; k = 30 . \tag{22}$$

Here units of length and of energy are measured in terms of the Lennard-Jones parameters σ and ε of the potential that acts between all effective monomers (bonded and nonbonded ones),

$$U_{\mathrm{LJ}}(r) = 4\varepsilon\left[\left(\frac{\sigma}{r}\right)^{12} - \left(\frac{\sigma}{r}\right)^{6}\right] + C, \quad r \leq 2r_{\mathrm{min}} , \tag{23}$$

the constant $C = 0.00775$ being chosen such that $U_{\mathrm{LJ}}(r = 2r_{\mathrm{min}}) = 0$, $r_{\mathrm{min}} = 2^{1/6}\sigma \approx 1.13$ being the position where $U_{\mathrm{LJ}}(r)$ has its minimum. For $r > 2r_{\mathrm{min}}$ the potential is then taken to be zero. For nearest neighbors along the chain, the superposition of the FENE and LJ potentials leads to a steep minimum at about $b_{\mathrm{min}} \approx 0.96$. This length scale preferred by the effective bond potential thus is clearly incompatible with the length scale preferred by the LJ potential ($r_{\mathrm{min}} \approx 1.13$), and hence formation of a simple crystal structure when the effective monomers are arranged on a face-centered cubic or body-centered cubic lattice is prevented. Of course, this polymer melt is still able to crystallize, if the chains all stretch out linearly and then are densely packed together in a parallel arrangement (*e.g.*, in planes perpendicular to those rodlike stretched chains the effective monomers could form a triangular lattice with the preferred spacing r_{min}). However, the liquid-to-solid transition from the melt, where the polymer has a random-walk–like configuration, to the crystal, where the polymer is stretched out, is hindered by huge kinetic barriers, and even for very short chains containing only $N = 10$ effective monomers crystallization has never been observed to occur in a simulation of this model. Only by preparing the system in the assumed crystalline ground state and heating it up it was possible [59] to estimate the melting temperature of the model roughly as $T_{\mathrm{m}} \approx 0.76$. On the other hand, cooling the melt slowly down at constant pressure ($p = 1$ in Lennard-Jones units) one observes a glass transition at about $T_{\mathrm{g}} \approx 0.41$ from the kink in the volume *vs.* temperature curve [60], similar to that one finds in experiment [61] (fig. 7). In the same temperature range, the static structure factor of the melt develops

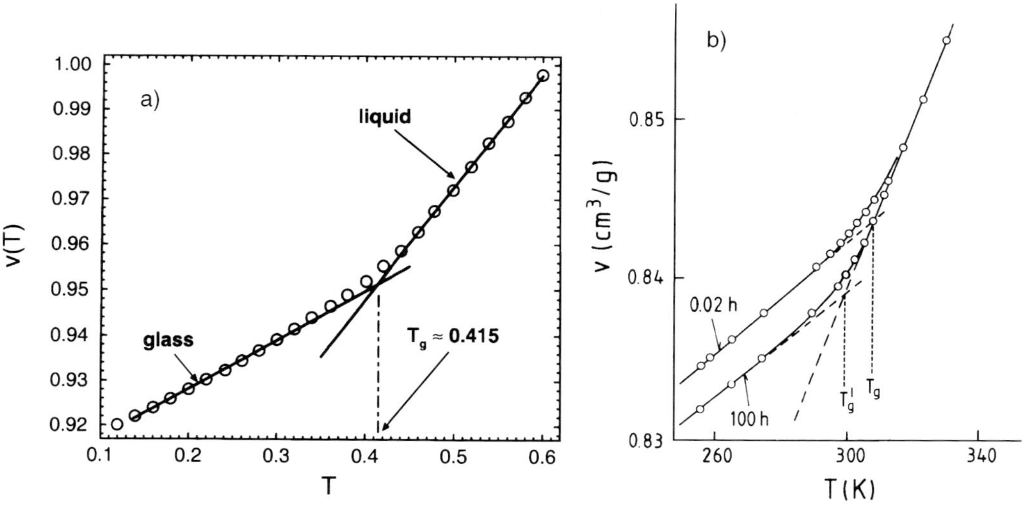

Fig. 7. – a) Volume $v(T)$ per effective monomer in a slow-cooling experiment, starting at pressure $p = 1$ from a well-equilibrated configuration at $T = 0.6$ (all temperatures being measured in units of the Lennard-Jones energy ε, putting $k_{\rm B} \equiv 1$), and lowering T every 500000 molecular-dynamics time steps δt by $\Delta T = 0.02$. From Binder *et al.* [60]. b) Specific volume v *vs.* temperature T for a poly(vinylacetate) polymer. The plot shows the effect of cooling velocity on the glass transition temperature: if the cooling takes 100 h rather than 0.02 h, the transition temperature is distinctly lower. Data extracted from fig. 1 of ref. [61].

an "amorphous halo" at $q \approx 6.9$, while $S(q \to 0)$ becomes very small (fig. 8 [62]), and the general shape of $S(q)$ for the model supercooled polymer melt (fig. 8) strikingly resembles experiment (fig. 2).

Figure 7b) also shows that the glass transition temperature $T_{\rm g}$ as estimated from the temperature of the kink in the $v(T)$ *vs.* T curve in a "slow cooling" experiment, where the temperature is lowered linearly with time, $T(t) = T_i - \gamma t$, T_i being some suitable initial high temperature far above $T_{\rm g}$, does depend on the cooling rate γ distinctly. Experiments [63] as well as simulations [38, 59, 63-65] show that this dependence of $T_{\rm g}$ on γ is in fact slower than logarithmic, and can be represented by

$$T_{\rm g}(\gamma) = T_{\rm VF} + \frac{E_{\rm VF}}{\{k_{\rm B} \ln[1/(\tau_\infty \gamma)]\}}, \tag{24}$$

if one assumes that the system falls out of equilibrium (*i.e.*, it freezes into the glass and thus the kink occurs) if the time constant of the cooling (γ^{-1}) becomes equal to the structural relaxation time τ which in eq. (24) was assumed to follow the Vogel-Fulcher relation, $\tau = \tau_\infty \exp[E_{\rm VF}/(T - T_{\rm VF})]$. Of course, since $v(T)$ is precisely linear in T neither in the glass regime nor in the fluid regime, and near $T_{\rm g}$ the "kink" anyway is rounded off (fig. 7), the data on $T_{\rm g}(\gamma)$ are never very precise, and typically only few decades of γ are available. Thus, eq. (24) can neither be taken as a proof for the Vogel-Fulcher relation

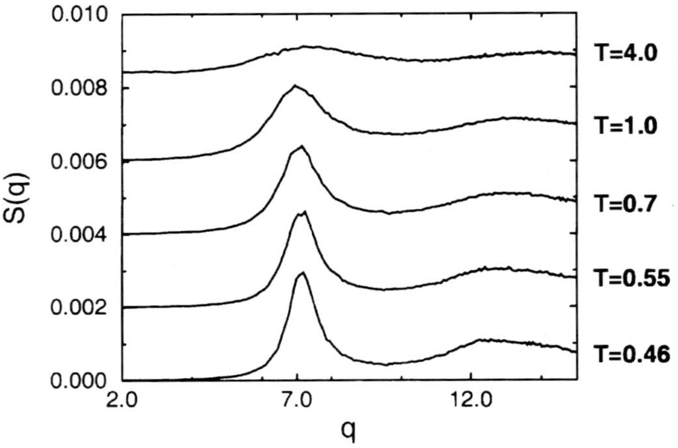

Fig. 8. – Structure factor $S(q)$ for the bead-spring model of flexible polymer chains, using 120 chains of length $N = 1$ at pressure $p = 1$ in a cubic box, for the model defined by eqs. (22), (23). Five temperatures are shown. Note that the zero of the ordinate axis is offset by 0.002, 0.004, 0.006 and 0.008 for $T = 0.55$, 0.71 and 4.0, respectively. From Bennemann *et al.* [62].

nor can it serve to yield reliable estimates for T_{VF} [38, 59, 63-65]. Nevertheless, it is of interest to roughly estimate what physical cooling rates the cooling rate of the Molecular Dynamics (MD) simulation in fig. 7a correspond to. Recall that MD simply means that Newton's equations of motion (m being the mass of the effective monomers; one chooses $m = 1$ in "MD units")

$$m\frac{\mathrm{d}^2}{\mathrm{d}t^2}\mathbf{r}_i = -\frac{\partial U(\{\mathbf{r}_i\})}{\partial \mathbf{r}_i} \tag{25}$$

are solved numerically for the simulated many-particle system, $U(\{\mathbf{r}_i\})$ containing all the interaction potentials of the chosen model. Integrating the system of equations of motions (eq. (25)) numerically (*e.g.*, with the help of the "velocity Verlet algorithm" [57, 58,62]), the trajectory of the system through the phase space is generated (in practice one often also introduces a coupling to external thermostats and/or barostats, not written in eq. (25), in order to realize the NVT or NpT ensemble of statistical mechanics rather than the microcanonical NVE ensemble). This numerical integration requires a finite timestep δt, which here [62] had to be chosen as $\delta t = 0.002\tau_{\mathrm{MD}}$, $\tau_{\mathrm{MD}} = \sigma(m/\varepsilon)^{1/2}$ being the natural unit of time with the chosen units. From the fact that one effective monomer represents several successive chemical monomers along a chain, we can assume that $\tau_{\mathrm{MD}} \approx 10^{-11}$ s, and $T = 1$ may correspond to 500 K: remember that melting of the crystalline polymer occurs at about $T = 0.76$, and crystalline polyethylene melts at $T_{\mathrm{m}} \approx 413$ K, while short alkanes melt at somewhat lower temperatures. Accepting these very rough estimates, we conclude that the data of fig. 7a correspond to a cooling rate of $\gamma = \Delta T/\Delta t \approx 10^9$ K/s (note $\Delta t = 500000\,\delta t$ here). Thus it is clear that the cooling rates of atomistic simulations are many orders of magnitude faster than those used in real experiments!

For those temperatures where the system is still well equilibrated (*i.e.*, $T > T_{\rm g}$) we can fix the temperature (and the volume) and follow the mean-square displacements of both the monomers, and the particles as a function of time (fig. 9). We focus here on the mean-square displacement of inner monomers $g_1(t)$ and of the center of mass $g_3(t)$, which are defined by

$$g_1(t) = \langle [\mathbf{r}_i(t) - \mathbf{r}_i(0)]^2 \rangle, \qquad g_3(t) = \langle [\mathbf{r}_{\rm CM}(t) - \mathbf{r}_{\rm CM}(0)]^2 \rangle. \tag{26}$$

Here $\mathbf{r}_i(t)$ is the spatial coordinate of an inner monomer (rather than a monomer at a chain end) of a polymer, while $\mathbf{r}_{\rm CM}(t)$ is the spatial coordinate of the center of mass of a chain at time t. The average $\langle \cdots \rangle$ in eq. (26) is meant to include both the standard thermal average and an average over all inner monomers (or all chains, respectively) in the simulated system, in order to gain statistics. From fig. 9 we recognize qualitative evidence for the "cage effect" at temperatures where the polymer melt is supercooled: there occurs a regime of intermediate times where $g_1(t)$ and $g_3(t)$ increase with time only very little and particles "rattle around" in the cage formed by their neighbors. The time it takes until effective monomers can escape from their cages gets longer and longer the lower the temperature. Therefore the time it takes before the asymptotic Einstein relation $g_1(t) = g_3(t) = 6Dt$ sets in also gets longer and longer, and in turn the diffusion constant D rapidly decreases with decreasing temperature (fig. 10). It is seen that the simulation data is compatible with a Vogel-Fulcher relation, with $T_{\rm VF} = 0.34$. As expected, this temperature occurs well below $T_{\rm g}$ (where the kink in the $v(T)$ *vs.* T curve was observed, fig. 7).

In order to test for the predictions of MCT, the intermediate incoherent scattering function $\phi_q^s(t)$ was examined,

$$\phi_q^s(t) = \frac{1}{\mathcal{N}} \sum_{n=1}^{\mathcal{N}} \langle \exp\{ i\mathbf{q} \cdot [\mathbf{r}_n(t) - \mathbf{r}_n(0)]\} \rangle, \tag{27}$$

where the sum is extended over all $\mathcal{N}$ effective monomers in the system. Typical data are shown in fig. 11 [66, 67]. One can clearly recognize that $\phi_q^s(t)$ decays in two steps (the decay towards the plateau is termed "β-relaxation", the final decay is termed "α-relaxation"). The final decay is well compatible with the Kohlrausch-William-Watts [22, 23] (KWW) relaxation (eq. (5)). While the initial decay is a simple Gaussian behavior, controlled by the average mean-square displacements $g_0(t)$, on the timescale of the β-relaxation the behavior starts to become non-Gaussian. When one compares the decay at different temperatures, one finds clear evidence for the time-temperature superposition principle [2], as far as the α-relaxation is concerned. The α-relaxation time τ_q is found to show a power law behavior over almost two decades in $T - T_{\rm c}$ with $T_{\rm c} = 0.45 \pm 0.01$,

$$\tau_q \propto (T - T_{\rm c})^{-\Delta_q}, \quad \Delta_q \approx 2.09\,. \tag{28}$$

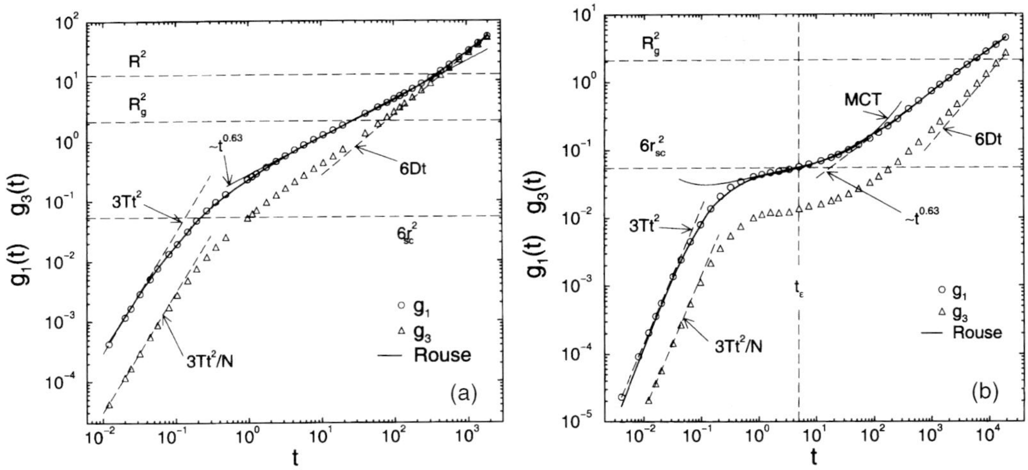

Fig. 9. – Mean-square displacements $g_1(t)$ of inner monomers in a chain (eq. (26)) and of the center of mass $g_3(t)$ of a chain (eq. (26)) plotted *vs.* time on a log-log plot, for $T = 1$ (a) and $T = 0.48$ (b). The thick solid line through $g_1(t)$ is computed from the Rouse model of polymer dynamics, using the correlation functions of the "Rouse modes" (eigenmodes in the coordinate system along the backbone of each chain) as observed in the simulation. The initial behavior for very short times (and extremely small displacements, over which the potentials due to the neighboring effective monomers are not yet felt!) is a free-flight ballistic motion, *i.e.* $g_1(t) = \langle v^2\rangle t^2 = 3Tt^2$ and $g_3(t) = 3Tt^2/N$, $\langle v^2\rangle$ being the mean-square thermal velocity of the particles and is indicated by dashed lines. Also the late-time diffusive behavior (Einstein relation $g_3(t) = 6Dt$ where D is the diffusion constant of the chains) is indicated by broken straight lines, while the Rouse-like power law behavior of the effective monomers at intermediate times ("anomalous diffusion" $g_1(t) \propto t^x$ with an effective exponent $x \approx 0.63 < 1$) is also shown. While at high temperatures ($T = 1$, case (a)) $g_1(t)$ crosses over smoothly from the ballistic regime into this Rouse-like regime, in the deeply supercooled melt ($T = 0.48$, case (b)) an intermediate regime occurs, where $g_1(t)$ becomes almost flat around a plateau at $6r_{\rm sc}^2$ (at a time-scale t_ε which characterizes the so-called "β-relaxation"). The length scale $r_{\rm sc}$ characterizes the typical excursions that an effective monomer can make which is confined in the "cage" of its nearest neighbors. In this intermediate regime, idealized MCT describes the simulation data reasonably well (for large t the thin curve shown would follow the so-called "von Schweidler" law, $g_1(t) \propto t^{0.75}$, which breaks down near $g_1(t) \approx 0.2$ where already a crossover to Rouse-like motion occurs). Note that the Einstein relation holds only if $g_3(t)$ exceeds the mean-square radius of gyration ($R_{\rm g}^2$), while also $g_1(t)$ follows the Einstein relation when $g_1(t) > R^2$, the mean-square end-to-end distance of the chains (with $R^2 = 6R_{\rm g}^2$ for Gaussian coils). From Baschnagel *et al.* [66].

Only for the lowest temperature, very close to $T_{\rm c}$, $T = 0.46$, clear deviations from eq. (28) were apparent, indicating that the divergence at $T_{\rm c}$ is rounded off. It turns out, however, that a Vogel-Fulcher law with $T_{\rm VF} \approx 0.34$ can be fitted to the data on τ_q about as well as the power law.

In eq. (28) an index q for Δ_q was used, because it was found that this exponent was distinctly q-dependent [69]; only for small q the exponent seems to set-

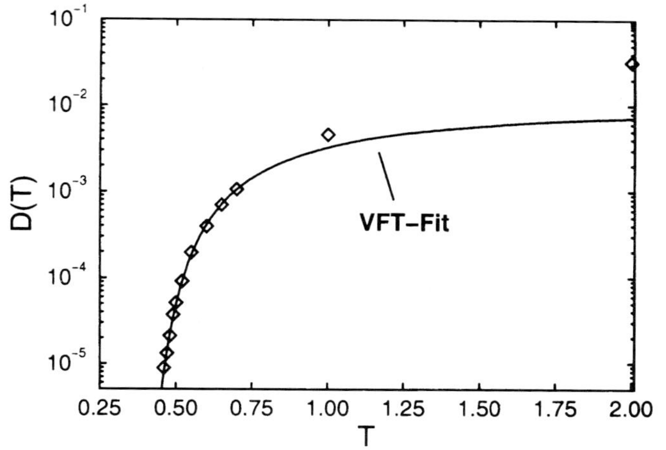

Fig. 10. – Self-diffusion constant $D(T)$ of the equilibrated supercooled polymer melt at pressure $p = 1$ plotted *vs.* temperature. For $T \leq 0.7$ the data can be fitted to a Vogel-Fulcher law, $D = D_0 \exp[-E_{\rm act}/(T - T_{\rm VF})]$, with $T_{\rm VF} \approx 0.34$. From Bennemann *et al.* [62].

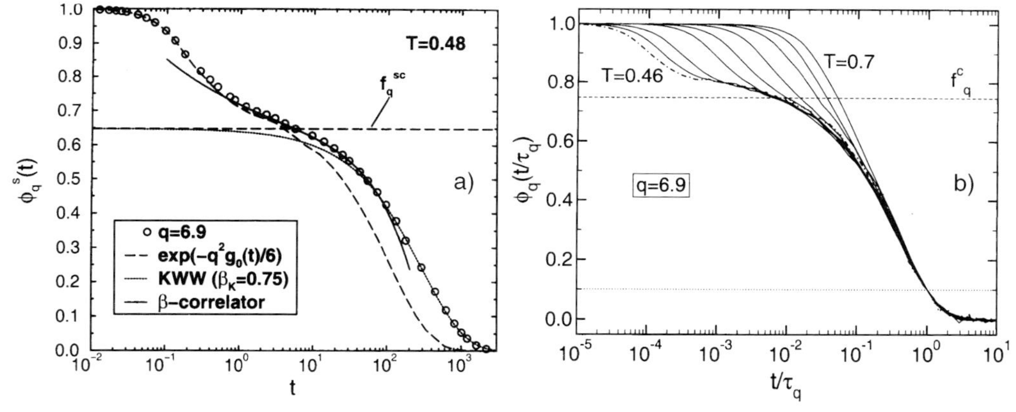

Fig. 11. – a) Comparison of simulation results for $\phi_q^s(t)$ (circles) for $T = 0.48$ and $q = 6.9$ (maximum of $S(q)$) with various theoretical approximations: a Gaussian approximation ($\phi_q^s(t) = \exp[-q^2 g_0(t)/6]$, $g_0(t)$ being the analog of $g_1(t)$ but including all monomers in the system), the MCT result for the β-relaxation regime, and the Kohlrausch-Williams-Watts (KWW) function [22,23] with $\beta_K = 0.75$. The nonergodicity parameter $f_q^{\rm sc}$ is indicated as a horizontal dashed line. From Baschnagel *et al.* [66]. (b) Testing the time-temperature-superposition principle of the α-relaxation for the coherent intermediate scattering function $\phi_q(t)$ at $q = 6.9$ (= maximum of $S(q)$). The temperatures investigated are: $T = 0.46, 0.47, 0.48, 0.50, 0.52, 0.6, 0.65$ and 0.7 (from left to right in the figure). The α-relaxation time τ_q was defined from the time it takes for this correlator to decay to 0.1, *i.e.* $\phi_q(t = \tau_q) = 0.1$, shown by the lower dotted horizontal straight line. The nonergodicity parameter f_q^c obtained from the analysis of the β-relaxation is shown as a dashed horizontal straight line as well. From Aichele and Baschnagel [67].

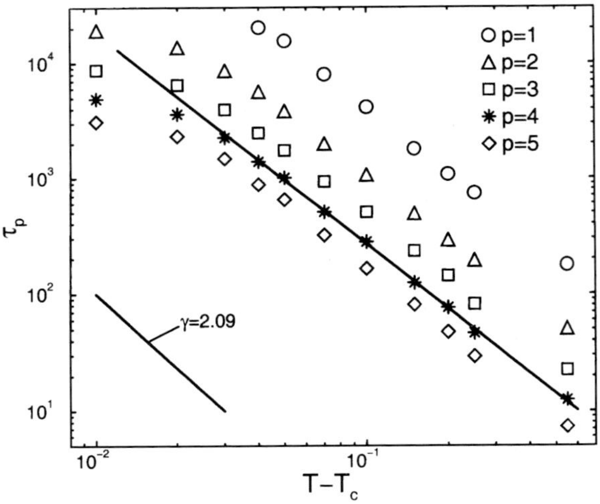

Fig. 12. – Log-log plot of the relaxation times τ_p of the first five Rouse modes plotted *versus* $T - T_c$, using $T_c = 0.45$, showing also a power law fit for $p = 4$ with $\Delta_0 = 1.83 \pm 0.02$. Within the error bars, this exponent provides a reasonable fit for all p shown. The straight line labelled $\gamma = 2.09$ indicates the exponent of eq. (28). From Bennemann *et al.* [68].

tle down at a q-independent value of about $\Delta_0 \approx 1.83$ (fig. 12). Here the relaxation times for the correlators $\langle \mathbf{X}_p(t) \cdot \mathbf{X}_p(0)\rangle / \langle \mathbf{X}_p(0) \cdot \mathbf{X}_p(0)\rangle$ for the "Rouse modes" $\{\mathbf{X}_p(t) = \sum_{n=1}^{N} \mathbf{r}_n(t) \cos[(n - 1/2)p\pi/N]/N\}$ are shown. While it is gratifying that many predictions of the "Rouse model" [70, 71] could be verified, *e.g.* one finds that $\tau_p D/R^2$ is essentially independent of temperature and depends on p proportional to $1/\sin^2(p\pi/2N)$ [68], it is disappointing that the simulations are unable to study the regime at and below T_c —this would already need considerably more computer resources than are available today. Therefore, we can see that the idealized MCT, which would imply a true divergence at T_c, is too simplified; but a significant test of the "extended" version of this theory (which predicts a smooth crossover from power law to Arrhenius-like behavior) is not possible. While the q-dependence of τ_q is not in accord with the theory either, this discrepancy between theory and simulations may well be due to the fact that data not close enough to T_c are analyzed. Of course, it is an inevitable problem of the idealized MCT that in order to confirm the predicted singularities one should study the region asymptotically close to T_c, but this in principle is not possible due to the rounding of the singular behavior. The extended version of the theory contains too many unknown parameters and therefore testing it with the present simulations is still elusive. Thus, although the theory is clearly valuable, as it presents a semi-quantitative picture of the cage effect, and also a unified qualitative description of the Rouse model and the cage effect has recently been achieved [72], there still remains a lot of work to be done.

1·3. *Testing the Gibbs-diMarzio theory with the bond fluctuation model.* – In the bond fluctuation model, a polymer chain is again represented by effective monomers connected by effective bonds, but now each effective monomer is an elementary cube at the simple cubic lattice, blocking all 8 sites at the corners of the cube from further occupation [35, 36]. The only nonbonded interaction between monomers is the excluded volume. The dynamics of the random conformational changes of the real polymers is represented in a crude way by attempting hops of randomly chosen monomers in randomly chosen lattice directions. If about one half of all lattice sites are occupied, the system behaves like a dense melt, and even short chains with chain length $N = 10$ already show the well-known random-walk–type properties of polymers, the gyration radius scales like $\sqrt{N}$, the single-chain static structure factor agrees with the Debye function [71], etc. Note that one allows bond lengths from 2 to $\sqrt{10}$ lattice spacings, and temperature enters only via a potential for the bond length [37-39]. Since real polymers show an increase of the persistence length and hence of their gyration radius with decreasing temperature, it is natural to energetically favor long bonds. So one chooses the potential in its minimum $U = 0$ for bond length $b = 3$, while $U = \varepsilon = 1$ for all other bond lengths [38]. This choice leads to "geometric frustration" [38] (fig. 13): each bond that reaches its ground state wastes a lot of free volume, since the four lattice sites between the adjoining effective monomers are blocked for further occupation by any other monomer. From the point of view of "packing" as many effective monomers as possible in a dense melt on the lattice, such bonds that waste lattice sites are clearly unfavorable: configurational entropy thus favors short bonds, which would not "waste" any lattice sites. Thus a conflict between entropy and energy is created, which is responsible for the glass transition. Note that when many bonds have already reached their ground state, it is quite likely that many of the remaining bonds are blocked in their motion. This mechanism is termed "geometric frustration". It is also clear that for this model a complicated (free) energy landscape results, depending on the geometrical arrangement of the chains and their blocked non-ground-state bonds on the lattice. In addition, one also expects that removal of "blockades" for bonds to reach their ground state will require large-scale cooperative rearrangements of many chains in their environment. Thus this model in a qualitative sense is clearly compatible with several of the basic ideas for the mechanism of glass transitions.

This model has the technical advantage that it can be equilibrated by the "slithering snake" algorithm [73-75]. In this type of Monte Carlo moves one randomly attempts to remove a bond from one chain end and attach it to the other chain end in a randomly chosen orientation. Although this algorithm does not correspond to any physically realistic dynamics of polymers, it is a perfectly admissible Monte Carlo move for studying equilibrium properties. Using this algorithm thermal equilibrium can be established at rather low temperature, such as $T = 0.16$, where after 10^7 Monte Carlo steps per monomer with the "random hopping" algorithm the autocorrelation of the end-to-end vector of the chains has not decayed yet below 90% of its starting value [76] (fig. 14). This correlation function is defined as

$$\phi_{\text{ete}} = \frac{\langle \mathbf{R}(t) \cdot \mathbf{R}(0) \rangle}{[\langle R^2 \rangle]} \tag{29}$$

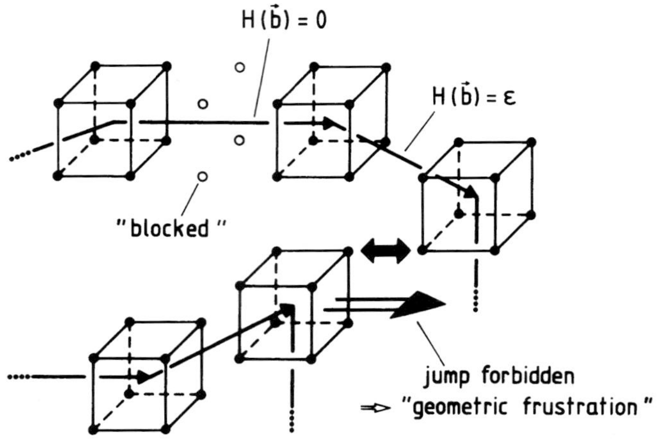

Fig. 13. – Sketch of a possible configuration of monomers belonging to two different chains in the bond fluctuation model of a polymer melt. For one monomer of the lower chain, an attempted move is indicated: this jump is forbidden, however, since it violates the excluded-volume constraint. The sketch also indicates the choice of a two-state energy function, namely $\mathcal{H}(\mathbf{b}) = 0$, if $\mathbf{b} = \mathbf{b}_{\min} = (0, 0, \pm 3)a$ (or a permutation thereof), where the lattice spacing a henceforth is chosen as $a \equiv 1$, and $\mathcal{H}(\mathbf{b}) = \varepsilon$ for all other bond vectors. Note that bonds which take a ground-state bond vector $\mathbf{b}_{\min}$ imply that four sites (shown as empty circles) are automatically blocked for further occupation. From Baschnagel *et al.* [38].

and we define an associated relaxation time via $\phi_{\text{ete}}(t = \tau_{\text{ete}}) = 0.2$. It is found that τ_{ete} is reasonably well compatible with the Vogel-Fulcher law, with $T_{\text{VF}} = 0.125$. Furthermore, when $\phi_{\text{ete}}(t)$ is replotted *vs.* the scaled time t/τ_{ete}, we find very good evidence once again for the time-temperature superposition principle (fig. 14b) and the KWW law, eq. (5), with $\beta_K \approx 0.69$. Similar results have been obtained for various other relaxation functions, including $S_q^s(t)$, the relaxation functions of the Rouse modes, etc. [76]. Various other criteria were used to extract several other relaxation times [76], and they all are compatible with the Vogel-Fulcher relation, eq. (1), and yield the same estimate for T_{VF} as τ_{ete}, within the statistical errors. However, also this model yields reasonably convincing evidence for the applicability of MCT [77, 78]. The estimate for T_{c} resulting from this analysis is $T_{\text{MC}} \approx 0.15$ [77]: Thus also in this model it is not possible to investigate the relaxation dynamics in the regime $T_{\text{VF}} < T \leq T_{\text{c}}$ by simulations. This is not really a surprise, however: estimating that the time unit of one attempted Monte Carlo step per monomer (MCS) corresponds to about 10^{-13} s [79], a run of 10^7 MCS would correspond to a physical time of 10^{-6} s, so also the Monte Carlo lattice model deals with times much shorter than are accessible in experiments.

Using the bond fluctuation model it was also possible to estimate how the glass transition temperature T_{g} depends on the chain length N [80]. The results are compatible with the law $T_{\text{g}}(\infty) - T_{\text{g}}(N) \propto 1/N$ found also experimentally [33], which is one of the most notable predictions of the theory of Gibbs and diMarzio [26]. Therefore, many

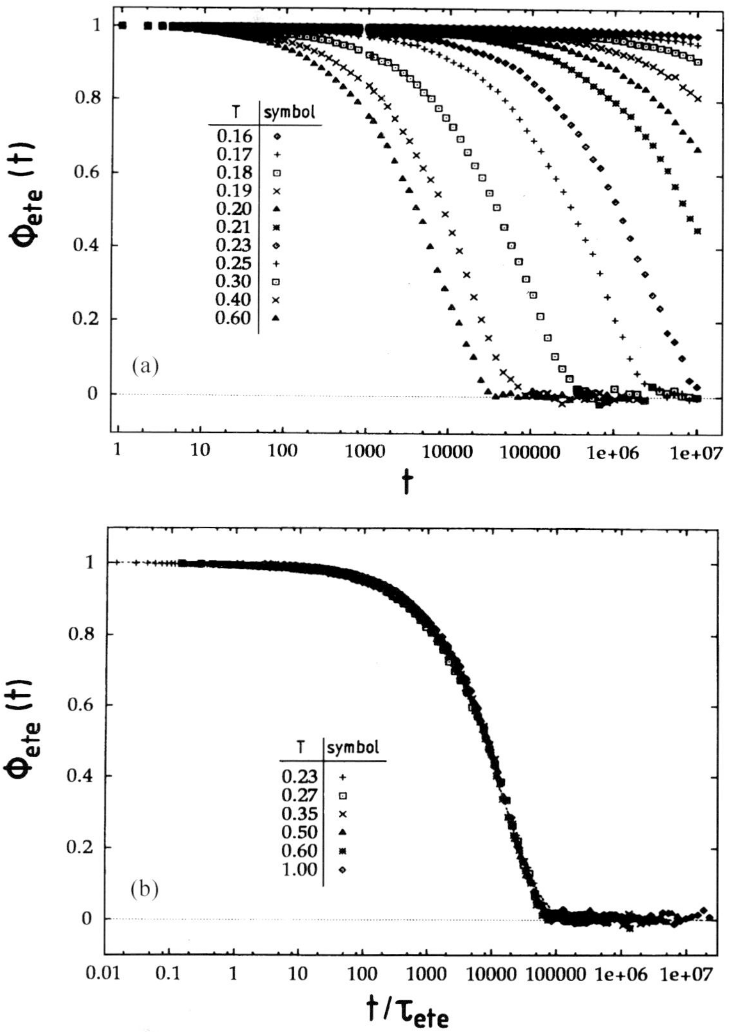

Fig. 14. – (a) Plot of the end-to-end vector correlation $\phi_{\text{ete}}(t)$ *vs.* the logarithm of time, for temperatures from $T = 0.16$ (the topmost curve) to $T = 0.60$ (lowest curve), for the three-dimensional bond fluctuation model. (b) Plot of $\phi_{\text{ete}}(t)$ *vs.* the logarithm of the rescaled time t/τ_{ete}. The dotted line underneath the data points is the KWW law, eq. (5). From Okun *et al.* [76].

experimentalists believe that this theory is valid. However, this conclusion is premature, as a study of the configurational entropy for the present lattice model shows (fig. 15) [83]. While the entropy does indeed decrease rather strongly with increasing values of the inverse temperature, starting out from an "athermal melt" (corresponding to infinite temperature), this decrease becomes slower when one approaches the vicinity of T_c,

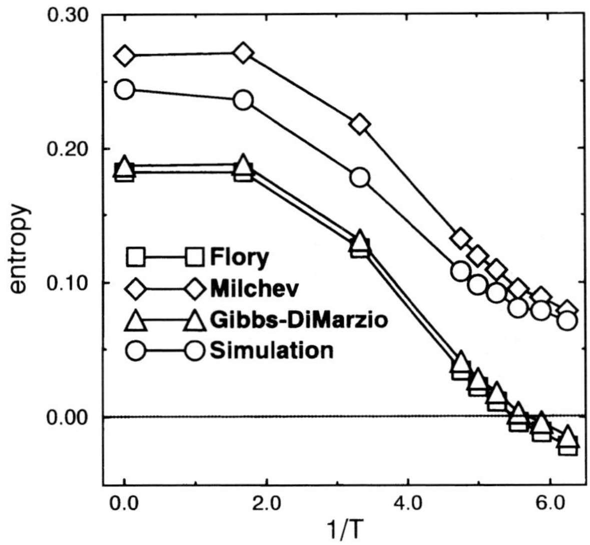

Fig. 15. – Comparison of the temperature dependence of the entropy per lattice site for the bond fluctuation model at a density $\phi = 0.5$ of occupied sites on the simple cubic lattice. Full dots show the simulation results, while the theoretical predictions of the theories of Gibbs and diMarzio [26], Flory [81] and Milchev [82] adapted to the present model are indicated by the open symbols. Recall that the estimates for T_c and T_{VF} for this model are $T_c \approx 0.15$ and $T_{VF} \approx 0.125$, respectively, while the Gibbs-diMarzio [26] and Flory [81] theories would predict a Kauzmann temperature $T_K \approx 0.18$, well in the fluid phase, and hence by no means are quantitatively reliable. From Wolfgardt *et al.* [83].

and the simulation data do not imply that the entropy vanishes (although they also cannot rule out a vanishing of the entropy at much lower temperatures than shown here). However, when one works out the theory of Gibbs and diMarizo [26] (or the related theory of Flory [81]) for the present lattice model, all the input parameters of the theory can also be extracted from the simulation, so there are no adjustable parameters whatsoever in the comparison of fig. 15.

Since the general trend of the actual entropy with temperature and the corresponding approximations of Flory [81] and Gibbs and diMarzio [26] is somewhat similar, apart from a constant shift along the ordinate axis, one suspects that the source of error in the analytical approximations is simply a bad approximation for the entropy of the athermal melt at $1/T \to 0$. In fact, this conjecture is corroborated by a somewhat different (but similarly plausible) approximation for the athermal melt entropy due to Milchev [82]. The latter theory [82] yields a slight overestimation of the simulation results at high temperatures, but is rather close to the simulation results at low temperatures. Since the theory of Milchev [82] (note that this work now is 20 years old, it was done long before the simulations!) yields a vanishing of the entropy at $T = 0$ only, while at low nonzero temperatures the entropy is small but positive, consistent with the general theoretical arguments of Stillinger [34], we consider it very likely that the bond fluctuation model actually does have a nonzero positive entropy at all nonzero temperatures. Of course, this does not rule out a transition where the $S(T)$ *vs.* T curve shows a kink singularity, as in the mean-field p-state Potts model with $p > 4$ (fig. 6): the Monte Carlo results in fig. 15 are not available at low enough temperatures to test this question.

1·4. *Some first conclusions; and what is necessary in order that a system be a good glass former?* – So far we have seen that glass-forming fluids exhibit a remarkable smooth increase of the structural relaxation time over many orders of magnitude, without an accompanying pronounced change of static structural properties, and hence the reason for this dramatic slowing-down is still unclear. The most appealing picture to explain the glass transition is the "random first-order transition" [54-56] scenario, as it is realized in the mean-field Potts glass (with $p > 4$ states) and related models: these models exhibit two transitions, a dynamical ergodic-to-nonergodic transition at a temperature T_{D} where no static quantity shows any anomaly, and a static transition at a temperature $T_0 < T_{\mathrm{D}}$, where the glass order parameter appears discontinuously, but there is no latent heat (the entropy has a kink only). The static transition then also presents the occurrence of an "entropy catastrophe" (at a Kauzmann temperature T_{K} the (configurational) entropy would become negative, but $T_0 > T_{\mathrm{K}}$ and hence this does not happen). The intriguing feature of this scenario is that the spin autocorrelation functions at the dynamic transition are described by precisely the same mode-coupling equations as those which had previously been proposed for structural glasses. This mode-coupling theory is in fact in rather good agreement both with various experiments (although also experiments exist which do not provide much evidence for MCT) and with simulations of various coarse-grained models. One such model, a bead-spring model for flexible polymers in the melt, has been described here in some detail. One important fact about structural glasses is that the dynamical transition (where in the ideal case the phase space splits in many valleys separated by infinitely high free-energy barriers) is rounded, *i.e.*, all what occurs is crossover from a power law divergence of the structural relaxation time to a much slower relaxation (tentatively this could be interpreted by "valleys" in phase space separated by barriers of a large but finite height, requiring thermally activated processes to overcome them). The question whether the static transition remains sharp or is also rounded is completely open: experiments and simulations cannot answer this question, since the expected transition temperature T_0 of this transition falls below the empirical glass transition temperature T_{g}, where the system falls out of equilibrium. There is one model calculation for a structural glass that claims to show such a static transition (by calculating a Kauzmann temperature $T_{\mathrm{K}} > 0$, where the configurational entropy of a lattice model of semiflexible polymers vanishes) due to Gibbs and diMarzio [26]. However, tests of this theory by Monte Carlo simulations of the bond fluctuation model show that this theory is very inaccurate; the predicted Kauzmann temperature is simply an artefact of inappropriate approximations.

While the simulations described here hence do not answer the question whether or not there is a true phase transition from a (metastable) undercooled fluid to a (metastable) ideal glass (the true stable solid is the crystalline state, of course), they give valuable insight into the mechanism that makes a fluid a good glass former (not all fluids are glass formers, *e.g.* fluids such as liquid argon do not form glasses, and also in simulations of simple Lennard-Jones fluids with a single kind of atoms crystallization cannot be avoided even for the rather fast cooling usually done in simulations). The simulations described here suggest that there must be some competing mechanisms in the system: *e.g.*, in the bead-spring model the preferred distances of beads along the chains and of nonbonded beads were different, preventing the formation of simple crystal structures which could be reached by small rearrangements of the bead positions. Similarly, in the bond fluctuation model the energetic preference for "long" bonds and the entropic preference for dense packing of effective monomers are in conflict with each other, leading to what has been termed "geometric frustration". It is plausible that related mechanisms are operative for real glass-forming fluids, too.

2. – Molten silicondioxide and its mixtures with other oxides: the case of strong glass formers

2˙1. *Fragile-vs.-strong glass formers.* – In sect. **1** it was emphasized that the generic behavior of the structural relaxation time τ is described by fig. 5, *i.e.* on a plot of $\ln \tau$ *vs.* inverse temperature $1/T$ we observe a concave function, compatible with the Vogel-Fulcher relation, eq. (1), and/or a crossover from the power law of mode-coupling theory (MCT) to a thermally activated (Arrhenius) behavior. However, a glance on the Angell plot (fig. 3) immediately shows that not all glasses do behave like this: the archetypical representative of a glass-forming fluid, molten SiO_2, and related materials show a simple Arrhenius behavior $\ln \tau \propto 1/T$, over the entire temperature range that is experimentally accessible. Such fluids are called "strong glass formers", while other materials that show pronounced curvature on the Angell plot are called "fragile glass formers" [84]. Does this distinctive behavior in the Angell plot mean that there exist different classes of glass-forming fluids, for which then a qualitatively different type of theoretical description is required? In order to answer this question, we look into the actual behavior of molten and glassy SiO_2 and related materials, namely its mixtures with Na_2O and Al_2O_3 more closely, presenting some recent extensive Molecular Dynamics (MD) results [12,65,85-101] in conjunction with pertinent experiments.

2˙2. *The BKS potential.* – In a MD simulation, as is well known, the statistical mechanics of condensed matter is reduced to averages along trajectories through the phase space of the chosen model system, generating these trajectories simply from Newton's equations of motion, using simple effective potentials for the interactions between the atoms. Thus, quantum-mechanical effects are neglected from the outset, and since in the considered systems the character of the bonding varies from ionic to covalent, it is not *a priori* clear that a sufficiently realistic description will be possible. However, it is perhaps a lucky accident that for SiO_2 a rather accurate description in terms of a simple pair potential is possible, proposed by van Beest *et al.* [102]. This potential was later generalized by Kramer *et al.* [103] to systems containing Na and Al. These potentials were derived from a parameterization of quantum-chemical calculations of small silica(-te) clusters, and the parameters were optimized to correctly reproduce some properties of bulk crystalline materials. Although some limitations of this potential to reproduce very small-scale structures were found [104-107], it describes both the properties of crystalline materials [108-111] and of glasses and melts [12, 65, 85-101] rather well.

This so-called BKS potential uses pseudo-Coulomb interactions between the ions and the Buckingham potential to describe the short-range part,

$$\phi_{\alpha\beta}(r) = \frac{q_\alpha q_\beta}{r}\, e^2 + A_{\alpha\beta}\exp[-B_{\alpha\beta} r] - \frac{C_{\alpha\beta}}{r^6}\,, \tag{30}$$

where $\alpha, \beta \in [\mathrm{Si}, \mathrm{O}, \mathrm{Na}, \mathrm{Al}]$, r being the distance between an ion of species α and an ion of species β, and e is the elementary charge. The values of the parameters $A_{\alpha,\beta}$, $B_{\alpha,\beta}$ and $C_{\alpha\beta}$ can be found in [102, 103]. The effective charges proposed for Si and O are $q_{\mathrm{Si}} = 2.4$, $q_{\mathrm{O}} = -1.2$, whereas for sodium and aluminium effective charges $q_{\mathrm{NA}} = 1.0$, $q_{\mathrm{Al}} = 1.9$ were proposed [103]. While for pure SiO_2 charge neutrality of the total system is fulfilled, since $q_{\mathrm{Si}} = 2q_{\mathrm{O}}$, this is not the case for mixtures such as $(Na_2O)x(SiO_2)$, $x = 2, 3, 5$ (abbreviated as NSx in the following), that will be considered below, or $(Al_2O_3)2(SiO_2)$. Therefore for Na and Al ions the above potential was slightly modified, by introducing somewhat *ad hoc* distance-dependent charges $q_\alpha(r)$ for Na and Al that ensure charge neutrality [91, 101]. *E.g.*, for Na the following function was used:

$$q_{\mathrm{Na}}(r) = \begin{cases} \widetilde{q}_{\mathrm{Na}}\{1 + \ln\left[C_{\mathrm{Na}}(r_{\mathrm{Na}} - r)^2 + 1\right]\}, & r > r_{\mathrm{Na}}\,, \\ \widetilde{q}_{\mathrm{Na}}\,, \quad r < r_{\mathrm{Na}}\,, \end{cases} \tag{31}$$

with $\widetilde{q}_{\mathrm{Na}} = 0.6 = -q_{\mathrm{O}}/2$, as it should be, and the parameter $r_{\mathrm{Na}} = 4.9\,\text{Å}$, $C_{\mathrm{Na}} = 0.926\,\text{Å}^{-2}$. Equation (31) for small distances reproduces almost perfectly the forces from the original potential, eq. (30), and furthermore many physical properties of the system are rather well described. The main merit of this potential is that purely with pairwise potentials with carefully chosen parameters in effect one simulates the presence of directional covalent bonds.

2·3. *Producing glasses by computer simulation.* – In spite of the convenient simplicity of this potential, MD simulations of this model are nevertheless rather demanding: the long range of the Coulomb interactions makes the use of Ewald summation techniques necessary, when we simulate a cubic box with periodic boundary conditions. In addition, the time step δt of the integration algorithm (the "velocity Verlet" algorithm is used [112, 113]) has to be rather small to ensure that the errors in the integration of Newton's equations of motion remains negligible, $\delta t = 1.6\,\mathrm{fs}$.

In a first set of simulations [65], simply the cooling schedule used in the industrial glass production has been imitated: one starts at a high initial temperature T_i and cools the system, linearly decreasing the temperature with increasing time $t\{T(t) = T_i - \gamma t\}$ until the desired temperature is reached. The difference from reality is that T_i is unphysically high (*e.g.*, $T_i = 7000\,\mathrm{K}$ [65]) and the cooling rate γ is many orders of magnitude larger than in the experiment (in [65] γ ranged from $4.44 \cdot 10^{12}\,\mathrm{K/s}$ up to $1.14 \cdot 10^{15}\,\mathrm{K/s}$, while in experiment $\gamma = 1\,\mathrm{K/s}$ is typical). In fact, nevertheless the configurations generated in this way look rather reasonable: amorphous SiO_2 is an irregular network of tetrahedra, such that in the center of each tetrahedron one has a Si atom, with four oxygens at the corners, and always two tetrahedra share an oxygen, so the stoichiometry works out, and no "dangling bonds" (oxygens bound to a single Si ion) occur in the network. The orientation of the tetrahedra relative to each other is rather random, however. This type of continuous random network appears for all cooling rates, and in the early literature on the subject it was even claimed that the obtained structure does not depend on the cooling rate. Of course, such a claim is foolish, and if one studies the problem more carefully, as Vollmayr *et al.* [65] did, one finds a distinct dependence of physical properties on the cooling rate: the distribution of the angle Θ between Si-O bonds in the tetrahedra sharpens up, as the cooling rate gets smaller, and the peak position moves towards the angle of an ideal regular tetrahedron. Also the enthalpy in the final state at zero temperature gets lower and lower as the cooling gets slower. Qualitatively, this is expected: when one cools fast, the system gets trapped in metastable states, and when one gives more time to the system to explore its phase space during the cooling, lower-lying states are found. But the functional form for this cooling rate dependence is not clear. In particular, for the density in the final state there are also appreciable statistical errors, and there is no way to predict the density of real glasses accurately from such methods. Particularly interesting are the properties of topological defects in the network structure, which one can identify from the so-called ring statistics: a ring is defined by moving from a silicon atom back to itself along covalent bonds via the oxygen atoms and neighboring silicon atoms, using each bond only once, and taking the smallest possible loop. In crystalline structures of SiO_2 even-membered rings ($n = 4, 6, 8$) occur in various periodic arrangements.

In a glass, however, also odd-membered rings occur, such as $n = 5$ and $n = 7$. The distribution of ring lengths is also strongly cooling rate dependent, and it is possible that in the real glass only rings with $n = 5, 6, 7, 8$ occur, while rings with $n = 3$ and $n = 4$ are artefacts of fast cooling. In the gauge theories of glasses [114] the "odd lines" which connect the centers of the odd-membered rings are the basic topological objects of the corresponding field-theoretic description (unfortunately, this approach does not lead to any testable predictions, at least to the knowledge of this author, and hence is not followed up here).

Since there is no real possibility to predict the density reliably, because of the cooling rate dependence, it was found to be more useful to simply fix the density in beforehand, at an experimentally reasonable value, thus performing the simulation at constant density rather than constant pressure. In practice a cubic box of edge length 48.37 Å was used, containing in total 8016 atoms. With this large number of atoms, the program is quite slow; in order to equilibrate the SiO_2 melt at $T = 2750\,\mathrm{K}$, one needs to run up to 20 ns, and this is only possible using rather powerful parallel computers such as CRAY T3E.

Could one have gained by simply running the programm for much smaller systems? The answer to this question is no, because of finite-size effects [85]! If one looks at static properties only, one does not see any effects in quantities such as partial structure factors, and one might think that as few as $N = 336$ atoms are enough. However, dynamic quantities such as the intermediate incoherent dynamic structure factor $F_s(q,t)$ show size effects very clearly [85], and only $N = 8016$ just is large enough! Thus, the computer simulation of glass-forming fluids is full of subtle pitfalls.

Anyway, careful equilibration of systems with $N = 8016$ atoms at $T = 2750\,\mathrm{K}$, which are then cooled (with a cooling rate of the order of $10^{12}\,\mathrm{K/s}$) to room temperature, does yield very useful results, as a comparison with numerous experimental data has shown [87-90, 92-94]. As an example, fig. 16 [88, 115] shows the static structure factor $S(q)$ and compares it to neutron scattering data. Here $S(q)$ is defined (for neutron scattering) as

$$S^{\mathrm{neu}}(q) = \left(N_{\mathrm{Si}}b_{\mathrm{Si}}^2 + N_{\mathrm{O}}b_{\mathrm{O}}^2\right)^{-1} \sum_{i,j} b_i b_j \left\langle \exp[i\mathbf{q}\cdot[\mathbf{r}_j(0) - \mathbf{r}_i(0)]]\right\rangle, \tag{32}$$

$b_\alpha(\alpha = \mathrm{Si}, \mathrm{O})$ being the (known) scattering lengths, $\mathbf{q}$ the scattering vector, and $\mathbf{r}_i(t)$ the position of the i-th particle at time t. Since there are no adjustable parameters whatsoever, the good agreement seen in fig. 16 is significant. Also, the temperature dependence of the specific heat [89] (fig. 17) is in good agreement with the corresponding experimental data [116-118]. Note that the specific heat per atom is not calculated directly from the simulation (by either taking the temperature derivative of the internal energy, $C_V = (\partial(\langle\mathcal{H}\rangle/N)\partial T)_V$, or from the fluctuation formula, $C_V = (\langle\mathcal{H}^2\rangle - \langle\mathcal{H}\rangle^2)/(Nk_{\mathrm{B}}T^2)$ $\mathcal{H}$ being the Hamiltonian of the system): both methods would yield the Dulong-Petit law, $C_V = 3k_{\mathrm{B}}$, as expected for all harmonic solids at low temperatures treated in the framework of classical statistical mechanics. Recall that MD is based on classical mechanics —how can we manage to get a result that complies with the third law, $C_V \to 0$ as $T \to 0$, which is a result of quantum-statistical mechanics? The answer is that for harmonic solids one can put in quantum mechanics "by hand" if one uses the density of states $g(\omega)$ for the harmonic frequencies ω together with the Bose-Einstein statistics, which yields the mean occupation number of energy $\hbar\omega$ as $[\exp(\hbar\omega/k_BT) - 1]^{-1}$, $\hbar$ being Planck's constant.

Thus

$$C_V = \frac{\hbar^2}{k_{\mathrm{B}}T^2} \int_0^\infty \frac{\omega^2 \exp(\hbar\omega/k_{\mathrm{B}}T)}{[\exp(\hbar\omega/k_{\mathrm{B}}T) - 1]^2}\, g(\omega)\mathrm{d}\omega\,. \tag{33}$$

The density of states $g(\omega)$ is calculated at $T = 30\,\mathrm{K}$, where it is a very good approximation to treat our classical frozen-in SiO_2 melt as a harmonic solid. One thus neglects the famous "two-level systems" (TLS), which are expected to dominate C_V at extremely low temperatures [120, 121], leading to $C_V \propto T$ rather than the Debye law $C_V \propto T^3$ that results from eq. (33).

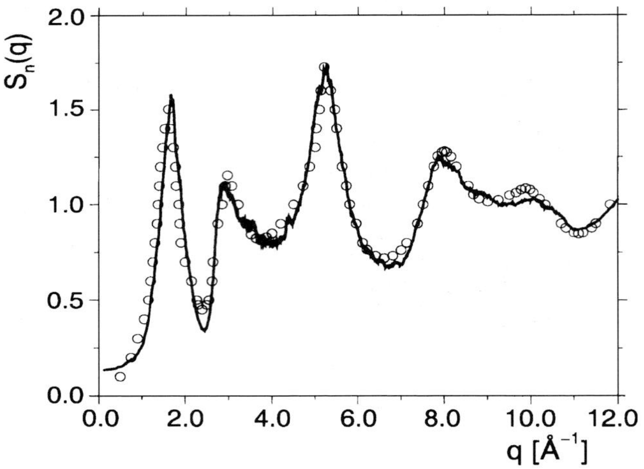

Fig. 16. – Static neutron structure factor of SiO_2 at room temperature ($T = 300\,\mathrm{K}$) plotted *vs.* wave number q. The full curve is the MD simulation, using the experimental neutron scattering lengths for Si and O atoms, while the symbols are the neutron scattering data of Price and Carpenter [115]. From Horbach and Kob [88].

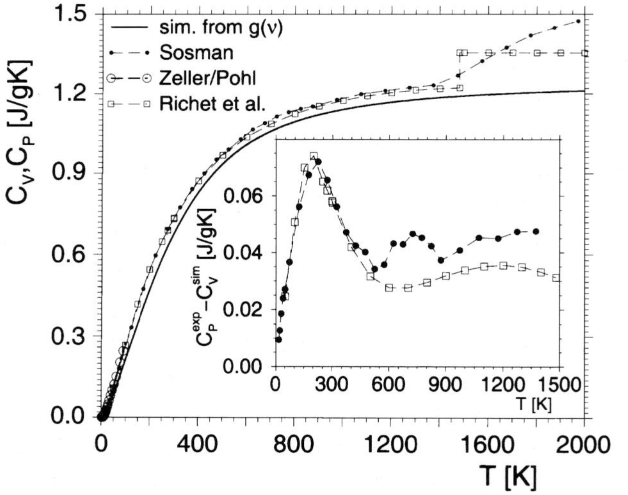

Fig. 17. – Temperature dependence of the specific heat at constant volume as predicted by the harmonic approximation (solid line). The symbols are experimental data for the specific heat at constance pressure from Sosman [116], from Zeller and Pohl [117], and from Richet *et al.* [118]. The inset shows the difference between the data of Richet *et al.* and of Sosman and the simulation data. From Horbach *et al.* [89].

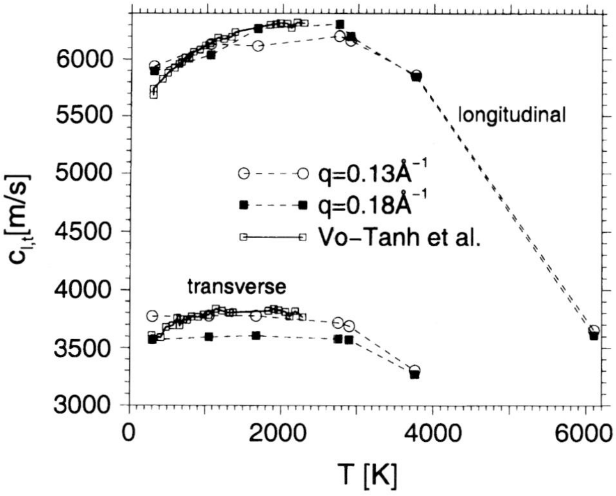

Fig. 18. – Temperature dependence of the (effective) longitudinal (C_ℓ) and transverse (C_t) sound velocities, determined from the maxima of the corresponding longitudinal and transverse current correlation functions at $q = 0.13\,Å^{-1}$ (filled squares). Also included are the experimental data of Vo-Tanh *et al.* [119] which were multiplied by the factor $(2.2/2.37)^{1/2}$ since the simulation was done at a density of $\rho_{\rm sim} = 2.37\,{\rm g/cm^3}$ while the experiment was done for $\rho_{\rm exp} = 2.2\,{\rm g/cm^3}$. From Horbach *et al.* [90].

One sees from fig. 17 that the harmonic approximation reproduces the experimental specific heat rather closely, the difference being only a few percent. Of course, the harmonic approximation cannot describe the step in the specific heat when the glass melts into the fluid. One also may question the comparison in fig. 17 since the data refer to C_p, the specific heat at constant pressure: but in SiO_2 the difference $C_p - C_V$ is much smaller than the discrepancies seen in fig. 17. Of course, it would be premature to expect that a simple potential model such as eq. (30) could reproduce all physical properties completely correctly, and hence small discrepancies as seen in fig. 17 are no surprise at all.

Figure 18 compares the temperature dependence of the (effective) longitudinal (C_ℓ) and transverse (C_t) sound velocities estimated from the simulation with corresponding experiments [119]. Note that strictly speaking there is no transverse sound (in the long-wavelength limit $q \to 0$) in any fluid, since fluids cannot resist a shear force: but sound-like propagating transverse modes can exist at nonzero q. Once more good agreement between the simulation [90] and corresponding experimental data [119] is found, although there do not occur any adjustable parameters whatsoever. Thus we conclude that the model, eq. (30), is good enough to merit a study of the slowing-down of SiO_2 melts as they are cooled down towards their glass transition.

2·4. *Dynamic properties of* SiO_2 *melts: diffusion and its mechanism, viscosity.* – Using once more the Einstein relation between the mean-square displacements of particles of

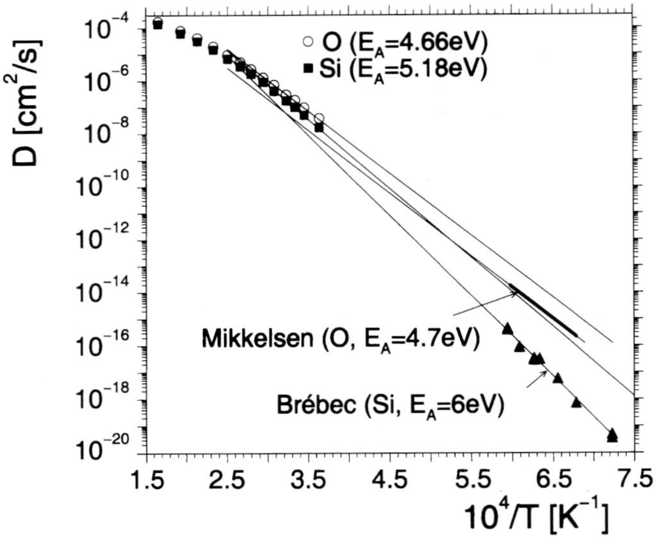

Fig. 19. – Plot of the self-diffusion constant D of silicon (Si) and oxygen (O) atoms in molten SiO_2 *vs.* inverse temperature. The open circles and full squares are MD results, and experimental data due to Mikkelsen [122] and Brebec *et al.* [123] are also included. The thin straight lines show simple Arrhenius behavior ($D \propto \exp[-E_{\mathrm{A}}/k_{\mathrm{B}}T]$) with various choices of the activation energy E_{A}, as indicated in the figure. The vertical broken lines indicate the experimental glass transition temperature, $T_{\mathrm{g}} = 1450\,\mathrm{K}$, as well as values for T_{g} that one obtains if one extrapolates the data from the simulations to low temperatures and then estimates T_{g} from the experimental value of the O diffusion constant ($D_{\mathrm{O}}(T = T_{\mathrm{g}}^{\mathrm{sim}}) = 10^{-16}\,\mathrm{cm}^2/\mathrm{s} \Rightarrow T_{\mathrm{g}}^{\mathrm{sim}} = 1381\,\mathrm{K}$) or the Si diffusion constant, respectively ($D_{\mathrm{Si}}(T = T_{\mathrm{g}}^{\mathrm{sim}}) = 5 \cdot 10^{-19}\,\mathrm{cm}^2\,\mathrm{s} \Rightarrow T_{\mathrm{g}}^{\mathrm{sim}} = 1303\,\mathrm{K}$). From Horbach and Kob [88].

species α and the respective self-diffusion constants D_α, $\lim_{t\to\infty}\langle[\mathbf{r}_i(t) - \mathbf{r}_j(0)]^2\rangle = 6Dt$, it is relatively straightforward to obtain both D_{O} and D_{Si} of SiO_2 melts at high enough temperatures ($T \geq 2750\,\mathrm{K}$) from the simulation (fig. 19 [88]). If we include experimental data [122, 123], we need to span 16 orders of magnitude, and it is clear that a wide gap exists between the lowest temperature accessible for the simulation ($T = 2750\,\mathrm{K}$) and the highest temperature for which experimental data exist. Nevertheless the activation energies E_A^{Si}, E_A^{O} that one can extract from the simulation are in rough agreement with their experimental counterparts.

It is also interesting to note that at very high temperatures there is distinct curvature in the $\ell n D$ *vs.* $1/T$ plot. As will be discussed later, in this high-temperature regime there is a region where mode-coupling theory (MCT) [4] works, *i.e.* one sees a crossover from a power law vanishing of $D (D \propto (T - T_{\mathrm{c}})^{\Delta}$, with $T_{\mathrm{c}} \approx 3300\,\mathrm{K}$ [88]), to the Arrhenius behavior emphasized in fig. 19.

Since we know that in the glass there is a rigid network of covalent bonds (silicon atoms in the centers of tetrahedra, oxygen atoms at the corners, shared between two neighboring tetrahedra), it is of interest to ask what is left from this network in the fluid. Defining "bonds" from a neighboring criterion (atoms being at the distance around the maximum of the pair correlation function $g_{\mathrm{SiO}}(r)$ up to the first minimum), we find that bonds exist also in the fluid, but they have a finite lifetime τ_{SiO}, they open up and close again. It is found that this breaking of bonds is responsible for the oxygen diffusion, since the product $D_{\mathrm{O}}\,\tau_{\mathrm{SiO}}$ is independent of temperature. When a bond breaks, typically one Si in the center of a tetrahedron has only 3 oxygen neighbors, while another Si nearby has 5 rather than 4 oxygen neighbors: the diffusion mechanism in the SiO_2 melt near T_{g} hence can be viewed as a kind of Frenkel defect pair mechanism as in crystals (there diffusion occurs via the creation of vacancy/interstitial pairs): defects in the Si coordination (3- and 5-fold coordinated silicon atoms) diffuse through the continuous random network. Of course, just as the Frenkel defect pair mechanism means that diffusion also occurs in a crystal, it is clear that diffusion also occurs in the glass. Therefore, it is not much of a surprise that some structural relaxation is still possible in the glass for $T < T_{\mathrm{g}}$, leading to the "aging" of physical properties [124].

Next we turn to the shear viscosity $\eta(T)$, which in the framework of a MD simulation can be extracted from the Kubo formula

$$\eta(T) = (k_{\mathrm{B}}TV)^{-1} \int_0^\infty \mathrm{d}t \left\langle \Pi_{\alpha\beta}(t)\Pi_{\alpha\beta}(0) \right\rangle, \quad \alpha \neq \beta (\alpha = x, y, z), \tag{34}$$

where $\Pi_{\alpha\beta}(t)$ is the pressure tensor (u_i^α being the α-component of the velocity of the particle i, and $\mathbf{F}_{ij}$ the force between particles i and j),

$$\Pi_{\alpha\beta}(t) = \sum_{i=1}^{N} m_i u_i^\alpha u_i^\beta + \sum_{i=1}^{N} \sum_{j>i} F_{ij}^\alpha r_{ij}^\beta . \tag{35}$$

Figure 20 shows corresponding results [88] together with the corresponding experimental data [125]. Again the simulations are roughly compatible with the corresponding experiment (but a gap in temperature where neither MD nor experimental results exist). And again we see that at high enough temperatures the plot $\ln\eta$ *vs.* $1/T$ is curved: as we will argue below, the system is compatible with MCT at $T > 3330\,\mathrm{K}$ (where $\eta(T = T_{\mathrm{c}}) \approx 8$ poise), and hence there is no difference in principle between strong and fragile glasses, but only a quantitative one: the larger $E_{\mathrm{A}}/k_{\mathrm{B}}T$ is, the steeper the slope of $\mathrm{d}(\ln\eta)/\mathrm{d}(T_{\mathrm{g}}/T)|_{T_{\mathrm{g}}=T}$, *i.e.* the more "fragile" the glass-forming fluid is, the closer together T_{c} and T_{g} must be. For SiO_2, $T_{\mathrm{c}} > 2T_{\mathrm{g}}$, T_{c} even exceeds the melting temperature, and hence experimentally no evidence whatsoever for MCT could be obtained.

2·5. *Evidence for MCT in* SiO_2 *melts.* – The assertion [86, 87, 93, 126, 127] that MCT works even for SiO_2, the prototype of a strong glass former, but at sufficiently high temperatures, has clearly been a completely unexpected and surprising finding of the simulations. We hence discuss here some of the key observations.

Figure 21 shows a plot of the incoherent intermediate scattering function $F_{\mathrm{s}}(q,t)$ *vs.* scaled time: one finds the typical two-step relaxation, just as for fragile glass formers

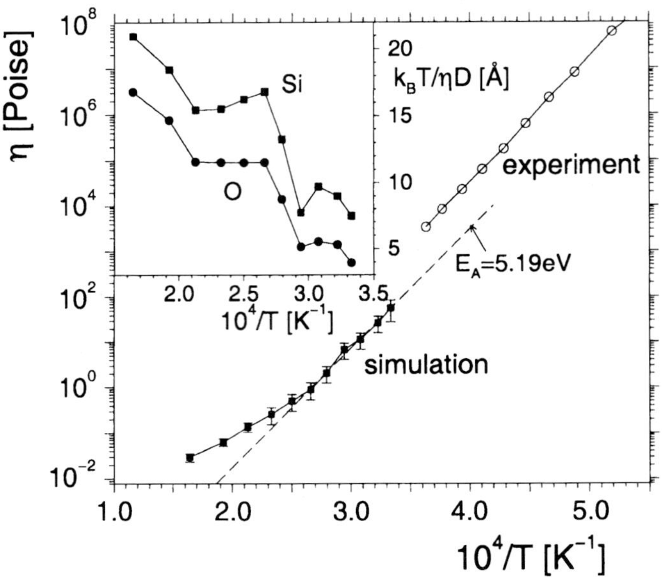

Fig. 20. – MD results for the viscosity of SiO_2 plotted *vs.* $1/T$. The dashed straight line indicates an Arrhenius fit with an activation energy $E_A = 5.19\,\text{eV}$. Experimental data of Urbain [125] are compatible with this value but would suggest a slightly different pre-exponential factor. Note that for SiO_2 the Stokes-Einstein relation $k_BT/(\eta D)$ = const holds neither for Si nor for O atoms. From Horbach and Kob [88].

(see fig. 11 in sect. **1**), the time-temperature superposition principle holds for the second step (the α-relaxation), and there also the stretched exponential decay (KWW relaxation function) is again found. However, the details of the onset of β-relaxation (the plateau-like region) are slightly different.

To provide a more stringent test of MCT, independent of "fitting" any parameters of the theory to the MD results, we note that in the β-relaxation regime MCT predicts a factorization property for all correlators [4, 30],

$$\phi(\mathbf{q}, t) = F_c(\mathbf{q}) + h(\mathbf{q})G(t), \tag{36}$$

$F_c(\mathbf{q})$ being the "nonergodicity parameter" for the considered quantity, $h(\mathbf{q})$ some time-independent amplitude function, and $G(t)$ depends on time (and temperature, of course) only. This property suggests to consider a function

$$R_\ell(t) \equiv \frac{[\phi(\mathbf{q}, t) - \phi(\mathbf{q}, t')]}{[\phi(\mathbf{q}, t'') - \phi(\mathbf{q}, t')]}; \tag{37}$$

if all times t, t', t'' lie in the β-regime where eq. (36) holds, the function $R_\ell(t)$ must be independent of q; *i.e.*, constructing $R_\ell(t)$ for various q all curves should superimpose in

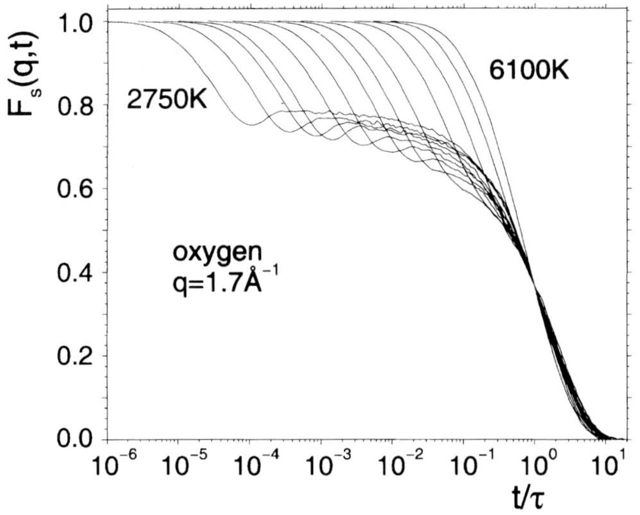

Fig. 21. – Incoherent intermediate scattering function $F_s(q,t)$ plotted *vs.* t/τ for Si atoms in SiO_2 melts at $q = 1.7\,\text{Å}^{-1}$, where the static structure factor $S(q)$ has its peak. Here τ is defined via $F_s(q, t = \tau) = 1/e$. Temperatures included a range from $T = 6100\,\text{K}$ down to $T = 2900\,\text{K}$. From Horbach *et al.* [86].

the β-regime! In fact, one finds that this property is true already for one decade in time for SiO_2; at $T = 4000\,\text{K}$ and at lower temperatures the factorization property holds for two decades and more [127].

Having identified the regime of times where eq. (36) holds and using the information on the function $G(t)$ that one can extract in this way, one can return to the original data on $\phi(\mathbf{q}, t)$ and extract estimates for $F_c(\mathbf{q})$ from the simulations. Using now static input from the simulations in the MCT equations, one can predict these nonergodicity parameters from the theory, without any adjustable parameters whatsoever!

Figure 22 shows the outcome of this test, using the coherent intermediate scattering function

$$F_{\alpha\beta}(q,t) = \frac{S^{\alpha\beta}_{\text{coh}}(q,t)}{S^{\alpha\beta}(q)}, \quad \alpha, \beta \in (\text{Si}, \text{O}), \tag{38}$$

where

$$S^{\alpha\beta}_{\text{coh}}(\mathbf{q},t) = \frac{1}{N_\alpha} \sum_{i\in\alpha} \sum_{j\in\beta} \big\langle \exp[i\mathbf{q} \cdot [\mathbf{r}_i(0) - \mathbf{r}_j(t)]] \big\rangle. \tag{39}$$

When one uses a simplified version of MCT (which uses two-body static correlators only, denoted as c_2-MCT in fig. 22), the agreement is qualitative only. However, using also the proper three-body correlators (denoted as c_3-MCT in fig. 22), there is almost perfect

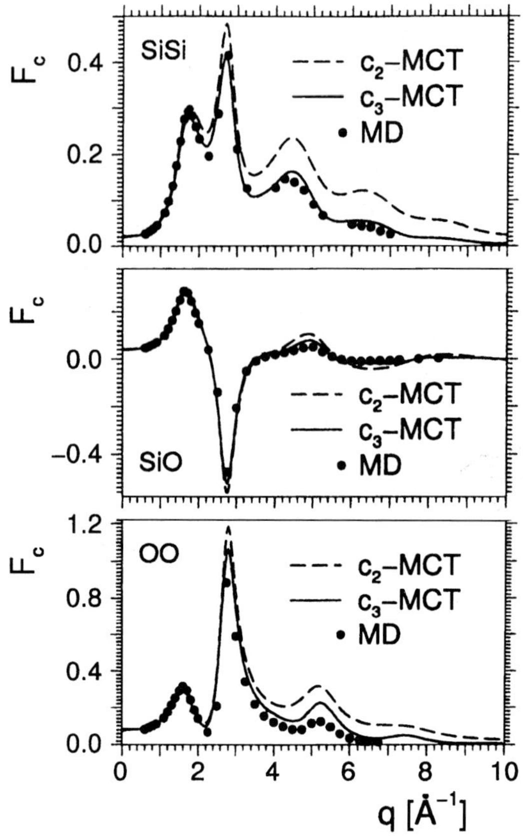

Fig. 22. – Wave vector dependence of the nonergodicity parameter $F_c^{\alpha\beta}(q)$ ($\alpha = (\mathrm{O}, \mathrm{Si})$) for the coherent intermediate scattering function $F(q,t)$ for SiO_2 melts. From Sciortino and Kob [126].

quantitative agreement! We stress that there is simply no other theory around that is able to make comparably significant and quantitative predictions for glass-forming fluids such as MCT does. Therefore, we have emphasized the test of MCT in these lectures.

2·6. Na_2O-SiO_2 *mixtures: breaking the continuous random network up.* – Most practically useful glasses (window glass, or a glass containing wine, etc.) are not at all pure SiO_2, but contain other oxides, such as Na_2O, Al_2O_3, CaO, etc. Some of these ingredients can be incorporated into the network of covalent bonds, even if the local topology gets changed (*e.g.*, Al_2O_3), and these oxides are then called "network formers"; other oxides (such as Na_2O) break the network locally up, creating "dangling Si-O bonds" (the so-called "network modifiers"). Here we briefly discuss the latter case.

As for the case of pure SiO_2, the validity of the chosen potentials (eqs. (30), (31)) can be checked by comparisons with suitable experimental data. As an example, fig. 23 shows the static neutron structure factor [128, 129] of $(Na_2O)(xSiO_2)$ with $x = 2$ (henceforth we shall use the abbreviation NSx for such sodium silicate melts). One can see that the overall agreement between simulation and experiment is reasonable. The "first sharp diffraction peak" at $q = 1.7\,\text{Å}^{-1}$ is a pronounced feature in pure SiO_2 melts as well (fig. 16); it arises from the tetrahedral network structure (the corresponding length scale $2\pi/1.7 \approx 3.7\,\text{Å}$ is the spatial extent of two connected SiO_4 tetrahedra). As expected, this feature is partly present in NSx also. The simulation also shows a weak "prepeak" at $q = 0.95\,\text{Å}^{-1}$, related to a superstructure formed by the Na and Si atoms: the length $(2\pi)/0.95 = 6.6\,\text{Å}$ is twice the mean distance of the nearest Na-Na or Na-Si neighbors, as one finds from an analysis of the partial radial distributions. Although this feature was not seen in the experiment of Misawa *et al.* [129], it was detected in a more recent experiment with better resolution [130], and it shows up very clearly in the partial structure factors (fig. 24) defined as

$$S_{\alpha\beta}(q) = \left(\frac{f_{\alpha\beta}}{N}\right) \sum_{i=1}^{N_\alpha} \sum_{j=1}^{N_\beta} \langle \exp[i\mathbf{q} \cdot (\mathbf{r}_i - \mathbf{r}_j)] \rangle, \quad \alpha, \beta \in \text{Si}, \text{Na}, \text{O}, \tag{40}$$

where $f_{\alpha\beta} = 1$ for $\alpha = \beta$, $f_{\alpha\beta} = 1/2$ for $\alpha \neq \beta$, and the sum over i extends only over atoms of species α, the sum over j only over species β. By recording all the $S_{\alpha\beta}(q)$ the simulation can go beyond the experiment, since for NSx the individual $S_{\alpha\beta}(q)$'s cannot be measured! Although the prepeak is so pronounced for $S_{\text{NaNa}}(\text{q})$ (fig. 24 [101]), it is difficult to see in the total neutron scattering function (fig. 23), since the mixed structure factors $S_{\text{NaSi}}(q)$ and $S_{\text{NaO}}(q)$ have a pronounced minimum there. However, it turns out that this structural feature is very important: it describes the length scale characterizing Na-rich channels that form a percolating network distinct from the network of SiO_4 tetrahedra that surrounds it. While this network of SiO_4 tetrahedra is relatively rigid, there is a high mobility of the Na atoms in these channels [95]. Computer simulations yield rather direct evidence both for the "dangling Si-O bonds" and for the channel picture [91, 95-97, 100], and can thus explain how it is possible that the diffusion constant of the Na ions has a much smaller activation energy (fig. 25). In contrast, for mixtures of Al_2O_3 and SiO_2 the behavior is completely different, as expected, since Al ions are also incorporated into the network, and hence the activation energies of all three types of atoms (Si, O, Al) in this case are similar (fig. 26) [101].

2·7. *The case of strong glass formers: a summary.* – Motivated by the distinctive behavior of various glass-forming materials in the so-called "Angell plot", fig. 3, where "strong glass formers" exhibit a simple Arrhenius law for the divergence of the structural relaxation time for all accessible temperatures, we have presented here extensive MD results on an archetypical "strong" glass former, namely molten SiO_2. While for "fragile glass formers" (such as polymer melts or glass-forming small molecule fluids, for instance) two characteristic temperatures exist, namely i) the critical temperature T_c of MCT (for $T > T_c$ the increase of the structural relaxation time approximately complies with a critical power law, $\tau \propto (T - T_c)^{-\Delta}$) and ii) the Kauzmann temperature T_K, where the (extrapolated) entropy difference ΔS between the undercooled fluid and the crystal

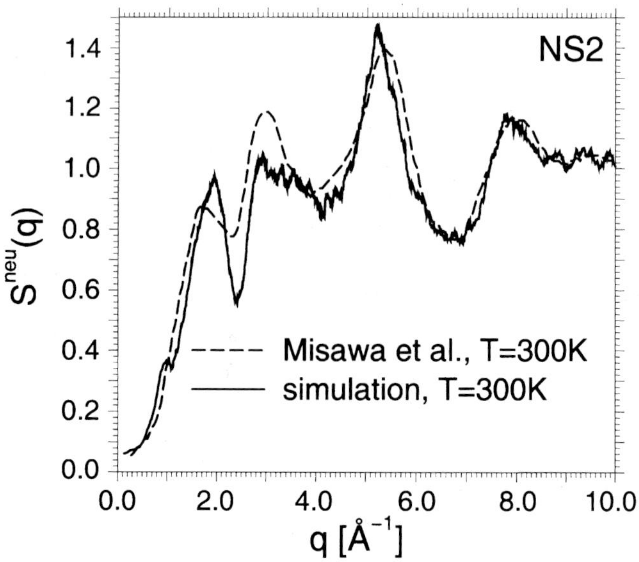

Fig. 23. – Static structure factor of sodium disilicate at $T = 300\,\mathrm{K}$ plotted *vs.* wave number. The full curve shows the MD results, where the experimental neutron scattering lengths for Si, O and Na atoms were used, so there is no adjustable parameter whatsoever. The broken curve shows the experimental data of Misawa *et al.* [129]. From Horbach and Kob [128].

would vanish, it has not been obvious *a priori* whether any of these temperatures T_c, T_K make sense for strong glass formers such as SiO_2; hence, the general usefulness of concepts such as the "random first-order transition" scenario is also called into question.

However, the MD simulations based on the simple BKS potential yielded clear evidence that MCT is useful for "strong glass formers" such as SiO_2, too: it was found, however, that T_c occurs at a very high temperature, not normally accessible in experiments on real materials. Thus it is suggested that the distinction between SiO_2 (and related "strong" glass formers) and "fragile" glass formers is not so fundamental as previously thought; different materials simply differ in their quantitative details, but a general unified description applies for the glass transition of all materials, at least qualitatively.

Fortunately, the MD simulations also show that the BKS potential (and its extension to include ions such as Na, Al, etc. into the description) allows a fairly faithful description of many physical properties of these materials (including diffusion constants, specific heat, static structure factor, sound velocities, viscosity, etc.). Also subtle phenomena may be modelled successfully, such as the effect of other ions which either may be incorporated into the network (*e.g.*, the Al ions in Al_2O_3-SiO_2 mixtures) or which modify the network by breaking it locally (*e.g.*, Na ions in Na_2O-SiO_2 mixtures). In both cases, "prepeaks" are seen in suitable static structure factors, which then show that on intermediate (mesoscopic) length scales chemical inhomogeneous structures form. These length scales should not be confused with the possible growth of an intermediate length

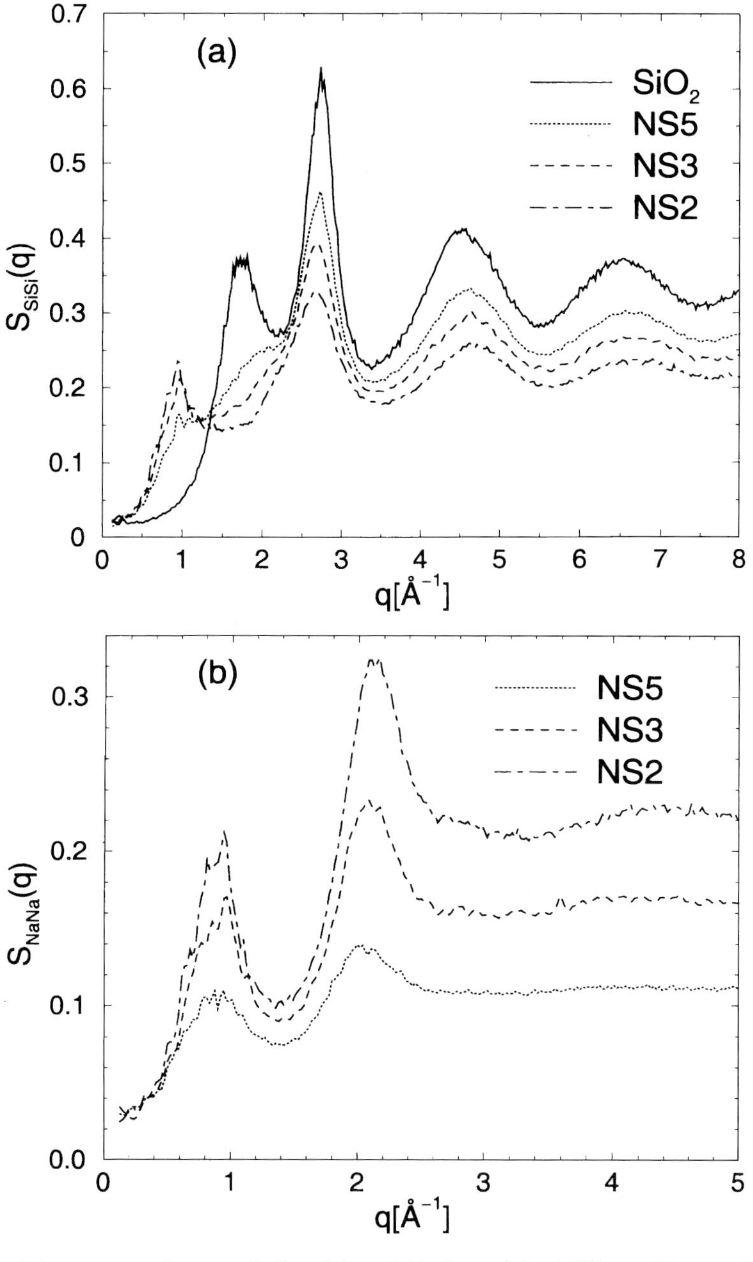

Fig. 24. – Partial structure factors a) $S_{\mathrm{SiSi}}(q)$ and b) $S_{\mathrm{NaNa}}(q)$ of NSx at $T = 300\,\mathrm{K}$ plotted *vs.* q. In part a), the result for pure $\mathrm{SiO_2}$ is included, in order to show that there the prepeak at $q \approx 0.9\,\mathrm{\AA}^{-1}$ is completely absent. From Winkler *et al.* [101].

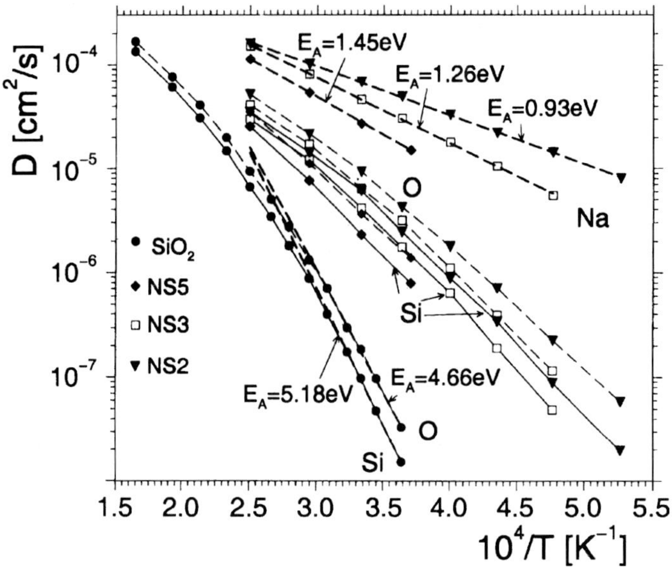

Fig. 25. – Arrhenius plot of the self-diffusion coefficients of Si, O and Na in SiO_2 and sodium silicate melts plotted *vs.* inverse temperature. Straight lines indicate the Arrhenius relations, $D_\alpha \propto \exp[-E_{\mathrm{A}}^{\alpha}/k_{\mathrm{B}}T]$, and the activation energies of the various ions are quoted in the figure. From Winkler *et al.* [101].

scale that might occur as a general characteristic of the approach to the glass transition, of course, but rather reflect a (nonuniversal!) microphase separation in these multi-component glasses. This interplay between chemical structural inhomogeneity and the resulting dynamic processes must be carefully analyzed, when one attempts a more detailed theoretical description of the glass transition in these materials.

3. – Search for a characteristic length of the glass transition via confinement effects

3˙1. *Motivation.* – As has already been discussed in subsect. **1**˙1, one does not understand well whether or not there exists an "order parameter" that distinguishes a glass from the corresponding supercooled fluid, and hence it is difficult to either verify or disprove the idea (eqs. (2)-(4)) that the dramatic slowing-down of the structural relaxation time results from the growth of a characteristic glass correlation length, as the glass transition is approached. In the absence of a good understanding of an "order parameter" it is clear that one also does not know which correlation function to study, which would be sensitive to the onset of glass-like ordering. It is clear, however, that one does not expect to learn about glass-like correlations from simple elastic-scattering experiments: just as in the case of spin glasses, where elastic magnetic scattering of neutrons contains information on ferromagnetic or antiferromagnetic ordering only, while the correlation function

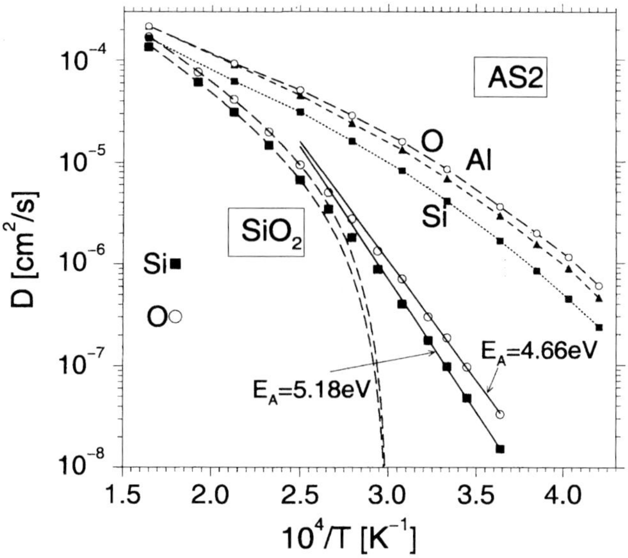

Fig. 26. – Arrhenius plot of the self-diffusion coefficients of SiO_2 and Al for the $(Al_2O_3) \cdot 2(SiO_2)$ melts. In the case of pure SiO_2, the curves shown represent fits according to MCT, $D \propto (1 - T/T_c)^{\Delta}$, $T_c \approx 3330\,\mathrm{K}$, $\Delta = 2$ (broken curves). Full straight lines (for SiO_2) are fits to an experimental Arrhenius variation, as shown in the previous figure. Broken curves for aluminium disilicate are guides to the eye only. From Winkler *et al.* [101].

sensitive to spin glass ordering is a higher-order correlation (such as $[\langle S_i S_j \rangle_T^2]_{\mathrm{av}}$). Thus the lack of any evidence for a growing length from scattering experiments near the glass transition is no proof that such a length does not exist, and indirect probes of growing glass correlation lengths should be considered.

In the context of computer simulations, one widely usable test for large correlation lengths that is viable for very different types of orderings (including the case of spin glasses alluded to above) is the study of finite-size effects, based on the idea of finite-size scaling [131-133]. The basic idea is that any observable of interest will have only a weak dependence on the linear dimension L of the system under consideration as long as $L \gg \xi$, the correlation length that is of interest for the considered problem; however, strong finite-size effects are observed in the inverse limit, when the correlation length ξ of the corresponding bulk system would exceed the linear dimension L. In computer simulations, such "intrinsic" finite-size effects are conveniently disentangled from surface effects by the use of periodic boundary conditions [132,133]. Thus, it has been a tempting idea to apply this approach for systems near a glass transition as well (and as we shall see in sect. **4**, for the infinite range Potts glass related finite-size effects indeed are very pronounced! [134]). Therefore it is somewhat disappointing that a study of finite-size effects for the bond fluctuation model (subsect. **1**˙3) did not reveal any indication of size effects in a temperature regime where the diffusion constant decreases by four orders of

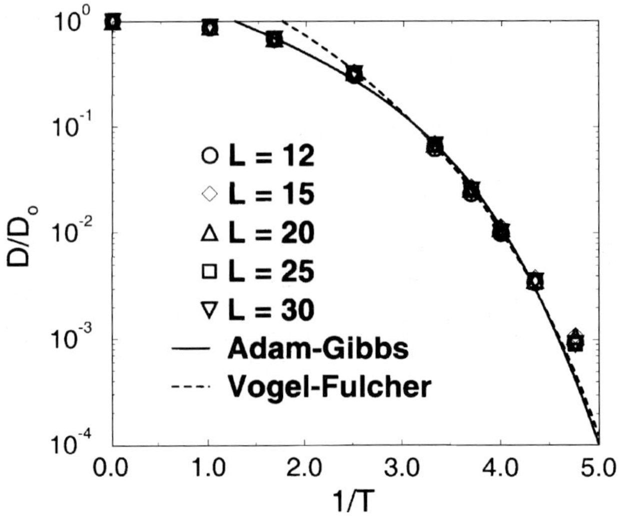

Fig. 27. – Self-diffusion coefficient of the three-dimensional bond fluctuation model (with energetic preference for bonds of length 3 lattice spacings, see subsect. **1**˙3) plotted *vs.* inverse temperature, using $L \times L \times L$ lattices with periodic boundary conditions for several choices of L, in order to check for finite-size effects. The solid and the dotted line are fits by the Vogel-Fulcher relation ($D/D_0 = \exp[-E_A/(T - T_{VF})]$, $T_{VF} = 0.125$) and by the Adam-Gibbs relation ($D/D_0 = \exp[-E_A/Ts(T)]$, $s(T)$ being the configurational entropy per lattice site of the model as obtained in fig. 15), respectively (D_0 and E_A are adjustable fit parameters). From Binder *et al.* [135].

magnitude [135] (fig. 27). Since we would have expected that the diffusion constant starts to show a size dependence when $\xi \approx L/2$, one concludes from fig. 27 that for $T \geq 0.2$ in this model ξ is still much smaller than 6 lattice spacings.

Since finite-size scaling obviously is not useful to estimate correlation lengths when the latter are very small [132, 133], we turn to an alternative method: perturbing the system by surface effects. Consider a glass-forming fluid confined in a cylindrical pore or in a thin film of width D: depending on the boundary conditions at the walls, the local relaxation time will either be enhanced or reduced (fig. 28), and the range over which this change exists before bulk behavior is reached can be taken to measure the desired glass correlation length ξ (fig. 28) [136]. One also sees that the range over which the relaxation time is enhanced near a rough surface is of the same order as the range over which the relaxation time is reduced near a smooth surface, and hence it is plausible that a correlation length can be extracted that is not specific for a particular perturbation at the surface, but a universal property of the considered glass-forming fluid. Hence we shall focus on the study of such surface and confinement effects on glass-forming fluids in this section.

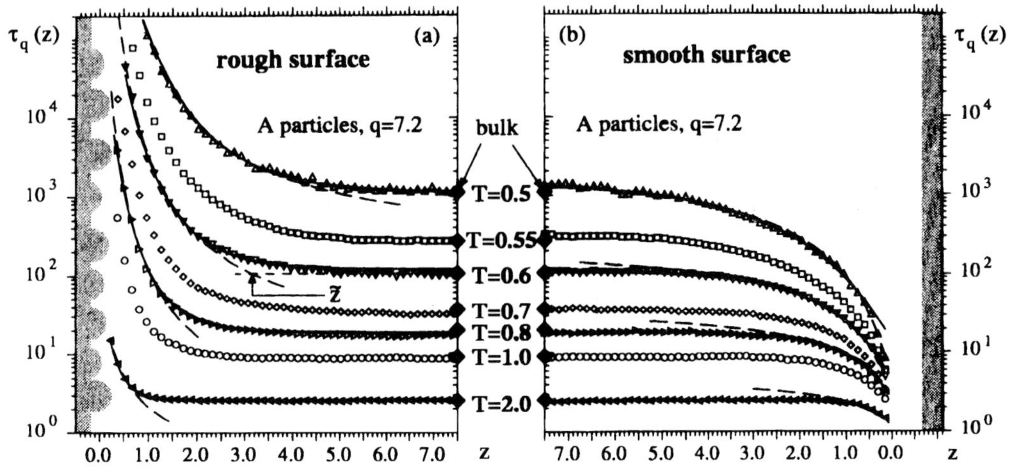

Fig. 28. – Structural relaxation time $\tau_q(z)$, defined from $F_s(q,z,t=\tau_q)=1/e$ where $F_s(q,z,t,)=N_\alpha^{-1}\sum\langle\exp[i\mathbf{q}\cdot(\mathbf{r}_j(t)-\mathbf{r}_j(0))]\delta(z_j(0)-z)\rangle$ and the wave vector $\mathbf{q}$ is oriented parallel to the surface, plotted as a function of the distance z from the wall, for (a) rough and (b) smooth surfaces at different temperatures. These results were obtained from a MD simulation of a binary Lennard-Jones mixture with $N_A=2400$ A-particles and $N_B=600$ B-particles (cf. text). The large diamonds indicate the corresponding values of the relaxation times in the bulk. The full and broken curves indicate various fits from which estimates for the glass correlation length are extracted (cf. text). From Scheidler *et al.* [136].

3˙2. *Polymer melts confined between smooth repulsive walls.* – Here we return to the bead-spring model of flexible glass-forming polymer chains [62, 66] that was already considered in subsect. **1**˙2. The chains are placed between two hard impenetrable walls at $z_{\rm wall}=\pm D/2$, at which a purely repulsive wall potential acts, $U_{\rm wall}(z)=(\sigma/z)^9$, where z is the distance of a monomer from the wall, $z=|z_i-z_{\rm wall}|$. The simulations were done using 200 chains of length $N=10$, adjusting the lateral dimensions (along the walls) such that the normal pressure p_N was kept fixed, $p_N=1$ (in Lennard-Jones units) [137-140]. Figure 29 shows typical results for the density profiles across the film [138]. One sees that at low temperatures near the hard wall the density profile develops pronounced oscillations. Despite this "layering" the polymer melt retains a fluid structure very similar to that of a melt in the bulk in the lateral directions parallel to the walls, as is evidenced by the radial distribution function $g(z,r)$ when $\mathbf{r}$ is oriented parallel to the walls [138]. In addition, $g(z,r)$ is only weakly dependent on z, despite the strong layering effect.

We now turn to the dynamic behavior of the confined polymer film, focusing attention first to the average mean-square displacements of the monomers (fig. 30 [137]). It is seen that qualitatively the behavior of $g_1(t)$ for the thin films is the same as for the bulk; but it is clearly seen that the average motions in the thin film are the faster the thinner the film.

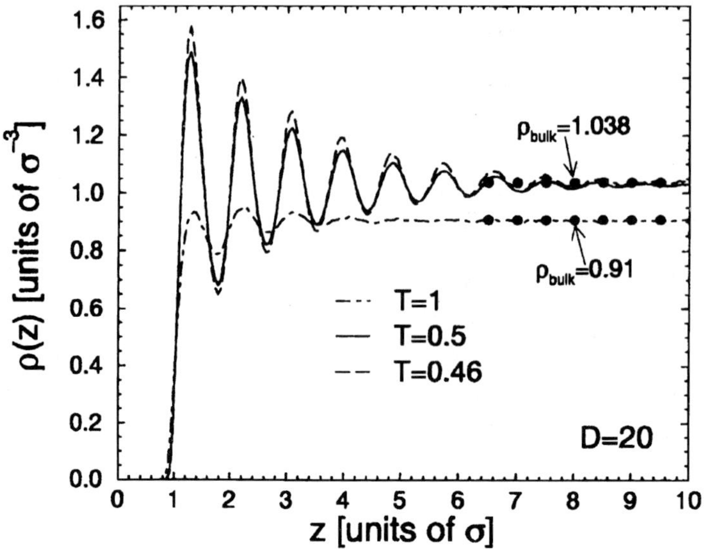

Fig. 29. – Temperature dependence of the monomer density profile $\rho(z)$ for the bead-spring model of flexible polymer chains confined between flat repulsive walls in a thin film of thickness $D = 20$ (corresponding to $14\,R_{\rm g}$, $R_{\rm g}$ being the gyration radius of the chains; unit of length is the Lennard-Jones diameter $\sigma \equiv 1$). Here z denotes the position of a monomer from the (left) wall. Since the profiles are symmetric with respect to the middle of the film, only one half is shown. The temperatures are characteristic of the high-temperature state ($T = 1$) and the supercooled state ($T = 0.46$). The filled circles on horizontal straight lines indicate the bulk densities $\rho_{\rm bulk}(T = 1, p = 1) = 0.91$ and $\rho_{\rm bulk}(T = 0.46,\ p = 1) = 1.038$, respectively. From Varnik *et al.* [137].

To quantify this observation one may define a relaxation time τ by requiring $g_i(t = \tau) = 1$, since then the escape of the monomers from the cage has occurred (fig. 30). This relaxation time $\tau(T)$ hence can be taken as a measure of the α-relaxation time of the thin film, and hence we expect that $\tau(T) \propto (T - T_{\rm c}(D))^{-\gamma(D)}$, if MCT holds as in the bulk. It turns out that all data are compatible with $\gamma(D) = 2.5$ for $D = 5$ for all quantities that were studied, fig. 31. But, as was found in the bulk [11, 62], one can alternatively fit $\tau(T)$ to the Vogel-Fulcher relation (see sect. **1**). Figure 32 shows a plot of $T_{\rm c}(D)$ and $T_{\rm VF}(D)$ *vs.* D [137]. Qualitatively, this behavior is similar to corresponding experimental data (fig. 33) [141].

In order to address the question to what extent these data can contribute to the problem of a glass correlation length, we consider spatially resolved mean-square displacements (fig. 34). One can clearly see that monomers at small distance from the wall diffuse much faster than those in the bulk; the fact that the average mean-square displacement of all monomers in a thin film is the larger the thinner the film hence has the obvious interpretation that the surface region adjacent to the walls is relatively the more important the thinner the film is.

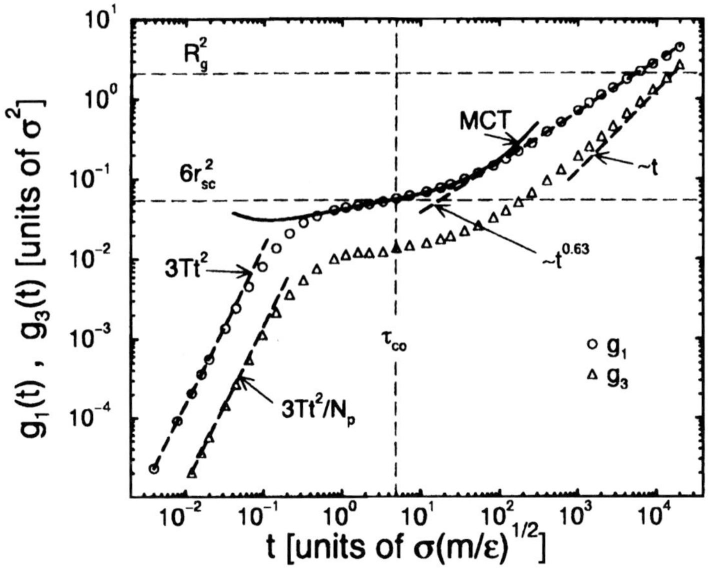

Fig. 30. – Log-log plot of mean-square displacement of inner monomers of the chains *vs.* time t for a temperature $T = 0.5$ (the bulk critical temperature being $T_c = 0.45$ for this model, see subsect. **1**˙2). The figure compares bulk data with the displacements measured parallel to the walls in films of different thickness, as indicated. The film data were multiplied by 3/2 to account for the different number of spatial directions used to calculate $g_1(t)$ (*i.e.*, three directions for the bulk, but only two for the thin films). The bulk mean-square end-to-end distance R_e^2, gyration radius R_g^2 and the plateau value $6r_{sc}^2$ of a MCT analysis are shown as horizontal dashed lines. The solid lines indicate the power laws expected in different time regimes (ballistic behavior $g_1(t) \propto t^2$ at short times, Rouse-like at intermediate times, $g_1(t) \propto t^x$ with $x \approx 0.63$, and diffusive at late times, $g_1(t) \propto t$). From Varnik *et al.* [137].

Defining now a z-dependent local relaxation time $\tau(z)$ from the condition $g_1(t = \tau(z), z) = 1$, one obtains fig. 35. One sees that at high temperatures there is a wide region around the film center, where τ is independent of z. Contrary to that, at low T the presence of the walls is "felt" also closer to the film center. Thus, fig. 35 is the counterpart to the results for the binary Lennard-Jones mixture, already shown in fig. 28.

3˙3. *The glass correlation length and its temperature dependence.* – Before we turn to the analysis of the results as shown in figs. 28 and 35, we explain the model used in fig. 28 in more detail. These results refer to a binary (A, B) mixture of particles interacting via Lennard-Jones (LJ) potentials, $V_{\alpha\beta}(r) = 4\varepsilon_{\alpha\beta}[(\sigma_{\alpha_\beta}/r)^{12} - (\sigma_{\alpha\beta}/r)^6]$ with $\alpha, \beta \in (A, B)$, cut-off radii $r^c_{\alpha\beta} = 2.5\,\sigma_{\alpha\beta}$, and interaction parameter $\varepsilon_{AA} = 1$, $\sigma_{AA} = 1$, $\varepsilon_{AB} = 1.5$, $\sigma_{\alpha\beta} = 0.8$ and $\varepsilon_{BB} = 0.88$. Time is here measured in units of $(m\sigma_{AA}^2/48\varepsilon_{AA})^{1/2}$, m being the mass of the particles. Previous simulations of this model fluid in the bulk have shown that it is a very good glass former, and the mode-coupling temperature (at a bulk density $\rho = 1.2$, for $N_A = 2400$, $N_B = 600$) is at about $T_c \approx 0.435$ [12, 142].

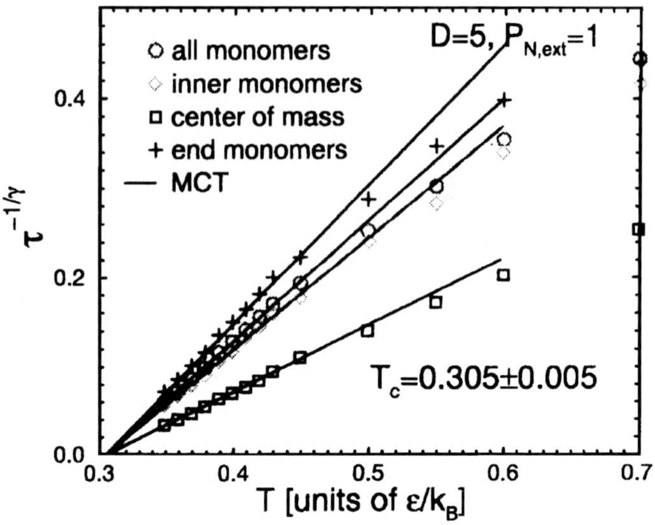

Fig. 31. – Plot of $\tau^{-1/\gamma}$ *vs.* temperature, for a film of thickness $D = 5$, using $\gamma = 2.5$, $T_c = 0.305$, and analyzing various mean-square displacements: of inner monomers, end monomers, the chain's center of mass, and all monomers averaged together. The straight lines indicate a power law behavior $\tau \propto (T - T_c(D))^{-\gamma}$, where $T_c(D)$ is independent of the type of quantity considered. From Varnik *et al.* [137].

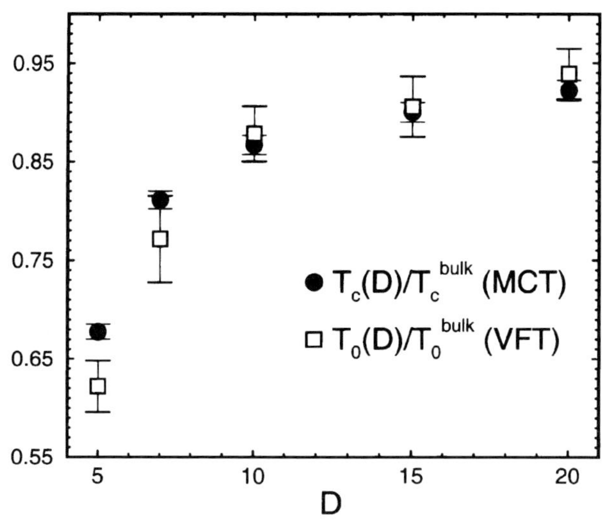

Fig. 32. – Ratios $T_c(D)/T_c^{\rm bulk}$ and $T_0(D)/T_0^{\rm bulk}$ *vs.* film thickness D. $T_0(D)$ denotes here the Vogel-Fulcher temperatures. Note that the error bars of $T_0(D)$ are larger than those of $T_c(D)$, since $T_0(D)$ is much lower than $T_c(D)$, but data on $\tau(T)$ are only available for $T \geq T_c(D)$: therefore the results for $T_c(D)$ are more reliable. From Varnik *et al.* [137].

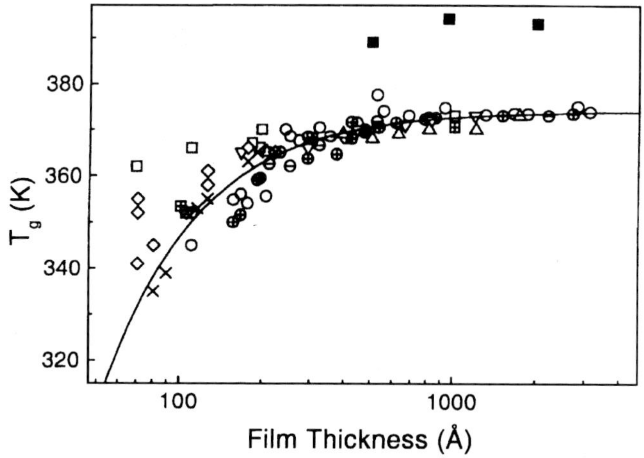

Fig. 33. – Experimental glass transition temperatures T_g plotted *vs.* film thickness for supported polystyrene films (upper surface of the film against air, lower against silicon). The various symbols indicate different experimental techniques and different molecular weights, ranging from $M_w = 120000$ to $M_w = 290000$. From Forrest and Jones [141].

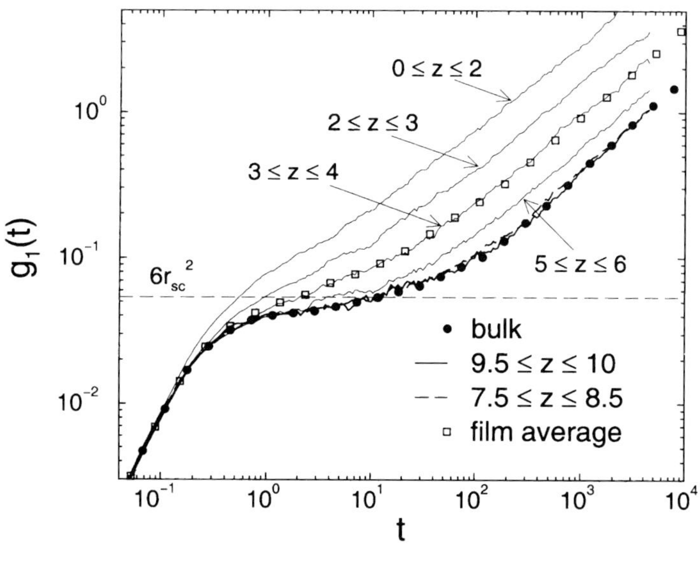

Fig. 34. – Layer-resolved mean-square displacements of the innermost monomers $g_1(t, z)$ *vs.* time for a film of thickness $D = 20$. Here z denotes the distances from the left wall. The displacements are calculated parallel to the wall and multiplied by $3/2$ (to put them on the same scale as the bulk data (full dots) which are averaged over three spatial dimensions instead of over only two for the film). The $g_1(z, t)$ data (indicated by lines) are averages over all the monomers between some $z_{\min} \le z \le z_{\max}$, which remain in the specified interval of z for all times shown in the figure. From Varnik *et al.* [137].

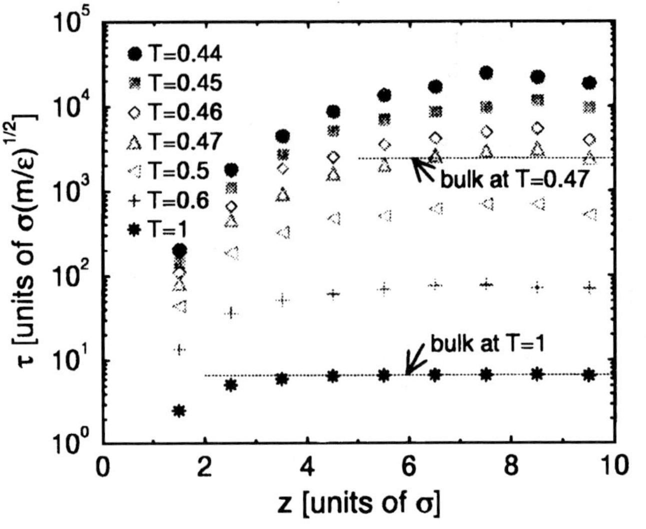

Fig. 35. – Relaxation time $\tau(g_0 = 1, z)$ *vs.* distance from the wall for a film of thickness $D = 20$. Here z stands for the distance of the center of a layer (bin) from the left wall; the thickness of a bin is $\Delta z = 1$. In the calculation of the local mean-square displacements of all monomers, $g_0(t, z)$, which underlies the definition of τ, only those monomers are considered which remain in the specified layer for all times $t' \le t$. From Varnik *et al.* [137].

As anticipated in fig. 28, two types of walls were used, a rough one and a smooth one. The rough wall was realized by freezing a slice of thickness $2.5\sigma_{AA}$ of the LJ liquid and applying the same LJ interaction (plus an additional hard core potential to prevent the liquid particles from penetrating into the wall). As a geometry, a film with area $L \times L$ and thickness D ($L = 12.88$, $D = 15.0$) was used, containing in total 3000 particles, as in the simulations of bulk behavior [142].

To study a fluid confined by smooth walls an external potential of the form $V_{\alpha\mathrm{w}}(z) = (4/45)\pi\,\rho\sigma_{AB}^3\;\varepsilon_{\alpha\mathrm{w}}[\sigma_{AB}/(z - z_\mathrm{w})]^9$ with $\varepsilon_{A\mathrm{w}} = 1.0$, $\varepsilon_{B\mathrm{w}} = 3.0$ was chosen, with $z_\mathrm{w} = -0.65$ and $z_\mathrm{w} = 15.65$, to obtain a film of width $D = 15.0$ [136].

While for the rough wall by construction there is only a very minor perturbation of the static structure of the fluid, for the smooth wall strong layering effects occur. To avoid layering in this case as well, the potential energy of the system was modified by adding a term that coupled directly to the deviation from a constant density profile, *i.e.* configurations whose profile was not constant were energetically disfavored. In this way, an essentially constant density profile was obtained.

Since one finds [143, 144] that the "local" incoherent intermediate scattering function

$$F_s(\mathbf{q}, z, t) = N_\alpha^{-1} \sum_{j=1}^{N_\alpha} \big\langle \exp[i\mathbf{q}\cdot[\mathbf{r}_j(t) - \mathbf{r}_j(0)]]\delta[z_j(0) - z]\big\rangle, \quad \alpha \in \mathrm{A}, \mathrm{B} \tag{41}$$

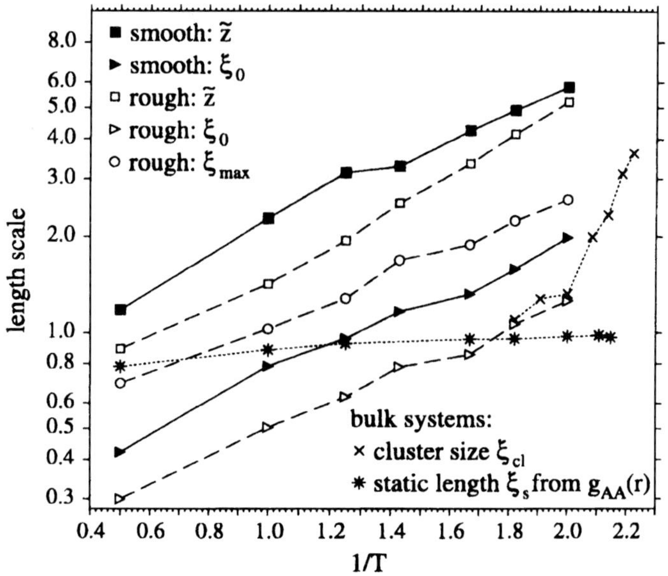

Fig. 36. – Different dynamic length scales (see text for definition) in confined systems as a function of inverse temperature in a logarithmic plot. A comparison with some static and dynamic length scales in bulk systems is also made. Data for the size of stringlike clusters of fast particles ξ_{cl} is taken from Donati *et al.* [145]. From Scheidler *et al.* [136].

where z is the distance from the wall and $\mathbf{q}$ is oriented parallel to the wall, shows a two-step decay similar to the behavior of $F_s(\mathbf{q}, t)$ in the bulk (see sects. **1**, **2**), one can characterize the α-relaxation time $\tau_q(z)$ simply by requiring $F_s(\mathbf{q}, zt = \tau_q) = 1/e$ [136]. This local structural relaxation time was already shown in fig. 28.

Several estimates for a glass correlation length were obtained from data such as those shown in fig. 28 [136,144]. Close to the surface, the z-dependence of $\tau_q(z)$ can be described by the empirical Ansatz [143]

$$\tau_q(z) = f_q(T) \exp\left[\frac{\pm\Delta_q(T)}{(z - z_p)}\right], \tag{42}$$

with three parameters $f_q(T)$, $z_p = -0.5 \pm 0.15$, and $\Delta_q(T)$. In eq. (42) and in the following the positive (negative) sign corresponds to systems with rough (smooth) surfaces. The z-range for which this function (dashed curves in fig. 28) works well increases with decreasing T. The location of the crossover to bulk behavior in the center (see data and fits for $T = 0.6$ in fig. 28) can be used to define a characteristic length scale $\tilde{z}(T)$ which is included in fig. 36.

An alternative Ansatz for the z-dependence of $\tau_q(z)$ is a function depending on

$\exp[-z/\xi_0]$, with a characteristic length scale $\xi_0(T)$. The functional form

$$\ln\left[\left(\frac{\tau_q(z)}{\tau_{q,\infty}}\right)^{\pm 1}\right] = A(T)\exp\left[\frac{-z}{\xi_0(T)}\right] \tag{43}$$

does indeed describe the z-dependence for all values of z (solid curves in fig. 28). Here $\tau_{q,\infty}$ is the relaxation time of the system in the bulk.

The length scales $\tilde{z}(T)$ and $\xi_0(T)$ are obtained directly from $\tau_q(z)$. A further length scale $\xi_{\max}$ was extracted in ref. [144] and is also included in fig. 36.

Figure 36 shows that all lengths $\tilde{z}(T)$, $\xi_{\max}$ are compatible with a simple Arrhenius behavior, $\xi \propto \exp[E/T]$, with an activation energy $E \approx 1.1$ that depends neither on the definition of the length scale nor on the type of wall. These findings give evidence that the length scales do indeed characterize a length scale *intrinsic* to the system. However, within the investigated temperature range the growth of this scale is rather small: *e.g.*, ξ_0 grows only by a factor of 3.5 between $T = 2.0$ and $T = 0.5$, while bulk relaxation times increase by several orders of magnitude. There is also no evidence for a divergence of this length scale at a nonzero temperature.

Figure 36 also includes the length scale $\xi_{\rm cl}$ that was proposed by Donati *et al.* [145], who used the van Hove autocorrelation function to identify the 5% most mobile particles in the system. These particles are not distributed at random, but instead form temporary string-like clusters ("dynamical heterogeneities"), the size of which grows with decreasing temperature.

As an example for a static length scale, the envelope of the radial pair distribution function $g_{AA}(r) - 1$ is fitted to $\exp[-r/\xi_s]$ to define a static length scale ξ_s. This length scale is only very weakly temperature-dependent (fig. 36), and hence clearly not related to the glass transition.

Some of these dynamic length scales shown in fig. 36 have also been extracted for the polymer model of the previous subsection, using fig. 35 [140], and the same conclusion as above emerges: all length scales are rather small, and increase only rather slowly as the temperature is lowered: so no evidence for a divergence of a length scale at nonzero temperature emerges.

3˙4. *Discussion.* – While the attempt to extract a glass correlation length from finite-size effects in systems of varying linear dimensions with periodic boundary conditions was completely unsuccessful, the study of surface effects on thin films of glass-forming fluids does allow to clearly extract an intrinsic length that grows with decreasing temperature like $\xi \propto \exp[E/T]$. However, the activation energy E and hence the length ξ itself stay rather small in the temperature region of interest, and it is not clear whether or not this length ξ is relevant for the dramatic slowing-down observed at the glass transition.

A very interesting and different approach to extract a dynamic length was proposed by Bennemann *et al.* [146]. For the model described in sects. **1**˙2, **3**˙2, the correlation function of displacements of particles ($\mu_i(t) = |\mathbf{r}_i(t+\Delta t) - \mathbf{r}_i(t)|$ is used to calculate a "susceptibility" which reaches a maximum $\kappa_{\max}$ at a time $\Delta t_{\max}$) is calculated, and it is suggested that $\kappa_{\max}$ diverges at the critical temperature of MCT. However, the error bars of this quantity $\kappa_{\max}$ are rather large, and it is likely that $\kappa_{\max}$ actually remains finite at T_c, as the structural relaxation time does.

In any case, the studies of glassy behavior of undercooled fluids confined in pores or in thin films have given a lot of valuable insight helpful for the interpretation of experiments, such as the shift of the glass transition temperatures (figs. 32, 33).

4. – The infinite-range Potts glass with $p > 4$ states: A scenario for the structural glass transition?

4.1. *Introduction.* – Already in sect. **1** the scenario of the "random first-order transition" was mentioned, namely that models with infinite-range interactions exist which indeed show both a dynamical ergodic-to-nonergodic transition at a higher temperature and a static transition, where a nonzero glass order parameter appears discontinuously although the model lacks a latent heat, at a lower temperature (fig. 6). In the present section, we shall study one model that belongs to this class of models more closely, namely the 10-state Potts glass. The purpose of this study is twofold: one aim is to elucidate the question on how these two transitions show up in simulations of finite-sized systems; also the simulations of realistic models of undercooled fluids undergoing a structural glass transition deal with systems of finite size, and hence it is of some importance to clarify size effects on such glass transitions. The second question that shall be asked is what is left from this scenario with the two transitions when one considers a model with short-range interactions rather than the mean-field limit of infinite-range interactions.

4.2. *A Monte Carlo study of the infinite-range Potts glass with $p = 10$ states.* – We return here to the model that was already specified in eqs. (6)-(8), choosing $\widetilde{J}_0=3-p=-7$ as a special case, so there is no tendency for ferromagnetic order in the model [47, 147].

At this point it is convenient to introduce a vector notation via the so-called "simplex representation" of the Potts spins [148]: the p states are represented by $(p-1)$-dimensional vectors $\mathbf{S}^{(\lambda)}$ pointing to the corners of a p-simplex (note that for $p = 3$ this would be a triangle, for $p = 4$ this would be a tetrahedron, etc.). Let us then consider two copies of exactly the same system, with identical bond configuration (sometimes these two copies are denoted as "real replicas" [3,47,149], but this should not at all be confused with the replicas of the "replica method" for handling the quenched averaging analytically [3]). The states of the i-th Potts spin in copies 1, 2 are then denoted as $\mathbf{S}_{i,1}$ and $\mathbf{S}_{i,2}$. Note that two such vectors satisfy the normalization condition $\mathbf{S}^{(\lambda)}\cdot\mathbf{S}^{(\lambda')} = p\delta_{\lambda\lambda'} - 1$.

Now the order parameter defined as a tensorial quantity $q^{\mu\nu}$ [42-47, 149] is

$$q^{\mu\nu} = N^{-1}\sum_{i=1}^{N}(\mathbf{S}_{i,1})^{\mu}(\mathbf{S}_{i,2})^{\nu}\,, \quad \mu,\nu = 1,2,\ldots,p-1\,, \tag{44}$$

μ, ν denoting the Cartesian components in this $(p-1)$-dimensional space. In practice, however, only the scalar quantity q is considered,

$$q = \left[\sum_{\mu,\nu=1}^{p-1} (q^{\mu\nu})^2 \right]^{1/2}, \tag{45}$$

as well as the distribution function of this glass order parameter, $P(q)$. Note that the "spin glass susceptibility" $\chi_{\rm SG}$ simply is related to the second moment of $P(q)$,

$$\chi_{\rm SG} \equiv \frac{N[\langle q^2 \rangle]_{\rm av}}{(p-1)} = N \int_0^1 \frac{q^2 P(q) {\rm d}q}{(p-1)}. \tag{46}$$

Remember that $\langle \cdots \rangle$ here means the standard canonical averaging (as is realized, for instance, by Monte Carlo sampling with the Metropolis algorithm), while $[\cdots]_{\rm av}$ means an averaging over the quenched bond disorder [3].

Note that the Ising spin glass ($p = 2$) [48] is still a special case of the model, eqs. (6)-(8), and that a second-order freezing transition occurs at $T_{\rm s} = 1$, where the spin glass susceptibility $\chi_{\rm SG}$ diverges according to a Curie-Weiss law, and the order parameter for $T < T_{\rm s}$ rises continuously,

$$\chi_{\rm SG} \propto \left(\frac{T}{T_{\rm s} - 1} \right)^{-\gamma_{\rm SG}}, \quad [\langle q \rangle]_{\rm av} \propto \left(\frac{1-T}{T_{\rm s}} \right)^{\beta_{\rm SG}}, \tag{47}$$

with exponents $\gamma_{\rm SG} = 1$, $\beta_{\rm SG} = 1$ in mean-field theory [3, 48]. In dynamic versions of the model the spin autocorrelation function, which for general p is defined as

$$C(t) = [N(p-1)]^{-1} \sum_{i=1}^{N} \left[\langle \mathbf{S}_i(t') \cdot \mathbf{S}_i(t'+t) \rangle \right]_{\rm av}, \tag{48}$$

exhibits critical slowing-down as $T_{\rm s}$ is approached from above, *i.e.* $C(t)$ shows an asymptotic decay described by $C(t) \propto \exp[-t/\tau]$ with

$$\tau \propto \left(\frac{T}{T_{\rm s} - 1} \right)^{-\Delta_{\rm SG}}, \quad \Delta_{\rm SG} = 2. \tag{49}$$

The same behavior occurs within mean-field theory for all $p \leq 4$ [3], while a very different behavior occurs for $p > 4$ [42-46], as sketched in fig. 6: a first-order phase transition from a "paramagnetic phase" to a "spin glass phase" occurs where the glass order parameter $q_{\rm EA}$ appears discontinuously, but there is no latent heat (fig. 6). This behavior is possible since $P(q)$ is found from the one-step replica symmetry breaking method [43, 46] as

$$P(q) = w(T)\delta(q - q_{\rm EA}) + [1 - w(T)]\delta(q), \quad T < T_0, \tag{50}$$

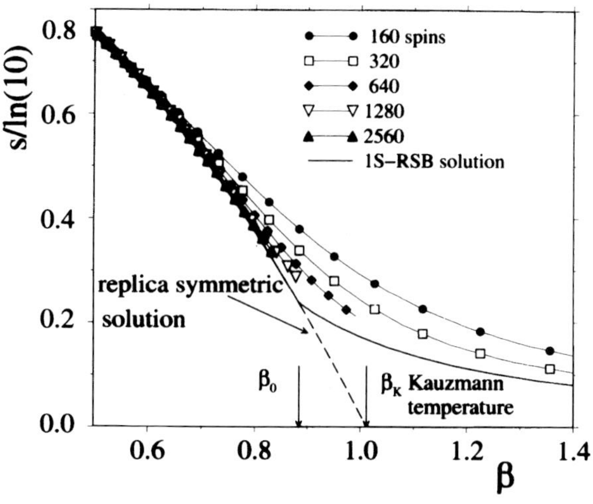

Fig. 37. – Entropy s per spin, normalized by its high-temperature value, plotted *vs.* inverse temperature for different system sizes (curves with symbols). The bold dashed curve and the bold solid curve are the replica-symmetric solution [42] and the one-step replica-symmetry breaking solution [46] of the replica theory of the Pott glass in the thermodynamic limit, $N \to \infty$. Vertical arrows indicate the static inverse transition temperature $\beta_0 = 1/T_0$ and the inverse $\beta_K = 1/T_{\rm K}$ of the "Kauzmann temperature" where the entropy of the replica-symmetric solution vanishes ($T_{\rm K} \approx 0.9885$ here). From Brangian *et al.* [47].

where the weight $w(T)$ of the ordered phase vanishes continuously as the temperature T_0 of the static phase transition is approached, $w(T) \propto 1 - T/T_0$. For the numerical Monte Carlo work [47, 147], arbitrarily the choice $p = 10$ for the number of Potts states was made, since then $q_{\rm EA}(T)$ at the transition temperature T_0 is already fairly large ($q_{\rm EA}(T_0) \approx 0.452$ [46]) and, in addition, T_0 is appreciably different from $T_{\rm s} = 1$, where the (extrapolated) inverse spin glass susceptibility $\chi_{\rm SG}^{-1}$ would vanish (fig. 6), namely $T_0 \approx 1.1312$ [46]. Unfortunately, the dynamical transition temperature $T_{\rm D}$ for this model is still very close to the static one, namely $T_{\rm D} = 1.142$ [46]. Hence, it is not possible to attempt to probe the complexity for this model by Monte Carlo methods (remember that the complexity counts the number of states in the temperature region $T_0 < T \leq T_{\rm D}$ and vanishes continuously as $T \to T_0^+$, while it is zero both for $T \leq T_0$ and $T > T_{\rm D}$, see sect. **1**). The ordinary (configurational!) entropy of the Potts glass can be studied by Monte Carlo methods, see fig. 37.

Simulations were carried out for $N = 160$ (using 500 samples of random bond configurations $\{J_{ij}\}$), 320 (200 samples), 640 (100 samples), 1280 (100 samples) and 2560 (up to 50 samples). In spite of the use of "parallel tempering" techniques [150], sizes $N \geq 640$ could not be equilibrated for $T \leq 1$, and therefore these curves for $N \geq 640$ in fig. 37 "stop" for $\beta \geq 1$. Once can see that the data superimpose for $\beta = 1/T \leq 0.7$ on the curve representing the (presumably) exact result for $N \to \infty$, but for β in the transition regime there occurs a very pronounced finite-size rounding. Thus, the data

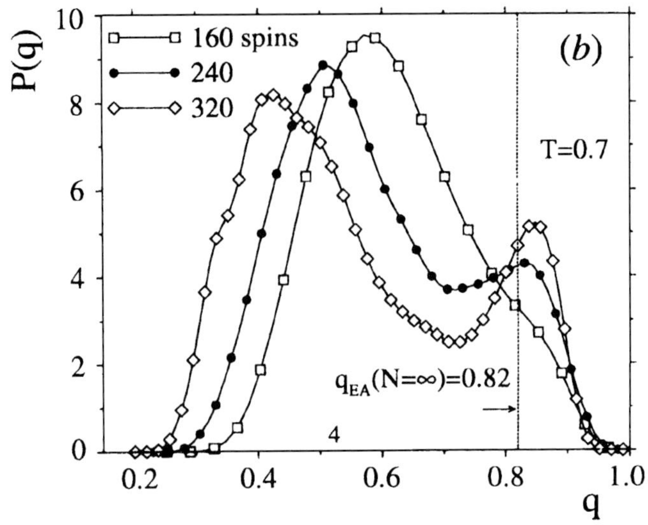

Fig. 38. – Order parameter distribution $P(q)$ *vs.* q for $T = 0.7$ and the three system sizes $N = 160$, 240 and 320. The asymptotic value of the order parameter q_{EA} (from [46]) is included as a vertical line. From Brangian *et al.* [47].

are suggestive of a convergence to the one-step replica symmetry breaking solution for $\beta > \beta_0$, but they certainly cannot prove it. Similar strong finite-size effects are found for other static properties (internal energy, spin glass susceptibility, and the glass order parameter distribution and its moments) as well [47]. As an example, fig. 38 shows $P(q)$ for $T = 0.7$. Only for the largest system ($N = 320$) is a clear evidence for a double-peak distribution apparent. Of course, we do expect that for any finite N the two delta-functions of eq. (50) are rounded into peaks of finite height and finite width, and the peak at $q = 0$ must be shifted to a nonzero positive value (cf. eq. (45), which is non-negative). However, it was rather unexpected that such strong finite-size effects occur here, although one can conclude in general that any phase transition must be rounded in a finite system [131-133]; but the surprising finding of [47, 147] is that for the random first-order phase transition in the 10-state Potts glass the rounding is particularly pronounced, and a theoretical description of the rounding for such unconventional phase transitions [42-47] is still lacking [147].

Also the dynamical transition is strongly rounded (fig. 39). Although it is established from the analytical work [44-46] that $C(t)$ must develop a two-step relaxation at T slightly above T_{D}, with a plateau at $C(t) \approx q_{\mathrm{EA}}(T_{\mathrm{D}})$ developing at intermediate times, the lifetime of this plateau diverging as $\tau \propto (T - T_{\mathrm{D}})^{-\Delta}$ as $T \to T_{\mathrm{D}}$, for $N \leq 1280$ at $T = T_{\mathrm{D}}$ no plateau is as yet seen (fig. 39). Only for significantly lower temperatures (*e.g.*, for $T = 0.7$ the dynamical behavior is still accessible for $N = 160$ Potts spins [47]) does one find the typical signatures of mode-coupling theory (pronounced two-step relaxation, time-temperature superposition holds for the final decay, etc.) [47]. But already the

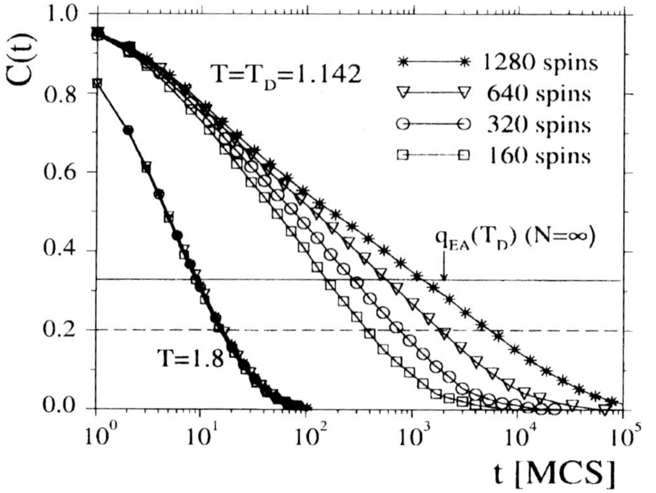

Fig. 39. – Time dependence of the autocorrelation function $C(t)$ of the Potts spins (eq. (48)) for $T = 1.8$ and for $T = T_\mathrm{D} = 1.142$ for several values of N. The solid line shows the theoretical value of the Edwards-Anderson order parameter $q_{\mathrm{EA}}(T_\mathrm{D})$ for $N \to \infty$ [46]. The dashed line shows the value that was used to define the relaxation time τ. From Brangian *et al.* [134].

"raw data" for the relaxation time τ, defined from (fig. 39) $C(t = \tau) = 0.2$, indicate a power law divergence at an effective transition temperature that converges towards the correct value, $T_\mathrm{D} = 1.142$ (fig. 40). This tentative extrapolation is corroborated by a dynamic finite-size scaling analysis (fig. 41). It implies a divergence $\tau \propto N^{z^*}$ at $T = T_\mathrm{D}$ [134], with $z^* \approx 1.5$. For N finite, however, T_D still has the significance of a crossover temperature where a rounded transition occurs from a power law divergence $\tau \propto (T - T_\mathrm{D})^{-\Delta}$ for $T > T_\mathrm{D}$ to an Arrhenius behavior $\tau \propto \exp[E_A(N)/k_\mathrm{B}T]$, with an activation energy $E_A(N)$ that strongly increases with N (fig. 42). Unfortunately, the range of N where data are available for τ at $T < T_\mathrm{D}$ does not suffice to make any firm statement on the functional form of $E_A(N)$.

4·3. *Monte Carlo study of the short-range Potts glass with $p = 10$ states.* – Figure 42 indeed is qualitatively similar to the behavior of real glasses (cf. figs. 3, 5, 20): does this finding imply that the Potts glass indeed provides a good scenario of what happens when the structural glass transition of an undercooled fluid is approached? Of course, for discussing real systems it is relevant to consider the behavior of systems with short-range forces in the thermodynamic limit, rather than finite mean-field type systems, as studied in the previous section. Thus, we consider now the Potts glass with nearest-neighbor random interaction J_{ij} on the simple cubic lattice, rather than the mean-field Hamiltonian, eqs. (6)-(8). Thus, the Hamiltonian for the short-range model is

(51) $$\mathcal{H} = -\sum_{\langle i,j \rangle} J_{ij}(p\delta_{\sigma_i\sigma_j} - 1),$$

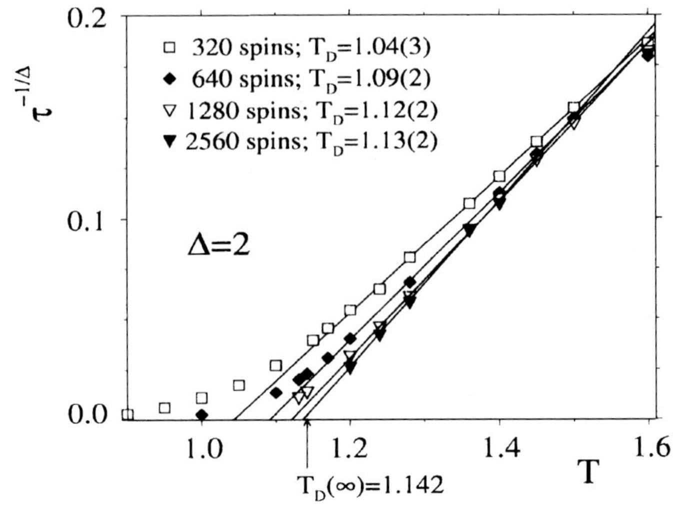

Fig. 40. – The temperature dependence of $\tau^{-1/\Delta}$ for different sizes, using $\Delta = 2.0$ as a trial value. The bold straight lines are fits to a proper subset of points. The resulting extrapolated values for $T_D(N)$ are quoted in the figure. From Brangian *et al.* [47].

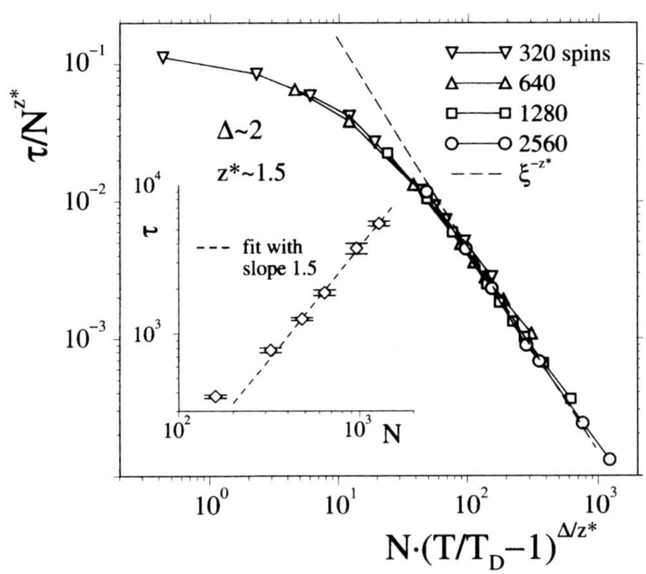

Fig. 41. – A log-log plot of the scaled relaxation time τ/N^{z^*} *vs.* the scaled distance $N(T/T_D - 1)^{\Delta/z^*}$ from the dynamical transition temperature T_D, choosing $z^* = 1.5$ and $\Delta/z^* = 1.3$. The inset is a log-log plot of $\tau(T = T_D)$ *vs.* N. From Brangian *et al.* [47].

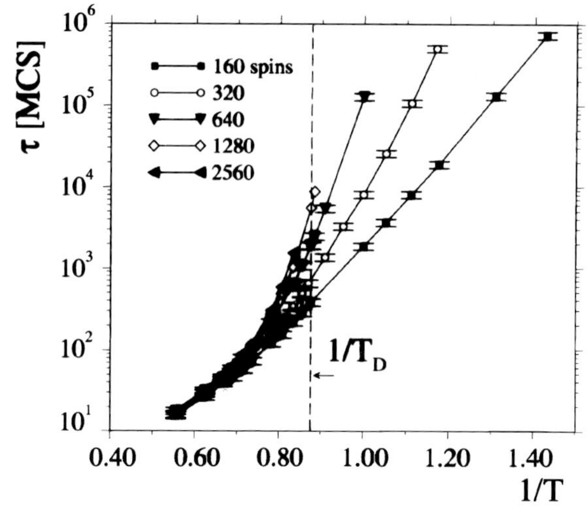

Fig. 42. – Plot of the relaxation time τ *vs.* inverse temperature for different system sizes. The broken vertical line indicates the location of the dynamical transition. From Brangian *et al.* [47].

where the Potts spins σ_i on the lattice sites i of the simple cubic lattice take p discrete values, $\sigma_i \in \{1, 2, \dots, p\}$, and the sum extends over all nearest-neighbor pairs $\langle i, j\rangle$ on the lattice once. For the sake of computational convenience, we now do not use the Gaussian distribution, eq. (7), but rather a bimodal distribution

$$P(J_{ij}) = x\delta(J_{ij} - J) + (1 - x)\delta(J_{ij} + J); \tag{52}$$

data for a Gaussian distribution also have been taken [49, 151], but will not be presented here. In both cases the first two moments of $P(J_{ij})$ are chosen to be $J_0 = [J_{ij}]_{\rm av} = -1$ and $(\Delta J)^2 = [J_{ij}^2]_{\rm av} - [J_{ij}]_{\rm av}^2 = +1$. Note that for $p = 10$ a sufficiently negative J_0 is needed to avoid ferromagnetic order and indeed it was found [151] that neither the magnetization nor the magnetic susceptibility show any sign of ferromagnetic order. For eq. (52) this choice means $J = \sqrt{2}$ and $x = (1-1/\sqrt{2})/2 \approx 0.146$. Monte Carlo runs were carried out with the standard heat bath algorithm [152] over times up to 10^8 Monte Carlo steps per spin (MCS), for lattices of sizes $L = 6$, $L = 10$ and $L = 16$, using periodic boundary conditions. The average over the quenched disorder $[\cdots]_{\rm av}$ was realized by averaging over 100 samples for $L = 6$ and 10 and over 50 samples for $L = 16$.

Turning now to the static properties of this model, it is found that the system stays disordered down to the lowest temperatures, and finite-size effects are insignificant. As an example, fig. 43 presents the spin glass susceptibility, eq. (46): it increases only weakly as the temperature is lowered, and for $T \leq 2.5$ it bends over to reach a saturation value at low temperatures. Note that the internal energy reaches its ground-state value at about the same temperatures [49, 151]: there it is clear that low enough temperatures have been reached. This behavior is very different from the behavior of the infinite-range model

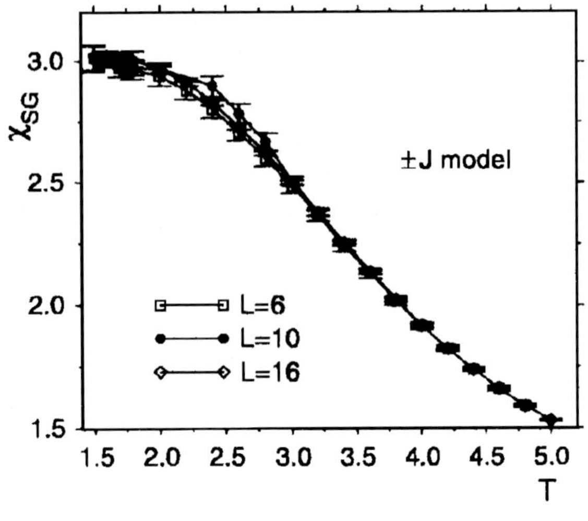

Fig. 43. – Temperature dependence of the spin glass susceptibility χ_{SG} for the 10-state nearest-neighbor Potts glass on the simple cubic lattice. From Brangian *et al.* [49].

(since $[\langle q^2 \rangle]_{av}$ is nonzero for $T < T_0$, χ_{SG} (eq. (46)) diverges to infinity for $T < T_0$ linear in N, and the actual simulation data are compatible with this behavior [47]). The dynamic correlation function $C(t)$ does show a two-step relaxation behavior at low temperatures (fig. 44), but the associated "lifetime" of the plateaus seen there only increases according to simple Arrhenius laws [49,151]. The conclusion is that both the static and the dynamic transition are completely wiped out in this model!

It has been argued, however, that the 10-state Potts glass with the particular choice of interactions described above is a rather unfavorable case [55], and the other models with short-range interactions that exhibit in the mean-field limit the random first-order transition scenario with the two glass transitions (a dynamic and a static one) should resemble the mean-field behavior more closely. In fact, one could imagine also a Potts glass model with a finite but a large range of interaction, and then indeed more pronounced precursor effects to the two glass transitions would be expected. It is maintained, however, that for any finite range of the interaction one expects qualitatively the same behavior as in the nearest-neighbor model, namely the transitions at T_D and at T_0 are ultimately rounded off and replaced by smooth, nonsingular crossovers.

4·4. *Concluding remarks.* – While the phase transition that mean-field theory predicts for the Ising spin glass [48] persists also for systems with short-range interactions [3, 9], though some detailed properties are rather different, the results presented in this section suggest a completely different conclusion for the 10-state Potts glass: both the dynamic and the static transition are completely wiped out, and almost no trace of the beautiful mean-field scenario (fig. 6) is left. Since the Ising spin glass is equivalent to a Potts glass with $p = 2$ states [3, 9], the question arises about what happens for intermediate

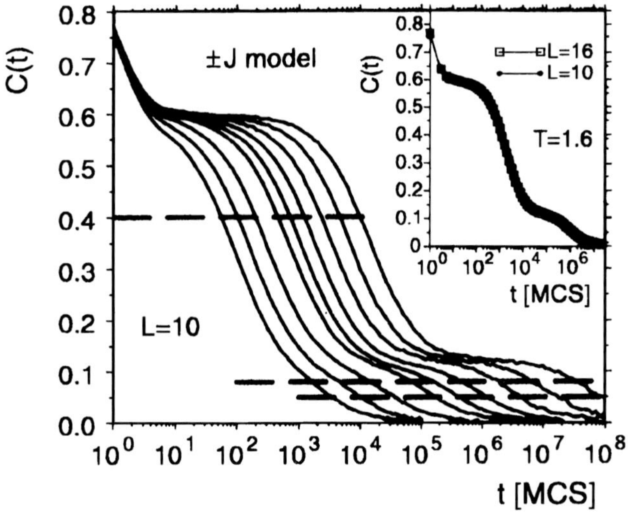

Fig. 44. – Time dependence of the spin autocorrelation function $C(t)$ for temperatures (left to right) $T = 2.4$, 2.2, 2.0, 1.8, 1.7, 1.6, 1.5, 1.4 and 1.3 ($L = 10$). The horizontal dashed lines are used to define relaxation times τ_i. In the inset $C(t)$ for $L = 10$ and $L = 16$ is compared at $T = 1.6$. From Brangian *et al.* [49].

values of p. While for the case $p = 3$ it was suggested that the system in $d = 3$ is exactly at its lower critical dimensionality for a spin glass transition of second order [149], implying an exponentially strong divergence of χ_{SG} as $T \to 0$, namely $\ln \chi_{SG} \propto T^{-2}$ [149], nothing is known about the behavior of Potts glasses with $p = 4, 5, \dots, 9$ states. Also extensive studies of the other models that yield in the mean-field limit the random first-order transition scenario, such as Ising spin glasses with p-spin interaction for $p \geq 3$, but using short-range interactions, would be illuminating. It is possible that for such generalized spin glass models with short-range interactions several "universality classes" with different types of behavior at the freezing transition exist. Thus, while we maintain that for the 10-state Potts glass with short-range forces there are no phase transitions, if the mean interaction is sufficiently negative (for positive mean interaction a transition to standard ferromagnetic order is always possible, of course), only a gradual slowing down at low temperatures occurs (fig. 44), and one cannot rule out different behaviors for other models. Thus, there is a clear need for further work in this area.

At this point we also comment again on the evidence for a static glass transition claimed from the Kauzmann paradox (fig. 5). For the nearest-neighbor Potts glass, it is found that the specific heat is almost quantitatively fitted by the simple Schottky-like behavior [151] $c(T) \propto T^{-2} \exp[-pJ/k_B T]$, with a maximum occurring at $T \approx 4.5$, and for $T \leq 1.5$ $c(T)$ is exponentially small. As a consequence, it immediately follows that the entropy $s(T)$ decreases from its high-temperature value (which is $s(T \to \infty) = (\ln 10)/k_B$) very strongly for $T < 4.5$, and becomes also exponentially small for $T \leq 1.5$. Any linear extrapolation of $s(T)$ with T for that model hence inevitably would also produce a Kauzmann temperature T_K below which the extrapolated entropy would become negative. We feel that this example again shows that such evidence from extrapolations needs to be considered with some reservations. While the attempts to extend the replica formalism of these generalized mean-field spin glasses [42-46, 52, 53, 153, 154] to structural glasses [54, 55, 155-157] are clearly very interesting, one must keep in mind that in physical systems the interactions are short-range, and a true dynamic transition where the phase space splits in infinitely many valleys separated by infinitely high barriers cannot occur, rather all free energy barriers remain finite, and while long-lived states may cause slow relaxation, ergodicity is not broken. Also numerical tests of replica theory for the structural glass transition need to be considered with great caution, since all motions of the particles that are still considered in a given valley are small-scale oscillations. However, in our opinion also small-scale structural rearrangements of particles (such as those which reduce to two-level systems at low temperature in the glass) need to be admitted in each "valley", rather than counting each such state resulting from a localized small-scale structural rearrangement (which also is separated from its neighboring states only by a barrier of a few k_BT) as separate valleys. If this is not done, the resulting entropy associated with all these states differing often only by these small-scale configurational changes is not the complexity but rather the total configurational entropy, comparable to the entropy of the Potts glass in fig. 37. Just as the latter is smooth at the dynamical transition and nonzero at the static transition, also the configurational entropy extracted in the numerical studies of various models for the structural glass transition is nonzero at temperatures above the critical temperature T_c of mode-coupling theory, and varies quite smoothly when T_c is crossed. Therefore, it is a clear misinterpretation to associate this quantity with the complexity, as it is sometimes done.

5. – Final comments

In these lectures we have described computer simulation studies of undercooled fluids and glasses and of related models, such as the Potts glass, in order to test some of the theoretical concepts on the glass transition. We have shown that the mode-coupling theory (MCT) does provide a meaningful description, for a variety of models, from bead-spring models of polymers to the BKS model of SiO_2. The success of MCT for the latter system is particularly noteworthy, since it has been common belief for some time that MCT cannot describe the so-called "strong glass formers". The simulations have shown, however, that even for the latter systems there is a regime at high temperatures where MCT works. However, in all systems the success of MCT is restricted to intermediate regimes of length, time, and temperature: usually small-scale atomistic structural effects are not included due to the coarse-grained nature of the basic model that is used as an input; also only times much larger than a corresponding microscopic time scale (of order 10^{-13} s) are described. On the other hand, at T_c where a crossover from (idealized) MCT to activated dynamics occurs, typically the structural relaxation time is of the order of 10^{-10} to 10^{-7} s: thus idealized MCT can describe only a fairly restricted window of relaxation phenomena, a few decades in time. Whether the extended form of MCT is a qualitatively correct or quantitatively accurate description of the remaining about 10 decades, over which the structural relaxation time rises in between T_c and T_g, is an open question —simulations could not access this regime so far, and the experimental evidence is not very conclusive either.

At this point we mention also a very fundamental objection that has been raised against MCT. As is well known, MCT is based on an input of static quantities, and at least a slight variation of these quantities is needed when the control parameter to approach the glass transition is varied, otherwise the dynamic correlator would not change, and no glass transition could occur. However, there exist models where the static correlations are completely trivial, and nevertheless, a dynamic transition is found to occur [158-161]. It is unclear how these models fit into the picture!

Another theory that has been considered here in some detail is the theory of Gibbs and diMarzio, which is broadly considered as a representative case of the entropy theories of the glass transition. The evidence from simulations shows that the Gibbs-diMarzio theory clearly is unreliable, since it is based on very inaccurate and unjustified approximations. While it is clear —both from the experiment and the simulation— that the entropy of an undercooled fluid decreases strongly as the glass transition is approached, the physical significance of the Kauzmann temperature (found from an extrapolation via the vanishing of the "configurational entropy") remains doubtful. Also the configurational entropy estimated in experiment and simulation should not be confused with the complexity, which is the entropy associated with the infinitely many states that appear when a genuine dynamic transition from ergodic to nonergodic behavior has taken place. The latter situation occurs in various mean-field models, but the extent to which such concepts are to be carried over to systems with short-range forces remains unclear.

Finally, also the concept of a growing characteristic length associated with the glass transition has been tested. While finite-size effects in systems with periodic boundary conditions [132] did not give any evidence for this length [135], it could be extracted from an analysis of wall effects in confined systems (which is a topic of great interest in its own right of course [162], but this was not discussed here). While it was found that such lengths do occur and do increase when the temperature is lowered, it is not clear that they are in any fundamental way responsible for the slowing-down occurring at the glass transition.

We also should emphasize that lack of space has prevented us from properly discussing many phenomenological concepts on the subject which are intuitively appealing (such as the "energy landscape picture", the "free volume theory", etc.). The concepts are potentially useful, but since they make predictions of a lesser extent which can be put to a quantitative test in simulations these concepts have been outside the scope of these lecture notes.

* * *

The material presented here is based on collaborative work with J. BASCHNAGEL, CH. BENNEMANN, C. BRANGIAN, B. DÜNWEG, J. HORBACH, W. KOB, K. OKUN,

W. PAUL, P. SCHEIDLER, T. STÜHN, F. VARNIK, K. VOLLMAYR, H.-P. WITTMANN, and M. WOLFGARDT. It is a pleasure to thank them for a fruitful and stimulating interaction, as well as the Deutsche Forschungsgemeinschaft (DFG) for financial support (under grant No. SFB 262/D1, D2). Work on silica and its mixtures was also supported by SCHOTT GLAS and the Bundesministerium für Forschung und Bildung (BMBF) under grant No. 03N6015. Finally, generous grants of computing time on CRAY-T3E from HLRS Stuttgart and NIC Jülich are acknowledged.

REFERENCES

[1] ZALLEN R., *The Physics of Amorphous Solids* (Wiley, New York) 1983.
[2] JÄCKLE J., *Rep. Prog. Phys.*, **49** (1986) 171.
[3] BINDER K. and YOUNG A. P., *Rev. Mod. Phys.*, **58** (1986) 801.
[4] GÖTZE W., in *Liquids, Freezing and the Glass Transition*, Les Houches, Session LI 1989, edited by HANSEN J. P., LEVESQUE D. and ZINN-JUSTIN J. (North-Holland, Amsterdam) 1991, p. 287.
[5] ZARZYCKI J. (Editor), *Materials Science and Technology*, Vol. **9** (VCH Publishing, Weinheim) 1991.
[6] FELTZ A., *Amorphous Inorganic Materials and Glasses* (VCH Publishing, Weinheim) 1993.
[7] GUTZOW I. and SCHMELZER J., *The Vitreous State: Thermodynamics, Structure, Rheology, and Crystallization* (Springer, Berlin) 1995.
[8] DEBENEDETTI P. G., *Metastable Liquids* (Princeton University Press, Princeton) 1997.
[9] YOUNG A. P. (Editor), *Spin Glasses and Random Fields* (World Scientific, Singapore) 1998.
[10] DONTH E., *The Glass Transition: Relaxation Dynamics in Liquids and Disordered Materials* (Springer, Berlin) 2001.
[11] BINDER K., BASCHNAGEL J. and PAUL W., *Prog. Polym. Sci.*, **28** (2003) 115.
[12] KOB W., preprint cond-mat/0212344.
[13] ARBE A., RICHTER D., COLMENERD J. and FARAGO B., *Phys. Rev. E*, **54** (1996) 3853.
[14] ANGELL C. A. and SICHINA W., *Ann. N.Y. Acad. Sci.*, **279** (1976) 53.
[15] VOGEL W., *Phys. Z.*, **22** (1921) 645.
[16] FULCHER G. S., *J. Amer. Chem. Soc.*, **72** (1925) 3701.
[17] COLBY R. H., *Phys. Rev. E*, **61** (2000) 1783.
[18] DONTH E., *Relaxations and Thermodynamics of Polymers: Glass Transition* (Akademie-Verlag, Berlin) 1992.
[19] FISCHER E. W., DONTH E. and STEFFEN W., *Phys. Rev. Lett.*, **68** (1992) 2344.
[20] DONATI C. and JÄCKLE J., *J. Phys. Condens. Matter*, **8** (1996) 2733.
[21] KOKSHENEV V. B., *Phys. Rev. E*, **57** (1998) 1187.
[22] KOHLRAUSCH F., *Poggendorffs Ann. Phys. Chem.*, **8** (1876) 332.
[23] WILLIAMS G. and WATTS D. C., *Trans. Faraday Soc.*, **66** (1980) 80.
[24] KIVELSON D., KIVELSON S. A., ZHAO X., NUSSINOV Z. and TARJUS G., *Physica A*, **219** (1995) 27.
[25] GREST G. S. and COHEN M. H., *Adv. Chem. Phys.*, **48** (1981) 455.
[26] GIBBS J. H. and DIMARZIO E. A., *J. Chem. Phys.*, **28** (1958) 373.
[27] ADAM G. and GIBBS J. H., *J. Chem. Phys.*, **43** (1965) 139.
[28] KAUZMANN W., *Chem. Rev.*, **43** (1948) 219.
[29] GÖTZE W., *J. Phys. Condens. Matter*, **11** (1999) A1.

[30] SCHILLING R., *Disorder Effects on Relaxation Processes*, edited by RICHERT R. and BLUMEN A. (Springer, Berlin) 1994.
[31] EDWARDS S. F. and ANDERSON P. W., *J. Phys. F*, **5** (1975) 965.
[32] GÖTZE W. and SJÖGREN L., *Z. Phys. B*, **65** (1987) 415; *J. Phys. C*, **21** (1988) 3407.
[33] PEZZIN G., ZILIO-GRANDE F. and SANMARTIN P., *Eur. Polym. J.*, **6** (1970) 1053.
[34] STILLINGER F., *J. Chem. Phys.*, **88** (1988) 7818.
[35] CARMESIN I. and KREMER K., *Macromolecules*, **21** (1988) 2819.
[36] DEUTSCH H. P. and BINDER K., *J. Chem. Phys.*, **94** (1991) 2294.
[37] WITTMANN H.-P., KREMER K. and BINDER K., *J. Chem. Phys.*, **96** (1992) 6291.
[38] BASCHNAGEL J., BINDER K. and WITTMANN H.-P., *J. Phys. Condens. Matter*, **5** (1993) 1597.
[39] PAUL W. and BASCHNAGEL J., *Monte Carlo and Molecular Dynamics Simulations in Polymer Science*, edited by BINDER K. (Oxford University Press, New York) 1995, p. 307.
[40] WITTMANN H.-P., *J. Chem. Phys.*, **95** (1991) 8449.
[41] WEBER H., PAUL W., KOB W. and BINDER K., *Phys. Rev. Lett.*, **78** (1997) 2136.
[42] ELDERFIELD D. and SHERRINGTON D., *J. Phys. C*, **16** (1983) L491, L971, L1169.
[43] GROSS D. J., KANTER I. and SOMPOLINSKY H., *Phys. Rev. Lett.*, **55** (1985) 304.
[44] KIRKPATRICK T. R. and THIRUMALAI D., *Phys. Rev. B*, **37** (1988) 5342.
[45] THIRUMALAI D. and KIRKPATRICK T. R., *Phys. Rev. B*, **38** (1988) 4881.
[46] DE SANTIS E., PARISI G. and RITORT F., *J. Phys. A*, **28** (1995) 3025.
[47] BRANGIAN C., KOB W. and BINDER K., *J. Phys. A*, **95** (2002) 191.
[48] SHERRINGTON D. and KIRKPATRICK S., *Phys. Rev. Lett.*, **35** (1975) 1972.
[49] BRANGIAN C., KOB W. and BINDER K., *Europhys. Lett.*, **59** (2002) 526.
[50] PALMER R. G., *Adv. Phys.*, **31** (1982) 669.
[51] VAN ENTER A. C. D. and VAN HEMMEN L., *Phys. Rev. A*, **29** (1984) 355.
[52] KIRKPATRICK S. and THIRUMALAI D., *Phys. Rev. Lett.*, **58** (1987) 2091.
[53] CRISANTI A. and SOMMERS H. J., *Z. Phys. B*, **87** (1992) 341.
[54] MÉZARD M. and PARISI G., *J. Phys. Condens. Matter*, **12** (2000) 6655.
[55] EASTWOOD M. P. and WOLYNES P. G., *Europhys. Lett.*, **60** (2002) 587.
[56] PARISI G., *Physica A*, **280** (2000) 115; MÉZARD M., *Physica A*, **306** (2002) 25.
[57] BINDER K. (Editor), *Monte Carlo and Molecular Dynamics Simulations in Polymer Science* (Oxford University Press, New York) 1995.
[58] KREMER K. and GREST G. S., *J. Chem. Phys.*, **92** (1990) 5057.
[59] BUCHHOLZ J., PAUL W., VARNIK F. and BINDER K., *J. Chem. Phys.*, **117** (2002) 7364.
[60] BINDER K., BENNEMANN C., BASCHNAGEL J. and PAUL W., *Anomalous Diffusion, from Basics to Applications*, edited by KUTNER R., PEKALSKI A. and SZNAJD-WERON K. (Springer, Berlin) 1998, p. 124.
[61] MC KENNA G. B., *Comprehensive Polymer Science*, Vol. **2**, edited by BOOTH C. and PRICE C. (Pergamon Press, Oxford) 1989, p. 311.
[62] BENNEMANN C., PAUL W., BINDER K. and DÜNWEG B., *Phys. Rev. E*, **57** (1998) 843.
[63] BRÜNING R. and SAMWER K., *Phys. Rev. B*, **46** (1996) 11318.
[64] VOLLMAYR K., KOB W. and BINDER K., *J. Chem. Phys.*, **105** (1996) 4714.
[65] VOLLMAYR K., KOB W. and BINDER K., *Phys. Rev. B*, **54** (1996) 15808.
[66] BASCHNAGEL J., BENNEMANN C., PAUL W. and BINDER K., *J. Phys. Condens. Matter*, **12** (2000) 6365.
[67] AICHELE M. and BASCHNAGEL J., *Eur. Phys. J. E*, **5** (2001) 245.
[68] BENNEMANN C., BASCHNAGEL J., PAUL W. and BINDER K., *Comput. Theor. Polym. Sci.*, **9** (1999) 217.
[69] BENNEMANN C., PAUL W., BASCHNAGEL J. and BINDER K., *J. Phys. Condens. Matter*, **11** (1999) 2179.

[70] Rouse P., *J. Chem. Phys.*, **21** (1953) 127.
[71] Doi M. and Edwards S. F., *The Theory of Polymer Dynamics* (Clarendon Press, Oxford) 1986.
[72] Chong S. H. and Fuchs M., *Phys. Rev. Lett.*, **88** (2002) 185702.
[73] Kron A. K., *Polymer Sci. USSR*, **7** (1965) 1361.
[74] Wall T. F. and Mandel F., *J. Chem. Phys.*, **63** (1975) 4592.
[75] Wolfgardt M., Baschnagel J. and Binder K., *J. Phys. (Paris)*, **5** (1995) 1035.
[76] Okun K., Wolfgardt M., Baschnagel J. and Binder K., *Macromolecules*, **30** (1997) 3075.
[77] Baschnagel J., *Phys. Rev. B*, **49** (1994) 135.
[78] Baschnagel J. and Fuchs M., *J. Phys. Condens. Matter*, **7** (1995) 6761.
[79] Tries V., Paul W., Baschnagel J. and Binder K., *J. Chem. Phys.*, **106** (1997) 738.
[80] Lobe B. and Baschnagel J., *J. Chem. Phys.*, **101** (1994) 1616.
[81] Flory P. J., *Proc. R. Soc. London, Ser. A*, **234** (1956) 60.
[82] Milchev A., *C.R. Bulg. Acad. Sci.*, **36** (1983) 1415.
[83] Wolfgardt M., Baschnagel J., Paul W. and Binder K., *Phys. Rev. E*, **54** (1996) 1535.
[84] Angell C. A., *Relaxation in Complex Systems*, edited by Ngai K. L. and Wright G. B. (US Dept. Commerce, Springfield) 1985, p. 1.
[85] Horbach J., Kob W., Binder K. and Angell C. A., *Phys. Rev. E*, **54** (1996) R5897.
[86] Horbach J., Kob W. and Binder K., *Philos. Mag. B*, **77** (1998) 297.
[87] Horbach J. and Kob W., *J. Noncryst. Solids*, **235-237** (1998) 320.
[88] Horbach J. and Kob W., *Phys. Rev. B*, **60** (1999) 3169.
[89] Horbach J., Kob W. and Binder K., *J. Phys. Chem. B*, **103** (1999) 4104.
[90] Horbach J., Kob W. and Binder K., *Eur. Phys. J. B*, **19** (2001) 531.
[91] Horbach J., Kob W. and Binder K., *Chem. Geol.*, **174** (2001) 87.
[92] Scheidler P., Kob W., Latz A., Horbach J. and Binder K., *Phys. Rev. B*, **63** (2001) 104204.
[93] Horbach J. and Kob W., *Phys. Rev. E*, **64** (2001) 041503.
[94] Horbach J. and Kob W., *J. Phys. Condens. Matter*, **14** (2002) 9237.
[95] Horbach J., Kob W. and Binder K., *Phys. Rev. Lett.*, **88** (2002) 125502.
[96] Horbach J., Winkler A., Kob W. and Binder K., *High Performance Computing in Science and Engineering '02*, edited by Krause E. and Jäger W. (Springer, Berlin) 2002, p. 109.
[97] Binder K., Horbach J., Kob W. and Winkler A., *Comput. Sci. Engin.*, **5** (2003) 60.
[98] Binder K., Horbach J., Kob W. and Varnik F., *J. Phys. Condens. Matter*, **16** (2004) S429.
[99] Binder K., Horbach J., Kob W. and Winkler A., preprint.
[100] Horbach J., Winkler A., Kob W. and Binder K., *Mat. Res. Symp. Proc.*, **754** (2003) 147.
[101] Winkler A., Horbach J., Kob W. and Binder K., *J. Chem. Phys.*, **120** (2004) 384.
[102] van Beest B. H. W., Kramer G. J. and van Santen R. A., *Phys. Rev. Lett.*, **64** (1990) 1955.
[103] Kramer G. J., de Man A. J. M. and van Santen R. A., *J. Am. Chem. Soc.*, **64** (1991) 6435.
[104] Benoit M., Ispas S., Jund P. and Julien R., *Eur. Phys. J. B*, **13** (2000) 631.
[105] Mischler C., Kob W. and Binder K., *Comput. Phys. Commun.*, **147** (2002) 222.
[106] Ispas S., Benoit M., Jund P. and Jullien R., *J. Noncryst. Solids*, **307-310** (2202) 946.
[107] Benoit M. and Kob W., *Europhys. Lett.*, **60** (2002) 269.

[108] TSE J. S. and KLUG D. D., *J. Chem. Phys.*, **95** (1991) 9176; *Phys. Rev. Lett.*, **67** (1991) 3559.

[109] TSE J. S., KLUG D. D. and TE PAGE Y., *Phys. Rev. B*, **46** (1992) 5993.

[110] MÜSER M. H., *J. Chem. Phys.*, **114** (2001) 6364.

[111] MÜSER M. H. and BINDER K., *Phys. Chem. Miner.*, **28** (2001) 746.

[112] ALLEN M. P. and TILDESLEY D., *Computer Simulation of Liquids* (Clarendon Press, Oxford) 1987.

[113] BINDER K. and CICCOTTI G. (Editors), *Monte Carlo and Molecular Dynamics of Condensed Matter Systems, Conf. Proc.*, Vol. **49** (Editrice Compositori, Bologna) 1996.

[114] RIVIER N., *Topological Disorder in Condensed Matter*, edited by YONEZAWA F. and NINOMIYA T. (Springer, Berlin) 1983, p. 13.

[115] PRICE D. L. and CARPENTER J. M., *J. Noncryst. Solids*, **92** (1987) 153.

[116] SOSMAN R. B., *The Properties of Silica* (Chemical Catalog Co., New York) 1927.

[117] ZELLER R. C. and POHL R. O., *Phys. Rev. B*, **4** (1971) 2029.

[118] RICHET P., BOTTINGA Y., DENIELOU D., PETITET J. P. and TEGUI C., *Geochim. Cosmochim. Acta*, **46** (1982) 2639.

[119] POLIAN A., VO-TANH D. and RICHET P., *Europhys. Lett.*, **57** (2002) 375.

[120] ANDERSON P. W., HALPERIN B. I. and VARMA C. M., *Philos. Mag.*, **25** (1972) 1.

[121] PHILLIPS W. A., *J. Low Temp. Phys.*, **7** (1972) 351.

[122] MIKKELSEN J. C., *Appl. Phys. Lett.*, **45** (1984) 1187.

[123] BREBEC G., SEGUIN R., SELLA C., BEVENOT J. and MARTIN J. C., *Acta Metall.*, **28** (1970) 327.

[124] STRUIK L. C. E., *Physical Aging in Amorphous Polymers and Other Materials* (Elsevier, Amsterdam) 1978.

[125] URBAIN G., *Geochim. Cosmochim. Acta*, **46** (1982) 1061.

[126] SCIORTINO F. and KOB W., *Phys. Rev. Lett.*, **86** (2001) 327.

[127] KOB W., HORBACH J. and BINDER K., *AIP Conf. Proc.*, **469** (1999) 441.

[128] HORBACH J. and KOB W., *Philos. Mag. B*, **79** (1999) 1981.

[129] MISAWA M., PRICE D. L. and SUZUKI K., *J. Noncryst. Solids*, **37** (1980) 35.

[130] MEYER A., SCHOBER H. and DINGWELL D. B., *Europhys. Lett.*, **59** (2002) 708.

[131] FISHER M. E., *Critical Phenomena*, edited by GREEN M. S. (Academic Press, New York) 1971, p. 1.

[132] BINDER K., *Computational Methods in Field Theory*, edited by GAUSTERER H. and LANG C. B. (Springer, Berlin) 1992, p. 59.

[133] BINDER K., *Rep. Prog. Phys.*, **60** (1997) 487.

[134] BRANGIAN C., KOB W. and BINDER K., *Europhys. Lett.*, **53** (2001) 756.

[135] BINDER K., BASCHNAGEL J., BÖHMER S., PAUL W. and WOLFGARDT M., *Non-Crystalline and Nanoscale Materials*, edited by RIVAS R. and LOPEZ-QUINTELA M. A. (World Scientific, Singapore) 1998, p. 252.

[136] SCHEIDLER P., KOB W. and BINDER K., *Europhys. Lett.*, **93** (2002) 701.

[137] VARNIK F., BASCHNAGEL J. and BINDER K., *Phys. Rev.*, **65** (2002) 021507.

[138] VARNIK F., BASCHNAGEL J. and BINDER K., *J. Chem. Phys.*, **113** (2000) 4444.

[139] VARNIK F., BASCHNAGEL J. and BINDER K., *Eur. Phys. J. E*, **8** (2002) 175.

[140] VARNIK F. *et al.*, preprint.

[141] FORREST J. A. and JONES R. A. L., *Polymer Surfaces, Interfaces and Thin Films*, edited by KARIM A. and KUMAR S. (World Scientific, Singapore) 2000, p. 251.

[142] KOB W. and ANDERSEN H.-C., *Phys. Rev. E*, **51** (1995) 4626.

[143] SCHEIDLER P., KOB W. and BINDER K., *Europhys. Lett.*, **52** (2000) 277.

[144] SCHEIDLER P., KOB W., BINDER K. and PARISI G., *Philos. Mag. B*, **82** (2002) 283.

[145] DONATI C., DOUGLAS J. F., KOB W., PLIMPTON S. J., POOLE P. H. and GLOTZER S. C., *Phys. Rev. Lett.*, **80** (1998) 2338.
[146] BENNEMANN C., DONATI C., BASCHNAGEL J. and GLOTZER S. C., *Nature*, **399** (1999) 246.
[147] BRANGIAN C., KOB W. and BINDER K., *Comput. Phys. Commun.*, **146** (2002) 9.
[148] ZIA R. K. P. and WALLACE D. J., *J. Phys. A*, **8** (1975) 1495.
[149] SCHEUCHER M., REGER J. D., BINDER K. and YOUNG A. P., *Phys. Rev. B*, **42** (1990) 6881.
[150] HUKUSHIMA K. and NEMOTO K., *J. Phys. Soc. Jpn.*, **64** (1996) 1604.
[151] BRANGIAN C., KOB W. and BINDER K., *J. Phys. Condens. Matter*, **36** (2003) 10847.
[152] LANDAU D. P. and BINDER K., *A Guide to Monte Carlo Simulation in Statistical Physics* (Cambridge University Press, Cambridge) 2000.
[153] KIRKPATRICK T. R. and THIRUMALAI D., *Phys. Rev. B*, **36** (1987) 5388.
[154] KIRKPATRICK T. R. and WOLYNES P. G., *Phys. Rev. B*, **36** (1987) 8552.
[155] MÉZARD M. and PARISI G., *J. Phys. A*, **29** (1996) 6515.
[156] MÉZARD M. and PARISI G., *J. Phys. Lett.*, **82** (1999) 747.
[157] MÉZARD M. and PARISI G., *J. Chem. Phys.*, **111** (1999) 1076.
[158] KOB W. and ANDERSEN H. C., *Phys. Rev. E*, **48** (1993) 4364.
[159] KAWASAKI K. and KIM B., *J. Phys. Condens. Matter*, **14** (2002) 2265.
[160] SCHILLING R. and SZAMEL G., *Europhys. Lett.*, **61** (2003) 2007.
[161] SCHILLING R. and SZAMEL G., *J. Phys. Condens. Matter*, **15** (2003) S967.
[162] ARNDT M., STANNARIUS R., GROOTHUES H., HEMPEL H. and KREMER F., *Phys. Rev. Lett.*, **79** (1997) 2077.

Study of the liquid-to-glass and glass-to-glass transitions in dense $L64$ copolymer micellar solution by scattering experiments

Sow-Hsin Chen and Wei-Ren Chen

Department of Nuclear Engineering, Massachusetts Institute of Technology Cambridge, MA 02139-4307, USA

F. Mallamace

Dipartimento di Fisica e Istituto Nazionale per la Fisica della Materia Università di Messina - Messina, Italy

Recent mode coupling theory (MCT) calculations show that if a short-range attractive interaction is added to the pure hard-sphere system, one may observe a new type of glass originating from the clustering effect (the attractive glass), as a result of the attractive interaction. This is in addition to the known glass-forming mechanism due to the cage effect in the hard-sphere system (the repulsive glass). The calculations also indicate that, if the range of attraction is sufficiently short compared to the diameter of the particle, within a certain interval of volume fractions where the two glass-forming mechanisms nearly balance each other, varying the external control parameter, the effective temperature, makes the glass-to-liquid-to-glass re-entrance and the glass-to-glass transitions possible. Here we present experimental evidence of both transitions, obtained from small-angle neutron scattering and photon correlation measurements taken from dense $L64$ copolymer micellar solutions in heavy water. Varying the temperature in certain predicted volume fraction range triggers a sharp transition between these two different types of glass. In particular, according to MCT, there is an end point (called A_3 singularity) of this glass-to-glass transition line, beyond which the long-time dynamics of the two glasses become identical. Our findings confirm this theoretical prediction.

Surprisingly, although the Debye-Waller factors, the long-time limit of the coherent intermediate scattering functions, of these two glasses obtained from photon correlation measurements indeed become identical at the predicted volume fraction, they exhibit distinctly different intermediate time relaxation. Furthermore, our experimental results obtained from volume fractions beyond the end point are characterized by the same features as the repulsive glass obtained before the end point. A complete phase diagram giving the boundaries of the structural arrest transitions for $L64$ micellar system is given for the first time.

1. – Introduction

Observation of gelation phenomena in colloids, such as in protein gels or in gel-like emulsion-polymer mixtures, is a common experience in our daily life [1]. The gel state of a soft matter is characterized by an apparent manifestation of a drastic slowing-down of the local structural relaxation dynamics around the typical particle in the system, which results in creation of a barrier for a dense, supercooled system to reach its true lowest free energy state, an ordered crystalline solid. Instead, a nonergodic state or a "structurally arrested" state, in which the local particle configuration is deviated from the thermodynamical equilibrium state, is formed due to the short-range interaction potential between the particles. Although the structurally arrested state is an ubiquitous state of matter in our environment, traditionally this state of matter is rather ill-characterized physically: It is amorphous but classified neither as a gas, a liquid nor as a crystal. Fortunately, recent progress [2] in the mode coupling theory (MCT) calculations opens up a possibility to gain a deeper physical insight into the detailed slow dynamical behavior of the system trapped in these structurally arrested states.

In the MCT, the structural arrest transition which causes the colloidal system to transform from a liquid state to an amorphous solid state is viewed as a result of a kinetic glass transition (KGT) predicted by the solution of the MCT bifurcation equation. The KGT has been studied extensively both experimentally and theoretically for a class of systems which can be modelled as a hard-sphere system in the past two decades [3-5].

Many colloids can be modelled as a system of spherical particles interacting via a specific form of pairwise state-dependent effective potential. Therefore, the predictions of the MCT can be tested unambiguously by comparing various experimental results obtained from a colloidal system with results calculated explicitly using a specific inter-particle potential. Initially, a hard-sphere system with a purely repulsive interaction potential was used to model the colloidal system. At low volume fractions, the behavior of the colloidal system is fluid-like. As the volume fraction increases, the test-particle time correlation function (self-intermediate scattering function, self-ISF) and the density-density correlation function (the intermediate scattering function, ISF) of the particles exhibit a two-stage relaxation process. The initial decay of the self-ISF corresponds to the rattling of a typical particle confined within a transient cage formed by its neighbors, followed by a slow decay, resulting from the relaxation of the cage itself and the escape of the trapped particle by re-arranging its nearest-neighbor configuration. The latter process leads to a possibility of particle diffusion through coupling to a structural relaxation process. If a system can be manipulated so as to avoid the known crystallization at the volume fraction of 0.495, for example by artificially creating a polidispersity in sizes of a few percent, at a critical volume fraction ϕ_c, which was predicted to be 0.516 (but determined to be 0.58 experimentally), the KGT can be observed [6]. At the KGT, both the particle diffusion and the long-time density fluctuations at all wavelengths freeze and the system undergoes an ergodic-to-nonergodic transition. Although this characteristic feature has been confirmed by computer molecular-dynamics simulations [7] and laboratory experiments [6, 8] involving several different glass-formers, there have been some anomalous dynamical observations, which cannot be interpreted in terms of the theory based on the hard-sphere potential alone [9].

A more complete picture of the KGT is obtained by modeling the inter-particle potential more accurately. Recent MCT calculations [10-12] show that if a system is characterized by a hard core plus an additional short-range attractive interaction, for example, by an adhesive hard-sphere system (AHS), a different structural arrest scenario emerges. Theoretically, the phase behavior of the AHS is characterized by an effective temperature $T^* = k_B T/u$, the volume fraction of the particles ϕ and the fractional attractive well width $\varepsilon = \Delta/R$, where $-u$ is the depth of the attractive square well, Δ the width of the well and R the diameter of the particle. In this case, for a given ε, aside from the volume fraction, as in the case of a pure hard-sphere system, a second external control parameter, the effective temperature T^*, is introduced into the description of the phase behavior of the system and the loss of ergodicity can take place either by increasing the volume fraction or by changing the effective temperature.

Figure 1 gives part of the predicted phase diagram based on the AHS with $\varepsilon = 0.03$. In this case, it is possible for the system to undergo a re-entrant (glass-to-liquid-to-glass) transition (for example at $\phi = 0.532$ in fig. 1) by varying the effective temperature. At high effective temperatures and at sufficiently high volume fractions, the system evolves into the well-known structurally arrested (glassy) state, called a "repulsive glass", as a result of the cage effect, a manifestation of the excluded-volume effect due to the existence of the hard core. However, at relatively low effective temperatures, an "attractive glass" can form in which motion of the typical particle is hindered instead by the cluster formation with neighboring particles due to the short-range attractive interaction. Moreover, in an AHS, aside from the hard-core diameter, an additional length scale, the range of the attractive well, should come into play (parameter ε). With this insight, we may divide spherical colloidal systems into two categories: the one-length-scale hard-sphere system, in which the glass formation is dictated by the cage effect alone; and the two-length-scale AHS, in which the two glass-forming mechanisms may coexist and compete with each other, depending on the values of the controlled parameters. The MCT calculations show that, with sufficiently small ε, variation of the two control parameters, T^* and ϕ, allows the transition between these two distinct forms of glass. Thus there is a branch of the KGT line across which transitions between the attractive glass and the repulsive glass are predicted. Of particular interest is the occurrence of an A_3 singularity (see fig. 1) at which point the glass-to-glass transition line terminates. MCT suggests that

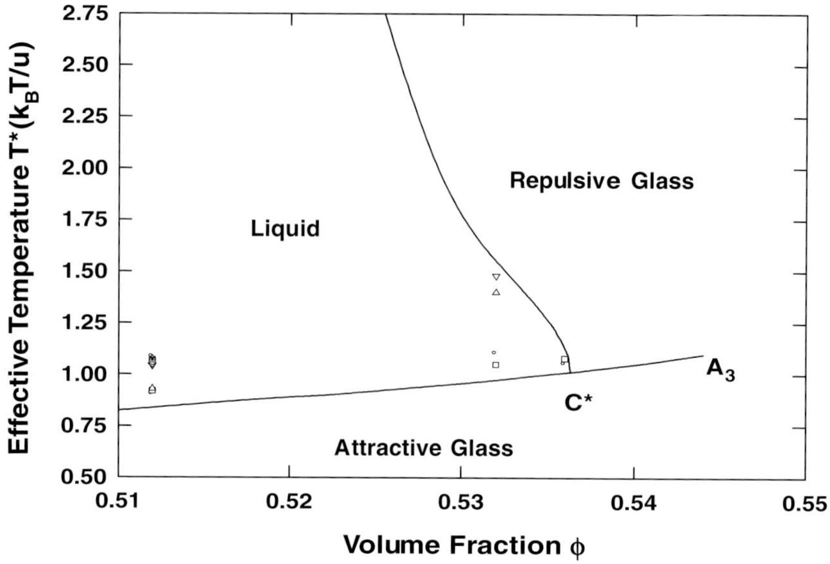

Fig. 1. – Theoretical phase diagram predicted by MCT calculations using the short-range attractive square-well potential with $\varepsilon = 0.03$. The calculations predict the attractive glass-to-liquid-to-repulsive glass re-entrant transition, the attractive glass-to-repulsive glass transition and the end point of the glass-glass transition line. Symbols (open circles, squares and triangles) represent the effective temperature T^*, determined by fitting the experimental SANS data taken in the liquid state with the method explained in the text, for different volume fractions.

the long-time dynamics of the two distinct structurally arrested states become identical at and beyond this point.

Several ongoing experimental investigations have confirmed some of the theoretical predictions, such as the re-entrant glass-to-liquid-to-glass transition phenomenon [13, 14] and the logarithmic relaxation of the glassy dynamics in the liquid states in the vicinity of the A_3 singularity [15]. They are considered to be the signatures of the glassy dynamics in the two-length-scale system. Yet, except for some recent experimental reports on a dense micellar system [16, 17], a detailed investigation of the glass-to-glass transition is just emerging, and the physical insight into the slow dynamics near its associated A_3 singularity still needs to be clarified.

The study of KGT by neutron scattering method is traditionally dominated by the use of the inelastic neutron scattering (INS) and the spin-echo spectroscopy techniques. However, by analyzing an extensive set of small-angle neutron scattering (SANS) intensity distribution obtained from $L64$ micellar solutions, which exposes all the characteristics of the AHS from our previous works [16, 17], we demonstrated that SANS is also a suitable tool to investigate the boundary of the KGT: By analyzing SANS intensity distributions, the difference in the structure factor before and after the KGT can be visualized vividly. Supplemental to the photon correlation spectroscopy (PCS), it provides rapid measurements of the peak of the structure factor over a wide temperature and volume fraction ranges, which is more difficult to do with INS or spin-echo spectroscopy. By this novel method, we successfully confirmed the predicted re-entrant phenomenon, the glass-to-glass transition and, most importantly, the existence of the A_3 singularity, in quantitative agreement with predictions of MCT. Furthermore, we are able, using SANS technique, to map out the complete structural arrest phase boundaries for the $L64$ micellar system for the first time.

2. – Experiment

The micellar system we study is a triblock copolymer $L64$, one member of the Pluronic family, used extensively in industrial applications. After the necessary purification procedure [18] to remove hydrophobic impurities, the polymer is dissolved in deuterated water (D_2O). Pluronic is made of polyethylene oxide (PEO) and polypropylene oxide (PPO). The chemical formula of $L64$ is $(PEO)_{13}(PPO)_{30}(PEO)_{13}$, having a molecular weight of 2990 Dalton. At low temperatures, both PEO and PPO are hydrophilic, so that $L64$ chains readily dissolve in water, and the polymers exist as unimers. As the temperature increases there is a decrease in the probability of hydrogen-bond formation between water and polymer molecules, and PPO tends to become less hydrophilic faster than PEO. This creates an inbalance of hyrophilicity between the end-blocks and the middle-block of the polymer molecule and the copolymer molecules acquire surfactant properties in the aqueous environment and self-assemble to form micelles. Thus the micellar formation is initiated at a well-defined Critical Micellar Temperature-Concentration (CMT-CMC) line. As the temperature further increases, water becomes progressively a poor solvent to both PPO and PEO chains, and the effective micelle-micelle interaction becomes attractive. The evidence for the increased short-range micellar attraction as a function of temperature comes from the existence of a lower consolute critical point at $C = 5.0\,\text{wt}\%$ and $T = 330.9\,\text{K}$ [19] and a percolation line detected by a jump of zero-shear viscosity of more than two orders of magnitude [20]. We have explored the dynamically arrested states and their structure in this micellar system, as a function of temperature at high volume fractions. During the experiments the temperature stability was controlled to within $\pm 0.1\,\text{K}$.

SANS measurements were performed at the $NG7$, 40 m SANS spectrometer in the NIST Center for Neutron Research, and at SAND station at the Intense Pulsed Neutron Source (IPNS) in Argonne National Laboratory. At $NG7$, incident monochromatic neutrons of wavelength $\lambda = 5\,\text{Å}$ with $\Delta\lambda/\lambda = 10\%$ were used. Sample-to-detector distance was fixed at 6 m, covering the magnitude of wave vector transfer (k) range from $0.008\,\text{Å}^{-1}$ to $0.3\,\text{Å}^{-1}$. At IPNS, high-energy spallation neutrons are generated by bombarding a heavy-metal target with repetitive pulses of 500 MeV protons. After moderation of these high-energy neutrons by a solid methane moderator, a pulse of white neutrons was selected with an effective wavelength range from 1.5 Å to 4 Å. In SAND all these neutrons are utilized by encoding their individual time of flight and their scattering angles determined by their detected position at a $2D$ area detector. The $2D$ area detector has an active area of $40 \times 40\,\mathrm{cm}^2$ and the sample to detector distance is 2 m. This configuration allows a maximum scattering angle of about 10°. The reliable k-range covered in the measurements were from $0.004\,\text{Å}^{-1}$ to $0.6\,\text{Å}^{-1}$. k-resolution functions of both of these SANS spectrometers are Gaussian and well characterized. It is essential that we apply the resolution broadening to the theoretical cross-section when fitting the intensity data. Sample liquid was contained in a flat quartz cell with 1 mm path length. The measured intensity was corrected for background and empty cell contributions and normalized by a reference scattering intensity of a polymer sample of known cross-section.

The Photon Correlation Spectroscopy (PCS) measurements were made at a scattering angle $\theta = 90°$, using a continuous-wave solid-state laser (Verdi-Coherent) operating at 50 mW ($\lambda = 5120\,\text{Å}$) and an optical scattering cell of diameter 1 cm in a refractive index matching bath. The intensity data were also corrected for turbidity and multiple scattering effects. PCS data have been taken using a digital correlator with a logarithmic sampling time scale which allows an accurate description of both the short- and the long-time regions of the Intermediate Scattering Function (ISF), from 1 μs up to 100 s. Following the method used for colloidal hard spheres [8], we measure the correlation function

$$g^{(2)}(k,\tau) = \frac{\langle\langle I(k,0)I(k,\tau)\rangle\rangle}{\langle\langle I(k,0)\rangle\rangle^2}, \tag{1}$$

where $I(k,\tau)$ is the intensity of light scattered at wave vector k and at a delay time τ. The first bracket denotes the time average, and the second bracket denotes the positional average over different parts of the sample. Since the system shows a structural arrest transition, and therefore a nonergodic behavior, particular care has been taken in averaging over many different positions in the sample for the measurement of the long-time part of the time correlation function. For each sample and each temperature we have performed more than 200 positional average measurements, observing a large scattering area corresponding to three or more independent Fourier components and changing the position of the sample in order to observe different scattering volumes. The ISF $g^{(1)}(k,\tau)$ can be obtained from eq. (1) in a straightforward way using the Siegert relation.

3. – SANS data analysis

The absolute intensity (in unit of cm^{-1}) of small-angle neutron scattering from a system of mono-dispersed micelles can be expressed by the following formula:

$$I(k) = cN\left[\sum_i b_i - \rho_{\mathrm{w}} v_{\mathrm{p}}\right]^2 \overline{P}(k)S(k), \tag{2}$$

where c is the concentration of polymer (number of polymers/cm^3), N the aggregation number of polymers in a micelle, $\sum_i b_i$ the sum of coherent scattering lengths of atoms comprising a polymer molecule, ρ_{w} the scattering length density of D_2O, v_{p} the molecular volume of the polymer. $\overline{P}(k)$ is the normalized intra-particle structure factor calculated by the modified cap-and-gown model [16] and $S(k)$ the inter-micellar structure factor of a spherical particle system interacting via an attractive square well potential with a repulsive hard core. $S(k)$ is calculated analytically by solving the Ornstein-Zernike (OZ) equation in Percus-Yevick (PY) approximation, for this square-well potential, to the first order in a series of small ε expansion [21]. The detailed analytical expression of $S(k)$ is given in the following section [16]. It is important to note that the $\overline{P}(k)$ in the modified cap-and-gown model is a function of the aggregation number N only and $S(k)$ is the function of the aggregation number N, the volume fraction of micelles ϕ, the fractional well width parameter ε and the effective temperature T^*. To fit SANS data in an absolute intensity scale, it is essential to take into account the incoherent background and the resolution function of the instrument used. The main point to emphasize here is that an absolute SANS intensity distribution in the liquid state can be calculated by the above eq. (2) unambiguously with the four parameters: N, ϕ, ε and T^*.

3·1. *The analytical expression for the inter-micellar structure factor* $S(k)$. – A square-well potential with a repulsive core of diameter R' and the real-particle diameter R is used to model a hard sphere with an adhesive surface layer. The pairwise interaction potential is defined as

$$V(r) = \begin{cases} +\infty, & \text{for } 0 < r < R', \\ -u, & \text{for } R' < r < R, \\ 0, & \text{for } r > R. \end{cases} \tag{3}$$

We define here $\varepsilon = (R - R')/R$ as the fractional well width parameter and $T^* = k_B T/u$ as the effective temperature. Then the inter-micellar structure factor is a function of four parameters: the real-particle diameter R, the volume fraction ϕ, the fractional well width parameter ε and the effective temperature T^*. It should be noted that the real-particle diameter R is tied uniquely to the aggregation number N and the volume fraction ϕ through the relation $\phi = c\pi R^3/6N$.

The Ornstein-Zernike (OZ) equation in Percus-Yevick (PY) approximation for this square-well potential can be solved analytically to the first order in ε expansion [21]. The result of the structure factor is given as follows:

$$\begin{aligned} \frac{1}{S(Q)} - 1 = {} & 24\varphi\left[\alpha f_2(Q) + \beta f_3(Q) + \frac{1}{2}\phi\alpha f_5(Q)\right] + 4\phi^2\lambda^2\varepsilon^2\left[f_2(\varepsilon Q) - \frac{1}{2}f_3(\varepsilon Q)\right] + \\ & + 2\phi^2\lambda^2\left[f_1(Q) - \varepsilon^2 f_1(\varepsilon Q)\right] - \frac{2\phi\lambda}{\varepsilon}\left[f_1(Q) - (1-\varepsilon)^2 f_1((1-\varepsilon)Q)\right] - \\ & - 24\phi\left[f_2(Q) - (1-\varepsilon)^3 f_2((1-\varepsilon)Q)\right], \end{aligned} \tag{4}$$

where

$$
\begin{aligned}
Q &= kR,\\
\phi &= \frac{c\pi R^3}{6N},\\
\alpha &= \frac{(1+2\phi-\mu)^2}{(1-\phi)^4},\\
\beta &= -\frac{3\phi(2+\phi)^2 - 2\mu\left(1+7\phi+\phi^2\right)+\mu^2(2+\phi)}{2(1-\phi)^4},\\
\mu &= \lambda\phi(1-\phi),\\
\lambda &= \frac{6\left(\Delta-\sqrt{\Delta^2-\Gamma}\right)}{\phi},\\
\Delta &= \tau + \frac{\phi}{(1-\phi)} = \frac{1}{12\varepsilon}\exp\left[-\frac{u}{k_{\mathrm{B}}T}\right] + \frac{\phi}{(1-\phi)},\\
\Gamma &= \frac{\phi(1+\phi/2)}{3(1-\phi)^2},\\
f_1(x) &= \frac{1-\cos x}{x^2},\\
f_2(x) &= \frac{\sin x - x\cos x}{x^3},\\
f_3(x) &= \frac{2x\sin x - \left(x^2-2\right)\cos x - 2}{x^4},\\
f_5(x) &= \frac{\left(4x^3-24x\right)\sin x - \left(x^4-12x^2+24\right)\cos x + 24}{x^6}.
\end{aligned} \tag{5}
$$

This formula is accurate in the k region around the first diffraction peak for $\varepsilon < 0.05$ [22]. It should be noted that the real volume of a micelle is significantly larger than the volume of N dry polymer chains. It consists of the volume of the N dry polymer chains plus the volumes of the associated hydration water per polymer chain in the corona region. The real volume of the micelle, rather than the polymer volume, determines the thermodynamic quantities such as the osmotic compressibility and the scattering intensities. It further determines the dynamical properties, such as diffusivity and viscosity, of the polymeric micellar solution [21]. Obtaining the real volume of a micelle in a self-consistent way is not a trivial matter. It requires the knowledge of the detailed microstructure such as the hydration number per polymer chain. This information of the associated solvent molecules is not available from the theoretical consideration but can be extracted accurately from the analysis of the SANS intensity distribution through the volume fraction ϕ and the aggregation number N.

To illustrate the model fitting, a SANS intensity distribution obtained from a 48.5 wt.% micellar solution at 333 K in an absolute scale is shown in the upper panel of fig. 2 as an example. Symbols give the experimental data and the dash line gives the model fitting taking into account the resolution correction. The same data plotted in a

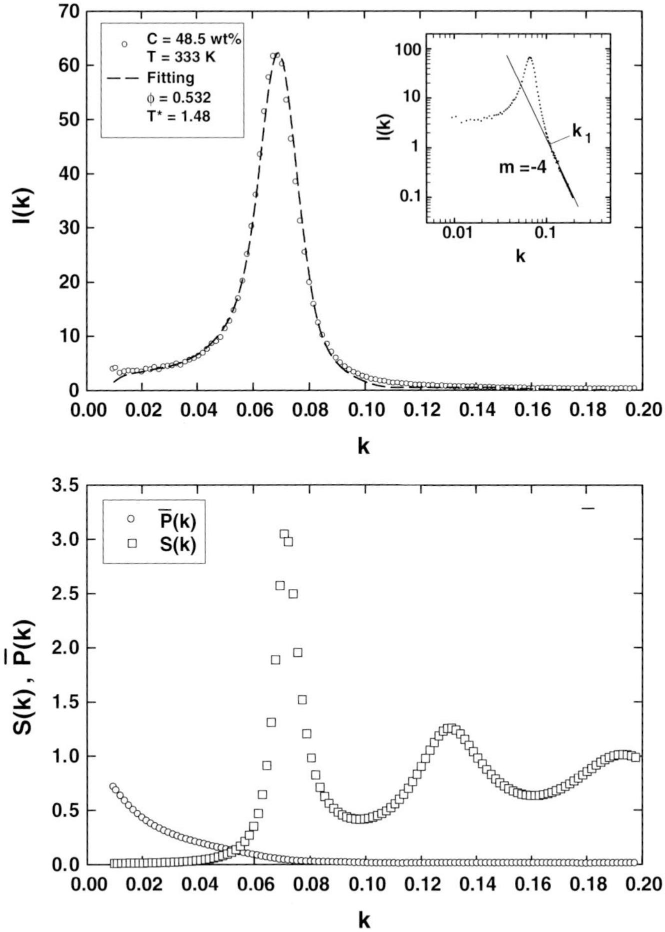

Fig. 2. – The upper panel shows the typical SANS intensity distribution in an absolute scale and its model fit taking into account the effect of resolution function. Symbols represent the experimental data and the dashed line the theory convoluted with the resolution function. The inset gives the same data but plotted in a log-log scale. It can be seen that Porod's law is satisfied at large k, as is expected in a two-phase system with a sharp interface. The lower panel gives the normalized intra-particle structure factor $\overline{P}(k)$ (circles) and the inter-particle structure factor $S(k)$ (squares) used to fit the data in the upper panel. The $S(k)$ is calculated by solving the OZ equation with a square-well inter-micellar potential and the $\overline{P}(k)$ is calculated using the modified cap-and-gown model [16] as the polymer segmental distribution function in a micelle. The observed SANS data is the product of these two functions and therefore it is clear that the interaction peak in the SANS data is primarily due to the first diffraction peak in the inter-micellar structure factor.

log-log scale is given in the inset. It can be seen that for sufficiently large k, the SANS absolute intensity decreases as k^{-4}, in agreement with Porod's law as expected in a two-phase system with a sharp interface. The lower panel gives the normalized intra-particle structure factor $\overline{P}(k)$ (circles) and the inter-particle structure factor $S(k)$ (squares) for this case. The observed SANS data is proportional to the product of these two functions. The fact that the first diffraction peak of $S(k)$ occurs at a relatively smooth tail part of $\overline{P}(k)$ implies that the interaction peak in the SANS intensity distribution is primarily reflecting the width of the first diffraction peak of $S(k)$.

4. – Scaling plot of SANS intensity distribution

At high enough polymer concentration, the SANS intensity distribution from the $L64/D_2O$ micellar system generally consists of a single, sharp interaction peak. Since there is a single peak in the observed SANS intensity distribution, we can assume that the system is characterized by a single length scale $\Lambda = 1/k_{\max}$, where $k_{\max}$ is the peak position of the intensity distribution. It is well known that the absolute intensity in a two-phase system (the micelles and the solvent) is given by a 3D Fourier transform of the Debye correlation function $\Gamma(r)$ and the Debye correlation function. The Debye correlation function is a two-point correlation function of the local scattering length density in the system. In this case it must be of the form $\Gamma(r/\Lambda)$ since Λ is the unique length scale in the system. Therefore

$$I(k) = \langle \eta^2 \rangle \int_o^\infty \mathrm{d}r 4\pi r^2 j_0(kr) \Gamma\left(\frac{r}{\Lambda}\right), \tag{6}$$

where $\langle \eta^2 \rangle$ is the so-called invariant of the intensity distribution.

By making a transformation of variables

$$x = \frac{r}{\Lambda} = k_{\max} r,$$
$$y = \frac{k}{k_{\max}},$$

it is straightforward to show that

$$\frac{k_{\max}^3 I(k)}{\langle \eta^2 \rangle} = \int_0^\infty \mathrm{d}x 4\pi x^2 j_0(xy) \Gamma(x). \tag{7}$$

From the above equation, it can be seen that the scaled intensity $k_{\max}^3 I(k)/\langle \eta^2 \rangle$ is a unique function of the scaled magnitude of the scattering vector, $y = k/k_{\max}$. Therefore, if we plot the scaled intensity distributions at different temperatures as a function of y, they should collapse into one single master curve in a single-phase amorphous state.

More specifically, taking fig. 2 as an example, the physical meaning of the scale intensity $k_{\max}^3 I(k)/\langle \eta^2 \rangle$ can be interpreted as follows:

The invariant $\langle \eta^2 \rangle$ is mathematically defined as

$$\langle \eta^2 \rangle = \frac{1}{2\pi^2} \int_0^\infty k^2 I(k) \mathrm{d}k. \tag{8}$$

From fig. 2 one can see that the minimum wave vector signaling the onset of Porod's law is at $k_1 \approx 0.09\,\text{Å}^{-1}$. According to the inset of fig. 2, Porod's law $I(k) = Ak^{-4}$ is valid, when $k \geqq k_1 = 0.09\,\text{Å}^{-1}$. $I(k_1) = 1\,\text{cm}^{-1} = 10^{-8}\,\text{Å}^{-1}$, thus $A = 6.5 \times 10^{-13}\,\text{Å}^{-5}$. Therefore, the integral in $\langle \eta^2 \rangle$ can be divided into two parts and the scale intensity $k_{\text{max}}^3 I(k)/\langle \eta^2 \rangle$ can be rewritten as

$$\begin{aligned} \frac{k_{\text{max}}^3 I(k)}{\langle \eta^2 \rangle} &= \frac{k_{\text{max}}^3 I(k)}{(1/2\pi^2) \int_0^\infty k^2 I(k) \mathrm{d}k} \\ &= \frac{k_{\text{max}}^3 I(k)}{(1/2\pi^2) \left[\int_0^{k_1} k^2 I(k) \mathrm{d}k + \int_{k_1}^\infty k^2 I(k) \mathrm{d}k \right]} \\ &= \frac{k_{\text{max}}^3 I(k)}{(1/2\pi^2) \left[\int_0^{k_1} k^2 I(k) \mathrm{d}k + \int_{k_1}^\infty k^2 (6.5 \times 10^{-13}/k^4) \mathrm{d}k \right]} \\ &= \frac{k_{\text{max}}^3 I(k)}{(1/2\pi^2) \int_0^{k_1} k^2 I(k) \mathrm{d}k + 3.3 \times 10^{-13}}. \end{aligned} \tag{9}$$

Furthermore, $\langle \eta^2 \rangle$ can be evaluated numerically and the result is $0.7 \times 10^{-12}\,\text{Å}^{-4}$. Since the contribution from Porod's part is only $3.3 \times 10^{-13}\,\text{Å}^{-4}$, which is about 7.0% of $\langle \eta^2 \rangle$, $k_{\text{max}}^3 I(k)/\langle \eta^2 \rangle$ can thus be approximated as

$$\frac{k_{\text{max}}^3 I(k)}{\langle \eta^2 \rangle} \sim \frac{k_{\text{max}}^3 I(k)}{(1/2\pi^2) \int_0^{k_1} k^2 I(k) \mathrm{d}k}. \tag{10}$$

As we see from fig. 2, $\overline{P}(k)$ varies slowly within the k range of the first diffraction peak of $S(k)$, and is negligibly small beyond, the scale intensity $k_{\text{max}}^3 I(k)/\langle \eta^2 \rangle$ can thus be well approximated by substituting eq. (2) into eq. (10) and canceling the common smooth factor $\overline{P}(k)$ from the numerator and the denominator of the equation to obtain the result

$$\frac{k_{\text{max}}^3 I(k)}{\langle \eta^2 \rangle} \sim \frac{k_{\text{max}}^3 S(k)}{(1/2\pi^2) \int_0^{k_1} k^2 S(k) \mathrm{d}k}, \quad \text{for } k \leq k_1. \tag{11}$$

Therefore, without losing generality, the scale intensity is proportional to the interparticle structure factor $S(k)$ in the region of its first diffraction peak. Examples and discussion of the scaling plots are given in the next section.

5. – Results and discussion

5·1. *Intermediate scattering function measured by photon correlation spectroscopy.* – To test the phase behavior predicted by MCT calculations for an AHS system, ISFs were measured for a set of volume fractions within the interval $0.525 < \phi < 0.546$, where the re-entrant phenomenon of the glass-to-liquid-to-glass transition, glass-to-glass transition and the A_3 singular point were predicted for the case of $\varepsilon = 0.03$, as shown in fig. 1. Figures 3a and b show ISFs obtained by PCS for two volume fractions, $\phi = 0.525$ and $\phi = 0.535$, respectively, at different temperatures. As can be seen, on increasing the temperature, the system, starting from a liquid state at a lower temperature, where the ISF decays to zero at long time, approaches a KGT, characterized by a diverging α relaxation time. At KGT the ISF tends to a finite plateau at long time ($F(k, t \to \infty) = f_k > 0$). Just before such an ergodic-to-nonergodic transition takes place, the ISF measured in an ergodic state exhibits a logarithmic relaxation (indicated by a straight-line fit) at an intermediate time followed by a power law decay before the final α relaxation sets in. This is in agreement with the MCT prediction. The non-ergodic state ($T = 300\,\mathrm{K}$, for $\phi = 0.535$), as indicated by the ISF having a finite plateau at long time, represents the attractive glass for which the Debye-Waller factor (DWF, the height of the plateau) is about $f_k = 0.5$. As discussed above, an important prediction of the MCT calculation for an AHS is a suggestion that there exists an attractive branch of KGT line near the cusp singularity separating two different glass phases [10, 23]. Motivated by this prediction, we extended the study of the micellar system to higher temperatures. Starting from a glass state, upon increasing the temperature further, the measured ISF reveals a surprising re-entry from the glass to the liquid state, as seen by the ISF again decaying to zero at higher temperatures.

The ISFs measure at $\phi = 0.538$ and $\phi = 0.540$, where the glass-to-glass transition is predicted, are given in figs. 3c and d. In the two-length-scale AHS, aside from the volume fraction ϕ, the glass transition can be triggered by varying the effective temperature T^* as well. Due to the fact that the depth of the potential well $-u$ is temperature dependent and increases on heating to a certain extent, the effective temperature T^* actually decreases as temperature rises for a certain interval of temperatures. Variation of temperature thus makes the transition between two distinct types of glass possible. Because the long-time limit of the ISF (DWF) reflects the degree of localization of the density fluctuation, these two different types of glass, if they have different degree of local disorder, can be identified by their different values of DWF. From figs. 3c and d it is obvious that all the ISFs can be grouped into two distinct sets of curves having two different values of DWF —one at $f_k = 0.5$ (attractive glass) and the other at $f_k = 0.4$ (repulsive glass). The two panels, 3c and 3d, confirm the predicted glass-to-glass transition.

Figure 3e shows the ISFs measured at $\phi = 0.544$, which, according to MCT, is the end point of the predicted glass-to-glass transition line and therefore the long-time limit of ISF of the two glasses should become identical at this point. Our measured ISFs verify this prediction, showing two nearly identical values of DWF, for the two glassy states, with an average value of $f_k \sim 0.45$. Although the DWFs of these two types of glasses

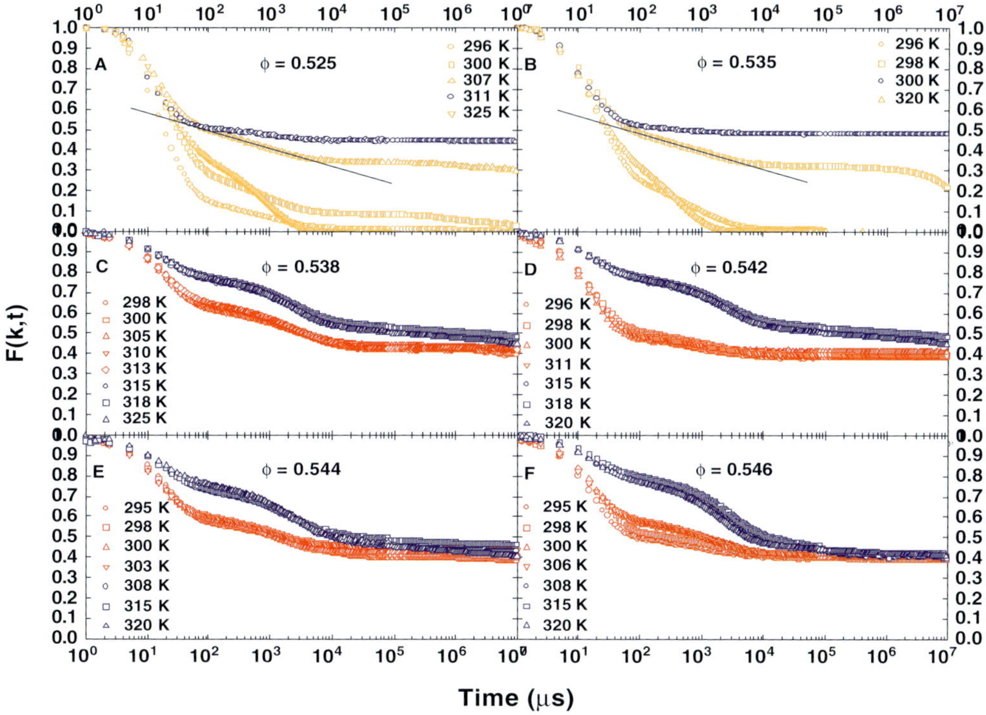

Fig. 3. – a) and b) represent ISFs measured at $\phi = 0.525$ and $\phi = 0.535$, respectively, where the liquid-to-attractive glass transition is found, as a function of temperature. In the liquid state, the long-time limit of ISF (DWF), f_k, is zero, while in the attractive glass state f_k is about 0.5. The structural arrest transition is thus characterized by a discontinuous change of f_k, called a bifurcation transition. The observed occurrence of a region of the logarithmic time dependence, preceding the plateau region for the system in the ergodic state just before the transition, are highlighted by a straight line in the linear-log plots, respectively. c) and d) represent ISFs measured at $\phi = 0.538$ and $\phi = 0.542$. According to MCT, in this volume fraction range, there is a possibility to observe glass-to-glass transition by varying the effective temperature T^*. Since the depth of the potential well $-u$ is temperature dependent and increases on heating, the effective temperature T^* actually decreases as temperature rises, making the transition from the repulsive glass to the attractive glass possible. By comparing the long-time limit of the ISFs, the two different types of the glasses can be identified by their respective DWFs, $f_k^{\mathrm{A}} \sim 0.5$ and $f_k^{\mathrm{R}} \sim 0.4$. The reason for observing two different values of DWFs can be interpreted as the different degrees of localization of the density fluctuation having the wave number k, for the two types of glasses. e) gives ISFs measured at $\phi = 0.544$, where, according to MCT, the long-time limits of ISF of the two glasses become identical. In our measurements, the long-time limit of ISFs of the attractive glass gradually decreases from 0.5, and gets closer to 0.4, the long-time limit of ISFs of the repulsive glass. Although the DWFs of these two types of glasses are almost identical, it is essential to recognize that there is a significant difference between the dynamics of their intermediate time relaxations. f) gives ISFs measured at $\phi = 0.546$. At this volume fraction, the long-time limit of ISFs of the attractive glass approaches closely 0.4, which is the long-time limit of ISFs of the repulsive glass. It indicates that, beyond the volume fraction 0.544, the two glasses merge into a single repulsive glass. Judging from e) and f), the experimentally determined volume fraction for A_3 singularity lies somewhere between $\phi = 0.544$ and $\phi = 0.546$.

are identical, it is important to recognize that there is a significant difference between the dynamics of their intermediate time relaxations (the β relaxation region).

The ISFs measured at $\phi = 0.546$, which is a volume fraction beyond the A_3 point, is shown in fig. 3f. This figure shows similar features as in fig. 3e, with a merged DWF of $f_k \sim 0.4$. It strongly hints that beyond the volume fraction $\phi = 0.544$, the system exists in the repulsive glass state. It is interesting to note here a critical point-like characteristic of the A_3 point.

5·2. *Fitting results of SANS data.* – Small-angle neutron scattering experiments were performed on Pluronic $L64/D_2O$ micellar solutions at different concentrations and temperatures. These experiments were carried out using $NG7$ SANS spectrometer at the NIST Center for Neutron Research and the SAND station at IPNS, Argonne National Laboratory. Part of the experimental results as well as their theoretical fits are given in fig. 4. Symbols are experimental data and lines are the fits. The fits are in an absolute intensity scale taking into account the effects of the instrument resolution and the incoherent background. It can be seen that for samples at both volume fractions, $\phi = 0.532$ and $\phi = 0.536$, the peak intensity is generally higher for the sample at the higher-temperature liquid phase. Furthermore, the position of the peak shifts toward smaller k as temperature increases. This is due to the enhanced self-association of the micelle (larger aggregation number) as the consequence of the increased hydrophobicity of the polymer segments at higher temperatures. It is important to note here that, in spite of the micellar growth as temperature increases, SANS data analyses show that the volume fraction of the micelles remains constant, for a given weight fraction of $L64$, at all temperatures studied.

Figure 5 summarizes the variation of the fitted effective temperature parameter T^*, as a function of temperature T, for several volume fractions studied. It can be seen clearly that, as T increases, T^* increases also. There are no data shown in the glass region, which appears as gaps in the plots, due to lack of an appropriate theory for the structure factor of a system in a non-ergodic state, showing the effect of ageing process.

5·3. *Scaling plots of SANS data.* – Figure 6 shows a series of SANS intensity distributions and their associated scaling plots for the 45 wt% concentration sample at different temperatures. The volume fraction of micelles for this polymer concentration is determined by SANS data analysis to be at $\phi = 0.522$, independent of temperature. In view of the fact that all $2D$ SANS patterns show an isotropic ring, we conclude that the sample stays amorphous in the entire temperature range studied at this concentration. Looking carefully at the SANS intensity distributions as a function of temperature given in the top panel, one already sees clues showing the kinetic glass transition: Judging from the width of the peak, the intensity distributions at different temperatures can be categorized into two groups —a sharp one and a broad one. The situation becomes clearer by analyzing the intensity distributions using the scaling plots. SANS intensity distribution $I(k)$ is proportional to the product $\overline{P}(k)S(k)$, where $\overline{P}(k)$ is the decaying Gaussian-like peak centered around $k = 0$. Thus the above product results in a single peak, reflecting

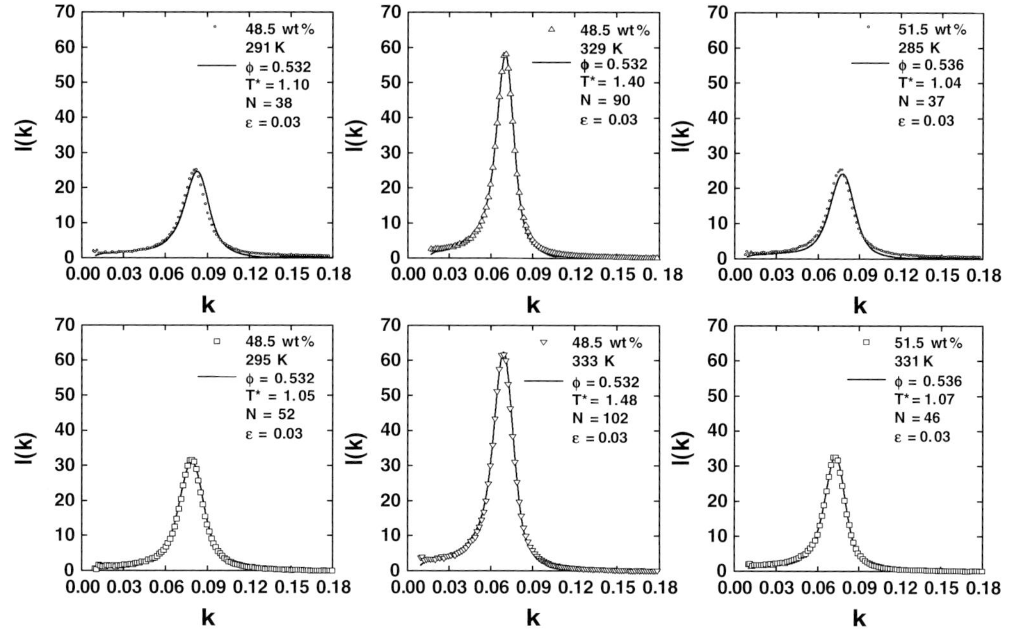

Fig. 4. – The theoretical fits to SANS data taken at $NG7$ SANS Spectrometer at the NIST Center for Neutron Research, for Pluronic $L64$ micellar solutions at various concentrations and temperatures. Symbols are experimental data and lines are the fits. The fits in absolute intensity scale take into account the effects of the resolution and the incoherent scattering backgrounds. The scattering intensities increase at the higher temperature liquid phase as compared to lower temperature liquid phase and the positions of the peak shifts toward smaller k due to the enhanced self-association (larger aggregation number) as the consequence of increased hydrophobicity of the polymer segments at higher temperatures. It is important to note that from these fits we obtain unique values of four parameters: volume fraction ϕ, fractional well width ε, the effective temperature $T^* = k_{\mathrm{B}}T/u$ and the aggregation number of the micelle N.

the position and the height of the first diffraction peak of $S(k)$, as explained in sect. **4**. It is well known that the position of the first diffraction peak of $S(k)$ reflects the mean inter-particle separation of the system and the height of the peak reflects the degree of local order surrounding a typical particle. Thus the height and the width of the scaling peak can be used to visualize the degree of local order surrounding a given particle in an amorphous state. Therefore, a sudden sharpening at a given temperature signals the onset of the liquid-to-amorphous solid transition. As one can see from the two lower panels, there are two distinct degrees of disorder, that depends on temperature. While the narrower peak, which is resolution limited, represents the glassy state, the broader peak, which is much broader than the resolution, represents the liquid state with a broader distribution of the inter-particle distances. This figure indicates that the system shows a re-entrant liquid-to-attractive glass-to-liquid transition as temperature increases. The sharpness of the scaling peaks which are resolution limited indicates that the nearest-

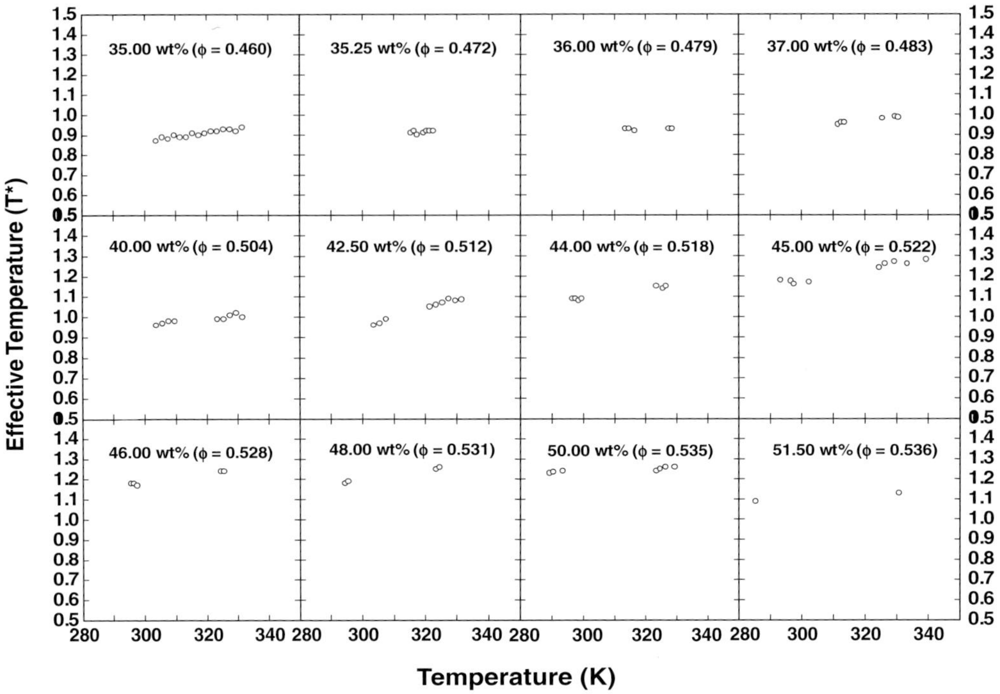

Fig. 5. – The effective temperature T^* of liquid state as a function of temperature, obtained by fitting SANS intensity distributions from different concentrations. For the indicated concentrations, T^* generally increases as temperature increases except in the glass region which appears as a gap in the figure.

neighbor distance in the glassy state (ranging from 306 to 321 K) is more uniform than that in the liquid state (288 K to 303 K, 325 K to 343 K). The inset gives the peak height of the scaling plots as a function of temperature. The variation of the peak heights indicates the re-entrant liquid-to-attractive glass-to-liquid transition.

SANS intensity distributions and their scaling plots for a sample at 48.5 wt%, or $\phi = 0.532$, at a series of temperatures ranging from 291 K to 340 K are given in fig. 7. The scenario is similar to the previous example: A temperature-dependent degree of disorder again characterizes the system. As the temperature rises, the system experiences a liquid-to-glass-to-liquid transition. However, in addition to all the similarities, when the temperature increases to 340 K, the system is driven into another glassy state peaked at $k = 0.082\,\text{Å}^{-1}$. From the scaling plots given in the bottom two panels, one can tell the differences between these two disordered glassy states. While the narrowest peak (340 K) is resolution limited, the slightly broader peak (298 K to 322 K) is also nearly resolution limited, but lower in the scaled intensity. Since the difference in local structure of different states is reflected in their scaling plots (see discussion in sect. **4**), we conclude that the degree of disorder is different for these two amorphous states. It can be interpreted

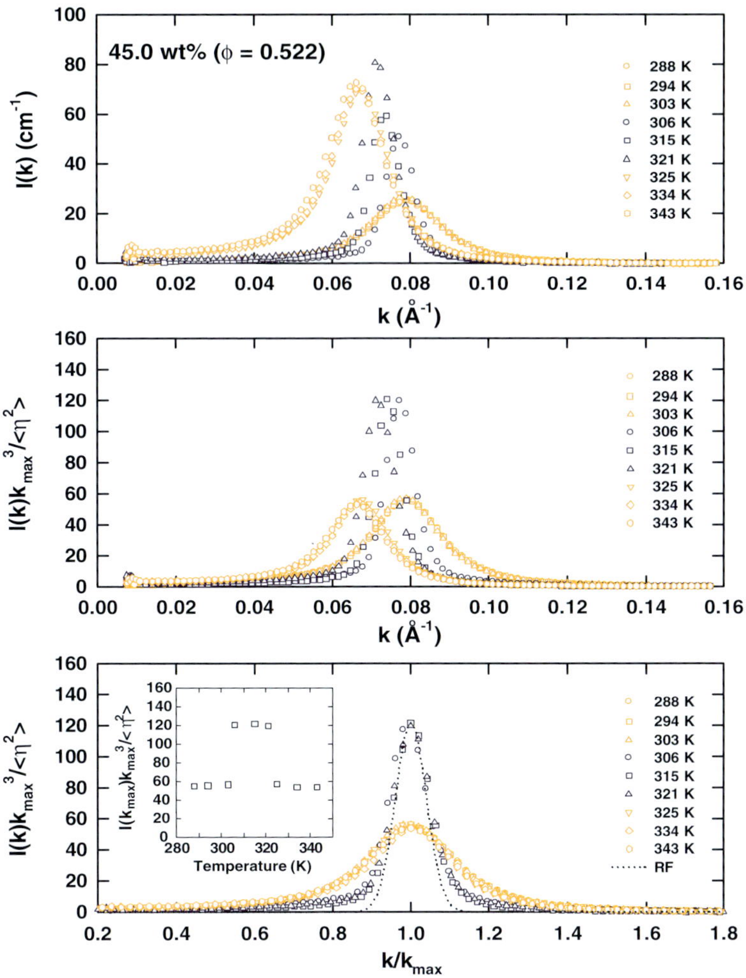

Fig. 6. – SANS intensity distributions and their scaling plots at $\phi = 0.522$, where the liquid-to-glass-to-liquid transition is observed, at a temperatures range spanning from 288 K to 343 K. The top panel gives the SANS intensity distributions as a function of k. The broader peaks (288 K to 303 K, 325 K to 343 K) represent the liquid states and the narrower ones (306 K to 321 K) represent the glassy state. The scattering intensities increase at the higher-temperature liquid phase (the left-hand side peak, 325 K to 343 K) as compared to lower-temperature liquid phase (the right-hand side peak, 288 K to 303 K) and the positions of the peak shifts toward smaller k due to the enhanced self-association (larger aggregation number) as the consequence of increased hydrophobicity of the polymer segments at higher temperatures. The scaling plots of SANS intensity distributions are given in the bottom two panels. It is seen clearly that there are two distinct degrees of disorder (judging from the width of the peak), which depend on temperature. While the narrower peak, which is resolution limited, represents the glassy state, the broader peak, which is much broader than the resolution, represents the liquid state with a broader distribution of the inter-particle distances. The inset gives the peak height of the scaling plots as a function of temperature. The variation of the peak heights indicates the re-entrant liquid-to-attractive glass-to-liquid transition.

that, by varying the temperature, the system shows a liquid-to-glass-to-liquid-to-glass transition.

Figure 8 gives the same plots taken at concentration of 51.5 wt%, or $\phi = 0.536$, for different temperatures. Generally it delivers the same message as fig. 7: it shows the re-entrant transition. However, it can be seen that the transition temperatures between different states for these two different volume fractions are different, especially at the high-temperature region. For example, for the case of $\phi = 0.536$, the high-temperature liquid state is only observed at 331 K. When the temperature increases to 333 K, the system transits to the repulsive glass state. In order to make this point clearer, the peak heights of the scaling plots as a function of temperature are given in the insets of the bottom panels of fig. 7 and fig. 8. The transition temperatures between different amorphous states can be visualized clearly from them. These two figures give a firm evidence of the re-entrant glass-to-liquid-to-glass transition which is in good agreement with the prediction of MCT for an AHS system. We would like to mention here that the KGT boundaries determined by SANS agree with the results obtained from the latest specific-heat measurements [24].

According to the MCT calculations for an AHS with sufficiently short-range attraction, a glass-to-glass transition is predicted [10]. Although the transitions between different amorphous glassy states are not uncommon in pure substances such as H_2O, Si, Ge *et al.* [25], yet there is no detailed investigation of the glass-to-glass transition in a micellar system so far, except for some recent reports [16, 17]. Combining with the PCS (see figs. 3c and d), a concrete evidence of such a transition in a micellar system was revealed by SANS. From SANS intensity distributions, taken at $\phi = 0.538$, given in fig. 9, one can see that the much broader peaks (liquid state) disappear and variation of temperature triggers the transition between the two amorphous solid states with different degrees of disorder. By increasing temperature, the variation of the peak heights of the scaling plots given in the insect shows a re-entrant repulsive glass-to-attractive glass-to-repulsive glass transition.

Perhaps the most important prediction of the AHS system is the existence of the end point in the glass-to-glass transition line, the so-called A_3 singularity. Comparing with the critical point in the equilibrium states, a certain degree of similarity between them can be found. Therefore, it is intriguing to speculate the extent to which one can draw the analogy between the A_3 singularity and the ordinary equilibrium critical point. In the case of $\varepsilon = 0.03$, the volume fraction of the A_3 point is predicted to be $\phi(A_3) = 0.544$. SANS intensity distributions and their associated scaling plots are shown in fig. 10. As one can see, all the scaling intensity curves collapse into one single master curve, indicating the fact that the first diffraction peaks of the structure factor for all states are identical. It suggests that the local structures of the two glasses are identical. The inset given in the bottom panel shows that all the scaling peaks have identical height (about 140), indicating that the two glasses indeed have the same degree of local order.

Increasing ϕ further to 0.546 (fig. 11), all the scaled intensities are again characterized by a unique temperature-independent length scale and collapse into one single master curve, independent of temperature, showing an identical local structure of the two glasses.

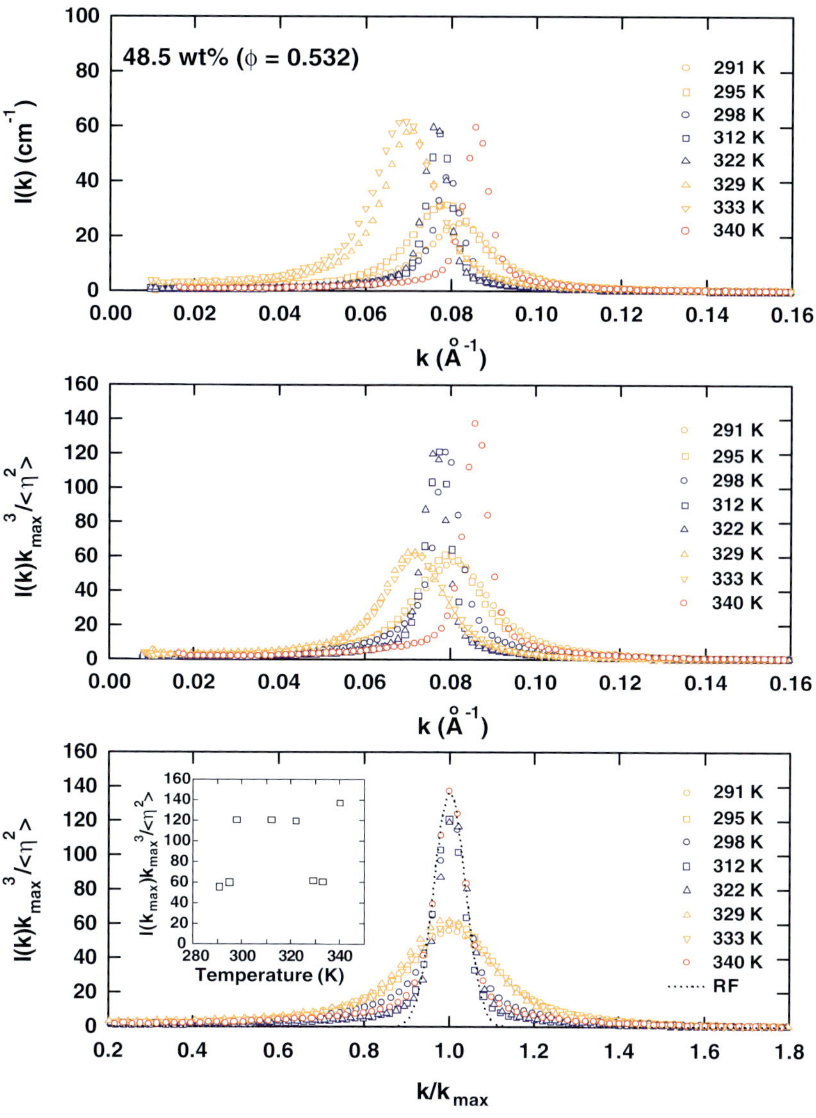

Fig. 7. – SANS intensity distributions and the scaling plots at $\phi = 0.532$, at a temperature range spanning from 290 K to 340 K. The top panel gives the SANS intensity distributions as a function of k. It shows features similar to those of the top panel of fig. 6: As temperature increases, a liquid-to-glass-to-liquid transition is observed. However, as the temperature is increased to 340 K, another disordered state peaked at $k = 0.082\,\text{Å}^{-1}$ is observed. The other two panels show the scaling plots of the SANS intensity distributions. A temperature-dependent degree of disorder characterizes the system. While the narrowest peak (340 K) is resolution limited, the slightly broader peak (298 K to 322 K) is also nearly resolution limited, but lower in intensity. From the inset given in the bottom panel, the liquid-to-glass-to-liquid-to-glass re-entrant transition can be seen clearly.

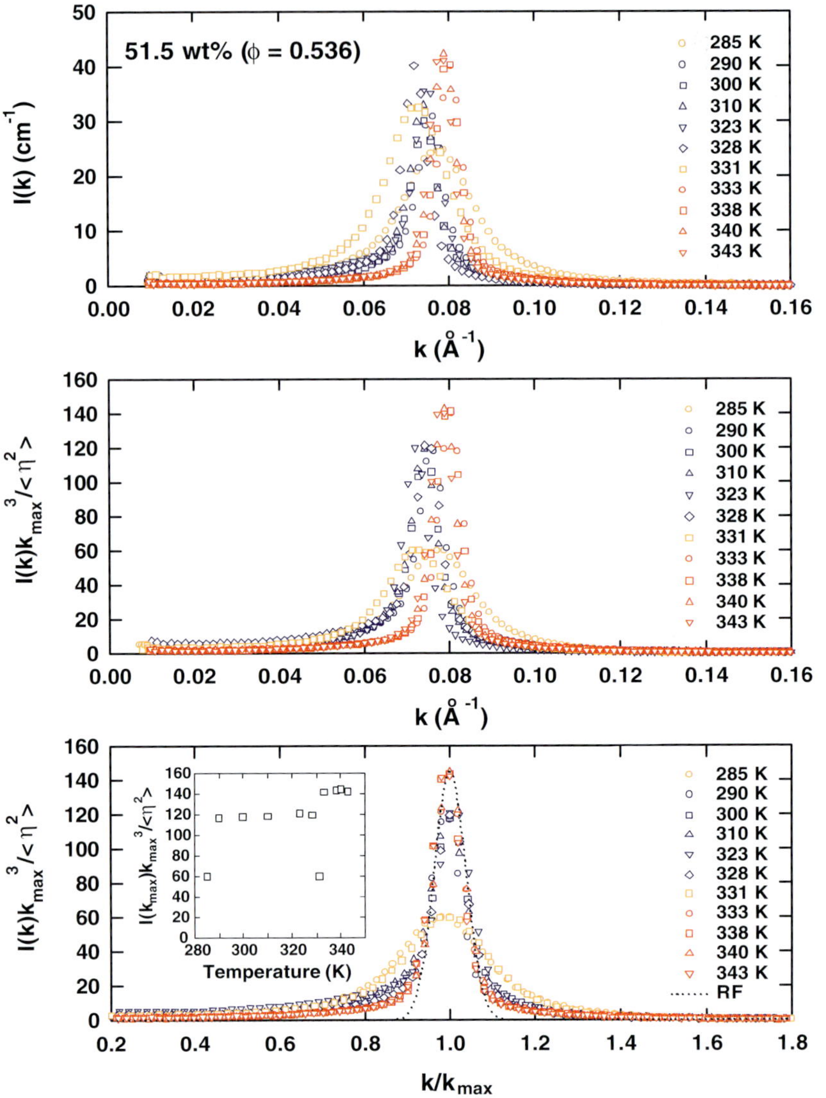

Fig. 8. – SANS intensity distributions and the scaling plots at $\phi = 0.536$, at a temperature range spanning from 285 K to 343 K. It exhibits all the features found in fig. 7. However, the high-temperature liquid state is only observed at 331 K, when the temperature increases to 333 K, the system transits to the repulsive glass state. Figures 7 and 8 together give a convincing evidence of the predicted re-entrant glass-to-liquid-to-glass transition. The only difference between them is that the transition temperature between different amorphous states are different. The inset shows the re-entrant transition more clearly.

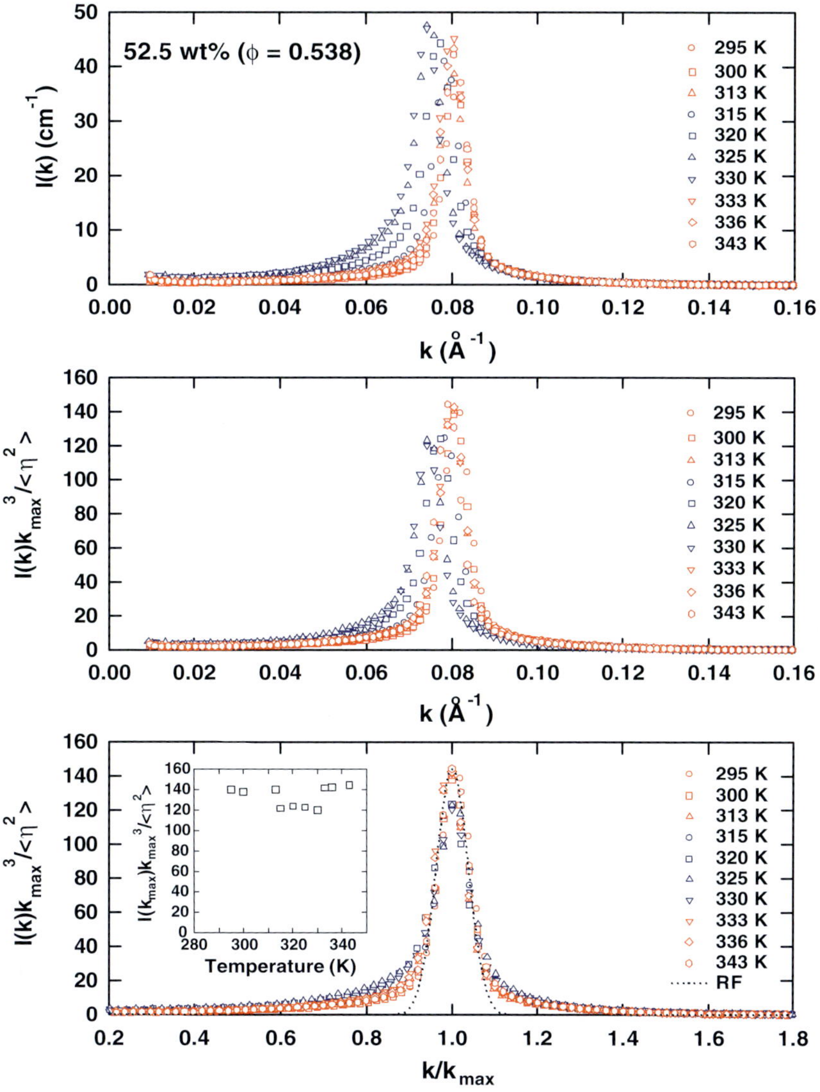

Fig. 9. – SANS intensity distributions and the scaling plots at $\phi = 0.538$, at a temperature range spanning from 295 K to 343 K. From the SANS intensity distributions given in the top panel, it can be seen that the much broader peaks (liquid state) disappear. Variation of temperature triggers the transition between the two amorphous solid states with different degrees of disorder. The variation of the peak heights of the scaling plots as a function of temperature given in the inset shows a re-entrant repulsive glass-to-attractive glass-to-repulsive glass transition for the first time.

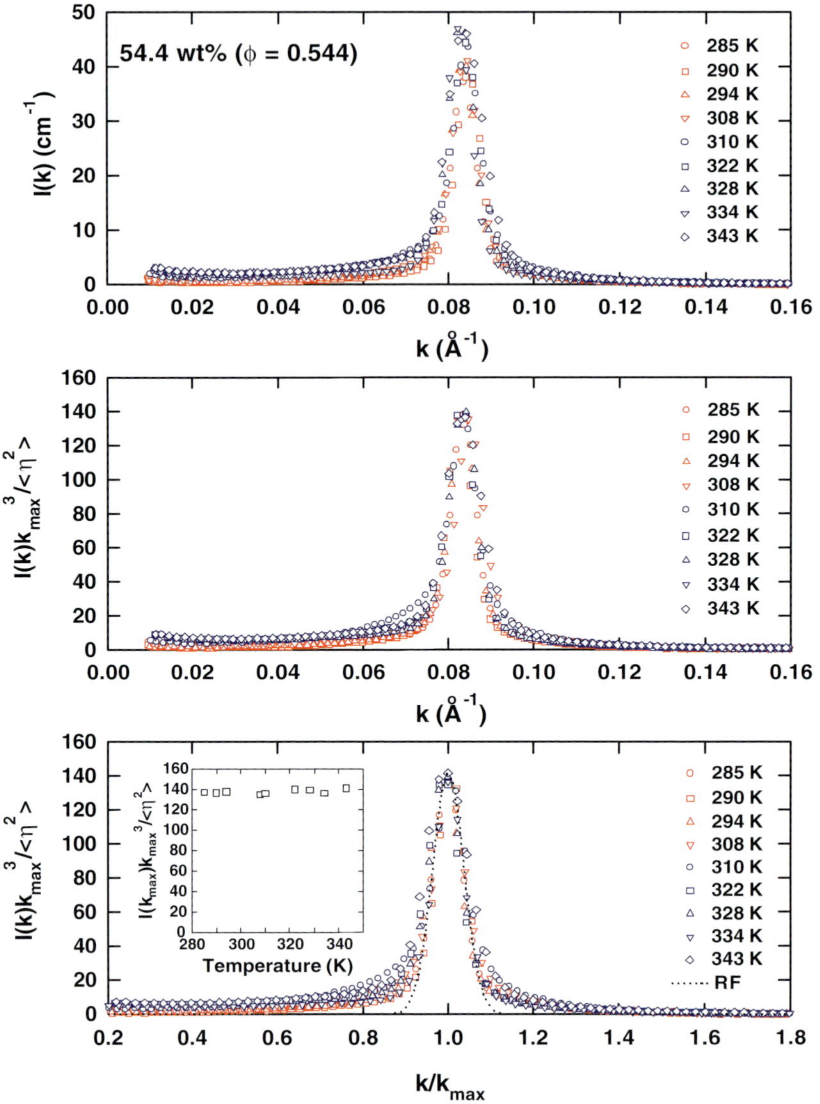

Fig. 10. – SANS intensity distributions and the scaling plots of SANS intensities at $\phi = 0.544$, which is predicted to be the volume fraction where the A_3 singularity point is located, at a temperature range spanning from 285 K to 343 K. From the top panel, it can be seen that all the peaks of the SANS intensity distributions are located at the same k value ($0.0836\,\text{Å}^{-1}$). Furthermore, from the scaling plots given in the bottom two panels, it can be seen that all the scaled intensities collapse into one single master curve. It suggests that the two different types of glass become identical in their local structure at this volume fraction. The inset given in the bottom panel shows that all the scaling peaks have an identical height (about 140), indicating that the two glasses indeed have the same degree of local order.

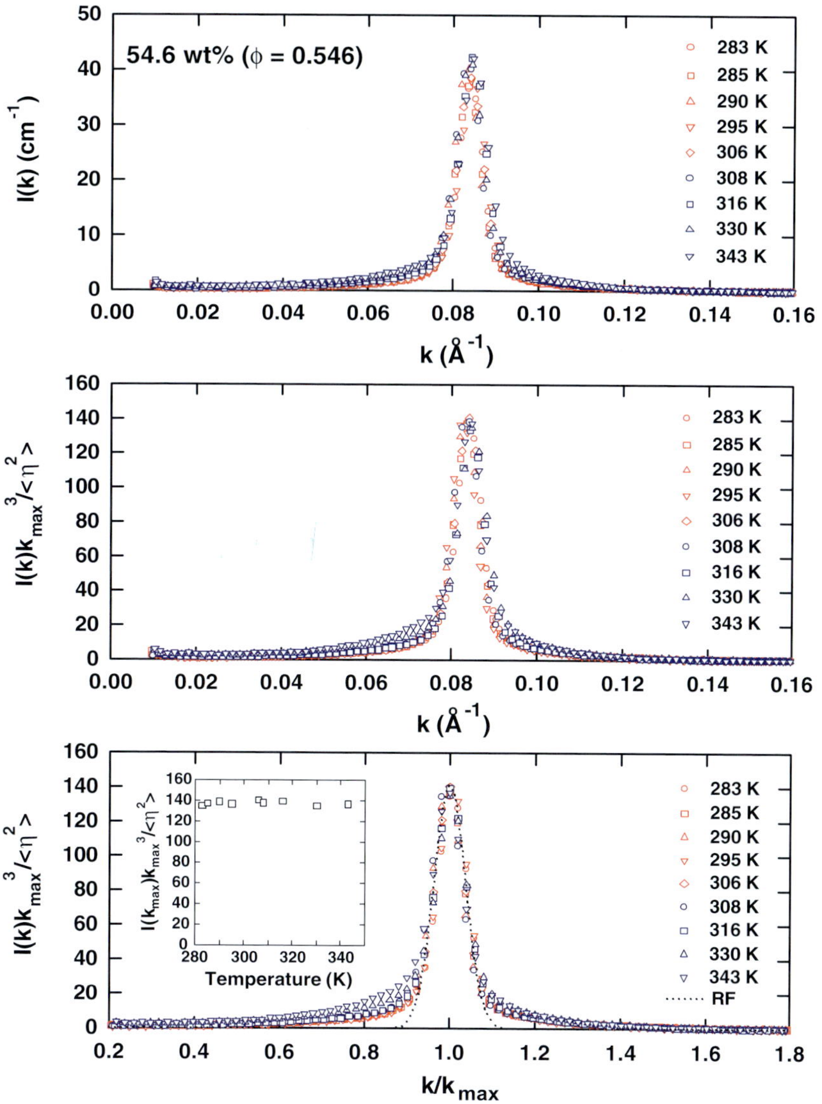

Fig. 11. – SANS intensity distributions and the scaling plots of SANS intensities at $\phi = 0.546$, a volume fraction beyond the A_3 point, at a temperature range spanning from 283 K to 343 K. Like fig. 10, all the scaled intensities are again characterized by a unique length scale and collapse into one single master curve, independent of temperature, showing identical local structure of these two glasses.

This is proof that the MCT predictions are accurate.

The phase diagram shown in fig. 12 summarizes the essential results of the extensive SANS and PCS data analysis. It contains the known equilibrium, liquid-to-hexagonal liquid crystalline phase boundary (solid line) [26], the experimentally determined KGT

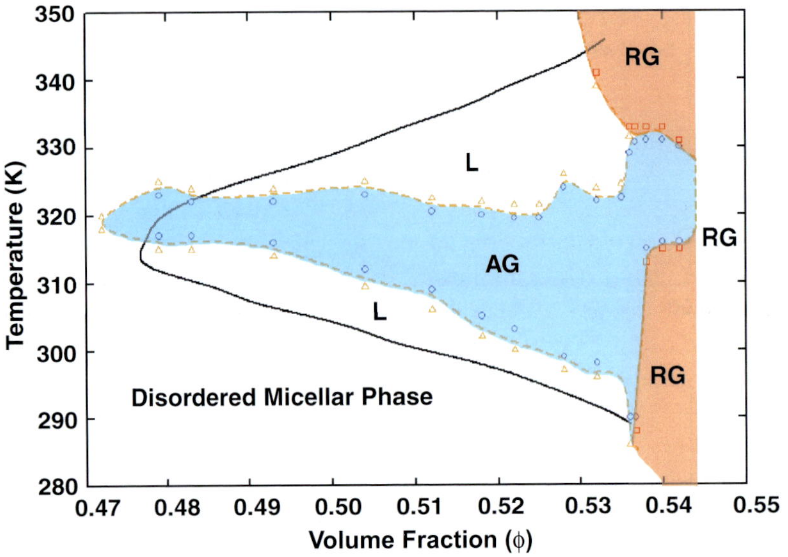

Fig. 12. – The experimental phase diagram of $L64/D_2O$ micellar system. The solid line represents the equilibrium phase boundary of disorder micellar liquid states and the hexagonal liquid crystalline states [26]. The dash line gives the kinetic glass transition boundary which is determined by SANS and PCS. The symbols represent the phase points where parts of the experimental data were taken. The triangles represents liquid state (L), the circles the attractive glass (AG), the squares the repulsive glass (RG). This figure gives several important pieces of information about this system: Within the region where the true ground state is the hexagonal liquid crystalline phase, only the metastable attractive glass is observed. The repulsive glass is found in the region where the volume fraction is larger than 0.536. It is interesting to see that there is a pocket of the attractive glass imbedded in between two separate repulsive glass regions spanning the volume fraction range between 0.536 and 0.544. Furthermore, it is important to note that the two different glasses become identical in their local structures and the long-time dynamics when the volume fraction exceeds $\phi(A_3) = 0.544$.

lines (dashed lines) and the phase points where parts of the experimental data are taken (symbols). This figure gives several important pieces of information about this system: First, only the metastable attractive glass is observed within the region where the true lowest free energy state is the hexagonal liquid crystalline phase. Next, the repulsive glass only exists in the region where the volume fraction is larger than 0.536. It is interesting to see that there is a pocket of the attractive glass imbedded in between two separate repulsive glass regions spanning the volume fraction range between 0.536 and 0.544, where the re-entrant glass-to-glass transition is observed. Furthermore, from fig. 3, fig. 10 and fig. 11, it is important to note that the two different glasses become identical in their local structures and the long-time dynamics when the volume fraction exceeds $\phi(A_3) = 0.544$. Furthermore, judging from the DWF and the peak height of the scaled intensity, the merged identical glassy state is the repulsive glass.

6. – Conclusions

In summary, we used photon correlation spectroscopy to verify the fact that the $L64/D_2O$ micellar system follows the overall structural arrest transition behavior predicted by the MCT, calculated using a square well potential with a short-range attraction relative to the micellar size. In particular, we show experimentally the existence of a glass-to-glass transition line which starts at point C^*, where the two glass phases and the liquid phase co-exist, and ends at point A_3, where the two glass phases merge [17]. Our SANS experiment further shows that while the local structures of the attractive and the repulsive glasses are in general different, they become identical at the predicted volume fraction of the A_3 singularity, independent of temperature. However, our PCS results indicate that the relaxation of the two glasses are different in the intermediate time region even at the A_3 point. The central result of this article is the use of the SANS method to pinpoint the exact volume fractions where the C^* point and the A_3 singularity are located [10] and furthermore, to verify that for volume fractions beyond the A_3 point, the system remains in the repulsive glass phase. We are able to map out the whole structural arrest transition boundaries in the $L64/D_2O$ micellar system using SANS method alone [27]. From SANS experiments, we observe a significant difference in the local structure when crossing the KGT boundaries at various temperatures for all the volume fractions studied. It is our conjecture that our SANS data are reflecting the ageing effect of the sample in the time scale of our measurements. We are in the process of investigating the time evolution of SANS intensity distributions at the present time.

* * *

The research at MIT is supported by a grant from Materials Science Division of US DOE, DE-FG02-90ER45429. The research in Messina is supported by INFM-PRA98 and MURST-PRIN2000. We are indebted to Dr. E. FRATINI of Department of Chemistry, University of Florence, Italy, for purifying $L64$ samples. We are grateful to NIST CNR and IPNS (ANL) respectively for granting the beam time of $NG7$ SANS and of SAND station and to Dr. C. J. GLINKA and Dr. P. THIYAGARAJAN for technical assistances.

REFERENCES

[1] EVERETT D. H., *Basic Principles of Colloid Science* (Royal Society of Chemistry, Letchworth, UK) 1988.
[2] DAWSON K., *Curr. Opin. Coll. Int. Sci.*, **7** (2002) 218.
[3] GÖTZE W., in *Liquids, Freezing and the Glass Transition*, edited by HANSEN J. P., LEVESQUE D. and ZINN-JUSTIN J. (North Holland, Amsterdam) 1991, p. 287.
[4] GÖTZE W. and SJÖGREN L., *Rep. Prog. Phys.*, **55** (1992) 241.
[5] GÖTZE W., *J. Phys. Condens. Matter*, **11** (1999) A1.
[6] PUSEY P. N. and VAN MEGAN W., *Phys. Rev. Lett.*, **59** (1987) 2083.
[7] FUCHS M., *Transport Theor. Stat. Phys.*, **24** (1995) 855.
[8] VAN MEGAN W. and UNDERWOOD S. M., *Phys. Rev. E*, **49** (1994) 4206.
[9] BARTSCH E., ANTONIETTI M., SCHUPP W. and SILLESCU H., *J. Chem. Phys.*, **97** (1992) 3950.

[10] DAWSON K., FOFFI G., FUCHS M., GÖTZE W., SCIORTINO F., SPERL M., TARTAGLIA P., VOIGTMANN TH. and ZACCARELLI E., *Phys. Rev. E*, **63** (2001) 011401.

[11] BERGENHOLTZ J. and FUCHS M., *Phys. Rev. E*, **59** (1999) 5706.

[12] BERGENHOLTZ J. and FUCHS M., *J. Phys. Condens. Matter*, **11** (1999) 10171.

[13] ECKERT T. and BARTSCH E., *Phys. Rev. Lett.*, **89** (2002) 125701.

[14] PHAM K. N., PUERTAS A. M., BERGENHOLTZ J., EGELHAAF S. U., MOUSSAID A., PUSEY P. N., SCHOFIELD A. B., CATES M. E., FUCHS M. and POON W. C. K., *Science*, **296** (2002) 104.

[15] MALLAMACE F., GAMBADAURO P., MICALI N., TARTAGLIA P., LIAO C. and CHEN S. H., *Phys. Rev. Lett.*, **84** (2000) 5431.

[16] WEI-REN CHEN, SOW-HSIN CHEN and FRANCESCO MALLAMACE, *Phys. Rev. E*, **66** (2002) 021403.

[17] SOW-HSIN CHEN, WEI-REN CHEN and FRANCESCO MALLAMACE, *Science*, **300** (2003) 619.

[18] CHEN S. H., LIAO C., FRATINI E., BAGLIONI P. and MALLAMACE F., *Colloid Surf. A*, **183-185** (2001) 95.

[19] LIAO C., CHOI S. M., MALLAMACE F. and CHEN S. H., *J. Appl. Crystallogr.*, **33** (2000) 677.

[20] LOBRY L., MICALI N., MALLAMACE F., LIAO C. and CHEN S. H., *Phys. Rev. E*, **60** (1999) 7076.

[21] LIU Y. C., CHEN S. H. and HUANG J. S., *Phys. Rev. E*, **54** (1996) 1698.

[22] FOFFI G., ZACCARELLI E., SCIORTINO F., TARTAGLIA P. and DAWSON K. A., *J. Stat. Phys.*, **100** (2000) 363.

[23] FABBIAN L., GÖTZE W., SCIORTINO F., TARTAGLIA P. and THIERY F., *Phys. Rev. E*, **59** (1999) R1347.

[24] MALLAMACE F., FERRARI C., MAZZAGLIA A., SALVETTI P., TOMBARI E. and CHEN S. H., to be published.

[25] PONYATOVSKY E. G. and BARKALOV O. I., *Mater. Sci. Rep.*, **8** (1992) 147.

[26] ZHANG K. Z., LINDMAN B. and COPPOLA L., *Langmuir*, **11** (1995) 538.

[27] WEI-REN CHEN, MALLAMACE F., GLINKA C. J., FRATINI E. and SOW-HSIN CHEN, *Phys. Rev. E*, **68** (2003) 041402.

Statistical Mechanics of granular media and glassy systems

A. Coniglio, A. de Candia, A. Fierro, M. Nicodemi,
M. Pica Ciamarra and M. Tarzia

Dipartimento di Fisica, Università di Napoli "Federico II"
INFM, Unità di Napoli, Complesso Universitario Monte Sant'Angelo
Via Cinthia, I-80126, Napoli, Italy

1. – Introduction

This paper describes some recent developments in the study of glasses and dense granular materials. These systems exhibit deep similarities in their "jamming" behaviors, but although the idea of a unified description is emerging, the precise nature of jamming in non-thermal systems, such as granular media, and the origin of its close connections to glassy phenomena in thermal ones are still open and very important issues.

Here we study glassy systems and granular media with the aid of simple models. We first discuss the glass transition within usual Statistical Mechanics. Then, by use of the Statistical Mechanics of powders introduced by Edwards, we describe the phase diagram of granular media and, in a quantitative way, its connections with glasses.

An important conceptual open problem concerning non-thermal systems, such as granular media, is the absence of an established theoretical framework where they might be described. Edwards [1] proposed a solution to such a problem by introducing the hypothesis that time averages of a system exploring its mechanically stable states subject to some external drive (*e.g.*, "tapping") coincide with suitable ensemble averages over its "jammed states". We discuss here some recent results validating and generalizing Edwards' proposal. In particular, within this approach we show that mean-field models for granular media undergo a phase transition from a (supercooled) "fluid" phase to a "glassy" phase, when their crystallization transition is avoided. The nature of such a "glassy" phase results to be the same found in mean-field models for glass formers: a discontinuous one-step replica symmetry-breaking phase preceded by a dynamical freezing point. These results are supported by Monte Carlo (MC) "tap dynamics" simulations which, in the region of low MC shaking amplitudes, show a pronounced jamming similar to the one found in experiments on granular media. As an application of Edwards' approach to powders we also briefly discuss segregation phenomena in these systems.

Section **2** shortly describes the phenomenology of the glass transition, and some theoretical approaches to that problem used in the past decades. Section **3** introduces a "lattice glass model", *i.e.* a lattice model suitable to study the glass transition. We present results for the thermodynamics and the dynamics both at the mean-field level (on the random graph) and on the three-dimensional cubic lattice (via Monte Carlo simulations). Section **4** gives an introduction to the phenomenology of granular materials relevant to the present purpose and discuss some correspondence with other systems "frozen" at $T = 0$, such as glasses in their inherent structures. Section **5** briefly reviews the essential lines in Edwards' Statistical Mechanics of powders. In sects. **6** to **7** we describe the results obtained in the study of hard spheres lattice models of granular materials, in order to check and apply Edwards' Statistical Mechanics. Finally, we draw some conclusions. In the Appendix, the method to derive a replica symmetry-breaking solution on the random graph, recently introduced by Mézard and Parisi [2], is reported.

2. – Phenomenology of the glass transition

One of the most intriguing questions in theoretical physics, today, is whether the glass is a new state of matter, or just a liquid that flows too slowly to observe it. If one cools a liquid below the melting temperature, avoiding crystallization, one observes a dramatic increase of the viscosity, until eventually, below some value $T_{\rm g}$, the liquid appears like a disordered solid, what is commonly called a glass (for a review see [3-5]). The glass transition $T_{\rm g}$, at which the liquid freezes, depends slightly on the cooling rate, being lower for a slower rate. It is usual anyway to define $T_{\rm g}$ roughly as the temperature at which the viscosity becomes greater than about 10^{12} poise, and the structural relaxation time greater than about 10^2 seconds.

Fifty-five years ago, Kauzmann [6] pointed out that, if the equilibrium curves of the entropy of the liquid, measured above $T_{\rm g}$, are extrapolated below the glass transition, one finds a temperature $T_{\rm K}$ below which the entropy of the liquid would become smaller than that of the crystal at the same temperature, which is clearly a paradox. As a consequence, if this extrapolation is valid, one is forced to admit that even for an infinitely slow cooling process, in which the liquid can reach equilibrium at all temperatures, the liquid phase cannot extend below the temperature $T_{\rm K}$. This fact suggested that, at temperature $T_{\rm K}$, there could be a thermodynamical transition to a new state of matter, the *ideal glass transition*, masked by the fact that, before getting to temperature $T_{\rm K}$, the liquid falls out of equilibrium.

This picture is also supported by the viscosity measurement, which for a large class of liquids, called *fragile liquids*, can be well fitted by the Vogel-Fulcher-Tamman law

$$\eta \propto \exp\left[\frac{A}{T - T_0}\right], \tag{1}$$

where $T_0 < T_g$. Experimentally, the temperature T_0 is very close to the Kauzmann temperature T_K, so that they may be in fact the same temperature. The temperature $T_0 \equiv T_K$ would mark therefore a divergence of the viscosity, which is proportional to the relaxation time of the liquid, and a breaking of ergodicity, that could be connected with a thermodynamical transition.

Together with the increasing of experimental data, it became evident, in the past decades, that a satisfactory theory of the glass transition should take into account not only the behavior of the system for temperatures near or below the glass transition, but also in a large interval above it. Several experimental findings signalled that the dynamics of supercooled liquids was in many respect different from that of a simple liquid, starting approximately near a temperature located halfway between the melting temperature T_m and the glass transition T_g, where the viscosity is about 10^2 poise, and the structural relaxation time about 10^{-8} seconds. Below such temperature, the dynamics of the supercooled liquid shows some peculiar features, among which the most apparent is the presence of a two-step relaxation. From dielectric measurement, Johari and Goldstein [7] distinguished a first short time decay in the relaxation functions, called β-relaxation, followed by a *plateau* and a much slower decay, called α-relaxation, that corresponds to structural relaxation and final decay to equilibrium. While the structural relaxation time τ_α, which is proportional to viscosity, diverges somewhere below the glass transition, the time of β-relaxation remains finite at all temperatures. As now has become clear, at low temperatures the particles begin to feel more strongly the presence of the other particles that surround them, and that act as a sort of cage. The β-relaxation describes the motion of the particles within the cage formed by the neighbors, while α-relaxation describes the process by which this cage is disrupted, structural relaxation takes place, and final equilibrium is achieved.

Furthermore, while at high temperatures relaxation is well approximated by a Debye model, where the structural relaxation functions decay exponentially with time, in supercooled liquid at low temperatures one observes a phenomenon called *stretching*. The long time tail of relaxation functions can be rather approximated by the *stretched exponential*

$$\phi(t) \propto \exp\left[-\left(\frac{t}{\tau_\alpha}\right)^\beta\right], \tag{2}$$

where $0 < \beta < 1$ is an exponent nearly independent of temperature (not to be confused with β-relaxation). This form of decay can be interpreted as the result of many processes, occurring over many time scales. The exponent β measures the enlargement of the resonance peak in the relaxation spectra over a frequency window larger than that of a simple Debye model.

One of the first attempts to explain the phenomenon of the glass transition, was that of Adams, Gibbs and Di Marzio [8]. They related the transport properties in the liquid, diffusion and viscosity, to large-scale *cooperative rearrangements* inside the liquid. At low temperature, the excess entropy of the equilibrium liquid, with respect to that of the crystal, becomes small. This means that the number of configurations accessible to the liquid is reduced. For any given temperature, there is a minimum-sized region which can undergo a cooperative rearrangement, and transform into a different configuration. The reduction of accessible configurations must be connected to a variation of the size of these minimum-sized regions. At temperature T_{K} this size diverges, and the configurational entropy vanishes. Thus the theory predicts a second-order phase transition at T_{K}, below which the configurational entropy remains zero, instead of taking negative values which have no significance.

Another important step, in the comprehension of glassy systems, came from the "mode-coupling theory" [9]. This theory gives a description of the so-called *cage effect*, that is the trapping of a tagged particle in the cage formed by its neighbors. For an isotropic system the equation of motion of the autocorrelation function $\phi_k(t)$ of wave vector k can be written as

$$\ddot{\phi}_k(t) + \nu_k \dot{\phi}_k(t) + \Omega_k^2 \phi_k(t) + \Omega_k^2 \int_0^t \mathrm{d}s \, m_k(t-s) \dot{\phi}_k(s) = 0, \tag{3}$$

where Ω_k is a microscopic frequency, which can be computed from the structure factor $S(k)$ via the relation $\Omega_k^2 = k_{\mathrm{B}} T k^2 / m S(k)$, where m is the mass of a molecule.

In the approximation of the so-called *idealized* version of the mode coupling theory, the kernel $m_k(t)$ is expressed by a quadratic form of the correlation functions $\phi_k(t)$, that is

$$m_k(t) = \sum_{k'+k''=k} V(k; k', k'') \phi_{k'}(t) \phi_{k''}(t), \tag{4}$$

where the vertices $V(k; k', k'')$ can be computed from the static structure factor $S(k)$. Within this approximation one therefore arrives at a closed set of coupled equations for $\phi_k(t)$. The temperature dependence of Ω_k, ν_k, and $V(k; k', k'')$, that are all static quantities, is assumed to be smooth throughout the whole temperature range, an assumption that is supported by experiments and computer simulations. Any singularity in the solutions of the equations of motion will be due to their non-linearity, and not to a singularity in the input parameters.

In the simplest version of the theory, one finds that there exists a critical temperature T_{c} where the structural relaxation time τ_α, or the inverse of the diffusion constant D^{-1}, diverges with a power law. In the *extended version* of the mode-coupling theory, one introduces the so-called *hopping processes*, by which particles can overcome the walls of the cage in an activated way, that make the structural relaxation time finite even for $T < T_{\mathrm{c}}$. The theory predicts that the Debye-Waller factor, that is the plateau in the correlation function, has a discontinuity at T_{c}, jumping from zero for $T > T_{\mathrm{c}}$ to a critical value f_{c} for $T \leq T_{\mathrm{c}}$, and also the form of the relaxation functions in the intermediate regime $\tau_\beta \ll t \ll \tau_\alpha$, and in the α-relaxation regime, where it predicts a stretched exponential decay. So the mode-coupling theory explains to a large extent the observed anomalies of the liquid dynamics for temperatures well above $T_{\rm g}$, and rests upon a solid basis of experimental confirms. It gives indeed a fairly accurate description of the experimental data for simple liquids, but also for complex liquids like some polymers.

Around the mid-eighties, the theory of the glass transition was therefore "wavering", between a thermodynamical approach, given by the phenomenological theory of Adams, Gibbs and Di Marzio, and a purely dynamical explanation of the structural arrest, represented by the mode-coupling theory. A "unification" of these different points of view came from the theory of spin glasses. It was found by Kirkpatrick, Thirumalai and Wolynes [10], that a class of mean-field spin models called p-spin glasses (for a review see [11]), with quenched disorder and p-body interactions, had a phenomenology pretty similar to that of structural glasses. They undergo a purely dynamical transition at some temperature $T_{\rm c}$, where correlation functions show a singularity of the same kind of that described by the mode-coupling theory. Below $T_{\rm c}$, phase space breaks into disconnected ergodic components, called "pure states" or "metastable states", separated by free-energy barriers that diverge with the number of particles. The logarithm of the number of pure states, divided by the number of particles, is called "complexity", and plays the role of the excess entropy, or configurational entropy, of Adams and Gibbs. At temperature $T_{\rm c}$ thermodynamical quantities do not show any singularity. By lowering the temperature, the complexity becomes smaller, until at some temperature $T_{\rm K}$ it vanishes. At that temperature the p-spin model undergoes a thermodynamical transition, of second order from the thermodynamical point of view, but with a discontinuous order parameter. Below $T_{\rm K}$, the system is in a different thermodynamical phase, where there is a "1-step replica symmetry breaking" [12]. These models provide therefore a mean-field description of structural glasses, where metastable states, differently than in real liquids, have an infinite lifetime. Hopping processes, that restore ergodicity in the temperature range $T_{\rm K} < T < T_{\rm c}$, are indeed absent in mean field. However, if one observes the system for times smaller than the time typical of hopping processes, the analogy between p-spin models and structural glasses is striking. For instance, both systems show a divergence of the so-called "dynamical susceptibility" at low temperatures [13], as predicted by the effective potential approach [14]. Moreover, the fluctuation-dissipation ratio shows the same pattern [15], typical of "1-step replica symmetry breaking" systems.

One interesting approach to the description of glassy behavior is through the study of inherent structures [16, 17], that is minima of the potential energy of the system. Phase space can be partitioned in basins, associating to each point the inherent structure that is obtained through a steepest-descent minimization. The logarithm of the number of inherent structures is usually called "configurational entropy". Recent studies have shown that the mode coupling temperature marks a topological change in the phase space visited by the system: for temperatures $T > T_{\rm c}$ the nearest saddle has some unstable directions, while for $T < T_{\rm c}$ it is a local minimum [18]. At temperature below the glass transition temperature $T_{\rm g}$ the system is out of equilibrium. In this situation response functions exhibit a strong slowing down as the waiting time (*i.e.* the time elapsed since the quench below T_g) increases, a phenomenon called physical aging. It has been shown that the (inverse) effective temperature of an aging system, measured by means of the fluctuation-dissipation ratio, coincides with the derivative of the configurational entropy with respect to the energy of the inherent structure [19].

3. – A lattice glass model

3·1. *The model.* – As explained in the previous section, the p-spin model is a kind of "mean-field version" of structural glasses, and it provides an analytically tractable description of glassy behavior, if one neglects the so-called "activated processes", or "hopping processes", that restore ergodicity, in real glass-forming liquids, below the mode-coupling temperature. On the other hand, when studied in finite dimensions the model loses many of the analogies with the glass-forming liquids. Moreover, the p-spin model is physically very different from a structural glass. It is a spin system with quenched disorder and p-body interaction and it has no crystalline phase.

For this reason, in the last years there were different attempts to devise a model for the glass transition physically more appealing, and that could be studied also in finite dimension. Recently it came out that frustration (and not quenched disorder) is the essential characteristic to have a glassy behavior. In this line of research, Biroli and Mézard have introduced a lattice glass model [20] that shows in three dimensions, by means of numerical simulations, a behavior typical of glass-forming liquids. Moreover the mean-field solution on a random graph (see Appendix) shows a scenario analogous to p-spin models, with one-step replica symmetry breaking. The model is constituted by lattice variables $n_i = 0, 1$. Frustration is achieved through a p-body potential, with $3 \leq p \leq z + 1$, where z is the coordination number of the lattice, with the nearest neighbors. The potential assures that each particle cannot have more than $p - 2$ nearest neighbors. In order to reduce the tendency to crystallize, the authors considered a mixture of two types of particles. The model has been further studied in [21]. Very similar behaviors are found in [22].

More recently we have introduced a three-dimensional lattice gas model with a simple geometrical interpretation [23] and two-body interactions. The model is the naive lattice gas model with the only difference that we let the center of mass of a particle assume different positions inside a cell. Here we review the properties of this model [23].

More precisely we partition the space in regular cells, such that not more that one particle can have its center of mass inside the cell. A coarse-grained discrete internal degree of freedom, which can assume a finite number q of states, describes the position and the orientation of a particle inside the cell. The interaction between nearest neighbors depends explicitly on the internal degree of freedom. The model is therefore described by the following Hamiltonian:

$$\mathcal{H} = \sum_{\langle ij \rangle} n_i n_j \phi_{ij}(\sigma_i, \sigma_j) - \mu \sum_i n_i\,, \tag{5}$$

where $n_i = 0, 1$ whether the i-th cell is occupied by a particle or not, $\sigma_i = 1, \ldots, q$

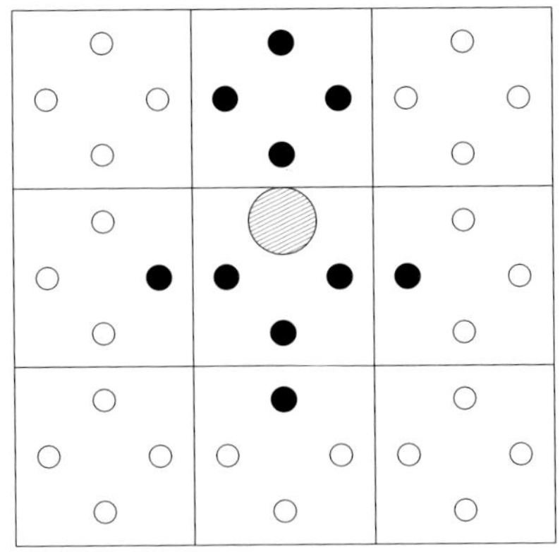

Fig. 1. – The model in two dimensions: the space is partitioned in square cells, and each cell can be occupied by at most one particle in one of four positions (small circles). A particle in a given position (big shaded circle) forbids the presence of another particle in the positions colored in black.

represents the internal position of the particle, and $\phi_{ij}(\sigma_i, \sigma_j)$ is the interaction energy between two particles in cells i and j, with internal positions σ_i and σ_j.

Is it clear that, by choosing a sufficiently large number q of internal positions, and an opportune interaction matrix $\phi_{ij}(\sigma_i, \sigma_j)$, one can approximate as closely as desired any model defined in the continuum, like for example a Lennard-Jones liquid. On the other hand, it is plausible that a few number of internal states may be enough to catch the fundamental characteristics of dynamics and thermodynamics of glass-forming liquids.

In [23] we have studied a particularly simple realization of the model described by eq. (5). In two dimensions, we partition the space in square cells, and subdivide each cell into four internal positions. When a cell is occupied by a particle in a given position, a hard-core repulsion forbids the presence of another particle in some of the internal states of the neighboring cells (see fig. 1). Therefore in this case $q = 4$, and the interaction $\phi_{ij}(\sigma_i, \sigma_j)$ is zero if the positions σ_i and σ_j are "compatible", infinite otherwise. The system is frustrated since each particle can have at most three of its four nearest neighbors occupied. This choice can be interpreted as a coarse-grained version of a hard-sphere system. In three dimensions, one partitions the space into cubic cells, and considers six internal positions instead of four.

For every spatial dimension d, the model has a crystalline ground state on lattices with sides multiples of $2d + 1$ lattice spacings, with density $2d/(2d + 1)$, where density means the fraction of occupied cells. For example, on a cubic lattice with periodic boundary conditions, the ground state can be found as follows: consider the cell with coordinates (x, y, z), and evaluate the number $a = (x + 2y + 3z \bmod 7)$: if $a = 0$ leave the cell empty; if $a = 1, 2, 3$ put a particle in the negative x, y or z direction, respectively; if $a = 4, 5, 6$ put a particle in the positive z, y or x direction, respectively. In fig. 2 we show the ground state of a two-dimensional 5×5 lattice, with density $4/5$.

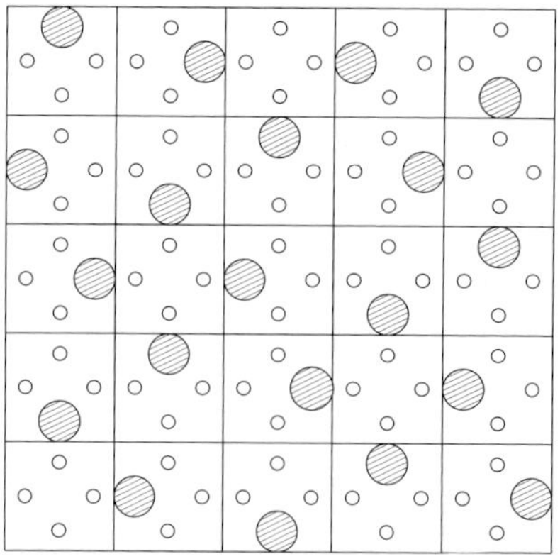

Fig. 2. – Ground state of a two-dimensional 5×5 lattice, with density 4/5.

3·2. *Numerical simulation.* – Numerical simulations show that the model reproduces the behavior of glass-forming liquids, and that its dynamics compare well, in many respects, with the predictions of MCT. We have simulated the system in three dimensions, by means of Monte Carlo canonical (fixed density) and grand-canonical (variable density) dynamics [23].

The nice feature of this model is that it has no tendency to crystallize. Consequently clear results on the high-density region are obtained without the introduction of a mixture, usually required to avoid crystallization. A compression experiment on a system of

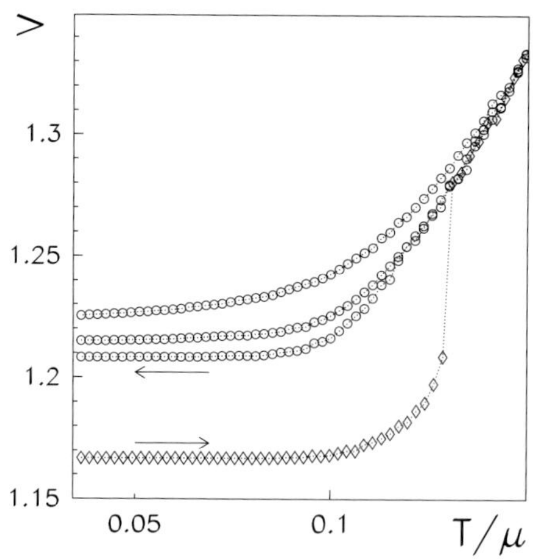

Fig. 3. – Specific volume (inverse density) as a function of temperature, for a three-dimensional system of size 28^3. Circles: the system is cooled starting from high temperature, for cooling rates (from top to bottom) $\dot{T}/T = -10^{-4}, -10^{-5}, -10^{-7}$. Diamonds: the system is heated starting from the crystalline ground state, with heating rate $\dot{T}/T = 10^7$.

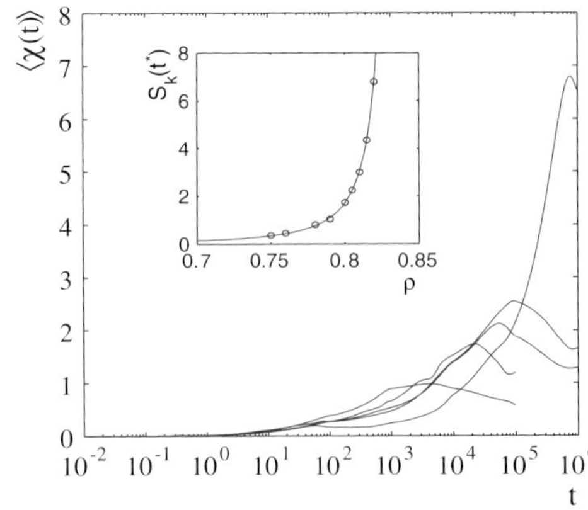

Fig. 4. – Non-linear susceptibility for density 0.78, 0.80, 0.805, 0.810, 0.820. Inset: the maximum $\chi(t^*)$ as a function of density. The fitting function is a power law $\chi(t^*) = 0.003(0.839 - \rho)^{-1.9}$.

size 28^3 is shown in fig. 3. A grand-canonical dynamics is performed, and temperature is slowly decreased, starting from some high value. The results are shown as open circles in fig. 3, for various cooling rates.

The final high-density state is strongly dependent on the cooling rate, as observed in glass-forming liquids. On the other hand, no tendency to crystallization is observed: even at the slowest cooling rate ($\dot{T}/T = -10^{-7}$) no transition to the crystalline state is observed. We have also prepared the system in crystalline states at very low temperature, and heated it up slowly (diamonds in fig. 3). The ordered state becomes unstable at $T_{\rm m} = 0.150\ \mu$, where the system falls into the liquid phase.

To avoid crystallization most continuum and discrete simulations of glass-forming liquids consider mixtures of particles with different properties (or introduce in the Hamiltonian some "*ad hoc*" terms [24]). This makes the models more complicated, and introduces in the relaxation processes two time scales that entangle with those associated with the cage effect. This is why the absence of crystallization makes this model suitable for the study of the glass transition.

After having equilibrated the system at a certain density, we switch to canonical dynamics. These simulations are performed on lattices of size 15^3. The mean square displacement and the relaxation of the self-overlap show that the model reproduces the cage effect and the presence of two characteristic time scales in the relaxation process [23]. The first one is associated to the vibration of a particle inside the cage formed by the surrounding particles, and gives rise to the first decay of the relaxation functions. The second time scale is associated to the global relaxation of the system, and gives rise to the final decay of the correlation functions.

The dynamical non-linear susceptibility associated with the self-overlap parameter,

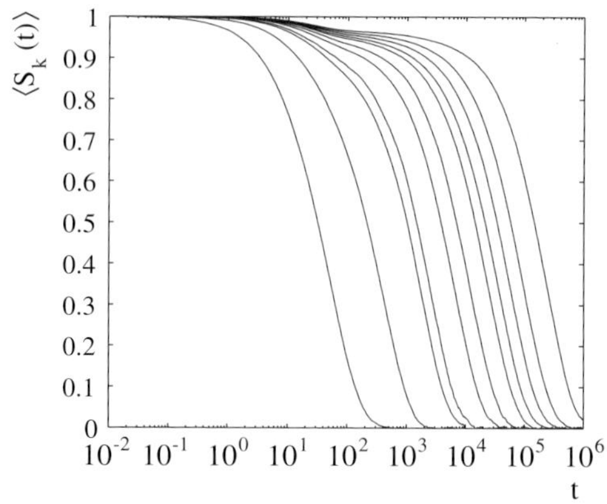

Fig. 5. – Relaxation functions for the self-part of the density-density fluctuations for $\mathbf{k} = (\pi, 0, 0)$ and densities $\rho = 0.6$, 0.7, 0.75, 0.76, 0.78, 0.79, 0.8, 0.805, 0.81, 0.815, 0.82.

which is defined as

$$\chi(t) = N\big[\langle q(t)^2\rangle - \langle q(t)\rangle^2\big], \tag{6}$$

is shown in fig. 4. The susceptibility displays a maximum $\chi(t^*)$ at time t^*. The growth of $\chi(t^*)$ with the density implies that the dynamics becomes more heterogeneous at higher density. We are able to fit $\chi(t^*)$ as a function of density with a power law $(\rho'_c - \rho)^{-\gamma'}$, in which $\rho'_c = 0.839$ and $\gamma' = 1.9$ (inset of fig. 4).

We have studied the self-part of the autocorrelation function of the density fluctuations, defined as

$$S_{\mathbf{k}}(t) = \frac{1}{N}\left\langle \sum_i e^{i\mathbf{k}\cdot(\mathbf{r}_i(t'+t)-\mathbf{r}_i(t'))}\right\rangle, \tag{7}$$

where $\mathbf{r}_i(t)$ is the position of the i-th particle in units of lattice constants, considering the internal positions as shifted of 1/4 of a lattice spacing with respect to the center of the cell. Due to the discreteness and periodicity of the lattice, the wave vector must have the form $\mathbf{k} = (2\pi/L)(n_x, n_y, n_z)$, where n_x, n_y and n_z are integers between 0 and $L/2$. We have chosen the wave vector $\mathbf{k} = (\pi, 0, 0)$, which corresponds to a wavelength equal to two lattice spacings, because for that $\mathbf{k}$ density fluctuations $\langle|\rho_{\mathbf{k}}(t')|^2\rangle$ are largest, where $\rho_{\mathbf{k}}(t')$ is the Fourier transform of the density. This means that this mode is the most important in the dynamics of the system. The results for various densities are plotted in fig. 5. Also the relaxation of the density-density fluctuations reproduces the cage effect and the presence of two characteristic time scales in the relaxation process.

We have then studied the behavior of the diffusion coefficient and of the inverse relaxation time (fig. 6). They vanish at the same density $\rho_c = 0.844$, with power laws

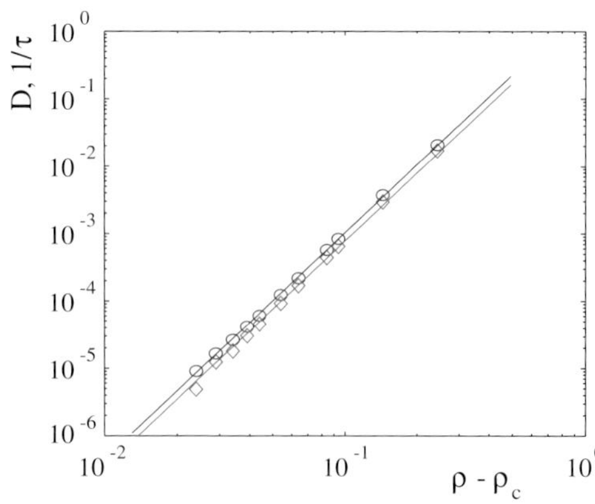

Fig. 6. – Diffusion coefficient D (circles) and inverse relaxation time $1/\tau$ (diamonds) as a function of $\rho - \rho_c$, with $\rho_c = 0.844$. The straight lines are power laws $\propto (\rho - \rho_c)^\gamma$ with $\gamma = 3.53$.

$(\rho_c - \rho)^\gamma$ with the same exponent $\gamma = 3.53$. The critical density ρ_c is compatible, within errors, with the density ρ'_c at which the dynamical non-linear susceptibility diverges.

We have compared the relaxation functions of the model with the predictions of MCT. The theory predicts that, near the transition, the intermediate time behavior ($t_0 \ll t \ll t_\alpha$, with t_0 the microscopic time and t_α the relaxation time) of the relaxation functions can be fitted by

$$\phi(t) = f + hc_\sigma g_\pm\left(\frac{t}{t_\sigma}\right), \tag{8}$$

with f (non-ergodicity parameter) and h constants, $c_\sigma = |\sigma|^{1/2}$, $t_\sigma = |\sigma|^{-1/2a}$, and the separation parameter σ is negative in the liquid phase and positive in the glassy one. The transition can be driven by temperature, in which case $\sigma \propto T_c - T$, or by density, in which case $\sigma \propto \rho - \rho_c$. At short times, $t_0 \ll t \ll t_\sigma$, the universal function has the asymptotic behavior $g_\pm(t/t_\sigma) \sim (t/t_\sigma)^{-a}$. At longer times, $t_\sigma \ll t \ll t_\alpha$, one has a functional form in the glassy phase ($\sigma > 0$), $g_+(t/t_\sigma) \sim$ constant, and a distinct one in the liquid phase ($\sigma < 0$), $g_-(t/t_\sigma) \sim -B(t/t_\sigma)^b$. The parameters a, b are related as follows:

$$\frac{\Gamma^2(1-a)}{\Gamma(1-2a)} = \frac{\Gamma^2(1+b)}{\Gamma(1+2b)} = \lambda, \tag{9}$$

where λ is the so-called exponent parameter.

We have tried to fit the intermediate timescale of the correlation functions with the functional form of eq. (8), as shown in fig. 7 (inset). In agreement with the prediction of MCT, the value of the exponent parameter λ, (and therefore of the exponents a, b) is constant within errors. The values extracted from the fit are: $\lambda = 0.584 \pm 0.005$, $a = 0.370 \pm 0.002$, $b = 0.84 \pm 0.01$.

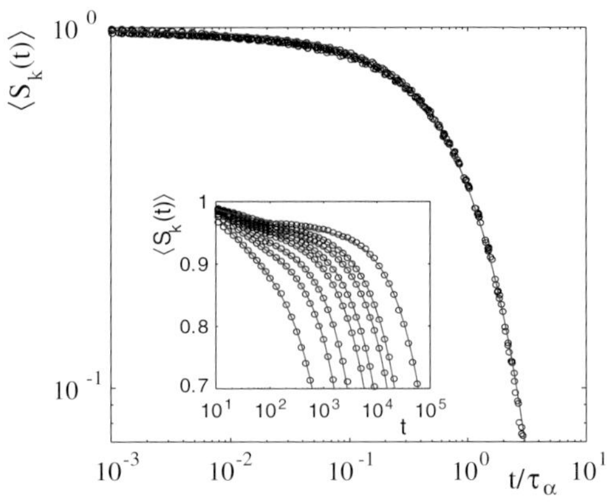

Fig. 7. – Time-density superposition principle for the correlation function, for densities $\rho = 0.75$, 0.76, 0.78, 0.79, 0.8, 0.805, 0.81, 0.815, 0.82. The fitting curve is a stretched exponential $\propto \exp[-(t/\tau)^\beta]$ with $\beta = 0.86$. Inset: fit of relaxation functions with the prediction of the mode coupling theory (eq. (8)). Densities $\rho = 0.76, 0.78, 0.79, 0.8, 0.805, 0.81, 0.815, 0.82$.

According to MCT, the exponent γ that characterizes the divergence of the relaxation times and of the inverse diffusivity should be the same, and equal to

$$\gamma = \frac{1}{2a} + \frac{1}{2b}\,. \tag{10}$$

As we said above, we found indeed that relaxation times and inverse diffusivity diverge with the same exponent $\gamma = 3.53$. On the other hand, using the values of a and b obtained by the fits of the intermediate part of the relaxation functions, and using eq. (10), one finds $\gamma = 1.94 \pm 0.02$. This part of the prediction is therefore not verified.

For what concerns the long-time regime ($t > t_\alpha$), the theory predicts that in the liquid phase, that is for temperature $T > T_c$ or density $\rho < \rho_c$, the correlation functions decay with a stretched exponential form, $c(t) \propto \exp[-(t/\tau_\alpha)^\beta]$, the exponent β being temperature (or density) independent. This prediction is called "time-temperature superposition principle", but applies as well for density driven transitions.

In fig. 7 we verify the prediction of MCT for densities between 0.75 and 0.82. In agreement with MCT the correlation functions collapse on the same stretched exponential when plotted *vs.* t/τ_α. In the present case the value of the exponent is $\beta = 0.86$.

3˙3. *Mean-field analysis.* – We have studied the model in mean field on the random graph as explained in the Appendix. In our case each site of the lattice is subdivided in $k+1$ internal positions corresponding to the internal degree of freedom and each site is connected to $k+1$ sites randomly chosen. The hard-core interaction between nearest-neighbor particles is shown in fig. 8. For large number of sites N, such a graph locally looks like a portion of a Cayley tree, but it displays loops of length of order $\ln N$, where N is the total number of sites. Locally the problem is unfrustrated, but the presence

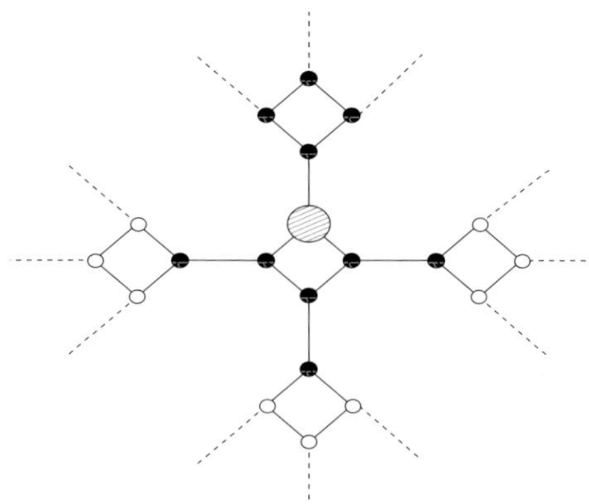

Fig. 8. – The model on a tree-like graph (here with $k = 3$): each site is subdivided in $k+1$ positions (little circles), and connected to $k+1$ randomly chosen neighbors. A particle in a given position (big shaded circle) forbids the presence of another particle in the positions colored in black.

of loops insures the existence of frustration which is a fundamental characteristic of glassy systems.

Here we generalize the procedure described in the Appendix. Let us consider a branch ending on a certain site i, connected to k sites $j \in \{1, \ldots, k\}$. In order to take into account the internal degrees of freedom, we have to introduce two local fields on each site. The recursion relations for the local fields for the iteration process are

$$e^{\beta a_i} = e^{\beta\mu}\left(\prod_{j=1}^{k}\frac{1+e^{\beta h_j}}{1+e^{\beta a_j}}\right)\left(1+\sum_{j=1}^{k}\frac{1}{1+e^{\beta h_j}}\right), \tag{11a}$$

$$e^{\beta h_i} = e^{\beta\mu}\left(\prod_{j=1}^{k}\frac{1+e^{\beta h_j}}{1+e^{\beta a_j}}\right)\left(\sum_{j=1}^{k}\frac{1}{1+e^{\beta h_j}}\right), \tag{11b}$$

while the free energy shift in the merging process (see fig. 9) is given by

$$e^{-\beta\Delta F} = \prod_{j=1}^{k}\left(1+e^{\beta a_j}\right). \tag{12}$$

According to eq. (A.3) to evaluate the total free energy density we have to compute the free energy shifts due to a site and link addition [2]:

$$e^{-\beta\Delta F^{(1)}} = \prod_{j=1}^{k+1}\left(1+e^{\beta a_j}\right) + e^{\beta\mu}\sum_{p=1}^{k+1}\prod_{j\neq p}\left(1+e^{\beta h_j}\right), \tag{13a}$$

$$e^{-\beta\Delta F^{(2)}} = 1+e^{\beta a_1}+e^{\beta a_2}+e^{\beta(h_1+h_2)}. \tag{13b}$$

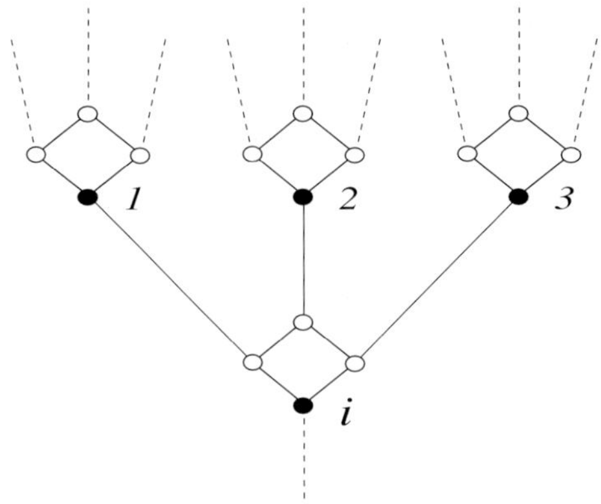

Fig. 9. – Merging of the branches $j = 1, \ldots, k$ (here with $k = 3$) onto the site i: the "external" positions are those colored in black, the others are the "internal" ones.

The Replica Symmetric (RS) solution of the problem corresponds to the case in which there is only one pure state and the local fields do not fluctuate from site to site ($h_i = h$, $a_i = a\ \forall i$). This homogeneous solution corresponds to the liquid phase characterized by translational invariance. This phase is expected to be stable for high values of the temperature T (or small values of the chemical potential μ).

In this case the cavity method [12] reduces to the simple Bethe-Peierls approximation. The local fields a and h are given by the fixed point of eqs. (11). The free energy can be easily computed by eqs. (13). From the local fields we can also evaluate the density ρ:

$$\rho = \frac{e^{\beta\mu}(k+1)(1+e^{\beta h})^k}{(1+e^{\beta a})^{k+1} + e^{\beta\mu}(k+1)(1+e^{\beta h})^k}\,. \tag{14}$$

Finally, since the internal energy per lattice site of the system is given by $E = -\mu\rho$, we obtain the following expression for the entropy per lattice site: $S = -\beta(F + \mu\rho)$.

As in the model studied by Biroli and Mézard [20], the entropy becomes negative for temperature below a certain value $T_{\mathrm{s}} = 0$ (see fig. 10). A similar behavior is observed for every $k > 1$. This behavior indicates that the homogeneous solution is not appropriate to describe the high density region.

As explained in the Appendix, it is also possible to find a crystalline (replica symmetric) solution of the problem where the fields do not fluctuate from site to site, but are different in different sites (breakdown of translational invariance). This is done via the introduction of three sub-lattices and three different couples of local fields. The merging is done taking into account the structure of the crystalline phase, described above (see fig. 2). The solution of the resulting equations appears discontinue at a spinodal point T_{sp}. It becomes thermodynamically stable when the corresponding free energy crosses the liquid one, at the melting temperature T_{m}. At that temperature we observe a first-

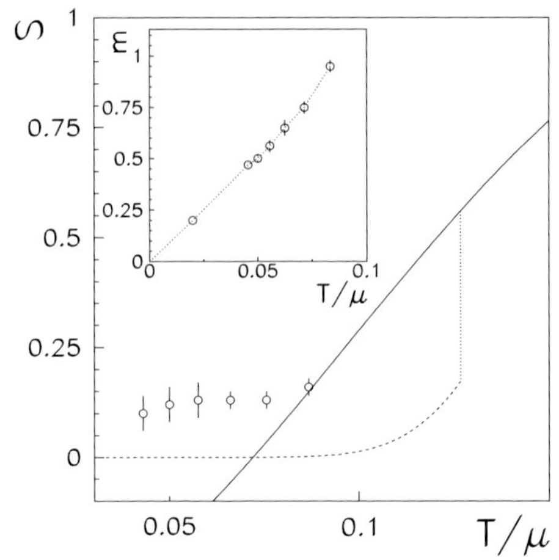

Fig. 10. – Entropy per site as a function of the temperature, on the random regular graph with $k = 5$. The notation is the same as in fig. 11. Inset: parameter m as a function of the temperature. The line is a guide for the eye.

order phase transition characterized by a spontaneous breakdown of the translational invariance, accompanied by a discontinuous change of the density and of the entropy per site. Increasing the chemical potential the density in the crystalline phase approaches quickly the maximum value $(k+1)/(k+2)$ and the entropy per site approaches quickly zero. For $k = 5$ we find the following values for the spinodal and the melting temperature: $T_{\rm sp} = 0.1633\mu$ and $T_{\rm m} = 0.1444\mu$. The corresponding specific volume and entropy per site are shown as dashed line in figs. 10 and 11.

As described in the Appendix, we have looked for a one-step RSB solution in the high density region [2,25]. We introduce a distribution probability $\mathcal{P}(a,h)$ defined as the probability that the fields a_i and h_i of a certain site i equal a and h (we work in the "factorized case" in which the probability distribution of the local fields on the single site is the same for all the sites of the lattices). Using the cavity method in the one-step replica symmetry broken ansatz we obtain the following integral self-consistency equation for the probability distribution of the local fields:

$$\mathcal{P}(a,h) = \mathcal{C} \int \prod_{j=1}^{k} \left[\mathrm{d}a_j \, \mathrm{d}h_j \, \mathcal{P}\big(a_j, h_j\big) \right] \delta(a - a_i)\delta(h - h_i) \exp[-\beta m \Delta F], \tag{15}$$

where $\mathcal{C}$ is a normalization constant, a_i, h_i and ΔF are functions of a_j and h_j via the recursion relations of eqs. (11) and (12), and the real parameter $m \in [0,1]$ is the usual one-step RSB parameter.

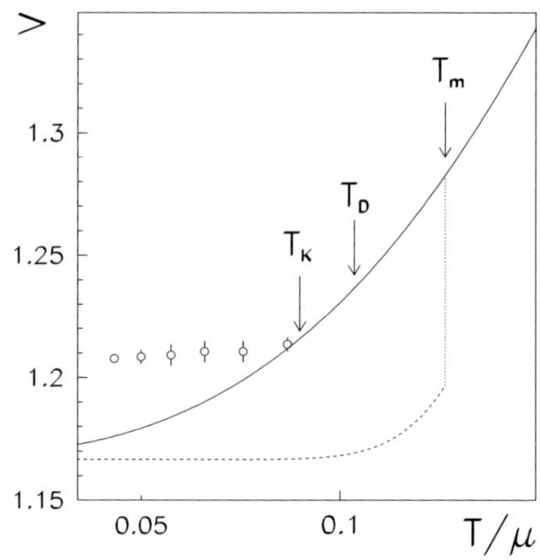

Fig. 11. – Specific volume (inverse density) as a function of temperature, on the random regular graph with $k = 5$. Solid line: replica symmetric liquid phase; broken line: crystalline phase; open circles: glassy phase in the 1-step RSB approximation. Arrows mark the various temperatures cited in the text.

As explained in the Appendix, the total free energy is now given by

$$F[m] = -\frac{1}{\beta m}\left\{ \log \int \prod_{j=1}^{k+1} \mathrm{d}a_j\, \mathrm{d}h_j\, \mathcal{P}(a_j,h_j) e^{-\beta m \Delta F^{(1)}} - \right. \\ \left. -\frac{k+1}{2} \log \int \prod_{j=1}^{2} \mathrm{d}a_j\, \mathrm{d}h_j\, \mathcal{P}(a_j,h_j) e^{-\beta m \Delta F^{(2)}} \right\}, \tag{16}$$

where $\Delta F^{(1)}$ and $\Delta F^{(2)}$ depend on a_j and h_j via eqs. (13). The parameter m is fixed by the maximization of the free energy with respect to it.

At $T = 0$ eq. (15) is easily solved. Indeed from eqs. (11) in the $\beta \to \infty$ limit we notice that the local fields on each site can only assume the following finite set of values:

$$a_i = (1-k)\mu, \dots, 0, \mu\,, \\ h_i = \begin{cases} a_i\,, & \text{if } a_i < \mu\,, \\ 0, \mu\,, & \text{if } a_i = \mu\,. \end{cases} \tag{17}$$

The probability distribution of the local fields is then the sum of $k+2$ delta-function $\mathcal{P}(a,h) = \sum_{r=1}^{k+2} p_r \delta(a - a_r)\delta(h - h_r)$. The self consistency eq. (15) gives rise to $k+2$ algebraic equations for the weights of the delta-functions p_r. For every value of $\beta m \mu$ we solve the equations for the coefficients. Finally we find that the maximum value of the free energy for $k = 5$ is $F = -0.8265\mu$, obtained for $\beta m \mu = 8.078$. Since $\beta m \mu$ stays finite for $\beta \to \infty$, m must go to zero at zero temperature, as it happens for the p-spin model.

At temperature $T > 0$ we have solved eq. (15) numerically with an iterative procedure and we have found a scenario identical to the one of p-spin models and discontinuous spin glasses. The starting distribution $\mathcal{P}(a, h)$ is the one that solves the equation at $T = 0$, that we have previously described. We have slowly increased the temperature, and for each value of the temperature we have applied iteratively eq. (15) until convergence.

For high values of the temperature we find that $\mathcal{P}(a, h) = \delta(a - \overline{a})\delta(h - \overline{h})$, where $\overline{a}$ and $\overline{h}$ are the values of the local fields in the liquid phase. Lowering the temperature we find first a dynamical transition at a certain temperature T_{D}, where a nontrivial solution of the 1-step RSB equation appears, signaling the existence of many pure states. This solution becomes thermodynamically stable below a static transition temperature T_{s}, where the maximum of the free energy as a function of m is at $m = 1$. For $k = 5$ we find that $T_{\mathrm{D}} = 0.105\mu$ and $T_{\mathrm{s}} = 0.087\mu$. The results for $k = 5$ are shown in figs. 10 and 11, and they are very similar to those found for other lattice glass models [20,22,26]. Figure 10 shows also the behavior of the parameter m as a function of the temperature, which turns out to be very similar to the one found for the p-spin model.

3'4. *Conclusions.* – In conclusion, we have studied a lattice glass model with two-body interactions. Monte Carlo simulations in three dimensions show that the model, despite the presence of a crystalline state, has no tendency to crystallize. This feature allows the study of the supercooled regime without introducing a mixture, as in many models of glass. The model has a clear glassy behavior, and reproduces remarkably well the cage effect and the presence of two different time scales in the relaxation process.

Moreover in mean field on a random regular graph we find the scenario typical of p-spin glasses, recently reproduced by other models of glass. Further investigation is required to understand whether in finite dimensions the static transition observed in mean field is still present.

In the next section we discuss another important class of systems showing "jamming" phenomena, *i.e.* granular media. In particular, by use of Edwards' approach, we will explain similarities and differences with glass formers.

4. – A few experimental facts about granular media

Granular media are large conglomerations of discrete macroscopic particles. Ubiquitous in the world around us, in many respect they behave differently from any of the other familiar forms of matter —solids, liquids, or gases (for a review see [27]). The static and dynamic properties of granular media, which are strongly dissipative systems, are not affected by temperature, because thermal fluctuations are usually negligible in comparison to the potential energy. In sand the potential energy of a grain raised a distance equal to its own diameter can be 10^{12} times $k_{\mathrm{B}}T$ at room temperature; therefore the temperature of the external bath can be considered zero and these media called *non-thermal*. As the system cannot explore its phase space, unless perturbed by external forces (such as shaking or tapping), it is frozen, at rest, in its mechanically stable microstates ("inherent states").

Even though granular media may form crystalline packings, in most cases they are found at rest in disordered configurations, characterized by "fluid" like distribution functions. Actually, gently shaken granular media exhibit a strong form of "jamming" [28-30], *i.e.* an exceedingly slow dynamics, which shows deep connections to "freezing" phenomena observed in many thermal systems such as glass formers [31,32]. It was discovered by the Chicago group [28] that under tapping the density of granular media tends to increase very slowly as a function of the number of shakes, in a stretched exponential [30] or even logarithmic [28] way. These [28] early observations triggered an intense theoretical and experimental research activity which showed in these systems typical "aging phenomena" along with huge relaxation times diverging *à la* Arrhenius or Vogel and Fulcher [29, 30], similarly to glass formers in the freezing region.

Although the idea of a unified description of these phenomena is emerging [32], the precise nature of jamming in non-thermal systems and the origin of its close connections to glassy phenomena in thermal ones are still important and open issues [33]. Here, we discuss these topics in the framework of the Statistical Mechanics of powders introduced by Edwards' [1,34]: we derive the phase diagram of granular media and explain in a quantitative way their similarities and differences with glass formers. The introduction of a Statistical Mechanics for powders is grounded on observations from experiments [28, 30] and simulations [35, 34] that, as much as in thermal systems, their macroscopic properties at stationarity are characterized by a few control parameters and their macrostates correspond to a huge number of microstates.

As an application of the Statistical Mechanics of powders, we also consider the intriguing phenomenon of segregation: in the presence of shaking a granular system is not randomized, but its components tend to separate [36, 37]. An example is the so-called "Brazil nut" effect (BNE) where, under shaking, large particles rise to the top and small particles move to the bottom of the container. Several mechanisms have been proposed to explain these phenomena, as for instance "percolation" [38], geometrical reorganization [39,40], convection [41], "condensation" [42] or inertia [43]. Interestingly, by changing grains sizes or mass ratio a crossover towards a "reverse Brazil nut" effect (RBNE) was more recently discovered [42] where small particles segregates to the top and large particles to the bottom. On the whole the criteria to predict segregation in a mixture, an issue of large practical relevance, are still largely unknown [36].

Via the study of a binary hard-sphere model under gravity subject to sequences of "taps" we show that a Statistical Mechanics description of segregation is indeed possible, and that two "configurational temperatures" are needed to characterize the macroscopic status of the system.

5. – Approaches *à la* Edwards to a Statistical Mechanics for granular media

In the Statistical Mechanics of powders introduced by Edwards [1] it is postulated that the system at rest (*i.e.* not in the "fluidized" regime) can be described by suitable ensemble averages over its "mechanically stable" states (called "inherent states"). Edwards proposed a method to individuate the probability, P_r, to find the system in its inherent state r, under the assumption that these mechanically stable states have the same *a priori* probability to occur. The knowledge of P_r has the conceptual advantage to substitute *time* with *ensemble averages*, allowing the description of the system properties by use of few basic theoretical concepts, as in thermodynamics. A possible approach to find P_r is as follows [1,34]. P_r is obtained as the maximum of the entropy,

$$S = -\sum_r P_r \ln P_r \tag{18}$$

with the macroscopic constraint that the system energy, $E = \sum_r P_r E_r$, is given. This assumption leads to the Gibbs result:

$$P_r \propto e^{-\beta_{\text{conf}} E_r}, \tag{19}$$

where β_{conf} is a *Lagrange multiplier*, called *inverse configurational temperature*, enforcing the above constraint on the energy:

$$\beta_{\text{conf}} = \frac{\partial S_{\text{conf}}}{\partial E}, \qquad S_{\text{conf}} = \ln \Omega_{\text{IS}}(E). \tag{20}$$

Here, $\Omega_{\text{IS}}(E)$ is the number of inherent states with energy E. Thus, summarizing, the system at rest has $T_{\text{bath}} = 0$ and $T_{\text{conf}} = \beta_{\text{conf}}^{-1} \neq 0$.

These basic considerations, to be validated by experiments or simulations, settle a theoretical Statistical Mechanics framework to describe granular media. Consider, for definiteness, a system of monodisperse hard spheres of mass m. In the system whole configuration space Ω_{Tot}, we can write Edwards' generalized partition function as

$$Z = \sum_{r \in \Omega_{\text{Tot}}} \exp[-\mathcal{H}_{\text{HC}} - \beta_{\text{conf}}\, mgH] \cdot \Pi_r\,, \tag{21}$$

where $\mathcal{H}_{\text{HC}}$ is the hard-core interaction between grains, mgH is the gravity contribution to the energy (H is particles height), and the factor Π_r is a projector on the space of "mechanically stable" states (*i.e.* inherent states space) Ω_{IS}: if $r \in \Omega_{\text{IS}}$ then $\Pi_r = 1$ else $\Pi_r = 0$. Usual Statistical Mechanics, where β_{conf}^{-1} is identified with T_{bath}, is recovered in the case where $\Pi_r = 1\ \forall r$.

In order to test Edwards' proposal, one introduces a dynamics (such as "tapping") that allows the system to explore the inherent states space. Using this dynamics one has first to check that at stationarity the properties of the system do not depend on the particular parameters of the dynamics. Then one must check that time averages obtained using such a dynamics compare well with ensamble averages over the distribution equation (19).

5$\cdot$1. *Configurational entropy of granular media and glasses frozen at* $T = 0$. – Edwards' proposal could also be relevant for a supercooled liquid quenched at very low temperature (that can be considered zero). In this case the system remains blocked in its "inherent structures", the local minima of the potential energy in the particle configuration space [44,45,17,19], because the thermal fluctuations are small in comparison to the barriers which separate two different minima (by analogy with the glass terminology we call "inherent states" the mechanically stable states in granular materials).

In glasses, in the inherent structure approach [44,45,17,19], the configurational entropy (and consequently the configurational temperature) associated to the number of inherent structures corresponding to a given energy, E, can be defined too. In this context two different ways were essentially used to allow the system explore its inherent structure space. One way is by quenching the system over and over from an equilibrium temperature, T, to zero temperature [44,17,19]. Using this procedure, Sciortino *et al.* [17] found that the configurational temperature numerically coincides with the equilibrium temperature T, provided that T is low enough. Another way to visit the inherent structures is by letting the system aging in contact with an almost zero bath temperature, $T_{\rm bath}$. During the aging process an effective temperature, $T_{\rm dyn}$, can be defined via the off-equilibrium extension of the fluctuation-dissipation ratio [46]. It happens that in mean-field models [47,48] this effective temperature coincides with the configurational temperature, $T_{\rm conf}$. This was also shown in ref. [49], in the limit $T_{\rm bath} \to 0$, for a class of finite-dimensional systems. The connection between Edwards' approach for granular media and the results in glass theory has been pointed out in [50,49,35,51,34].

A procedure, called "tap dynamics", similar to that used in the compaction of real granular materials was instead used in [31,34,51-55]. In particular in refs. [51,34] each tap consists in raising the bath temperature to a value T_Γ and, after a lapse of time τ_0, quenching it back to zero. By cyclically repeating the process the system explores the space of the inherent states. Once the stationary state is reached a temperature, $T_{\rm fd}$, is defined via the equilibrium fluctuation-dissipation relation, and if Edwards' assumption applies, $T_{\rm fd}$ coincides with the configurational temperature. This was verified for different finite-dimensional models [51,34]. Many other studies confirming Edwards' approach were also presented [52-54,56].

In the following sections we give a comprehensive view of the results obtained in [51,34], where the validity of the statistical-mechanics approach is shown for three schematic models of granular media and glasses. Finally a mean-field theory for the monodisperse hard-sphere system under gravity is presented. The very good agreement with the tap dynamics simulation results shows how powerful could be such an approach in order to clarify some of the not yet completely understood properties of granular media, and the similarity between granular media and glasses.

6. – Models for granular media

6$\cdot$1. *A monodisperse hard-sphere system under gravity.* – The simplest model for granular media we considered is a system of hard-spheres of equal diameter $a_0 = 1$,

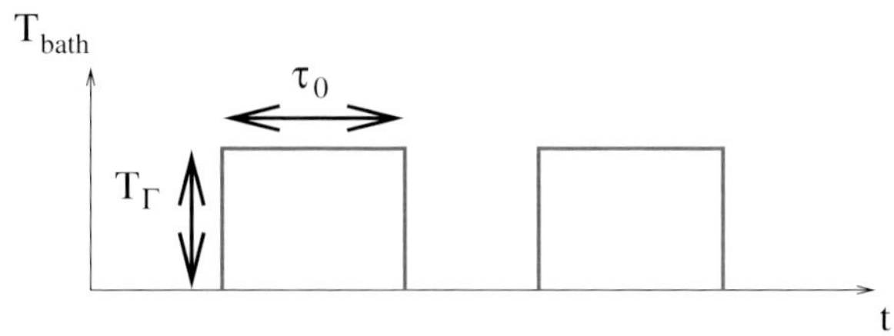

Fig. 12. – Our lattice models for granular media are subject to a Monte Carlo dynamics made of "tap" sequences. A "tap" is a period of time, of length τ_0 (the tap duration), during which the system evolves at a finite bath temperature T_Γ (the tap amplitude); after each "tap" the system evolves at $T_\Gamma = 0$ and reaches an inherent state.

subjected to gravity. We have studied this model on a lattice, constraining the centers of mass of the spheres on the sites of a cubic lattice (see upper inset in fig. 15). We use periodic boundary conditions on the horizontal directions.

The Hamiltonian of the system is

$$\mathcal{H} = \mathcal{H}_{\rm hc}(\{n_i\}) + gm \sum_i n_i z_i \,, \tag{22}$$

where the height of site i is z_i, $g = 1$ is the gravity, $m = 1$ the grains mass, n_i the usual occupancy variable and $\mathcal{H}_{\rm hc}(\{n_i\})$ a hard-core interaction term that prevents the overlapping of nearest-neighbor grains (this term can be written as $\mathcal{H}_{\rm hc}(\{n_i\}) = J \sum_{\langle ij \rangle} n_i n_j$, where the limit $J \to \infty$ is taken). Note that this Hamiltonian is similar to that defining the lattice glass model previously studied (eq. (5)).

The grains are subject to a dynamics made of a sequence of Monte Carlo "taps" (see fig. 12): a single "tap" is a period of time, of length τ_0 (the tap duration), where particles can diffuse laterally with probability $p_{\rm l} = 1/2$, upwards with probability $p_{\rm up} \in [0, 1/2]$, and downwards with probability $1 - p_{\rm up}$. When the "tap" is off grains can only move downwards (*i.e.* $p_{\rm up} = 0$). In this case the system reached a blocked state (*i.e.* an "inherent state") where no grain can move downwards without violating the hard-core repulsion. After each tap we evolve the system with $p_{\rm up} = 0$ until it reaches a blocked configuration. The parameter $p_{\rm up}$ has an effect equivalent to keep the system in contact (for a time τ_0) with a bath temperature $T_\Gamma = mga_0/\ln[(1 - p_{\rm up})/p_{\rm up}]$ (called the "tap amplitude"). The properties of the system are measured when this is in a blocked state. Time averages, therefore, are averages over the blocked configurations reached with this dynamics. Time t is measured in units of taps applied to the system.

Under such a tap dynamics the systems reaches a stationary state where the Statistical Mechanics approach to granular media can be tested, and particularly Edwards hypothesis can be verified by comparing time averages to ensemble averages over the distribution of eq. (19).

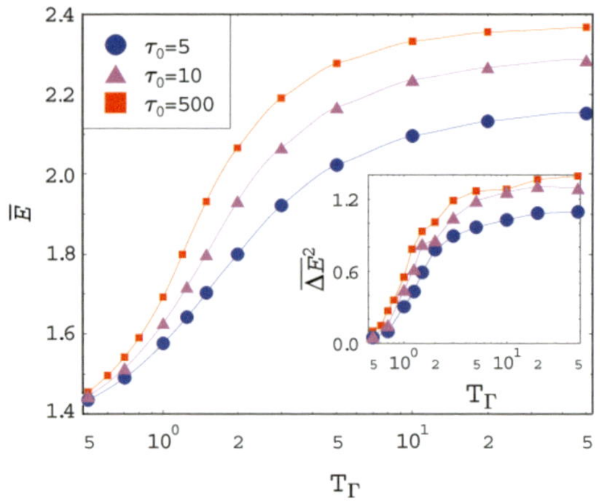

Fig. 13. – The time average of the energy, $\overline{E}$, and (inset) its fluctuations, $\overline{\Delta E}^2$, recorded at stationarity during a tap dynamics, as a function of the tap amplitude, T_Γ, in the 3D lattice monodisperse hard-sphere model. Different curves correspond to sequences of tap with different values of the duration of each single tap, τ_0.

6·2. *The results obtained under the tap dynamics.* – During the tap dynamics, in the stationary state, the time average of the energy, $\overline{E}$, and its fluctuations, $\overline{\Delta E^2}$, are calculated. Figure 13 shows $\overline{E}$ (main frame) and $\overline{\Delta E}^2$ (inset) as a function the tap amplitude, T_Γ (for several values of the tap duration, τ_0). Since sequences of taps, with same T_Γ and different τ_0, give different values of $\overline{E}$ and $\overline{\Delta E}^2$, it is apparent that T_Γ is not the right thermodynamic parameter. On the other hand, if the stationary states are indeed characterized by a *single* thermodynamic parameter the curves corresponding to different tap sequences (*i.e.* different T_Γ and τ_0) should collapse onto a single master function, when $\overline{\Delta E}^2$ is parametrically plotted as a function of $\overline{E}$. This is the case in the present model, where the data collapse is in fact found and shown in fig. 14. Such is a prediction could be easily checked in real granular materials (where one could consider the density which is easier to measure than the energy).

A technique to derive from raw data the thermodynamic parameter $\beta_{\rm fd}$ conjugated to E (apart from an integration constant, β_0), is through the usual *equilibrium* Fluctuation-Dissipation relation:

$$-\frac{\partial \overline{E}}{\partial \beta_{\rm fd}} = \overline{\Delta E^2}\,. \tag{23}$$

By integrating eq. (23), $\beta_{\rm fd} - \beta_0$ can be expressed as a function of $\overline{E}$ or (for a fixed value of τ_0) as a function of β_Γ: $\beta_{\rm fd} - \beta_0 \equiv g(\beta_\Gamma)$. The constant β_0 can be determined as explained in [34].

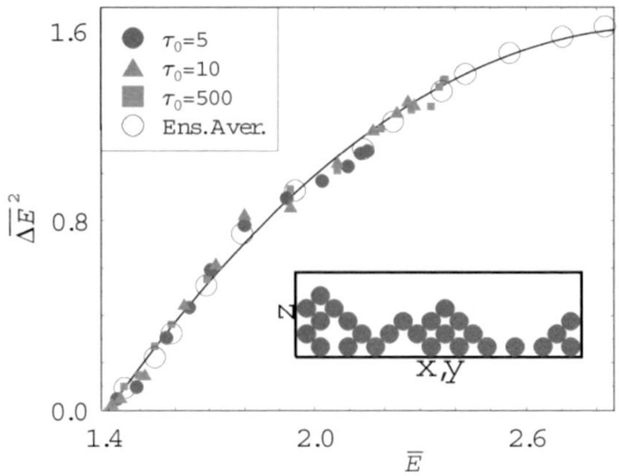

Fig. 14. – Time averages of energy fluctuations $\overline{\Delta E^2}$ plotted as a function of the time average of energy $\overline{E}$. ●, ▲ and ■ are time averages obtained with different tap dynamics. ○ are ensamble avereges obtained according to eq. (19). The collapse of the data obtained with different dynamics shows that the system stationary states are characterized by a *single* thermodynamic parameter. The agreement with the ensamble averages shows the success of Edwards' approach to describe the system macroscopic properties.

6˙3. *Edwards' averages.* – In subsect. **6**˙2 we have found that the fluctuations of the energy in the stationary state depend only on the energy, E, and not on the past history. If all macroscopic quantities depend only on the energy, E, or on its conjugate thermodynamic parameter, $\beta_{\rm fd}$, the stationary state can be genuinely considered a "thermodynamic state". In this case one can attempt to construct an equilibrium statistical mechanics, as originally suggested by Edwards [1].

To test Edwards' hypothesis one has to compare the time average of the energy, $E(\beta_{\rm fd})$, recorded during the taps sequences, with the ensemble average, $\langle E\rangle(\beta_{\rm conf})$, over the distribution equation (19). Therefore we have independently calculated the average $\langle E\rangle$, as a function of $\beta_{\rm conf}$.

In particular we have constructed an effective Hamiltonian, $\mathcal{H}_{\rm eff} = \mathcal{H}+\mathcal{H}_\Pi$, by adding the term $\mathcal{H}_\Pi = -\ln\Pi$ to eq. (22), which is zero if the configuration is an inherent state, and infinite otherwise. The canonical distribution for this Hamiltonian gives a weight, $e^{-\beta_{\rm conf}\mathcal{H}_{\rm eff}}$, which coincides with the weight in the distribution of eq. (19), for each accessible configuration. By using standard Monte Carlo simulations, we have calculated $\langle E\rangle(\beta_{\rm conf})$. Figure 15 outlines a very good agreement between $\langle E\rangle(\beta_{\rm conf})$ and $\overline{E}(\beta_{\rm fd})$ (notice that there are no adjustable parameters).

In order to check the scenario in further details, we have also calculated the system density on the bottom layer, $\rho_{\rm b}$, and the density self-overlap function Q: $Q = 1/N\sum_{i=1}^N n_i^{(1)}n_i^{(2)}$, where the sum is over all the lattice sites and $n^{(1)}$ and $n^{(2)}$ refer to two different replicas of the systems, which are at stationarity under the same tap dynamics. We have verified that, when plotted as a function of $T_{\rm fd}$, both $\rho_{\rm b}$ and Q scale on a single master function (see fig. 16).

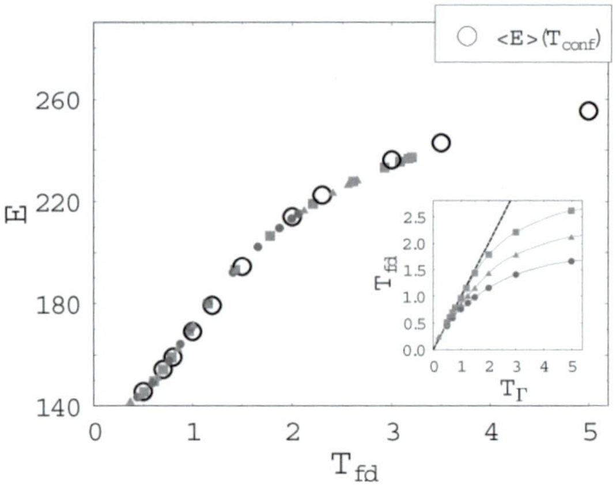

Fig. 15. – Main frame: The time average $\overline{E}$ and the ensemble average over the distribution equation (19) $\langle E \rangle$, plotted respectively as a function of T_{fd} and T_{conf} (in units mga_0), in the 3D monodisperse hard-sphere system under gravity described in the text. Symbols are as in fig. 14. Time averages over the tap dynamics and Edwards' ensemble averages then coincide. Lower inset: The temperature $T_{\mathrm{fd}} \equiv \beta_{\mathrm{fd}}^{-1}$ defined by eq. (23) as a function of T_Γ (in units mga_0) for $\tau_0 = 500, 10, 5$ MCS (from top to bottom). The straight line is the function $T_{\mathrm{fd}} = T_\Gamma$.

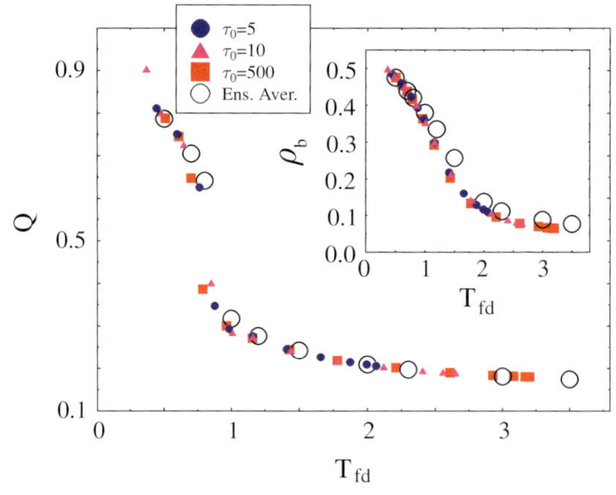

Fig. 16. – The density self-overlap function, Q, and (upper inset) the system density on the bottom layer, ρ_b, plotted as a function of T_{fd} (in units mga_0), compared with the ensemble averages over the distribution equation (19) (the black empty circles), plotted as a function of T_{conf} (in units mga_0), in the 3D monodisperse hard-sphere system under gravity. Also for this system there is a very good agreement between the independently calculated time averages over the tap dynamics and the statistical mechanics ensemble averages *à la* Edwards.

As described in subsect. **6**˙3, on the same master function collapse Edwards' averages as a function of $\beta_{\rm conf}$. Figure 15 shows a very good agreement between $\langle E\rangle(T_{\rm conf})$ and $\overline{E}(T_{\rm fd})$. The same agreement is found for the other quoted observables, $\rho_{\rm b}$ and Q (see fig. 16). In fig. 15 (inset) we also show the dependence of the configurational temperature $T_{\rm conf}$ on the parameters of the tap dynamics T_Γ and τ_0.

6˙4. *The frustrated lattice gas model.* – We have also tested Edwards' scenario in the "frustrated lattice gas" model. This is a lattice model described by the following Hamiltonian:

$$-H = \sum_{\langle ij\rangle} n_i n_j \phi_{i,j}(S_i, S_j) + \mu \sum_i n_i - g \sum_i z_i n_i \,, \tag{24}$$

where the sum $\sum_{\langle ij\rangle}$ is over nearest-neighbor sites, $n_i = 0, 1$ are occupation variables, μ is the particle chemical potential, g the gravity acceleration and z_i the height of the site i. The function ϕ defines the interaction between neighbors: $\phi_{i,j}(S_i, S_j) = J(\epsilon_{ij} S_i S_j - 1)$ where J is a coupling constant, $S_i = \pm 1$ are Ising variables describing grain orientations and, in the frustrated lattice gas model, ϵ_{ij} are quenched and random variables, equal to ± 1 with equal probability. This model, for $g = 0$, reproduces the Ising spin glass in the limit $\mu \to \infty$ (*i.e.* when all sites are occupied, $n_i \equiv 1$).

In the other limit, $J \to \infty$, the model describes a frustrated lattice gas with properties recalling those of a "frustrated" liquid. Note the similarity between the Hamiltonian of this model (eq. (24)) and the Hamiltonian of the lattice glass model previously studied (eq. (5)). The frustrated loops in the model are the same of the spin glass model and correspond to those loops in a liquid which, due to geometrical hindrance, cannot be fully occupied by the particles. In 3D [57-59], the model has a maximum density $\rho_{\max} \simeq 0.68$, and a transition at $\rho_{\rm c} \simeq 0.62$ where the non-linear spin susceptibility diverges.

In [34] we have studied the 3D cubic frustrated lattice gas model with J finite. We have simulated the dynamics of the model, that was driven by a sequence of taps (see fig. 12), and the static properties, exploring the inherent states space by using a Monte Carlo procedure based on an effective Hamiltonian (which gives to each inherent state the same weight given by eq. (19)). As in the case of the hard-sphere system, the results [34] show that Edwards' proposal successfully apply to this model too.

6˙5. *Relaxation during MC "tap" dynamics.* – The MC tap "dynamics", in both the frustrated lattice gas and hard-sphere models, exhibits a rich structure in agreement with experimental findings [28, 30]. In the region of small tap amplitudes, the system gets "jammed" and "memory" phenomena are observed, along with "irreversibility" effects [28, 31]. In particular, it is interesting to consider correlation functions such as $C(t, t_w) = B(t, t_w)/B(t_w, t_w)$, where $B(t, t_w) = \sum_i [\langle n_i(t + t_w) n_i(t_w)\rangle - \langle n_i(t + t_w)\rangle\langle n_i(t_w)\rangle]$. In the high-$T_\Gamma$ region, $C(t, t_w)$ has a time translation invariant (TTI) behavior, *i.e.* $C(t, t_w) = C(t)$. Asymptotically $C(t)$ can be well fitted by stretched exponentials: $C(t) = C_0 \exp[-(t/\tau)^\beta]$. The exponent β becomes significantly lower than 1 at low amplitudes. The above fit defines the relaxation time $\tau(T_\Gamma)$: the growth of τ

by decreasing T_Γ is well approximated by an Arrhenius or Vogel-Tamman-Fulcher law (as early found in [31, 34]), resembling the slowing down of glass formers close to the glass transition, a result also recently experimentally reported in granular media [29,30]: $\tau \simeq \tau_0 \exp[E_0/(T_\Gamma - T_\Gamma^{\rm K})]$. The divergence point, $T_\Gamma^{\rm K}$ (which in simulations is difficult to precisely locate; here it is between 0 and 1), of τ is interpreted as the numerical location of the point of dynamical arrest of the system, where an "ideal" transition to a glassy phase occurs. By quenching the system at low values of T_Γ, the TTI character of relaxation is lost and logarithmic aging behaviors are found. For slow quenches the hard-spheres model is able, anyway, to attain its crystal phase.

Another quantity of interest is grains mean square displacement, $R^2(t)$, defining for $t \to \infty$ the diffusion coefficient $D(\rho) = R^2(t)/t$. As much as in glass formers, D goes to zero around a high density $\rho_{\rm c}$, signaling a localization transition in which particles are confined in local cages and the macroscopic diffusion-like processes are suppressed [31]. This phenomenon may also be described in a different way: $\rho_{\rm c}$ is the density above which it becomes impossible to obtain a macroscopic rearrangement of particles positions without increasing the system volume, *i.e.* the density at which macroscopic shear in the system is impossible without dilatancy. This remark outlines the close correspondence with the phenomenon of the Reynolds transition in granular media [31].

7. – A mean-field study of hard spheres under gravity on a random graph

Summarising, we have discussed schematic models for granular media shown to be well described by Edwards' assumption. Since the exact calculation of Edwards' partition function, Z, for the above model of monodisperse hard spheres on a cubic lattice is hardly feasible, we now discuss a mean-field theory [60] based on a random graph version of such a lattice which is sketched in fig. 17. More specifically we consider a 3D lattice box with H horizontal layers (*i.e.*, $z \in \{1, \ldots, H\}$) occupied by hard spheres. Each layer is a random graph of given connectivity, $k-1$ (we take $k=4$). Each site in layer z is also connected to its homologous site in $z-1$ and $z+1$ (the total connectivity is thus $k+1$). The Hamiltonian is the one of eq. (5) plus a chemical potential term to control the overall density. Hard-core repulsion prevents two connected sites to be occupied at the same time. In the present lattice model we adopt a simple definition of "mechanical stability": a grain is "stable" if it has a grain underneath. The operator Π_r has thus a simple expression: $\Pi_r = \lim_{K\to\infty} \exp[-K\mathcal{H}_{\rm Edw}]$, where $\mathcal{H}_{\rm Edw} = \sum_i \delta_{n_i(z),1}\delta_{n_i(z-1),0}\delta_{n_i(z-2),0}$ (for clarity, we have shown the z dependence in $n_i(z)$).

Using the "cavity method" [2] as shown in the Appendix, the phase diagram is found. At low $N_{\rm s}$ or high $T_{\rm conf}$ a fluid-like phase is found, characterized by a homogeneous Replica Symmetric (RS) solution, in which only one pure state exists and the local fields are the same for all the sites of the lattice (translational invariance). For a given $N_{\rm s}$, by lowering $T_{\rm conf}$ (see figs. 18 and 19), a phase transition to a crystal phase (an RS solution with no space translation invariance) is found at $T_{\rm m}$ [61]. Notice that the fluid phase still exists below $T_{\rm m}$ as a metastable phase corresponding to a supercooled fluid found when crystallization is avoided.

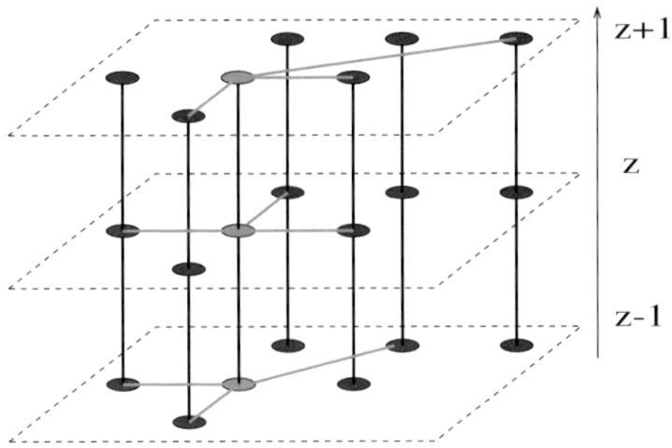

Fig. 17. – In our mean-field approximation, the hard spheres describing the system grains are located on a Bethe lattice, sketched in the figure, where each horizontal layer is a random graph of given connectivity. Homologous sites on neighboring layers are also linked and the overall connectivity, c, of the vertices is $c \equiv k + 1 = 5$.

Within the one-step Replica Symmetry Breaking (1RSB) ansatz of the cavity method [2, 62], a non-trivial solution appears for the first time at a given temperature $T_{\rm D}(N_{\rm s})$, signaling the existence of an exponentially high number of pure states. In mean-field theory $T_{\rm D}$ is interpreted as the location of a purely dynamical transition as in mode-coupling theory, but in real systems it might correspond just to a crossover in the dynamics (see [48, 20, 63] and references therein). The 1RSB solution becomes stable

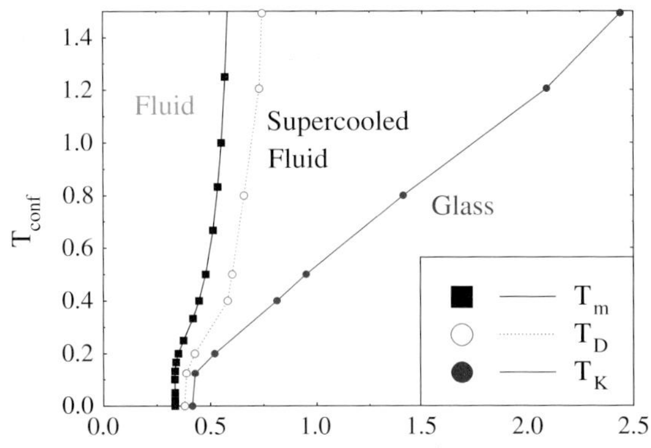

Fig. 18. – The system mean-field phase diagram is plotted in the plane of its two control parameters $(T_{\rm conf}, N_{\rm s})$: $T_{\rm conf}$ is Edwards' "configurational temperature" and $N_{\rm s}$ the average number of grains per unit surface in the box. At low $N_{\rm s}$ or high $T_{\rm conf}$, the system is found in a fluid phase. The fluid forms a crystal below a melting transition line $T_{\rm m}(N_{\rm s})$. When crystallization is avoided, the "supercooled" (*i.e.* metastable) fluid has a thermodynamic phase transition, at a point $T_{\rm K}(N_{\rm s})$, to a replica symmetry-breaking "glassy" phase with the same structure found in mean-field theory of glass formers. In between $T_{\rm m}(N_{\rm s})$ and $T_{\rm K}(N_{\rm s})$ a dynamical freezing point, $T_{\rm D}(N_{\rm s})$, is located, where the system characteristic time scales diverge.

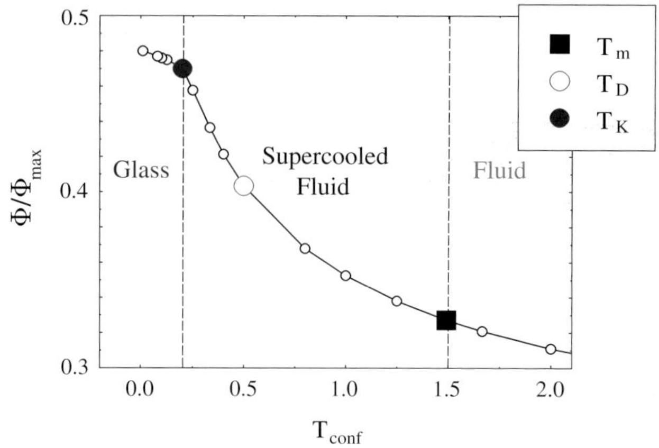

Fig. 19. – For a system with a given number of grains (*i.e.* a given N_s), the overall number density, $\Phi \equiv N_s/2\langle z\rangle$ ($\langle z\rangle$ is the average height), calculated in mean-field approximation is plotted as a function of T_{conf}; $\Phi(T_{conf})$ has a shape very similar to the one observed in the "reversible regime" of tap experiments and MC simulations of the cubic lattice model for $\Phi(T_\Gamma)$. The location of the glass transition, T_K (filled circle), corresponds to a cusp in the function $\Phi(T_{conf})$. The passage from the fluid to supercooled fluid is T_m (filled square). The dynamical crossover point T_D is found around the flex of $\Phi(T_{conf})$ and well corresponds to the position of a characteristic shaking amplitude Γ^* found in experiments and simulations where the "irreversible" and "reversible" regimes approximately meet.

at a lower point T_K, where a thermodynamic transition from the supercooled fluid to a 1RSB glassy phase takes place (see fig. 18) in a scenario *à la* Kauzmann with a vanishing complexity of pure states (which stays finite for $T_K < T < T_D$).

The results of these calculations, summarized in the phase diagram of fig. 18, are further illustrated in fig. 19: in a system with a given number of grains (*i.e.* a given N_s), the overall number density, Φ, is plotted as a function of T_{conf} (here by definition $\Phi \equiv N_s/2\langle z\rangle$, where $\langle z\rangle$ is the average height). The shown curve, $\Phi(T_{conf})$, is the equilibrium function here calculated. It has a shape very similar to the one observed in tap experiments [28, 30], or in MC simulations on the cubic lattice (see also [31]), where the density is plotted as a function of the shaking amplitude Γ (along the so-called "reversible branch").

Summarizing, in the present mean-field scenario of a granular medium with N_s particles per surface, in general, at high T_{conf} (*i.e.* high shaking amplitudes) a fluid phase is located (see fig. 18). By lowering T_{conf}, a phase transition to a crystal phase is found at T_m. However, when crystallization is avoided, the fluid phase still exists below T_m as a metastable phase corresponding to a supercooled fluid. At a lower point, T_D, an exponentially high number of new metastable states appears, interpreted, at a mean-field level, as the location of a purely dynamical transition, which in real system is thought to correspond just to a dynamical crossover. Finally, at an even lower point, T_K, the supercooled fluid has a genuinely thermodynamics discontinuous phase transition to glassy

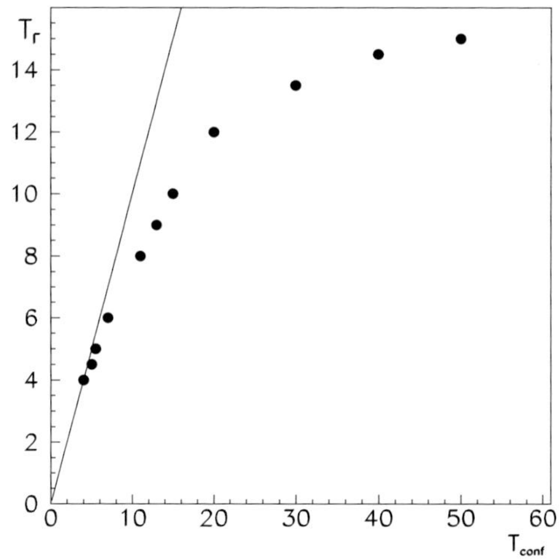

Fig. 20. – T_Γ as a function of $T_{\rm conf}$ for a tap duration $\tau_0 = 100$.

state. The structure of the glass transition of the present model for granular media, obtained in the framework of Edwards' theory, is the same found in the glass transition of the p-spin glass and in other mean-field models for glass formers [48, 20].

We have compared these mean-field results with simulations of the 3D model of hard spheres under the tap dynamics. In particular we have considered $N = 288$ grains confined in a 3D box of linear horizontal size $L = 12$ and height $H = 20$ (*i.e.* $N_{\rm s} = 2$). In fig. 20, T_Γ is plotted as a function of $T_{\rm conf}$ [64, 65], for a tap duration $\tau_0 = 100$; again we find that T_Γ and $T_{\rm conf}$ coincide at low values. For instance, we have compared the density profile $\sigma(z) \equiv N(z)/L^2$ (where $N(z)$ is the number of particles on the layer at height z) obtained at $T_\Gamma = 3$ (fig. 21) with the one obtained in the mean field at $T_{\rm conf} = 2.778$. As we can see the agreement between the simulations and the theoretical predictions obtained in mean-field approximation is very good. At $T_\Gamma = 3$ the system is in the crystal phase (as confirmed by the calculation of the staggered density), and the density profile is the one predicted by the mean-field approximation.

8. – A hard-sphere binary mixture under gravity

In order to study segregation and to test Edwards' proposal to a more complicate system, here we consider a hard-sphere binary system made of two species 1 (small) and 2 (large) with grain diameters a_0 and $\sqrt{2}a_0$, under gravity on a cubic lattice of spacing $a_0 = 1$. We set the units such that the two kinds of grain have masses $m_1 = 1$ and $m_2 = 2$, and the gravity acceleration is $g = 1$. The hard-core potential $\mathcal{H}_{\rm hc}$ is such that two large nearest-neighbor particles cannot overlap. This implies that only couples of small particles can be nearest neighbors on the lattice. The system overall Hamiltonian is

$$\mathcal{H} = \mathcal{H}_{\rm hc} + m_1 g H_1 + m_2 g H_2 \,, \tag{25}$$

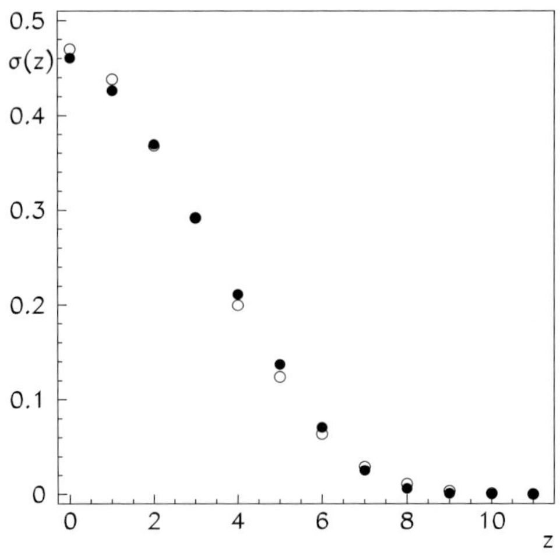

Fig. 21. – The density profile $\sigma(z) \equiv N(z)/L^2$ (where $N(z)$ is the number of particles on the layer at height z) obtained using the tap dynamics with tap amplitude $T_\Gamma = 3$ (the black circles in figure), compared to the one obtained in the mean-field crystal phase at $T_{\rm conf} = 2.778$ (the empty circles in figure). In both cases $N_{\rm s} = 2$.

where $H_1 = \sum_i^{(1)} z_i$ and $H_2 = \sum_i^{(2)} z_i$, the height of site i is z_i and the two sums are over all particles of species 1 and 2, respectively. In the above units, the gravitational energies in a given configuration are thus $E_1 = H_1$ and $E_2 = 2H_2$.

As before, grains are confined in a box of linear size L with periodic boundary conditions in the horizontal directions. $N_1 = 120$ grains of type 1 and $N_2 = 40$ grains of type 2 are initially prepared in a random loose stable pack. Under the tap dynamics the system approaches a stationary state for each value of the tap parameters T_Γ and τ_0 used. In fig. 22, we plot as a function of T_Γ (for several values of τ_0) the asymptotic value of the *vertical* segregation parameter, *i.e.* the difference of the average heights of the small and large grains at stationarity, $\Delta h(T_\Gamma, \tau_0) \equiv h_1 - h_2$. Here h_1 and h_2 are the average of H_1/N_1 and H_2/N_2 over the tap dynamics in the stationary state.

The results given in the main panel of fig. 22 apparently show that T_Γ is not a right thermodynamic parameter, since sequences of taps with different τ_0 give different values for the system observables. However, if the stationary states corresponding to different tap dynamics (*i.e.* different T_Γ and τ_0) are indeed characterized by a single thermodynamic parameter, the curves of fig. 22 should collapse onto a universal master function when $\Delta h(T_\Gamma, \tau_0)$ is parametrically plotted as a function of another macroscopic observable such as the average energy, $e(T_\Gamma, \tau_0) = (E_1 + E_2)/N$ (N is the total number of particles). This collapse of data is clearly not observed here, as it is apparent in the inset of fig. 22. We show, instead, that two macroscopic quantities may be sufficient to characterize uniquely the stationary state of the system. These two quantities are, for instance, the energy e and the height difference Δh. Of course since $e = ah_1 + 2bh_2$ (where $a = N_1/N$ and $b = N_2/N$) and $\Delta h = h_1 - h_2$, we can also choose h_1 and h_2 to characterize

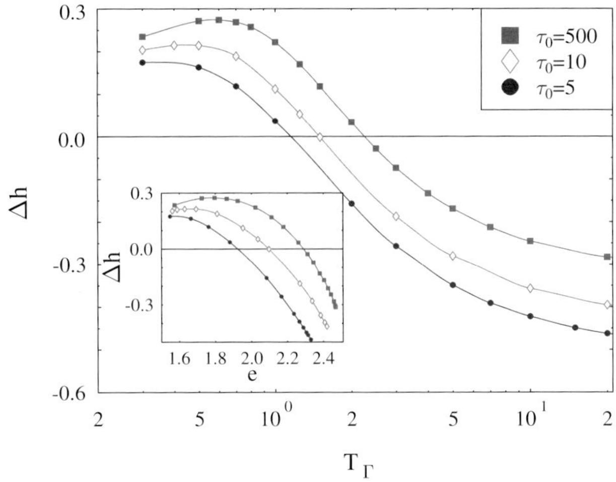

Fig. 22. – Main frame: The difference of the average heights of small and large grains, $\Delta h = h_1 - h_2$, measured at stationarity in the binary hard-spheres mixture under gravity, is plotted as a function of tap amplitude, T_Γ (in units mga_0). The three sets of points correspond to the shown tap durations, τ_0. At high T_Γ larger grains are found above the smaller, *i.e.* $\Delta h < 0$, as in the Brazil Nut Effect (BNE). Below a $T^*_\Gamma(\tau_0)$ the opposite is found (Reverse Brazil Nut Effect, RBNE). Inset: The Δh data of the main frame are plotted as a function of the corresponding average energy, e. The three sets of data do not collapse onto a single master function, showing that a single macroscopic observable, such as e, does not characterize the system status.

the stationary state. Namely, we show that any macroscopic quantity A, averaged over the tap dynamics in the stationary state, is only dependent on h_1 and h_2, *i.e.* $A = A(h_1, h_2)$. We have checked that this is the case for several independent observables, such as the number of contacts between large particles, N_c, the density of small and large particles on the bottom layer, ρ_1^b and ρ_2^b, and others. In particular, as shown in figs. 23 and 24, we find with good approximation that: $N_c \simeq N_c(e) = N_c(ah_1+bh_2)$, $\rho_2^b \simeq \rho_2^b(h_2)$, $\rho_1^b \simeq \rho_1^b(h_1)$. Therefore we need both h_1 and h_2 to characterize unambiguously the state of the system; namely all the observables assume the same values in a stationary state characterized by the same values of h_1 and h_2, independently of the previous history (*i.e.* in our case independently of the particular tapping parameters T_Γ and τ_0).

These findings imply that in this case an extension of Edwards' approach is required, where at least *two* thermodynamic parameters have to be included. As before, this can be obtained by assuming that the microscopic distribution is given by the principle of maximum entropy with the constraint that the average gravitational energies of the two species $E_1 = \sum_r P_r E_{1r}$ and $E_2 = \sum_r P_r E_{2r}$ are independently fixed. This gives *two* Lagrange multipliers:

$$\beta_1 = \frac{\partial \ln \Omega_{\rm IS}(E_1, E_2)}{\partial E_1} \qquad \beta_2 = \frac{\partial \ln \Omega_{\rm IS}(E_1, E_2)}{\partial E_2},$$

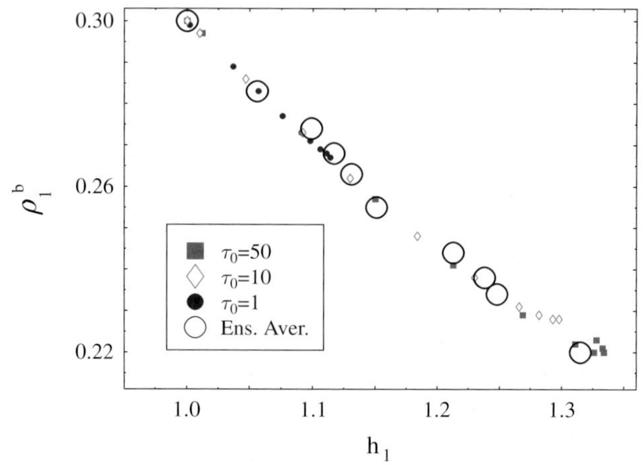

Fig. 23. – The average density of small grains on the box bottom layer, ρ_1^{b}, measured at stationarity as a function of the height of small particles, h_1, in the binary hard-spheres mixture under gravity. Data corresponding to different T_Γ and τ_0 approximately scale on a single master function. The empty circles are the corresponding values obtained by the ensemble average with the two temperatures Gibbs measure proposed in the text.

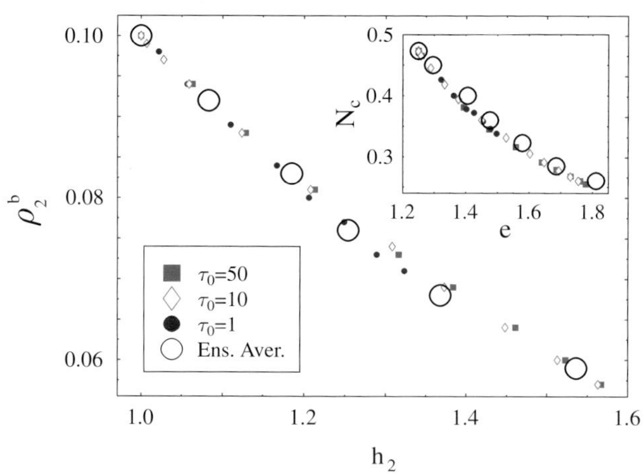

Fig. 24. – Main frame: The average density of large grains on the box bottom layer, ρ_2^{b}, obtained for different T_Γ and τ_0, scales almost on a single master function when plotted as a function of the large grains height, h_2. Upper inset: The average number of contacts between large grains per particle, N_{c}, obtained for different T_Γ and τ_0, scales on a single master function when plotted as a function of the system energy, e.

where $\Omega_{\mathrm{IS}}(E_1, E_2)$ is the number of inherent states with E_1, E_2.

The hypothesis that the ensemble distribution at stationarity is the above can be tested as we have already previously shown. We have to check that the time average of any quantity, $A(h_1, h_2)$, as recorded during the taps sequences in a stationary state characterized by given values h_1 and h_2, coincides with the ensemble average, $\langle A\rangle(h_1, h_2)$, over the generalized version of distribution equation (19). To this aim, we have calculated the ensemble averages $\langle N_{\mathrm{c}}\rangle$, $\langle\rho_2^{\mathrm{b}}\rangle$, $\langle\rho_1^{\mathrm{b}}\rangle$ for different values of β_1 and β_2; we have expressed parametrically $\langle N_{\mathrm{c}}\rangle$, $\langle\rho_2^{\mathrm{b}}\rangle$, $\langle\rho_1^{\mathrm{b}}\rangle$, as function of the average of h_1 and h_2, and compared them with the corresponding quantities, N_{c}, ρ_1^{b} and ρ_2^{b}, averaged over the tap dynamics. The two sets of data are plotted in figs. 23 and 24 showing a good agreement (notice, there are no adjustable parameters). In order to calculate the ensemble averages, we simulate the model with $\mathcal{H}$ from eq. (25) where we impose the constraint that the only accessible states are the inherent states, as already described in subsect. **6**˙3.

Finally, it is worth to recall that the crossover observed in vertical segregation from Brazil Nut Effect (BNE) to Reverse Brazil Nut Effect (RBNE), found here, corresponds to the drift of the minimum of Edwards' generalized free energy, $F = -\ln Z$, from negative to positive values of Δh and, in general, no phase transitions are associated to it. Instead, in the same framework, phenomena of horizontal segregation are shown to be typically related to true phase transitions in the system, leading to phase separation mechanisms (see ref. [34]).

In [66] the mean-field approximation (see sect. **7**) is applied to the binary mixture, giving a possible interpretation of the segregation phenomena observed in the model eq. (25) [67], and experimentally in granular media.

9. – Conclusions

Summarising, in the first part of this review we have introduced a "lattice glass model" of particles interacting with a hard-core potential. Monte Carlo simulations show that the model reproduces the behavior of glass-forming liquids, and that its dynamics compare well, in many respects, with MCT. The analytical solution of the model in mean field shows a scenario analogous to p-spin glass models, with a 1-step replica symmetry breaking glass transition.

In the second part, we have shown that the same kind of lattice models in the presence of gravity can be used to describe granular matter. More specifically, we have shown, by Monte Carlo simulations of "tap" dynamics, that Edwards' approach to granular media at rest is well grounded in these models. The system stationary states are indeed independent of the sample history as in a "thermodynamics" system, and can be described in terms of a distribution function characterized by a few control parameters (such as configurational temperatures). By use of Edwards' approach, we have derived, by analytical calculations at a mean-field level, the phase diagram of these systems. In particular, we discovered that "jamming" corresponds to a phase transition from a "fluid" to a "glassy" phase, observed when crystallization is avoided. Interestingly, the nature of such a "glassy" phase turns out to be the same found in mean-field models for glass formers. In the same framework, we have also briefly discussed segregation patterns observed in a hard-sphere binary model under gravity subject to sequences of taps.

In practice, even though the general validity of Edwards' approach to "frozen" systems has just begun to be assessed, it turns out that a first reference framework is emerging to understand the physics of granular media and their deep connections with other "jamming" systems such as glass formers.

Appendix

The cavity method

The cavity method [2] is a technique that allows to calculate the properties of ground states of systems defined on the Bethe lattice. The solution obtained with the cavity method can describe (depending on its symmetry) the liquid phase (low-density phase with translational invariance), the crystalline phase (high-density phase without translational invariance), and the glassy phase (high-density phase with translational invariance). This section introduces the cavity method and shows how it applies to a system of hard spheres.

Here we use the cavity method on the Bethe lattice (or "random regular graph"), which is a random graph with fixed connectivity $k+1$. For large number of sites N, such graph locally looks like a portion of a Cayley tree, but it displays loops of length of order $\ln N$. The system is locally unfrustrated, but the presence of loops ensures the existence of frustration, which is a fundamental characteristic of glassy systems. It is useful to use this kind of graph because there is a simple argument that allows to derive a general expression for the free energy of the system. Let us introduce (following [2]), $G_{(N,j)}$, which is a random graph with $N-j$ sites having connectivity $k+1$ and j site having connectivity k (cavity sites). $G_{(N,0)}$, therefore, is a Bethe lattice with N sites. Now let us consider the graph $G_{(N,2(k+1))}$. The addition of $k+1$ links connecting pairs of cavity sites transforms $G_{(N,2(k+1))}$ in $G_{(N,0)}$. On the other hand, the addition of two sites, each of them connected to $k+1$ cavity sites, transforms $G_{(N,2(k+1))}$ in $G_{(N+2,0)}$. To summarize:

$$G_{(N,0)} \underset{\text{addition}}{\overset{k+1\text{links}}{\longleftarrow}} G_{(N,2(k+1))} \underset{\text{addition}}{\overset{2\text{sites}}{\longrightarrow}} G_{(N+2,0)}. \tag{A.1}$$

It follows that the difference in free energy between a Bethe lattice with N sites and a Bethe lattice with $N+2$ sites is related to the average free energy shift for a site addition ($\Delta F^{(1)}$) and for a link addition ($\Delta F^{(2)}$). Precisely

$$F_{N+2} - F_N = 2\Delta F^{(1)} - (k+1)\Delta F^{(2)}. \tag{A.2}$$

Assuming that in the thermodynamic limit the free energy is linear in N, one obtains the following expression for the free energy density:

$$F = \Delta F^{(1)} - \frac{k+1}{2}\Delta F^{(2)}. \tag{A.3}$$

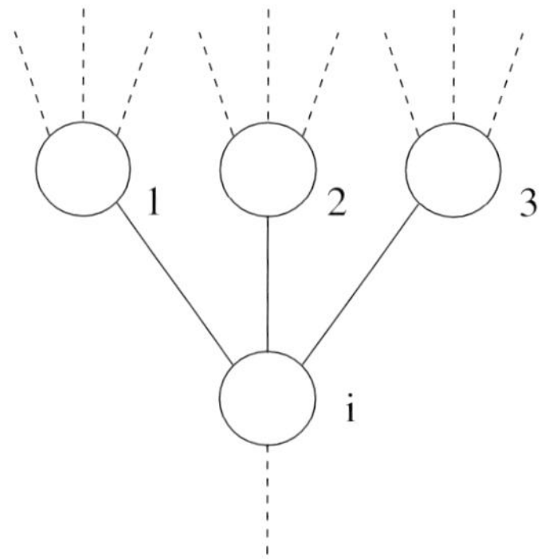

Fig. 25. – The merging of $k = 3$ sites into the site i.

The cavity method offers a general procedure to calculate the free energy shifts for a site and for a link addition. Here we present the solution for a system of hard spheres, which is described by the following Hamiltonian:

$$\mathcal{H} = J \sum_{\langle i,j \rangle} n_i n_j \,, \tag{A.4}$$

in the limit $J \to \infty$. The sum is over all the pairs of nearest-neighbors sites. The Hamiltonian realizes a hard-core interaction between particles which forbids the occupation of two nearest-neighbors sites.

Let us consider a branch ending on a certain site i connected to k sites $j \in \{1, \dots, k\}$ (see fig. 25, where $k = 3$). We suppose that the effect of the whole lattice (except for the site i) on the site j is described by an effective field h_j. Consequently the Hamiltonian of the system, having introduced the site i, is

$$H = J n_i \sum_{j=1}^{k} n_j - \sum_{j=1}^{k} n_j h_j - \mu n_i \,, \tag{A.5}$$

while the partition function is given by

$$Z = \sum_{\{n_i, n_j\}} \exp[-\beta H] = \sum_{n_i=0}^{1} \exp[\beta \mu n_i] \prod_{j=1}^{k} \sum_{n_j=0}^{1} \exp\big[-\beta J n_i n_j + \beta h_j n_j\big]. \tag{A.6}$$

If we look for two functions $a(h_j)$ and $c(h_j)$ satisfying

$$\sum_{n_j=0}^{1} \exp\big[-\beta J n_i n_j + \beta h_j n_j\big] = a(h_j) \exp\big[\beta c(h_j) n_i\big], \tag{A.7}$$

we find, considering the effect of the hard-sphere Hamiltonian (*i.e.* that $n_i n_j = 0$), $a(h_j) = 1 + \exp[\beta h_j]$ and $a(h_j) \exp[\beta c(h_j)] = 1$. Therefore, the partition function becomes

$$Z = \prod_{j=1}^{k} a(h_j) \sum_{n_i=0}^{1} \exp\left[\sum_{j=1}^{k} (\beta c(h_j) + \beta\mu) n_i\right] = \prod_{j=1}^{k} a(h_j) \sum_{n_i=0}^{1} \exp\left[\beta h_i n_i\right], \tag{A.8}$$

where

$$\beta h_i = \mu + \beta \sum_{j=1}^{k} c(h_j) = \beta\mu + \sum_{j=1}^{k} \log \frac{1}{1 + \exp\left[\beta h_j\right]} . \tag{A.9}$$

Equation (A.9) relates the effective field h_i on site i to the fields h_j on sites j. Equation (A.8) shows that in the merging process of the sites j into the site i there is a free energy shift given by

$$e^{-\beta \Delta F} = \prod_{j=1}^{k} a(h_j) = \prod_{j=1}^{k} (1 + e^{\beta h_j}). \tag{A.10}$$

According to eq. (A.3) to evaluate the total free energy density we have to compute the free energy shifts due to a site ($\Delta F^{(1)}$) and a link ($\Delta F^{(2)}$) addition. These are given by [2]

$$e^{-\beta \Delta F^{(1)}} = \prod_{j=1}^{k+1} (1 + e^{\beta h_j}) + e^{\beta\mu}, \tag{A.11}$$

$$e^{-\beta \Delta F^{(2)}} = 1 + e^{\beta h_1} + e^{\beta h_2}. \tag{A.12}$$

Others thermodynamic quantities, like density,

$$\rho = -\frac{1}{\beta}\frac{\partial F}{\partial \mu} = \frac{e^{\beta\mu}}{(1 + e^{\beta h})^{k+1} + e^{\beta\mu}} , \tag{A.13}$$

internal energy per lattice site,

$$E = -\mu\rho , \tag{A.14}$$

and entropy,

$$S = -\beta(F + \mu\rho), \tag{A.15}$$

are also easily obtained.

To summarize, in order to obtain the static properties of the model one has to solve eq. (A.9) to find the value of the local cavity field h_i on each site i at each temperature $\beta\mu$. Equation (A.9) is actually a system of N equations (where N is the number of lattice sites) that depends on the realization of the Bethe lattice. In the following we solve eq. (A.9) by making some simple hypothesis.

A'1. *Liquid solution.* – To obtain the liquid solution one supposes that there is only one pure state (one minimum of the free energy) and that the local cavity field h_i is the same on all lattice sites ($h_i = h\ \forall i$). This assumption, which corresponds to the Replica Symmetric (RS) solution of the problem, gives a solution characterized by translational invariance.

In this case the local field h is given by the fixed point of eq. (A.9). The results of the replica symmetric solution for $k = e$ are shown in fig. 26. The entropy of this solution becomes negative at $\beta\mu \simeq 5.50$ ($\rho \simeq 0.42$): the RS solution is not appropriate to describe the high-density region. The hypothesis underlying the RS approach, *i.e.* the absence of correlation between the various cavity sites, fails in the high-density phase because of the packing of the system in some disordered state.

A'2. *Crystalline solution.* – It is also possible to find a replica symmetric crystalline solution of the problem. To find this solution, we use our knowledge about the properties of the crystal. Since in the crystal each empty (occupied) site is surrounded by occupied (empty) sites, two sub-lattices emerge. Therefore, we introduce two non-fluctuating fields h_a and h_b which describe the two sub-lattices, respectively. The relation between the cavity fields of different sites (eq. (A.9)) becomes

$$\beta h_{a,b} = \beta\mu + k \log \frac{1}{1+\exp[\beta h_{b,a}]} \,. \tag{A.16}$$

The free energy shift due to a site addition becomes

$$\Delta F^{(1)} = \frac{\Delta F_a^{(1)} + \Delta F_b^{(1)}}{2} \,, \tag{A.17}$$

where

$$e^{-\beta\Delta F_{a,b}^{(1)}} = \left(1+e^{\beta h_{b,a}}\right)^{k+1} + e^{\beta\mu}\,; \qquad e^{-\beta\Delta F_{a,b}^{(2)}} = 1 + e_a^h + e_b^h \,, \tag{A.18}$$

and the density is $\rho = \rho_a + \rho_b$, where the density of each sub-lattice is

$$\rho_{a,b} = \frac{e^{\beta\mu}}{\left(1+e^{\beta h_{a,b}}\right)^{k+1} + e^{\beta\mu}} \,. \tag{A.19}$$

The results for the crystalline solution are shown in fig. 26. The free energy of the crystal F_{crystal} becomes smaller than that of the liquid F_{liquid} at the melting temperature $\beta_{\text{m}}\mu \simeq 1.46$. A peculiarity of the hard-sphere systems is that $F_{\text{crystal}} = F_{\text{liquid}}$ for $\beta \geq \beta_{\text{m}}$. Therefore the density

$$\rho = -\frac{1}{\beta}\frac{\partial}{\partial\mu} \min(F_{\text{crystal}}, F_{\text{liquid}}) \tag{A.20}$$

is a continuous function, *i.e.* the system has a second-order liquid-solid transition.

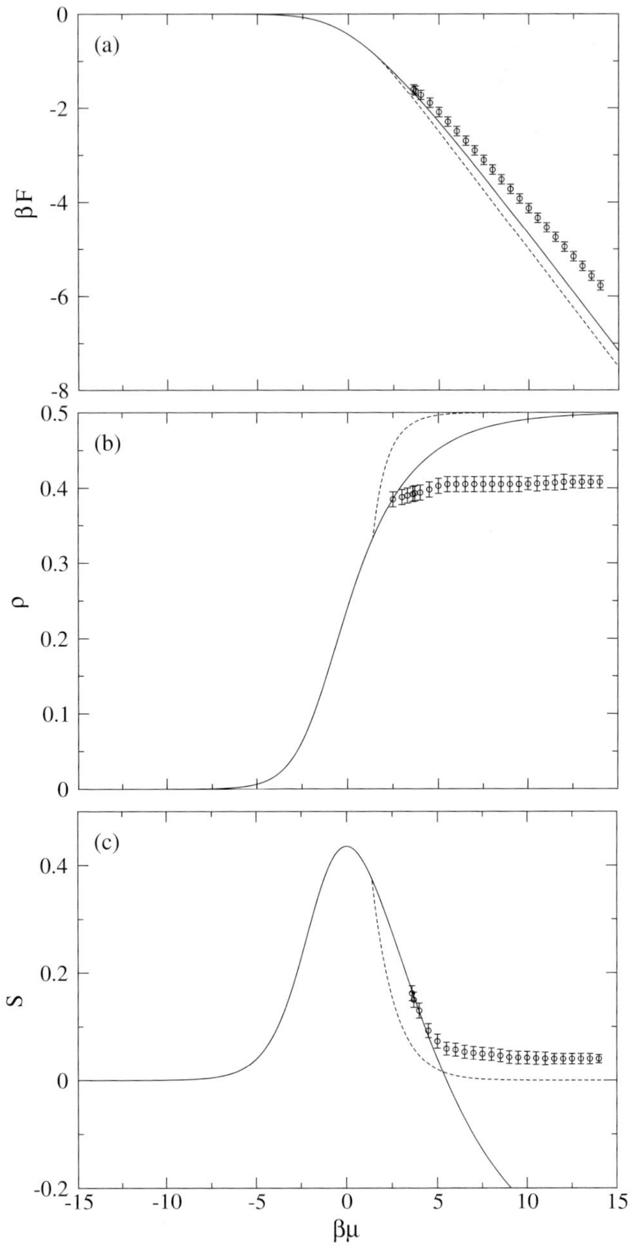

Fig. 26. – Liquid (plain line), crystalline (dashed line), and glassy (circles) solutions for the equilibrium properties of a system of hard spheres as obtained with the cavity method on a Bethe lattice with $k = 3$. (a): free energy density. (b): density. (c): entropy. The liquid phase becomes unstable at $\beta\mu \simeq 5.50$ ($\rho \simeq 0.42$), when its entropy becomes negative.

A.3. *Glassy solution.* – The glassy phase of the system is a high-density translational invariant phase. This phase is not described well by the RS liquid solution since this predicts an entropy vanishing at $\rho > 0.42$. The RS solution fails because it does not take into account the existence of several local minima of the free energy (pure states), that we know to exist at low temperature. Therefore a solution at the level of one-step Replica Symmetry Breaking (RSB) must be searched.

To take into account the existence of many metastable states, we allow the local fields to fluctuate, *i.e.* we introduce a probability distribution $P_i(h)$ defined as the probability that the field h_i on the site i equals h. Since we are looking for a solution that is translational invariant we work in the so-called "factorized case", *i.e.* we assume that the probability distribution of the local fields on the single site is the same for all the site of the lattices (*i.e.* $P_i(h_i) = P(h_i)\ \forall i$).

Using the cavity method in the one-step replica symmetry broken ansatz eq. (A.9) becomes the following integral self-consistency equation for the probability distribution of the local fields $P(h)$ [2]:

$$P(h) = \mathcal{C} \int \prod_{j=1}^{k} [\mathrm{d}h_j P(h_j)]\, \delta(h - h_j) \exp[-\beta m \Delta F], \tag{A.21}$$

where $\mathcal{C}$ is a normalization constant, h_i and ΔF are functions of h_j via the recursion relations of eqs. (A.9) and (A.10), and the real parameter $m \in [0,1]$ is the usual one-step RSB parameter.

To evaluate the free energy one introduces the functional

$$\Phi[m] = -\frac{1}{\beta m}\left\{ \log \int \prod_{j=1}^{k+1} \mathrm{d}h_j P(h_j) e^{-\beta m \Delta F^{(1)}} - \right. \\ \left. -\frac{k+1}{2} \log \int \prod_{j=1}^{2} \mathrm{d}h_j P(h_j) e^{-\beta m \Delta F^{(2)}} \right\}, \tag{A.22}$$

where $\Delta F^{(1)}$ and $\Delta F^{(2)}$ depend on h_j via eqs. (A.11) and (A.12). The parameter m is fixed by the maximization of the free energy functional with respect to it. This is typical in the replica approach since m turns out to be the breakpoint in Parisi's order parameter function at the 1-step RSB level. For a spin glass it has been rigorously proved that in the $k \to \infty$ limit $\Phi[m]$ is a lower bound to the correct free energy, so it is natural to find the preferred value of m by the maximization of $\Phi[m]$.

At $T = 0$ eq. (A.21) is easily solved. Indeed from eq. (A.9) in the $\beta \to \infty$ limit we notice that the local field on each lattice site can only assume the finite set of values $h = (-k+1)\mu, (-k+2)\mu, \ldots, 0, \mu$. This result is obtained considering that in the $\beta \to \infty$ limit $\prod_{j=1}^{k}(1 + e^{\beta h_j})^{-1} = \exp[-\beta\mu \sum_j \theta(h_j)]$, where θ is the Heaviside function. The probability distribution of the local fields is then the sum of $k+1$ delta:

$$P(h, T=0) = \sum_{r=1}^{k+1} p_r \delta(h - h_r) \quad \text{where } h_r = (-k+r)\mu\,. \tag{A.23}$$

From eq. (A.21) we obtain $k+2$ algebraic equations for the weights of the delta p_r and for the normalization constant $\mathcal{C}$.

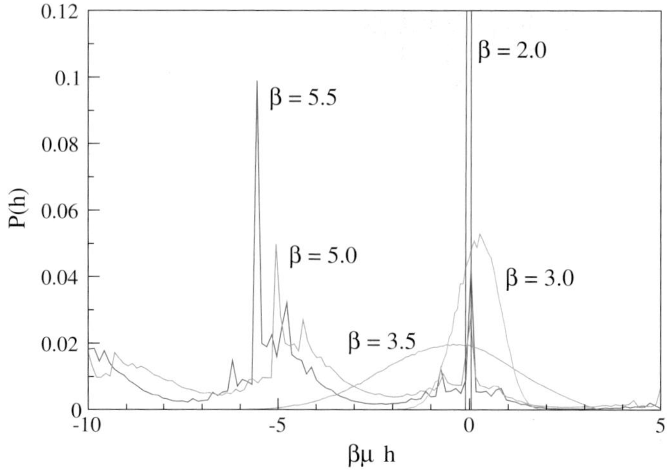

Fig. 27. – Probability distribution function $P(h)$ of the local field at temperatures $\beta = 2.0$, 3.0, 3.5, 5.0, 5.5. The critical temperature is $\beta_c \simeq 3.6$, while the dynamical temperature is $\beta_d \simeq 3$. For $\beta > \beta_d$ $P(h)$ has support on the whole range $[-k\beta : \beta]$ and is characterized by the presence of several peaks. For $\beta < \beta_d$, on the contrary, $P(h)$ has a single peak and support in a sub-interval of $[-k\beta : \beta]$.

Equation (A.21) is then solved with a population dynamics algorithm for all temperatures. The first step is to construct a population of N fields extracted from $P(h, T = 0)$. Then one increments the temperature by intervals $\mathrm{d}T$, and at each value of the temperature T and of the replica symmetry breaking parameter m one finds the distribution $P(h, T, m)$ that solves eq. (A.21). This is done by updating the distribution until convergence with the following procedure. Using eq. (A.9) and eq. (A.10) we can generate a new field h^n and the corresponding free energy shift ΔF^n starting with $k+1$ fields extracted from the distribution $P(h)$. We repeat this procedure M times generating the fields $h^1, \ldots, h^M$ and the free energy shifts $\Delta F^1, \Delta F^2, \ldots, \Delta F^M$. Then we substitute each field h^n to a randomly chosen field of the population with probability $\exp[-\beta m \Delta F^n]$.

Having obtained $P(h, T, m)$ (see fig. 27) we find $F[m, T]$ and, by its maximization, $m^*(T)$. From $P(h, T, m^*)$, to conclude, one obtains the values of the thermodynamic properties in the glassy phase. These are shown in fig. 26. Since the glassy solution needs $m \leq 1$ [2], for $m^*(\beta)$ one defines the critical temperature β_c with $m^*(\beta_c) = 1$. At β_c the complexity $\Sigma = \log \mathcal{N}$, where $\mathcal{N}$ is the number of metastable states, vanishes. In the case under study, *i.e.* a system of hard spheres with $k = 3$, we find $\beta_c \simeq 3.6$.

The glassy solution also predicts the existence of a dynamical transition temperature $T_d > T_c$. By looking at the shape of $P(h, T > T_c, m = 1)$, in fact, one notices that there is a temperature T_d such that for $T > T_d$ $P(h)$ has the shape typical of the liquid solution, *i.e.* a delta (or a Gaussian-like shape because of numerical errors). For $T < T_d$, on the contrary, $P(h)$ is characterized by the presence of several peaks that become delta functions in the limit $T \to 0$. For the system of hard spheres we are considering $T_d \simeq 0.3$.

The presence of several peaks in $P(h)$ for $T < T_d$ is related to the presence of many metastable states. We expect (at least in the thermodynamic limit) the system to get strapped in one of these states, and the dynamics to halt. Therefore this transition temperature is considered to be the analogous of the critical temperature corresponding to an ergodicity breaking transition predicted by the mode coupling theory.

* * *

This work has been partially supported by MURST-PRIN 2002, MIUR-FIRB 2002, and by the INFM Parallel Computing Initiative.

REFERENCES

[1] EDWARDS S. F. and OAKESHOTT R. B. S., *Physica A*, **157** (1989) 1080; MEHTA A. and EDWARDS S. F., *Physica A*, **157** (1989) 1091; EDWARDS S. F., in *Disorder in Condensed Matter Physics* (Oxford Science Pubs) 1991, p. 148; in *Granular Matter: an Interdisciplinary Approach*, edited by A. MEHTA (Springer-Verlag, New York) 1994.
[2] MÉZARD M. and PARISI G., *Eur. Phys. J. B*, **20** (2001) 217; *J. Stat. Phys.*, **111** (2003) 1.
[3] BOUCHAUD J. P., CUGLIANDOLO L. F., KURCHAN J. and MEZARD M., in *Spin-Glasses and Random Fields*, edited by A. P. YOUNG (World Scientific, Singapore) 1997.
[4] DEBENEDETTI P. G., *Metastable Liquids: Concept and Principles* (Princeton University Press) 1996.
[5] DONTH E., *The Glass Transition: Relaxation Dynamics in Liquids and Disordered Materials* (Springer-Verlag) 2001.
[6] KAUZMANN W., *Chem. Rev.*, **43** (1948) 219.
[7] JOHARI P. and GOLDSTEIN M., *J. Chem. Phys.*, **53** (1970) 2372; **55** (1971) 4245.
[8] GIBBS J. H. and DI MARZIO E. A., *J. Chem. Phys.*, **28** (1958) 373; ADAMS G. and GIBBS J. H., *J. Chem. Phys.*, **43** (1965) 139.
[9] BENGTZELIUS U., GÖTZE W. and SJÖLANDER A., *J. Phys. C*, **17** (1984) 5919; GÖTZE W., in *Liquids, Freezing, and Glass Transition*, edited by J. P. HANSEN, D. LEVESQUE and J. ZINN-JUSTIN (Elsevier) 1991; GÖTZE W., *Science*, **267** (1995) 1924.
[10] KIRKPATRICK T. R. and WOLYNES P. G., *Phys. Rev. A*, **34** (1986) 1045; KIRKPATRICK T. R. and THIRUMALAI D., *Phys. Rev. Lett.*, **58** (1987) 2091; KIRKPATRICK T. R., THIRUMALAI D. and WOLYNES P. G., *Phys. Rev. A*, **40** (1989) 1045.
[11] VINCENT E., HAMMANN M., OCIO M., BOUCHAUD J. P. and CUGLIANDOLO L. F., *Proceedings of the Sitges conference*, edited by E. RUBI (Springer-Verlag) 1997, cond-mat/9607224.
[12] MÉZARD M., PARISI G. and VIRASORO M. A., *Spin Glass Theory and Beyond* (World Scientific) 1987.
[13] FRANZ S., DONATI C., PARISI G. and GLOTZER S. C., *Philos. Mag. B*, **79** (1999) 1827.
[14] FRANZ S. and PARISI G., *J. Phys. I*, **5** (1995) 1401; *Phys. Rev. Lett.*, **79** (1997) 2486; *Physica A*, **261** (1998) 317.
[15] PARISI G., *Phys. Rev. Lett.*, **78** (1997) 4581; KOB W. and BARRAT J. L., *Phys. Rev. Lett.*, **79** (1997) 3660.
[16] STILLINGER F. H. and WEBER T. A., *Phys. Rev. A*, **25** (1982) 978; SASTRY S., DEBENEDETTI P. G. and STILLINGER F. H., *Nature*, **393** (1998) 554.
[17] SCIORTINO F., KOB W. and TARTAGLIA P., *Phys. Rev. Lett.*, **83** (1999) 3214.
[18] CAVAGNA A., *Europhys. Lett.*, **53** (2001) 490.
[19] KOB W., SCIORTINO F. and TARTAGLIA P., *Europhys. Lett.*, **49** (2000) 590; SCIORTINO F. and TARTAGLIA P., *Phys. Rev. Lett.*, **86** (2001) 107.
[20] BIROLI G. and MÉZARD M., *Phys. Rev. Lett.*, **88** (2002) 025501.

[21] Lawlor A., Reagan D., McCullagh G. D., De Gregorio P., Tartaglia P. and Dawson K. A., *Phys. Rev. Lett.*, **89** (2003) 245503.

[22] Hartmann L. and Weigt M., *Europhys. Lett.*, **62** (2003) 533.

[23] Pica Ciamarra M., Tarzia M., de Candia A. and Coniglio A., *Phys. Rev. E*, **67** (2003) 057105; Pica Ciamarra M., Tarzia M., de Candia A. and Coniglio A., *Phys. Rev. E*, **68** (2003) 066111.

[24] Di Leonardo R., Angelani L., Parisi G. and Ruocco G., *Phys. Rev. Lett.*, **84** (2000) 6054.

[25] Monasson R., *J. Phys. A*, **31** (1998) 513.

[26] Franz S., Leone M., Ricci-Tersenghi F. and Zecchina R., *Phys. Rev. Lett.*, **87** (2001) 127209.

[27] Jaeger H. M., Nagel S. R. and Behringer R. P., *Rev. Mod. Phys.*, **68** (1996) 1259.

[28] Knight J. B., Fandrich C. G., Lau C. N., Jaeger H. M. and Nagel S. R., *Phys. Rev. E*, **51** (1995) 3957; Nowak E. R., Knight J. B., Ben-Naim E., Jaeger H. M. and Nagel S. R., *Phys. Rev. E*, **57** (1998) 1971; Nowak E. R., Knight J. B., Povinelli M., Jaeger H. M. and Nagel S. R., *Powder Technol.*, **94** (1997) 79.

[29] D'Anna G. and Gremaud G., *Nature*, **413** (2001) 407.

[30] Philippe P. and Bideau D., *Europhys. Lett.*, **60** (2002) 677.

[31] Nicodemi M., Coniglio A. and Herrmann H. J., *Phys. Rev. E*, **55** (1997) 3962; *J. Phys. A*, **30** (1997) L379; Coniglio A. and Herrmann H. J., *Physica A*, **225** (1996) 1.

[32] Liu A. J. and Nagel S. R., *Nature*, **396** (1998) 21.

[33] O'Hern C. S., Langer S. A., Liu A. J. and Nagel S. R., *Phys. Rev. Lett.*, **86** (2001) 111; O'Hern C. S., Silbert L. E., Liu A. J. and Nagel S. R., *Phys. Rev. E*, **68** (2003) 011306.

[34] Coniglio A., Fierro A. and Nicodemi M., *Physica A*, **302** (2001) 193; Fierro A., Nicodemi M. and Coniglio A., *Europhys. Lett.*, **59** (2002) 642; *Phys. Rev. E*, **66** (2002) 061301; Nicodemi M., Fierro A. and Coniglio A., *Europhys. Lett.*, **60** (2002) 684.

[35] Makse H. A. and Kurchan J., *Nature*, **415** (2002) 614.

[36] Ottino J. M. and Khakhar D. V., *Ann. Rev. Fluid Mech.*, **32** (2000) 55; Jaeger H. M., Nagel S. R. and Behringer R. P., *Rev. Mod. Phys.*, **68** (1996) 1259; Bridgewater J., *Chem. Eng. Sci.*, **50** (1994) 4081.

[37] Makse H. A., Havlin S., King P. R. and Stanley H. E., *Nature*, **386** (1997) 379; Shinbrot T., Alexander A. and Muzzio F. J., *Nature*, **397** (1999) 675; Möbius M. E., Lauderdale B. E., Nagel S. R. and Jaeger H. M., *Nature*, **414** (2001) 270.

[38] Rosato T., Prinze F., Standburg K. J. and Swendsen R., *Phys. Rev. Lett.*, **58** (1987) 1038.

[39] Bridgewater J., *Powder Technol.*, **15** (1976) 215; Williams J. C., *Powder Technol.*, **15** (1976) 245.

[40] Duran J., Rajchenbach J. and Clement E., *Phys. Rev. Lett.*, **70** (1993) 2431.

[41] Knight J., Jaeger H. and Nagel S., *Phys. Rev. Lett.*, **70** (1993) 3728.

[42] Hong D. C., Quinn P. V. and Luding S., *Phys. Rev. Lett.*, **86** (2001) 3423; Both J. A. and Hong D. C., *Phys. Rev. Lett.*, **88** (2002) 124301.

[43] Shinbrot T. and Muzzio F., *Phys. Rev. Lett.*, **81** (1988) 4365.

[44] Stillinger F. H. and Weber T. A., *Phys. Rev. A*, **25** (1982) 978; *Science*, **225** (1984) 983; Stillinger F. H., *Science*, **267** (1995) 1935; Sastry S., Debenedetti P. G. and Stillinger F. H., *Nature*, **393** (1998) 554.

[45] Coluzzi B., Parisi G. and Verrocchio P., *Phys. Rev. Lett.*, **84** (2000) 306.

[46] Cugliandolo L. F., Kurchan J. and Peliti L., *Phys. Rev. E*, **55** (1997) 3898.

[47] Monasson R., *Phys. Rev. Lett.*, **75** (1995) 2847; Nieuwenhuizen Th. M., *Phys. Rev. E*, **61** (2000) 267; Franz S. and Virasoro M. A., *J. Phys. A*, **33** (2000) 891; Crisanti A. and Ritort F., *J. Chem. Phys.*, **113** (2000) 10615.
[48] Cugliandolo L. and Kurchan J., *Phys. Rev. Lett.*, **71** (1993) 173; Kurchan J., cond-mat/9812347; Kurchan J., in *Jamming and Rheology: Constrained Dynamics on Microscopic and Macroscopic Scales*, edited by A. J. Liu and S. R. Nagel (Taylor and Francis, London) 2001.
[49] Barrat A., Kurchan J., Loreto V. and Sellitto M., *Phys. Rev. Lett.*, **85** (2000) 5034; *Phys. Rev. E*, **63** (2001) 051301.
[50] Nicodemi M., *Phys. Rev. Lett.*, **82** (1999) 3734.
[51] Coniglio A. and Nicodemi M., *Physica A*, **296** (2001) 451.
[52] Brey J. J., Prados A. and Sánchez-Rey B., *Physica A*, **275** (2000) 310; Prados A., Brey J. J. and Sánchez-Rey B., *Physica A*, **284** (2000) 277.
[53] Dean D. S. and Lefèvre A., *Phys. Rev. Lett.*, **86** (2001) 5639; *J. Phys. A: Math. Gen.*, **34** (2001) L213; Dean D. S. and Lefèvre A., in cond-mat/0106220.
[54] Berg J., Franz S. and Sellitto M., *Eur. Phys. J. B*, **26** (2002) 349.
[55] Berg J. and Mehta A., *Europhys. Lett.*, **56** (2001) 784.
[56] Berthier L., Cugliandolo L. F. and Iguain J. L., *Phys. Rev. E*, **63** (2001) 051302.
[57] Coniglio A., in *"Frustration and connectivity in glass forming systems and granular materials"*, *Physics of Complex Systems*, in *Proceedings of the International Summer School of Physics "Enrico Fermi", Course CXXXIV*, edited by F. Mallamace and H. E. Stanley (IOS Press, Amsterdam) 1997.
[58] Coniglio A., de Candia A., Fierro A. and Nicodemi M., *J. Phys. Cond. Mat.*, **11** (1999) A167.
[59] de Candia A. and Coniglio A., *Phys. Rev. E*, **65** (2002) 16132.
[60] Coniglio A., de Candia A., Fierro A., Nicodemi M. and Tarzia M., *Europhys. Lett.*, **66** (2004) 531.
[61] In the present model the "melting" transition is continuous. This pathology can be cured as explained in [60] or in [22].
[62] Since the glassy phase is expected to be translational invariant, we work in the factorized case in which the probability distributions at a given height are equal for all the sites of the layer.
[63] Toninelli C., Biroli G. and Fisher D. S., cond-mat/0306746.
[64] The system has reached the stationary state for all values of T_Γ here considered, in the sense that the time translation invariance is observed in two-time autocorrelation functions.
[65] We have simulated the Hamiltonian equation (22) under Edwards' constraint with Monte Carlo simulations and we have evaluated the overall number density, Φ (which in this case equals $1/\langle z\rangle$), as a function of the configurational temperature, T_{conf}, $\Phi \equiv f_1(T_{\text{conf}})$. From the tap dynamics we have also obtained $\Phi \equiv f_2(T_\Gamma)$, then we have calculated $T_\Gamma \equiv f(T_{\text{conf}})$, where $f(x) \equiv f_2^{-1}(f_1(x))$.
[66] Coniglio A., Fierro A., Nicodemi M. and Tarzia M., cond-mat/0405486.
[67] Fierro A., Coniglio A. and Nicodemi M., *J. Phys. Condens. Matter*, **15** (2003) S1095.

Force chains in granular packings

R. Cruz Hidalgo, H. J. Herrmann and E. J. R. Parteli

Institute for Computer Applications 1, University of Stuttgart
Pfaffenwaldring 27 70569, Stuttgart, Germany

F. Kun

Department of Theoretical Physics, University of Debrecen
P.O. Box 5, H-4010, Debrecen, Hungary

1. – Introduction

Granular materials present a wide spectrum of different basic characteristics and phenomena which are still poorly understood, although well known for a long time. Their unusual behaviour makes them very distinguishable from liquids and solids. A classical example is the formation of a pile of sand when grains are poured onto a table. Unlike water that would spread out on the surface, the final state is a conical heap that grows through sequences of avalanches on the incline, until it reaches a stable configuration whose slope is characterized by a steepest angle, which is called the *angle of repose* θ_r. This angle is a material property of the granular system [1]. For a sand pile, θ_r is typically between 25° and 35°, while if the grains are glass beads, it is about 10°–15°, depending mostly on the shape and the surface roughness of the grains, and the humidity of the air.

One of the pioneers of the research on granular media was Osborne Reynolds [2]. He introduced the concept of dilatancy, an effect we encounter while walking on a wet beach: When we press our foot strongly into the sand, a dry halo appears around the foot. Naively one would expect that the pressure of the foot would rather produce a hole in the sand which would fill with water, but the opposite occurs. The physical reason for this effect is that the grains were so densely packed that one needs to separate them before deformation occurs.

Individual grains are typically rigid and can have different densities, shapes, sizes and surface roughnesses. Between the grains there can exist interactions characterized by local rotations and point-like contacts, as we are going to see in the following. A consequence of particular interest of the friction between the particles is that granular matter can be set in many stable static configurations. It is due to energy dissipation during the interactions between the grains that sand poured on the floor immediately comes to rest and looses all its kinetic energy. Dissipation introduces a characteristic length scale into a granular system, namely the distance over which the energy decays by a factor $1/e$, since the multiplicative restitution coefficient implies an exponential relaxation [3].

Because of dissipation, every experiment in which a granular system moves must constantly be furnished with energy. This can be done for instance i) by gravity, as in mass flows (*e.g.* in a hopper, a pipe or an hourglass) [4, 5] or in surface flows (like heaps, rotating drums and inclined chutes) [6-8]; ii) by using a loudspeaker or a vibrating conveyor belt [9, 10]; iii) for dense systems like packings, energy can be inputed by shearing the material under gravity for instance in shear cells, which are a very important device to characterize soils [11-13]; iv) the input of energy in granular systems can be done also by means of hydrodynamics (by a surrounding fluid). In the industry, many processes push a fluid through a granular bed (system called a "fluidized bed") [14-16]; in the hydrology, the formation of river beds on beaches and sedimentation are of major importance; the wind in the desert also acts like a fluid moving the grains and forming beautiful dunes. In fluid sand, dissipation plays an essential role in the evolution of density instabilities, which are responsible for the appearance of clusters with higher concentrations of particles [17, 18].

In fact, the density is the most important parameter characterizing the behaviour of a granular system. While loose granular matter can flow like a fluid, dense granular packings behave like a solid. Bagnold was the first to introduce the concept of viscosity η and "granular temperature" $T_{\rm g}$ for flowing sand in 1954 [1], and empirically found the law named after him that η is proportional to the shear velocity $\dot{\epsilon}$ itself, and not a constant as is the case for Newtonian fluids like water or air. In the 1980s, a kinetic gas theory was developed by Savage and Jenkins [19], and Haff [20]. It describes agitated granular media in the framework of fluid dynamics, the only important addition being a dissipation term in the energy flux equations.

Granular materials in densely packed forms are relevant for a wide variety of technical processes. In soil mechanics, as we mentioned above, one of the major problems is the prediction of properties of sheared granular media. Granular geomaterials like sand and silts are good examples of deformation-rate-insensitive plastic solids. The classical description for such deformations, which are mostly irreversible, is given by the so-called non-associate Mohr-Coulomb plasticity theory [21, 22].

The non-trivial macroscopic properties observed in granular materials are intimately related to the microscopic nature of the force distributions in the systems, which are formed through the contacts between the grains. In the next sections, we focus our attention to this topic.

2. – Friction and contact dynamics of grains in densely packed form

Granular matter means a system of particles whose size (or diameter) d is larger than $50\,\mu$m. For $d < 10\,\mu$m, thermal agitation is important, and Brownian motion can be seen. For this reason, granular systems are also called systems with zero temperature [23]. Unlike in the atomic regime, where the particles (atoms) interact by overlapping their various orbitals and sharing electrons, the contact between the grains is essentially point-like. Friction between the grains is thus characterized by gigantic local forces. This behaviour leads to important effects, for instance, in the wearing of glass beads, which makes the particles rougher, producing "glass dust", and thus increasing the specific area of the granular medium [24]. On the other hand, if the particles are made of aluminum, plastic deformation can occur. As a consequence of these interactions, collisions between the particles are typically inelastic, and energy can be dissipated in the form of heat or acoustic emissions.

The friction between two interacting surfaces has been studied for a long time, but still no universally accepted explanation of the basic laws of friction exists. Amonton ennounced two basic laws of friction in 1699, that were already known by Leonardo da Vinci and are still teached in the present: the first law is that friction is proportional to the normal $F_{\rm n}$ of the interfaces in contact, *i.e.* $F_{\rm fr} = \mu F_{\rm n}$, and the second law is that friction is independent of the area of contact. Later, Coulomb observed that friction was furthermore independent of the relative sliding velocity of the surfaces in contact. However, it has been observed that friction presents features much more complex than this. Bowden and Tabor in 1950 [25] suggested that the existence of friction is related to the roughnesses of the surface areas in contact. This means that the "real" area of contact is in general much smaller than the "apparent" area. Understanding the mechanism of friction in granular systems is a subject of extensive theoretical research [26, 27].

The static distribution of stresses in granular materials has been studied for more than one hundred years by researchers in different areas like applied mechanics and geotechnical and chemical engineerings. Janssen [28] studied in 1895 the structure of the forces inside a pipe filled with granular material. The schematic model is depicted in fig. 1. We want to know the force distribution of granular matter in a silo. They assumed that the lateral forces (per unit area) exerted by the grains on the wall are proportional to the vertical stresses:

$$F_{\rm n} = kP, \tag{1}$$

where $P(h)$ is the pressure at the height h, and k is a phenomenological coefficient. The friction between the grains and the vertical walls can be written according to Amonton's law [29]: $F_{\rm fr} = \mu F_{\rm n} = \mu k P$, where μ is the related coefficient of friction. Thus, for an horizontal slice of area πR^2 and height Δh, the equilibrium condition reads

$$\pi R^2 \rho g \Delta h = \pi R^2 \frac{\mathrm{d}P}{\mathrm{d}h}\Delta h + 2\pi R \Delta h \mu k P \Longrightarrow$$
$$\Longrightarrow \rho g = P' + \frac{2\mu k}{R}P,$$

where ρ is the density of the column of granular matter. This expression can be written in a simpler form by introducing the characteristic length $\lambda = R/2\mu k$, which leads to

$$P = \lambda \rho g - \lambda P',$$

which is a first-order differential equation that has as solution pressure profiles of the form

$$P(h) = c \exp\left[-\frac{h}{\lambda}\right] + \lambda \rho g. \tag{2}$$

The value of the constant c can be determined by using the convention that $P(0) = 0$ at the top of the silo, from which it follows that $c = -\lambda\rho g$, and the final equation reads as

$$P(h) = \lambda \rho g \left(1 - \mathrm{e}^{-h/\lambda}\right). \tag{3}$$

Equation (3) has a very important consequence: As we can see from fig. 2, for a pipe filled with grains, the pressure at the base of the container does not rise indefinitely, but saturates on a maximum value $c = \lambda\rho g$, independent of the height. Note that this is very different from the case of a normal fluid, where $P(h) = \rho g h$. The physical origin of this behaviour lies in the way that forces are transmitted in granular media, from grain to grain at their contact points. In a packing of grains, this gives rise to a complex network of force lines that can branch at the contacts, and stress is transmitted laterally. Thus, a normal stress is exerted by the grains on the walls of the container, which support all the extra weight of the material.

On industrial scale, the pipe of fig. 1 would correspond to a silo. Real silos, however, present a funnel on the bottom, leading to a small orifice of diameter D through which the granular material flows out with an essentially constant velocity, as in the case of an hourglass. If D is too small the silo will clog due to the formation of so-called arches. A schematic picture of a single arch is shown in fig. 3 as a simple geometrical model, which is classical in structural mechanics particularly used to describe the static of bridges. The equilibrium condition is the result of the interplay between gravity g and the contacts with the walls of the container. The forces transmitted among the grains of mass m and diameter $d = 2R$ can be decomposed in components F_x and F_z. We now determine the form of the arch $z(x)$. Since all the force added to the arch along x goes in the z-direction, F_x equals the normal force that reaches the walls and is constant along x. The force transmition through the grains in the direction of the vector $\mathrm{d}\vec{s}$ as shown in fig. 3 follows

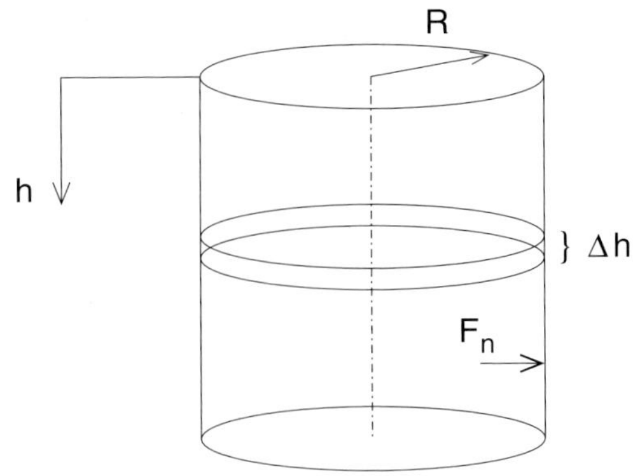

Fig. 1. – Schematic representation of the pipe considered in the model by Janssen discussed in the text.

the equation $d(\partial\vec{F}/\partial s) = m\vec{g}$ [30]. In the case of an isolated arch (or bridge), illustrated in fig. 3(a), we have

$$\frac{\mathrm{d}F_x}{\mathrm{d}s} = 0; \qquad \frac{\mathrm{d}F_z}{\mathrm{d}s} = \frac{mg}{d},$$

from which it follows that $F_x = \text{const}$, and $F_z = (mgs/d) + \text{const}$, or simply $F_z = mgs/d$, since $F_z(0) = 0$, where the origin is at the top of the arch. Thus, we can write

$$\frac{\mathrm{d}x}{\mathrm{d}z} = \frac{F_x}{F_z} = \frac{k}{s}, \quad \text{where } k = \frac{F_x d}{mg} = \text{const}.$$

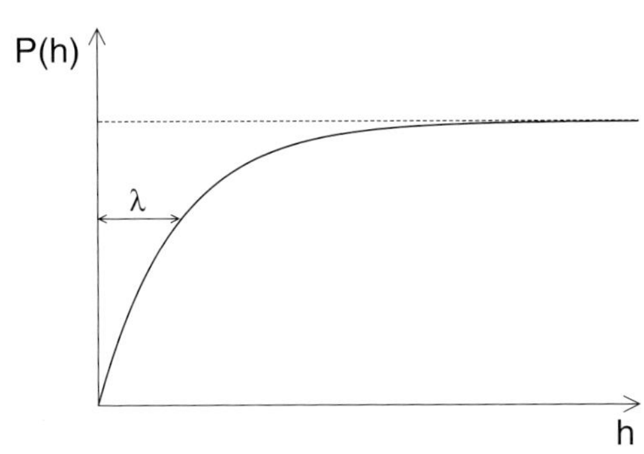

Fig. 2. – Qualitative behaviour of the pressure $P(h)$ exerted by granular matter on the walls of the pipe at height h. The curve represented by the full line corresponds to eq. (3).

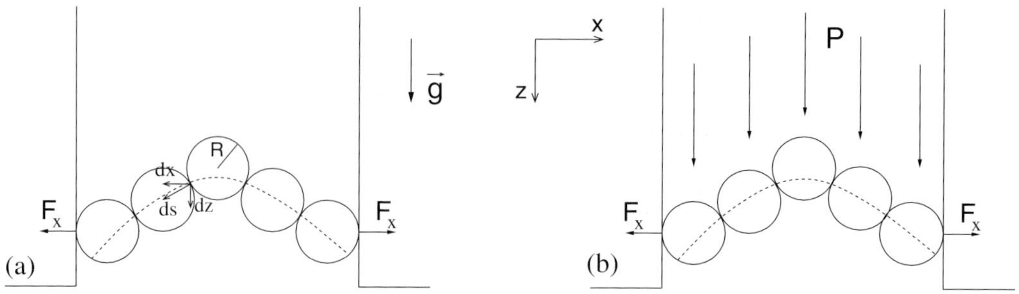

Fig. 3. – Schematic diagram of an arch clogging the outlet of a silo. In (a) we consider a single arch, and in (b) we take into account the pressure P exerted by the material above it.

But $\mathrm{d}s^2 = \mathrm{d}x^2 + \mathrm{d}z^2 = \mathrm{d}z^2(1 + (k^2/s^2))$. Therefore we can write

$$x = \int_0^x \frac{\mathrm{d}x}{\mathrm{d}z}\mathrm{d}z = \int_0^x \frac{k}{s}\frac{\mathrm{d}s}{\sqrt{1+(k^2/s^2)}} = k\int_0^x \frac{\mathrm{d}u}{u\sqrt{1+(1/u^2)}} = k\int_0^x \frac{\mathrm{d}u}{\sqrt{1+u^2}} = k\,\sinh^{-1}\frac{s}{k},$$

where we did the variable change $u = s/k$. Inverting the equation above, it follows that

$$\sinh\frac{x}{k} = \frac{\mathrm{d}z}{\mathrm{d}x} \Longrightarrow z = k\cosh\frac{x}{k}. \tag{4}$$

We now have to take into account the interaction between the many arches present in the material. Each arch has to support a pressure P resulting from the weight of the material on the top, as shown in fig. 3(b). For simplicity, we assume that the pressure is applied uniformly along the arch, such that the vertical force added at each interval $\mathrm{d}x$ is in fact $\mathrm{d}F_z = P\,\mathrm{d}x$. Since the force that acts on the arch does not vary along the x-direction, we now replace the parametrization s by integrating over the variable x. Thus, we have at position x

$$z = \int_0^z \frac{\mathrm{d}z}{\mathrm{d}x}\mathrm{d}x = \int_0^z \frac{x}{k}\mathrm{d}x \Longrightarrow z = \frac{2}{k}x^2, \tag{5}$$

which is a parabolic shape. The clogging of silos is an enormous industrial problem and many costly accidents have occurred.

The force distribution in granular materials is a subject of intense theoretical and experimental research [31, 32]. Great effort has been devoted in the last years to the understanding of the global behaviour of granular media in terms of microscopic phenomena which occur at the level of discrete particles [33-39]. The force network inside a granular packing is very heterogeneous, with certain chains along which the stresses are particularly intense. Particles ("rattlers") lying between lines ("cages") of the force network do not support any load, and can even be removed from the packing without changing its mechanical properties. These structures of force lines can be clearly visualized by means

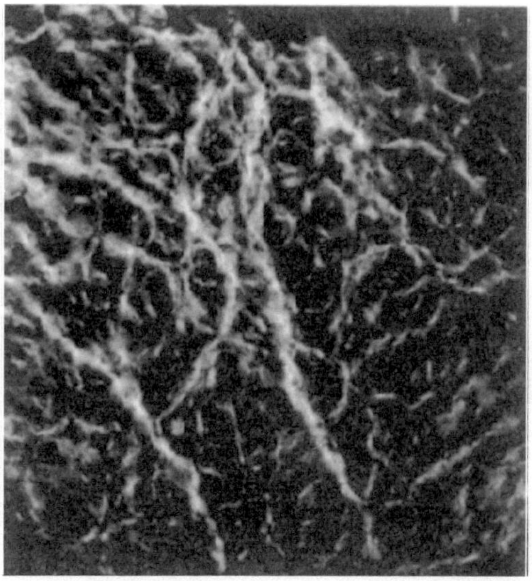

Fig. 4. – Force chains in a granular material as viewed between two crossed-circular polarizers. The particles are 3 mm Pyrex spheres surrounded by a mixture of water and glycerol that matches the index of refraction of the Pyrex [33].

of a simple experimental setup constituted of grains made of photoelastic material, immersed in a liquid mixture between two crossed polarizers. When the packing is loaded, the grains that are stressed rotate their optical axis and light up, forming a beautiful pattern, as shown in fig. 4. It is observed that, as the external stress is increased, more and more force lines appear. Indeed, each force line undergoes an erratic transformation before reaching a stationary state when the load is high enough, regime in which all the grains light up equally [33].

The distribution of stresses in a granular medium as a network of force chains is associated with the non-trivial behaviour of the normal pressures on the floor below a heap of sand [23, 40]. Along the principal axis, there is a "channeling" of forces that percolate through the material in such a manner that two peaks of loads on the bottom appear. This distribution will be different if the sand pile is for instance compressed from the top, since the actual force network will be modified and new force lines can appear. During the compression of a granular material, the creation and restructuring of percolating force chains are associated with relative displacements of the grains, which rearrange their positions until a new stable configuration of the system is achieved. In the next section, we present a model for the evolution of these percolating force chains, that very well captures the restructuring processes at the level of the particles.

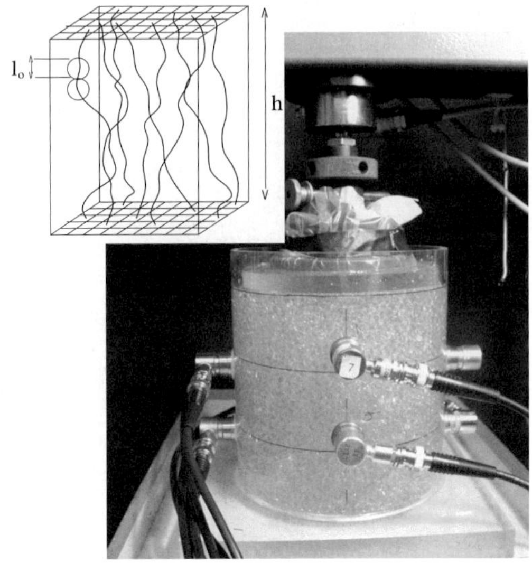

Fig. 5. – Experimental setup and sketch of the array of force chains for modeling the uniaxial compression of the granular media.

3. – Evolution of percolating force chains in compressed granular media

To investigate the evolution of the force chains percolating through a granular system submitted to a uniaxial compression, experiments have been performed in collaboration with the group of Dr. Christian Grosse, at the *Institute of Construction Materials* (IWB) of the University of Stuttgart. In the experiments, the granular material consisted on glass beads of 5 mm diamater, which were immersed in water inside a cylindrical container made of PMMA with a thickness of 5 mm and a diameter of 140 mm (fig. 5). By moving the traverse at a constant speed of some mm/minute, a uniaxial compression at the top level of the glass beads was carried out under strain controlled conditions. During the compression, the acoustic waves emitted due to sudden relative displacements of the beads were recorded by eight sensors placed at the container wall, as shown in fig. 5. An eight channel (two with 10 MHz and six with 1 MHz) transient recorder was used as an analogue-digital converter to enable the storing of the acoustic emission waveforms and a signal-based data (with a sampling rate of 1 ms or 100 ns). Typically several hundred signals were recorded during the experiment. By averaging over the event size distributions detected by each recorder independently, the influence of the position of the recorders was reduced. The inset of fig. 6 shows a typical averaged time series of the automatically extracted peak amplitudes of the burst signals. The main plot of this figure shows the distribution $D(s)$ of the events with size s. The straight line represents a power law fit with exponent $\delta = 1.15 \pm 0.05$, which is followed by the experimental data over two orders of magnitude.

A model for the hardening of the individual force lines during compression has been

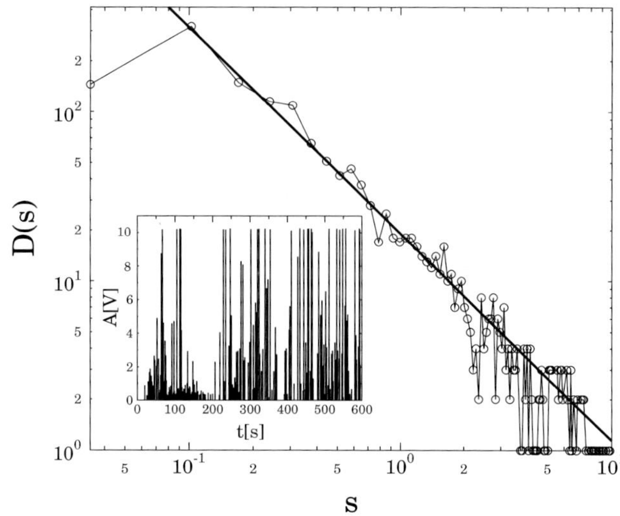

Fig. 6. – Statistics of the acoustic signals recorded by the sensors placed at the container wall (see fig. 5). The size distribution corresponding to the averaged time series shown in the inset is fitted by a power law with exponent $\delta = 1.15 \pm 0.05$.

recently proposed [41], based on an analogy to fiber models, which are used in rupture mechanics to describe the failure of fiber-reinforced composites. Fiber bundles mean composites constituted by parallel fibers of identical elastic properties but stochastically distributed breaking thresholds. A fiber fails during the loading process when the local load on it exceeds its breaking threshold. Fiber failures are followed by a redistribution of load on the remaining intact fibers according to the range of interaction in the system. If an infinite range of interaction between the fibers is considered (mean-field approach), then the load of a broken fiber is redistributed among all the intact remaining fibers. The so-called Continuous Damage Model introduced recently [42] is particularly suited to model granular materials since it captures gradual stiffness changes in the system.

In the model of compressed granular packings, force lines formed by particles are represented by an array of lines organized in a square lattice, as illustrated in fig. 5. Let us consider a system with N of such lines with identical stiffness (or Young modulus) E_0 but with random failure thresholds d_i, $i = 1, 2, \ldots, N$. The failure strength d_i of individual lines is an independent random variable uniformly distributed with probability density $p(d/d_c)$ and an associated cumulative probability $P(d/d_c) = \int_0^d p(x/d_c)\mathrm{d}x$, where d_c is the characteristic strength of the force lines [43]. During the compression process, when the local load on a line exceeds its threshold value d, the line undergoes a sudden restructuring, becoming stiffer and straighter. The stiffness of the broken line *increases* after restructuring in a multiplicative manner:

$$E_i \longrightarrow a \times E_i, \quad a > 1. \tag{6}$$

This is why the model is also called an "inverted fiber bundle model": The factor a is no longer in the range $0 < a < 1$, which would imply a *reduction* of the stiffness of the broken fiber by the factor a [44].

Once a fiber has failed, its damage threshold d_i can either be kept constant for the further failures (quenched disorder), or new failure thresholds of the same distribution can be chosen (annealed disorder). The damage law of the model is illustrated in fig. 7. For modeling the breakings of force chains in granular packings, we choose the second type of disorder shown in fig. 7(b), as it is able to model microscopic rearrangements in the system (redistribution of the grains in the packing). Therefore, after each rearrangement, the force chain gets a new threshold value d from a distribution of the same functional form. On the other side, the characteristic strength d_c of the distribution is increased in a multiplicative way ($d_c \to d_0 \times d_c$), so that after k rearrangement events, the disorder distribution takes the form

$$P_k\left(\frac{d}{d_c}\right) = P_0\left(\frac{d}{d_0^k d_c}\right), \tag{7}$$

where we set d_c to be initially equal to unity. After failure of a line, a certain amount of discharged load has to be taken by the other fibers. For the load redistribution we assume an infinite range of interaction among the force chains (mean-field approach); furthermore, equal strain condition is imposed, which implies that stiffer lines of the system carry more load. At a strain ϵ, the load of the line i that has failed $k(i)$ times reads as

$$f_i(\epsilon) = E_0\, a^{k(i)}\, \epsilon, \tag{8}$$

where $E_0\, a^{k(i)}$ is the actual stiffness of the line i. An important parameter of the model is the maximum number of failures, $k_{\max}$, allowed for a single line. In the fiber bundle model, the stiffness of a breaking fiber is decreased after each failure $k = 1, 2, \ldots$, until $k_{\max}$ failures happen and the fiber cannot carry any more load. For the granular packings, when the particles are elastic, $a^{k_{\max}}$ is a finite value related to the stiffness E_p of an individual particle as $E_p = a^{k_{\max}} E_0$. In this case, after all the lines reach the allowed restructuring events number $k_{\max}$, the system will have the same constitutive behaviour as a single contact. The maximum value of possible restructurings, $k_{\max}$, is proportional to the number of contacts, and therefore to h/l_0 (see fig. 5). In the following, the initial stiffness E_0 of the force lines will be set to unity, so that the stiffness of the line i after k failures reads as $a^{k(i)}$.

The statistics of the microscopic rearrangements in a set of compressed grains due to an external load as defined above can be described by the probability $pb_k(\epsilon)$ that, during the loading of a specimen, an arbitrarily chosen force line at a strain ϵ breaks precisely

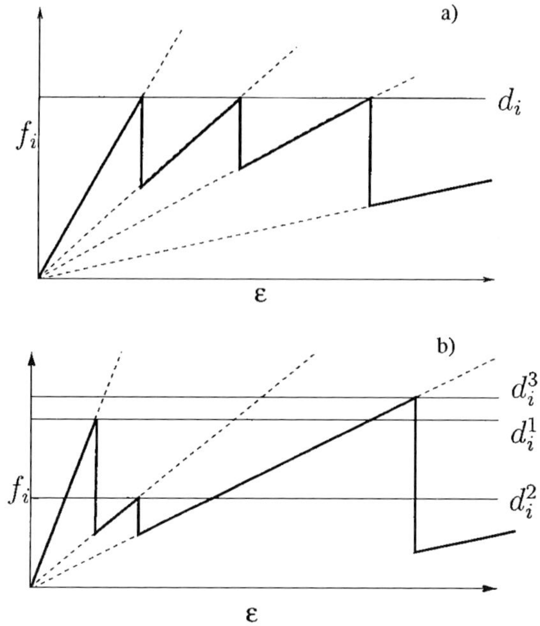

Fig. 7. – Damage law for the restructuring of force chains according to the continuous damage model with multiple failures for (a) quenched and (b) annealed disorder. The horizontal lines indicate the corresponding damage thresholds d_i.

k-times. For the *quenched disorder* illustrated in fig. 7(a), we have

$$pb_k(\epsilon) = \begin{cases} 1 - P_0(\epsilon), & k = 0, \\ P_k\left(a^{k-1}\epsilon\right) - P_k\left(a^k\epsilon\right), & 1 \leq k \leq k_{\max} - 1, \\ P_{k_{\max}}\left(a^{k_{\max}-1}\epsilon\right), & k = k_{\max}, \end{cases} \tag{9}$$

whereas for the *annealed disorder* (fig. 7(b)), which is used to model the compression of the granular packings considered here, $pb_k(\epsilon)$ can be cast in the following form:

$$pb_k(\epsilon) = \begin{cases} \left[1 - P_k\left(a^k\epsilon\right)\right] \prod_{j=0}^{k-1} P_j\left(a^j\epsilon\right), & 0 \leq k \leq k_{\max} - 1, \\ \prod_{j=0}^{k_{\max}-1} P_j\left(a^j\epsilon\right), & k = k_{\max}. \end{cases} \tag{10}$$

It can be easily seen that the probability equations above fulfill the normalization condition $\sum_{k=0}^{k_{\max}} pb_k(\epsilon) = 1$. With these equations, average quantities of the ensemble of force chains during a loading process can be calculated, as for instance the average load or stress σ on a line at a given strain ϵ:

$$\sigma = \epsilon \left[\sum_{k=0}^{k_{\max}} a^k p b_k(\epsilon) \right], \tag{11}$$

which provides the macroscopic constitutive behaviour of the model. The expression in the brackets of eq. (11) represents the macroscopic effective stiffness (or Young modulus) of the sample. For example, after one restructuring event, the above defined stress σ is written as

$$\sigma = \epsilon[1 - P_0(\epsilon)] + a\epsilon P_0(\epsilon), \tag{12}$$

where $P_0(\epsilon)$ and $1 - P_0(\epsilon)$ are the fraction of failed and intact lines, respectively. The total load carried by the intact (failed) lines is thus represented by the first (second) term of eq. (12). Note that in the Fiber Bundle Model, eq. (12) represents the constitutive law for the case in which fibers are allowed to fail only once ($k_{\max} = 1$). Indeed, if we set $a = 0$ in this equation (*i.e.* broken fibers do not carry any load), the "dry" Fiber Bundle Model [45-48] is recovered, while the case in which $a = 0.5$ corresponds to the so-called "micromechanical model of fiber reinforced ceramic matrix composites" (CMCs), which has been extensively studied in the literature [49-51]. When fibers are allowed to fail more than once, we have to distinguish between quenched and annealed disorder. As mentioned before, here we will consider annealed disorder, since it models the microscopic redistributions of the grains in the packing during compression.

After two restructuring events, eq. (11) can be calculated by using the probability law (10) with $k = 2$, which gives

$$\sigma = \epsilon[1 - P_0(\epsilon)] + a\epsilon P_0(\epsilon)[1 - P_1(a\epsilon)] + a^2\epsilon P_0(\epsilon)P_1(a\epsilon), \tag{13}$$

where $P_0(\epsilon)[1 - P_1(a\epsilon)]$ is the fraction of lines that failed only once, and $P_0(\epsilon)P_1(a\epsilon)$ is the fraction of lines that failed already twice. Finally, after $k_{\max}$ restructuring events, the constitutive equation (11) is given by

$$\sigma = \sum_{k=0}^{k_{\max}-1} a^k \epsilon \left[1 - P_k(a^k\epsilon)\right] \prod_{j=0}^{k-1} P_j(a^j\epsilon) + a^{k_{\max}}\epsilon \prod_{k=0}^{k_{\max}-1} P_k(a^k\epsilon). \tag{14}$$

The probability distributions P_k, P_j above follow the relation $P_k(a^k\epsilon) = P_0(\epsilon/(d_0)^k)$ ($d_c = 1$), which determines the evolution of the system with the restructuring events. Therefore, the ratio $\tau = a/d_0$ is a very important parameter of the model; it decides whether the force chain becomes more fragile ($\tau > 1$) or more ductile ($\tau < 1$) as a result of restructuring. For the case in which $a = d_0$, we have

$$\tau = 1 \Longrightarrow P_k(a^k\epsilon) = P_0(\epsilon). \tag{15}$$

For this particular regime, and in the limit of $k_{\max} \to \infty$ where the particles are hard, eq. (14) can be written as

$$\sigma(\epsilon) = \sum_{k=0}^{k_{\max}-1} a^k \epsilon[1 - P_0(\epsilon)][P_0(\epsilon)]^k + a^{k_{\max}} \epsilon[P_0(\epsilon)]^{k_{\max}} \tag{16}$$
$$= \epsilon[1 - P_0(\epsilon)] \sum_{k=0}^{k_{\max}-1} [aP_0(\epsilon)]^k + \epsilon[aP_0(\epsilon)]^{k_{\max}}.$$

It can be seen from eq. (16) that if the maximum number $k_{\max}$ of possible restructuring events goes to infinity, the stress σ has finite values *only* for $aP_0(\epsilon) < 1$. In this case the second term goes to zero, and the constitutive equation takes the form

$$\sigma(\epsilon) = \epsilon[1 - P_0(\epsilon)] \frac{1}{1 - aP_0(\epsilon)}, \quad aP_0(\epsilon) < 1. \tag{17}$$

It follows that the stress σ diverges when ϵ approaches a critical value ϵ_c, where $P_0(\epsilon_c) = 1/a$. Expanding $P_0(\epsilon)$ into a Taylor series at ϵ_c as $P_0(\epsilon) = P_0(\epsilon_c) + p_0(\epsilon_c)(\epsilon_c - \epsilon) + \cdots$, where $p_0(\epsilon) = \mathrm{d}P_0/\mathrm{d}\epsilon$ is the distribution density corresponding to $P_0(\epsilon)$, and substituting it into eq. (17), the behaviour of σ in the vicinity of ϵ_c reads as

$$\sigma(\epsilon) \approx \frac{1}{ap_0(\epsilon_c)(\epsilon_c - \epsilon)} \sim (\epsilon_c - \epsilon)^{-1}. \tag{18}$$

It means that the stress σ exhibits a power law divergence when ϵ approaches the critical value ϵ_c. The *critical exponent* can be calculated as

$$\nu = \lim_{\epsilon \to \epsilon_c} \frac{\ln \sigma(\epsilon)}{\ln(\epsilon_c - \epsilon)} = -1. \tag{19}$$

The value of the exponent is universal; it does not depend on the form of the disorder distribution P_0. On the other side, the value of ϵ_c does depend on the choice of P_0. For a uniform distribution $P_0(\epsilon) = \epsilon$, we have $\sigma(\epsilon) = \epsilon[1-\epsilon](1/1-a\epsilon) \to 1-a\epsilon_c = 0 \Rightarrow \epsilon_c = 1/a$, while for a cumulative *Weibull distribution*

$$P_k(d) = 1 - \exp\left[-\left(\frac{d}{{d_0}^k}\right)^m\right], \tag{20}$$

which has been proved to be a good empirical statistical distribution for modeling solid strength in material science, the critical value ϵ_c can be calculated, again for $\tau = 1 \Rightarrow a = d_0$, as follows:

$$P_k(a^k\epsilon) = P_0(\epsilon) = 1 - \mathrm{e}^{-\epsilon^m} \Longrightarrow \tag{21}$$
$$\Longrightarrow \sigma(\epsilon) = \epsilon\big[1 - \big(1 - \mathrm{e}^{-\epsilon^m}\big)\big] \frac{1}{1 - a\big(1 - \mathrm{e}^{-\epsilon^m}\big)} \Longrightarrow \epsilon_c = \left(\ln \frac{a}{a-1}\right)^{1/m}.$$

Experiments and discrete element simulations [33,39,52] have revealed that the number of effective force chains increases during the compression process, until it reaches a saturation value. To incorporate this phenomenological aspect in the model, we write the number of force chains as $N(\epsilon) = N_0 G(\epsilon)$, where N_0 is the saturated number of chains, and the profile $G(\epsilon)$ has the property $G(\epsilon) \to 1$ with increasing ϵ. The number of force lines $\mathrm{d}N$ emerging due to an infinitesimal deformation increment from ϵ to $\epsilon + \mathrm{d}\epsilon$ is given by

$$\mathrm{d}N = N_0 g(\epsilon)\mathrm{d}\epsilon, \tag{22}$$

where $g(\epsilon) = \mathrm{d}G(\epsilon)/\mathrm{d}\epsilon$. A more general constitutive equation of compressed granular systems can be obtained by taking into account the whole loading history of the system, as well as possible non-linearities of the contacts between the grains. For this, we write the load $f_i(\epsilon)$ introduced in eq. (8) in a different way: $f_i(\epsilon - \epsilon_0) = E_0 a^{k(i)}(\epsilon - \epsilon_0)^\alpha$, where now $\alpha \neq 1$, and the local strain $\epsilon - \epsilon_0$ is different from the externally imposed strain value ϵ, since it also depends on the initial strain ϵ_0. By introducing the density function $g(\epsilon_0)$ as defined above into eq. (11), the average load or stress σ on a line reads as

$$\sigma(\epsilon) = \sum_{k=0}^{k_{\max}} \int_0^\epsilon (\epsilon - \epsilon_0)^\alpha a^k \left[pb_k(\epsilon - \epsilon_0)\right] g(\epsilon_0)\mathrm{d}\epsilon_0, \tag{23}$$

where E_0 is again set to be unity. The probabilities $pb_k(\epsilon - \epsilon_0)$ can be calculated with eq. (10). It follows that the macroscopic constitutive equation of the compressed granular system, again with the condition $\tau = 1$, can be cast into the following expression:

$$\begin{aligned}\sigma(\epsilon) = \sum_{i=0}^{k_{\max}} \int_0^\epsilon a^i (\epsilon - \epsilon_0)^\alpha P_0^i(\epsilon - \epsilon_0)[1 - P_0(\epsilon - \epsilon_0)] g(\epsilon_0)\mathrm{d}\epsilon_0 \,+ \\ + \int_0^\epsilon a^{k_{\max}} (\epsilon - \epsilon_0)^\alpha P_0^{k_{\max}}(\epsilon - \epsilon_0) g(\epsilon_0)\mathrm{d}\epsilon_0,\end{aligned} \tag{24}$$

which can be numerically calculated by choosing the correct parameters and the appropriated form of $g(\epsilon_0)$. Figure 8 shows how eq. (24) can be fitted to experimental data obtained with the apparatus of fig. 5, with the measured force presented as a function of the relative displacement of the cylinder top. The parameters corresponding to the plot shown in this figure are: $E_0 = 4600\,\mathrm{N/m^2}$, $d_0 = 3\,\mathrm{N}$ and $k_{\max} = 70$. It has been assumed that the new force chains emerge following an exponential law $g(\epsilon) = \mathrm{e}^{-\epsilon/\beta}$, with $\beta = 0.01$. A power law $F \sim \epsilon^3$ is also shown. The emergence of gradual hardening force chains due to rearrangements of grains inside the container is responsible on the microscopic level for the strong non-linearity observed macroscopically. If we consider the number N of force lines as fixed, analytic results can be obtained also for the statistics of the restructuring events. Restructuring occurs during the compression process whenever the local load on a force line exceeds its threshold value. Since loading is performed under strain-controlled conditions, there is no load redistribution among existing

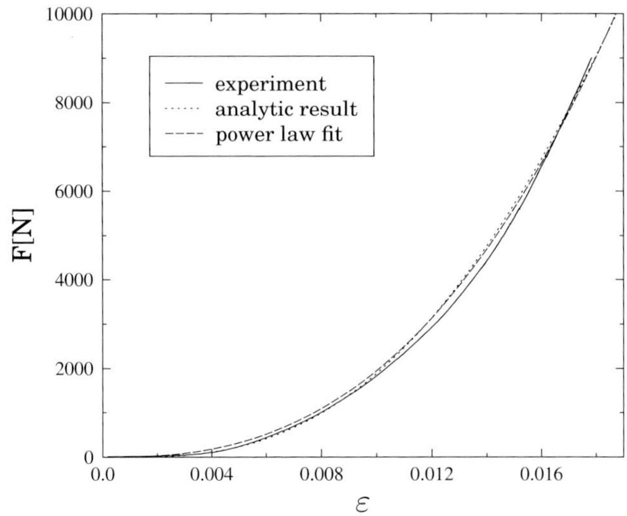

Fig. 8. – The constitutive behaviour measured experimentally is shown as a full line, while the analytic solution obtained with eq. (24) is represented by the dashed line. A power law with exponent 3 is also presented (dotted line). See text for details.

force lines. If the new threshold value assigned to the force line after rearrangement is smaller than the local load, the actual line undergoes successive restructurings until it gets stabilized. The number of steps to reach the stable state defines the size s of the restructuring event. The number $n_{k,s}$ of restructuring events of size s starting in force chains which have already suffered k restructurings reads as

$$\frac{n_{k,s}(\epsilon)}{N_0} = \prod_{j=1}^{k-1} P_j(a^j\epsilon)\frac{\mathrm{d}P_k(a^k\epsilon)}{\mathrm{d}\epsilon} \times \prod_{r=1}^{s-1} P_r(a^r\epsilon)[1-P_s(a^s\epsilon)],$$

for $s \leq k_{\max} - k - 1$, and

$$\frac{n_{k,s}(\epsilon)}{N_0} = \prod_{j=1}^{k-1} P_j(a^j\epsilon)\frac{\mathrm{d}P_j(a^j\epsilon)}{\mathrm{d}\epsilon} \times \prod_{r=1}^{s-1} P_r(a^r\epsilon), \quad \text{for } s = k_{\max} - k.$$

The conditional probability is normalized by the total number of elements N_0, and is shown in fig. 9 for several values of k, with $s = 2$. For $\tau = 1$ we have $P_k(a^k\epsilon) = P_0(\epsilon)$, and the expression for $n_{k,s}(\epsilon)$ can be written as follows:

$$(25) \qquad \frac{n_{k,s}(\epsilon)}{N_0} = \begin{cases} \dfrac{\mathrm{d}P_0(\epsilon)}{\mathrm{d}\epsilon}{P_0}^{s+k-1}(\epsilon)[1-P_0(\epsilon)], & s \leq k_{\max}-k-1, \\ {P_0}^{k_{\max}-1}(\epsilon)\dfrac{\mathrm{d}P_0(\epsilon)}{\mathrm{d}\epsilon}, & s = k_{\max}-k. \end{cases}$$

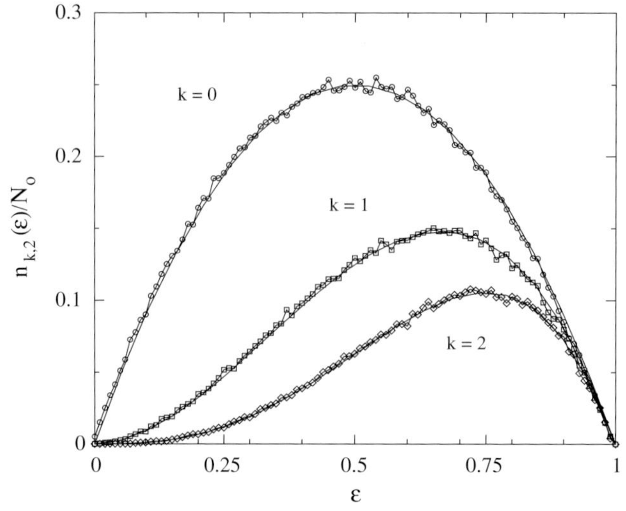

Fig. 9. – Normalized number of restructuring events of size $s = 2$ starting in force chains which have already failured k times. The circles represent the conditional probability obtained experimentally, while the analytic results are shown as full lines.

The number $D(s)$ of events of size s can be determined by integrating over the entire loading history and summing over all possible k values:

$$D(s) = \sum_{k=0}^{k_{\max}-s-1} \int_0^{\epsilon_c} \frac{n_{k,s}(\epsilon)}{N_0}\,\mathrm{d}\epsilon + \int_0^{\epsilon_c} \frac{n_{k_{\max}-s,s}(\epsilon)}{N_0}\,\mathrm{d}\epsilon. \tag{26}$$

Substituting expression (25) into the equation above, we get

$$\begin{aligned} D(s) &= \sum_{k=0}^{k_{\max}-s-1} \int_0^{\epsilon_c} \frac{\mathrm{d}P_0(\epsilon)}{\mathrm{d}\epsilon} {P_0}^{s+k-1}(\epsilon)[1-P_0(\epsilon)]d\epsilon + \int_0^{\epsilon_c} {P_0}^{k_{\max}-1}(\epsilon)\frac{\mathrm{d}P_0(\epsilon)}{\mathrm{d}\epsilon}\mathrm{d}\epsilon \\ &= \sum_{k=0}^{k_{\max}-s-1} \int_0^1 {P_0}^{s+k-1}[1-P_0]\mathrm{d}P_0 + \int_0^1 {P_0}^{k_{\max}-1}\,\mathrm{d}P_0 \Longrightarrow \\ &\Longrightarrow D(s) = s^{-1}, \quad \text{where } 1 \le s \le k_{\max}, \end{aligned} \tag{27}$$

i.e. the distribution of microscopic restructuring events exhibits a universal power law behaviour with an exponent 1, which is completely independent of the disorder distribution. Numerical simulations revealed that this behaviour holds even if the gradual creation of force chains is taken into account, *i.e.* when the system is described by eq. (24). In fig. 10 we show the statistics of restructuring events obtained by Monte Carlo simulations performed with the assumption $N(\epsilon) = N_0(1 - \mathrm{e}^{-\epsilon/\beta})$. A sequence of local events of different sizes, that occurred during the loading process, is shown in the inset of this

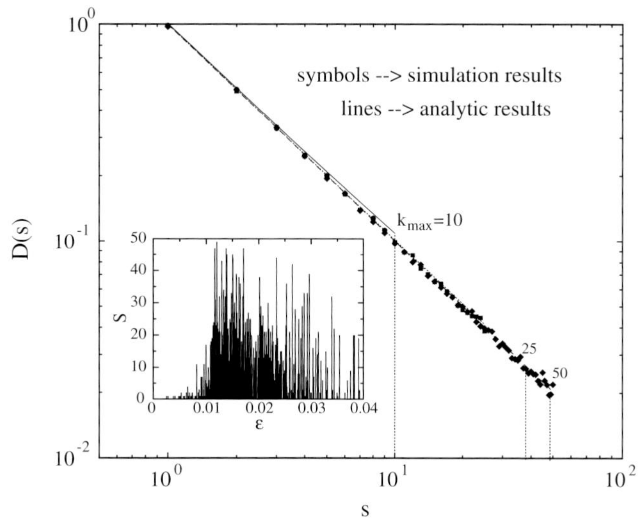

Fig. 10. – The inset shows local events of size s obtained from a Monte Carlo simulation of strain controlled loading. The distribution $D(s)$ is presented in a double logarithmic plot in the main figure.

figure, while the corresponding distribution $D(s)$ is presented in the main plot. As we can see from the comparison of this figure with fig. 6, the theoretical results on the statistics of the restructuring events are in a very good quantitative agreement with the experimental findings. It is important to emphasize that the functional form of $D(s)$, as well as the value of the exponent of the power law (27), have their origin mainly in the local behaviour of the restructuring processes, as there is no load redistribution among intact lines. The excellent agreement of the results presented here with experimental data indicates that this is probably the microscopic mechanism responsible for the power law statistics of acoustic signals observed experimentally.

4. – Conclusions

In this lecture, we have seen several aspects of the non-trivial behaviour of granular materials. A short review of the research on grains was presented, and we concentrated on the phenomenology of granular media in the static form. The last section of these lecture notes was dedicated to the model of percolating force chains in compressed granular packings, recently published [41]. Other sections could be added for the study of grains in the fluidized form, but this is rather the subject for another lecture. It has been shown that many open problems on granular media remain to be investigated in the future. We hope very much that this pedagogical enterprise will help to motivate young researchers to take up some of these challenges.

* * *

We acknowledge S. McNamara for a critical reading in this manuscript and helpful comments. E. J. R. Parteli acknowledges a fellowship from CAPES, Brasília/Brazil.

REFERENCES

[1] Bagnold R. A., *The Physics of Blown Sand and Desert Dunes* (Chapman & Hall, London) 1956.
[2] Reynolds O., *Philos. Mag. Ser. 5*, **469** (1885) 20.
[3] Herrmann H. J., *Physica A*, **313** (2002) 188.
[4] Måløy K. J., Pennec T. Le, Flekkøy E. G., Bideau D., Ammi M., Messager J. C., Wu X. L. and Hansen A., *Granular flow in hoppers and tubes*, in *Physics of Dry Granular Media*, edited by Herrmann H. J., Hovi J.-P. and Luding S. (Kluwer Academic Publishers) 1998, pp. 509-531.
[5] Nakahara A., *1/f noise in pipe flow*, in *Physics of Dry Granular Media*, edited by Herrmann H. J., Hovi J.-P. and Luding S. (Kluwer Academic Publishers) 1998, pp. 533-538.
[6] Schinner A., Scherer M., Rehberg I. and Kassner K., *Rotation and reptation*, in *Physics of Dry Granular Media*, edited by Herrmann H. J., Hovi J.-P. and Luding S. (Kluwer Academic Publishers) 1998, pp. 639-644.
[7] Jaeger H. M., Liu C. H., Nagel S. R. and Witten T. A., *Europhys. Lett.*, **11** (1990) 619.
[8] Lemiaux P.-A. and Durian D. J., *Phys. Rev. Lett.*, **85** (2000) 4273.
[9] Jaeger H. M. and Nagel S. R., *Science*, **255** (1992) 1523.
[10] Jaeger H. M., Nagel S. R. and Behringer R. P., *Rev. Mod. Phys.*, **68** (1996) 1259.
[11] Savage S. B., *J. Fluid Mech.*, **241** (1992) 109.
[12] Wang D. G. and Campbell C. S., *J. Fluid Mech.*, **224** (1992) 527.
[13] Thompson P. A. and Grest G. S., *Phys. Rev. Lett.*, **67** (1991) 13.
[14] Tsimring L. S., Ramaswamy R. and Sherman P., *Phys. Rev. E*, **60** (1990) 7126.
[15] Aranson I. S. and Tsimring L. S., *Phys. Rev. E*, **65** (2002) 061303.
[16] Argentina M., Clerc M. G. and Soto R., *Phys. Rev. Lett.*, **89** (2002) 044301.
[17] Savage S. B. and Jeffrey D. J., *J. Fluid Mech.*, **110** (1981) 255.
[18] McNamara S. and Young W. R., *Phys. Rev. E*, **53** (1996) 5089.
[19] Jenkins J. T. and Savage S. B., *J. Fluid Mech.*, **130** (1983) 187.
[20] Haff P. K., *J. Fluid Mech.*, **134** (1983) 401.
[21] Mohr O., *Zeit. d. Vereines Dtsch. Eng.*, **44** (1900) 1.
[22] Vardoulakis I., *Behavior of Granular Materials*, in *Lemaitre Handbook of Materials Behavior Models* (Academic Press) 2001, pp. 1093-1106.
[23] de Gennes P. G., *Rev. Mod. Phys.*, **71** (1999) S374.
[24] Restagno F., Ursini C., Gayvallet H. and Charlaix É., *Phys. Rev. E*, **66** (2002) 021304.
[25] Bowden F. P. and Tabor D., *The Friction and Lubrication of Solids - Part I.* (Clarendon Press, Oxford) 1950.
[26] Luding S., *Collisions & contacts between two particles*, in *Physics of Dry Granular Media*, edited by Herrmann H. J., Hovi J.-P. and Luding S. (Kluwer Academic Publishers) 1998, pp. 285-304.
[27] Hayakawa H., *Phys. Rev. E*, **60** (1999) 4500.
[28] Janssen H. A., *Zeit. d. Vereines Dtsch. Eng.*, **39** (1895) 1045.

[29] BOWDEN P. and TABOR D., *Friction: An Introduction* (Doubleday, New York) 1973.
[30] EDWARDS S. F. and MOUNTFIELD C. C., *Physica A*, **226** (1996) 1; 12; 25.
[31] ASTE T., DI MATTEO T. and D'AGLIANO E. G., *J. Phys. Condens. Matter*, **14** (2002) 2391.
[32] MUETH D. M., JAEGER H. M. and NAGEL S. R., *Phys. Rev. E*, **57** (1998) 3164.
[33] LIU C. H., NAGEL S. R., SCHECTER D. A., COPPERSMITH S. N., MAJUMDAR S., NARAYAN O. and WITTEN T. A., *Science*, **269** (1995) 513.
[34] TRAVERS T., BIDEAU D., GERVOIS A., TROADEC J.-P. and MESSAGER J.-C., *J. Phys. A*, **19** (1986) L1033.
[35] MILLER B., O'HERN C. and BEHRINGER R. P., *Phys. Rev. Lett.*, **77** (1996) 3110.
[36] ROUX S., STAUFFER D. and HERRMANN H. J., *J. Phys. (Paris)*, **48** (1987) 341.
[37] STAUFFER D., HERRMANN H. J. and ROUX S., *J. Phys. (Paris)*, **48** (1987) 347.
[38] RINTOUL M. D. and TORQUATO S., *Phys. Rev. Lett.*, **77** (1996) 4198.
[39] MAKSE D., JOHNSON D. L. and SCHWARTZ L. M., *Phys. Rev. Lett.*, **84** (2000) 4160.
[40] SMID J. and NOVOSAD J., *I. Chem. E. Symp. Ser.*, **63:D3/V** (1981) 1.
[41] HIDALGO R. C., GROSSE C. U., KUN F., REINHARDT H. W. and HERRMANN H. J., *Phys. Rev. Lett.*, **89** (2002) 205501.
[42] KUN F., ZAPPERI S. and HERRMANN H. J., *Eur. Phys. J. B*, **17** (2000) 269.
[43] CHAKRABARTI B. K. and BENGUIGUI L. G., *Statistical Physics of Fracture and Breakdown in Disordered Systems* (Clarendon Press, Oxford) 1997.
[44] HIDALGO R. C., KUN F. and HERRMANN H. J., *Phys. Rev. E*, **64** (2001) 066122.
[45] DANIELS H. E., *Proc. R. Soc. London, Ser. A*, **183** (1944) 405.
[46] COLEMAN B. D., *J. Appl. Phys.*, **29** (1958) 968.
[47] SORNETTE D., *J. Phys. A*, **22** (1989) L243.
[48] SORNETTE D., *J. Phys. (Paris)*, **50** (1989) S. 745.
[49] CURTIN W. A., *J. Am. Ceram. Soc.*, **75** (1991) 2837.
[50] CURTIN W. A., *J. Mech. Phys. Solids*, **41** (1993) 217.
[51] ZHOU S. J. and CURTIN W. A., *Acta Metall. Mater.*, **43** (1995) 3093.
[52] STEENBERG B. and BERGKVIST L., *Cellulose Chem. Technol.*, **14** (1980) 797.

Pattern formation and shocks in granular gases

Harry L. Swinney and E. C. Rericha

Center for Nonlinear Dynamics and Department of Physics
University of Texas at Austin - Austin, Texas 78712 USA

1. – Introduction

Granular materials include sand, sugar, crushed coal, cereals, pills, cosmetics, and asteroids. The transport, mixing, and segregation of granular materials is important in the pharmaceutical, mining, agricultural, metal, food, and energy industries. Even in the chemical industry, the majority of the products are in granular rather than liquid form [1]. Thus a large engineering literature has developed on "powders and particulates". However, a basic understanding of the physical mechanisms underlying the collective behavior of particles in a granular medium is lacking. Granular media can exhibit both solid and fluid properties (*e.g.*, one can walk on a beach or pour the sand from a bucket), but granular media are less well understood than solids and fluids. While fluids are processed in industry with high efficiency, the efficiency of handling (crushing, mixing, separation) of granular materials is estimated to be well below optimum [1].

The scientific study of granular systems has a long history, including a discussion by Galileo in his *Dialogues* and a 1831 study by Faraday of convective motion of grains in heaps in vertically oscillated granular layers [2]. In recent years there has been a resurgence of interest in granular media among physicists, thanks to Pierre-Giles de Gennes, who recommended in the early 1980s to young French scientists that granular matter was an interesting subject worthy of study [3-6]. In the past two decades there has been an explosion of interest in granular systems. A search in INSPEC on the word "granular" followed by "system, medium, matter, flow, or gas" yields 20 papers in the three year period 1980–82, while a decade later the number of papers in a three year period jumped to 112, and in 2000–2002 there were 691 papers on the subject. The growth is not so dramatic for literature searches using terms often used in industry (powder, particles, particulates), but most of that work is empirical, and much of that literature concerns pastes, soil, and fine particles (powders). We make no attempt to review the enormous literature on granular/particulate matter but list a few reviews [7-13], books [14-18], and several journals: *Powder Technology*, *Granular Matter*, *Particle Science and Technology*, and *Advanced Powder Technology*.

The key property distinguishing collisions of granular particles from collisions of atoms in an ordinary gas is dissipation: collisions between macroscopic grains are inelastic —in the absence of external forcing, the particles in a granular medium all come to rest. A measure of this dissipation is the coefficient of restitution e, which is the ratio of the relative normal velocity of two particles after a collision to the relative normal velocity before collision. Some representative values of e for a relative normal velocity of $0.1\,\mathrm{m/s}$ are 0.96 for hardened bronze, 0.85 for aluminum, and 0.3 for lead [19]; the values depend on the particle bulk and surface properties in a complicated way (see, *e.g.*, [20]).

This chapter concerns rapid granular flows, which is called the collisional regime to distinguish it from the quasi-static regime where particles are at rest or nearly so. Much of the granular literature concerns the quasi-static regime where inertia is not significant and chains of particles in contact bear most of the load. Understanding the development and evolution of these force chains and the role of steric hindrance is often the focus of the research on sand piles and other granular systems at rest or nearly at rest. The collisional and quasi-static regimes are each difficult, but the intermediate regime with some particles moving rapidly while others are at rest is even more difficult. For an example, see the contribution in this book on the formation of craters in a granular medium [21,22].

We will consider situations where only contact forces are important. For small particles (less than $\approx 50\,\mu\mathrm{m}$), electrostatic and van der Waals forces become important. Further, air friction can be significant, but for particles greater than about $1\,\mathrm{mm}$ in diameter, air friction is often negligible if the velocity is not too large.

An oft-studied example of a granular medium in the collisional regime is a collection of particles in a vertically oscillating container. Section **2** describes spatial patterns formed by such a system. Since the number of particles in an experiment can be small (less than one million), Newton's laws for the motion of these particles can be directly implemented on a personal computer. Section **3** discusses such Molecular Dynamics (MD) simulations for hard, spherical particles that are characterized by a restitution coefficient and a frictional coefficient. Most theoretical analyses of granular flows examine frictionless (smooth) inelastic spheres, but there exist no frictionless macroscopic particles, just as there are no elastic particles. Molecular dynamics simulations show that realistic models of rapid granular motions must include friction —friction provides another mode of dissipation and also results in reduced grain mobility and a higher particle collision rate [23]. Molecular dynamics simulations including friction describe experimental observations on oscillating granular media very well, as we shall describe in sect. **3**, but simulations without friction fail to capture even qualitatively some important aspects of the observations.

Studies of oscillated granular layers have revealed localized structures, "oscillons", which are stable for a range of container oscillation frequencies and amplitudes. Section **4** describes the properties of oscillons and shows how they can be considered as the basic building blocks of some extended spatial patterns.

The continuum approach to granular flows is introduced in sect. **5**, where equations of motion derived by Jenkins and Richman for a dilute dissipative fluid are presented. The Jenkins-Richman equations are similar to the Navier-Stokes equations for a fluid but are modified to include dissipative effects arising from collisions between the grains. However, the equations do not include the effect of friction in grain-grain collisions or grain-container collisions. As will be discussed, the application of continuum equations to granular flows must be considered with caution because, unlike an ordinary fluid, in granular flows there is not a large separation between the microscopic and macroscopic length and time scales.

In a granular fluid the speed of propagation of pressure fluctuations (the sound speed) is only a few centimeters/second, which is very small compared to the 330 m/s sound speed in air. (The granular sound speed here refers to particles in vacuum; there is no air and the only particles are the dissipative grains.) Hence the average streaming speed of granular flows generally exceeds the sound speed in the flow, and shock waves form whenever the flow encounters an obstacle. These shock waves are the subject of sect. **6**, where observations from experiments are compared with molecular dynamics and continuum simulations.

We will conclude with a discussion in sect. **7** that mentions some of the many issues open in the understanding of systems of dissipative particles.

2. – Patterns in vertically oscillated granular layers

Consider a layer of non-cohesive grains in a container oscillating sinusoidally with frequency f and dimensionless acceleration amplitude $\Gamma = 4\pi^2 f^2 A/g$, where $2A$ is the peak-to-peak amplitude of the displacement of the container and g is the gravitational acceleration. We consider patterns that arise spontaneously, not from sidewall forcing or from interstitial gas, but from correlations induced by multiple collisions between the grains and between the grains and the container bottom. To minimize sidewall effects, the horizontal dimensions of the container are made large compared to the layer depth h. The layer is illuminated from the side, as fig. 1 illustrates [24, 25].

For $\Gamma > 1$, on each cycle the layer loses contact with the container, flies in the air, and then collides with the container. However, the layer remains compact and flat until $\Gamma \approx 2.5$, where a standing wave pattern spontaneously forms. A square pattern forms at low frequencies and a stripe pattern at high frequencies, as figs. 2(a) and (b), respectively, illustrate [26, 27]. The pattern is subharmonic, repeating every 2τ, where $\tau = 1/f$ is the container oscillation period; thus a ridge in a striped pattern at an instant of time becomes a valley one container oscillation period later.

A heuristic argument for the critical value of Γ for the onset of instability of a flat oscillating layer was given in [25]. The authors argued that a flat layer becomes unstable

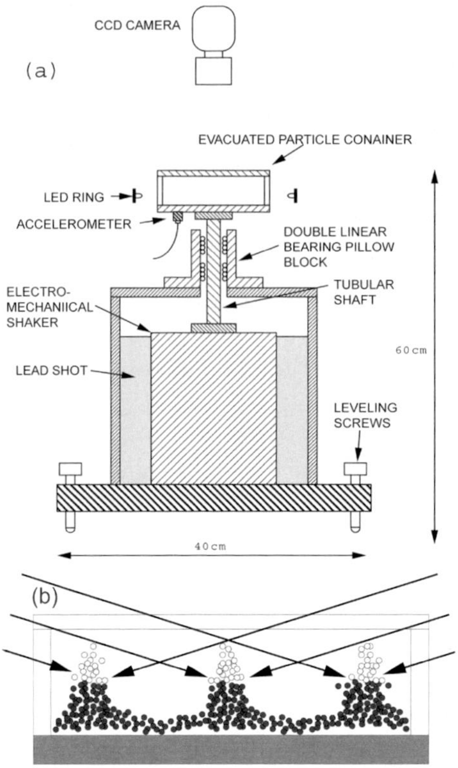

Fig. 1. – (a) Electromechanical shaker system for studying pattern formation in vertically oscillating layers. (b) Spatial patterns are illuminated from the side by light incident at low angles. When viewed from above, high regions are bright and low regions are dark. From [24].

when the collision occurs at the plate's lowest point. Then a time $\tau/2$ is taken for the layer to free fall from its highest point through a distance $2A$ to collide with the plate, so that $2A = (1/2)g(\tau/2)^2 = (1/8)g\tau^2$, which gives $\Gamma_c = 2.5$, in accord with the observed critical acceleration for the onset of patterns.

The onset of patterns (squares or stripes) at $\Gamma \approx 2.5$ is quite robust, independent of layer depth (depths from about a monolayer to about 25 particle diameters σ), container shape, and particle properties (size, restitution coefficient, surface roughness, material). Most studies of patterns have been conducted for particles of uniform size, but studies with a range of particle sizes have found that the onset remains sharp for size distributions ranging up to about 30% [28]. Square and stripe patterns have also been observed to form at $\Gamma \approx 2.5$ for irregular particles such as rice grains and grass seed [28].

The phase diagram (fig. 3) shows the stability regions for different patterns as a function of Γ and dimensionless frequency f^*, where $f^* = \sqrt{h/g}$ and h is the depth of the layer at rest [29]. The transitions are well defined and are only weakly dependent on f^*. Except for the transition from a flat layer to squares, the hysteresis is small or

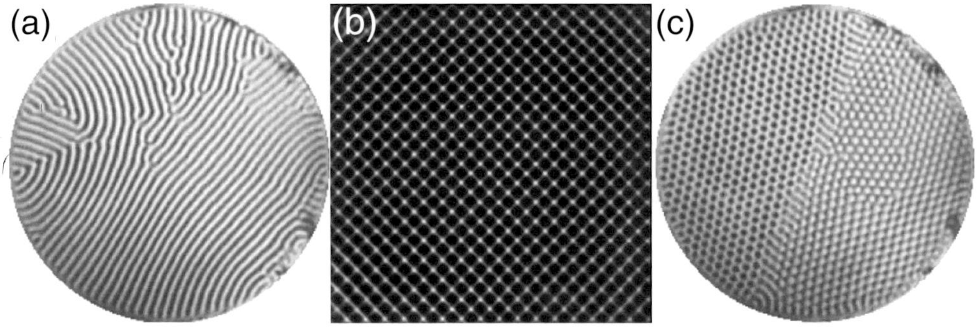

Fig. 2. – Patterns in oscillating granular layers: (a) stripes, (b) squares, and (c) hexagons (with different phases on the left and right). The patterns oscillate at $f/2$, where f is the container oscillation frequency. Here Γ, f, and h/σ (ratio of the depth of the layer at rest to the diameter of the particles) are given by (a) 3.3, 67 Hz, 7, (b) 2.9, 25 Hz, 4, and (c) and (d) 4.0, 67 Hz, 7. In (a) and (c) the diameter of the container is 770σ and in (b) the container is square, $1100\sigma \times 1100\sigma$. The particles are bronze spheres 0.165 mm diameter. (a) and (c) are from [26] and (b) is from [27].

perhaps zero. We know of no argument for the square to stripe transition as a function of frequency with Γ fixed, but we note that the transition has been found to occur at $f^* \approx 1/3$ [24].

Some insight into the dynamics can be gained from consideration of the one-dimensional motion of a single completely inelastic ball on a vertically oscillating plate [30]. Such a model cannot of course describe the 2D spatial patterns that form in granular layers on an oscillating plate, but it does help in understanding the transitions in behavior as a function of Γ [26, 29]. The inelastic ball motion, illustrated in fig. 4, models the center-of-mass motion of a granular layer at small Γ, where the layer remains fairly compact and is highly dissipative as a consequence of multiple collisions. We will refer to the inelastic ball as "the layer", meaning the motion of the center of mass of the granular layer.

For $\Gamma > 1$, the layer leaves the plate at the point in each cycle when the plate acceleration exceeds $-g$. The layer continues in free flight until it later collides with the plate. In the regime with squares or stripes oscillating at $f/2$, the layer leaves and hits the plate every cycle, as fig. 4(a) illustrates. At $\Gamma \approx 4.0$, there is a bifurcation in the dynamics, illustrated by fig. 4(b): successive trajectories now have different flight durations, with a short trajectory initiated by a collision of the layer with the plate whose acceleration greater than g, followed by a long trajectory initiated by a collision of the layer with the plate whose acceleration is less than g. This bifurcation corresponds to a value of Γ close to that for the onset of hexagonal spatial patterns [26], pictured in fig. 2(c). While square or stripe patterns have the same appearance whether an image is obtained at a time t or $t + \tau$, hexagonal patterns at these two times are different. The two phases are both present and are separated by a phase discontinuity in fig. 2(c) —the

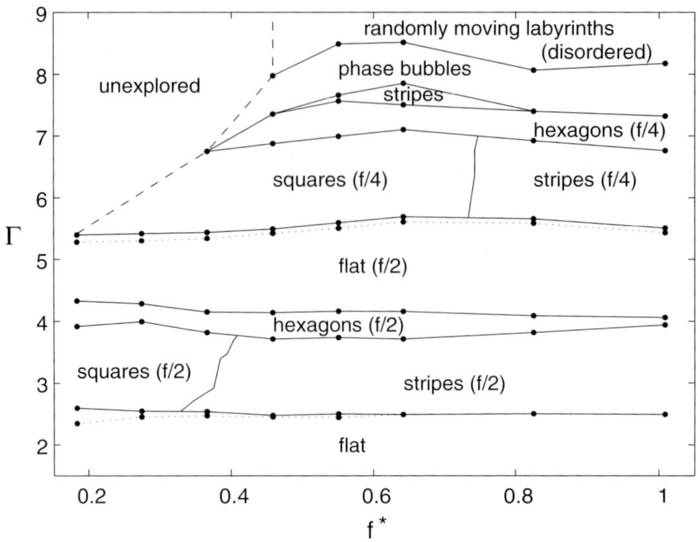

Fig. 3. – Phase diagram for granular patterns observed in a vertically oscillated container, as a function of the dimensionless acceleration Γ and dimensionless frequency $f^* = f\sqrt{h/g}$. The transitions from a flat layer to squares are hysteretic: solid lines denote the transition for increasing Γ while dotted lines denote decreasing Γ. (Bronze spheres, $\sigma = 0.165$ mm; layer depth, 5.0σ; container diameter, 770σ.) From [29].

pattern on the right consists of a hexagonal array of dots, while the hexagonal pattern on the left is cellular; the two patterns will be switched one period, τ, later.

Another bifurcation in the dynamics occurs for $\Gamma > 4.5$, where the layer flight duration exceeds τ, and the layer hits the container every other cycle. Now the velocity of the layer relative to the plate at the instant of collision goes to zero (at about $\Gamma = 4.6$), and the layer makes a soft landing; not enough momentum is transferred from the vertical to horizontal direction to form patterns. When Γ is increased above about 5.4, there is again a transition from a flat layer to a pattern, stripes at low frequencies and squares at high frequencies. However, now the pattern period is $f/4$ instead of $f/2$. Further increase in Γ leads to another bifurcation from trajectories with a single period to successive trajectories with different flight durations (fig. 4(d)). Again this bifurcation in the dynamics of a ball corresponds to the bifurcation of the patterns from squares or stripes to hexagons.

At larger Γ the layer becomes sufficiently dilated so that the inelastic single ball model no longer provides a useful description of the dynamics. But even the $f/3$ and $f/6$ regimes of the single ball model ($8 < \Gamma < 11$) have been observed transiently in molecular dynamics simulations and laboratory experiments [29].

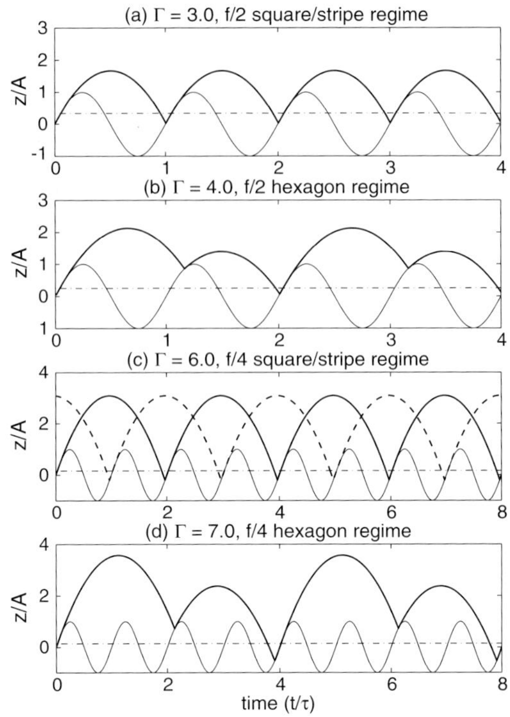

Fig. 4. – Trajectory of a completely inelastic ball on an oscillating plate. This is a model for the motion of the center of mass of a granular layer. The sinusoidal curve is the trajectory of the plate. The ball leaves the plate when the acceleration of the plate becomes $-g$, that is, when the dot-dashed line intersects the trajectory of the ball. If the ball collides with the plate above the dot-dashed line, as in (b) and (d), it leaves the plate immediately. From [29].

3. – Molecular Dynamics simulations

Simulations of granular media using rigid particles, soft particles, and Monte Carlo methods have been conducted by many researchers since the 1980s [31, 7, 15]. We consider a granular layer modeled as a collection of hard spheres that interact only through instantaneous binary collisions. Between successive collisions the particles move only under the influence of gravity. This is known as an Event Driven (ED) type of Molecular Dynamics (MD) simulation [32]. Linear and angular momentum are conserved in collisions, while energy is dissipated through collisions and surface friction. The interaction between particles is described by the coefficient of restitution e, the coefficient of sliding friction μ, and the rotation coefficient of restitution β. (The same values of e, μ and β are used for ball-wall collisions as for ball-ball collisions.) This collision model was developed by Walton [33] and used in hard sphere MD simulations conducted by Bizon [25] and Moon [29], as we shall describe.

The restitution coefficient is often taken to be a constant in MD simulations, but a constant value of e can lead to "inelastic collapse", where particles undergo an infinite number of collisions in a finite amount of time [34-36]. However, physically e must approach unity as the relative normal velocity v_n approaches zero (the elastic limit). The simulations presented here assume a form for e that goes to unity in the $v_n \to 0$ limit: $e(v_n) = 1 - B{v_n}^{3/4}$ for $v_n < v_0$, where v_0 is a crossover velocity, and $e(v_n) = e_0 =$ *constant* for $v_n > v_0$. The approach of e toward unity for low v_n avoids inelastic collapse, while the constant value e_0 at high v_n is computationally efficient. The MD results are insensitive to the form assumed for $e(v_n)$ for $v_n < v_0$, as long as $e(v_n)$ increases to unity as v_n approaches zero [25].

In a collision the tangential impulse is given by μ times the normal impulse, with a cutoff corresponding to the crossover from a sliding contact to a rolling contact. The crossover ratio of the relative surface velocity after collision to that before the collision is given by β, which we fix at -0.35, as in [33]. The parameters e_0 and μ are set to 0.7 and 0.5, respectively, values obtained by fitting MD simulation results to laboratory observations for square patterns in layers of lead particles at one set of conditions [25].

Results from simulation and experiment are compared in fig. 5 for seven values of (f^*, Γ). At every point in the phase diagram in fig. 3, the results from the MD simulation agree with experiment, not only in the form of the pattern but also in the wavelength of the pattern. The dispersion relation relating the wavelength λ to f^* for layers of varying depths collapse onto a single curve, $\lambda/h = 1+1.1{f^*}^{-1.32\pm0.03}$, when the container velocity exceeds a critical value, $v_{\mathrm{gm}} \approx 3\sqrt{\sigma g}$, where v_{gm} corresponds to a transition in the grain mobility (gm): for $v > v_{\mathrm{gm}}$, there is a hydrodynamic-like horizontal sloshing motion of the layer, while for $v < v_{\mathrm{gm}}$, the grains are essentially immoblile and the stripe pattern apparently arises from a bending of the granular layer [24].

Most MD simulations and theoretical analyses of granular media consider frictionless spherical particles [37-39]. However, MD simulations comparing the behavior of frictional and frictionless particles indicate that the effect of friction cannot be mimicked by increasing the dissipation (decreasing e) [23]; thus friction is not merely an additional mechanism of dissipation. Even a small amount of friction increases the overall dissipation significantly, not because the frictional dissipation is significant in each collision, but because the friction reduces the grain mobility and increases the overall collision rate. MD simulations with frictional particles yield square and hexagonal patterns like those observed in experiments (fig. 5), while simulations without friction do not yield square or hexagonal patterns, even if the restitution coefficient is decreased to compensate for the absence of friction [23]. Simulations of frictionless particles do yield stripe patterns, but the critical Γ for the onset of patterns is smaller ($\Gamma_c \approx 1.9$) than for frictional particles ($\Gamma_c \approx 2.5$), and the stripes formed by frictionless particles are less robust than those formed by particles with friction [23].

4. – Localized structures and lattice dynamics

Localized stable standing wave structures dubbed "oscillons" can occur in oscillating granular layers when Γ is decreased slightly below the value corresponding to the onset of

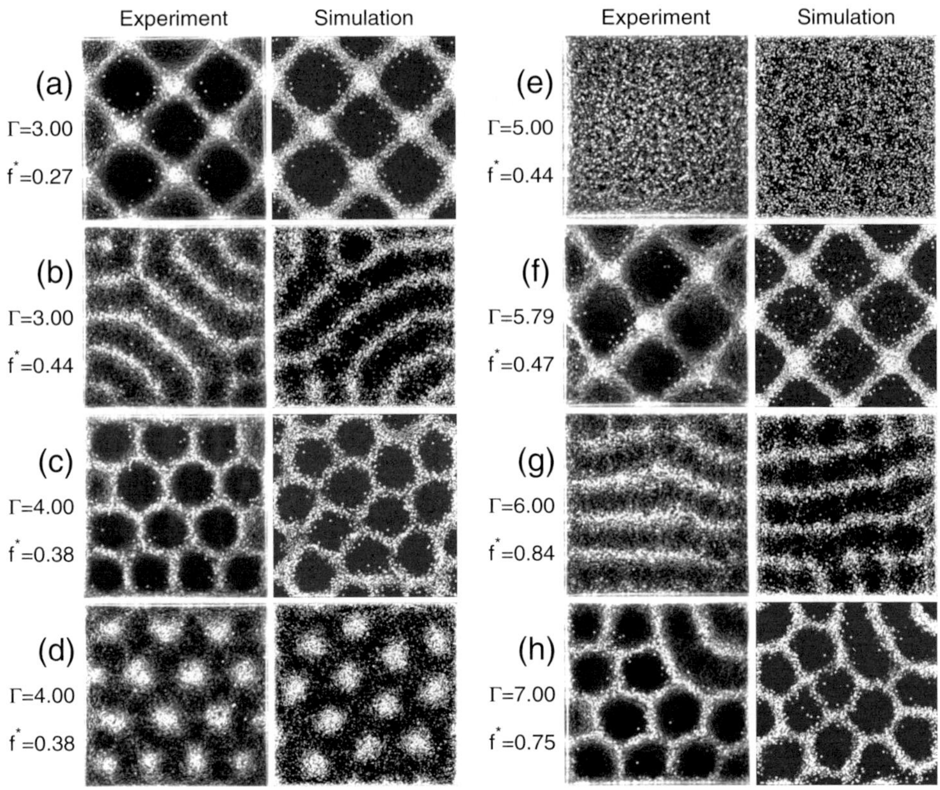

Fig. 5. – Standing-wave granular patterns in laboratory experiments and molecular dynamics simulations for the same number of particles, 60 000, which fill a $100\sigma \times 100\sigma$ container to a depth of 5.4 layers: (a) squares, (b) stripes, (c) and (d) alternating phases of hexagons, (e) flat layers, (f) squares, (g) stripes, and (h) hexagons. Patterns (a)-(e) oscillate at $f/2$, (f)-(h) at $f/4$. The experiment used lead particles with $\sigma = 0.55$ mm. From [25].

squares with increasing Γ [28,40]. Top and side views of oscillons are shown in fig. 6. An oscillon is a small, circularly symmetric excitation that oscillates at $f/2$; during one cycle of the container, it is a peak; on the next cycle it is a crater. Unlike solitons, oscillons are stationary (nonpropagating). Oscillons form with equal probability at all locations in the container, and they live indefinitely. If the container acceleration were increased slowly from rest to a value just below the onset of squares, no oscillons would appear, but an oscillon can be formed by a finite amplitude perturbation (a puff of air or a poke with a rod). When oscillons are obtained by decreasing Γ from the regime with square patterns, the number of oscillons that form is not fixed; as many as fifty oscillons were observed in the experiment in fig. 6, but no oscillons occurred if Γ was quasi-statically decreased [41].

Oscillons of like phase show a repulsive interaction that has a range not much larger than the diameter of an oscillon, while oscillons of opposite phase that are closer than

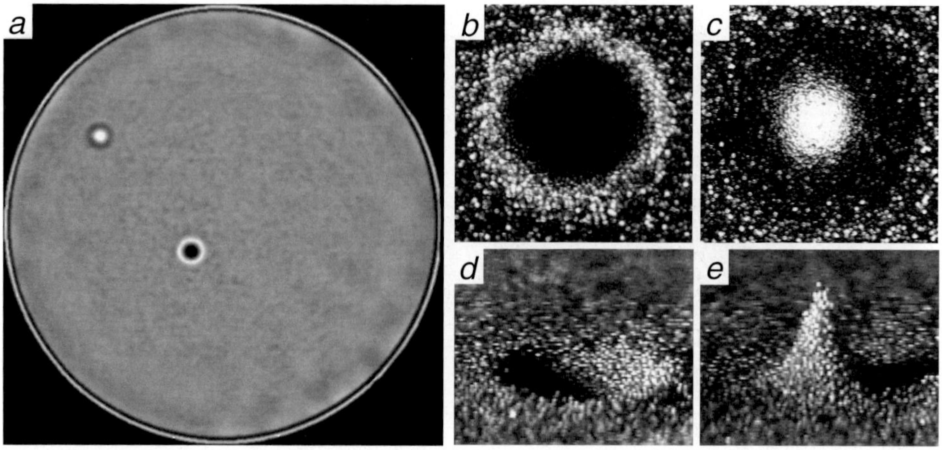

Fig. 6. – (a) Snapshot of a container with two oscillons (viewed from above); one is a peak (upper left) and the other is a crater (near the center). (b) and (d), Close-ups of an oscillon crater viewed from the top and from the side, respectively. (c) and (e) Close-ups of an oscillon peak viewed from the top and from the side, respectively. Individual bronze spheres ($\sigma = 0.165$ mm) are discernible in (b)-(e). ($f = 25$ Hz, $\Gamma = 2.45$, layer depth $h = 17\sigma$.) From [28].

about 1.4 oscillon diameters attract and form a stable dipole structure, as shown in fig. 7(a) [28, 40], and more complex structures like the tetramer pictured in fig. 7(b) and the polymer chain in pictured in fig. 7(c).

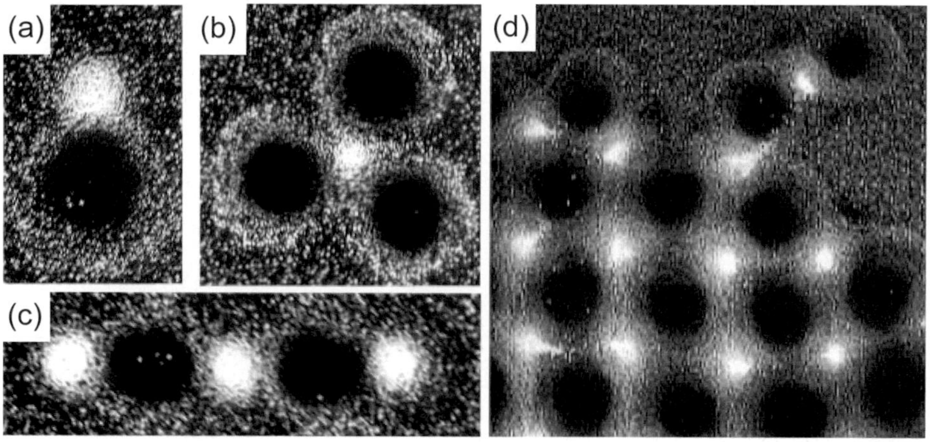

Fig. 7. – (a) Dimer formed by two bound oscillons of opposite phase; one period of oscillation of the container later the white peak will have become a crater and the black will have become a peak. (b) Tetramer formed of four oscillons. (c) Polymer chain of five oscillons. (d) A square lattice grows by nucleating oscillons. Individual bronze spheres ($\sigma = 0.165$ mm) are discernible in (a)-(c). From [28].

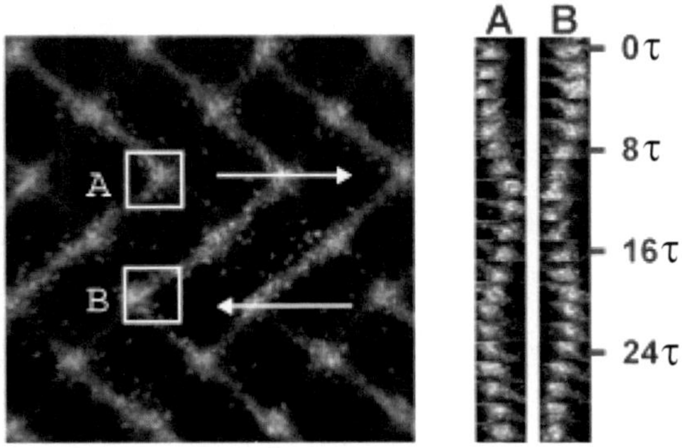

Fig. 8. – Left: close-up snapshot of a granular lattice ($\Gamma = 2.9$, $f = 25\,\text{Hz}$, $h = 4\sigma$). Right: time evolution of the peaks in the boxes A and B in the left-hand image. The peaks oscillate out of phase with a frequency about twenty times smaller than $f = 1/\tau$. From [27].

Can oscillons be considered as building blocks ("atoms") of the square lattice that forms with an increase in Γ? This view is suggested by the observation of the formation of a square lattice as Γ is slowly increased (see fig. 7(d)): an oscillon seeds a square lattice by spawning oscillons, which adjust to form a square array. This observation suggests that the granular lattice could be modeled as a system of coupled oscillon atoms, each of which is comprised of hundreds of particles that are colliding hundreds of times during each oscillation cycle. Thus the lattice approach is much simpler than the full description of all the particle motions and simpler than a continuum fluid description of the granular medium. As we will now describe, the lattice picture is supported by an analysis of the dynamics of the lattice.

Close examination of the center of mass of a peak in a square lattice reveals harmonic motion about the equilibrium position of the peak for a wide range of Γ and f [27]; such a lattice oscillation is illustrated in fig. 8. One test of the conjecture that the lattice of peaks can be modeled by a lattice of balls connected with springs is to compare the dispersion relation for the two lattices. A time sequence of images of the granular pattern was Fourier-transformed in space and then in time to obtain the frequencies of oscillations of lattice modes with different wave vectors. The frequency f_{L} of the lattice modes as a function of k (the magnitude of the wave vector) was found to be well described by the dispersion relation for balls connected by springs, $f_{\mathrm{L}} = f_{\mathrm{BZ}}|\sin(ka/(2\sqrt{2})|$, where f_{BZ} is the frequency at the edge of the Brillouin zone and a the lattice spacing [27]. The lattice oscillation frequency f_{L} is typically an order of magnitude smaller than f but depends on the plate acceleration Γ and frequency f.

In a crystalline solid, defects form when the amplitude of oscillation of atoms about their equilibrium position in the lattice becomes large. To investigate the possible formation of defects in a granular lattice, the plate frequency f was modulated at the lattice frequency f_{L}. For sufficiently large modulation amplitude, defects formed, breaking the long-range order of the square lattice. It was found that the amplitude of the lattice

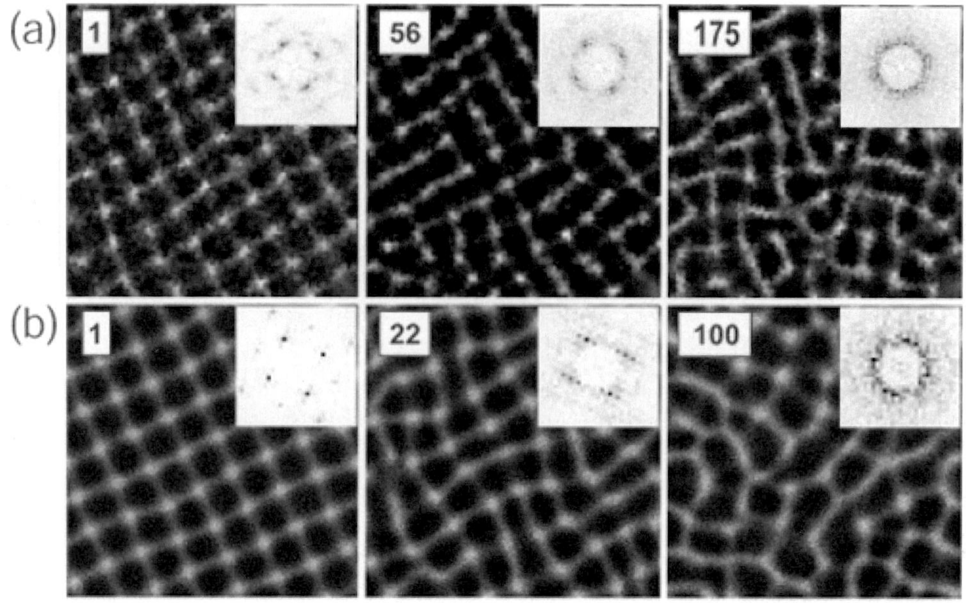

Fig. 9. – Defect creation and melting of a square granular pattern in a vertically oscillating granular layer: (a) experiment and (b) molecular dynamics simulation. The insets show Fourier transforms of the spatial patterns. In the experiment the plate oscillation frequency ($f = 32\,\mathrm{Hz}$) was modulated at the natural oscillation frequency of the lattice ($2\,\mathrm{Hz}$), and at $t = \tau$ graphite powder was added to the layer of bronze spheres. By $t = 56\tau$ defects had formed, and by $t = 175\tau$ the lattice had melted. In the MD simulation the friction coefficient was reduced from $\mu = 0.5$ to zero at $t = \tau$; by $t = 22\tau$ defects had formed, and by $t = 100\tau$ the lattice had melted. ($\Gamma = 2.9$.) From [27].

oscillations could be increased further by adding a lubricant (graphite powder) to reduce the friction between particles. The result was that the granular lattice melted [27]: the spatial Fourier transform became a circular ring (about $k = 0$) rather than sharp peaks, as fig. 9 illustrates.

Lattice oscillations and defect formation have also been studied in MD simulations, where friction can be easily varied or set to zero. First a square lattice was simulated with $\mu = 0.5$, as described previously and illustrated in the first panel of fig. 9(b). Then μ was set to zero, and defects were observed to form quickly as the lattice oscillation amplitude increased. Finally the lattice melted, as illustrated by the last panel of fig. 9(b).

Studies of melting in two-dimensional solids have shown that melting occurs when the Lindemann ratio, $\gamma = \langle |u_m - u_n|^2 \rangle / a^2$, exceeds 0.1 [42, 43]. Here u corresponds to the displacements of atoms from equilibrium lattice sites, a is the lattice constant, and the average is taken over all nearest neighbors m and n. The Lindemann ratio was computed for the granular lattice in the MD simulation, and it was found that when the friction coefficient μ was decreased, γ increased. Further, when γ reached the value 0.1 (which happened for $\mu = 0.1$ for the conditions of the simulation), the granular lattice melted, in accord with the result for crystalline solids [27].

5. – Continuum description

Section **3** introduced Molecular Dynamics simulations as a useful tool in describing granular flows. This technique models the system on a microscopic level, evolving individual particle trajectories using Newton's laws and computing the effects of each collision. Averaging over many collisions and particle trajectories gives the macroscopic behavior of the flow. A complementary method for understanding granular flows is to model the macroscopic motion directly by a continuum field theory that describes the bulk motion of the flow in terms of the density, velocity and temperature fields. Unlike MD simulations, the continuum approach is not limited by particle number. A personal computer currently contains enough memory for useful MD simulations of laboratory experiments. However, industrial processes contain billions of particles, far outside the abilities of MD simulations. Another reason that a continuum approach is attractive is that it could exploit tools such as stability analysis, amplitude equations, and perturbation theory, which have been developed through more than a century of research on the the Navier-Stokes equations and other partial differential equations.

Granular flows present many difficulties in developing a continuum theory [44-46]. Continuum theory requires a separation of length and time scales: variations over space should be small and occur over long distances, so that the behavior of local collections of individual particles can be averaged and replaced with small fluid elements. Changes in time for the flow should occur for times long compared to the mean time between particle collision so that particles moving between fluid elements do not affect the average values in a fluid element. Unfortunately, inelastic collisions between particles create an inherent lack of scale separation [45, 11]. Sufficient separation of scales may only be present for granular flows in the specific circumstances of low density and low dissipation [46,45,11].

The derivation of the continuum equations from kinetic theory makes assumptions about the underlying statistics of granular flows, assumptions which have not been verified by MD simulations. For instance, the velocity distribution function is assumed to have a steady-state functional form that is nearly Gaussian. Since granular flows are dissipative, a steady-state distribution function can only be achieved in the presence of forcing. Granular experiments have yielded velocity distributions that depend on the forcing characteristics and experimental geometry [47-50]. Also, most derivations of continuum equations assume Boltzmann's molecular chaos (particle velocities before collisions are uncorrelated), but strong velocity correlations have been found in MD simulations [51].

Despite the reservations regarding a continuum approach in granular media, observations of granular media have revealed many phenomena similar to those observed in continuum systems. For example, the stripe patterns shown in fig. 10(a) look like those in vertically oscillated liquid layers [52] (fig. 10(b)), chemical reaction-diffusion systems [53], Rayleigh-Bénard convection in fluids [54, 55] (fig. 10(c)), and liquid crystals [56].

Not only are the patterns similar for granular and continuum systems, but also some

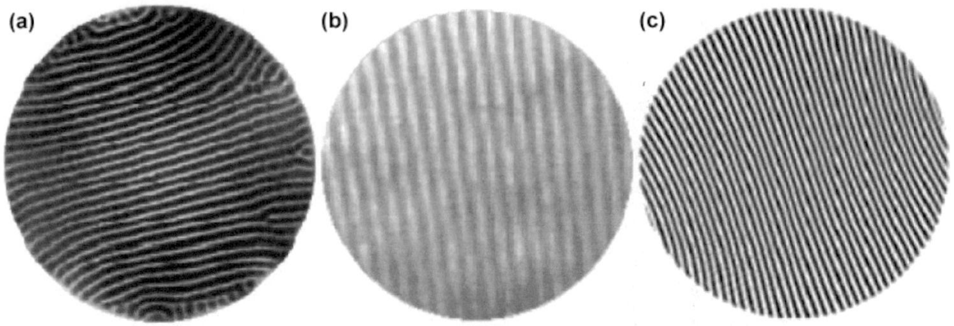

Fig. 10. – Forced granular materials produce qualitatively similar patterns as forced fluids: (a) stripe pattern formed by a vertically oscillated granular layer [26], (b) stripe pattern formed by a vertically oscillated layer of water [52], (c) stripe pattern formed in thermal convection of a fluid (CO_2) [55].

the same pattern instabilities have been observed. For example, when the wave number of parallel convection rolls (stripes) in a Rayleigh-Bénard convection becomes small, an instability leads to the formation of cross rolls with a larger wave number that are perpendicular to the original rolls [57, 58]; the same instability has been observed for stripes in oscillated granular layers, as fig. 11 illustrates. The cross rolls invade the region of small wave number stripes such that, after sufficient time, the region contains a pattern of straight stripes perpendicular to the original pattern and with a larger wave number.

Granular stripe patterns also exhibit a skew varicose instability like that in convection roll patterns (fig. 11) [57, 59-61]. When the local wave number becomes too large, an initially straight pattern of stripes will develop a distortion which evolves into a dislocation defect. The defect propagates away; destroying one of the stripes and decreasing the local wave number of the pattern. The stability of the stripe pattern in fluid convection is well described by amplitude equations derived from the Navier-Stokes equations for fluids [58]. That the granular pattern shows the same behavior strongly suggests a continuum description for the vibrated system is applicable.

Aspects of the phase diagram for granular patterns (fig. 3) have been reproduced by amplitude equation models. For example, a phenomenological continuum model requiring that the mass of the layer is conserved locally, produces stripe, square, and oscillon-like patterns similar to those found in experiment [62, 63]. A continuum, shallow-water–like model of the granular layer captures the patterns and yields a dispersion relation which agrees with experiment [64]. The success of these and other models [65-68] provides further motivation for considering continuum equations derived for a granular gas.

Additional evidence for the applicability of continuum theory to granular media is provided by a recent study of noise in vertically oscillating granular layers. In the Rayleigh-Bénard system below the onset of convection, thermal noise has been found to drive

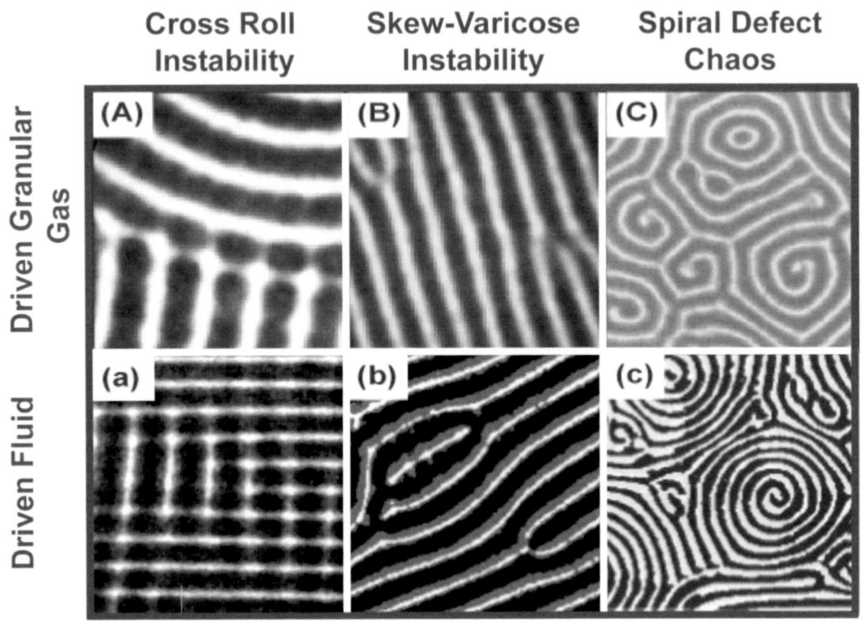

Fig. 11. – Instabilities of patterns found in oscillating granular layers and Rayleigh-Bénard convection in a fluid. Cross roll instability in stripes: (A) Vibrated granular layer [57] and (a) Rayleigh-Bénard convection [59]. Skew varicose instability in stripes: (B) Granular layer [57] and (b) Rayleigh-Bénard convection [60]. Spiral defect chaos in: (C) vibrated granular layer [61] and (c) Rayleigh-Bénard convection [60].

noisy transient disordered waves with a characteristic length scale. The intensity and coherence of these modes increases as the transition from conduction to convection is approached [69]. This behavior is well described by the fluctuating hydrodynamic theory of Swift and Hohenberg [70]. Remarkably, the same noisy incoherent modes are observed just below the transition from a flat vertically oscillating granular layer to a square pattern [71]. The Swift-Hohenberg continuum theory describes the observations for the granular system very well, even though the noise is not thermal noise, which is many orders of magnitude too small; apparently the noise arises from the fluctuations due to the small number of particles [71].

Fired by the promise of quantitative predictive power and encouraged by the qualitative similarity of granular flows to fluid flows, researchers have proposed various continuum descriptions for rapid granular flows [72-78]. This section focuses on one such description [75] and compares results from it to MD simulations and a granular flow experiment.

Jenkins and Richmann derived a set of inelastic continuum equations in a manner similar to the derivation of the Navier-Stokes equations [75]. This approach begins with the single-particle distribution function $f^{(1)}(\mathbf{r}, t)$, which gives the probability of finding a particle at a position $\mathbf{r}$ with a velocity $\mathbf{v}$ at a given time t. Integrating $f^{(1)}$ over all possible velocities gives the local number density, $n(\mathbf{r}, t)$. The ensemble-averaged value of any particle property ψ is determined by

$$\langle\psi\rangle = \frac{1}{n}\int \psi(\mathbf{v}) f^{(1)}(\mathbf{v}, \mathbf{r}, t) \mathrm{d}\mathbf{v}. \tag{5.1}$$

The Boltzmann equation describes how $f^{(1)}$ changes in time. Particles can move in and out of volume elements due to streaming motion; particle velocities can change in response to external forces $\mathbf{F}$; or particles can be scattered out of elements by collisions. The time rate of change for $f^{(1)}$ is given by

$$\frac{\partial f^{(1)}}{\partial t} + \mathbf{v}\cdot\nabla_{\mathbf{r}} f^{(1)} + \mathbf{F}\cdot\nabla_{\mathbf{v}} f^{(1)} = \Theta(f^{(1)}), \tag{5.2}$$

where $\Theta(f^{(1)})$ is the collision operator. Collisions are considered to be binary, frictionless, and inelastic with a constant coefficient of restitution e_0 [75]. Integrating eq. (5.2) yields the balance law for the number density,

$$\frac{\partial n}{\partial t} + \nabla\cdot(n\mathbf{u}) = 0, \tag{5.3}$$

where $\mathbf{u}(\mathbf{r}, t) = (1/n)\int \mathbf{v} f^{(1)}(\mathbf{r}, \mathbf{v}, t)\mathrm{d}\mathbf{v}$ is the local average velocity. Multiplying by the velocity and then integrating gives the balance law for momentum,

$$n\left(\frac{\partial \mathbf{u}}{\partial t} + \mathbf{u}\cdot\nabla\mathbf{u}\right) = \nabla\cdot\underline{\mathbf{P}} - ng\hat{\mathbf{z}}\,, \tag{5.4}$$

Finally, multiplying by $\mathbf{v}^2$ and integrating gives the balance law for the energy, where the *granular temperature* T is proportional to the average kinetic energy of the random motion of particles,

$$T = \frac{1}{3}(\langle\mathbf{v}^2\rangle - \langle\mathbf{v}\rangle^2), \tag{5.5}$$

$$\frac{3}{2}n\left(\frac{\partial T}{\partial t} + \mathbf{u}\cdot\nabla T\right) = -\nabla\cdot\mathbf{q} + \underline{\mathbf{P}} : \underline{\mathbf{E}} - \gamma. \tag{5.6}$$

The granular temperature T is many orders of magnitude greater than the Boltzmann temperature: thermal fluctuations are negligible ($mg\sigma \gg k_{\mathrm{B}}T_{\mathrm{B}}$).

A series of approximations is required in order to derive the form of the pressure tensor $\underline{\mathbf{P}}$, the velocity gradient tensor $\underline{\mathbf{E}}$, and the heat flux $\mathbf{q}$. One assumes that $f^{(1)}$ is nearly Gaussian, that spatial derivatives of n, $\mathbf{u}$, and T are small, and that $(1 - e_0)$ is small. With these assumptions, the components of the velocity gradient tensor $\underline{\mathbf{E}}$ are given by: $E_{ij} = (1/2)(\partial_j u_i + \partial_i u_j)$. The components of the stress tensor $\underline{\mathbf{P}}$ are given by the constitutive relation

$$P_{ij} = \left[-p + \left(\lambda - \frac{2}{3}\mu\right)E_{kk}\right]\delta_{ij} + 2\mu E_{ij}, \tag{5.7}$$

and the heat flux is given by Fourier's law

$$\mathbf{q} = -\kappa \nabla T. \tag{5.8}$$

The transport coefficients are fully determined and are the same as for a dense gas of hard spheres. The bulk viscosity is given by

$$\lambda = \frac{8}{3\sqrt{\pi}} n\sigma T^{1/2} G(\nu), \tag{5.9}$$

the shear viscosity by

$$\mu = \frac{\sqrt{\pi}}{6} n\sigma T^{1/2} \left[\frac{5}{16} \frac{1}{G(\nu)} + 1 + \frac{4}{5}\left(1 + \frac{12}{\pi}\right) G(\nu) \right], \tag{5.10}$$

and the thermal conductivity by

$$\kappa = \frac{15\sqrt{\pi}}{16} n\sigma T^{1/2} \left[\frac{5}{24} \frac{1}{G(\nu)} + 1 + \frac{6}{5}\left(1 + \frac{32}{9\pi}\right) G(\nu) \right], \tag{5.11}$$

where

$$G(\nu) = \nu g_0(\nu), \tag{5.12}$$

and the radial distribution function at contact, g_0, is [79]

$$g_0(\nu) = \left[1 - \left(\frac{\nu}{\nu_{\max}} \right)^{(4/3)\nu_{\max}} \right]^{-1}, \tag{5.13}$$

where ν is the volume fraction of the flow and $\nu_{\max} = 0.65$ is the 3-dimensional random close-packed volume fraction.

The only difference between these equations and those for an elastic gas is γ in eq. (5.6), which accounts for the temperature loss due to inelastic collisions:

$$\gamma = \frac{12}{\sqrt{\pi}} (1 - e_0^2) \frac{nT^{3/2}}{\sigma} G(\nu). \tag{5.14}$$

The system is closed by an equation of state, proposed by Goldshtein *et al.* in [79]:

$$p = nT\left[1 + 2(1 + e_0)G(\nu)\right]. \tag{5.15}$$

Direct experimental verification of the inelastic continuum equations has been slow in coming due to the complexity of solving the equations and also due to difficulties in finding an appropriate experimental system [7]. The presence of strong gradients in granular materials [45, 11] adds additional difficulty to solving continuum equations. For instance, simulations of the vertically vibrated layer find that the temperature varies by three orders of magnitude throughout the cycle [80]. Thus, unlike most Navier-Stokes simulations, the transport coefficients (λ, μ and κ) cannot be treated as constants, but must be recomputed at each grid point at every time step. Additionally, a complete set of boundary conditions for granular flows is still not established and this remains an active area of research [81-86]. Without the correct boundary conditions, numerical solutions can be unstable and are not guaranteed to converge to a correct solution in the bulk. For a good discussion on the difficulties in determining the correct boundary conditions, see Goldhirsch's review paper [11].

In a 1990 review paper [7], Campbell made a resounding call for granular flow experiments to make quantitative tests of the inelastic continuum approach. The application of new technologies such as particle tracking in two and three dimensions is now making these measurements feasible, but granular flow experiments still present technical challenges. Plugs develop in pipe flow [87,88], wall effects dominate in quasi–two-dimensional experiments [89], and detailed bulk flow measurements are difficult to make in fully three-dimensional experiments [90].

The distinguishing feature of granular flows is that inelastic collisions dissipate energy. Without an external source of energy, the granular temperature decays to zero with all particles coming to rest. Experimentally, energy can only be put into the flow through the boundaries. Shock waves serve as a mechanism to deliver energy from the boundary to the bulk of the flow. Studying the balance between the transfer of energy by shock waves and the energy dissipation through inelastic collisions is important in understanding granular flows [12]. In the next section we present two studies of shock waves as a test for the inelastic continuum equations.

6. – Shock waves in granular materials

The sound speed (the speed a pressure wave travels) in a granular medium is typically much smaller than the streaming velocity. Hence shocks are common in granular media. For example, imagine pouring sand out of a bucket. Gravity accelerates the flow downward, creating an average velocity U that reaches 100 cm/s after the sand has fallen by only 5 cm. In contrast, the sound speed c in the granular gas becomes small, typically 10 cm/s, as multiple particle collisions cool the gas, reducing the random velocities of the particles. The simple act of turning over a bucket full of sand can easily generate a supersonic flow with Mach number $= c/U = 10$. (A flow with Mach number greater than unity is supersonic.)

The sound speed in a gas can be determined from thermodynamic relations

$$c = \sqrt{\left(\frac{\partial P}{\partial \rho}\right)_S} = \sqrt{\frac{c_p}{c_v}\left(\frac{\partial P}{\partial \rho}\right)_T}, \tag{6.1}$$

where c is the speed of sound, ρ is the local density, S is the entropy, c_p is the specific heat at constant pressure, and c_v is the specific heat at constant volume. For a dense inelastic gas, c is given by [91]

$$c = \sqrt{T\chi\left(1 + \frac{2}{3}\chi + \frac{\nu}{\chi}\frac{\partial\chi}{\partial\nu}\right)}, \tag{6.2}$$

where $\chi = 1 + 2(1+e)G(\nu)$.

A shock forms when a supersonic flow encounters an obstacle. The shock separates two regions of the flow, an undisturbed region that is unaware of the obstacle, and a compressed region which has adjusted to fit the boundary conditions at the obstacle. The compressed region has a higher temperature and smaller velocity than the undisturbed region. In an ideal fluid with no viscosity, heat conduction, or dissipation, a shock is a zero-width surface of discontinuity. In a non-ideal fluid the shock has a finite width on the order of a particle's mean free path in the fluid [92].

When a fluid with velocity $U > c$ impinges perpendicularly onto an obstacle, a *normal* shock forms and propagates in the $-U$ direction. Subsection **6**˙2 will discuss the formation and propagation of normal shocks in a vertically vibrated granular layer where the flow fields and thus the shocks are highly time dependent [80]. If, instead, the fluid velocity and the obstacle are not perpendicular, an *oblique* shock forms and propagates into the flow at an angle and with a speed determined by the local flow values. Subsection **6**˙1 describes an oblique shock formed in a steady-state laboratory flow past a wedge [89].

6˙1. *Steady state flow past an obstacle.* – We now describe an experimental study of shocks in a time-independent flow, and then the experimental results will be compared to two simulations: an event-driven molecular dynamics simulation, similar to those discussed in sect. **3**, and a two-dimensional, finite-element solution of the inelastic continuum equations presented in sect. **5**.

In the experiment, stainless steel spheres (particle diameter $\sigma = 1.2\,\mathrm{mm}$) fell under gravity past a wedge sandwiched between two glass plates separated by 1.6σ. The particles were initially distributed uniformly on a conveyor belt. As the conveyor turned, particles fell off into a hopper that guided the particles into the cell formed of the closely spaced plates; the wedge was located a distance of 42σ below the top of the cell. The positions and velocities of the particles were determined from high speed images of the falling particles, and data from many thousands of particles were averaged to obtain the time-independent velocity, volume fraction, and temperature fields. The average free stream speed of sound determined from the measurements for flow incident on the wedge was $0.09\,\mathrm{m/s}$. The flow entered the top of the cell with a Mach number of 7 and accelerated under gravity to a Mach number of 12 at the tip of the wedge.

The horizontal-velocity field measured in the experiment is shown in fig. 12(a). A shock separates the undisturbed region, where the horizontal velocity is nearly zero, from the compressed region, whose stream lines follow the flow around the obstacle. Because of gravity and inelasticity, the shock does not extend out at a constant angle but curves towards the wedge.

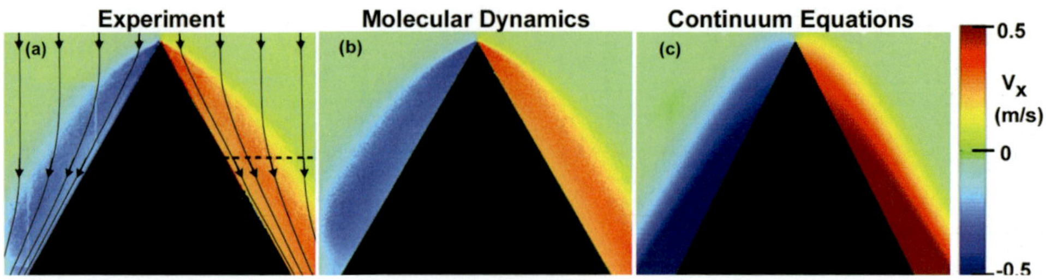

Fig. 12. – Horizontal component of the velocity field of a granular flow incident downward on a wedge, determined by three methods: (a) experiment, (b) MD simulations, and (c) integration of inelastic continuum equations. Each picture shows a region 130σ by 104σ. The solid lines with arrows denote streamlines. Quantitative comparisons along the dashed line in (a) are shown in figs. 14 and 15. From [89].

At the bottom of the wedge the compressed gas expands in a fan-like structure as the volume available to the flow increases (fig. 13). In an expansion fan the density and temperature decrease and the Mach number increases. The expansion fan is a smooth transition radiating from the bottom corner of the wedge.

The flow was computed numerically in a three-dimensional MD simulation (fig. 12(b)) and in a two-dimensional finite-difference simulation of the inelastic continuum equations (fig. 12(c)). The two simulations yield results for the horizontal component of velocity in qualitative accord with experiment: a shock forms at the tip of the obstacle, and behind the shock the flow is compressed, has a higher temperature, and lower mean velocity. Quantitative comparisons among the methods are plotted for values of the fields along

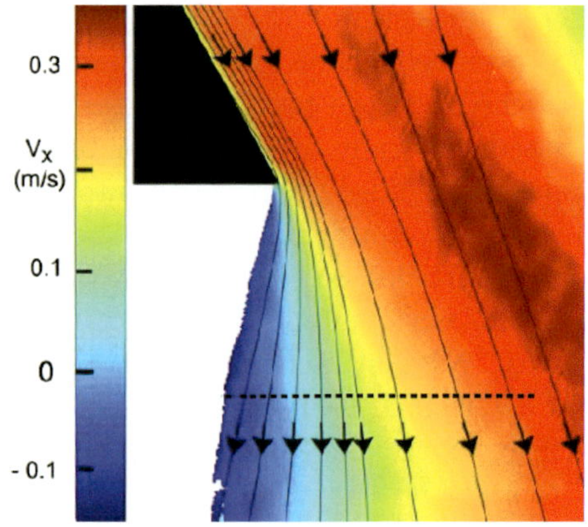

Fig. 13. – The horizontal velocity field measured for the expansion fan that formed when the supersonic granular flow reached the bottom of the wedge. The solid lines indicate selected streamlines. The total height of the region shown is 55σ. The white region below the wedge had too few particles for the velocity to be determined. From [89].

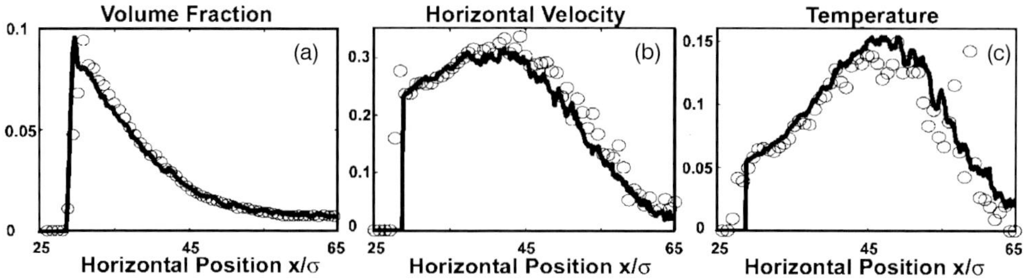

Fig. 14. – Shock profiles for granular flow past a wedge measured in an experiment (circles) are compared with results from molecular dynamics (solid lines): (a) volume fraction, (b) horizontal component of the velocity, and (c) temperature. The profiles are taken along the dashed line in fig. 12. From [89].

the dashed line shown in fig. 12(a).

Three parameters were adjusted in the MD simulation to achieve the agreement with the experiment shown in fig. 14. The same coefficient of restitution $e_0 = 0.97$ and friction coefficient $\mu = 0.15$ were used to model interparticle and particle-wall collisions. The initial conditions of the experiment were modeled by placing particles into the top of the cell at a constant rate. Incoming particles were placed randomly at the top of the cell with a mean downward velocity measured from the experiment, and fluctuations were chosen from a Gaussian distribution determined by the measured temperature. An additional parameter α, defined as the ratio of temperature perpendicular to the wall to that parallel to the wall, was set to 0.8. These parameters, which were not measured in the experiment, were adjusted to provide agreement in the full flow fields, including the free-stream velocity.

Results from the MD simulation are compared with experiment in fig. 14 for the volume fraction, horizontal velocity component, and temperature. The agreement is quite good with a root mean square difference between experiment and simulation of less than 2% for the volume fraction and velocity fields and 10% for the temperature field.

The simple geometry and steady-state behavior of the experiment provided a good system for testing the inelastic continuum equations. However, a full three-dimensional simulation of the experiment was found to be time prohibitive. Instead, the equations were solved on a two-dimensional grid; consequently the simulation could not capture the interaction of particles with the confining glass side walls. Frictional collisions with the side walls strongly affected the flow in the experiment and in the fully three-dimensional MD simulations. The average downward acceleration of a single-particle falling between the two glass plates in the experiment was $8.9\,\mathrm{m/s^2}$, while the same particle falling outside the cell accelerated with the expected $9.8\,\mathrm{m/s^2}$.

The continuum equations were numerically solved by a second-order accurate, finite-difference method. The only fit parameter in the equations was the coefficient of restitution, which was set to the same value of $e_0(0.97)$ used in the MD simulation. Boundary conditions at the inlet were determined by the experiment and at the outlet were free.

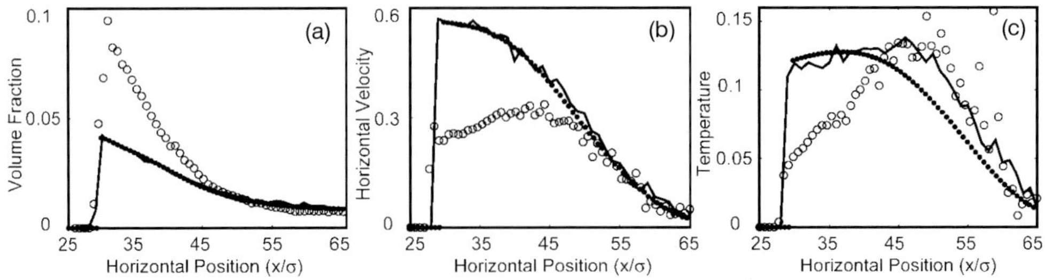

Fig. 15. – Comparison of shock profiles for granular flow past a wedge obtained from molecular dynamics (solid lines) and inelastic continuum equations (dotted line), assuming no friction. (a) Volume fraction, (b) horizontal velocity profile, and (c) temperature along the dashed line in fig. 12(a). Experimental measurements (open circles) show similar qualitative behavior but disagree quantitatively. The difference between the simulations and the experiment is due to wall friction. From [89].

Slip velocity boundary conditions were used along the wedge boundary. The heat flux at the wedge was taken to be proportional to the local ν and $T^{3/2}$ [85]. Euler time stepping was used to increment the simulation until the flow reached a steady state where the horizontally averaged mass flux was constant to 0.01.

Experiment and continuum simulation showed similar behavior, but the shape of the curves differed and the magnitudes of the fields disagreed by as much as a factor of two (fig. 15). This disagreement was attributed to the frictional drag of the confining side walls in the experiment. A three-dimensional simulation of the inelastic continuum equations with viscous boundary conditions along the side walls should agree better with the experiment [12].

Molecular dynamics simulations were done with wall drag neglected for comparison with the two-dimensional simulation of the continuum equations. The two simulations agreed remarkably well in all regions of the flow except within 5σ of the wedge tip. Near the tip of the wedge, the two simulations disagreed due to different boundary conditions. The agreement between the two simulations in the bulk of the flow confirms the applicability of the continuum description for granular flows. The disagreement with the experiment emphasizes once again the importance of including friction in a continuum description, both in the equations and in the boundary conditions.

6˙2. *Shock waves in a vertically oscillating layer.* – Section **2** described the patterns that form when a layer of granular materials is vertically oscillated. Shocks play an important role in this system [80]. Each time the layer collides with the plate a shock forms and propagates through the layer, transmitting energy upward through the layer to the surface. The striking similarity of the granular patterns to their counterparts in continuum systems strongly suggests that a continuum description of granular flows could prove useful. If this is true, the continuum description must also capture the shock dynamics in the bulk of the layer.

Here we compare inelastic, frictionless, fully three-dimensional MD and continuum simulations. In order to focus on the formation and propagation of the shock wave, pattern formation is intentionally suppressed by considering a container smaller than one wavelength of the pattern in either horizontal direction. We focus on one point in the phase space shown in fig. 3: $\Gamma = 3$ and $f^* = 0.42$ for a layer with depth $h = 9\sigma$. For this set of parameter values, a vibrated layer in a larger cell would have a $f/2$ stripe pattern.

In both simulations periodic boundary conditions were used in the two horizontal directions and impermeable boundary conditions, $u_z = 0$, were applied at the plate. The additional boundary conditions required for the continuum simulations were taken from the MD simulation. In the MD simulation, the vertical derivatives at the plate were negligible throughout most of the cycle. For simplicity, the continuum simulation assumed $\partial u_x/\partial z = 0$, $\partial u_y/\partial z = 0$, and $\partial T/\partial z = 0$ at all times in the cycle.

The evolution of the shock wave throughout a plate cycle is shown in figs. 16 and 17. The dynamics of the cycle occurs in the time interval between $ft = 0$ and one cycle later, $ft = 1$.

At $ft = 0$ the container is at its minimum height. The layer, having been thrown off the plate in the previous cycle, now falls towards the plate. Inelasticity has dissipated most of the energy so that the layer's temperature is nearly zero. The Mach number of the layer with respect to the plate is much greater than one. The MD and continuum simulation show similar behavior in the v and T fields.

At $ft = 15$ the layer begins to collide with the plate. A shock wave forms, separating the region near the plate where ν and T increases from the undisturbed region still falling towards the plate.

At $ft = 0.22$ the shock wave is moving through the layer. The compressed region continues to grow. Collisions between particles in this high density region cause the layer to cool behind the shock, creating a lower temperature near the plate.

At $ft = 0.40$ the shock has propagated through the layer and into the very dilute region above the layer. At this time, the plate is approaching its maximum height and the layer begins to leave the plate as the downward plate acceleration exceeds g. The layer continues to cool behind the shock, setting the stage for the next oscillation.

The MD and continuum simulations show good agreement throughout the cycle, despite the presence of large spatial gradients and a strong time dependence. In the dilute regimes above and below the layer, numerical solutions of the inelastic continuum equations are unstable unless artificial dissipation is added [80], following the example from numerical solutions of Knudsen gases [93]. The effect of the extra dissipation is most pronounced in the falling layer and accounts for the disagreement between MD and continuum at the top of the cell.

Inelastic collisions between particles are the distinguishing characteristics of a granular gas, but few studies have examined how granular flow properties depend on the restitution coefficient. Simulations for the oscillating layer were modified to study the effect of varying e_0 on the propagation of the shock. The initial conditions for this numerical experiment were taken for a layer with $e_0 = 0.99$, using the same parameters as in the above discussion. At $ft = 0.33$, when the center-of-mass layer was near its maximum

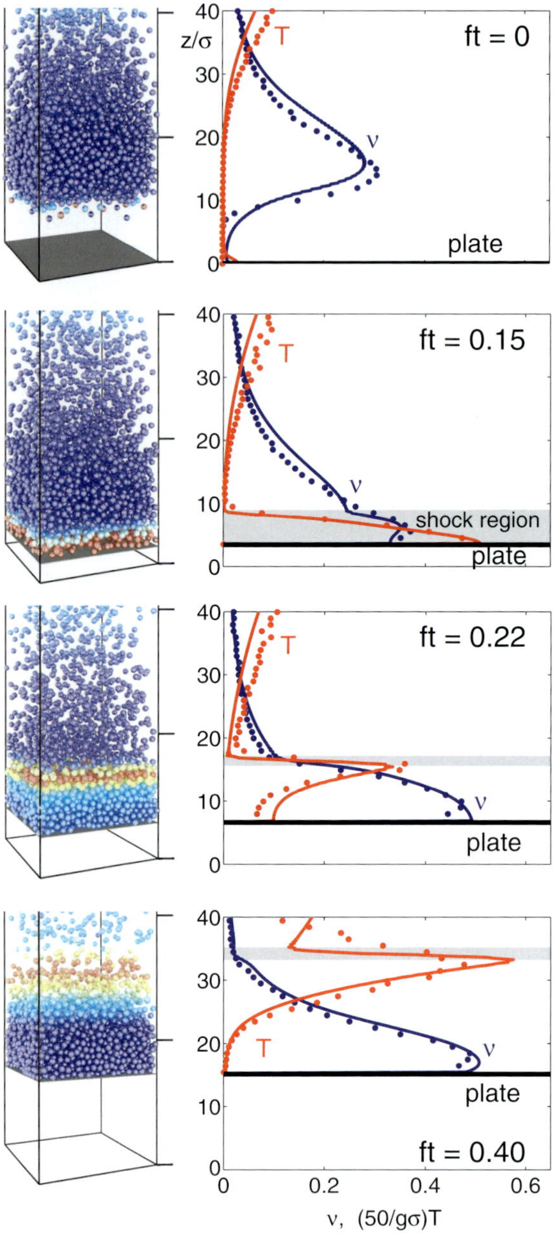

Fig. 16. – Dimensionless temperature $T/g\sigma$ and volume fraction ν as functions of the dimensionless height z/σ at four times ft in the oscillation cycle. For each time, a snap shot from the MD simulation is shown in the left column, with individual particles color coded according to temperature: high T in red, low T in blue, and the bottom plate of the container shaded solid gray. The right column shows horizontally averaged ν (blue) and $T/g\sigma$ (red) for the same four times. The plate is shown as a horizontal black solid line, results from MD simulation are shown as dots, and continuum results are solid lines. From [80].

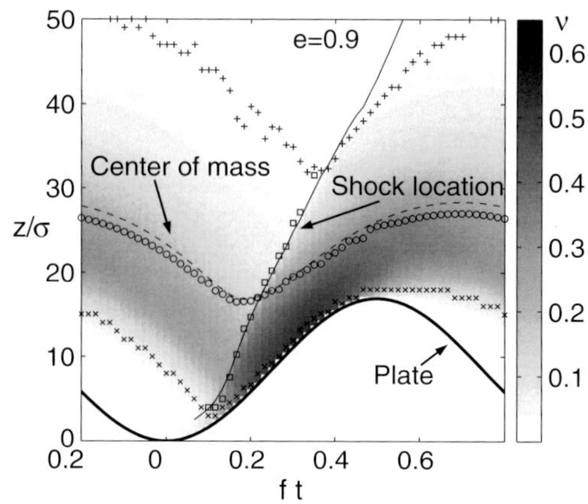

Fig. 17. – Location of the shock (solid line for continuum, squares for MD) and the center of mass of the layer (dashed line for continuum, circles for MD) as a function of time ft during one cycle of the plate (thick solid line) for particles with $e_0 = 0.90$. The plot is shaded according to the volume fraction from the continuum simulation, so that high volume fraction is dark and low volume fraction is light. The "top" and the "bottom" of the layer from MD (when the volume fraction drops to less than 4% of that for random close packed particles) are shown as +'s and ×'s respectively. The material below the shock is compressed as compared to the region above the shock, as can be seen from the shading. From [80].

height above the plate, the coefficient of restitution was suddenly changed, which changed the subsequent evolution of the shock.

As before, when the layer hits the plate it compresses and forms a shock that propagates through the layer. The smaller the value of e_0, the faster the layer cools and compacts; for small e_0, the layer remains very compact throughout the cycle and leaves the plate almost as a solid body. For higher values of e_0, the layer dilates quickly after each collision with the plate. The maximum height of the center of mass in a cycle increases with increasing e_0.

The speed of the shock (eq. (6.2)) depends on both the temperature of the flow and on its density. Since the density and temperature of the flow change throughout the cycle, so does the shock speed as the shock propagates through the layer. The behavior of the average speed of the shock as a function of inelasticity is shown in fig. 18. For small values of e_0 the shock speed asymptotes to a fixed value of $47\sqrt{g\sigma}$. The shock speed monotonically increases with increasing e_0. The special case of elastic particles appears to match with the limit of $e \to 1$, suggesting that there is no qualitative difference in shock propagation through elastic and inelastic gases.

For both problems we have considered —granular flow past a wedge and a vertically oscillating granular layer— numerical solutions of the inelastic continuum equations of Jenkins and Richman agree well with MD simulations for frictionless particles. The continuum equations were derived for a weakly dissipative, low density, frictionless granular

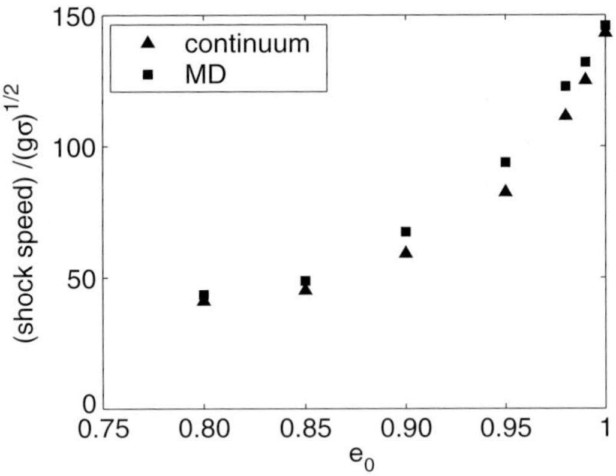

Fig. 18. – Average dimensionless shock speed, $v_{\mathrm{shock}}/\sqrt{g\sigma}$, in the reference frame of the plate. v_{shock} is calculated as the average speed of the shock from when the shock is formed until it leaves the layer. From [80].

gas, assuming small gradients in the flow fields in both space and time. Nevertheless, the equations capture the evolution of a shock through a dense, inelastic oscillating layer, and qualitatively capture the properties of a shock formed in flow past a wedge. With more research on boundary conditions and the incorporation of friction, these continuum equations show great promise.

7. – Discussion

More than one thousand papers have been published on granular materials since de Gennes brought the subject to the attention of physicists, and Bak's work (1987) on self-organized criticality stimulated interest in sand piles [94]. However, much remains to be done to achieve a level of understanding of granular media comparable to that for fluids and crystalline solids. Experiments and simulations have investigated a wide range of problems including the angle of repose [95] and internal structure of sand piles [96,97], shear forces in Couette-Taylor flows [98,90], convection due to temperature gradients [99-101] and due to buoyancy [102], flows in a rotating drum [103, 104], and chute flows [105-107, 31]. Much of the research has concerned granular media as a solid where particles are in continuous contact, while this chapter has concerned rapid granular flows (the "collisional regime") where inertial effects are important and force chains do not play a major role. We have further limited the considerations to particles that interact only on contact.

Understanding flows of grains that interact only on contact would seem at first to involve a straightforward application of Newton's laws. However, the energy loss in collisions complicates the application of standard statistical methods. We have focused on two systems that hint at the rich variety of phenomena exhibited by granular media in the collisional regime: a vertically vibrated granular layer and supersonic granular flow past an obstacle. These two problems were chosen because they are amenable to direct comparison of experiment, molecular dynamics simulations, and continuum theory.

A vertically vibrated granular layer spontaneously forms spatially extended patterns. Experiments and MD simulations reveal that the collective motion of grains arises due to dissipative collisions between particles, and does not require mediation by an interstitial gas or side walls. For square granular patterns, an approach intermediate between molecular dynamics and continuum models has been found to describe the dynamics of the lattice pattern: a collection of particles that form a peak (an oscillon) is like an atom in a crystalline lattice. The modes of the granular lattice obey the dispersion relation for a two-dimensional lattice, and the granular lattice even forms defects and melts in the same way as a two-dimensional crystal of atoms (sect. **2**).

The spatial patterns formed by oscillating granular layers exhibit marked similarities to those observed in continuum nonequilibrium systems such as convecting fluids and oscillating liquids (sect. **5**). Further, the cross-roll and skew-varicose instabilities observed in thermal convection in a fluid and interpreted in terms of the hydrodynamic equations (more specifically, the Boussinesq equations) have also been observed in oscillated granular layers. Various amplitude equation models have been found to describe granular patterns and their instabilities. Even the subtle effects of noise on the transition from conduction to convection in fluids have been found also in oscillating granular layers near the onset of the transition from a flat layer to a square pattern.

The striking similarities of granular patterns to those found in nonequilibrium continuum systems and in experiments on granular flows under shear and in rotating drums suggest that granular gases may be describable by continuum theory. Inspired by these observations, researchers have proposed many continuum descriptions. The descriptions differ in the particle properties included in the collision model; for instance, collision models can be frictionless [108, 75] or can account for friction between particles [109, 110]. Equations of motion obtained by Goldshtein and Shapiro include, in addition to the terms in the Navier-Stokes equation, a term accounting for heat transport by density gradients [76]. A Model presented by Bocquet *et al.* [90] includes corrections to the viscosity due to velocity correlations. None of these models has been definitively established.

We have compared predictions of continuum equations derived by Jenkins and Richman with experiments on shocks in vibrating layers and flow past an obstacle. For both geometries, numerical solutions of the inelastic continuum equations agree well with results from MD simulations of smooth (frictionless) inelastic spheres. However, comparisons of the continuum equations with experiment and with MD simulations for particles with friction have demonstrated the crucial role of friction in granular flows. For continuum equations to achieve quantitative predictive power, the effects of friction between particles and between particles and boundaries must be included.

Derivations of granular hydrodynamic equations have thus far assumed weak dissipation and small particle volume fractions. Future work should extend theory to higher densities and larger dissipation. Because inelastic collisions dissipate temperature, granular flows frequently coexist with solid-like sand piles. A major challenge is the development of theory that bridges the gap from the collisional regime to the quasi-static regime where particles are always in contact.

As continuum equations of motion become better established, it will become possible to exploit the power of the continuum description. Continuum models for larger and denser systems may reveal new phenomena. Continuum simulations are better suited to time-dependent flows than MD simulations. Linear and nonlinear stability analyses could provide insight into bifurcations, just as a century of stability analyses of the Navier-Stokes equations has given insights into diverse fluid flow phenomena. Stability analyses have been conducted for simplified continuum models (see, *e.g.*, [111]), but thus far no stability analysis has been conducted for a realistic set of granular hydrodynamic equations. Further, given a set of granular equations like the Navier-Stokes equations, it should be possible to derive amplitude equations, which can yield a better understanding of instabilities in granular flows.

The experiments and theory presented in this chapter involve spheres of uniform size, while industrial applications usually involve a wide range of particle sizes and shapes. Experiments and simulations on flows with two particle sizes show an additional richness to granular flow phenomena such as size segregation [23, 112-115], nonequipartition of energy [116, 117], and increased normal stress [117]. Much theoretical and experimental work is needed on systems of particles with a range of sizes and shapes.

Air friction is usually neglected in simulations and in the interpretation of experiments, whether or not the experiments are conducted in vacuum. In contrast, in granular systems in industry, air friction and buoyancy are often important, although the air effects can be negligible for large particles (say greater than 1 mm). The interaction of the interstitial fluid and sand leads to the development of sand dunes [118] and sand ripples [119, 120] and the formation of heaps in vibrated layers [121]. The inclusion of interstitial flow in continuum theories for granular materials is another challenge for future research on granular materials.

In conclusion, much remains to be done to establish a fundamental understanding granular flows. Future studies should seek more examples of granular flows amenable to experiment, molecular dynamics simulations, and continuum theory. A concerted attack from three approaches should lead to a better understanding of outstanding issues concerning boundary conditions, the incorporation of particle friction and velocity correlations into theory, and the extension of theory to higher particle volume fractions and higher dissipation.

* * *

In the past two decades granular media have been studied by many scientists. For convenience, our examples have been taken largely from research in the Center for Nonlinear Dynamics at The University of Texas at Austin (see `http://chaos.utexas.edu`). We thank particularly CHRIS BIZON, JONATHAN BOUGIE, FRANCISCO MELO, DANIEL GOLDMAN, W. D. MCCORMICK, SUNG JOON MOON, MARK SHATTUCK, JACK SWIFT, and PAUL UMBANHOWAR. This research was supported by the Engineering Research Program of the Office of Basic Energy Sciences of the U.S. Department of Energy and the Texas Advanced Research Program.

REFERENCES

[1] ENNIS B. J., GREEN J. and DAVIES R., *Chem. Engin. Prog.*, **90** (1994) 32.

[2] FARADAY M., *Philos. Trans. R. Soc. London*, **121** (1831) 299.

[3] EVESQUE P. and RAJCHENBACH J., *Phys. Rev. Lett.*, **62** (1989) 44.

[4] FAUVE S., DOUADY S. and LAROCHE C., *Physica Scripta*, **T29** (1989) 250.

[5] DOUADY S., FAUVE S. and LAROCHE C., *Europhys. Lett.*, **8** (1989) 621.

[6] DE GENNES P. G., *Rev. Mod. Phys.*, **71** (1999) S374.

[7] CAMPBELL C. S., *Annu. Rev. Fluid Mech.*, **2** (1990) 57.

[8] JAEGER H. M. and NAGEL S. R., *Science*, **255** (1992) 1523.

[9] JAEGER H. M., NAGEL S. R. and BEHRINGER R. P., *Physics Today*, **49**, no. 4 (1996) 32.

[10] RAJCHENBACH J., *Adv. Phys.*, **49** (2000) 229.

[11] GOLDHIRSCH I., *Annu. Rev. Fluid. Mech.*, **35** (2003) 267.

[12] GOLDSHTEIN A., ALEXEEV A. and SHAPIRO M., in *Granular Gas Dynamics*, edited by T. POSCHEL and N. BRILLIANTOV (Springer, Berlin) 2003, pp. 188–225.

[13] XU H., LOUGE M. and REEVES A., *Continuum Mech. Thermodyn.*, **15** (2003) 321.

[14] BAGNOLD R. A., *Proc. Royal Soc. London*, **49** (1954) 225.

[15] BIDEAU D. and HANSEN A. (Editors), *Disorder and Granular Media* (Elsevier, Amsterdam) 1993.

[16] DURAN J., *Sands, Powders, and Grains: An Introduction to the Physics of Granular Materials* (Springer, Berlin) 2000.

[17] POSCHEL T., *Granular Cases, Lecture Notes in Physics* (Cambridge University Press, Cambridge) 2003.

[18] POSCHEL T. and BRILLIANTOV N. (Editors), *Granular Glass Dynamics* (Springer, Berlin) 2003.

[19] GOLDSMITH W., *Impact* (Edward Arnold Ltd., London) 1960.

[20] LORENZ A., TUOZZOLO C. and LOUGE M., *Exp. Mech.*, **37** (1997) 292.

[21] PICA CIAMARRA M., LARA A., LEE A., GOLDMAN D. and SWINNEY H., in this volume, p. 469.

[22] PICA CIAMARRA M., LARA A., LEE A., GOLDMAN D., VISHIK I. and SWINNEY H., *Phys. Rev. Lett.*, **92** (2004) 194301.

[23] MOON S. J., GOLDMAN D. I., SWIFT J. B. and SWINNEY H. L., *Phys. Rev. E*, **69** (2004) 03131.

[24] UMBANHOWAR P. and SWINNEY H. L., *Physica A*, **288** (2000) 344.

[25] BIZON C., SHATTUCK M. D., SWIFT J. B., MCCORMICK W. D. and SWINNEY H. L., *Phys. Rev. Lett.*, **80** (1998) 57.

[26] MELO F., UMBANHOWAR P. B. and SWINNEY H. L., *Phys. Rev. Lett.*, **75** (1995) 3838.

[27] GOLDMAN D. I., MOON S. J., SWIFT J. B. and SWINNEY H. L., *Phys. Rev. Lett.*, **90** (2003) 104302.

[28] UMBANHOWAR P., MELO F. and SWINNEY H. L., *Nature*, **382** (1996) 793.

[29] MOON S. J., SHATTUCK M. D., BIZON C., GOLDMAN D. I., SWIFT J. B. and SWINNEY H. L., *Phys. Rev. E*, **65** (2002) 011301.

[30] MEHTA A. and LUCK J. M., *Phys. Rev. Lett.*, **65** (1990) 393.

[31] CAMPBELL C. S. and BRENNEN C. E., *Trans. ASME E: J. Appl. Mech.*, **52** (1985) 172.

[32] RAPAPORT D., *The Art of Molecular Dynamics Simulation* (Cambridge University Press, Cambridge) 1995.

[33] WALTON O. R., in *Particulate Two-Phase Flow*, edited by M. C. ROCO (Butterworth-Heinemann, Boston) 1993, pp. 884–911.

[34] MCNAMARA S. and YOUNG W. R., *Phys. Fluids A*, **4** (1992) 496.

[35] LUDING S., CLÉMENT E., RAJCHENBACH J. and DURAN J., *Europhys. Lett.*, **36** (1996) 247.
[36] GOLDMAN D., SHATTUCK M. D., BIZON C., MCCORMICK W. D., SWIFT J. B. and SWINNEY H. L., *Phys. Rev. E*, **57** (1998) 4831.
[37] ARGENTINA M., CLERC M. and SOTO R., *Phys. Rev. Lett.*, **89** (2002) 044301.
[38] NIE X., BEN-NAIM E. and CHEN S., *Phys. Rev. Lett.*, **89** (2002) 204301.
[39] KHAIN E. and MEERSON B., *Phys. Rev. E*, **67** (2003) 021306.
[40] UMBANHOWAR P. B., MELO F. and SWINNEY H. L., *Physica A*, **249** (1998) 1.
[41] UMBANHOWAR P. B., *Wave Patterns in Vibrated Granular Layers*, Ph.D. thesis, University of Texas at Austin (1996).
[42] BEDANOV V. M., GADIYAK G. V. and LOZOVIK Y. F., *Phys. Lett. A*, **109** (1985) 289.
[43] ZHENG X. H. and EARNSHAW J. C., *Europhys. Lett.*, **41** (1998) 635.
[44] ZHOU T. and KADANOFF L. P., *Phys. Rev. E*, **54** (1996) 623.
[45] TAN M. L. and GOLDHIRSCH I., *Phys. Rev. Lett.*, **81** (1998) 3022.
[46] KADANOFF L. P., *Rev. Mod. Phys.*, **71** (1999) 435.
[47] OLAFSEN J. and URBACH J. S., *Phys. Rev. E*, **60** (1999) 4268.
[48] KUDROLLI A. and HENRY J., *Phys. Rev. E*, **62** (2000) 1489.
[49] VEJE C. T., HOWELL D. W. and BEHRINGER R. P., *Phys. Rev. E*, **59** (1999) 739.
[50] ROUYER F. and MENON N., *Phys. Rev. Lett.*, **85** (2000) 3676.
[51] MOON S. J., SHATTUCK M. D. and SWIFT J. B., *Phys. Rev. E*, **64** (2001) 31303.
[52] KUDROLLI A. and GOLLUB J. P., *Physica D*, **97** (1996) 133.
[53] OUYANG Q. and SWINNEY H. L., *Nature*, **352** (1991) 610.
[54] BEHRINGER R., *Rev. Mod. Phys.*, **57** (1985) 657.
[55] PLAPP B., private communication.
[56] DUBOIS-VIOLETTE E., DUNRAND G., GUYON E., MANNEVILLE P. and PIERANSKI P., in *Liquid Crystals*, edited by L. LIEBERT (Academic Press, New York) 1978, p. 147.
[57] DE BRUYN J. R., BIZON C., SHATTUCK M. D., GOLDMAN D., SWIFT J. B. and SWINNEY H. L., *Phys. Rev. Lett.*, **81** (1998) 1421.
[58] CROSS M. C. and HOHENBERG P. C., *Rev. Mod. Phys.*, **65** (1993) 851.
[59] BUSSE F. H. and WHITEHEAD J. A., *J. Fluid Mech.*, **47** (1971) 305.
[60] ASSENHEIMER M. and STEINBERG V., *Phys. Rev. Lett.*, **70** (1993) 3888.
[61] DE BRUYN J. R., LEWIS B. C., SHATTUCK M. D., SWIFT J. B. and SWINNEY H. L., *Phys. Rev. E*, **63** (2001) 041305.
[62] TSIMRING L. and ARONSON I., *Phys. Rev. Lett.*, **79** (1997) 213.
[63] CERDA E., MELO F. and RICA S., *Phys. Rev. Lett.*, **79** (1998) 4570.
[64] EGGERS J. and RIECKE H., *Phys. Rev. E*, **59** (1999) 4476.
[65] SHINBROT T., *Nature*, **389** (1997) 574.
[66] SAKAGUCHI H. and BRAND H. R., *J. Phys. II*, **7** (1997) 1325.
[67] ROTHMAN D. H., *Phys. Rev. E*, **57** (1998) R1239.
[68] VENKATARAMANI S. C. and OTT E., *Phys. Rev. Lett.*, **80** (1998) 3498.
[69] OH J. and ALHERS G., *Phys. Rev. Lett.*, **91** (2003) 094501.
[70] SWIFT J. and HOHENBERG P., *Phys. Rev. A*, **15** (1977) 319.
[71] GOLDMAN D. I., SWIFT J. B. and SWINNEY H. L., *Phys. Rev. Lett.*, **92** (2004) 174302.
[72] HAFF P. K., *J. Fluid Mech.*, **134** (1983) 401.
[73] AHMADI G. and SHAHINPOOR M., *Int. J. Non-linear Mechanics*, **19** (1983) 177.
[74] LUN C. K. K., *J. Appl. Mech.*, **54** (1987) 47.
[75] JENKINS J. T. and RICHMAN M. W., *Arch. Rat. Mech. Anal.*, **87** (1985) 355.
[76] GOLDSHTEIN A. and SHAPIRO M., *J. Fluid Mech.*, **282** (1995) 75.
[77] SELA N., GOLDHIRSCH I. and NOSKOWICZ S. H., *Phys. Fluids*, **8** (1996) 2337.
[78] BREY J. J., DUFTY J. W. and SANTOS A., *J. Stat. Phys.*, **87** (1997) 1051.

[79] GOLDSHTEIN A., SHAPIRO M. and GUTFINGER C., *J. Fluid Mech.*, **327** (1996) 117.
[80] BOUGIE J., MOON S. J., SWIFT J. B. and SWINNEY H. L., *Phys. Rev. E*, **66** (2002) 1.
[81] SILBERT L., GREST G. and PLIMPTION S., *Phys. Fluids*, **14** (2002) 2637.
[82] SOTO R. and MANSOUR M., *Physica A*, **327** (2003) 88.
[83] JENKINS J. T., *Trans. ASME*, **59** (1992) 120.
[84] JENKINS J. and ASKARI E., *Chaos*, **9** (1999) 654.
[85] JENKINS J. T. and LOUGE M. Y., *Phys. Fluids*, **9** (1997) 2835.
[86] BREY J., RUIZ-MONTERO M. and MORENO F., *Phys. Rev. E*, **62** (2000) 5339.
[87] AIDER J. L., SOMMIER N., RAAFAT T. and HULIN J. P., *Phys. Rev. E*, **59** (1998) 778.
[88] PÖSCHEL T., *J. Phys. II France*, **3** (1993) 27.
[89] RERICHA E. C., BIZON C., SHATTUCK M. D. and SWINNEY H. L., *Phys. Rev. Lett.*, **88** (2002) 014302.
[90] BOCQUET L., LOSERT W., SCHALK D., LUBENSKY T. and GOLLUB J., *Phys. Rev. E*, **65** (2001) 011307.
[91] SAVAGE S. B., *J. Fluid Mech.*, **194** (1988) 457.
[92] ANDERSON J. D., *Modern Compressible Flow with Historical Perspective* (McGraw-Hill, Boston) 1990.
[93] ROACHE P., *Computational Fluid Dynamics* (Hermosa Publishers, Albuquerque) 1976.
[94] BAK P., TANG C. and WIESENFELD K., *Phys. Rev. Lett.*, **59** (1987) 381.
[95] JAEGER H. M., LIU C. and NAGEL S. R., *Phys. Rev. Lett.*, **62** (1989) 40.
[96] MEHTA A. and BARKER G. C., *Rep. Prog. Phys.*, **57** (1994) 383.
[97] VANEL L., HOWELL D., CLARK D., BEHRINGER R. and CLEMENT E., *Phys. Rev. E*, **60** (1999) R5040.
[98] HOWELL D., BEHRINGER R. P. and VEJE C., *Phys. Rev. Lett.*, **82** (1998) 5241.
[99] HAYAKAWA H., YUE S. and HONG D. C., *Phys. Rev. Lett.*, **75** (1995) 2328.
[100] RAMIREZ R., RISSO D. and CORDERO P., *Phys. Rev. Lett.*, **85** (2000) 1230.
[101] EHRICHS E. E., JAEGER H. M., KARCZMAR G. S., KNIGHT J. B., KUPERMAN V. Y. and NAGEL S. R., *Science*, **267** (1995) 1632.
[102] CORDERO P., RAMIREZ R. and RISSO D., *Physica A*, **327** (2003) 82.
[103] SHEN A., *Physics of Fluids*, **14** (2002) 462.
[104] HILL K., GIOIA G. and TOTA V., *Phys. Rev. Lett.*, **91** (2003) 064302.
[105] SANDERS B. E. and ACKERMANN N. L., *J. Eng. Mech.*, **117** (1991) 2396.
[106] NOTT P. and JACKSON R., *J. Fluid Mech.*, **241** (1992) 125.
[107] FORTERRE Y. and POULIQUEN O., *Phys. Rev. Lett.*, **86** (2001) 5886.
[108] LUN C. K. K., SAVAGE S. B., JEFFREY D. J. and CHEPURNIY N., *J. Fluid Mech.*, **140** (1983) 223.
[109] LUN C. K. K., *J. Fluid Mech.*, **233** (1991) 539.
[110] LUDING S., HUTHMANN M., MCNAMARA S. and ZIPPELIUS A., *Phys. Rev. E*, **58** (1998) 3416.
[111] BIZON C., SHATTUCK M. D., SWIFT J. B. and SWINNEY H. L., *Phys. Rev. E*, **60** (1999) 4340.
[112] HILL K., JAIN N. and OTTINO J., *Phys. Rev. E*, **64** (2001) 011302.
[113] WILLIAMS J. C., *Powder Technol.*, **15** (1976) 245.
[114] ROSATO A., STRANDBURG K. J., PRINZ F. and SWENDSEN R. H., *Phys. Rev. Lett.*, **58** (1987) 1038.
[115] METCALFE G. and SHATTUCK M., *Physica A*, **133** (1996) 709.
[116] FEITOSA K. and MENON N., *Phys. Rev. Lett.*, **88** (2002) 198301.
[117] ALAM M. and LUDING S., *J. Fluid Mech.*, **476** (2003) 69.
[118] BAGNOLD R. A., *The Physics of Blown Sand and Desert Dunes* (Methuen and Co. Ltd., London) 1954.

[119] DURAN J., *Phys. Rev. Lett.*, **84** (2000) 5126.
[120] VITTORI G. and BLONDEAUX P., *J. Fluid Mech.*, **218** (1990) 19.
[121] PAK H. K., VAN DOORN E. and BEHRINGER R. P., *Phys. Rev. Lett.*, **74** (1995) 4643.

Constraint optimization and statistical mechanics

G. PARISI

Dipartimento di Fisica, INFM, SMC and INFN
Università di Roma "La Sapienza" - P.le A. Moro 2, 00185 Rome, Italy

1. – Introduction

Statistical mechanics of disordered systems and constraint optimization of random problems have many points in common. In these lectures I will describe some of the results that have been obtained by applying the techniques developed in the study of disordered systems to optimization theory. The aim of these studies is twofold:

- Statistical mechanics techniques are very sophisticated and powerful: using them it is possible to obtain very relevant heuristic and eventually exact results for optimization theory. Better algorithms can also be found.

- Applying the techniques of statistical mechanics in a rather different setting needs the invention of new theoretical tools that can be carried back to the study of systems of physical interest.

This cross-fertilization process between different fields usually produces very interesting results.

These lectures are organized as follows. In the next section I will present some general considerations on the relation between optimization theory and statistical mechanics. In sect. **3** I will describe a well-understood problem, *i.e.* bipartite matching, for which many analytic and numeric results are available. In sect. **4** I will introduce some of the basic ideas and notations in constraint optimization and I will stress the relations with statistical mechanics and with the theory of phase transitions. In sect. **5** I will recall the main properties of random lattices and of the Bethe (cavity) approximation, that will be heavily used in the rest of these lectures. In sect. **6** I will present the problem of graph coloring and I will derive the appropriate cavity equations in the colorable phase. In sect. **7** I will sketch the survey method that allows us to compute in a (hopefully) exact way the phase transition between the colorable and the uncolorable phase. Finally in the last section I will present some conclusions and perspectives.

2. – General considerations

In statistical mechanics [1] one is interested in computing the partition function as a function of temperature, the partition function being defined as

$$Z(\beta) = \sum_{\mathcal{C}} \exp[-\beta H(\mathcal{C})], \tag{1}$$

where $\beta = 1/(kT)$, T being the temperature and k being equal to two thirds of the Boltzmann-Drude constant. The variable $\mathcal{C}$ denotes a generic element of the configuration space. The quantity $H(\mathcal{C})$ is the Hamiltonian of the system. Starting from $Z(\beta)$ we can reconstruct the other thermodynamic quantities.

In optimization problems we are interested in finding out the configuration $\mathcal{C}^*$ (or the configurations, as far as the ground state may be degenerate) that minimizes the function $H(\mathcal{C})$, that in this contest is called the cost function. We are also interested in knowing the minimal cost, *i.e.*

$$H^* \equiv H(\mathcal{C}^*). \tag{2}$$

It is trivial to realize that optimization is a particular case of statistical mechanics [2]. Let us denote by $E(T)$ and $S(T)$ the expectation value of the energy and the entropy as a function of the temperature T. One finds that only the zero-temperature behaviour is relevant for optimization:

$$H^* = E(0) \tag{3}$$

and the number of optimizing configurations is just equal to $\exp[S(0)]$.

Often in optimization we are interested in knowing something more (*e.g.* their number) about the nearly optimizing configurations, *i.e.* those configurations such that $H(\mathcal{C}) = H^* + \epsilon$. This information may be gathered by studying the system at small temperatures.

In both cases, statistical mechanics and optimization theory, we are interested in studying what happens in the thermodynamic limit, *i.e.* when the number N of variables (the dimensions of the configuration space) goes to infinity, or is very large (finite-size effects are more important in optimization theory than in statistical mechanics).

While optimization theory could be considered only from an abstract point of view as the zero-temperature limit of statistical mechanics, in reality it differs from statistical mechanics in many crucial aspects, *e.g.* goals, problems and techniques. Optimization theory is really a different science, with many points of contact with statistical mechanics.

Optimization problems, where the Hamiltonian has a simple form, are not usually the most interesting (on the contrary an incredibly large amount of work has been done in statistical mechanics on the Ising problems). In optimization theory it is natural to consider an ensemble of Hamiltonians and to find out the properties of a generic Hamiltonian that belongs to this ensemble.

In other words we have a Hamiltonian that is characterized by a set of parameters (collectively denoted by J) and we have a probability distribution $\mu(J)$ on this parameter space. One would like to compute the ensemble average(1), *e.g.*

$$E_{\mathrm{AV}}(T) = \int \mathrm{d}\mu(J) E_J(T) \equiv \overline{E_J(T)}\,, \tag{4}$$

where the overline denotes the average of the ensemble of Hamiltonians. Eventually we are going to set $T = 0$. This approach has been developed in statistical mechanics for studying systems with quenched disorder (*e.g.* spin glasses)(2). Only the recent development of these techniques allows us to use statistical mechanics tools to study optimization theory.

In general we are interested in computing the probability distribution $P(E)$ of the zero-temperature energy E over the ensemble

$$P(E) = \overline{\delta(E_J(0) - E)}\,. \tag{5}$$

In the thermodynamics limit, where the number N of variables goes to infinity, if E is well normalized in this limit, we expect that its probability distribution becomes a delta-function, according to the general principle that its intensive quantity does not fluctuate in the thermodynamic limit.

The interest of computer scientists is not limited to the computation of the ensemble average: they live in the real world and they are extremely interested in finding an efficient algorithm to compute for a given $H_J(\mathcal{C})$ (*i.e.* for an instance of the problem) the configuration $\mathcal{C}^*$ that minimizes it. Of course the configuration $\mathcal{C}^*$ can always be found (at least in the case of a finite configuration space) by computing $H(\mathcal{C})$ for all possible choices of the configurations, but this algorithm is very very slow(3). One would like to find out the most efficient algorithms and to correlate their performances to the properties of the ensemble of problems.

A given algorithm can be tested by checking how it performs for different instances: for each problem there are well-known benchmarks that correspond to different types of ensembles. We can also do a more theoretical analysis. For a given algorithm we can

---

(1) Sometimes the ensemble is defined in a loose way, *e.g.* problems that arise from practical instances, *e.g.* chips placements on a computer board, or register allocation.
(2) Many of the techniques developed for disordered systems can be used also for the study of structural disordered systems where no disorder is present in the Hamiltonian.
(3) This algorithm usually takes a time proportional to $\exp[AN]$.

define the time t_J as the time (*i.e.* the number of operations) it takes to find the ground state of the Hamiltonian H_J. We can define the average time (in a logarithmic scale) as

$$\ln(t_{ln}) = \overline{\ln(t_J)}\,. \tag{6}$$

The introduction of the logarithm is very important: very often the linearly averaged time

$$t_L = \overline{t_J} \tag{7}$$

is much greater than t_{AV} because it may be dominated by rare configurations that give an extremely large contribution([4]). We can also define both the median and the most likely time (*i.e.* the time where the probability distribution of time has a maximum).

In many cases the logarithmically averaged time, the median and the most likely times behave in quite a similar way when the number N of variables becomes large: in the best of the possible worlds

$$\frac{\ln(t_J)}{\overline{\ln(t_J)}} \tag{8}$$

does not fluctuate in the thermodynamics limit or it is with probability one a quantity of order 1.

On the contrary very often the worst case time, defined as

$$t_{\mathrm{WC}} = \max_J(t_J), \tag{9}$$

is much larger that the other times [3]. The most likely time (or the logarithmically averaged time) is the most interesting, unfortunately the computation of the worst case time is the easiest one from the theoretical point of view (in empirical studies, *i.e.* when testing the algorithm with real instances, the situation is just the opposite).

It is rather difficult to study analytically the most likely time: if the algorithm is interpreted as a physical system, we must study its time evolution toward equilibrium. In spite of the spectacular progresses that have been done in recent years, a general theory of the approach to equilibrium is lacking and this is one of the most difficult problems of statistical mechanics. Fortunately present-days theory allows us to do some predictions [4, 5].

3. – A well-understood problem: bipartite matching

Let us be specific and let us consider an example of optimization problem for which we have a good understanding. Let us consider the so-called bipartite matching.

([4]) This is the same argument that implies that the annealed free energy and the quenched free energy are very different: $\overline{\ln(Z_J)}$ is very different from $\ln(\overline{Z_J})$.

The problem is defined as follows: we have N cities and N wells that must be connected one with the other: in other words there is one and only one well $\pi(i)$ that is connected to (matches) the city i ([5]). In other words $\pi(i)$ is a permutations and the configuration space of all possible matchings contains $N!$ elements.

The Hamiltonian $H_d(\pi)$ is given by

$$H_d(\pi) = \sum_{i=1,N} d_{i,\pi(i)}\,, \tag{10}$$

where $d_{i,k}$ is an $N \times N$ matrix expressing the cost of establishing a connection from i to k and it characterizes an instance of the problem.

In spite of the fact that the configuration space contains $N!$ elements, there are algorithms that for any d compute the minimum energy configuration $\mathcal{C}^*(d)$ and the corresponding energy $(E^*(d))$ in a time that is less than N^3 multiplied by an appropriate constant. The algorithmic simplicity of the model is related to the fact that it can be recast as a linear programming problem. Indeed we can introduce the N^2 variables $n_{i,k} = 0, 1$ such that

$$\sum_{i=1,N} n_{i,k} = \sum_{k=1,N} n_{i,k} = 1. \tag{11}$$

It is obvious that there is a one-to-one correspondence of these variables with the permutation and the Hamiltonian may be rewritten as

$$H_d(n) = \sum_{i,k=1,N} d_{i,k} n_{i,k}\,. \tag{12}$$

Up to this point we have not gained too much. However we can enlarge the configuration space by considering real variables $s_{i,k}$ that satisfy both the constraint eq. (11) (*i.e.* $\sum_{i=1,N} s_{i,k} = \sum_{k=1,N} s_{i,k} = 1$) and the bound

$$0 \le s_{i,k} \le 1. \tag{13}$$

The problem is more general and can be solved using linear programming tools in a very efficient way. The surprise is that the minimum of

$$H_d(s) = \sum_{i,k=1,N} d_{i,k} s_{i,k} \tag{14}$$

happens always when $s_{i,k} = 0, 1$, as can be easily proved. In this way, by enlarging the configuration space, we are able to find a fast solution of the original problem.

([5]) In simple matching we have N cities that must be pairwise matched: here the configuration space contains $(N-1)!!$ elements.

From the algorithmic point of view the situation is quite good. A different set of questions arise if we consider the ensemble average of the value of H at the mimima. In this case we have to specify the ensemble by specifying the probability distribution of the matrix d.

A well-studied case is when the matrix elements $d_{i,k}$ are independent identically distributed (i.i.d.) variables: here we have to assign the probability of a single matrix element $P(d_{i,k})$. Simple results are obtained when

$$P(d_{i,k}) = \exp[-d_{i,k}], \tag{15}$$

i.e. $d_{i,k} = -\ln(r_{i,k})$, where $r_{i,k}$ is a random number, with flat probability distribution in the interval $[0-1]$.

It was firstly conjectured [6] and then proved rigorously [7] that

$$\overline{H(d)} = \sum_{n=1,N} \frac{1}{n^2}\,. \tag{16}$$

The proof of this very simple result is a *tour de force* in combinatorial optimization [7] and it is rather unlikely that similar results are valid for different choices of the function $P(d)$.

The situation becomes simpler if we consider the case where N goes to infinity. Using the techniques of statistical mechanics [8] we can prove that

$$\lim_{N\to\infty} \overline{H(d)} = \zeta(2) = \frac{\pi^2}{6} \tag{17}$$

for all the probability distributions such that $P(0) = 1$. Moreover, if

$$P(d) = 1 - Ad + O(d^2), \tag{18}$$

one finds that [8-10]

$$\overline{H(d)} = \zeta(2) - \frac{2(1-A)\zeta(3)+1}{N} + O(N^{-2}). \tag{19}$$

This result is in agreement with the previous formula for the exponential probability where $A = 1$. The computation of the leading term when $N \to \infty$ is nowadays a real mathematical theorem [11] (that was obtained 15 years after the original result derived using the usual non-rigorous methods of statistical mechanics), while the computation of the subleading term (the one proportional to $1/N$) was only done using statistical mechanics techniques.

The computation of the subleading term is a highly non-trivial computation; the first two attempts gave wrong results (a factor two [9] and a Jacobian [10] were forgotten), however now we have the correct versions [10]. It has been verified with at least 3 decimal digits. The computation of the term $-2(1-A)\zeta(3)/N$ is rather simple. The real difficulty is the term $-1/N$ (in the second computation with the missing Jacobian this term was $-\pi^2/(12N)$), however its correctness can be checked with the exact result for the exponential probability.

A similar problem is the travelling salesman: here we have to find the least expensive tour that travels through all the cities (cities and wells are identified). The matrix d represents the cost for going from city i to k and is usually taken to be symmetric. The configuration space is restricted to permutations that contain only one cycle (*i.e.* cyclic permutations) and the Hamiltonian has always the same form (*i.e.* eq. (10)).

It is amusing to note that, when we restrict the configuration space, the nature of the problem changes dramatically and it becomes much more difficult. The travelling-salesman problem is NP complete; this statement implies (if a well-known conjecture, worth one million dollars, is correct) that there is no algorithm that can find a solution of the travelling-salesman problem in time that is bounded by a polynomial in N in the worst case([6]). However, it is empirically found (and I believe it has also been proved) that, in the typical case of random distribution of the elements of the matrix d, a standard simple algorithm takes a time that does not increase faster than N^4. This example shows that there is a dramatic difference among the time in the typical case and in the worst case. Unfortunately, while it is relatively easy to measure the scaling of the typical time as a function of N, the results for the worst cases are very difficult to be obtained numerically.

These examples are particularly interesting because they tell us that the scenario may radically change by an apparently small change in the definition of the problem. However, from the point of view of analytic computations the results are only slightly more difficult in the travelling-salesman problem. If $P(0)$ (defined in eq. (15)) is equal to one, we find that in the large-N limit

$$\overline{H(d)} = 2.041. \tag{20}$$

Apparently there are no serious difficulties in computing the $1/N$ corrections to the travelling-salesman problem and the $1/N^2$ corrections to the bipartite matching, however the computations become rather long and they have not been done.

Summarizing, statistical-mechanics techniques, based on the replica approach or the cavity approach, are able to provide the exact solution to these models in the thermodynamic limit: they can be used to compute also the leading finite-size correction (that is a highly non-trivial test).

These methods are working very well when there is no structure in the ensemble of Hamiltonians (*i.e.* the d's are independent identically distributed variables). A different situation may arise in other cases. For example, let us define a three-dimensional bipartite

---

([6]) NP does not means non-polynomial, but Polynomial on a Non-standard machine [3], *e.g.* a computer that has an unbounded number of nodes that work in parallel; however it is quite likely that this misinterpretation of the name is not far from reality.

matching problem: we consider a cube of size 1, we extract randomly n points x_i and y_i inside the cube and we construct the matrix d using the Euclidean distance among the points:

$$d_{i,k} = |x_i - y_k|. \tag{21}$$

In this way the distances are correlated random variables (*e.g.* they satisfy the triangular inequality) and the exact computation of the minimal energy state is at least as difficult as the computation of the free energy of the three-dimensional Ising model. Both problems have a three-dimensional structure that is absent in models where the distances are uncorrelated.

The situation where the distances are uncorrelated plays the role of a mean-field theory and the results in finite dimensions can be approximately obtained by doing a perturbative expansion around the uncorrelated mean-field case [12]. The results are rather good (there are still some points that are not so well understood [13]), however the description of these techniques would lead us too far from the main goal of these lectures.

4. – Constraint optimization

4·1. *General considerations.* – Constraint optimization is a particular case of combinatorial optimizations and will be the main subject of these lectures. Let us consider a simple case: a configuration of our system is composed by a number N of variables σ_i that may take q values (*e.g.* from 1 to q). These values usually have some meaning in real life, but this does not necessarily concern us. An instance of the problems is characterized by M functions $f_k[\sigma]$ $(= 1, M)$; each function takes only the values 0 or 1 and depends on a small number of σ variables.

Let us consider the following example with $N = 4$ and $M = 3$:

$$\begin{aligned} c_1[\sigma] &= \theta(\sigma_1\sigma_2 - \sigma_3\sigma_4), \\ c_2[\sigma] &= \theta(\sigma_1\sigma_3 - \sigma_2\sigma_4), \\ c_3[\sigma] &= \theta(\sigma_1\sigma_4 - \sigma_2\sigma_3), \end{aligned} \tag{22}$$

where the function $\theta(x)$ is 1 if the argument is non-negative and it is 0 if the argument is negative. The function we want to minimize is

$$H[\sigma] = \sum_{k=1,M} c_k[\sigma]. \tag{23}$$

In particular we are interested in knowing if there is a minimum with $H[\sigma] = 0$. If this happens all the functions c_k must be zero. It is quite evident that imposing the condition $H[\sigma] = 0$ is equivalent to finding the solution to the following inequalities:

$$\begin{aligned} \sigma_3\sigma_4 &> \sigma_1\sigma_2\,, \\ \sigma_2\sigma_4 &> \sigma_1\sigma_3\,, \\ \sigma_2\sigma_3 &> \sigma_1\sigma_4\,. \end{aligned} \tag{24}$$

In other words each function imposes a constraint and the function H is zero if all the constraints are satisfied. If this happens, the minimal total energy is zero and we are in the satisfiable case. On the contrary, if it is not possible to satisfy all the constraints, the minimal total energy is different from zero and we stay in the unsatisfiable case. In this case the minimum value of H is the minimum number of constraints that have to be violated. It is clear that for each set of inequalities of the previous kind there are arithmetic techniques to find out if there is a solution and, if any, to count their number.

4·2. *The thermodynamic limit.* – Given N and M we can easily define an ensemble as all the possible different sets of M inequalities of the type

$$\sigma_{i_1(k)}\sigma_{i_2(k)} > \sigma_{i_3(k)}\sigma_{i_4(k)}\,. \tag{25}$$

The interesting limit is when N goes to infinity with

$$M = N\alpha\,, \tag{26}$$

α being a parameter. Hand-waving arguments suggest that for small α it should be possible to satisfy all the constraints, while for very large α also in the best case most of the constraints will not be satisfied. It is believed that in this and in many other similar problems there is a phase transition from a satisfiable to a non-satisfiable phase [14]. More precisely, let us define the energy density

$$e(\alpha, N) = \frac{\overline{H^*}}{N}\,, \tag{27}$$

where the average is done over all possible constraints over the set of N variables. The results should depend on q but we have not indicated this dependence.

Usual arguments of statistical mechanics imply that the following limit is well defined:

$$e(\alpha) = \lim_{N\to\infty} e(\alpha, N) \tag{28}$$

and that sample-to-sample fluctuations vanish when N goes to infinity:

$$\frac{\overline{(H^*)^2} - \left(\overline{H^*}\right)^2}{N^2} \longrightarrow 0. \tag{29}$$

According to the previous discussion there must be a phase transition at a critical value of α_c, such that

$$e(\alpha) = 0 \quad \text{for } \alpha \leq \alpha_c\,, \qquad e(\alpha) > 0 \quad \text{for } \alpha > \alpha_c\,. \tag{30}$$

A simple argument shows that $e(\alpha)$ is a continuous differentiable function, so that the satisfiability-unsatisfiability transition cannot be a first-order transition in the thermodynamic sense.

The goal of the statistical mechanics approach is to compute α_c and eventually $e(\alpha)$. We would also like to compute the entropy density, that is related to the number of zero-energy configurations, but we will not address this point in this lecture. We will see that in order to compute α_c we will need a generalization of the entropy, *i.e.* the complexity $\Sigma(\alpha)$ that is the exact equivalent for this problem of the configurational entropy used in the study of glasses.

This random inequality model has never been studied, as far as I know, however the computation of α_c should be relatively straightforward, for not too large values of q. We will consider in the next section a much more studied problem: the coloring of a random graph with q different colors.

5. – An intermezzo on random graphs

The definition of random graphs has been discussed in Havlin's lectures, however it is convenient to recall here the main properties [15].

There are many variants of random graphs: fixed local coordination number, Poisson-distributed local coordination number, bipartite graphs.... They have the same main topological structures in the limit where the number (N) of nodes goes to infinity.

We start by defining the random Poisson graph in the following way: given N nodes we consider the ensemble of *all* possible graphs with $M = \alpha N$ edges (or links). A random Poisson graph is a generic element of this ensemble.

The first quantity we can consider for a given graph is the local coordination number z_i, *i.e.* the number of nodes that are connected to the node i. The average coordination number z is the average over the graph of the z_i:

$$z = \frac{\sum_{i=1,N} z_i}{N}\,. \tag{31}$$

In this case it is evident that

$$z = 2\alpha\,. \tag{32}$$

It takes a little more work to show that in the thermodynamic limit ($N \to \infty$) the probability distribution of the local coordination number is a Poisson distribution with average z.

In a similar construction two random points i and k are connected with a probability that is equal to $z/(N-1)$. Here it is trivial to show that the probability distribution of the z_i is Poisson, with average z. The total number of links is just $zN/2$, apart from corrections proportional to $\sqrt{N}$. The two Poisson ensembles, *i.e.* fixed total number of links and fluctuating total number of links, cannot be distinguished locally for large N and most of the properties are the same.

Random lattices with fixed coordination number z can be easily defined; the ensemble is just given by all the graphs with $z_i = z$ and a random graph is just a generic element of this ensemble.

One of the most important facts about these graphs is that they are locally a tree, *i.e.* they are locally cycleless. In other words, if we take a generic point i and we consider the subgraph composed by those points that are at a distance less than d on the graph(7), this subgraph is a tree with probability one when N goes to infinity at fixed d. For finite N this probability is very near to 1 as soon as

$$\ln(N) > A(z)\, d, \tag{33}$$

$A(z)$ being an appropriate function. For large N this probability is given by $1-O(1/N)$.

If $z > 1$ the nodes percolate and a finite fraction of the graph belongs to a single giant connected component. Cycles (or loops) do exist on this graph, but they have typically a length proportional to $\ln(N)$. Also the diameter of the graph, *i.e.* the maximum distance between two points of the same connected component, is proportional to $\ln(N)$. The absence of small loops is crucial because we can study the problem locally on a tree and we have eventually to take care of the large loops (that cannot be seen locally) in a self-consistent way, *i.e.* as a boundary condition at infinity. This problem will be studied explicitly in the next section for the ferromagnetic Ising model.

5·1. *The Bethe approximation.* – Random graphs are sometimes called Bethe lattices, because a spin model on such a graph can be solved exactly using the Bethe approximation. Let us recall the Bethe approximation for the two-dimensional Ising model.

In the standard mean-field approximation, one writes a variational principle assuming that all the spins are not correlated [1]; at the end of the computation one finds that the magnetization satisfies the well-known equation

$$m = \text{th}(\beta J z m) \tag{34}$$

where $z = 4$ on a square lattice ($z = 2d$ in d dimensions) and J is the spin coupling ($J > 0$ for a ferromagnetic model). This well-studied equation predicts that the critical point (*i.e.* the point where the magnetization vanishes) is $\beta_c = 1/z$. This result is not

(7) The distance between two nodes i and k is the minimum number of links that we have to traverse in going from i to k.

very exciting in two dimensions (where $\beta_c \approx 0.44$) and it is very bad in one dimension (where $\beta_c = \infty$). On the other hand, it becomes more and more correct when $d \to \infty$.

A better approximation can be obtained if we look at the system locally and we compute the magnetization of a given spin (σ) as a function of the magnetization of the nearby spins (τ_i, $i = 1, 4$). If we assume that the spins τ are uncorrelated, but have magnetization m, we obtain that the magnetization of the spin σ (let us call it m_0) is given by

$$m_0 = \sum_\tau P_m[\tau] \mathrm{th}\left(\beta J \sum_{i=1,4} \tau_1\right), \tag{35}$$

where

$$P_m[\tau] = \prod_{i=1,4} P_m(\tau_i), \qquad P_m(\tau) = \frac{1+m}{2}\delta_{\tau,1} + \frac{1-m}{2}\delta_{\tau,-1}\,. \tag{36}$$

The sum over all the 2^4 possible values of τ can be easily done.

If we impose the self-consistent condition

$$m_0(m) = m, \tag{37}$$

we find an equation that enables us to compute the value of the magnetization m.

This approximation remains unnamed (as far as I know) because with a little more work we can get the better and simpler Bethe approximation. The drawback of the previous approximation is that the spins τ cannot be uncorrelated because they interact with the same spin σ: the effect of this correlation can be taken into account and this leads to the Bethe approximation.

Let us consider the system where the spin σ has been removed. There is a cavity in the system and the spins τ are on the border of this cavity [16]. We assume that in this situation these spins are uncorrelated and they have a magnetization m_C. When we add the spin σ, we find that the probability distribution of this spin is proportional to

$$\sum_\tau P_{m_C}[\tau) \exp\left[\beta J \sigma \sum_{i=1,4} \tau_i\right]. \tag{38}$$

The magnetization of the spin σ can be computed and after some simple algebra we get

$$m = \mathrm{th}\{z \,\mathrm{arth}[\mathrm{th}(\beta J) m_C]\}, \tag{39}$$

with $z = 4$.

This seems to be a minor progress because we do not know m_C. However, we are very near the final result. We can remove one of the spins τ_i and form a larger cavity (two spins removed). If in the same vein we assume that the spins on the border of the cavity are uncorrelated and they have the same magnetization m_C, we obtain

$$m_C = \text{th}\{(z-1)\text{arth}[\text{th}(\beta J)m_C]\}. \tag{40}$$

Solving this last equation we can find the value of m_C and using the previous equation we can find the value of m.

It is rather satisfactory that in 1 dimension ($z = 2$) the cavity equations become

$$m_C = \text{th}(\beta J)m_C\,. \tag{41}$$

This equation for finite β has no non-zero solutions, as it should be.

The internal energy can be computed in a similar way: we get that the energy density per link is given by

$$E_{\text{link}} = \frac{\text{th}(\beta J) + m_C^2}{1 + \text{th}(\beta J)m_C^2} \tag{42}$$

and we can obtain the free energy by integrating the internal energy as a function of β.

In a more sophisticated treatment we write the free energy as a function of m_C:

$$\frac{\beta F(m_C)}{N} = F_{\text{site}}(m_C) - \frac{z}{2}F_{\text{link}}(m_C), \tag{43}$$

where $F_{\text{link}}(m_C)$ and $F_{\text{site}}(m_C)$ are appropriate functions [17]. This free energy is variational, in other words the equation

$$\frac{\partial F}{\partial m_C} = 0 \tag{44}$$

coincides with the cavity equation (40). However for lack of space we will not discuss this interesting approach [17, 18].

5·2. *Bethe lattices and replica symmetry breaking.* – It should be clear now why the Bethe approximation is correct for random lattices. If we remove a node of a random lattice, the nearby nodes (that were at distance 2 before) are now at a very large distance, *i.e.* $O(\ln(N))$. In this case we can write

$$\langle \tau_{i_1}\tau_{i_2}\rangle \approx m_{i_1}m_{i_2} \tag{45}$$

and everything seems easy.

This is actually easy in the ferromagnetic case where in the absence of magnetic field at low temperature the magnetization may take only two values ($\pm m$). In more complex cases, (*e.g.* antiferromagnets) there are many different possible values of the magnetization because there are many equilibrium states and everything becomes complex (as it should) because the cavity equations become equations for the probability distribution of the magnetizations [19].

This case has long been studied in the literature and for historical reasons it is usually said that the replica symmetry is spontaneously broken [19,20]. Fortunately for the aims of this lecture we need only a very simple form of replica symmetry breaking (we are interested in the zero-temperature case) and we are not going to describe the general formalism.

6. – Coloring a graph

6˙1. *Basic definitions.* – For a given graph G we would like to know if using q colors the graph can be colored in such a way that adjacent nodes have different colors. It is convenient to introduce the Hamiltonian

$$H_G = \sum_{i,k} A(i,k)\delta_{\sigma_i,\sigma_i}\,, \tag{46}$$

where $A_G(i,k)$ is the adjacency matrix (*i.e.* 1 if the two nodes are connected, 0 elsewhere) and the variable σ may take values that go from 1 to q. This Hamiltonian describes the antiferromagnetic Potts model with q states. The graph G is colorable if and only if the ground state of this Hamiltonian (E_G) is zero.

For large N on a random graph we expect that the energy density

$$\frac{E_G}{N} = e(z) \tag{47}$$

does not depend on G: it should depend only on the average coordination number z of the graph. It can be proved that

$$\begin{aligned} &e(z) = 0 \quad \text{for } z < 1,\\ &e(z) \propto \sqrt{z} \quad \text{for } z \longrightarrow \infty. \end{aligned} \tag{48}$$

Consequently there must be a phase transition at z_c between the colorable phase $e(z) = 0$ and the uncolorable phase $e(z) \neq 0$. Although it is possible to compute the function $e(z)$ for all z, in this lecture we address only the simpler problem of computing the value of z_c [21].

For $q = 2$ $z_c = 1$, as can be intuitively understood: loops with an odd number of sites cannot be coloured with two colours and for $z > 1$ there are many large loops that have an even or an odd number of sites with equal probability. The $q = 2$ case is an antiferromagnetic Ising model on a random graph, *i.e.* a standard spin glass [22, 18].

6˙2. *The cavity equations.* – Let us start with the basic definitions. Let us consider a legal coloring (*i.e.* all adjacent nodes have different colors). We take a node i and we consider the subgraph of nodes at distance d from a given node. Let us call $\mathcal{B}(i,d)$ the interior of this graph. With probability one (when N goes to infinity) this graph is a tree and the nodes at distance less than or equal to d are the leaves of this tree (there may be also other leaves at shorter distance). In the future we shall assume that this graph is a tree and all the statements we shall do will be valid only with probability one when N goes to infinity.

We ask the following questions:

- Are there other legal colorings of the graph that coincide with the original coloring outside $\mathcal{B}(i,d)$ and differ inside $\mathcal{B}(i,d)$? (Let us call the set of all these colorings $\mathcal{C}(i,d)$.)

- Which is the list of colors that the node i may have in one of the colorings belonging to $\mathcal{C}(i,d)$? (Let us call this list $L(i,d)$. This list depends on the legal configuration σ, however for lightening the notation we will not indicate this dependence in this section.)

The cardinality of $L(i,d)$ increases with d so that $L(i,d)$ must have a limit when d goes to infinity (still remaining in the region $d \ll A(z)\ln(N)$). We call this limit $L(i)$. We have not to forget that $L(i)$ is fuzzy for finite N, but it becomes sharper and sharper when N goes to infinity. In other words, $L(i)$ is the list of all the possible colors that the site i may have if we change only the colors of the nearby nodes and we do not change the colors of far-away nodes [23, 24].

Let us study what happens on a graph where the site i has been removed. We denote by k a node adjacent to i and we call $L(k;i,d)$ the list of the possible colors of the node k, while the colors outside $\mathcal{B}(i,d)$ are fixed. The various nodes k do not interact directly and in this situation they can have a color independently of the color of the other nodes.

In this situation it is evident that $L(i,d+1)$ can be written as a function of all the $L(k;i,d)$. In order to write explicit equations it is convenient to indicate by an overbar ($\overline{L}$) the list of the forbidden colors([8]). Let us indicate by $F(L)$ the list of colors that must be forbidden at the nodes that are adjacent to a node where the list of allowed colors is L. Barring the case where the list L is empty it is easy to obtain that:

- $F(L) = L$ if L contains only one element,

- $F(L) = \emptyset$ ($\emptyset$ being the empty set) if L contains more than one element.

With these preliminaries we have that

$$L(i,d+1) = \overline{\bigcup_k F(L(k;i,d))}, \tag{49}$$

([8]) The list of forbidden colors is just the list of colors that do not belong to the list of possible colors. From the point of view of set theory the set of forbidden colors is just the complement of the set of possible colors. Using this notation we obviously have $\overline{\overline{L}} = L$.

where k runs over all the nodes adjacent to i, and the union is done in a set theoretical sense.

The previous formula can be easily transcribed into words. We have to consider all the neighbours (k) of the node i; if a neighbour may be colored in two ways, it imposes no constraint, if it can be colored in only one way, it forbids the node i to have its color. Considering all nearby nodes we construct the list of the forbidden colors, and the list of the allowed colors is just composed by those colors that are not forbidden.

The previous formula is exact (with probability one, as usual). Also the next formula is valid with probability one:

$$L(i,d) = \overline{\bigcup_k F(L(k;i,d))}\,, \tag{50}$$

because in most of the cases $L(i,d+1) = L(i,d)$.

If we do the limit $d \to \infty$ in any of the two previous formulae, we get

$$L(i) = \overline{\bigcup_k F(L(k;i))}\,. \tag{51}$$

Of course everything is true with probability 1 when $N \to \infty$.

A further simplification in the previous formulae may be obtained if we associate to a list L a variable ω that takes values from 0 to q defined as follows:

- The variable ω is equal to i if the list contains only the i-th color.

- The variable ω is equal to 0 if the list contains more than one color.

Let us call this mapping $\Omega(L)$.

In the nutshell we have introduced an extra color, white, and we say that a site is white if it can be colored in more than two ways without changing the colors of the far-away sites [23-25]. The rationale for introducing the variable ω is that $F(L)$ depends only on $\Omega(L)$. The previous equations induce equations for the variables ω, *i.e.*

$$\omega(i) = \Omega\left(\overline{\bigcup_k F(\omega(k;i))}\right). \tag{52}$$

In the same way as before we have to consider all the neighbours (k) of the node i; if a neighbour is white, it imposes no constraint, if it is colored it forbids the node i to have its color. Considering all nearby nodes we construct the list of the forbidden colors. If more than one color is *not* forbidden, the node is white, if only one color is *not* forbidden, the node has this color.

The previous equation is just the generalization of eq. (39) where we have the colors, white included, instead of the magnetizations. We have discrete, not continuous variables, because we are interested in the ground state, not in the behaviour at finite temperature.

We have to write down the equivalent of the cavity equations, eq. (39), *i.e.* for the quantities $\omega(i;l)$. They are given by

$$\omega(i;l) = \Omega\left(\overline{\bigcup_{k\neq l} F(\omega(k;i))}\right). \tag{53}$$

We can also say that $\omega(i;l)$ is the information that is transmitted from the node i to the node l and it is computed using the information transmitted by all the nearby nodes, with the exclusion of l.

The previous equations are called the belief equations (sometimes the strong belief equations, in order to distinguish them from the weak belief equations that are valid at non-zero temperature). We can associate to any legal coloring a solution (or a quasi-solution) of the belief equations in a constructive way. Sometimes the solution of the belief equations is called a whitening [23, 24], because some nodes that were colored in the starting legal configuration become white.

The reader should notice that at this stage we can only say that the belief equations should be satisfied in a fraction of nodes that goes to 1 when N goes to infinity. However, it is possible that the total number of nodes where the beliefs equations are not satisfied remains finite or goes to infinity with N and for this reason in general we can only say that these equations have quasi-solutions, not true solutions [23, 24].

7. – Analytic computations

7·1. *An interpretation.* – The main reason for introducing the whitening is that each legal coloring has many other colorings nearby that differ only by the change of the colors of a small number of nodes. The number of these legal colorings that can be reached by a given coloring by making this kind of moves is usually exponentially large. On the other hand, two colorings that differ only by the change of the colors of a small number of nodes correspond to the same whitening.

Let us be more precise. For each coloring let us write

$$\omega(i|\{\sigma\}) = \omega(i, d|\{\sigma\}) \quad \text{for } 1 \ll d \ll A(z)\ln(N), \tag{54}$$

where we have indicated in an explicit way that the whitening $\omega(i|\{\sigma\})$ depends on the legal coloring $\{\sigma\}$.

We say that two whitenings are equivalent if they differ in a fraction of nodes less than $\epsilon(N)$, where $\epsilon(N)$ is a function that goes to zero when $N \to \infty$. For finite N everything is fuzzy and depends on the precise choice of d and ϵ. Let us suppose that for large N with probability one two whitenings are equivalent or they differ in a large fraction of nodes.

Generally speaking we have three possibilities.

- For all the legal configurations the corresponding whitenings $(\omega(\{\sigma\}))$ are such that $\omega(i, \{\sigma\}) = 0$, *i.e.* all nodes are white.

- For a generic legal configuration the corresponding whitening is non-trivial. *i.e.* for a finite fraction of the nodes $\omega(i,\{\sigma\}) = 0$.
- The graph is not colorable and there are no legal configurations.

In the second case we would like to know how many whitenings there are, how they differ and which are their properties, *e.g.* how many sites are colored. We shall see in the next section how these properties may be computed analytically and as byproduct we will find the value of z (z_c) that separates the colorable phase from the uncolorable phase.

In the case where there are many whitenings one can argue that the set of all the legal configurations breaks in a large number of different disconnected regions that are called with many different names in the physical literature [27, 18, 26] (states, valleys, clusters, lumps...). Roughly speaking, the set of all the legal configurations can be naturally decomposed into clusters of proximate configurations, while configurations belonging to different clusters (or regions) are not close. The precise definition of these regions is rather complex [4]; roughly speaking, we could say that two legal configurations belong to the same region if they are in some sense adjacent, *i.e.* they belong to a different region if their Hamming distance is greater than ϵN. In this way the precise definition of these regions depends on ϵ, however it can be argued that there is an interval in ϵ where the definition is non-trivial and is independent of the value of ϵ(9). It is usually assumed that each whitening is associated to a different cluster of legal solutions.

7·2. *Surveys.* – Let us consider the case (that we suppose to be non-empty) where there are a large number of non-equivalent whitenings and we want to study the properties of this ensemble. To this end it is convenient to introduce the probability that for a generic whitening ω of the ensemble we have that $\omega(i) = c$; we will denote this probability $P_i(c)$. We obviously have

$$\sum_{c=0,q} P_i(c) = 1. \tag{55}$$

The quantities $P_i(c)$ generalize the physical concept of magnetization. In the statistical analysis of the ensemble of configurations of an Ising model, the local variables may have only two values (± 1) and

$$P(\pm 1) = \frac{1 \pm m_i}{2}, \tag{56}$$

where m_i is the magnetization at the site i. Here $q+1$ colors are possible (white included) and the magnetization is a $(q+1)$-dimensional vector, normalized to 1.

In order to do some computation we have to consider also the two colors probabilities $P_{i,l}(c_1, c_2)$. We will assume a factorization hypothesis: for points i and l that are far away

(9) For a rigorous definition of these regions see [28-30].

on the graph the probability $P_{i,l}(c_1, c_2)$ factorizes into the product of two independent probabilities:

$$P_{i,l}(c_1, c_2) = P_i(c_1)P_l(c_2), \tag{57}$$

neglecting corrections that go to zero (in probability) when N goes to infinity. This hypothesis is not innocent: there are many cases where it is not correct; however we cannot discuss here this important and subtle point.

A similar construction can be done with the cavity coloring and in this way we define the probabilities $P_{i;k}(c)$, where k is a neighbour of i. These probabilities are called surveys (and they are denoted by $\vec{s}(i;k)$) because they quantify the probability distribution of the messages sent from the node i to the node k.

Under the previous hypothesis the surveys satisfy simple equations. Let us see one of them, *e.g.* the one that relates $P_i(c)$ to the $P_{k;i}(c)$. The final equations are simple, but the notation may become easily heavy, so let us write everything in an explicit way in the case where the point i has three neighbours k_1, k_2, k_3. The generalization is intuitive.

The formulae of the previous section tell us which is the color of the site i if we know the colors $c_1 = \omega(k_1; i)$, $c_2 = \omega(k_2; i)$ and $c_3 = \omega(k_3; i)$. This relation can be written as

$$c = T_3(c_1, c_2, c_3). \tag{58}$$

Therefore the probability distribution of c can be written as a function of the probability distribution of the c_i, that are supposed to factorize. We thus get

$$P_i(c) = \sum_{c_1,c_2,c_3} P_{k_1;i}(c_1)P_{k_2;i}(c_2)P_{k_3;i}(c_3)\delta_{c,T_3(c_1,c_2,c_3)}\,. \tag{59}$$

Similar formulae can be written for computing the surveys $P_{k_1;i}$ as a function of other surveys, *e.g.*

$$P_{i;k_1}(c) = \sum_{c_2,c_3} P_{k_2;i}(c_2)P_{k_3;i}(c_3)\delta_{c,T_2(c_2,c_3)}. \tag{60}$$

In this way we obtain the so-called survey propagation equations [27, 31, 32].

The survey propagation equations always have a trivial solution corresponding to all sites white: $P_i(0) = 1$ for all i. Depending on the graph there can also be non-trivial solutions of the survey equations. Let us assume that if such a solution exists, it is unique (also this point should be checked). In the next section we will find the statistical properties of this solution and we will identify the values of z where such a solution is present.

7'3. *A high-level statistical analysis.* – We are near the end of our trip. Let us look at eq. (60). The quantity $P_{i;k_1}$ depends on $P_{k_2;i}$ and $P_{k_3;i}$: we indicate this relation by $P_{i;k_1} = \mathcal{T}_2[P_{k_2;i}, P_{k_3;i}]$. However, if we neglect the large loops, the quantities $P_{k_2;i}$ and $P_{k_3;i}$ do *not* depend on $P_{i;k_1}$.

If we consider the whole graph we can define the probability $\mathcal{P}[P]$, *i.e.* the probability that a given node has a probability $P(c)$. If we consider only points with two neighbours we have

$$\mathcal{P}[P] = \int \mathrm{d}\mathcal{P}[P_1]\mathrm{d}\mathcal{P}[P_2]\delta[P - \mathcal{T}_2(P_1, P_2)]. \tag{61}$$

If we sum over all the possible coordination numbers with the Poisson distribution, the final equation is

$$\mathcal{P}[P] = \sum_n \frac{z^n}{n!} \int \mathrm{d}\mathcal{P}[P_1] \dots \mathrm{d}\mathcal{P}[P_n]\delta[P - \mathcal{T}_n(P_1 \dots P_n)]. \tag{62}$$

We arrived at an integral equation for the probabilities of the local probabilities (*i.e.* the surveys). This integral equation looks formidable and it is unlikely that it has an analytic solution. Fortunately the solution of integral equations can often be computed numerically without too many difficulties on present-day computers and this is what happens in this case.

One finds that there is a range $z_d < z < z_U$ where the previous integral equation has a non-trivial solution and its properties can be computed [32]. The fact that the survey equation has a non-trivial solution does not imply that there are whitenings that correspond to legal configurations, so that at this stage we cannot compute the critical value of z. This problem will be solved in the next section.

7'4. *The complexity.* – In the same way as the entropy counts the number of legal colorings, the complexity counts the number of different whitenings; more precisely for a given graph we write

$$\text{number of whitenings} = \exp[\Sigma_G], \tag{63}$$

where Σ_G is the complexity.

We assume that for large N all graphs with the same average coordination number (z) have the same complexity density:

$$\Sigma_G \approx N\Sigma(z). \tag{64}$$

There is a simple way to compute the complexity. It consists in counting the variation in the number of whitenings when we modify the graph. To this end it is convenient to consider the complexity as a function of N and of M (*i.e.* the total number of edges). Asymptotically we should have

$$\Sigma(M, N) = N\Sigma\left(\frac{2M}{N}\right). \tag{65}$$

There are two possible modifications we shall consider: adding one edge or adding one site (with the related edges).

Let us suppose that we add an edge between the nodes i and k. Only those whitenings in which the color of two nodes is different may be extended to the larger graph. We thus find

$$
\begin{aligned}
(66) \qquad & \text{number of whitenings}\,(N, M+1) = \\
& = \text{number of whitenings}\,(N, M)\left(1 - \sum_{c=1,q} \omega_i(c)\omega_k(c)\right),
\end{aligned}
$$

$$
\Sigma(N, M+1) = \Sigma(N, M) + \ln\left(1 - \sum_{c=1,q} \omega_i(c)\omega_k(c)\right) \equiv \Sigma(N, M) + \Delta\Sigma_{\text{edge}}(z).
$$

In the same way, if we add a site (and automatically z links in the average) we get that the complexity increases by an addictive factor $\Delta\Sigma_{\text{site}}$ that can be easily computed (it is equal to the logarithm of the probability that the total number of different non-white colors of the nodes to which the new added node is linked is less than q).

However, if we want to change N by one unit at fixed z, we have to add one site and only $z/2$ edges. Putting everything together we find that

$$
(67) \qquad \Sigma(z) = \Delta\Sigma_{\text{site}}(z) - \frac{z}{2}\Delta\Sigma_{\text{edge}}(z).
$$

We know the probability distribution of the variables ω from the previous analysis. We can now get the results for the complexity. It is shown in fig. 1 for the case of the three colorings.

The complexity jumps from 0 to a finite value at $z_d = 4.42$ and it decreases with increasing z and becomes eventually negative at $z = 4.69$. A negative value of Σ implies a number of whitenings less than one and it is interpreted as the signal that there are no whitenings (and no legal configurations) with probability 1. Indeed an explicit computation shows that in the region where the complexity is negative a correct computation of the energy $e(z)$ gives a non-zero (positive) result. The value where the complexity becomes zero is thus identified as the colorability threshold $z_c = 4.69$. We have thus obtained [21] the result we wanted for $q = 3$. Similar results may be obtained for higher values of q [21].

8. – Conclusions

The methods we have used in this lecture have been developed in the study of disordered systems (*i.e.* spin glasses) firstly in the context of infinite-range models (*i.e.* in the limit where $z \to \infty$). They have been extended to finite-range models first at finite temperature and later at zero temperature (where notable simplifications are present). The concept of complexity emerged in the study of the spin glasses and it was also introduced in the study of glasses under the name of configurational entropy. The behaviour of the

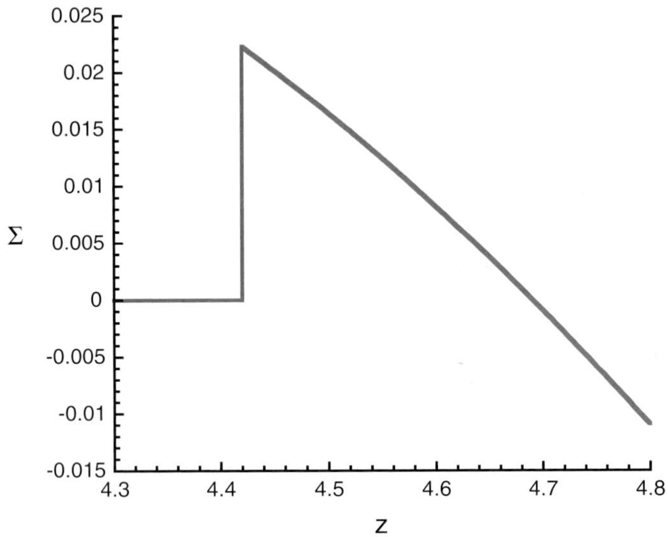

Fig. 1. – The complexity *vs.* the average connectivity z for $q = 3$ (from [21]).

complexity as a function of z is very similar to what is supposed to happen in glasses as a function of β.

These methods have been successfully applied to combinatorial optimization in the case of the K satisfiability where our goal is to find a set of N Boolean variables that satisfy αN logical clauses (of the OR type) containing K variables. Also in this case the satisfiability threshold [33, 34] can be computed (for example, we found [27] that for $K = 3$, $\alpha_c = 4.267$).

The research on these subjects is quite active. There are many problems that are open.

- The extension to other models. This is interesting *per se*; moreover surprises may be present.

- Verification of the self-consistency of the different hypotheses done and the identification of effective strategies in the case where they fail [35, 36].

- The construction of effective algorithms for finding a solution of the optimization problem. A first algorithm has been proposed [27, 31, 37] and it has been later improved by adding a simple form of backtracking [38]. A goal is to produce an algorithm that for large N finds a solution with probability one on a random graph in a polynomial time as soon as $z < z_c$ (*i.e.* in a computer time that is bounded by $N^{\lambda(z)}$). Finding this algorithm is interesting from the theoretical point of view (it is not clear at all if such an algorithm does exist) and it may have practical applications.

– Last but not least, everything that we said up to now was derived using physicists style. One should be able to transform the results derived in this way into rigorous theorems. It is very interesting that after a very long effort Talagrand [39], using some crucial results of Guerra [40], has recently been able to prove that a similar, but more complex, construction gives the correct results in the case of infinite-range spin glasses (*i.e.* the Sherrington-Kirkpatrick model) that was the starting point of the whole approach. It is relieving to know that the foundations of this elaborate building are sound. Some rigorous results have also been obtained for models with z [41].

The whole field is developing very fast and it is likely that in a few years we should have a much clearer and more extended picture.

REFERENCES

[1] See, for example: Parisi G., *Statistical Field Theory* (Academic Press, New York) 1987.
[2] Martin O. C., Monasson R. and Zecchina R., *Theor. Comput. Sci.*, **265** (2001) 2.
[3] Garey M. R. and Johnson D. S., *Computers and Intractability* (Freeman, New York) 1979.
[4] Parisi G., *Glasses, replicas and all that*, cond-mat/0301157 (2003).
[5] Cugliandolo T. F., *Dynamics of glassy systems*, cond-mat/0210312 (2002).
[6] Parisi G., *A conjecture on random bipartite matching*, cond-mat/9801176 (1998).
[7] Dotsenko V., *J. Phys. A*, **33** (2000) 2015; Linusson S. and Wästlund J., *A Proof of Parisi's Conjecture on the Random Assignment Problem*, preprint (2003); Nair C., Prabhakar B. and Sharma M., *A Proof of the Conjecture due to Parisi for the Finite Random Assignment Problem*, preprint (2003).
[8] Mézard M. and Parisi G., *J. Phys. (Paris) Lett.*, **46** (1985) L-771; *Eur. Phys. Lett.*, **2** (1986) 913.
[9] Brunetti R., Krauth W., Mézard M. and Parisi G., *Eur. Phys. Lett.*, **14** (1991) 295.
[10] Parisi G. and Ratiéville M., *Eur. Phys. J. B*, **29** (2002) 457.
[11] Aldous D. J., *Random Struct. Algorith.*, **18** (2001) 381.
[12] Mézard M. and Parisi G., *J. Phys. (Paris)*, **49** (1988) 2019.
[13] Houdayer J., Boutet de Monvel J. H. and Martin O. C., *Eur. Phys. J. B*, **6** (1998) 383.
[14] Dubois O., Monasson R., Selman B. and Zecchina R., *Phase Transitions in Combinatorial Problems, Theoret. Comp. Sci.*, **265** (2001); Biroli G., Cocco S. and Monasson R., *Physica A*, **306** (2002) 381.
[15] Erdös P. and Rènyi A., *Publ. Math. (Debrecen)*, **6** (1959) 290.
[16] Thouless D. J., Anderson P. A. and Palmer R. G., *Philos. Mag.*, **35** (1977) 593.
[17] Katsura S., Inawashiro S. and Fujiki S., *Physica A*, **99** (1979) 193.
[18] Mézard M. and Parisi G., *Eur. Phys. J. B*, **20** (2001) 217.
[19] Mézard M., Parisi G. and Virasoro M. A., *Spin Glass Theory and Beyond* (World Scientific, Singapore) 1997.
[20] Parisi G., *Field Theory, Disorder and Simulations* (World Scientific, Singapore) 1992.
[21] Mulet R., Pagnani A., Weigt M. and Zecchina R., *Phys. Rev. Lett.*, **89** (2002) 268701; Braunstein A., Mulet R., Pagnani A., Weigt M. and Zecchina R., *Phys. Rev. E*, **68** (2003) 036702.

[22] BAILLIE C., JOHNSTON D., MARINARI E. and NAITZA C., *J. Math. Phys.*, **29** (1996) 6683.
[23] PARISI G., cs.CC/0212047, *On local equilibrium equations for clustering states* (2002).
[24] PARISI G., *On the probabilistic approach to the random satisfiability problem*, cs.CC/0308010 (2003).
[25] BRAUNSTEIN A., MEZARD M., WEIGT M. and ZECCHINA R., cond-mat/0212451, *Constraint Satisfaction by Survey Propagation* (2002).
[26] MÉZARD M. and PARISI G., *J. Stat. Phys.*, **111** (2003) 1.
[27] MÉZARD M., PARISI G. and ZECCHINA R., *Science*, **297** (2002) 812.
[28] TALAGRAND M., *Prob. Theory Relat. Fields*, **117** (2000) 303.
[29] DUBOIS O. and MANDLER J., *FOCS* (2002) 769.
[30] COCCO S., DUBOIS O., MANDLER J. and MONASSON R., *Phys. Rev. Lett.*, **90** (2003) 047205.
[31] MÉZARD M. and ZECCHINA R., *Phys. Rev. E*, **66** (2002) 056126.
[32] PARISI G., *On the survey-propagation equations for the random K-satisfiability problem*, cs.CC/0212009 (2002).
[33] KIRKPATRICK S. and SELMAN B., *Science*, **264** (1994) 1297.
[34] DUBOIS O., BOUFKHAD Y. and MANDLER J., *Typical random 3-SAT formulae and the satisfiability threshold*, in *Proceedings of the 11th ACM-SIAM Symposiun on Discrete Algorithms*.
[35] MONTANARI A. and RICCI-TERSENGHI F., *Phys. Rev. Lett.*, **90** (2003) 017203.
[36] MONTANARI A., PARISI G. and RICCI TERSENGHI F., *Instability of one-step replica-symmetry-broken phase in satisfiability problems*, `cond-mat/0307543` (2003).
[37] PARISI G., *Some remarks on the survey decimation algorithm for K-satisfiability*, cs.CC/0301015 (2003).
[38] PARISI G., *A backtracking survey propagation algorithm for K-satisfiability*, cond-mat/0308510 (2003).
[39] TALAGRAND M., *Spin Glasses. A Challenge for Mathematicians. Mean-field Models and Cavity Method* (Springer-Verlag Berlin) 2003 and private comunication.
[40] GUERRA F., *Comm. Math. Phys.*, **233** (2002) 1; GUERRA F. and TONINELLI F. L., *Comm. Math. Phys.*, **230** (2002) 71.
[41] FRANZ S. and LEONE M., *J. Stat. Phys.*, **111** (2003) 535.

Introduction to nonextensive statistical mechanics and thermodynamics(*)

C. TSALLIS

Centro Brasileiro de Pesquisas Fisicas
Xavier Sigaud 150, 22290 - 180 Rio de Janeiro, RJ, Brazil
Santa Fe Institute - 1399 Hyde Park Road, Santa Fe, NM 87501, USA

F. BALDOVIN

Centro Brasileiro de Pesquisas Fisicas
Xavier Sigaud 150, 22290 - 180 Rio de Janeiro, RJ, Brazil

R. CERBINO

Istituto Nazionale per la Fisica della Materia
Dipartimento di Fisica, Università degli Studi di Milano
Via Celoria 16, 20133 Milano, Italy

P. PIEROBON

Hahn-Meitner Institut, Abteilung Theorie
Glienicker Strasse 100, D-14109 Berlin, Germany and
Fachbereich Physik, Freie Universität Berlin
Arnimallee 14, 14195 Berlin, Germany

(*) The present manuscript reports the content of the lecture delivered by C. Tsallis, and is based on the corresponding notes prepared by F. Baldovin, R. Cerbino and P. Pierobon, students at the School.

1. – Introduction

Entropy emerges as a classical thermodynamical concept in the 19th century with Clausius but it is only due to the work of Boltzmann and Gibbs that the idea of entropy becomes a cornerstone of statistical mechanics. As a result we have that the entropy S of a system is given by the so-called Boltzmann-Gibbs (BG) entropy

$$S_{\mathrm{BG}} = -k \sum_{i=1}^{W} p_i \ln p_i \tag{1}$$

with the normalization condition

$$\sum_{i=1}^{W} p_i = 1. \tag{2}$$

Here p_i is the probability for the system to be in the i-th microstate, and k is an arbitrary constant that, in the framework of thermodynamics, is taken to be the Boltzmann constant ($k_{\mathrm{B}} = 1.38 \times 10^{-23}$ J/K). Without loss of generality one can also arbitrarily assume $k = 1$. If every microstate has the same probability $p_i = 1/W$ (*equiprobability* assumption), one obtains the famous *Boltzmann principle*

$$S_{\mathrm{BG}} = k \ln W\,. \tag{3}$$

It can be easily shown that entropy (1) is *nonnegative*, *concave*, *extensive* and *stable* [1] (or *experimentally robust*). By extensive we mean the fact that, if A and B are two *independent* systems in the sense that $p_{ij}^{A+B} = p_i^A p_j^B$, then we straightforwardly verify that

$$S_{\mathrm{BG}}(A+B) = S_{\mathrm{BG}}(A) + S_{\mathrm{BG}}(B). \tag{4}$$

Stability will be addressed later on. One might naturally expect that the form (1) of S_{BG} would be rigorously *derived* from microscopic dynamics. However, the difficulty of performing such a program can be seen from the fact that still today this has not yet been accomplished from first principles. Consequently (1) is in practice a *postulate*. To better realize this point, let us place it on some historical background.

Albert Einstein says in 1910 [2]:

"In order to calculate W, one needs a *complete* (molecular-mechanical) theory of the system under consideration. Therefore it is dubious whether the Boltzmann principle has any meaning without a *complete* molecular mechanical theory or some other theory which describes the elementary processes. $S = k \ln W$ + *constant* seems without content, from a phenomenological point of view, without giving in addition such an *Elementartheorie*."

In his famous book *Thermodynamics*, Enrico Fermi says in 1936 [3]:

"The entropy of a system composed of several parts is very often equal to the sum of the entropies of all the parts. This is true if the energy of the system is the sum of the energies of all the parts and if the work performed by the system during a transformation is equal to the sum of the amounts of work performed by all the parts. Notice that these conditions are not quite obvious and that in some cases they may not be fulfilled."

Laszlo Tisza says in 1961 [4]:

"The situation is different for the additivity postulate $P\ a2$, the validity of which cannot be inferred from general principles. We have to require that the interaction energy between thermodynamic systems be negligible. This assumption is closely related to the homogeneity postulate $P\ d1$. From the molecular point of view, additivity and homogeneity can be expected to be reasonable approximations for systems containing many particles, provided that the intramolecular forces have a short range character."

Peter Landsberg says in 1978 [5]:

"The presence of long-range forces causes important amendments to thermodynamics, some of which are not fully investigated as yet."

If we put all this together, as well as many other similar statements available in the literature, we may conclude that physical entropies different from the BG one could exist which would be the appropriate ones for anomalous systems. Among the anomalies that we may focus on we include i) metaequilibrium (metastable) states in large systems involving long-range forces between particles; ii) metaequilibrium states in small systems, *i.e.*, whose number of particles is relatively small, say up to 100-200 particles; iii) glassy systems; iv) some classes of dissipative systems; v) mesoscopic systems with non-Markovian memory, and others which, in one way or another, might violate the usual simple ergodicity. Such systems might have a multifractal, scale-free or hierarchical structure in their phase space.

In this spirit, an entropy, S_q, which generalizes $S_{\rm BG}$, has been proposed in 1988 [6] as the basis for generalizing BG statistical mechanics. The entropy S_q (with $S_1 = S_{\rm BG}$) depends on the index q, a real number to be determined *a priori* from the microscopic dynamics. This entropy seems to describe a large number of natural and artificial systems quite well. As we shall see, the property chosen to be generalized is extensivity, *i.e.*, eq. (4). In this lecture we will introduce, through a *metaphor*, the form of S_q, and will then describe its properties and applications as they have emerged during the last 15 years.

A clarification might be worthy. Why introducing S_q through a metaphor, why not *deducing* it? If we knew how to deduce $S_{\rm BG}$ from first principles for those systems (*e.g.*, short-range-interacting Hamiltonian systems) whose microscopic dynamics ultimately leads to ergodicity, we could try to generalize along that path. But this procedure is still unknown, the form (1) being adopted, as we already mentioned, at the level of a *postulate*. It is clear that we are then obliged to do the same for any generalization of it. Indeed, there is no logical/deductive way to generalize any set of postulates that are useful for theoretical physics. The only way to do that is precisely through some kind of metaphor.

A statement through which we can measure the difficulty of (rigorously) making the main features of BG statistical mechanics to descend from (nonlinear) dynamics is that

of the mathematician Floris Takens. He said in 1991 [7]:
"The values of p_i are determined by the following dogma: if the energy of the system in the ith state is E_i and if the temperature of the system is T then: $p_i = \exp\{-E_i/kT\}/Z(T)$, where $Z(T) = \sum_i \exp\{-E_i/kT\}$, (this last constant is taken so that $\sum_i p_i = 1$). This choice of p_i is called *Gibbs distribution*. We shall give no justification for this dogma; even a physicist like Ruelle disposes of this question as 'deep and incompletely clarified'."

It is a tradition in mathematics to use the word "dogma" when no theorem is available. Perplexing as it might be for some readers, no theorem is available which establishes, on detailed microscopic dynamical grounds, the necessary and sufficient conditions for being valid the use of the celebrated BG factor. We may say that, at the bottom line, this factor is ubiquitously used in theoretical sciences because seemingly "it works" extremely well in an enormous amount of cases. It just happens that more and more systems (basically the so-called *complex systems*) are being identified nowadays where that statistical factor seems to *not* "work"!

2. – Mathematical properties

2·1. *A metaphor.* – The simplest ordinary differential equation one might think of is

$$\frac{\mathrm{d}y}{\mathrm{d}x} = 0\,, \tag{5}$$

whose solution (with initial condition $y(0) = 1$) is $y = 1$. The next simplest differential equation might be thought to be

$$\frac{\mathrm{d}y}{\mathrm{d}x} = 1\,, \tag{6}$$

whose solution, with the same initial condition, is $y = 1 + x$. The next one in increasing complexity that we might wish to consider is

$$\frac{\mathrm{d}y}{\mathrm{d}x} = y\,, \tag{7}$$

whose solution is $y = e^x$. Its inverse function is

$$y = \ln x\,, \tag{8}$$

which has the same functional form of the Boltzmann-Gibbs entropy (3), and satisfies the well-known additivity property

$$\ln(x_A x_B) = \ln x_A + \ln x_B. \tag{9}$$

A question that might be put is: can we unify all three cases (5), (6), (7) considered above? A trivial positive answer would be to consider $\mathrm{d}y/\mathrm{d}x = a + by$, and play with

(a, b). Can we unify with *only one* parameter? The answer still is positive, but this time *out of linearity*, namely with

$$\frac{\mathrm{d}y}{\mathrm{d}x} = y^q \quad (q \in \mathcal{R}), \tag{10}$$

which, for $q \to -\infty$, $q = 0$ and $q = 1$, reproduces respectively the differential equations (5), (6) and (7). The solution of (10) is given by the *q-exponential* function

$$y = [1 + (1-q)x]^{1/(1-q)} \equiv e_q^x \quad (e_1^x = e^x), \tag{11}$$

whose inverse is the *q-logarithm* function

$$y = \frac{x^{1-q} - 1}{1-q} \equiv \ln_q x \quad (\ln_1 x = \ln x). \tag{12}$$

This function satisfies the *pseudo-additivity* property

$$\ln_q(x_A x_B) = \ln_q x_A + \ln_q x_B + (1-q)(\ln_q x_A)(\ln_q x_B). \tag{13}$$

2·2. *The nonextensive entropy* S_q. – We can rewrite eq. (1) in a slightly different form, namely (with $k = 1$)

$$S_{\mathrm{BG}} = -\sum_{i=1}^{W} p_i \ln p_i = \sum_{i=1}^{W} p_i \ln \frac{1}{p_i} = \left\langle \ln \frac{1}{p_i} \right\rangle, \tag{14}$$

where $\langle \ldots \rangle \equiv \sum_{i=1}^{W} (\ldots) p_i$. The quantity $\ln(1/p_i)$ is sometimes called *surprise* or *unexpectedness*. Indeed, $p_i = 1$ corresponds to certainty, hence zero surprise if the *expected* event does occur; on the other hand, $p_i \to 0$ corresponds to nearly impossibility, hence infinite surprise if the *unexpected* event does occur. If we introduce the *q-surprise* (or *q-unexpectedness*) as $ln_q(1/p_i)$, it is kind of natural to define the following q-entropy:

$$S_q \equiv \left\langle \ln_q \frac{1}{p_i} \right\rangle = \sum_{i=1}^{W} p_i \ln_q \frac{1}{p_i} = \frac{1 - \sum_{i=1}^{W} p_i^q}{q-1}. \tag{15}$$

In the limit $q \to 1$ one has $p_i^q = p e^{(q-1)\ln p_i} \sim p_i[1 + (q-1)\ln p_i]$, and the entropy S_q coincides with the Boltzmann-Gibbs one, *i.e.*, $S_1 = S_{\mathrm{BG}}$. Assuming equiprobability (*i.e.*, $p_i = 1/W$) one obtains straightforwardly

$$S = \frac{W^{1-q} - 1}{1-q} = \ln_q W. \tag{16}$$

Consequently, it is clear that S_q is a generalization of and not an alternative to the Boltzmann-Gibbs entropy. The pseudo-additivity of the q-logarithm immediately implies (for the following formula we restore arbitrary k)

$$\frac{S_q(A+B)}{k} = \frac{S_q(A)}{k} + \frac{S_q(B)}{k} + (1-q)\frac{S_q(A)}{k}\frac{S_q(B)}{k}, \tag{17}$$

if A and B are two *independent* systems (*i.e.*, $p_{ij}^{A+B} = p_i^A p_j^B$). It follows that $q = 1$, $q < 1$ and $q > 1$, respectively, correspond to the *extensive*, *superextensive* and *subextensive* cases. It is from this property that the corresponding generalization of the BG statistical mechanics is often referred to as *nonextensive statistical mechanics*.

Equation (17) is true under the hypothesis of independency between A and B. But if they are correlated in some special, strong way, it may exist q^* such that

$$S_{q^*}(A+B) = S_{q^*}(A) + S_{q^*}(B), \tag{18}$$

thus recovering extensivity, but of a *different* entropy, not the usual one! Let us illustrate this interesting point through two examples:

i) A system of N nearly independent elements yields $W(N) \sim \mu^N$ (with $\mu > 1$) (*e.g.*, $\mu = 2$ for a coin, $\mu = 6$ for a dice). Its entropy S_q is given by

$$S_q(N) = \ln_q W(N) \sim \frac{\mu^{N(1-q)} - 1}{1-q} \tag{19}$$

and extensivity is obtained *if and only if* $q = 1$. In other words, $S_1(N) \sim N \ln \mu \propto N$. This is the usual case, discussed in any textbook.

ii) A system whose elements are correlated at all scales might correspond to $W(N) \sim N^\rho$ (with $\rho > 0$). Its entropy S_q is given by

$$S_q(N) = \ln_q W(N) \sim \frac{N^{\rho(1-q)} - 1}{1-q} \tag{20}$$

and extensivity is obtained *if and only if* $q = q^* \equiv 1 - (1/\rho) < 1$. In other words, $S_{q^*}(N) \propto N$. It is allowed to think that such possibility could please as much as the usual one (present example i)) somebody —like Clausius— wearing the "glasses" of classical thermodynamics! Indeed, it does not seem that Clausius had in mind any specific functional form for the concept of entropy he introduced, but he was surely expecting it to be proportional to N for large N. We see, in both examples that we have just analyzed, that it can be so. In particular, the entropy S_q is nonextensive for independent systems, but *can be extensive for systems whose elements are highly correlated.*

2·3. *The 1988 derivation of the entropy* S_q. – Let us point out that the 1988 derivation [6] of the entropy S_q was slightly different. If we consider $0 < p_i < 1$ and $q > 0$ we have that $p_i^q \gtrless p_i$ if $q \lessgtr 1$ and $p_i^q = p_i$ if $q = 1$. Many physical phenomena seem to be controlled either by the rare events or the common events. Similarly to what occurs in multifractals, one could take such effect into account by choosing $q < 1$ (hence *low probabilities are enhanced and high probabilities are depressed*; indeed, $\lim_{p_i \to 0} p_i^q/p_i \to \infty$) or $q > 1$ (hence *the other way around*; indeed, $\lim_{p_i \to 0} p_i^q/p_i = 0$). The present index q is completely different from the multifractal index (currently noted q_{M} in the nonextensive statistical literature to avoid confusion, where M stands for *multifractal*), but the type of bias is obviously analogous. What happens then if we introduce this type of *bias* into the entropy itself?

At variance with the concept of energy, the entropy should *not* depend on the physical support of the information, hence it must be invariant under permutations of the events. The most natural choice is then to make it depend on $\sum_{i=1}^W p_i^q$. The simplest dependence being the linear one, one might propose $S_q = A + B\sum_{i=1}^W p_i^q$. If some event has $p_i = 1$ we have certainty, and we consequently expect $S_q = 0$, hence $A + B = 0$, hence $S(\{p_i\}) = A(1 - \sum_{i=1}^W p_i^q)$. We impose now $S_1 = S_{\mathrm{BG}}$ and, taking into account $p_i^q = p_i e^{(q-1)\ln p_i} \sim p_i[1 + (q-1)\ln p_i]$, we obtain $S_1 \sim -A(q-1)\sum_{i=1}^W p_i \ln p_i$. So we choose $A = 1/(q-1)$ and straightforwardly reobtain (15).

2·4. S_q *and the Shannon-Khinchin axioms.* – Shannon in 1948 and then Khinchin in 1953 gave quite similar sets of axioms about the form of the entropic functional [8, 9]. Under reasonable requests about entropy they obtained that the only functional form allowed is the Boltzmann-Gibbs entropy.

Shannon theorem:

1. $S(p_1, \ldots, p_W)$ continuous function with respect to all its arguments
2. $S(p_1 = p_2 = \cdots = p_W = \frac{1}{W})$ monotonically increases with W
3. $S(A+B) = S(A) + S(B)$ if $p_{ij}^{A+B} = p_i^A p_j^B$
4. $S(p_1, \ldots, p_W) = S(p_L, p_M) + p_L S(\frac{p_1}{p_L}, \ldots, \frac{p_{W_L}}{p_L}) + p_M S(\frac{p_{W_L+1}}{p_M}, \ldots, \frac{p_W}{p_M})$, where $W = W_L + W_M$, $p_L = \sum_{i=1}^{W_L} p_i$, $p_M = \sum_{i=W_L+1}^{W} p_i$ (hence $p_L + p_M = 1$)

if and only if $S(p_1, \ldots, p_W)$ is given by eq. (1).

Khinchin theorem:

1. $S(p_1, \ldots, p_W)$ continuous function with respect to all its arguments
2. $S(p_1 = p_2 = \cdots = p_W = \frac{1}{W})$ monotonically increases with W
3. $S(p_1, \ldots, p_W, 0) = S(p_1, \ldots, p_W)$
4. $S(A+B) = S(A) + S(B|A)$, $S(B|A)$ being the conditional entropy

if and only if $S(p_1, \ldots, p_W)$ is given by eq. (1).

The following generalizations of these theorems have been given by Santos in 1997 [10] and by Abe in 2000 [11]. The latter was first conjectured by Plastino and Plastino in 1996 [12] and 1999 [13].

Santos theorem:

1. $S(p_1, \ldots, p_W)$ continuous function with respect to all its arguments
2. $S(p_1 = p_2 = \cdots = p_W = \frac{1}{W})$ monotonically increases with W
3. $\frac{S(A+B)}{k} = \frac{S(A)}{k} + \frac{S(B)}{k} + (1-q)\frac{S_q(A)}{k}\frac{S_q(B)}{k}$, if $p_{ij}^{A+B} = p_i^A p_j^B$
4. $S(p_1, \ldots, p_W) = S(p_L, p_M) + p_L^q S(\frac{p_1}{p_L}, \ldots, \frac{p_{W_L}}{p_L}) + p_M^q S(\frac{p_{W_L+1}}{p_M}, \ldots, \frac{p_W}{p_M})$, where $W = W_L + W_M$, $p_L = \sum_{i=1}^{W_L} p_i$, $p_M = \sum_{i=W_L+1}^{W} p_i$ (hence $p_L + p_M = 1$)

if and only if

(21)
$$S(p_1, \ldots, p_W) = k\frac{1 - \sum_{i=1}^{W} p_i^q}{q-1} \,.$$

Abe theorem:

1. $S(p_1, \ldots, p_W)$ continuous function with respect to all its arguments
2. $S(p_1 = p_2 = \cdots = p_W = \frac{1}{W})$ monotonically increases with W
3. $S(p_1, \ldots, p_W, 0) = S(p_1, \ldots, p_W)$
4. $\frac{S(A+B)}{k} = \frac{S(A)}{k} + \frac{S(B|A)}{k} + (1-q)\frac{S_q(A)}{k}\frac{S(B|A)}{k}$, $S(B|A)$ being the conditional entropy

if and only if $S(p_1, \ldots, p_W)$ is given by eq. (21).

The Santos and the Abe theorems are clearly important. Indeed, they show that the entropy S_q is the *only* possible entropy that extends the Boltzmann-Gibbs entropy maintaining the basic properties but allowing, if $q \neq 1$, nonextensivity (of the form of Santos' third axiom, or Abe's fourth axiom).

2·5. *Other mathematical properties.*

Reaction under bias: It has been shown [14] that the Boltzmann-Gibbs entropy can be rewritten as

(22)
$$S_{\mathrm{BG}} = -\left[\frac{\mathrm{d}}{\mathrm{d}x}\sum_{i=1}^{W} p_i^x\right]_{x=1} \,.$$

This can be seen as a reaction to a translation of the bias x in the same way as differentiation can be seen as a reaction of a function under a (small) *translation* of the abscissa. Along the same line, the entropy S_q can be rewritten as

$$S_q = -\left[D_q \sum_{i=1}^{W} p_i^x\right]_{x=1}, \tag{23}$$

where

$$D_q h(x) \equiv \frac{h(qx) - h(x)}{qx - x} \quad \left(D_1 h(x) = \frac{\mathrm{d}h(x)}{\mathrm{d}x}\right) \tag{24}$$

is Jackson's 1909 generalized derivative, which can be seen as a reaction of a function under *dilatation* of the abscissa (or under a *finite* increment of the abscissa).

Concavity: If we consider two probability distributions $\{p_i\}$ and $\{p'_i\}$ for a given system ($i = 1, \ldots, W$), we can define the *convex sum* of the two probability distributions as

$$p''_i \equiv \mu p_i + (1-\mu)p'_i \quad (0 < \mu < 1). \tag{25}$$

An entropic functional $S(\{p_i\})$ is said *concave* if and only if for all μ and for all $\{p_i\}$ and $\{p'_i\}$

$$S(\{p''_i\}) \geq \mu S(\{p_i\}) + (1-\mu)S(\{p'_i\}). \tag{26}$$

By convexity we mean the same property where $\geq$ is replaced by $\leq$. It can be easily shown that the entropy S_q is concave (convex) for every $\{p_i\}$ and every $q > 0$ ($q < 0$). It is important to stress that this property implies, in the framework of statistical mechanics, thermodynamic stability, *i.e.*, stability of the system with regard to energetic perturbations. This means that the entropic functional is defined such that the stationary state (*e.g.*, thermodynamic equilibrium) makes it extreme (in the present case, maximum for $q > 0$ and minimum for $q < 0$). Any perturbation of $\{p_i\}$ which makes the entropy extreme is followed by a tendency toward $\{p_i\}$ once again. Moreover, such a property makes it possible for two systems at different temperature to equilibrate to a common temperature.

Stability or experimental robustness: An entropic functional $S(\{p_i\})$ is said *stable* or *experimentally robust* if and only if, for any given $\varepsilon > 0$, there exists $\delta_\varepsilon > 0$ such that, independently of W,

$$\sum_{i=1}^{W} |p_i - p'_i| \leq \delta_\varepsilon \Longrightarrow \left|\frac{S(\{p_i\}) - S(\{p'_i\})}{S_{\max}}\right| < \varepsilon\,. \tag{27}$$

This implies in particular that

$$\lim_{\varepsilon\to 0}\lim_{W\to\infty}\left|\frac{S(\{p_i\})-S(\{p_i'\})}{S_{\max}}\right| = \lim_{W\to\infty}\lim_{\varepsilon\to 0}\left|\frac{S(\{p_i\})-S(\{p_i'\})}{S_{\max}}\right| = 0. \tag{28}$$

Lesche [1] has argued that experimental robustness is a necessary requisite for an entropic functional to be a physical quantity because essentially assures that, under arbitrary small variations of the probabilities, the relative variation of entropy remains small. This property is to be not confused with the *thermodynamical stability*, considered above. It has been shown [15] that the entropy S_q exhibits, for any $q > 0$, this property.

2·6. *A remark on other possible generalizations of the Boltzmann-Gibbs entropy.* – There have been in the past other generalizations of the BG entropy. The *Renyi entropy* [16] is one of them and is defined as follows:

$$S_q^{\mathrm{R}} \equiv \frac{\ln \sum_{i=1}^{W} p_i^q}{1-q} = \frac{\ln[1+(1-q)S_q]}{1-q}\,. \tag{29}$$

Another entropy has been introduced by Landsberg and Vedral [17] and independently by Rajagopal and Abe [18]. It is sometimes called *normalized nonextensive entropy*, and is defined as follows:

$$S_q^{\mathrm{N}} \equiv S_q^{\mathrm{LVRA}} \equiv \frac{1-\left(1/\sum_{i=1}^{W} p_i^q\right)}{1-q} = \frac{S_q}{1+(1-q)S_q}\,. \tag{30}$$

A question arises naturally: Why not using one of these entropies (or even a different one such as the so-called *escort entropy* S_q^{E}, defined in [19, 20]), instead of S_q, for generalizing BG statistical mechanics? The answer appears to be quite straightforward. S_q^{R}, S_q^{LVRA} and S_q^{E} are not concave nor experimentally robust. Neither yield they a *finite* entropy production for unit time, in contrast with S_q, as we shall see later on. Moreover, these alternatives do not possess the suggestive structure that S_q exhibits associated with the Jackson generalized derivative. Consequently, *for thermodynamical purposes*, it seems nowadays quite natural to consider the entropy S_q as the best candidate for generalizing the Boltzmann-Gibbs entropy. It might be different for other purposes: for example, Renyi entropy is known to be useful for geometrically characterizing multifractals.

3. – Connection to thermodynamics

Dozens of entropic forms have been proposed during the last decades, but not all of them are necessarily related to the physics of nature. Statistical mechanics is more than the adoption of an entropy: the (meta)equilibrium probability distribution must not only optimize the entropy but also satisfy, in addition to the norm constraint, constraints on quantities such as the energy. Unfortunately, when the theoretical frame is generalized, it is not obvious which constraints are to be maintained and which ones are to be generalized, and in what manner. In this section we derive, following the lines of Gibbs, a thermodynamics for (meta)equilibrium distribution based on the entropy defined above. It should be stressed that the distribution derived in this way for $q \neq 1$ does *not* correspond to thermal equilibrium (as addressed within the BG formalism through the celebrated Boltzmann's *molecular chaos hypothesis*), but rather to a metaequilibrium or a stationary state, suitable to describe a large class of nonergodic systems.

3'1. *Canonical ensemble.* – For a system in thermal contact with a large reservoir, and in analogy with the path followed by Gibbs [6,19], we look for the distribution which optimizes the entropy S_q defined in eq. (21), with the normalization condition (2) and the following constraint on the energy [19]:

$$\frac{\sum_{i=1}^{W} p_i^q E_i}{\sum_{j=1}^{W} p_j^q} = U_q \,, \tag{31}$$

where

$$P_i \equiv \frac{p_i^q}{\sum_{j=1}^{W} p_j^q} \tag{32}$$

is called *escort distribution*, and $\{E_i\}$ are the eigenvalues of the system Hamiltonian with the chosen boundary conditions. Note that, in analogy with BG statistics, a constraint like $\sum_{i=1}^{W} p_i E_i = U$ would be more intuitive. This was indeed the first attempt [6], but though it correctly yields, as stationary (metaequilibrium) distribution, the q-exponential, it turns out to be inadequate for various reasons, including those related to Lévy-like superdiffusion, for which a diverging second moment exists.

Another natural choice [6, 21] would be to fix $\sum_{i=1}^{W} p_i^q E_i$ but, though this solves annoying divergences, it creates some new problems: the metaequilibrium distribution is *not* invariant under change of the zero level for the energy scale, $\sum_{i=1}^{W} p_i^q \neq 1$ implies that the constraint applied to a constant does *not* yield the same constant, and above all, the assumption $p_{ij}^{A+B} = p_i^A p_j^B$ and $E_{ij}^{A+B} = E_i^A + E_j^B$ does *not* yield $U_{ij}^{A+B} = U_i^A + U_j^B$, *i.e.*, the energy conservation principle is *not* the same in the microscopic and macroscopic worlds.

It is by now well established that the energy constraint must be imposed in the form (31), using the normalized escort distribution (32). A detailed discussion of this important point can be found in [22].

The optimization of S_q with the constraints (2) and (31) with Lagrange multiplier β yields

$$p_i = \frac{\left[1-(1-q)\beta_q(E_i - U_q)\right]^{1/(1-q)}}{\overline{Z}_q} = \frac{e_q^{-\beta_q(E_i-U_q)}}{\overline{Z}_q} \tag{33}$$

with

$$\overline{Z}_q \equiv \sum_{j=1}^{W} e_q^{-\beta_q(E_i - U_q)} \tag{34}$$

and

$$\beta_q \equiv \frac{\beta}{\sum_{j=1}^{W} p_j^q} . \tag{35}$$

It turns out that the metaequilibrium distribution can be written hidding the presence of U_q in a form which sometimes is more convenient when we want to use experimental or computational data:

$$p_i = \frac{\left[1 - (1-q)\beta_q' E_i\right]^{1/(1-q)}}{Z_q'} = \frac{e_q^{-\beta_q' E_i}}{Z_q'} \tag{36}$$

with

$$Z_q' \equiv \sum_{j=1}^{W} e_q^{-\beta_q' E_i} \tag{37}$$

and

$$\beta_q' \equiv \frac{\beta}{\sum_{j=1}^{W} p_j^q + (1-q)\beta U_q} . \tag{38}$$

It can be easily checked that i) for $q \to 1$, the BG weight is recovered, *i.e.*, $p_i = e^{-\beta E_i}/Z_1$, ii) for $q > 1$, a power law tail emerges, and iii) for $q < 1$, the formalism imposes a high-energy cut-off ($p_i = 0$) whenever the argument of the q-exponential function becomes negative.

Note that distribution (33) is generically not an exponential law, *i.e.*, it is generically *not* factorizable (under sum in the argument), and nevertheless is invariant under choice of the zero energy for the energy spectrum (this is one of the pleasant facts associated with the choice of energy constraint in terms of a normalized distribution like the escort one).

3·2. *Legendre structure.* – The Legendre-transformation structure of thermodynamics holds for every q (*i.e.*, it is q-invariant) and allows us to connect the theory developed so far to thermodynamics.

We verify that, for all values q,

$$\frac{1}{T} = \frac{\partial S_q}{\partial U_q} \quad \left(T \equiv \frac{1}{k\beta}\right). \tag{39}$$

Also, it can be proved that the free energy is given by

$$F_q \equiv U_q - TS_q = -\frac{1}{\beta}\ln_q Z_q\,, \tag{40}$$

where

$$\ln_q Z_q \equiv \frac{Z_q^{1-q}-1}{1-q} = \frac{\overline{Z}_q^{1-q}-1}{1-q} - \beta U_q\,, \tag{41}$$

and the internal energy is given by

$$U_q = -\frac{\partial}{\partial\beta}\ln_q Z_q\,. \tag{42}$$

Finally, the specific heat reads

$$C_q \equiv T\frac{\partial S_q}{\partial T} = \frac{\partial U_q}{\partial T} = -T\frac{\partial^2 F_q}{\partial T^2}\,. \tag{43}$$

In addition to the Legendre structure, many other theorems and properties are q-invariant, thus supporting the thesis that this is a right road for generalizing the BG theory. Let us briefly list some of them.

1. Boltzmann H-theorem (*macroscopic time irreversibility*):

$$q\frac{\mathrm{d}S_q}{\mathrm{d}t} \geq 0 \quad (\forall q). \tag{44}$$

   This inequality has been established under a variety of irreversible time evolution mesoscopic equations ([23, 24] and others), and is consistent with the second principle of thermodynamics [25], which turns out to be satisfied for all values of q.

2. Ehrenfest theorem (*correspondence principle between classical and quantum mechanics*): Given an observable $\widehat{\mathcal{O}}$ and the Hamiltonian $\widehat{\mathcal{H}}$ of the system, it can be shown (see [26] for unnormalized q-expectation values; for the normalized ones, the proof should follow along the same lines) that

$$\frac{\mathrm{d}\langle\widehat{\mathcal{O}}\rangle_q}{\mathrm{d}t} = \frac{i}{\hbar}\big\langle[\widehat{\mathcal{H}},\widehat{\mathcal{O}}]\big\rangle_q \quad (\forall q). \tag{45}$$

3. Factorization of the likelihood function (*thermodynamically independent systems*): The likelihood function satisfies [27-29]

$$W_q(\{p_i\}) \propto e_q^{S_q(\{p_i\})}\,. \tag{46}$$

Consequently, if A and B are two probabilistically independent systems, it can be verified that

$$W_q(A+B) = W_q(A)W_q(B) \quad (\forall q)\,, \tag{47}$$

as expected by Einstein [2].

4. Onsager reciprocity theorem (*microscopic time reversibility*): It has been shown [30-32] that the reciprocal linear coefficients satisfy

$$L_{jk} = L_{kj} \quad (\forall q), \tag{48}$$

thus satisfying the fourth principle of thermodynamics.

5. Kramers and Kronig relations (*causality*): They have been proved in [30] for all values of q.

6. Pesin theorem (*connection between sensitivity to the initial conditions and the entropy production per unit time*). We can define the *q-generalized Kolmogorov-Sinai entropy* as

$$K_q \equiv \lim_{t\to\infty} \lim_{W\to\infty} \lim_{N\to\infty} \frac{\langle S_q\rangle(t)}{t}\,, \tag{49}$$

where N is the number of initial conditions, W is the number of windows in the partition (*fine graining*) we have adopted, and t is the (discrete) time. Let us mention that the standard Kolmogorov-Sinai entropy is defined in a slightly different manner in the mathematical theory of nonlinear dynamical systems. See more details in [33] and references therein.

The *q-generalized Lyapunov coefficient* λ_q can be defined through the sensitivity to the initial conditions

$$\xi \equiv \lim_{\Delta x(0)\to 0} \frac{\Delta x(t)}{\Delta x(0)} = e_q^{\lambda_q t}\,, \tag{50}$$

where we have focused on a one-dimensional system (basically $x(t+1) = g(x(t))$, $g(x)$ being a nonlinear function, for example that of the logistic map). It was conjectured in 1997 [34], and recently proved for unimodal maps [35], that they are related through

$$K_q = \begin{cases} \lambda_q\,, & \text{if } \lambda_q > 0, \\ 0, & \text{otherwise.} \end{cases} \tag{51}$$

To be more explicit, we have $K_1 = \lambda_1$ if $\lambda_1 \geq 0$ (and $K_1 = 0$ if $\lambda_1 < 0$). But if we have $\lambda_1 = 0$, then we have a special value of q such that $K_q = \lambda_q$ if $\lambda_q \geq 0$ (and $K_q = 0$ if $\lambda_q < 0$).

Notice that the q-invariance of all the above properties is kind of natural. Indeed, their origins essentially lie in mechanics, and what we have generalized is *not* mechanics but only the concept of information upon it.

4. – Applications

The ideas related with nonextensive statistical mechanics have received an enormous amount of applications in a variety of disciplines including physics, chemistry, mathematics, geophysics, biology, medicine, economics, informatics, geography, engineering, linguistics and others. For description and details about these, we refer the reader to [22, 36] as well as to the bibliography in [37]. The *a priori* determination (from microscopic or mesoscopic dynamics) of the index q is illustrated for a sensible variety of systems in these references. This point obviously is a very relevant one, since otherwise the present theory would *not* be *complete*.

In the present brief introduction we shall address only two types of systems, namely a long-range-interacting many-body classical Hamiltonian, and the logistic-like class of one-dimensional dissipative maps. The first system is still under study (*i.e.*, it is only partially understood), but we present it here because it might constitute a direct application of the thermodynamics developed in sect. **3**. The second system is considerably better understood, and illustrates the various concepts which appear to be relevant in the present generalization of the BG ones.

4·1. *Long-range-interacting many-body classical Hamiltonians.* – To illustrate this type of system, let us first focus on the inertial XY ferromagnetic model, characterized by the following Hamiltonian [38, 39]:

$$\mathcal{H} = \sum_{i=1}^{N} \frac{p_i^2}{2} + \sum_{i \neq j} \frac{1 - \cos(\theta_i - \theta_j)}{r_{ij}^{\alpha}} \qquad (\alpha \geq 0), \tag{52}$$

where θ_i is the i-th angle and p_i the conjugate variable representing the angular momentum (or the rotational velocity since, without loss of generality, unit moment of inertia is assumed).

The summation in the potential is extended to all couples of spins (counted only once) and not restricted to first neighbors; for $d = 1$, $r_{ij} = 1, 2, 3, \ldots$; for $d = 2$, $r_{ij} = 1, \sqrt{2}, 2, \ldots$; for $d = 3$, $r_{ij} = 1, \sqrt{2}, \sqrt{3}, 2, \ldots$. The first-neighbor coupling constant has been assumed, without loss of generality, to be equal to unity. This model is an inertial version of the well-known XY ferromagnet. Although it does not make any relevant difference, we shall assume periodic boundary conditions, the distance to be considered between a given pair of sites being the smallest one through the $2d$ possibilities introduced by the periodicity of the lattice. Notice that the two-body potential term has been written in such a way as to have zero energy for the global fundamental state (corresponding to $p_i = 0$, $\forall i$, and all θ_i equal among them, and equal to say zero). The $\alpha \to \infty$ limit corresponds to only first-neighbor interactions, whereas the $\alpha = 0$ limit corresponds to infinite-range interactions (a typical Mean Field situation, frequently referred to as the HMF model [38]).

The quantity $\tilde{N} \equiv \sum_{i \neq j} r_{ij}^{-\alpha}$ corresponds essentially to the potential energy *per rotator*. This quantity, in the limit $N \to \infty$, converges to a finite value if $\alpha/d > 1$, and diverges like $N^{1-\alpha/d}$ if $0 \leq \alpha/d < 1$ (like $\ln N$ for $\alpha/d = 1$). In other words, the energy is extensive for $\alpha/d > 1$ and nonextensive otherwise. In the extensive case (here referred to as *short-range interactions*; also referred to as *integrable interactions* in the literature), the thermal equilibrium (stationary state attained in the $t \to \infty$ limit) is known to be the BG one (see [40]). The situation is much more subtle in the nonextensive case (*long-range interactions*). It is this situation that we focus on here. $\tilde{N}$ behaves like $\int_1^{N^{1/d}} dr\, r^{d-1} r^{-\alpha} \propto ((N^{1-\alpha/d} - 1)/(1-\alpha/d)) \sim ((N^{1-\alpha/d} - \alpha/d)/(1-\alpha/d))$. All these three equivalent quantities ($\tilde{N}$ or $(N^{1-\alpha/d} - 1)/(1-\alpha/d)$ or $(N^{1-\alpha/d} - \alpha/d)/(1-\alpha/d)$) are indistinctively used in the literature to scale the energy per particle of such long-range systems. In order to conform to the most usual writing, we shall from now on replace the Hamiltonian $\mathcal{H}$ by the following rescaled one:

$$\mathcal{H}' = \sum_{i=1}^{N} \frac{p_i^2}{2} + \frac{1}{\tilde{N}} \sum_{i \neq j} \frac{1 - \cos(\theta_i - \theta_j)}{r_{ij}^{\alpha}} \quad (\alpha \geq 0). \tag{53}$$

The molecular dynamical results associated with this Hamiltonian (now artificially transformed into an extensive one for *all* values of α/d) can be trivially transformed into those associated with Hamiltonian $\mathcal{H}$ by re-scaling time (see [39]).

Hamiltonian (53) exhibits in the microcanonical case (isolated system at fixed total energy U) a second-order phase transition at $u \equiv U/N = 0.75$. It has anomalies both above and below this critical point.

Above the critical point it has a Lyapunov spectrum which, in the $N \to \infty$ limit, approaches, for $0 \leq \alpha/d \leq 1$, zero as $N^{-\kappa}$, where $\kappa(\alpha/d)$ decreases from 1/3 to zero when α/d increases from zero to unity, and remains zero for $\alpha/d \geq 1$ [39, 41]. It has a Maxwellian distribution of velocities [42], and exhibits no aging [43]. Although it has no aging, the typical correlation functions depend on time as a q-exponential. Diffusion is shown to be of the normal type.

Below the critical point (*e.g.*, $u = 0.69$), for a nonzero-measure class of initial conditions, a longstanding quasistationary (or metastable) state precedes the arrival to the BG thermal equilibrium state. The duration of this quasistationary state appears to diverge with N like $\tilde{N}$ [42, 44]. During this anomalous state, there is aging (the correlation functions being well reproduced by q-exponentials once again), and the velocity distribution is not Maxwellian, but rather approaches a q-exponential function (with a cut-off at high velocities, as expected for any microcanonical system). Anomalous superdiffusion is shown to exist in this state. The mean kinetic energy ($\propto T$, where T is referred to as the *dynamical temperature*) slowly approaches the BG value from below, the relaxation function being once again a q-exponential one. During the anomalous aging state, the zeroth principle of thermodynamics and the basic laws of thermometry have been shown to hold as usual [45, 46]. The fact that such basic principles are preserved constitutes a major feature, pointing towards the applicability of thermostatistical arguments and methods to this highly nontrivial quasistationary state.

Although none of the above indications constitutes a proof that this long-range system obeys, in one way or another, nonextensive statistical mechanics, the set of so many consistent evidences may be considered as a very strong suggestion that so it is. Anyhow, work is in progress to verify closely this tempting possibility (see also [47]).

Similar observations are in progress for the Heisenberg version of the above Hamiltonian [48], as well as for an XY model including a local term which breaks the angular isotropy in such a way as to make the model to approach the Ising model [49].

Lennard-Jones small clusters (with N up to 14) have been numerically studied recently [50]. The distributions of the number of local minima of the potential energy with k neighboring saddle-points in the configurational phase space can, although not mentioned in the original paper [50], be quite well fitted with q-exponentials with $q = 2$. No explanation is still available for this suggestive fact. Qualitatively speaking, however, the fact that we are talking of very *small* clusters means that, despite the fact that the Lennard-Jones interaction is not a long-range one thermodynamically speaking (since $\alpha/d = 6/3 > 1$), all the atoms sensibly see each other, therefore fulfilling a nonextensive scenario.

4·2. *The logistic-like class of one-dimensional dissipative maps.* – Although low-dimensional systems are often an idealized representation of physical systems, they sometimes offer the rare opportunity to obtain analytical results that give profound insight in the comprehension of natural phenomena. This is certainly the case of the logistic-like class of one-dimensional dissipative maps, that may be described by the iteration rule

$$x_{t+1} = f(x_t) \equiv 1 - a|x_t|^z, \tag{54}$$

where $x \in [-1, 1]$, $t = 0, 1, \ldots$ is a discrete-time variable, $a \in [0, 2]$ is a control parameter, and $z > 1$ characterizes the universality class of the map (for $z = 2$ one obtains one of the possible forms of the very well-known logistic map). In particular, this class of maps captures the essential mechanisms of chaos for dissipative systems, like the period-doubling and the intermittency routes to chaos, and constitute a prototypical example that is reported in almost all textbooks in the area (see, *e.g.*, [51, 52]).

As previously reported, the usual BG formalism applies to situations where the dynamics of the system is sufficiently chaotic, *i.e.*, it is characterized by *positive* Lyapunov coefficients. A central result in chaos theory, valid for several classes of systems, is the identity between (the sum of positive) Lyapunov coefficients and the entropy growth (for Hamiltonian systems this result is called Pesin theorem). The failure of the classical BG formalism and the possible validity of the nonextensive one is related to the vanishing of the classical Lyapunov coefficients $\lambda \equiv \lambda_1$ and to their replacement by the generalized ones λ_q (see eq. (50)).

Reminding that the sensitivity to initial conditions of one-dimensional maps is associated to a single Lyapunov coefficient, the Lyapunov spectra of the logistic map ($z = 2$), as a function of the parameter a, are displayed in fig. 1, together with the attractor $x_\infty = \{x \in [-1,1] : x = \lim_{t\to\infty} x_t\}$. For a smaller than a critical value $a_c = 1.401155198\ldots$, a zero Lyapunov coefficient is associated to the pitchfork bifurcations (period-doubling); while for $a > a_c$ the Lyapunov coefficient vanishes for example in correspondence of the tangent bifurcations that generate the periodic windows inside the chaotic region. In ref. [53], using a renormalization-group (RG) analysis, it has been (exactly) proven that the nonextensive formalism describes the dynamics associated to these critical points. The sensitivity to initial conditions is in fact given by the q-exponential eq. (50), with $q = 5/3$ for pitchfork bifurcations and $q = 3/2$ for tangent bifurcations of any nonlinearity z, while λ_q depends on the order of the bifurcation. It is worthwhile to notice that these values are *not* deduced from fitting; instead, they are analytically calculated by means of the RG technique that describes the (universal) dynamics of these critical points.

Perhaps the most fascinating point of the logistic map is the *edge of chaos* $a = a_c$, that separates regular behavior from chaoticity. It is another point where the Lyapunov coefficient λ vanishes, so that no nontrivial information about the dynamics is attainable using the classical approach. Nonetheless, once again the RG approach reveals to be extremely powerful. Let us focus, for definiteness, on the case of the logistic map $z = 2$. Using the Feigenbaum-Coullet-Tresser RG transformation one can in fact show (see [54, 35] for details) that the dynamics can be described by a series of subsequences labelled by $k = 0, 1, \ldots$, characterized by the shifted iteration time $t_k(n) = (2k+1)2^{n-k} - 2k - 1$ (n is a natural number satisfying $n \geq k$), that are related to the bifurcation mechanism. For each of these subsequences, the sensitivity to initial conditions is given by the q-exponential eq. (50). The value of q (that is the same for all the subsequences) and $\lambda_q^{(k)}$ are deduced by one of Feigenbaum's universal constants $\alpha_F = 2.50290\ldots$ and are given by

$$(55) \qquad q = 1 - \frac{\ln 2}{\ln \alpha_F} = 0.2445\ldots \qquad \text{and} \qquad \lambda_q^{(k)} = \frac{\ln \alpha_F}{((2k+1)\ln 2)}\,.$$

In fig. 2(b) this function is drawn for the first subsequence ($k = 0$), together with the result of a numerical simulation. For comparison purposes, fig. 2(a) shows that when the map is fully chaotic ξ grows exponentially with the iteration time, with the Lyapunov coefficient $\lambda = \ln 2$ for $a = 2$.

For the edge of chaos it is also possible to proceed a step further and consider the entropy production associated to an ensemble of copies of the map, set initially out of equilibrium. Remarkably enough, if (and only if) we consider the entropy S_q precisely with $q = 0.2445\ldots$ for the definition of K_q (see eq. (49)), for all the subsequences we obtain a *linear* dependence of S_q with (shifted) time, *i.e.*, a generalized version of the Pesin identity:

$$(56) \qquad K_q^{(k)} = \lambda_q^{(k)}.$$

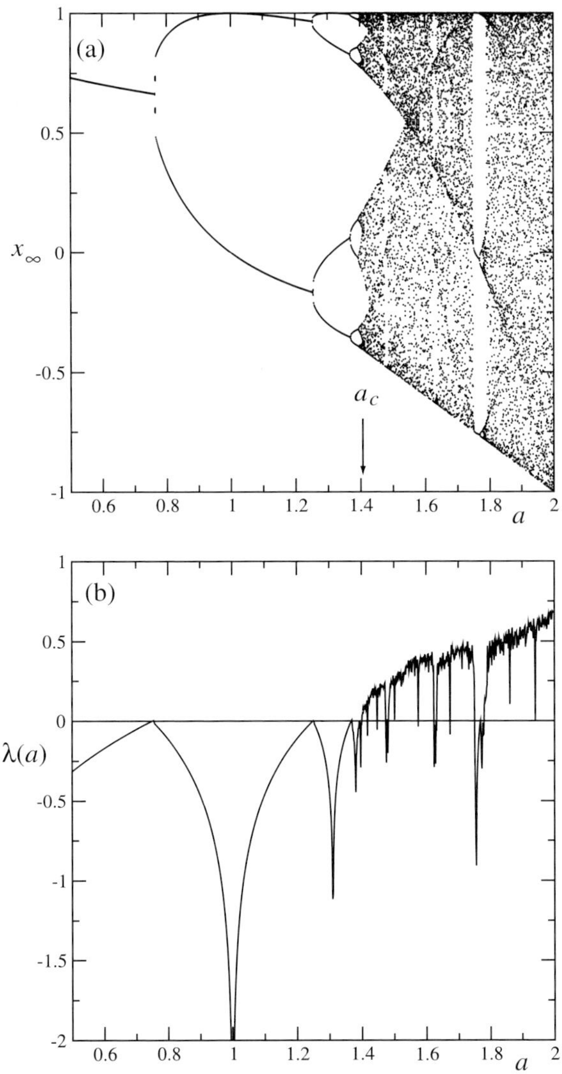

Fig. 1. – (a) Attractor of the logistic map ($z = 2$) as a function of a. The edge of chaos is at the critical value $a_c = 1.401155198\ldots$. (b) Lyapunov exponent λ as a function of a.

Figure 3(b) shows a numerical corroboration of this analytical result. The q-logarithm of the sensitivity to initial conditions plotted as a function of S_q displays in fact a 45° straight line for all iteration steps. Again, fig. 3(a) presents the analogous result obtained for the chaotic situation $a = 2$ using the BG entropy. The inset of fig. 3(b) gives an insight of fig. 2(b), showing that the linearity of the q-logarithm of ξ with the iteration time is valid for all the subsequences, once the shifted time t_k is used.

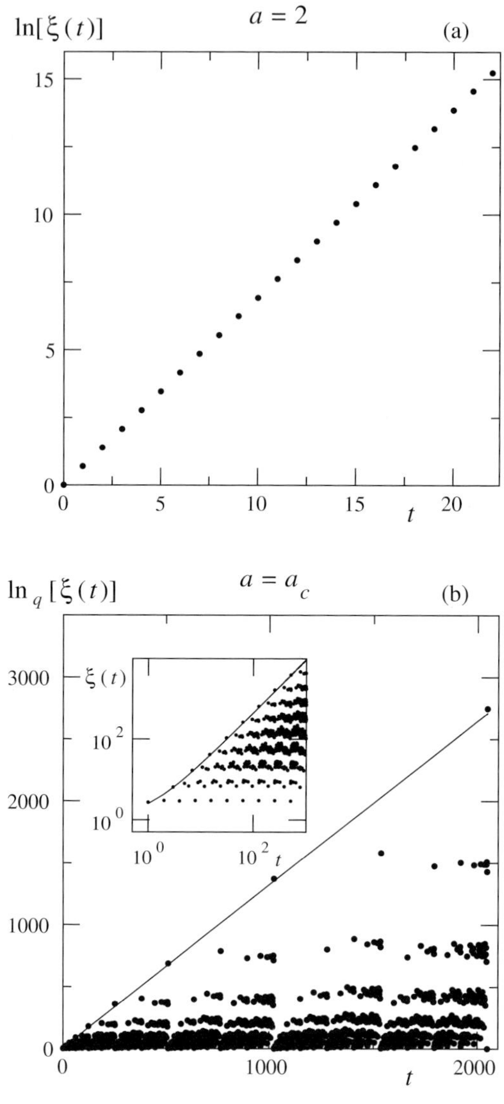

Fig. 2. – Sensitivity to initial conditions for the logistic map ($z = 2$). The dots represent $\xi(t)$ for two initial data started at $x_0 = 1/2$ and $x'_0 \sim 1/2 + 10^{-8}$ (a), and $x_0 = 0$ and $x'_0 \sim 10^{-8}$ (b). For $a = 2$ the log-linear plot displays a linear increase with a slope $\lambda = \ln 2$. For $a = a_c$ the q-log-linear plot displays a linear increase of the upper bound (sequence $k = 0$). The solid line is the function in eq. (50), with $q = 0.2445\ldots$ and $\lambda_q = \ln \alpha_F / \ln 2 = 1.3236\ldots$. In the inset of (b) the same data are represented in a log-log plot.

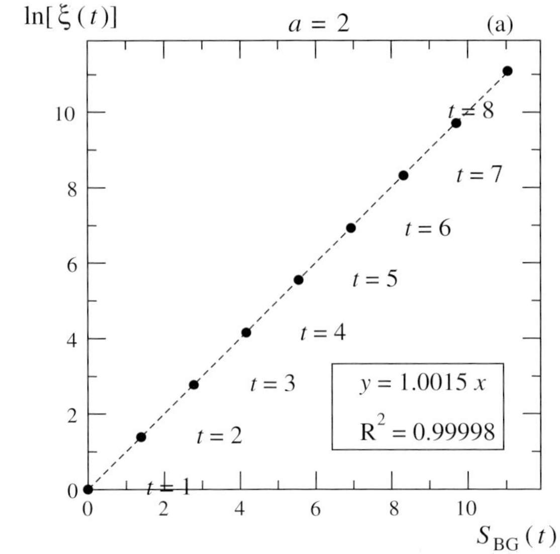

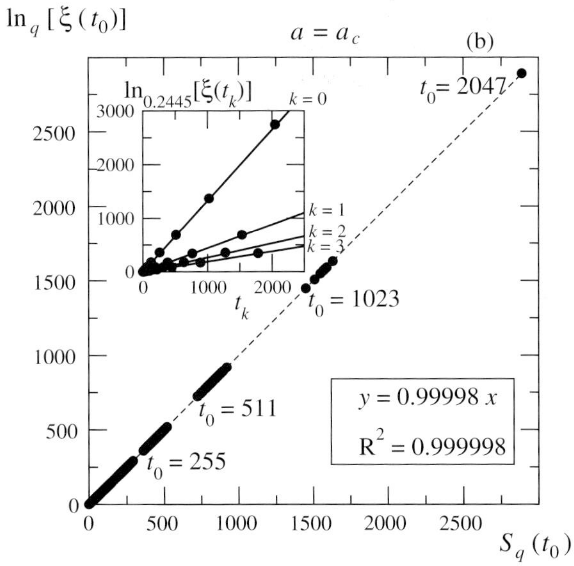

Fig. 3. – Numerical corroboration (full circles) of the generalized Pesin identity $K_q^{(k)} = \lambda_q^{(k)}$ for the logistic map. On the vertical axis we plot the q-logarithm of ξ (equal to $\lambda_q^{(k)} t_k$) and in the horizontal axis S_q (equal to $K_q^{(k)} t_k$). Dashed lines are linear fittings. (a) For $a = 2$ the identity is obtained using the BG formalism $q = 1$; while (b) at the edge of chaos $q = 0.2445\ldots$ must be used. Numerical data in (b) are obtained partitioning the interval $[-1, 1]$ into cells of equal size 10^{-9} and considering a uniform distribution of 10^5 points inside the interval $[0, 10^{-9}]$ as initial ensemble; ξ is calculated using, as inital conditions, the extremal points of this same interval. A similar setup gives the numerical results in (a). In the inset of (b) we plot the q-logarithm of ξ as a function of the shifted time $t_k = (2k+1)2^{n-k} - 2k - 1$. Full lines are from the analytical result eq. (50).

To conclude this illustration of low-dimensional nonextensivity, it is worthy to explicitly mention that the nontrivial value $q(z)$ (with $q(2) = 0.2445\ldots$) can be obtained from microscopic dynamics through at least *four* different procedures. These are: i) from the *sensitivity to the initial conditions*, as lengthily exposed above and in [34, 54, 55]; ii) from *multifractal geometry*, using

$$\frac{1}{1-q(z)} = \frac{1}{\alpha_{\min}(z)} - \frac{1}{\alpha_{\max}(z)} = \frac{(z-1)\ln\alpha_{\mathrm{F}}(z)}{\ln 2}, \tag{57}$$

whose details can be found in [56]; iii) from *entropy production per unit time*, as exposed above and in [33, 35]; and iv) from *relaxation associated with the Lebesgue measure shrinking*, as can be seen in [57] (see also [58]).

5. – Final remarks

Classical thermodynamics, valid for both classical and quantum systems, is essentially based on the following principles: 0th principle (*transitivity of the concept of thermal equilibrium*), 1st principle (*conservation of the energy*), 2nd principle (*macroscopic irreversibility*), 3rd principle (*vanishing entropy at vanishing temperature*), and 4th principle (*reciprocity of the linear nonequilibrium coefficients*). All these principles are since long known to be satisfied by Boltzmann-Gibbs statistical mechanics. However, a natural question arises: *Is BG statistical mechanics the only one capable of satisfying these basic principles?* The answer is *no*. Indeed, the present nonextensive statistical mechanics appears to also satisfy all these five principles (thermal equilibrium being generalized into stationary or quasistationary state or, generally speaking, *metaequilibrium*), as we have argued along the present review. The second principle in particular has received a new confirmation very recently [25].

The connections between the BG entropy and the BG exponential energy distribution are since long established through various standpoints, namely steepest descent, large numbers, microcanonical counting and variational principle. The corresponding q-generalization is equally available in the literature nowadays. Indeed, through *all these procedures*, the entropy S_q has been connected to the q-exponential energy distribution, in particular in a series of works by Abe and Rajagopal (see [22, 36] and references therein).

In addition to all this, S_q shares with the BG entropy concavity, stability, finiteness of the entropy production per unit time. Other well-known entropies, such as the Renyi one for instance, do not.

Summarizing, the dynamical scenario which emerges is that whenever ergodicity (or at least an ergodic sea) is present, one expects the BG concepts to be the adequate ones. But when ergodicity fails, very particularly when it does so in a special hierarchical (possibly multifractal) manner, one might expect the present nonextensive concepts to naturally take place. Furthermore, we conjecture that, in such cases, the visitation of phase space occurs through some kind of scale-free topology.

* * *

The present effort has benefited from partial financial support by CNPq, Pronex/MCT, Capes and Faperj (Brazilian agencies) and SIF (Italy).

REFERENCES

[1] LESCHE B., *J. Stat. Phys.*, **27** (1982) 419.
[2] EINSTEIN A., *Ann. Phys. (Leipzig)*, **33** (1910) 1275 (Translation: PAIS A., *Subtle is the Lord...* (Oxford University Press) 1982).
[3] FERMI E., *Thermodynamics* (Prentice-Hall, New York) 1937.
[4] TISZA L., *Ann. Phys. (N.Y.)*, **13** (1961) 1 (or in *Generalized Thermodynamics* (MIT Press, Cambridge) 1966, p. 123).
[5] LANDSBERG P. T., *Thermodynamics and Statistical Mechanics* (Oxford University Press, Oxford) 1978; also (Dover) 1990, p. 102.
[6] TSALLIS C., *J. Stat. Phys.*, **52** (1988) 479.
[7] TAKENS F., in *Structures in Dynamics - Finite Dimensional Deterministic Studies*, edited by BROER H. W., DUMORTIER F., VAN STRIEN S. J. and TAKENS F. (North-Holland, Amsterdam) 1991, p. 253.
[8] SHANNON C. E. and WEAVER W., *The Mathematical Theory of Communication* (Urbana University of Illinois Press, Urbana) 1962.
[9] KHINCHIN A. I., *Mathematical Foundations of Information Theory* (Dover, New York) 1957.
[10] SANTOS R. J. V., *J. Math. Phys.*, **38** (1997) 4104.
[11] ABE S., *Phys. Lett. A*, **271** (2000) 74.
[12] PLASTINO A. R. and PLASTINO A., *Condensed Matter Theories*, edited by LUDENA E., Vol. **11** (Nova Science Publishers, New York) 1996, p. 327.
[13] PLASTINO A. and PLASTINO A. R., in *Nonextensive Statistical Mechanics and Thermodynamics*, edited by SALINAS S. R. A. and TSALLIS C., *Braz. J. Phys.*, **29** (1999) 50.
[14] ABE S., *Phys. Lett. A*, **224** (1997) 326.
[15] ABE S., *Phys. Rev. E*, **66** (2002) 046134.
[16] RENYI A., *Proceedings of the 4th Berkeley Symposium* (University of California Press, Berkeley, Los Angeles) 1961, p. 547.
[17] LANDSBERG P. T. and VEDRAL V., *Phys. Lett. A*, **247** (1998) 211.
[18] RAJAGOPAL A. K. and ABE S., *Phys. Rev. Lett.*, **83** (1999) 1711.
[19] TSALLIS C., MENDES R. S. and PLASTINO A. R., *Physica A*, **261** (1998) 534.
[20] TSALLIS C. and BRIGATTI E., to appear in *Non additive entropy and non extensive statistical mechanics*, in *Continuum Mechanics and Thermodinamics*, Vol. **16**, edited by SUGIYAMA M.(Springer Verlag, Heidelberg) 2004.
[21] CURADO E. M. F. and TSALLIS C., *J. Phys. A*, **24** (1991) L69 (Corrigenda: **24** (1991) 3187 and **25** (1992) 1019).
[22] TSALLIS C., *What should a statistical mechanics satisfy to reflect nature?* in *Anomalous Distributions, Nonlinear Dynamics and Nonextensivity*, edited by SWINNEY H. L. and TSALLIS C., to be published in *Physica D* (2004).
[23] MARIZ A. M., *Phys. Lett. A*, **165** (1992) 409.
[24] RAMSHAW J. D., *Phys. Lett. A*, **175** (1993) 169.
[25] ABE S. and RAJAGOPAL A. K., *Phys. Rev. Lett.*, **91** (2003) 120601.
[26] PLASTINO A. R. and PLASTINO A., *Phys. Lett. A*, **177** (1993) 384.

[27] CACERES M. O. and TSALLIS C., private discussion (1993).
[28] CHAME A. and DE MELLO V. M., *J. Phys. A*, **27** (1994) 3663.
[29] TSALLIS C., *Chaos, Solitons Fractals*, **6** (1995) 539.
[30] RAJAGOPAL A. K., *Phys. Rev. Lett.*, **76** (1996) 3469.
[31] CACERES M. O., *Physica A*, **218** (1995) 471.
[32] CHAME A. and DE MELLO V. M., *Phys. Lett. A*, **228** (1997) 159.
[33] LATORA V., BARANGER M., RAPISARDA A. and TSALLIS C., *Phys. Lett. A*, **273** (2000) 97.
[34] TSALLIS C., PLASTINO A. R. and ZHENG W. M., *Chaos, Solitons Fractals*, **8** (1997) 885.
[35] BALDOVIN F. and ROBLEDO A., cond-mat/0304410, to be published in *Phys. Rev. E* (2004).
[36] GELL-MANN M. and TSALLIS C. (Editors), in *Nonextensive Entropy: Interdisciplinary Applications* (Oxford University Press) 2004.
[37] `http://tsallis.cat.cbpf.br/biblio.htm`
[38] ANTONI M. and RUFFO S., *Phys. Rev. E*, **52** (1995) 2361.
[39] ANTENEODO C. and TSALLIS C., *Phys. Rev. Lett.*, **80** (1998) 5313.
[40] FISHER M. E., *Arch. Rat. Mech. Anal.*, **17** (1964) 377; *J. Chem. Phys.*, **42** (1965) 3852; *J. Math. Phys.*, **6** (1965) 1643; FISHER M. E. and RUELLE D., *J. Math. Phys.*, **7** (1966) 260; FISHER M. E. and LEBOWITZ J. L., *Commun. Math. Phys.*, **19** (1970) 251.
[41] CAMPA A., GIANSANTI A., MORONI D. and TSALLIS C., *Phys. Lett. A*, **286** (2001) 251.
[42] LATORA V., RAPISARDA A. and TSALLIS C., *Phys. Rev. E*, **64** (2001) 056134.
[43] MONTEMURRO M. A., TAMARIT F. and ANTENEODO C., *Phys. Rev. E*, **67** (2003) 031106.
[44] CABRAL B. J. C. and TSALLIS C., *Phys. Rev. E*, **66** (2002) 065101(R).
[45] TSALLIS C., *Phys. Rev. E*, **69** (2004) 038101.
[46] MOYANO L. G., BALDOVIN F. and TSALLIS C., cond-mat/0305091.
[47] TSALLIS C., BORGES E. P. and BALDOVIN F., *Physica A*, **305** (2002) 1.
[48] NOBRE F. D. and TSALLIS C., *Phys. Rev. E*, **68** (2003) 036115.
[49] BORGES E. P., TSALLIS C., GIANSANTI A. and MORONI D., in *Tendencias da Fisica Estatistica no Brasil*, edited by TOME T. (Editora Livraria da Fisica, Sao Paulo) 2003, p. 84 (in Portuguese).
[50] DOYE J. P. K., *Phys. Rev. Lett.*, **88** (2002) 238701.
[51] BECK C. and SCHLOGL F., *Thermodynamics of Chaotic Systems* (Cambridge University Press, UK) 1993.
[52] OTT E., *Chaos in Dynamical Systems* (Cambridge University Press, UK) 1993.
[53] BALDOVIN F. and ROBLEDO A., *Europhys. Lett.*, **60** (2002) 518.
[54] BALDOVIN F. and ROBLEDO A., *Phys. Rev. E*, **66** (2002) 045104(R).
[55] COSTA U. M. S., LYRA M. L., PLASTINO A. R. and TSALLIS C., *Phys. Rev. E*, **56** (1997) 245.
[56] LYRA M. L. and TSALLIS C., *Phys. Rev. Lett.*, **80** (1998) 53; LYRA M. L., *Ann. Rev. Comp. Phys.*, edited by STAUFFER D. (World Scientific, Singapore) 1998, p. 31.
[57] BORGES E. P., TSALLIS C., ANANOS G. F. J. and OLIVEIRA P. M. C., *Phys. Rev. Lett.*, **89** (2002) 254103; see also WEINSTEIN Y. S., LLOYD S. and TSALLIS C., *Phys. Rev. Lett.*, **89** (2002) 214101 for a quantum illustration.
[58] DE MOURA F. A. B. F., TIRNAKLI U. and LYRA M. L., *Phys. Rev. E*, **62** (2000) 6361.

New developments in density-functional theory

M. H. COHEN

Department of Physics and Astronomy, Rutgers University
136 Frelinghuysen Rd. Piscataway, NJ 08854-8019, USA
Department of Chemistry, Princeton University
Washington Rd., Princeton NJ 08544-1009, USA

A. WASSERMAN

Department of Chemistry and Chemical Biology, Rutgers University
610 Taylor Rd. Piscataway, NJ 08854-8087, USA

1. – Introduction

The theme of the present Course is "The Physics of Complex Systems (New Advances and Perspectives)". The lectures reported in this paper concern new developments introduced into density-functional theory (DFT) by the present authors. How then does DFT relate to the physics of complex systems? Our answer proceeds from the observation that we humans perceive nature to be hierarchical. In biology there is a hierarchy of organization: molecules, organelles, cells, tissues, organs, organisms, species, etc., up through ecosystems to the biosphere. In physics there is a hierarchy in the scales of space and time: M-"theory", the standard model, nuclei, atoms, molecules, the mesoscale, the macroscale, and so on to the Universe. In a certain scale range —atoms, molecules, mesoscale, and macroscale— ordinary quantum mechanics (QM) provides the overarching principle of organization. Through QM, DFT deals with a *metasystem*, the collection of all systems in that scale range, simple or *complex*.

Density-functional theory has provided both a fertile basis for conceptual analysis of such systems and a practical basis for accurate numerical computation of their properties. Here, the primary focus is on what has come to be called conceptual DFT.

The lectures fall naturally into two parts. The first part, sect. **2**, is concerned with how DFT should be extended to hold for systems with noninteger numbers of electrons whereas it was introduced specifically for integer numbers. The goal is to provide a sound foundation for the theory of "atoms" in "molecules" and thence for chemical reactivity theory. The second part, sect. **3**, is concerned with constructing a density-functional theory of electron-hole excitation spectra for integer electron numbers with important consequences for our understanding of the band gaps of semiconductors and insulators and the Fermi surfaces of metals.

Section **2** begins in subsect. **2**˙1 with a very brief introduction to DFT. It is followed by an analysis in subsect. **2**˙2 of how the current literature has led to what we perceive to be an impass in extending DFT to noninteger numbers of electrons despite the importance and need of such an extension in that arguments can be put forward against both of the two extensions in the literature. Preparation for breaking that impass is carried out in subsect. **2**˙3 through a proof of the existence of density functionals continuous in electron number N and a derivation of the associated Kohn-Sham equations, thus constructing an N-continuous DFT (NCDFT). In subsect. **2**˙4 we introduce the conjecture that the hardness is negative for an isolated system in which noninteger N occurs through fractional occupation of a single Kohn-Sham HOMO spin state, removing the fundamental objection to NCDFT. It is then shown that the sign of the hardness is indefinite in general and that it is negative in the Hartree-Fock approximation, making the conjecture plausible. A novel form of the theory of "atoms" in "molecules" (AIM) is constructed in subsect. **2**˙5 based on NCDFT. For the AIM to be used as a basis for chemical reactivity theory, however, the hardness of the "atoms" within a saturated molecule must be positive. How this transformation from negative to positive hardness occurs is explained and demonstrated explicitly in the Hartree-Fock approximation. The analysis of sect. **2** is summarized in subsect. **2**˙6 and its implications discussed.

Section **3** is concerned with the relation between the true excitation spectrum of a system and the excitation spectrum derived from the Kohn-Sham (KS) energy-level separations derived in turn from the ground-state Kohn-Sham potential of the system for which the number of electrons is the same integer in both the ground and the excited state, as discussed in subsect. **3**˙1. In subsect. **3**˙2 we introduce the band-gap problem, the discrepancy between KS calculations and measured values of the band gap of semiconductors and insulators which became apparent in the 1980's. We describe the solution to the band-gap problem provided by direct computation of the quasi-particle band gap as well as the conventional interpretation that the differences between the computed KS gaps and the observed gaps are fundamental in character. However, as we report in subsect. **3**˙3, when the exchange energy was treated exactly, it was discovered that the discrepancy between the KS and observed gaps largely vanished for semiconductors and insulators with not too wide gaps, reopening the question of whether the discrepancy was fundamental or arose from errors in the density functional. Accordingly, in subsect. **3**˙4 we introduce time-dependent density-functional theory and in subsect. **3**˙5 Casida's formalism for the determination of the true excitation energies. We specialize Casida's formalism to apply explicitly to extended crystals in subsect. **3**˙6, and we prove in subsubsect. **3**˙6.3 that the Kohn-Sham gap and the exact gap are identical. Similarly, in subsubsect. **3**˙6.4, we show that the Kohn-Sham Fermi surface and the exact Fermi surface are identical.

2. – Revisiting N-continuous density-functional theory; chemical reactivity and "atoms" in "molecules"

2˙1. *Introduction to density-functional theory.* – In the late 20's, quantum mechanics was developed and, paraphrasing Dirac, all of chemistry and much of physics became understandable in principle. The barrier to that understanding was the difficulty of solving the many-electron Schrödinger equation

$$\mathcal{H}_N \Psi_N = E_N \Psi_N \,, \tag{2.1}$$

where $\mathcal{H}_N$ is the N-electron Hamiltonian, Ψ_N is an N-electron wave function antisymmetric in the electron coordinates and spin, and E_N the corresponding energy eigenvalue of $\mathcal{H}_N$. One way around the difficulty was proposed almost immediately: finding an approximately equivalent, manageable independent-electron problem. This was done first by Thomas [1] and by Fermi [2] independently, who introduced the eponymous Thomas-Fermi approximation in 1927. The Hartree approximation [3], more fully quantum-mechanical, emerged in 1928 and was followed by the Hartree-Fock approximation [4]. The latter has the great advantage of being derivable from the Rayleigh-Ritz variational principle, but the resulting potential in which the electrons moved independently contained a nonlocal, computationally cumbersome contribution from the exchange. Slater [5] argued in the early 50's for replacing the nonlocal exchange potential by a local potential and introduced a heuristic, parameter-dependent form for it. Shortly afterwards Sharp and Horton [6] pointed out that there was an optimal mapping from the nonlocal exchange potential to a local one via a variational argument. That potential is now called the optimized effective potential (OEP) [7].

At that point the stage was set for the emergence of density-functional theory, which occured in the mid 60's in two landmark papers by Hohenberg and Kohn [8] and by Kohn and Sham [9]. These papers established an exact correspondence between the actual interacting-electron problem presented by eq. (2.1) and an equivalent, noninteracting-electron problem, concluding a search lasting over 30 years. Slowly at first, but with subsequent exponential growth, the literature in the resulting field, DFT, has become vast. Those wishing entry into the field would be well advised to start with the monographs which have been written on the subject, two of which we refer to here [10, 11].

2˙1.1. The HK Density Functional. Hohenberg and Kohn [8] proved that for a nondegenerate ground state of N interacting electrons moving in an external potential $V_e(\mathbf{r})$, *e.g.* that provided by the nuclei, there is an invertible one-to-one mapping between the potential and the ground-state wave function Ψ_N^G up to a constant in the potential. The latter indeterminacy can be eliminated by restricting consideration to potentials $V_e(\mathbf{r})$

which vanish as $r \uparrow \infty$. They then proved that there is an invertible one-to-one mapping between $\Psi_N^{\rm G}$ and the ground-state electron density $n_{\rm G}(\mathbf{r})$,

$$n_{\rm G}(\mathbf{r}) = \left(\Psi_N^{\rm G}, \hat{n}(\mathbf{r})\Psi_N^{\rm G}\right). \tag{2.2}$$

In eq. (2.2), the superposed caret indicates an N-electron operator. Electron densities like those in eq. (2.2) are termed v-representable. The Hamiltonian $\mathcal{H}_N$ contains a universal part consisting of the kinetic-energy operator $\hat{T}$ and the electron-electron interaction-energy operator $\hat{U}$ as well as the nonuniversal operator $\hat{V}_{\rm e}$ for the energy of interaction of the electrons with an external potential $V_{\rm e}$, containing the electrostatic potential of the nuclei and potentials from any other sources external to the system,

$$\hat{\mathcal{H}}_N = \hat{T} + \hat{U} + \hat{V}_{\rm e}\,. \tag{2.3}$$

Let Ψ be the ground state of the N electrons interacting with a different potential $V_{\rm e}'$ and n the associated electron density. Then, according to the first Hohenberg-Kohn (HK) theorem, the energy can be considered a functional of the electron density, the density functional (DF)

$$E[n] = \left(\Psi_N, \hat{\mathcal{H}}_N \Psi_N\right) = E[\Psi_N] \tag{2.4}$$

because of the mapping (2.2) between Ψ_N and n. The functional (2.4) is thus defined only for densities derivable from ground states Ψ_N of an N-electron Hamiltonian containing some external potential different in general from that within $\hat{\mathcal{H}}_N$ in eq. (2.4). The Rayleigh-Ritz variational principle together with the first HK theorem implies that $E[n]$ is a minimum, $E_{\rm G}$, at the ground-state density $n_{\rm G}$, the second HK theorem,

$$E_{\rm G} = {\rm Min}_n\, E[n] = E[n_{\rm G}]. \tag{2.5}$$

2·1.2. The Kohn-Sham equations. Following Kohn and Sham [9] (KS), consider now a system of noninteracting electrons moving in an external potential $V_{\rm KS}$. Their ground state is

$$\Phi_N = \frac{1}{(N!)^{1/2}} {\rm Det}\{\psi_i\}, \quad i = 1 \ldots N. \tag{2.6}$$

The orthonormal spin-orbitals $\psi_i(x)$, $x = (\mathbf{r}, s)$ with s being spin, yield an electron density

$$n(\mathbf{r}) = \sum_{i,s} |\psi_i(x)|^2. \tag{2.7}$$

They satisfy the one-electron Schrödinger equation

$$\mathcal{H}_{\rm KS}\psi_i = \varepsilon_i\psi_i\,, \tag{2.8}$$

$$\mathcal{H}_{\rm KS} = \frac{p^2}{2m} + V_{\rm KS}\,, \tag{2.9}$$

$$\varepsilon_i \leq \varepsilon_{i+1}\,. \tag{2.10}$$

KS supposed that there exists a potential $V_{\rm KS}$ for which the ground-state density of the noninteracting electrons, eq. (2.7), is identical to that, eq. (2.2), of the interacting electrons in the specific external potential $V_{\rm e}$, and they found it by a variational argument.

There were two shortcomings of the KS analysis. First, there has to be a proof that such a potential exists, which was subsequently established through the proof of the adiabatic connection theorem [12]. Second, their variational argument requires that the ψ_i and ψ_i^* be varied arbitrarily, apart from the requirement of orthogonality. Such arbitrary variation of necessity generates electron densities which are not V-representable and for which the DF was not defined.

The Rayleigh-Ritz minimization of $(\Psi_N, \hat{\mathcal{H}}_N\Psi_N)$ can be carried out in two steps:

$$E_{\rm G} = {\rm Min}_n \left\{{\rm Inf}_{\Psi_N\to n}\left[E_N[\Psi_N] = \left(\Psi_N, \hat{\mathcal{H}}_N\Psi_N\right)\right]\right\}. \tag{2.11a}$$

The required generalization of the DF emerges naturally from eq. (2.11a),

$$E[n] = {\rm Inf}_{\Psi_N\to n}E_N[\Psi_N], \tag{2.11b}$$

the Levy [13]-Lieb [14] constrained search algorithm.

In eqs. (2.11a) and (2.11b), the infimum is taken over all states Ψ which yield the preset electron density n. The infimum is searched for, instead of the minimum, to allow for end points and cusps of $E_N[\Psi_N]$ in the restricted Hilbert space $\Psi_N \to n$. It was later proved that for any nonnegative, sufficiently well behaved $n(\mathbf{r})$ there is at least one Ψ so that the set $\{\Psi \to n\}$ is not empty [15]. In fact, the proof proceeds by showing that there is an infinity of non-interacting particle wave functions Φ which give rise to the preset density $n(\mathbf{r})$. That Φ which minimizes the noninteracting total kinetic energy can be chosen, leading to an alternative resolution of the first issue above, the Lagrange multiplier of the density becoming the desired potential.

With this foundation in place, one can proceed to identify $V_{\rm KS}$ by making explicit the components of the DF. Corresponding to eq. (2.3), the DF is

$$E[n] = T[n] + U[n] + \int {\rm d}\mathbf{r}V_{\rm e}(\mathbf{r})n(\mathbf{r}) + V_{NN}\,, \tag{2.12a}$$

where we have added the nuclear electrostatic interaction energy V_{NN} to the $\hat{\mathcal{H}}_N$ of eq. (2.3) to avoid divergences in the thermodynamic limit. The kinetic energy and electron-electron interaction functionals, $T[n]$ and $U[n]$, are obtained by taking the expectation values of $\hat{T}$ and $\hat{U}$, respectively, with the infimal Ψ of eq. (2.11a). The DF, eq. (2.12a), contains a universal part,

$$F[n] = T[n] + U[n], \tag{2.12b}$$

the same for all N-electron systems as well as a non universal part $V_e[n]+V_{NN}$ particular to the system under study, where

$$V_e[n] = \int d\mathbf{r} V_e(\mathbf{r}) n(\mathbf{r}). \tag{2.12c}$$

The terms in eq. (2.12a) are rearranged to facilitate derivation of V_{KS}:

$$E[n] = T_s[n] + U_H[n] + E_{xc}[n] + \int d\mathbf{r} V_e(\mathbf{r}) n(\mathbf{r}) + V_{NN}\,. \tag{2.13}$$

In eq. (2.13) $T_s[n]$ is the expectation value of $\hat{T}$ with the noninteracting wave function Φ,

$$T_s[n] = \sum_{i=1}^{N} \left(\psi_i, \frac{p^2}{2m} \psi_i \right). \tag{2.14}$$

It can be shown to be a functional of n through application of the Levy-Lieb procedure of eq. (2.11a) to the noninteracting case. $U_H[n]$ is the Hartree approximation to $(\Psi, \hat{U}\Psi)$,

$$U_H[n] = \int d\mathbf{r}\, d\mathbf{r}' \frac{n(\mathbf{r}) e^2 n(\mathbf{r}')}{|\mathbf{r} - \mathbf{r}'|}\,. \tag{2.15}$$

$E_{xc}[n]$ is the exchange-correlation functional,

$$E_{xc}[n] = T[n] - T_s[n] + U[n] - U_H[n]. \tag{2.16}$$

At a non-degenerate ground state of the interacting system, $E[n]$ is stationary (actually minimal) with respect to variations in n. In varying n, the equality of the n of the interacting and noninteracting systems is maintained; eq. (2.7) holds for the interacting n as well. Thus the variations in the interacting n are generated by the variations in the noninteracting ψ_i subject to the orthonormality constraint

$$(\psi_i, \psi_j) = \delta_{ij}\,. \tag{2.17}$$

Equations (2.8)–(2.10) follow immediately, with the Kohn-Sham potential V_{KS} given by

$$V_{KS}(\mathbf{r}) = V_e(\mathbf{r}) + V_H(\mathbf{r}) + V_{xc}(\mathbf{r}). \tag{2.18}$$

In eq. (2.18), $V_{\mathrm{H}}(\mathbf{r})$ is the Hartree potential

$$V_{\mathrm{H}}(\mathbf{r}) = \int \mathrm{d}\mathbf{r}' \frac{e^2}{|\mathbf{r}-\mathbf{r}'|} n(\mathbf{r}'), \tag{2.19}$$

and the exchange-correlation potential $V_{\mathrm{xc}}(\mathbf{r})$ is

$$V_{\mathrm{xc}}(\mathbf{r}) = \frac{\delta E_{\mathrm{xc}}[n]}{\delta n(\mathbf{r})} . \tag{2.20}$$

With this explicit form for V_{KS}, the independent-particle Schrödinger equations (2.8)–(2.10) become the Kohn-Sham equations.

Great effort has been devoted to finding, assessing, and succesfully utilizing workable approximations to E_{xc} and thence V_{xc}. The above brief exposition of the root structure of DFT is sufficient for our present purposes, however.

2˙2. *The status of DFT for noninteger N.*

2˙2.1. Why noninteger N is of interest. Through the original HK arguments or those of Levy and of Lieb, the density functional is defined only for integer N. Moreover, both imply (cf. eq. (2.11a)) that the DF depends explicitly on the integer number N, as well as implicitly through $n(\mathbf{r})$; that is, we should write the DF as $E[N, n(\mathbf{r})]$. That explicit dependence could be subsumed into the functional dependence of E on n, since $N = \int \mathrm{d}\mathbf{r}\; n(\mathbf{r})$, but making the distinction will prove useful later in subsect. **2**˙3. However, from the beginning of DFT in the initial paper of Hohenberg and Kohn, arguments were adduced which required knowledge of the DF for noninteger N. To illustrate why noninteger N is of interest, we give two examples.

In the 1964 HK paper [8], the problem of finding the ground-state energy of an N-electron system with N an integer, $E_{\mathrm{G}} = \mathrm{Min}_n\, E[n]$ subject to the constraint $N = \int \mathrm{d}\mathbf{r}\; n(\mathbf{r})$, was addressed by relaxing the constraint through a Legendre transformation from $E[n]$ to the grand potential $\Omega[n]$,

$$\Omega[n] = E[n] - \mu N, \tag{2.21}$$

and minimizing $\Omega[n]$. The Lagrange multiplier μ is the chemical potential, and the minimization of $\Omega[n]$ results in

$$\mu = \frac{\delta E[n]}{\delta n} = \frac{\mathrm{d}E(N)}{\mathrm{d}N} . \tag{2.22}$$

This relaxation of the constraint with the consequent arbitrary variation of n implies that $E[n]$ is presumed known for arbitrary, that is, noninteger N. This is seen clearly from the second equality in eq. (2.22), where $E(N)$ is the ground-state energy of N electrons, and the derivative can be regarded as defined by the usual limiting process

$$\frac{\mathrm{d}E}{\mathrm{d}N} = \lim_{\Delta N \to 0} \frac{E(N+\Delta N) - E(N)}{\Delta N} . \tag{2.23}$$

Clearly, if N is an integer, $N + \Delta N$ cannot be.

Our second example concerns chemical reactivity theory (CRT), which is of great fundamental importance in chemistry. In the 1980's Parr and collaborators [11,16] provided a deep basis for CRT via DFT. The subject has been termed "Conceptual DFT". To measure the extent of activity in this field, we note that a recent review by Geerlings, De Proft, and Langenaeker [17] has 812 references. Chemical potential μ, or equivalently electronegativity, and absolute hardness(1) [18,19] (or simply hardness) η are central concepts in CRT. Both are defined through differentiation of the total energy with respect to electron number, μ via eq. (2.22) and η via

$$\eta = \frac{\mathrm{d}\mu}{\mathrm{d}N} = \frac{\mathrm{d}^2 E(N)}{\mathrm{d}N^2}\,. \tag{2.24}$$

Both thus require knowledge of the density functional for noninteger N, which is true as well for other, more complex indices of chemical reactivity.

2·2.2. Existing interpolations of the DF. The required knowledge can in principle be obtained by interpolating the DF between integer values of N. Define the ray $g(\mathbf{r})$ in the function space of densities through

$$n(\mathbf{r}) = Ng(\mathbf{r}), \qquad \int \mathrm{d}\mathbf{r}\; g(\mathbf{r}) = 1, \tag{2.25}$$

as illustrated in fig. 1. Along the ray $g(\mathbf{r})$, $E[N, Ng(\mathbf{r})]$ is defined only at the integers $N = M$, where the rays $g(\mathbf{r})$ intersect the subspaces of densities $n(\mathbf{r})$ for which $\int \mathrm{d}\mathbf{r}\; n(\mathbf{r}) = M$, in fig. 1. The question we face is how to choose among the infinity of possible interpolations between these points.

Two interpolations exist in the literature. The first is *N-continuous DFT* (NCDFT). All proposed approximations to $E[N, n]$ contain no explicit N dependence ($E[N, n] \to E[n]$) and are continuous in n and therefore in N. Whether an exact NCDFT can be constructed is an open question which we address in subsect. **2·3**.

The second is the ground-state ensemble DF (EDF) proposed by Perdew *et al.* [20] in 1982. A consistent density-functional theory (EDFT) can be constructed based on the density matrix of the grand canonical ensemble, as shown by Mermin [21] for finite temperature. Perdew *et al.* took over this idea for the ground state. A simple energy minimization argument then shows that for $M < N < M+1$, *i.e.* $N = M + \nu$, $0 < \nu < 1$,

$$E[N, n] = (1-\nu)E[M, n_M] + \nu E[M+1, n_{M+1}], \tag{2.26}$$

$$n = (1-\nu)n_M + \nu n_{M+1}\,. \tag{2.27}$$

(1) For convenience, we have dropped in eq. (2.24) a factor of 1/2 present in the original definition.

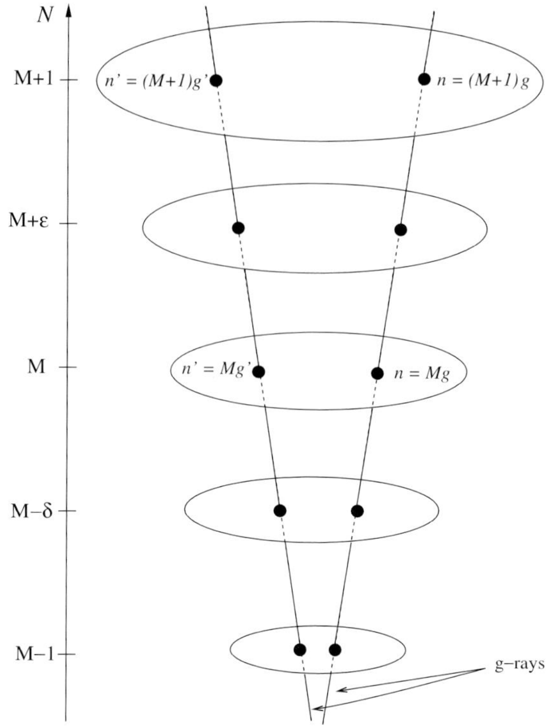

Fig. 1. – Illustration of g-rays in the space $\mathcal{N}$ of all positive-indefinite, integrable densities; $\int d\mathbf{r}\; n(\mathbf{r}) = N$, $n(\mathbf{r}) = Ng(\mathbf{r})$.

In eq. (2.26) n_M is an M-electron density, n_{M+1} is an $(M+1)$-electron density, and n is any density which can be formed by averaging two such densities. In this EDFT, all ground-state properties are piecewise linear in N. The DF (2.26) contains both explicit N dependence through ν and implicit dependence through eq. (2.27).

In both NCDFT and EDFT, ground-state properties have derivative discontinuities across integer N. In NCDFT, they arise from discontinuous changes in the occupation of the Kohn-Sham states ψ_i, that is, from the implicit N dependence, whereas in EDFT they arise from the explicit N dependence as well.

2·2.3. The argument against NCDFT. Perdew *et al.* [20] rejected NCDFT as a valid interpolation scheme because, they argued, it led to unphysical results. They considered the adiabatic separation to infinity of the atoms A and B of the heteronuclear diatomic molecule AB. The ground-state energy of AB, E_{AB}, decomposes into a sum of the ground-state energies of the individual atoms, $E_A + E_B$. In NCDFT the energy E_A is a continuous function of N_A and similarly for E_B subject only to the constraint $N_A + N_B = N_{AB}$. Carrying out the usual electronegativity equalization argument of CRT and supposing that the neutral atoms differ little in electronegativity, one finds for example

$$N_A \longrightarrow N_A^0 + \Delta N, \qquad N_B \longrightarrow N_B^0 - \Delta N, \tag{2.28a}$$

$$\Delta N = \frac{\mu_B - \mu_A}{\eta_A + \eta_B} \neq 0 \tag{2.28b}$$

by minimizing $E_A + E_B$ with respect to ΔN. More generally, NCDFT, according to this argument, implies that the separated atoms will have nonzero integer or fractional charges in the minimum energy configuration. This is unphysical on two grounds. First, a fractional charge on an isolated atom is impossible, and second, no known electron affinity is greater than any known ionization energy among the neutral atoms, ruling out integer charge transfer. A and B must remain neutral after separation. There can be no minima in $E[N]$ but only infima, in contradiction to what their analysis of NCDFT implies. They then showed that the cusps in $E(N)$ following from its piecewise linearity in EDFT introduces the required infima at $\Delta N = 0$. They proposed that the EDF provides a satisfactory interpolation of $E[N, n]$ between integers, a widely accepted conclusion.

2˙2.4. Objections to EDFT. However, there are three objections to EDFT [22]. First, the usual requirement for the use of an ensemble in a physical problem is that ensemble averages equal temporal averages of physical properties. Use of the ensemble of DFT would therefore require that the valence of an *isolated* system would fluctuate between two values on a time scale much shorter than any relevant observation time, an impossibility. Instead, the ensemble describes a physical admixture of noninteracting systems in two different, stable valence states, not the single dynamical system desired.

Second, the chemical potential has a staircase dependence on n,

$$\mu(N) = -A_M, \quad M \leq N < M + 1, \tag{2.29}$$

so that its derivative, the hardness, is a sum of δ-functions

$$\eta(N) = \frac{\mathrm{d}\mu}{\mathrm{d}N} = \sum_{M=0}^{\infty} (I_M - A_M)\delta(N - M). \tag{2.30}$$

Thus η vanishes between integers, turning the hardness into a concept useless for chemistry. In general, reactivities of species M towards adding electrons are identical to those of species $M + 1$ towards losing electrons in EDFT, which makes no chemical sense, cf. eq. (2.29).

Finally, it is not clear that all $n(\mathbf{r})$ of chemical interest can be represented in EDFT by eq. (2.27). For these reasons, we re-examine the separation argument against NCDFT in subsect. **2**˙4. First, however, we demonstrate that a well-defined NCDFT can be constructed, and we give the KS equations which follow from it in subsect. **2**˙3. Subsections **2**˙3 through **2**˙5 contain an extension of an analysis first reported in ref. [23].

2·3. *Defining the NCDF; Kohn-Sham equations.*

2·3.1. A separation argument for continuing the DF. In this subsection, we use the constrained search algorithm of Levy and Lieb to define the DF of NCDFT. Consider a molecule or larger finite system AB of arbitrary complexity. A and B are two parts into which it can be decomposed, both defined by their respective nuclear configurations. Separate the centers of gravity of the nuclei of A and of those of B adiabatically to infinity so that each nucleus of A becomes infinitely distant from all those of B. Consider all possible electron densities $n_{AB}(\mathbf{r})$ which decompose into a sum $n_A(\mathbf{r}) + n_B(\mathbf{r})$ of nonoverlapping densities associated with A and B for which

$$n_A(\mathbf{r})n_B(\mathbf{r}) \downarrow 0. \tag{2.31}$$

Apart from eq. (2.31), the only restriction on $n_{AB}(\mathbf{r})$ is that its integral be the integer N_{AB}. No such restriction applies to N_A and N_B individually, only that their sum be an integer,

$$N_A + N_B = N_{AB}\,. \tag{2.32}$$

Now the DF of AB, $E_{AB}[N_{AB}, n_{AB}]$, is defined via constrained search for all nonnegative n_{AB} and integer N_{AB} consistent with finite kinetic energy. But, because there are no infinite-range interactions present, E_{AB} must break up into two terms associated with A or B in the limit,

$$E_{AB}[N_{AB}, n_{AB}] \longrightarrow E_A[N_{AB}, n_A] + E_B[N_{AB}, n_B]. \tag{2.33}$$

If $E_{AB}[N_{AB}, n_{AB}]$ is continuous in n_{AB}, then $E_\alpha[N_{AB}, n_\alpha]$ $(\alpha = A, B)$ must be continuous in n_α and therefore in $N_\alpha = \int \mathrm{d}\mathbf{r} n_\alpha(\mathbf{r})$ for noninteger N_α. The possibility of the cancellation of discontinuities in one E_α by those in the other is ruled out by the nonoverlap condition (2.31) and the absence of infinite-range interactions.

Thus, from the constrained search algorithm and the supposition of continuity in n_{AB} for integer N_{AB}, we arrive at a density functional for A of the form $E_A[N_{AB}, n_A]$ for noninteger N_A which is continuous in N_A. Moreover, because of the absence of infinite-range interactions, $E_A[N_{AB}, n_A]$ must be insensitive to the presence of B and must be the same for all B's, as long as $N_{AB} \geq N_A$. That is

$$E_A[N_{AB}, n_A] \longrightarrow E_A[n_A]. \tag{2.34}$$

$E_A[n_A]$ is the DF of an isolated system with no explicit dependence on N_A. Since the choice of A, B, and AB is arbitrary, we have shown that the constrained search DF of any system contains no explicit N dependence and can be continued to noninteger N given sufficient knowledge of its dependence on n for integer N. The continuation of $V_\mathrm{e}[n]$ is trivial since it is linear in n, eq. (2.12c). What we have really accomplished via the separation argument is to show that the universal part of the DF, $F[n]$, can be defined through the constrained search algorithm for integer N for $n(\mathbf{r})$ with non-compact support, the pieces of which individually contain noninteger electron number. When these pieces separate to infinity, $F[n]$ is continued via a distributional argument to $n(\mathbf{r})$ with compact support and noninteger N. At this point, we have shown only that one particular continuation of the DF to noninteger N is possible. Why we choose this continuation will emerge more clearly in the course of the discussion of chemical reactivity and atoms in molecules in subsect. **2**·5.

2·3.2. Kohn-Sham equations for noninteger N. Having chosen the density functional, the next step is to construct Kohn-Sham equations. Consider first a noninteracting electron ensemble which has the same electron density as the interacting system:

$$n(\mathbf{r}) = \sum_{i=1,s}^{\infty} \nu_i |\psi_i(x)|^2 ; \tag{2.35}$$

$$\int \mathrm{d}\mathbf{r}\; n(\mathbf{r}) = \sum_{i=1}^{\infty} \nu_i = N = M + \omega, \quad 0 < \omega \leq 1, \tag{2.36}$$

with M an integer. We show in Appendix A that there is an infinite number of such ensembles. Choose the ensemble which minimizes the total energy of the noninteracting system (constrained search for noninteracting ensembles),

$$\mathcal{H}_{\mathrm{KS}}\psi_i = \varepsilon_i \psi_i , \tag{2.37}$$

$$\varepsilon_i \leq \varepsilon_{i+1} , \tag{2.38}$$

$$\nu_i = \begin{cases} 1, & i = 1 \ldots M, \\ \omega, & i = M+1, \\ 0, & i > M+1. \end{cases} \tag{2.39}$$

This choice allows for proof of a generalized adiabatic connection theorem, which we omit. Substitute eq. (2.31) into $E[n]$ and perform the same variational analysis which led to eqs. (2.8)-(2.10). The result is formally identical, but now, for later purposes, we separate out explicitly the exact-exchange contributions to E_{xc} and V_{xc}:

$$V_{\mathrm{KS}} = V_{\mathrm{e}} + V_{\mathrm{H}} + V_{\mathrm{x}} + V_{\mathrm{c}} ; \tag{2.40}$$

$$V_{\mathrm{x}} = \frac{\delta E_{\mathrm{x}}}{\delta n} , \qquad V_{\mathrm{c}} = \frac{\delta E_{\mathrm{c}}}{\delta n} ; \tag{2.41}$$

$$E_{\mathrm{x}} = -\frac{1}{2} \sum_{s} \int\!\!\int \mathrm{d}\mathbf{r}\, \mathrm{d}\mathbf{r}' \frac{\rho(\mathbf{r}s, \mathbf{r}'s)\rho(\mathbf{r}'s, \mathbf{r}s)e^2}{|\mathbf{r} - \mathbf{r}'|} ; \tag{2.42}$$

$$\rho(\mathbf{r}s, \mathbf{r}'s) = \sum_{i} \nu_i \psi_i(\mathbf{r}s) \psi_i^*(\mathbf{r}'s); \tag{2.43}$$

$$E_{\mathrm{c}} = E_{\mathrm{xc}} - E_{\mathrm{x}} . \tag{2.44}$$

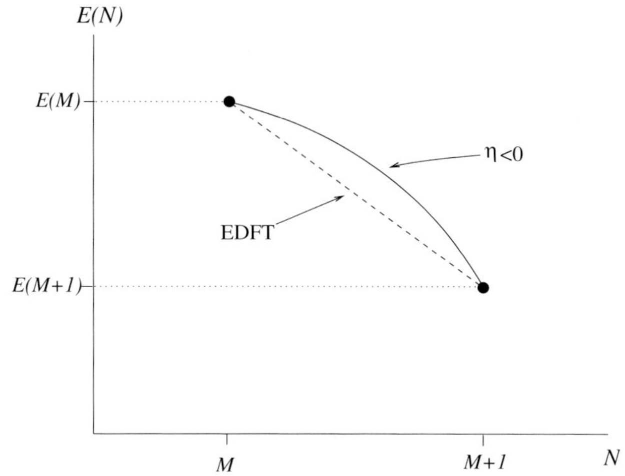

Fig. 2. – Ground state $E(N)$ as a function of noninteger electron number N for N between the integers M and $M+1$ for EDFT (dashed line) and the hypothetical case of negative hardness (solid line).

We now have what we need to address the Perdew *et al.* objection to NCDFT.

2˙4. *The hardness conjecture, part I.*

2˙4.1. The conjecture. The separation argument of Perdew *et al.*, which leads to their rejection of NCDFT implicitly assumes that a minimum exists in E_A+E_B as a function of ΔN. The condition for the existence of a minimum is that E_A+E_B be locally convex in ΔN. The condition for convexity is that the sum of the hardnesses be positive for all ΔN,

$$\eta_A + \eta_B > 0. \tag{2.45}$$

Suppose that $\eta < 0$ *for isolated finite systems instead, the hardness conjecture, part I.* The electronegativity equalization argument would then fail. There would be no minima in E_A+E_B at fractional ΔN. There would only be infima at $\Delta N = 0, \pm 1$, etc. because of cusps in $E[n]$ at integer N. Because all atomic ionization energies are larger than all atomic electron affinities, $\Delta N = 0$ is the global infimum. NCDFT would then yield the correct description of the dissociation of a heteronuclear diatomic molecule. The NCDF with negative η and the EDF are compared in fig. 2.

2˙4.2. Possibility. Could it possibly be true that the continuation of $E[n]$ to noninteger n via the constrained search definition for integer N and the separation argument of subsect. **2**˙3. would yield negative hardness for isolated finite systems? We address this question in two steps, first establishing possibility and then establishing plausibility.

We start by carrying out a formal analysis of η. The variational principle which generates the KS eqs. (2.37)-(2.44) guarantees that when N is changed, no contribution to $\mathrm{d}E/\mathrm{d}N$ comes from changes in the wave functions. Moreover, from eq. (2.39), as N changes, only the occupation number of the $(M+1)$st state, the HOMO, changes. It follows from the Janak theorem [24], therefore, that

$$\mu = \frac{\mathrm{d}E}{\mathrm{d}N} = \varepsilon_{M+1}\,. \tag{2.46}$$

The hardness consequently is

$$\eta = \frac{\mathrm{d}\varepsilon_{M+1}}{\mathrm{d}N}\,, \tag{2.47}$$

which, by the Hellmann-Feynman theorem [25], becomes

$$\eta = \left(\psi_{M+1}, \frac{\mathrm{d}V_{\mathrm{KS}}}{\mathrm{d}N}\psi_{M+1}\right). \tag{2.48}$$

Applying the chain rule results in

$$\frac{\mathrm{d}V_{\mathrm{KS}}(\mathbf{r})}{\mathrm{d}N} = \int \mathrm{d}\mathbf{r}'\frac{\delta V_{\mathrm{KS}}(\mathbf{r})}{\delta n(\mathbf{r}')}\frac{\mathrm{d}n(\mathbf{r}')}{\mathrm{d}N}\,. \tag{2.49}$$

The first factor in the integrand of eq. (2.49) is the Kohn-Sham interaction

$$f(\mathbf{r},\mathbf{r}') = \frac{\delta V_{\mathrm{KS}}(\mathbf{r})}{\delta n(\mathbf{r}')}\,. \tag{2.50a}$$

The second factor is the Fukui function

$$F(\mathbf{r}) = \frac{\mathrm{d}n(\mathbf{r})}{\mathrm{d}N} \tag{2.50b}$$

introduced by Parr and Yang [26,11] in 1984. Cohen *et al.* [27] showed subsequently that $F(\mathbf{r})$ is a screened version of the frontier electron density $n_{\mathrm{F}}(\mathbf{r})$,

$$F(\mathbf{r}) = \int \mathrm{d}\mathbf{r}' K^{-1}(\mathbf{r},\mathbf{r}')n_{\mathrm{F}}(\mathbf{r}'), \tag{2.51}$$

$$n_{\mathrm{F}}(\mathbf{r}) = \sum_s |\psi_{M+1}(\mathbf{r},s)|^2, \tag{2.52}$$

$$\hat{K} = \frac{\delta V_{\mathrm{e}}}{\delta V_{\mathrm{KS}}} = \hat{1} + \hat{\chi}_0\cdot\hat{f}. \tag{2.53}$$

In eq. (2.53), the static KS screening function $\hat{K}$ is expressed in terms of the isoelectronic KS susceptibility,

$$\chi_0(\mathbf{r},\mathbf{r}') = -\left.\frac{\delta n(\mathbf{r})}{\delta V_{\mathrm{KS}}(\mathbf{r}')}\right|_N\,. \tag{2.54}$$

Inserting eqs. (2.50a)–(2.53) into eq. (2.48), we obtain

$$\eta = \left(n_{\mathrm{F}}, \hat{f} \cdot \left[\hat{1} + \hat{\chi}_0 \cdot \hat{f}\right]^{-1} n_{\mathrm{F}}\right) \tag{2.55}$$

for η, which is readily transformed into our final expression for η,

$$\eta = \left(n_{\mathrm{F}}, \left[\hat{f}^{-1} + \hat{\chi}_0\right]^{-1} n_{\mathrm{F}}\right). \tag{2.56}$$

In parallel with eq. (2.18) for V_{KS}, $\hat{f}$ can be decomposed into Hartree and exchange-correlation contributions,

$$\hat{f} = \hat{f}_{\mathrm{H}} + \hat{f}_{\mathrm{xc}}\,. \tag{2.57}$$

$\hat{f}_{\mathrm{H}}$ is just the bare Coulomb interaction which is positive definite,

$$f_{\mathrm{H}}(\mathbf{r}, \mathbf{r}') = \frac{e^2}{|\mathbf{r} - \mathbf{r}'|}\,, \tag{2.58}$$

whereas the exchange-correlation contribution is negative. Thus $\hat{f}$ and $\hat{f}^{-1}$ are indefinite kernels. On the other hand, $\hat{\chi}_0$ is readily shown to be positive definite. Thus the sign of η depends on the relative importance of $\hat{f}_{\mathrm{xc}}$, $\hat{f}_{\mathrm{H}}$ and $\hat{\chi}_0$ in eq. (2.56), and it is *possible* that η can be negative.

2·4.3. Plausibility. It is possible, but is it plausible? We address the question of plausibility by considering the Hartree-Fock approximation (HFA). The total energy is, in the HFA,

$$E = T_s + V_{\mathrm{e}} + U_{\mathrm{H}} + E_{\mathrm{x}}\,, \tag{2.59}$$

with T_s given by eq. (2.14), V_{e} by eq. (2.12c), U_{H} by eq. (2.15), E_{x} by eqs. (2.42)-(2.43), and $n(\mathbf{r})$ given by eqs. (2.35)-(2.36). We note first that the only terms in E explicitly quadratic in ν_{M+1}, the only fractional occupation number, are the Coulomb interaction of ψ_{M+1} with itself in U_{H} and the exchange interaction of ψ_{M+1} with itself in E_{x}. These cancel exactly so that E is explicitly only linear in ν_{M+1}, not quadratic. First, keep the ψ_i fixed and calculate $E(N)$. Since $\nu_{M+1} = N - M$, this unrelaxed energy is linear in N, as shown in fig. 3. Correspondingly, the unrelaxed hardness vanishes. Next, relax the ψ_i to obtain the true ground-state energy. $E(N)$ then falls below the unrelaxed line by an amount which increases monotonically in N according to the variational principle. Also, according to the variational principle, at M the relaxed and unrelaxed μ's are identical. $E(N)$ is thus concave as shown in fig. 3, and we conclude that the hardness is negative in the HFA for fractional occupation of a single HOMO. As the correlation energy E_{c} is generally smaller in magnitude than E_{x}, and E_{x} in turn is generally smaller in magnitude than T_s, V_{e}, and U_{H} in finite systems, it is highly plausible that η remains negative for "real" systems when E_{c} is included.

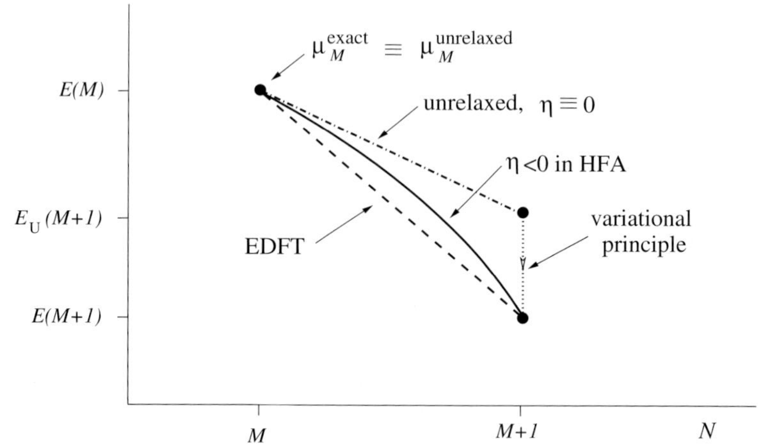

Fig. 3. – Three dependences of total energy $E(N)$ on electron number N, $M \leq N \leq M+1$, M an integer. The dashed line shows $E(N)$ in EDFT. The dashed-dotted and solid lines show $E(N)$ in the Hartree-Fock approximation. Only one member of the Kramers conjugate degenerate pair of HOMO's is fractionally occupied with occupation number ν_{M+1} in $[0,1]$. The dashed-dotted line shows the unrelaxed energy, whereas the solid line shows the exact energy, the concavity of which displays the negative hardness of an isolated system in the HFA. From ref. [23]

2·5. *Chemical reactivity theory and "atoms" in "molecules".*

2·5.1. Chemical reactivity theory in brief. The goal of CRT is to predict or interpret the reactive proclivities of atoms or groups of atoms in response to the chemical stimuli encountered in molecules or larger systems. The prediction or interpretation is to follow in CRT from the properties in isolation of the atoms or groups. The reactive proclivities have been quantified by various reactivity indices.

Before DFT the indices were largely empirical. Pauling introduced his empirical scale for electronegativity χ to quantify the tendency to donate or accept electrons [28]. Mulliken [29] pointed out on theoretical grounds that if χ were set equal to

$$\chi_m = \frac{1}{2}(I + A), \tag{2.60}$$

where I is the ionization energy and A is the electron affinity, there would be no transfer of charge between species having the same χ_m, and electrons would flow towards the species of larger χ_m. Moreover, he pointed out that disproportionation (that is, charge transfer between like systems) would not occur as along as $I - A > 0$. Interestingly, if one fits a parabola

$$E(N) = E(M) + \mu_d(N - M) + \frac{1}{2}\eta_d(N - M)^2, \tag{2.61}$$

to $E(M)$ and $E(M \pm 1)$, one obtains μ_d, the *discrete* chemical potential, as the average first difference

$$\mu_{\mathrm{d}} = \frac{1}{2}\big[(E(M+1) - E(M)) + (E(M) - E(M-1))\big] \\ = -\frac{1}{2}(I + A) = -\chi_m\,, \tag{2.62}$$

and the *discrete* hardness η_{d} as the second difference,

$$\eta_{\mathrm{d}} = E(M+1) + E(M-1) - 2E(M) = I - A. \tag{2.63}$$

For all known species, η_{d} is positive, presumably an as yet unproven consequence of the Pauli principle and the repulsive nature of the electron-electron interaction. Disproportionation does not occur.

The notions that $\mu = \mathrm{d}E/\mathrm{d}N$ [30] and the hardness $\eta = \mathrm{d}^2E/\mathrm{d}N^2$ [18,19] (apart from a factor of 2) were both introduced before DFT; as were hybridization [28]; the role of polarizability [19], Mulliken's charge and bond orders [31]; Pearson's hard-soft, acid-base principle [19]; the principle of maximum hardness [32]; and the frontier orbital concept of Fukui [33].

All of these concepts and indices have been given a deep foundation in DFT by Parr and his collaborators [11]. The earlier indices have been embedded in DFT, and powerful new ones have been added. These various indices have been studied quantitatively via approximate NCDFT [34] as well as conceptually [17]. Many empirical rules have been better understood thereby, *e.g.* rules for regio- and stereo-selectivity, the Woodward-Hoffman rules, aromaticity, etc. [17].

These reactivity indices fall into three categories: *global*, in which a derivative is taken of a global quantity such as total energy with respect to another global quantity such as number (the chemical potential and hardness are global indices); *local*, in which a derivative is taken of a local quantity such as the density with respect to a global quantity (the Fukui function $F(\mathbf{r})$ of eq. (2.50b) is a local index); and *nonlocal*, in which a functional derivative is taken of a local quantity with respect to another local quantity. $\hat{f}^{-1}$, cf. eq. (2.50a), and $\hat{\chi}_0$, cf. eq. (2.54), are such nonlocal quantities, but it is the usual susceptibility,

$$\chi(\mathbf{r},\mathbf{r}') = -\frac{\delta n(\mathbf{r})}{\delta V_{\mathrm{e}}(\mathbf{r}')}\,, \tag{2.64}$$

which has appeared in the chemical literature as a nonlocal reactivity index termed the softness kernel [35,36].

All of these indices, either directly or indirectly, require taking derivatives with respect to N. Even in $\chi(\mathbf{r},\mathbf{r}')$, the functional derivative can be taken at constant N or constant μ. Knowledge of the energy functional and the ground-state properties of the atom or larger group (conflated here into "atom") as functions of N is therefore essential to embedding CRT within DFT.

These "atoms" interact with the rest of a complex in the course of a chemical reaction or with the rest of the "molecule" in a stable configuration. Yet in CRT we wish to capture the effects of these environmental influences through use of the properties of the isolated species. Clearly what is needed is a theory of "atoms" in "molecules" (AIM) in a form suitable for use within CRT. We construct such a theory in the next subsection.

2·5.2. "Atoms" in "Molecules". Consider a "molecule" M with the ground-state electron density $n_M(\mathbf{r})$. Suppose that M can be partitioned into a set of components $\alpha = 1 \ldots A$, each containing a distinct subset of nuclei. M and the α can be of arbitrary complexity. The *AIM problem* can be stated as follows: What is the optimal decomposition of $n_M(\mathbf{r})$ into $\sum_\alpha n_\alpha(\mathbf{r})$, where $n_\alpha(\mathbf{r})$ is an electron density associated specifically with the component α isolated in the nuclear configuration it has in M?

There have been many proposed solutions in the literature. These have been critically assessed by Nalewajski and Parr [37] who, in the end, favor only the stockholder partition of Hirshfeld [38]. The latter, however, has $n_\alpha(\mathbf{r})$ proportional to the unperturbed ground-state density $n_\alpha^0(\mathbf{r})$ which at best captures only charge transfer to or from α and not the distortions of n_α^0 caused by polarization, hybridization, or bond formation; moreover, it is not an exact decomposition. By focusing on the density alone as has been traditional, the changes in spin density and spin orientation attendant to formation of the molecule cannot easily be captured. Accordingly, we have proposed a new solution to the AIM problem which leads to identification of the state of the isolated "atom" with respect to which the reactivity indices should be defined.

Consider a molecule M with the ground-state local density matrix

$$\rho_M(\mathbf{r})_{ss'} = N_M \int \mathrm{d}x_2 \ldots \mathrm{d}x_N \Psi(\mathbf{r}s, x_2 \ldots x_N)\Psi^*(\mathbf{r}s', x_2 \ldots x_N), \tag{2.65}$$

where N_M is the number of electrons in M and $\int \mathrm{d}x_2 \ldots \mathrm{d}x_N$ implies integration over space and sum over spin for the $N-1$ unspecified space-spin coordinates in M's ground-state wave function Ψ. Suppose that the nuclei of M can be partitioned into subsets $\alpha = 1 \ldots A$ which define the components of the molecule. The AIM problem can then be further restated as follows: What is the optimal *exact* decomposition of $\rho_M(\mathbf{r})_{ss'}$ into the sum $\sum_\alpha \rho_\alpha(\mathbf{r})_{ss'}$ of contributions $\rho_\alpha(\mathbf{r})_{ss'}$, where $\rho_\alpha(\mathbf{r})_{ss'}$ is a local density matrix of the part α isolated in the nuclear configuration it has in M,

$$\rho_M(\mathbf{r})_{ss'} = \sum_\alpha \rho_\alpha(\mathbf{r})_{ss'}\,? \tag{2.66}$$

A local spin-density matrix can be decomposed into an electron density $n(\mathbf{r})$ and a spin-density matrix $\mathcal{S}_{ss'}(\mathbf{r})$

$$\rho(\mathbf{r})_{ss'} = \frac{1}{2}n(\mathbf{r})\delta_{ss'} + \mathcal{S}(\mathbf{r})_{ss'}\,, \tag{2.67a}$$

$$n(\mathbf{r}) = \sum_s \rho(\mathbf{r})_{ss}\,. \tag{2.67b}$$

This formulation of the AIM problem differs from previous formulations in that the spin-density is included in the decomposition as well as the electron density and that optimality requires the decomposition to be exact, not approximate as in the stockholder model [37, 38]. The inclusion of spin-density is essential if we are to understand how the chemical behavior of an "atom" within a "molecule" relates to its chemical behavior in true isolation as expressed by, *e.g.*, its hardness even if that molecule is saturated, having vanishing spin density. We have seen in subsect. **2**˙4 the importance of spin in establishing the sign of the hardness.

Let $E_\alpha[n_\alpha]$ be the density functional of the α-th part of M defined for noninteger N_α by the continuation procedure of subsubsect. **2**˙3.1 for α isolated in the nuclear configuration it has in M. The arguments of subsubsect. **2**˙3.1 are readily generalized to lead to an analogous definition of a spin-density matrix functional $E_\alpha[\hat{\rho}_\alpha]$, leading in turn to the spin-dependent generalization of NCDFT to NCSDFT. Define $E_{\mathrm{I}}[\{\hat{\rho}_\alpha\}]$ as the sum of the set of isolated parts, $\alpha = 1 \dots A$, of M,

$$E_{\mathrm{I}}[\{\hat{\rho}_\alpha\}] = \sum_\alpha E_\alpha[\hat{\rho}_\alpha]. \tag{2.68}$$

We propose here as the solution of the AIM problem so formulated *that* set of $\rho_\alpha(\mathbf{r})_{ss'}$ which yields the infimal value of $E_{\mathrm{I}}[\{\hat{\rho}_\alpha\}]$ subject to the condition (2.66) and the implied condition

$$\sum_\alpha N_\alpha = N_M\,, \quad N_\alpha = \int \mathrm{d}\mathbf{r}\; n(\mathbf{r}). \tag{2.69}$$

It is convenient to express $\hat{\mathcal{S}}(\mathbf{r})$ in terms of the Pauli matrices $\hat{\boldsymbol{\sigma}} = (\hat{\sigma}_x, \hat{\sigma}_y, \hat{\sigma}_z)$

$$\hat{\mathcal{S}}(\mathbf{r}) = \mathbf{S}(\mathbf{r}) \cdot \hat{\boldsymbol{\sigma}}, \tag{2.70}$$

where $\mathbf{S}(\mathbf{r})$ is the vector spin density. Condition (2.66) then gets replaced by

$$n_M(\mathbf{r}) = \sum_\alpha n_\alpha(\mathbf{r}), \tag{2.71a}$$

$$\mathbf{S}_M(\mathbf{r}) = \sum_\alpha \mathbf{S}_\alpha(\mathbf{r}). \tag{2.71b}$$

Carrying out a Legendre transformation, we look for an infimum of the Legendre-transformed grand potential

$$\Omega[\{n_\alpha, \mathbf{S}_\alpha\}] = E_{\mathrm{I}}[\{n_\alpha, \mathbf{S}_\alpha\}] + \int \mathrm{d}\mathbf{r}\big[(V_{R,M} - \mu_M) n_\alpha(\mathbf{r}) - \mathrm{tr}\, \mathsf{H}_{R,M}(\mathbf{r}) \cdot \mathbf{S}_\alpha(\mathbf{r})\big]. \tag{2.72}$$

In eq. (2.72) $V_{R,M}$, $\mathsf{H}_{R,M}$, and μ_M are the Lagrange multipliers which guarantee that eqs. (2.71a), (2.71b), and (2.69), respectively, are satisfied. $V_{R,M}$ is unique up to a constant which must be absorbed by μ_M. Imposing the condition $V_{R,M}(\mathbf{r}) \to 0$, $r \uparrow \infty$, fixes $V_{R,M}(\mathbf{r})$ uniquely.

If a global minimum of Ω exists, that is if it is stationary with positive Hessian there, then a modified set of KS equations results,

$$\hat{\mathcal{H}}'_{\mathrm{KS},\alpha}\hat{\psi}_{i\alpha} = \varepsilon_{i\alpha}\hat{\psi}_{i\alpha}\,, \tag{2.73}$$

$$\hat{\mathcal{H}}'_{\mathrm{KS},\alpha} = \hat{\mathcal{H}}_{\mathrm{KS},\alpha} + V_{R,M}(\mathbf{r})\hat{I} - \mathsf{H}_{R,M}(\mathbf{r}) \cdot \hat{\boldsymbol{\sigma}}, \tag{2.74}$$

for the KS spin-orbitals $\hat{\psi}_{i\alpha}$. In addition to eq. (2.73), an *internal* electronegativity equalization condition holds,

$$\mu'_\alpha = \frac{\mathrm{d}E'_\alpha(N_\alpha)}{\mathrm{d}N_\alpha} = \mu_M, \quad \forall\alpha\,, \tag{2.75}$$

where $E'_\alpha(N_\alpha)$ is

$$E'_\alpha[n_\alpha, \mathbf{S}_\alpha] = E_\alpha[n_\alpha, \mathbf{S}_\alpha] + \int \mathrm{d}\mathbf{r}\big[V_{R,M}(\mathbf{r})n_\alpha(\mathbf{r}) - \mathrm{tr}\,\mathsf{H}_{R,M}(\mathbf{r}) \cdot \mathbf{S}_\alpha(\mathbf{r})\big] \tag{2.76}$$

evaluated at the $n_\alpha(\mathbf{r})$ and $\mathbf{S}_\alpha(\mathbf{r})$ constructed from the solution of eq. (2.73) at fixed N_α. The solution of eq. (2.73) is to be obtained self-consistently via the Car-Parinello method [39], or other optimization scheme, by identifying the interacting spin-density matrix with that derived from a suitable noninteracting ensemble

$$\rho_\alpha(\mathbf{r})_{ss'} = \sum_i \nu_{i\alpha}\psi_{i\alpha}(\mathbf{r}, s)\psi^*_{i\alpha}(\mathbf{r}, s'). \tag{2.77}$$

Such an identification is always possible. The occupation numbers $\nu_{i\alpha}$ will of course depend on $\rho_M(\mathbf{r})_{ss'}$.

$V_{R,M}(\mathbf{r})$ can be regarded as a reactivity potential. It is the previously undefined external potential of CRT introduced by Nalewajski [35] and by Berkowitz and Parr [36]. $\mathsf{H}_{R,M}(\mathbf{r})$ has no precedent in CRT. It is required for the generalization of the existing formalism of CRT to describe reactive proclivities in which spin plays a key role. μ_M becomes the internal chemical potential of the molecule.

A condition for the utility of this formulation of the AIM problem as a foundation for CRT is that the $\hat{\psi}_{i\alpha}$ be localized to the vicinity of fragment α. Since $V_{R,M}(\mathbf{r})$ and $\mathsf{H}_{R,M}(\mathbf{r})$ must in general extend throughout the entire molecule, it is possible that they delocalize the $\hat{\psi}_{i\alpha}$, causing them to extend well beyond the domain of the fragment α to which they belong. Whether or not this occurs can be established only numerically. If, however, charge transfer between parts is effected primarily by the internal electronegativity-equalization condition, eq. (2.75), $V_{R,M}(\mathbf{r})$ and $\mathsf{H}_{R,M}(\mathbf{r})$ may primarily cause entirely local hybridization and spin/charge polarization, inducing no delocalization. We have implicitly assumed this to be the case. Much of chemical interest could then be expected from the $n_\alpha(\mathbf{r})$ and $\mathbf{S}_\alpha(\mathbf{r})$: fractional charge transfer, covalent bond strengths, pairwise additivity, spin sensitivity, etc.

2˙5.3. The hardness conjecture, part II. In arriving at eq. (2.73), we have supposed that a minimum exists, one requirement for which is

$$\sum_{\alpha} \eta'_{\alpha} > 0, \tag{2.78}$$

where the modified hardness η'_{α} is

$$\eta'_{\alpha} = \frac{\mathrm{d}^2 E'_{\alpha}(N_{\alpha})}{\mathrm{d}N_{\alpha}^2}\,. \tag{2.79}$$

Equation (2.78) can be trivially satisfied if all $\eta'_{\alpha} > 0$. This appears to be in direct contradiction of the hardness conjecture, part I. However, the latter applies to truly isolated systems, whereas in the AIM problem the parts α are constrained to replicate collectively $n_M(\mathbf{r})$ and $\mathbf{S}_M(\mathbf{r})$ through the presence of $V_{R,M}(\mathbf{r})$ and $\mathsf{H}(\mathbf{r})$ in the Kohn-Sham equations. Limitations of space prevent further discussion of the general case of nonvanishing $\mathbf{S}_M(\mathbf{r})$. We now restrict ourselves to the simpler case of saturated molecules for which $\mathbf{S}_M(\mathbf{r}) \equiv 0$ and N_M is even. The vanishing of $\mathbf{S}_M(\mathbf{r})$ is guaranteed by noting that all $\psi_{i\alpha}(\mathbf{r}, s)$ possess Kramers degeneracy or, ignoring spin-orbit coupling, spin degeneracy, and that both members of a degenerate pair are identically occupied, whether fully or fractionally so, or unoccupied. The occupation numbers are

$$\begin{cases} \nu_{i\alpha} = 1, & 1 \le i \le M_{\alpha}\,, \\ \phantom{\nu_{i\alpha}} = \dfrac{1}{2}\omega, & i = M_{\alpha}+1,\ M_{\alpha}+2, \\ \phantom{\nu_{i\alpha}} = 0, & i > M_{\alpha}+2, \end{cases} \tag{2.80}$$

$$N_{\alpha} = \sum_i \nu_{i\alpha} = M_{\alpha} + \omega, \tag{2.81a}$$

$$0 \le \omega \le 2, \tag{2.81b}$$

$$M_{\alpha} \le N \le M_{\alpha} + 2. \tag{2.81c}$$

With eqs. (2.80) and (2.81) for the occupation numbers of the self-consistent solution of the KS eqs. (2.73), $\mathsf{H}_{R,M}(\mathbf{r})$ vanishes, also self-consistently.

We now introduce the second part of our *hardness conjecture*: An isolated system in interaction with a reactivity potential by which all spin-orbitals are forced to be doubly occupied, whether fractionally or not, or unoccupied, has *positive* hardness η'. In contrast to part I in which only one spin state of the HOMO is fractionally occupied, in part II both spin states of the HOMO are fractionally equioccupied.

We first reexamine the "possibility" argument of subsubsect. **2**˙4.2. The formal equation for the electron density, eq. (2.35), is unchanged, but eq. (2.36) is modified with M now even and $0 < \omega \le 2$. The Kohn-Sham equations (2.37) and (2.38) are unmodified,

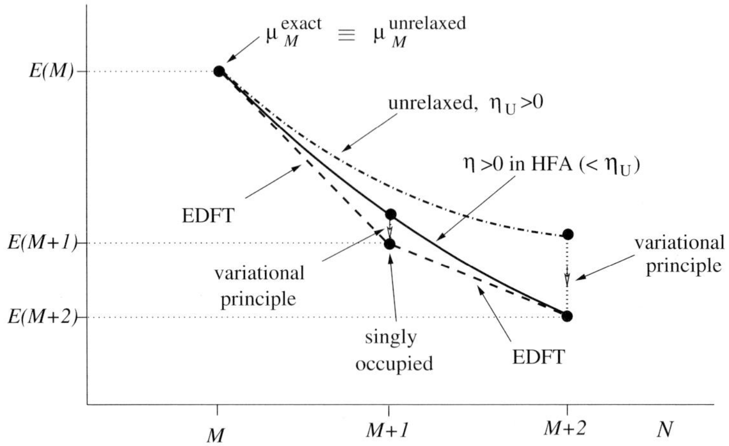

Fig. 4. – Same as for fig. 3 except that the range of N is from M to $M+2$, M even, and most important, both Kramers degenerate HOMO pairs are fractionally occupied with $0 \le \nu_{M+1} = \nu_{M+2} \le 1$. Note that $E(M+1)$ in the HFA must lie above the exact $E(M+1)$ of the EDFT because the exact ground state would have $\nu_{M+1} = 1$, whereas in this version of the HFA adapted to the AIM problem for saturated molecules, $\nu_{M+1} = \nu_{M+2} = 1/2$. The hardness of the exact HFA solution is now positive, as displayed by the convexity of the solid line. From ref. [23]

as are their generalizations, eqs. (2.73) and (2.74), but the occupation numbers change from those of eq. (2.39) to those of eqs. (2.80). Equation (2.56) for η remains formally unchanged, but the frontier density is now

$$n_{\mathrm{F}}(\mathbf{r}) = \frac{1}{2}\sum_s \left[|\psi_{M+1}(\mathbf{r},s)|^2 + |\psi_{M+2}(\mathbf{r},s)|^2\right]. \tag{2.82}$$

However, time reversal invariance imposes

$$\sum_s |\psi_{M+1}(\mathbf{r},s)|^2 = \sum_s |\psi_{M+2}(\mathbf{r},s)|^2 \,; \tag{2.83}$$

it follows that n_{F} is still given by eq. (2.52). Equation (2.56) holds for the η''s as well, except that all quantities within it are associated with $\mathcal{H}'_{\mathrm{KS}}$ instead of $\mathcal{H}_{\mathrm{KS}}$. Thus, it is *possible* that the η'_α are all positive and eq. (2.78) holds.

Once again, we establish *plausibility* by considering the HFA. The N_α-dependence of $E'_\alpha(N_\alpha)$ is carried entirely by its dependence on $\nu_{M+1,\alpha}$ and $\nu_{M+2,\alpha}$. Thus, if the $\hat{\psi}_{i\alpha}$ are kept fixed, the only quadratic term in ω_α (and therefore N_α) arises from the positive Coulomb repulsion between electrons in the degenerate HOMO's. Dropping the subscript α, the unrelaxed $E'(N)$ is consequently a parabola with constant, positive curvature passing through the exact HFA $E'(M)$. The fully relaxed $E'(N)$ falls progressively further below the unrelaxed parabola as N increases. However, it retains positive curvature, as

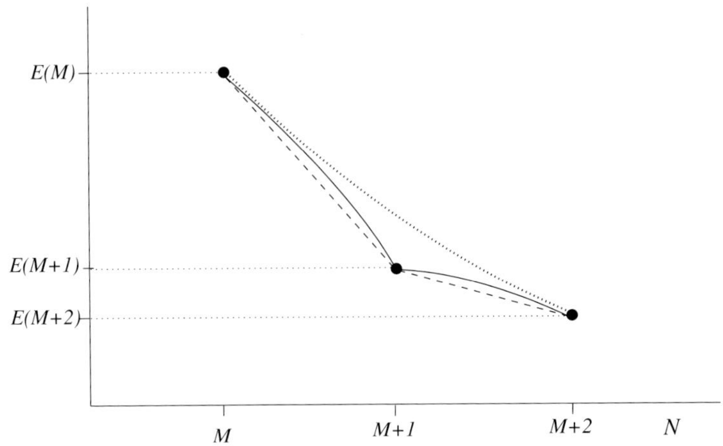

Fig. 5. – Comparison of the noninteger number dependence of the energy among the EDFT (dashed line), $\eta = 0$ hardness; fractional occupation of a single spin state of the LUMO (solid line), $\eta < 0$; and equal fractional occupation of both spin states of the LUMO (dotted line), $\eta > 0$.

$E'(M + 1)$ must lie above the ground-state energy for $M + 1$ since in the ground state only ψ_{M+1} is occupied, not ψ_{M+2}. Thus the η'_α are positive in the HFA. The above arguments are illustrated in fig. 4.

These same considerations hold for the isolated systems as well, so that $\eta_\alpha > 0$ for fractional double occupancy without imposition of the density constraint eq. (2.71a). Figure 5 shows the relationship of the N dependences of $E(N)$ for single and double fractional occupancy. There is an analogy here to the crossover between the Heitler-London picture of molecular bonding in which electrons move primarily around individual atoms ($\eta < 0$) and the Hund-Mulliken picture in which all the relevant states are molecular orbitals extended through the molecule, doubly occupied in a saturated molecule for which $\eta > 0$ if double occupancy is maintained as N changes by dN. There is a similar crossover in theories of magnetic materials from the Van Vleck [40] picture based on localized orbitals and the Stoner [41] or Slater [42] picture based on Bloch functions. An analogous crossover occurs in the Mott-Hubbard [43, 44] metal-insulator transition from Mott insulator to normal metal. However, the details of the transition from positive to negative η are complex and will be discussed elsewhere.

2·6. *Summary and discussion.* – From the beginnings of DFT and especially in the reformulation of CRT within the framework of DFT, knowledge of the density functional for noninteger N was implicitly required. The lack of explicit N dependence and the continuity in n of the standard approximations to the DF suggested simply interpolating them to noninteger N (NCDFT). Calculations of various reactivity indices have been made on that basis with perfectly sensible results [17]. Perdew *et al.*, on the other hand, have argued against NCDFT on the grounds that NCDFT yields fractional charges on the atoms of a heteronuclear diatomic molecule separated to infinity instead of the correct neutral atoms. To avoid this contradiction, they introduced ground-state ensemble-density-functional theory (EDFT). However, EDFT is unsatisfactory on physical grounds because it represents a mixture of noninteracting systems of differing valence and not a single system. It is also unsatisfactory on chemical grounds because it yields vanishing hardness and the identity of the electrophilic reactivities of a system with $M+1$ electrons with the nucleophilic reactivities of a system with M electrons.

The separation argument of Perdew *et al.*, implicitly assumes that the sum of the hardnesses of the two isolated atoms is positive. We have pointed out that this need not be the case and, with the possibility of negative hardness in mind, have reexamined NCDFT. We have first shown that the Levy-Lieb constrained-search algorithm for integer N implies, through a different separation argument, that the DF exists for noninteger N for isolated systems and contains no explicit N dependence. This DF is continuous in N if the DF for integer N is continuous in n. Though we are dealing in this way with a single interacting dynamical system, the electron density of that system can be represented exactly by the density of an ensemble of noninteracting electrons and thus Kohn-Sham equations can be derived.

Next, we put forward the first part of a two-part hardness conjecture: the hardness of a system in which only one member of a pair of Kramers degenerate KS states is occupied is negative. We first prove that the sign of the hardness of such a system is indeterminate. Then we add to the plausibility of the conjecture by showing it to hold in the Hartree-Fock approximation.

The NCDFT so formulated is then used as a basis for a novel "atoms" in "molecules" theory in which the electron spin-density matrix of the "molecule" is exactly partitioned into a sum of contributions from its parts by minimizing the sum of the isolated DF's of its parts subject to that constraint, eq. (2.71). For there to be a minimum, the hardnesses of the parts should be positive, setting the stage for part II of the hardness conjecture, that the hardness of a part of a saturated molecule is positive when both members of a Kramers degenerate pair of KS HOMO states are equally fractionally occupied, also shown to hold in the Hartree-Fock approximation. This sensitivity of hardness to spin parallels the transition from Heitler-London to Hund-Mulliken molecular bonding. It suggests that the concept of hardness be generalized explicitly to include spin or the corresponding index of the Kramers degenerate pair,

$$\eta_{ss'} = \frac{\mathrm{d}^2 E}{\mathrm{d}N_s\,\mathrm{d}N_{s'}}, \tag{2.84}$$

to facilitate the analysis of free-radical reactions, a subject for future research. The Lagrange multiplier of the constraint eq. (2.71a) has the form of a reactivity potential $V_{R,M}(\mathbf{r})$ which can be taken as the previously undefined external potential of chemical reactivity theory. That of the constraint (2.71b) is a new reactivity field.

We conclude that there exists an internally consistent density-functional theory in which the density functional is continuous in electron number. This theory can be made the basis of chemical reactivity theory via a theory of "atoms" in "molecules" provided both parts of our hardness conjecture are true. While the plausibility of the conjecture has been established here, formal proof is not possible with our present understanding of the mathematics of functionals. What is needed now is direct numerical computation of the hardnesses of systems of both types. Treating the exchange exactly would be essential in such calculations as the exact cancellation of the self-interaction within the Hartree potential by the exchange lies at the heart of our arguments about hardness.

Arguments such as those of Perdew *et al.* [20] and those of subsect. **2**˙3 can lead to valid inferences about the properties of isolated systems only if the separation process is such that the separated parts become dynamically independent (no interactions) and statistically independent (no entanglement). Dealing with entanglement is particularly important when degeneracy enters as it always does for noninteger N.

Both Perdew *et al.* in ref. [20] and the present authors in subsect. **2**˙3 have implicitly assumed that disentanglement occurs, without describing the mechanism. In both cases, the principal conclusion remains correct (explicit in subsect. **2**˙3, implicit in ref. [20]): the electron density is a linear superposition of non-overlapping parts, and the density functional is a simple sum of the density functionals of the parts. Disentanglement will be treated explicitly in a separate paper.

A recent analysis by Yang *et al.* [45] of degeneracy and fractional number obtains, without resorting to ensembles, the same piecewise linear dependence on noninteger number of electrons as that of EDFT. However, their analysis depends in an essential way on a separation argument for degenerate states which retains complete entanglement, holding for a set of states of measure zero while failing for the remaining states of finite measure. Detailed criticism of ref. [45] will be given elsewhere with an analysis of disentanglement needed to understand more deeply our separation argument for NCDFT and the transition from negative to positive η.

3. – Excitation spectra in DFT

3˙1. *Introduction.* – The Hohenberg-Kohn theorem established that the ground-state properties of a system with a non-degenerate ground state are unique functionals of the ground-state electron density. The Kohn-Sham equations give us a procedure for computing that density and thence those properties once the relevant functionals are known to sufficient accuracy. There are properties in addition which can be computed directly from the KS wave functions and eigenvalues, *e.g.* χ_0 of eq. (2.54). The orbitals can figure prominently in, for example, the indices of CRT, viz. the frontier density within the Fukui function, eqs. (2.50a)-(2.54). Consequently, it is natural to ask what is the physical meaning of the KS energies, ε_i. From the beginnings of DFT, it has been assumed that the exact excitation energies of a system are not given by differences of its KS eigenvalues, and there has been much discussion of their actual physical significance [20]. In this section, we explore the relationships between the KS energies and the excitation energies for extended systems. In contrast to sect. **2**, here we treat only systems with fixed electron number. The question of how to formulate DFT for noninteger N simply does not arise. Yet, the results of the analysis bear directly on the issues raised in sect. **2**.

3·2. *The band-gap problem.* – By the early 1980's, it was established that the ground-state properties could be calculated to satisfactory accuracy for atoms, molecules, metals, semiconductors, insulators, surfaces, and imperfections in crystals via the KS formulation of DFT using the LDA [46,47], generalizations of the LDA like the GGA [48], or various heuristic approximations to the $V_{\rm KS}$ or the DF [49].

By that time, it had been clearly demonstrated that the KS band structure derived from the ground-state $V_{\rm KS}$ yielded by those approximations, $\epsilon_n^{\rm KS}(\mathbf{k})$, gave band gaps much smaller than observed [46, 50]. Here, $\epsilon_n^{\rm KS}(\mathbf{k})$ is the KS energy within the n-th energy band at wave vector $\mathbf{k}$ in the first Brillouin zone. For example, Ge was incorrectly predicted to be a metal. This contrast between agreement with experiment for ground-state properties and disagreement for excitation energies focussed attention once again on the physical meaning of the KS energies ϵ_i. There were two possible interpretations of that disagreement. The first is that the KS band structure should have given the correct gaps, but, while adequate for the ground-state properties, the computational scheme used was not sufficiently accurate for the excitation energies. That computational scheme had two central elements, pseudopotential theory and the LDA. Hamann [50] had already shown that the discrepancy did not arise from the use of pseudopotentials by repeating the computations via an all-electron scheme. The issue of the accuracy of the LDA in this context was not addressed because a second interpretation had been favored from the introduction of KS theory in 1965.

It had been recognized, once Landau's Fermi-liquid theory had been formulated [51], that experiments which yielded information about electronic band structure were in fact measuring the Landau quasiparticle spectrum, $\epsilon_n^{\rm QP}(\mathbf{k})$. The second interpretation, then, is that as a matter of principle the KS spectrum and the quasiparticle spectrum are distinct,

$$\epsilon_n^{\rm KS}(\mathbf{k}) \neq \epsilon_n^{\rm QP}(\mathbf{k}). \tag{3.1}$$

Hybertsen and Louie [52] carried out a calculation of the quasiparticle band gaps in the GW approximation of Hedin [53] for silicon and diamond and found excellent agreement with experiment, supporting this second interpretation, eq. (3.1). Since then their method has been applied to a broad range of semiconductors and insulators, 29 in number in a recent compilation by Shirley and Louie [54] with band gaps ranging over two orders of magnitude, always agreing with experiment.

The QP band gap computations have been extended to the valence bands and the low-lying conduction bands of semiconductors by Rohlfing *et al.* [55], again with excellent agreement with photoemission [56] and inverse photoemission experiments [57], respectively.

Thus the gap problem appeared to be solved, and it is natural to ask what was missing from the KS band structure; that is, how it specifically differed from the QP band structure. The QP band structure is derived from a Dyson-like equation [58],

$$\left[\frac{p^2}{2m} + V_n + V_{\rm H} + \Sigma_1\left(\epsilon_n^{\rm QP}(\mathbf{k})\right)\right]\psi_{\mathbf{k}n}^{\rm QP} = \epsilon_n^{\rm QP}(\mathbf{k})\psi_{\mathbf{k}n}^{\rm QP}\,. \tag{3.2}$$

On the other hand, the KS band structure is derived from the KS equation

$$\left[\frac{p^2}{2m} + V_n + V_{\rm H} + V_{\rm xc}\right]\psi_{\mathbf{k}n}^{\rm KS} = \epsilon_n^{\rm KS}(\mathbf{k})\psi_{\mathbf{k}n}^{\rm KS}\,. \tag{3.3}$$

The $V_{\rm xc}$ in eq. (3.3) is replaced in eq. (3.2) by the real part of the self-energy operator Σ_1 evaluated at the quasi-particle energy itself. The mapping from the self-energy to $V_{\rm xc}$ derived by Sham and Schlüter [58] makes it highly unlikely that the two could be identical. Perdew and Levy [59], Sham and Schlüter [58], and many later authors have ascribed the difference primarily to the derivative discontinuities arising from generating $V_{\rm xc}$ within the conceptual framework provided by EDFT, arguing that

$$V_{\rm xc}^{N+1} \neq V_{\rm xc}^{N}\,. \tag{3.4}$$

where the superscript indicates the number of electrons. From the Janak theorem [24], they argue that

$$\epsilon_{\rm LUMO}^{\rm KS}\big|_{N+1} = -A_N\,. \tag{3.5}$$

From the derivative discontinuity argument, eq. (3.4), they argue that

$$\epsilon_{\rm LUMO}^{\rm KS}\big|_{N} \neq \epsilon_{\rm LUMO}^{\rm KS}\big|_{N+1}\,. \tag{3.6}$$

Similarly

$$\epsilon_{\rm HOMO}^{\rm KS}\big|_{N} = -I_N \tag{3.7}$$

follows from the Janak theorem. Now

$$E_{\rm G}^{\rm QP} = I_N - A_N \tag{3.8}$$

and

$$E_{\rm G}^{\rm KS} = \epsilon_{\rm LUMO}^{N} - \epsilon_{\rm HOMO}^{N} \tag{3.9}$$

hold by definition so that from eqs. (3.5)–(3.9), it follows that

$$E_{\rm G}^{\rm QP} = E_{\rm G}^{\rm KS} + \Delta_{\rm xc}\,, \quad \Delta_{\rm xc} = \epsilon_{\rm LUMO}^{\rm KS}\big|_{N+1} - \epsilon_{\rm LUMO}^{\rm KS}\big|_{N}\,, \tag{3.10}$$

with the nonzero correction Δ_{xc} arising from the derivative discontinuities in V_{xc}. Δ_{xc} is interpreted as a constant (**k**-independent) shift of the conduction band relative to the valence band [58, 59]. The implication is that LDA results for the valence band are satisfactory but that the LDA conduction band is too low by an amount Δ_{xc}. This contention is consistent with the results of the QP computations.

3·3. *EXX: Reopening the question.* – In 1996 Görling [60] published a feasible scheme for calculating the exchange part V_x of the local V_{xc} exactly, EEX. Städele *et al.* [61] have used the EEX to calculate E_G^{KS} for a series of narrow- and moderately wide-gap semiconductors. Their results for E_G^{KS} agreed essentially as well with experiments as the computed E_G^{QP} did and thus imply that Δ_{xc} is either small or zero. Using accepted methods to compute Δ_x, they found it to be positive and of the same size as that of the Δ_{xc} extracted from the LDA results. They infer that Δ_c is negative and largely cancels Δ_x for the materials studied.

Magyar *et al.* [62], using the same EEX code, have studied the rare-gas crystals and some ionic crystals, all very wide-gap materials. They find KS gaps which are significantly smaller than the observed gaps. However, the EEX method uses the same $E_c[n]$ as the LDA, a local approximation. For these tightly bound materials, the polarization potential, a highly nonlocal contribution to $E_c[n]$ becomes very important. It shifts both the conduction band and the valence band downward, but perhaps the latter shift is much greater because of a near resonant effect.

It has been clear for some time that one important error in the LDA or GGA [63] is the lack of cancellation of the self-interaction term in the Hartree contribution to V_{KS}. Heuristic self-interaction corrections have led to much better agreement between KS and observed energy gaps in a broad range of materials, particularly Mott insulators [63]. Inclusion of exact exchange in the EEX provides for precise cancellation of the self-interaction. Imperfect but much better cancellation than in the LDA or GGA is provided by hybrid functionals such as B3LYP [64] or PBE1PBE [65]. Kudin *et al.* [66], using the latter, have obtained the observed antiferromagnetic insulating ground state of UO_2 and a KS gap of 2.6 eV *vs.* the observed gap of 2.1 eV, whereas the LSDA and GGA yield a ferromagnetic metal.

Thus the question of the relation between E_G^{KS} and E_G^{QP} and, more generally, between $\epsilon_n^{KS}(\mathbf{k})$ and $\epsilon_n^{QP}(\mathbf{k})$ remains open. What is particularly confusing is that the QP corrections to the LDA band structures, which presumably fix up both exchange and correlation, and which yield agreement with experiment, show an upward shift to the conduction bands alone, consistent with the Δ_{xc} argument; whereas the EEX computations, which fix up the exchange interaction only, also give good E_G values for low to medium gaps without the need to invoke Δ_{xc}. One way to settle the issue is through the analysis of an exact theory of excitation spectra grounded in DFT. We devote the remainder of this part of the paper to that task. Among other results which contradict widely held beliefs, we shall show thereby that Δ_{xc} is identically zero.

3·4. *Time-dependent density-functional theory.* – In 1984 Runge and Gross [67] showed how to extend the conceptual structure of DFT into the time domain, formulating time-dependent DFT, TDDFT. Gross and Kohn [68] then developed a linear response theory, and Dreizler and Kohl introduced an improved definition of the energy functional entering the theory [69]. The construction of TDDFT was far more complex and subtle than that of DFT, with many pitfalls along the way. Instead of developing it here, we refer the reader to the excellent exposition of van Leeuwen [70]. All we need to know for our present purposes is that time-dependent KS equations can be derived.

In 2002, Kim and Görling applied the linear response theory to compute the dielectric response of bulk Si, using the EEX methodology but making the so-called adiabatic approximation to the functional [71], that is assuming the functional to be the same as for static problems. Their results agree with experiment essentially as well as the many-body computations of, *e.g.* Rohlfing and Louie [72] for GaAs, who use the GW approximation for the QP energies and compute the dynamic effects of electron-hole correlation by solving the Bethe-Salpeter equation. These results support our intention of carrying out an exact analysis of the electron-hole excitation spectrum via TDDFT.

The TDKS equations are

$$i\hbar\frac{\partial\psi_i}{\partial t} = \mathcal{H}_{\mathrm{KS}}\psi_i\,, \quad i = 1\ldots N. \tag{3.11}$$

Equation (2.9) still holds, but V_{KS} is now time dependent so that eq. (2.18) changes to

$$V_{\mathrm{KS}}(\mathbf{r},t) = V_{\mathrm{e}}(\mathbf{r},t) + V_{\mathrm{H}}(\mathbf{r},t) + V_{\mathrm{xc}}(\mathbf{r},t) \tag{3.12}$$

with V_{e} now time dependent, $V_{\mathrm{H}}(\mathbf{r},t)$ given by the time-dependent generalization of eq. (2.19), and $V_{\mathrm{xc}}(\mathbf{r},t)$ quite different in principle. It is now given by

$$V_{\mathrm{xc}}(\mathbf{r},t) = \frac{\delta A_{\mathrm{xc}}[n]}{\delta n(\mathbf{r},t)}\,, \tag{3.13}$$

where the exchange-correlation action A_{xc} is defined in subsect. 4.2 of ref. [70].

3·5. *Excitation energies via TDDFT.* – In TDDFT, knowing the electron density is sufficient for predicting or interpreting the result of any measurement of a physical property. All of the experiments mentioned or referred to implicitly in subsect. **3**·2 in connection with the band gap problem involve measurements of the linear response of a material to some time-dependent external perturbation. M. E. Casida has developed an explicit formal theory of the electron-hole excitation spectrum relevant to linear-response theory [73]. Casida was interested in molecular systems. As we need to discuss extended systems as well, we shall rederive Casida's ingenious formalism here with minor modifications of the analysis. We shall suppose that the perturbation is an external potential containing only finite wavelengths so that eqs. (2.9), (3.11) and (3.12) can provide a basis for getting at the excitation spectrum. Nevertheless, the results are of general validity for any perturbation. We restrict ourselves to finite systems no matter how large. For the treatment of crystals, we impose periodic boundary conditions with a finite but macroscopically large supercell. In both cases, the KS spectra are discrete. In the following, we set $\hbar = 1$.

A first-order change of the electron density, $n_1(\mathbf{r}, t)$, implies the existence of a corresponding first-order change in the functional representing the time-dependent expectation value of an operator $\hat{\mathcal{O}}$, $\mathcal{O}[n]$,

$$\mathcal{O}_1(t) = \int \mathrm{d}\mathbf{r}\,\mathrm{d}t' \frac{\delta \mathcal{O}([n], t)}{\delta n(\mathbf{r}, t')}\bigg|_{\mathrm{G}} n_1(\mathbf{r}, t'), \tag{3.14}$$

where the subscript G indicates that the derivative is to be evaluated at the ground-state density. In the KS framework, the density is given by

$$n(\mathbf{r}, t) = \sum_{i=1,s}^{N} |\psi_i(x, t)|^2. \tag{3.15}$$

Equation (3.15) can be rewritten as

$$n(\mathbf{r}, t) = \mathrm{tr}[\hat{n}(\mathbf{r})\hat{\rho}(t)], \tag{3.16}$$

where $\hat{n}(\mathbf{r})$ is the single-particle density operator $\delta(\mathbf{r} - \hat{\mathbf{r}})$ and $\hat{\rho}(t)$ is the one-particle density matrix. In the coordinate representation $\hat{\rho}(t)$ becomes

$$\rho(x, x'; t) = \sum_i \nu_i \psi_i(x, t)\psi_i^*(x', t). \tag{3.17}$$

In eq. (3.17), the ν_i are time independent, having the same values of 1 or 0 they have in the initial non-interacting N-particle Slater determinant at $t = t_0$. Those states for which ν_i is initially unity evolve with time according to

$$\psi_i(t) = \hat{U}(t, t_0)\psi_i(t_0), \quad i = 1 \ldots N, \tag{3.18}$$

$$\hat{U}(t, t_0) = \mathcal{T}\exp\left[\int_{t_0}^{t} \mathrm{d}t' \hat{\mathcal{H}}_{\mathrm{KS}}(t')\right], \tag{3.19}$$

where $\mathcal{T}$ is the time-ordering operator (later to the left). Those for which $\nu_i = 0$ can be taken as any orthonormal complement of the set (3.18) in the Hilbert space. From eq. (3.16) $n_1(\mathbf{r}, t)$ is given by

$$n_1(\mathbf{r}, t) = \mathrm{tr}[\hat{n}(\mathbf{r})\hat{\rho}_1(t)]. \tag{3.20}$$

Inserting eq. (3.20) into (3.14) and invoking time-translation invariance shows us that the excitation spectrum of the linear response of the system enters into measurements through the time dependence of $\hat{\rho}_1(t)$ and the functional derivative of $\mathcal{O}([n],t)$. But, following Casida [73, 74], it is sufficient to focus our attention on $\hat{\rho}_1(t)$ to derive Casida's powerful formalism for the analysis of excitation energies.

The equation of motion of $\hat{\rho}$,

$$i\dot{\rho} = [\mathcal{H}, \rho], \tag{3.21}$$

leads to

$$i\dot{\rho}_1 = [\mathcal{H}_1, \rho_0] + [\mathcal{H}_0, \rho_1]. \tag{3.22}$$

In eqs. (3.21) and (3.22), we have dropped the subscripts KS and the superposed caret for simplicity. We suppose the system to be in the ground state at $t_0 \downarrow -\infty$ and that an external perturbation $V_{1e}(t)$ is turned on adiabatically and vanishes sufficiently rapidly as t and $t_1 \uparrow \infty$.

Solving eq. (3.22) by the standard methods of first-order perturbation theory, introducing as a basis the complete set of ground-state Kohn-Sham eigenfunctions, and taking the Fourier transform results in

$$\rho_{ij}^{(1)}(\omega) = -\frac{\nu_{ji}\mathcal{H}_{ij}^{(1)}(\omega)}{\epsilon_{ij} - \omega^+}, \tag{3.23a}$$

$$\epsilon_{ij} = \epsilon_i - \epsilon_j\,, \tag{3.23b}$$

$$\nu_{ji} = \nu_j - \nu_i\,. \tag{3.23c}$$

The ground-state KS eigenvalues ϵ_i satisfy condition (2.10). The ground-state occupation numbers are

$$\nu_i = 1, \quad i \leq N, \tag{3.23d}$$

$$\nu_i = 0, \quad i > N, \tag{3.23e}$$

$$\omega^+ = \omega + i\alpha, \quad \alpha \downarrow 0. \tag{3.23f}$$

The first order change in $\mathcal{H}$ is just

$$\mathcal{H}^{(1)} = V_{\mathrm{e}}^{(1)} + V_{\mathrm{H}}^{(1)} + V_{\mathrm{xc}}^{(1)}\,, \tag{3.24a}$$

$$V_{\mathrm{H}}^{(1)}(\mathbf{r},t) + V_{\mathrm{xc}}^{(1)}(\mathbf{r},t) = \int_{-\infty}^{\infty} \mathrm{d}t' \int \mathrm{d}\mathbf{r}'\; f(\mathbf{r},t;\mathbf{r}',t') n_1(\mathbf{r}',t'), \tag{3.24b}$$

$$f(\mathbf{r},t;\mathbf{r}',t') = \frac{e^2}{|\mathbf{r}-\mathbf{r}'|}\delta(t-t') + f_{\mathrm{xc}}(\mathbf{r},t;\mathbf{r}',t'), \tag{3.24c}$$

$$f_{\mathrm{xc}}(\mathbf{r},t;\mathbf{r}',t') = \frac{\delta V_{\mathrm{xc}}(\mathbf{r},t)}{\delta n(\mathbf{r}',t')} = \frac{\delta^2 A_{\mathrm{xc}}}{\delta n(\mathbf{r}',t')\delta n(\mathbf{r},t)}\,. \tag{3.24d}$$

$f(\mathbf{r},t;\mathbf{r}',t')$ is retarded in t and t', it vanishes for $t'>t$; it is real; and it is time-translation invariant (temporal uniformity) so that $f(t,t') = f(t-t')$. As a consequence, $\mathcal{H}^{(1)}(\omega)$ becomes

$$\mathcal{H}^{(1)}(\omega) = V_{\rm e}^{(1)}(\omega) + \int \mathrm{d}\mathbf{r}' f(\mathbf{r},\mathbf{r}';\omega) n_1(\mathbf{r}',\omega), \tag{3.25}$$

and $f(\mathbf{r},\mathbf{r}';\omega)$ is not Hermitian for finite α,

$$f(\mathbf{r},\mathbf{r}';\omega)^\dagger = f(\mathbf{r},\mathbf{r}';-\omega). \tag{3.26}$$

In (3.26), the imaginary part of $-\omega$ equals α and is positive. However, for the discrete spectra we consider, $f(\mathbf{r},\mathbf{r}';\omega)$ becomes real as $\alpha \downarrow 0$. The perturbation $n_1(\mathbf{r},\omega)$ in the density is

$$n_1(\mathbf{r},\omega) = \sum_{k,l;s} \psi_k(x)\psi_l^*(x)\rho_{kl}^{(1)}(\omega). \tag{3.27}$$

Inserting eq. (3.27) into eq. (3.25) and the result into eq. (3.23a) yields

$$\rho_{ij}^{(1)}(\omega) + \frac{\nu_{ji}}{\epsilon_{ij}-\omega^+}\sum_{kl} K_{ij,kl}(\omega)\rho_{kl}^{(1)}(\omega) = -\frac{\nu_{ji}}{\epsilon_{ij}-\omega^+} V_{\rm e}^{(1)}(\omega)_{ij}\,, \tag{3.28a}$$

$$K_{ij,kl}(\omega) = \sum_{s,s'}\int \mathrm{d}\mathbf{r}\,\mathrm{d}\mathbf{r}'\psi_i^*(x)\psi_j(x) f(\mathbf{r},\mathbf{r}';\omega)\psi_k(x')\psi_l^*(x'). \tag{3.28b}$$

From eq. (3.21) or, more explicitly, eq. (3.28a), we see that $\rho_{ij}^{(1)}(\omega)$ vanishes if $\nu_i = \nu_j$, that is if i and j are both occupied or unoccupied KS states. Accordingly, we can restrict consideration to pairs ij and kl which are differently occupied, reserving the symbols i and k for unoccupied and j and l for occupied states so that $\nu_{ij} = -1$ and $\nu_{ji} = 1$. Next, we simplify by supposing that the ground state is time-reversal invariant so that the KS orbitals are real and $f(\mathbf{r},\mathbf{r}';\omega)$ is real and symmetric in $\mathbf{r}$ and $\mathbf{r}'$. Equation (3.28a) can then be rewritten as

$$\sum_{kl}\{[(\epsilon_{ij}-\omega)\delta_{ij,kl} + K_{ij,kl}(\omega)]\rho_{kl}^{(1)}(\omega) + K_{ij,kl}(\omega)\rho_{lk}^{(1)}(\omega)\} = -V_{\rm e}^{(1)}(\omega)_{ij}\,, \tag{3.29a}$$

$$\sum_{kl}\{[(\epsilon_{ij}+\omega)\delta_{ij,kl} + K_{ij,kl}(\omega)]\rho_{lk}^{(1)}(\omega) + K_{ij,kl}(\omega)\rho_{kl}^{(1)}(\omega)\} = -V_{\rm e}^{(1)}(\omega)_{ji}\,, \tag{3.29b}$$

because of the invariance of K with respect to interchange of i with j or k with l. We now introduce the convention that i or k will always be written before j or l, respectively. This permits eqs. (3.29a)-(3.29b) to be written in abstract supermatrix form

$$\left[\begin{pmatrix} \hat\epsilon + \hat K & \hat K \\ \hat K & \hat\epsilon + \hat K \end{pmatrix} - \omega \begin{pmatrix} \hat 1 & 0 \\ 0 & -\hat 1 \end{pmatrix}\right]\begin{pmatrix} \rho^{(1)} \\ \rho^{(1)T} \end{pmatrix} = -\begin{pmatrix} V_{\rm e}^{(1)} \\ V_{\rm e}^{(1)T} \end{pmatrix}. \tag{3.30}$$

Interchanging the x and z axes of quantization of the 2×2 supermatrices by reflection transforms eq. (3.30) to

$$\left[\begin{pmatrix} \hat{\epsilon} + 2\hat{K} & 0 \\ 0 & \hat{\epsilon} \end{pmatrix} - \omega \begin{pmatrix} 0 & \hat{1} \\ \hat{1} & 0 \end{pmatrix}\right] \begin{pmatrix} \rho_+ \\ \rho_- \end{pmatrix} = - \begin{pmatrix} V_+ \\ V_- \end{pmatrix}, \tag{3.31a}$$

$$\rho_\pm = \rho^{(1)} \pm \rho^{(1)T}, \qquad V_\pm = V_{\mathrm{e}}^{(1)} \pm V_{\mathrm{e}}^{(1)T}. \tag{3.31b}$$

Decoupling ρ_+ and ρ_- in eq. (3.31a) leads to

$$\left[(\hat{\epsilon} + 2\hat{K}) - \omega^2 \hat{\epsilon}^{-1}\right]\rho_+ = -V_+ + \omega \hat{\epsilon}^{-1} V_- , \tag{3.32a}$$

$$\left[\hat{\epsilon} - \omega^2 (\hat{\epsilon} + 2\hat{K})^{-1}\right]\rho_- = -V_- + \omega (\hat{\epsilon} + 2\hat{K})^{-1} V_+ . \tag{3.32b}$$

The poles in ω of ρ_+ and ρ_- are identical and occur when ω^2 is an eigenvalue of the operator $\hat{\epsilon}^2 + 2\hat{\epsilon}^{1/2}\hat{K}\hat{\epsilon}^{1/2}$ according to eqs. (3.32a)-(3.32b). Consequently, the poles of the entire $\rho^{(1)}$ are given by the same condition, via eq. (3.31b). The imaginary part of any response of the system to the perturbation $V_{\mathrm{e}}^{(1)}$ is non-zero only at the poles of $\rho^{(1)}$, where energy absorption occurs. Thus the squares of the exact excitation energies of the system in linear response to $V_{\mathrm{e}}^{(1)}$ are given by the eigenvalue spectrum of the operator $\hat{\epsilon}^2 + 2\hat{\epsilon}^{1/2}\hat{K}\hat{\epsilon}^{1/2}$. Writing this out explicitly results in the Casida equation for the excitation energies,

$$\sum_{kl} \left[\epsilon_{ij}^2 \delta_{ij,kl} + 2\epsilon_{ij}^{1/2} K_{ij,kl}(\omega) \epsilon_{kl}^{1/2}\right] F_{kl}^\lambda = \omega_\lambda^2 F_{ij}^\lambda , \tag{3.33}$$

where $\{\omega_\lambda\}$ is the set of exact excitation energies and $\{F^\lambda\}$ the corresponding eigenfunctions related to the normal modes of spontaneous linear fluctuations of the single-particle density matrix.

Casida's equation, (3.33), allows for the direct calculation of excitation energies, for which it has already found use [75]. In the following, however, we shall employ it to obtain several general results of fundamental importance to DFT.

3·6. *Extended systems: crystals.*

3·6.1. Adapting the Casida formalism.
For perfect crystals, the solutions of the Kohn-Sham equations take the form of Bloch functions $\psi_{\mathbf{k}n}$

$$\mathcal{H}_{\mathrm{KS}}\psi_{\mathbf{k}n} = \epsilon_n(\mathbf{k})\psi_{\mathbf{k}n} , \tag{3.34a}$$

$$\psi_{\mathbf{k}n}(\mathbf{r}) = N^{-1/2} e^{i\mathbf{k}\cdot\mathbf{r}} u_{\mathbf{k}n}(\mathbf{r}). \tag{3.34b}$$

The KS eigenvalues $\epsilon_n(\mathbf{k})$ form a set of energy bands with band index n, functions of the wave vector $\mathbf{k}$ within the first Brillouin zone. The factors $u_{\mathbf{k}n}(\mathbf{r})$ in eq. (3.34b) have the periodicity of the crystal and, for given $\mathbf{k}$, form a complete orthonormal set within a unit cell of the crystal. N is the number of unit cells within the volume on which periodic boundary conditions are imposed and is macroscopically large. Following the notation convention introduced after eqs. (3.37), we replace i by the symbol $\mathbf{k}_\mathrm{e}c$ for empty conduction band states and j by $\mathbf{k}_\mathrm{v}v$ for occupied valence and core band states. In this notation, eq. (3.33) becomes

$$\text{(3.35a)} \quad [\epsilon_{cv}(\mathbf{k}_\mathrm{e},\mathbf{k}_\mathrm{v})]^2 F_{cv}(\mathbf{k}_\mathrm{e},\mathbf{k}_\mathrm{v}) + 2\sum_{\mathbf{k}'_\mathrm{e},\mathbf{k}'_\mathrm{v},c',v'} \sqrt{\epsilon_{cv}(\mathbf{k}_\mathrm{e},\mathbf{k}_\mathrm{v})}K_{cv,c'v'}(\mathbf{k}_\mathrm{e},\mathbf{k}_\mathrm{v};\mathbf{k}'_\mathrm{e},\mathbf{k}'_\mathrm{v}) \times$$
$$\times \sqrt{\epsilon_{c'v'}(\mathbf{k}'_\mathrm{e},\mathbf{k}'_\mathrm{v})}F_{c'v'}(\mathbf{k}'_\mathrm{e},\mathbf{k}'_\mathrm{v}) = \omega^2 F_{cv}(\mathbf{k}_\mathrm{e},\mathbf{k}_\mathrm{v}),$$

$$\text{(3.35b)} \quad K(\mathbf{k}_\mathrm{e},\mathbf{k}_\mathrm{v};\mathbf{k}'_\mathrm{e},\mathbf{k}'_\mathrm{v}) =$$
$$= \int d\mathbf{r}\, d\mathbf{r}' \frac{e^{-i(\mathbf{k}_\mathrm{e}-\mathbf{k}_\mathrm{v})\cdot\mathbf{r}}}{N^2} u_{\mathbf{k}_\mathrm{e}c}(\mathbf{r})^* u_{\mathbf{k}_\mathrm{v}c}(\mathbf{r}) f(\mathbf{r},\mathbf{r}';\omega) u_{\mathbf{k}'_\mathrm{e}c'}(\mathbf{r}') u_{\mathbf{k}'_\mathrm{v}v'}(\mathbf{r}')^* e^{i(\mathbf{k}'_\mathrm{e}-\mathbf{k}'_\mathrm{v})\cdot\mathbf{r}'}.$$

To arrive at eqs. (3.35), we remember that the crystal has time-reversal symmetry, which permits transforming from the real KS eigenfunctions employed in deriving eq. (3.33) to the complex Bloch functions of eq. (3.34b) without modification of the formal structure of eq. (3.33).

It is readily established that $f(\mathbf{r},\mathbf{r}';\omega)$ is translation invariant; that is

$$\text{(3.36)} \quad f(\mathbf{r},\mathbf{r}';\omega) = f(\mathbf{r}+\mathbf{R}_\ell,\mathbf{r}'+\mathbf{R}_\ell),$$

where $\mathbf{R}_\ell$ is any translation vector of the crystal lattice. We now introduce the total center-of-mass crystal momentum $\mathbf{P}$,

$$\text{(3.37a)} \quad \mathbf{P} = \mathbf{k}_\mathrm{e} - \mathbf{k}_\mathrm{v} = \mathbf{k}_\mathrm{e} + \mathbf{k}_\mathrm{h}$$

and the relative crystal momentum $\boldsymbol{p}$,

$$\text{(3.37b)} \quad \boldsymbol{p} = \frac{1}{2}(\mathbf{k}_\mathrm{c} + \mathbf{k}_\mathrm{v}) = \frac{1}{2}(\mathbf{k}_\mathrm{e} - \mathbf{k}_\mathrm{h}),$$

of the excited electron-hole pair, where $\mathbf{k}_\mathrm{e}$ is the crystal momentum of the electron and $\mathbf{k}_\mathrm{h} = -\mathbf{k}_\mathrm{v}$ that of the hole. Because of the translation invariance of f, eq. (3.36), $\mathbf{P}$ is conserved, so that eq. (3.35a) can be rewritten with an obvious change of notation,

$$\text{(3.38)} \quad [\epsilon_{cv,\mathbf{P}}(\boldsymbol{p})]^2 F_{cv,\mathbf{P}}(\boldsymbol{p}) + 2\sum_{c'v',\boldsymbol{p}'} \sqrt{\epsilon_{cv,\mathbf{P}}(\boldsymbol{p})}K_{cv,c'v';\mathbf{P}}(\boldsymbol{p},\boldsymbol{p}') \times$$
$$\times \sqrt{\epsilon_{cv,\mathbf{P}}(\boldsymbol{p})}F_{c'v',\mathbf{P}}(\boldsymbol{p}') = \omega^2_{\mathbf{P}} F_{cv,\mathbf{P}}(\boldsymbol{p}).$$

The interaction matrix $K_{cv,c'v';\mathbf{P}}(\boldsymbol{p},\boldsymbol{p}')$ now has the form

$$K_{cv,c'v';\mathbf{P}}(\boldsymbol{p},\boldsymbol{p}') = \frac{1}{N}\int_{\Delta} \mathrm{d}\mathbf{r} \int \mathrm{d}\mathbf{r}' e^{-i\mathbf{P}\cdot(\mathbf{r}-\mathbf{r}')} u^*_{\mathbf{k}_{\mathrm{e}}c}(\mathbf{r}) u_{\mathbf{k}_{\mathrm{v}}v}(\mathbf{r}) \times \times f(\mathbf{r},\mathbf{r}';\omega) u_{\mathbf{k}'_{\mathrm{e}},c'}(\mathbf{r}') u_{\mathbf{k}'_{\mathrm{v}}v'}(\mathbf{r}'), \tag{3.39a}$$

$$\mathbf{k}_{\mathrm{e}} = \boldsymbol{p} + \frac{1}{2}\mathbf{P}, \qquad \mathbf{k}_{\mathrm{v}} = \boldsymbol{p} - \frac{1}{2}\mathbf{P}, \qquad \mathbf{k}'_{\mathrm{e}} = \boldsymbol{p}' + \frac{1}{2}\mathbf{P}, \qquad \mathbf{k}'_{\mathrm{v}} = \boldsymbol{p}' - \frac{1}{2}\mathbf{P}. \tag{3.39b}$$

Note particularly that the integral over $\mathbf{r}$ is over a single unit cell while that on $\mathbf{r}'$ is over the entire material. In consequence, $K_{cv,c'v';\mathbf{P}}(\boldsymbol{p},\boldsymbol{p}')$ is $\mathcal{O}(1/N)$, a fact central to the discussion of the band-gap in subsubsect. **3**˙6.3.

3˙6.2. The asymptotics of $f(\mathbf{r},\mathbf{r}';\omega)$. Gross and Kohn [68] have shown that

$$\chi^{-1}(\mathbf{r},t;\mathbf{r}',t') = \chi_0^{-1}(\mathbf{r},t;\mathbf{r}',t') + f(\mathbf{r},t;\mathbf{r}',t'), \tag{3.40}$$

where

$$\chi(t,t') = -\frac{\delta n(t)}{\delta V_{\mathrm{e}}(t')}; \qquad \chi_0 = -\frac{\delta n(t)}{\delta V_{\mathrm{KS}}(t')}, \tag{3.41}$$

using their formulation of TDDFT for v-representable densities. Explicit expressions for $\chi(\mathbf{r},\mathbf{r}';\omega)$ and $\chi_0(\mathbf{r},\mathbf{r}';\omega)$ can be obtained from first-order perturbation theory, for the former in terms of the interacting-electron eigenfunctions of the true Hamiltonian and for the latter in terms of the corresponding Kohn-Sham quantities derived from the ground-state Kohn-Sham potential. Translation invariance like that exhibited by f in eq. (3.36) leads to both becoming matrices of the form $M_{\boldsymbol{\kappa},\boldsymbol{\kappa}'}(\mathbf{q})$ upon Fourier transformation, where $\mathbf{q}$ is a wave-vector in the first Brillouin zone and $\boldsymbol{\kappa}$, $\boldsymbol{\kappa}'$ are wave vectors of the reciprocal lattice. We can now follow an argument initiated by Ambegaokar and Kohn in 1960 [76] and developed further by Pick *et al.* in 1970 [77] to obtain their asymptotic behavior. That argument concerned static response functions, but can be carried over without change to the dynamical response functions considered here. Because of orthogonality between ground and excited states for χ and occupied and unoccupied KS states for χ_0, their leading terms in $\mathbf{q}$ are

$$\begin{cases} M_{0,0}(\mathbf{q}) \longrightarrow \mathbf{q}\cdot B\cdot\mathbf{q}, \\ M_{0,\boldsymbol{\kappa}}(\mathbf{q}) \longrightarrow \mathbf{q}\cdot\mathbf{A}_{\boldsymbol{\kappa}}, \\ M_{\boldsymbol{\kappa},0}(\mathbf{q}) \longrightarrow \mathbf{A}^*_{\boldsymbol{\kappa}}\cdot\mathbf{q}, \\ M_{\boldsymbol{\kappa},\boldsymbol{\kappa}'}(\mathbf{q}) \longrightarrow M_{\boldsymbol{\kappa}\boldsymbol{\kappa}'}(0), \quad \boldsymbol{\kappa},\boldsymbol{\kappa}' \neq 0, \end{cases} \tag{3.42}$$

as $q \to 0$. Using standard methods for inverting the matrices, one obtains ($\hat{q}$ is a unit vector)

$$(3.43)\qquad \begin{cases} M^{-1}_{0,0}(\mathbf{q}) \longrightarrow \dfrac{1}{q^2} B'(\hat{q}), \\ M^{-1}_{0,\boldsymbol{\kappa}}(\mathbf{q}) \longrightarrow \dfrac{1}{q} A'(\hat{q}), \\ M^{-1}_{\boldsymbol{\kappa},0}(\mathbf{q}) \longrightarrow \dfrac{1}{q} A'^{*}(\hat{q}), \\ M^{-1}_{\boldsymbol{\kappa},\boldsymbol{\kappa}'}(\mathbf{q}) \longrightarrow M^{-1}_{\boldsymbol{\kappa},\boldsymbol{\kappa}'}(\hat{q}), \quad \boldsymbol{\kappa}, \boldsymbol{\kappa}' \neq 0. \end{cases}$$

Inverting the Fourier transform, one obtains from the dominant $M^{-1}_{00}(\mathbf{q})$ term

$$(3.44)\qquad M^{-1}(\mathbf{r}, \mathbf{r}') \sim \frac{D(\hat{R})}{\mathbf{R}}, \quad \mathbf{R} = \mathbf{r} - \mathbf{r}'.$$

It follows from (3.40) that

$$(3.45)\qquad f(\mathbf{r}, \mathbf{r}'; \omega) \sim \frac{U(\hat{R}, \omega)}{R}$$

is the asymptotic form of the dynamic Kohn-Sham interaction, a result previously obtained by Ghosez *et al.* for the static case [78]. Based on broad experience with the excitation spectra of crystals, we can anticipate that the effect on the spectrum of $U(\hat{R}, \omega)$ is repulsive for ω near and below the band gap and attractive for much higher frequencies.

3·6.3. Was there an energy gap problem? We now examine the relation between the Kohn-Sham energy gap, $E_{\mathrm{G}}^{\mathrm{KS}} = \mathrm{Min}_{\mathbf{P},\boldsymbol{p}}[\epsilon_{cv,\mathbf{P}}(\boldsymbol{p})]$ with c the lowest conduction band, and v the highest valence band, and the corresponding value of ω which would be the quasiparticle energy gap $E_{\mathrm{G}}^{\mathrm{QP}}$. We assume for simplicity that the relevant value of $\mathbf{P}$ is the same for both the KS and QP energy bands. We first project out all states other than those of the bands c and v, transforming eq. (3.38) to

$$(3.46)\qquad [\epsilon_{cv,\mathbf{P}}(\boldsymbol{p})]^2 F_{cv,\mathbf{P}}(\boldsymbol{p}) + \sum_{p'} 2\sqrt{\epsilon_{cv,\mathbf{P}}(\boldsymbol{p})}\mathcal{K}_{cv,\mathbf{P}}(\boldsymbol{p}, \boldsymbol{p}') \times \\ \times \sqrt{\epsilon_{cv,\mathbf{P}}(\boldsymbol{p}')} F_{cv,\mathbf{P}}(\boldsymbol{p}') = \omega_{\mathbf{P}}^2 F_{cv,\mathbf{P}}(\boldsymbol{p}),$$

$$(3.47)\qquad \mathcal{K}_{cv,\mathbf{P}}(\boldsymbol{p}, \boldsymbol{p}') = K_{cv,cv;\mathbf{P}}(\boldsymbol{p}, \boldsymbol{p}') + 2 \sum_{c'v', c''v'', \bar{p}, \bar{p}'} K_{cv,c'v';\mathbf{P}}(\boldsymbol{p}, \bar{\boldsymbol{p}}) \times \\ \times \sqrt{\varepsilon_{c'v',\mathbf{P}}(\bar{\boldsymbol{p}})} \left[\hat{\omega}^2 - \hat{\epsilon}^2 - 2\hat{\epsilon}^{1/2}\hat{K}\hat{\epsilon}^{1/2}\right]^{-1}_{c'v',c''v'',\bar{p}\bar{p}';\mathbf{P}} \times \\ \times \sqrt{\epsilon_{c''v'',\mathbf{P}}(\bar{\boldsymbol{p}}')} K_{c''v'',cv;\mathbf{P}}(\bar{\boldsymbol{p}}', \boldsymbol{p}')$$

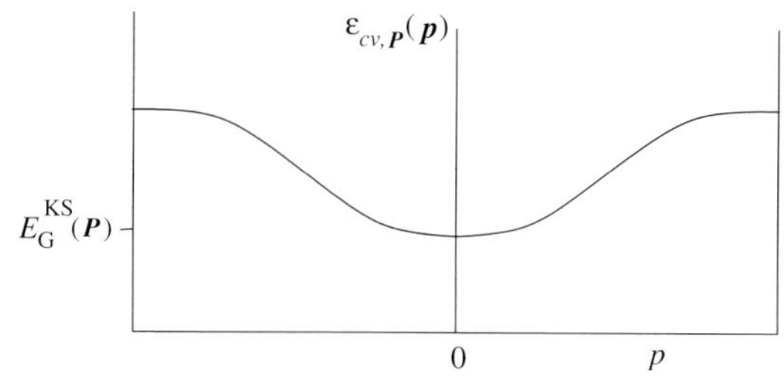

Fig. 6. – A schematic plot of the dispersion of the unperturbed KS excitation energy from the uppermost valence band to the lowest conduction band for given total pair momentum $\mathbf{P}$ along a particular direction in the first Brillouin zone of the relatice momentum $\boldsymbol{p}$, $\varepsilon_{cv,\mathbf{P}}(\boldsymbol{p})$. The minimum in $\varepsilon_{cv,\mathbf{P}}(\boldsymbol{p})$ is the KS energy gap $E_{\mathrm{G}}^{\mathrm{KS}}(\mathbf{P})$ for that total momentum.

In eq. (3.47), the sum over $c'v'$ and $c''v''$ excludes cv, and the operator $[\hat{\omega}^2 - \hat{\epsilon}^2 - 2\hat{\epsilon}^{1/2}\hat{K}\hat{\epsilon}^{1/2}]^{-1}$ operates entirely in the domain of $c'v'$ and $c''v''$, as indicated by the carets. We note that $\mathcal{K}_{cv,\mathbf{P}}(\boldsymbol{p},\boldsymbol{p}')$ in eqs. (3.46) and (3.47) remains $\mathcal{O}(1/N)$, just as $K_{cv,c'v';\mathbf{P}}(\boldsymbol{p},\boldsymbol{p}')$ is in eqs. (3.38) and (3.39a).

The formal structure of Casida's equation, (3.46), is identical to that of the analogous equation in the time-dependent Hartree approximation (TDHA). In 1957, Sawada *et al.* [79], and Brout [80] adapted Wentzel's scalar-meson pair theory of 1942 [81] to explore, inter alia, the interrelation between plasmons and the free electron-hole pair excitation of what was then called the electron gas and now is termed the uniform Fermi liquid. In 1959, Ehrenreich and Cohen showed that the analysis of refs. [79] and [80] was equivalent to the TDHA [82]. That analysis is readily generalized to TDDFT and the above version of the Casida equation, eq. (3.46).

We sketch in fig. 6, a representative variation of the KS free-electron-hole excitation energy $\epsilon_{cv,\mathbf{P}}(\boldsymbol{p})$ with relative momentum $\boldsymbol{p}$ along a symmetry direction in the first Brillouin zone. With careful definition of the bands c and v, this free electron-hole continuum will be bounded from above and below. The lower bound is the KS energy gap $E_{\mathrm{G}}^{\mathrm{KS}}(\mathbf{P})$. The electron-hole pairs interact via the interaction $\mathcal{K}_{cv,\mathbf{P}}(\boldsymbol{p},\boldsymbol{p}')$ of eq. (3.47) according to eq. (3.46). This interaction is of long range according to (3.45), and its effective sign depends on the eigenvalue ω. For low frequency it is attractive and generates excitons, collective modes which are split off below the continuum. For high frequencies, it is repulsive and generates plasmons, collective modes split off above the continuum if they are not degenerate with higher electron-hole continua. Finally, because $\mathcal{K}_{cv,\mathbf{P}}(\boldsymbol{p},\boldsymbol{p}')$ is $\mathcal{O}(1/N)$, the shifts of the bounds of the continua and, in particular, the band gap are also $\mathcal{O}(1/N)$ and disappear in the macroscopic limit if the number of collective modes M is of lower order than N, *i.e.* $M/N \downarrow 0$, $N \to \infty$.

More explicitly, an unperturbed pair excitation state at $\boldsymbol{p}_0$ in the continuum is transformed by the interaction into the state

(3.48a) $$F_{cv,\mathbf{P}}(\boldsymbol{p}) = \delta_{\boldsymbol{p},\boldsymbol{p}_0} + \mathcal{O}\left(\frac{M}{N}\right),$$

and its excitation energy changes to

$$\omega_{cv,\mathbf{P}}(\boldsymbol{p}) = \epsilon_{cv,\mathbf{P}}(\boldsymbol{p}) + \mathcal{O}\left(\frac{M}{N}\right). \tag{3.48b}$$

Equation (3.48b) implies that

$$E_{\mathrm{G}}^{\mathrm{QP}} = E_{\mathrm{G}}^{\mathrm{KS}} + \mathcal{O}\left(\frac{M}{N}\right) \tag{3.49}$$

is the relation between the QP and KS gaps. Thus the central question about that relation is how does M scale with N as $N \uparrow \infty$.

The most rapid increase of M with N occurs if the collective modes include a Rydberg series of Wannier excitons for which $M \uparrow \infty$ as $N \uparrow \infty$ and the continuum threshold is an upper accumulation point. If n is the principle quantum number of the excitonic state and a_n its radius, then

$$a_n \propto n^2, \tag{3.50a}$$

$$M_n \propto n^3, \tag{3.50b}$$

hold, where M_n is the number of excitons, *i.e.* collective modes, with principle quantum number $\leq n$. For finite N with quantization volume of linear scale $L \; \alpha \; N^{1/3}$, the transition from bound to continuum states occurs when $a_{n_{\mathrm{Max}}} \simeq L$, implying that $n_{\mathrm{Max}} \propto L^{1/2}$. This in turn implies that $M \propto L^{3/2} \propto N^{1/2}$ and, strikingly, that $M/N \sim N^{-1/2}$. This allows us to conclude from (3.49) that

$$E_{\mathrm{G}}^{\mathrm{QP}} \equiv E_{\mathrm{G}}^{\mathrm{KS}} \quad (N \uparrow \infty). \tag{3.51}$$

In words, the *exact* QP gap of a perfect crystalline insulator or semiconductor is *identical* to the KS gap computed from the *exact* ground-state KS potential. There never was a band-gap problem. The problem was instead insufficient accuracy of the LDA and the GGA for the computation of the energy gap. The Δ_{xc} of eq. (3.10) is in fact zero. That is $\epsilon_{\mathrm{LUMO}}^{\mathrm{KS}}|_{N+1}$ equals $\epsilon_{\mathrm{LUMO}}^{\mathrm{KS}}|_N$ to order $1/N$ for a macroscopic crystalline insulator or semiconductor. This was already understood at least implicitly in the 1930's in the context of the validity of Koopmans' theorem in the HFA [83, 84].

3·6.4. The Fermi surfaces of metals. In fig. 7a we sketch the Fermi surface (FS) of a metal with the first Brillouin zone as the boundary between the occupied and the unoccupied states in the partially occupied conduction band c. We show also the excitation of an electron from just inside the FS to just above it in fig. 7b. Recalling that $\mathbf{P} = \mathbf{k}_{\mathrm{e}} + \mathbf{k}_{\mathrm{h}}$, eq. (3.37a), and recognizing that $\mathbf{k}_{\mathrm{h}}$ now equals $-\mathbf{k}'_{\mathrm{e}}$, the wave vector of the empty state, we see that as the hole and electron approach the Fermi surface, $\mathbf{k}'_{\mathrm{e}} \to \mathbf{k}_{\mathrm{e}} = \mathbf{k}_{\mathrm{F}}$, $\boldsymbol{p} \to \mathbf{k}_{\mathrm{F}}$, and $\mathbf{P} \to 0$, where $\mathbf{k}_{\mathrm{F}}$ is a wave vector on the Fermi surface. These statements are true both for the KS and QP Fermi surfaces.

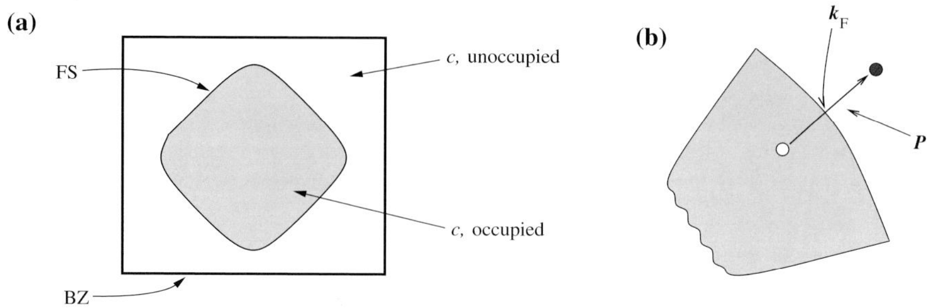

Fig. 7. – (a) A schematic illustration of the Fermi surface (FS) of a metal in the first Brillouin zone (BZ), dividing the occupied (shaded) from the unoccupied states (unshaded). (b) Creation of an electron-hole pair of center of mass momentum $\mathbf{P}$ by excitation from below to above the Fermi surface. As the hole approaches a point $\mathbf{k}_\mathrm{F}$ on the FS from inside and the electron does from outside, $\mathbf{P}$ goes to zero and the relative crystal momentum $\boldsymbol{p}$ goes to $\mathbf{k}_\mathrm{F}$.

We can take advantage of these observations to rewrite the Casida equation (3.46) for excitations at the FS:

$$(3.52)\quad [\epsilon_{cc,\mathbf{0}}(\boldsymbol{p})]^2 F_{cc,\mathbf{0}}(\boldsymbol{p}) + \sum_{p'} 2\sqrt{\epsilon_{cc,\mathbf{0}}(\boldsymbol{p})}\mathcal{K}_{cc,\mathbf{0}}(\boldsymbol{p},\boldsymbol{p}')\sqrt{\epsilon_{cc,\mathbf{0}}(\boldsymbol{p}')}F_{cc,\mathbf{0}}(\boldsymbol{p}') = \omega_\mathrm{F}^2 F_{cc,\mathbf{0}}(\boldsymbol{p}).$$

Now on the KS Fermi surface, $\boldsymbol{p} = \mathbf{k}_\mathrm{F}$, $\epsilon_{cc,\mathbf{0}}(\boldsymbol{p})$ vanishes. Thus, if we set

$$(3.53)\qquad F_{cc,\mathbf{0}}(\boldsymbol{p}) = \delta_{\boldsymbol{p},\mathbf{k}_\mathrm{F}}$$

in (3.52), we see that the QP excitation energy ω_F vanishes. In words, the exact KS Fermi surface generated by the ground-state Kohn-Sham potential is identical to the exact QP Fermi surface.

Note that the transformations leading from eqs. (3.36) to eqs. (3.37) involve inverting $\hat{\epsilon}$ to $\hat{\epsilon}^{-1}$. Equation (3.52) must therefore be understood to hold as the limit $P \to 0$ is approached, when $\epsilon_{cc,\mathbf{P}}(\boldsymbol{p}) \to 0$. The conclusion that the KS FS is identical to the QP FS has thus been reached by a distributional argument.

It is well known that even the LDA gives computed Fermi surfaces of metals in good agreement with the corresponding experimental determinations. A common rationalization for this greater success of the LDA for the Fermi surfaces of metals is that dynamical screening effects lead to shorter-range correlation holes in metals, enhancing the relevance of any local-density approximation.

APPENDIX A.

Noninteracting electron ensembles of given density

In this appendix we show that there is an infinite number of noninteracting ensembles which yield a preset electron density $n(\mathbf{r})$ for noninteger N,

$$n(\mathbf{r}) \geq 0, \qquad \int \mathrm{d}\mathbf{r}\; n(\mathbf{r}) = N. \tag{A.1}$$

Introduce the shape function $g(\mathbf{r})$,

$$n(\mathbf{r}) = N g(\mathbf{r}). \tag{A.2}$$

The density matrix of the grand canonical ensemble is taken as diagonal in electron number so that

$$n(\mathbf{r}) = \sum_M p_M n_M(\mathbf{r}), \tag{A.3}$$

and

$$N = \sum_M p_M M, \tag{A.4}$$

where p_M is the probability that a system of integer M electrons is present in the ensemble with density $n_M(\mathbf{r})$. Take the latter to be

$$n_M(\mathbf{r}) = M g(\mathbf{r}). \tag{A.5}$$

According to ref. [15], the set of M-electron Slater determinants Ψ_M which yield the density (A.5)

$$M g(\mathbf{r}) = \big(\Psi_M, \hat{n}_M(\mathbf{r}), \Psi_M\big) \tag{A.6}$$

is not empty. Inserting (A.5) into (A.3), and (A.4) into the result, yields (A.2). The only unspecified parameters are the p_M, and there is an infinite set of them which satisfy (A.4).

REFERENCES

[1] THOMAS L. H., *Proc. Cambridge Phil. Soc.*, **23** (1927) 542.
[2] FERMI E., *Rend. Accad. Naz. Lincei*, **6** (1927) 602; *Z. Phys.*, **48** (1928) 73.
[3] HARTREE D. R., *Proc. Camb. Phil. Soc.*, **24** (1928) 328.
[4] FOCK V. A., *Z. Phys.*, **15** (1930) 126.
[5] SLATER J. C., *Phys. Rev.*, **81** (1951) 385.
[6] SHARP R. T. and HORTON G. K., *Phys. Rev.*, **90** (1953) 317.

[7] Talman J. D. and Shadwick W. F., *Phys. Rev. A*, **46** (1976) 36; Sahni V., Gruenebaum J. and Perdew J. P., *Phys. Rev. B*, **26** (1982) 4371.
[8] Hohenberg P. and Kohn W., *Phys. Rev. B*, **136** (1964) 864.
[9] Kohn W. and Sham L. J., *Phys. Rev. A*, **140** (1965) 1133.
[10] Dreizler R. M. and Gross E. K. U., *Density Functional Theory* (Springer-Verlag, Berlin) 1990.
[11] Parr R. and Yang W., *Density Functional Theory of Atoms and Molecules* (Oxford University Press, New York) 1989.
[12] Harris J. and Jones R. O., *J. Phys. F*, **4** (1974) 1170; Gunnarsson O. and Lundqvist B. I., *Phys. Rev. B*, **13** (1976) 4274; Langreth D. C. and Perdew J. P., *Phys. Rev. B*, **15** (1977) 2884; Harris J., *Phys. Rev. A*, **29** (1984) 1648.
[13] Levy M., *Proc. Nat. Acad. Sci. USA*, **76** (1979) 6062; *Phys. Rev. A*, **26** (1982) 1200.
[14] Lieb E. H., in *Physics as Natural Philosophy*, edited by A. Shimony and H. Feshback (MIT Press, Cambridge) 1982, p. 111; *Int. J. Quant. Chem.*, **24** (1983) 243; in *Density Functional Methods in Physics*, edited by R. M. Dreizler and J. da Providencia, *NATO ASI Ser. B*, Vol. **123** (Plenum, New York) 1985, p. 31.
[15] Harriman J. E., *Phys. Rev. A*, **24** (1980) 680; Zumbach G. and Masche K., *Phys. Rev. A*, **28** (1983) 544; erratum, *Phys. Rev. A*, **29** (1984) 1585.
[16] Nalewajski R. F., Korchowiec J. and Michalak A., in *Topics in Current Chemistry*, **183** (1996) 25.
[17] Geerlings P., De Proft F. and Langenaeker W., *Chem. Rev.*, **103** (2003) 1793.
[18] Parr R. G. and Pearson R. G., *J. Am. Chem. Soc.*, **105** (1983) 7512.
[19] Pearson R. G., *J. Am. Chem. Soc.*, **85** (1963) 3533; *Science*, **157** (1966) 172; *Chem. Brit*, **3** (1967) 103; *Inorg. Chem.*, **27** (1988) 734.
[20] Perdew J. P., Parr R. G., Levy M. and Balduz J. R. jr., *Phys. Rev. Lett.*, **49** (1982) 1691; Perdew J. P., in *Density Functional Methods in Physics*, edited by R. M. Dreizler and J. da Providencia, *NATO ASI Ser. B*, Vol. **123** (Plenum, New York) 1985, p. 265.
[21] Mermin N. D., *Phys. Rev.*, **137** (1965) A1441.
[22] Cohen M. H., in *Topics in Current Chemistry*, **183** (1996) 143.
[23] Cohen M. H. and Wasserman A., *Israel J. Chem.*, **43** (2003) 219.
[24] Janak J. F., *Phys. Rev. B*, **18** (1978) 7165.
[25] Hellmann H., *Z. Physik*, **85** (1933) 180; Feynman R. P., *Phys. Rev.*, **56** (1939) 340.
[26] Parr R. G. and Yang W., *J. Am. Chem. Soc.*, **106** (1984) 4049.
[27] Cohen M. H., Ganduglia-Pirovano M. V. and Kudonovský J., *J. Chem. Phys.*, **101** (1994) 8988; **103** (1995) 3543.
[28] Pauling L., *The Nature of the Chemical Bond*, 3rd edition (Cornell, N.Y.) 1960.
[29] Mulliken R. S., *J. Chem. Phys.*, **2** (1934) 782.
[30] Pritchard H. O. and Sumner F. H., *Proc. R. Soc. London, Ser. A*, **235** (1956) 136.
[31] Mulliken R. S., *J. Chem. Phys.*, **23** (1955) 1833.
[32] Pearson R. G., *J. Chem. Educ.*, **64** (1987) 561; Parr R. G. and Chattaraj P. K., *J. Am. Chem. Soc.*, **113** (1991) 1854.
[33] Fukui K., Yonezawa T. and Shingu H., *J. Chem. Phys.*, **20** (1952) 722; Fukui K., Yonezawa T., Nagata C. and Shingu H., *J. Chem. Phys.*, **22** (1954) 1433; Fukui K., *Theory of Orientation and Stereoselection* (Springer-Verlag, Berlin) 1965; Fukui K., *Science*, **218** (1982) 747.
[34] Galvan M., del Pino A. and Joannopolous J., *Phys. Rev. Lett.*, **70** (1993) 21.
[35] Nalewajski R. F., *J. Chem. Phys.*, **78** (1984) 6112; **81** (1984) 2088; *J. Phys. Chem.*, **89** (1985) 2831.
[36] Berkowitz M. and Parr R. G., *J. Chem. Phys.*, **88** (1988) 2554.
[37] Nalewajski R. F. and Parr R. G., *Proc. Nat. Acad. Sci. USA*, **97** (2000) 8879.

[38] Hirshfeld F. L., *Theo. Chim. Acta*, **44** (1977) 129.
[39] Car R. and Parinello M., *Phys. Rev. Lett.*, **55** (1985) 2471.
[40] Van Vleck J. H., *Phys. Rev.*, **52** (1937) 1178.
[41] Stoner E. C., *Phil. Mag.*, **21** (1936) 145; *Proc. R. Soc. London, Ser. A*, **165** (1938) 372; **169** (1939) 339.
[42] Slater J. C., *Phys. Rev.*, **82** (1951) 538.
[43] Mott N. F., *Proc. Phys. Soc. A*, **62** (1949) 416; *Can. J. Phys.*, **34** (1956) 1356; *Phil. Mag.*, **6** (1961) 287; *J. Non-Cryst. Solids*, **1** (1968) 1.
[44] Hubbard J., *Proc. R. Soc. A*, **277** (1969) 237; **281** (1964) 401.
[45] Yang W., Zhang Y. and Ayers P. W., *Phys. Rev. Lett.*, **84** (2000) 5172.
[46] Yin M. T. and Cohen M. L., *Phys. Rev. B*, **26** (1982) 5668.
[47] Perdew J. P. and Zunger A., *Phys. Rev. B*, **23** (1981) 5048.
[48] Perdew J. P., Burke K. and Ernzerhof M., *Phys. Rev. Lett.*, **77** (1996) 3865.
[49] Koch W. and Holthausen M. C., *A Chemist's Guide to Density Functional Theory* (Wiley-VCH, Germany) 2000, p. 78.
[50] Hamann D. R., *Phys. Rev. Lett.*, **42** (1979) 662; Lieb E. H., *Int. J. Quant. Chem.*, **24** (1983) 243.
[51] Landau L., *J. Exp. Theoret. Phys. (USSR)*, **30** (1956) 1058; **32** (1957) 59; **35** (1958) 97.
[52] Hybertsen M. S. and Louie S. G., *Phys. Rev. B*, **34** (1986) 2920.
[53] Hedin L., *Phys. Rev.*, **139** (1965) 796.
[54] Louie S. G., *First-Principles Theory of Electron Excitation Energies in Solids, Surfaces, and Defects* in *Topics in Computational Materials Science*, edited by C. Y. Fong (World Scientific, Singapore) 1997, p. 96.
[55] Rohlfing M., Krüger P. and Pollmann J., *Phys. Rev. B*, **48** (1993) 17791.
[56] Chaing T. C. *et al.*, *Phys. Rev. B*, **21** (1980) 3513; **27** (1983) 4770.
[57] Ortega J. E. and Himpsel F. J., *Phys. Rev. B*, **47** (1993) 2130.
[58] Sham L. J. and Schlüter M., *Phys. Rev. Lett.*, **51** (1983) 1888.
[59] Perdew J. P. and Levy M., *Phys. Rev. Lett.*, **51** (1884) 1983.
[60] Görling A., *Phys. Rev. Lett.*, **53** (1996) 7024.
[61] Städele M. *et al.*, *Phys. Rev. Lett.*, **79** (1997) 2089; *Phys. Rev. B*, **59** (1999) 10031.
[62] Magyar R., Fleszar A. and Gross E. K. U., *Phys. Rev. B* (in press).
[63] Zunger A., Perdew J. P. and Oliver G. L., *Solid State Commun.*, **37** (1981) 249; Perdew J. P. and Zunger A., *Phys. Rev. B*, **23** (1981) 5048; Svane A. and Gunnarsson O., *Phys. Rev. Lett.*, **65** (1990) 1148; Svane A., *Phys. Rev. Lett.*, **68** (1994) 1900.
[64] Stevens P. J., Devlin J. F., Chabalowski C. F. and Frisch M. J., *J. Phys. Chem.*, **98** (1994) 11623.
[65] Adamo C. and Barone V., *J. Chem. Phys.*, **110** (1999) 6158.
[66] Kudin K. N., Scuseria G. E. and Martin R. L., *Phys. Rev. Lett.*, **89** (2002) 266402.
[67] Runge E. and Gross E. K. U., *Phys. Rev. Lett.*, **52** (1984) 997.
[68] Gross E. K. U. and Kohn W., *Phys. Rev. Lett.*, **55** (1985) 2850.
[69] Dreizler R. M. and Kohl H., *Phys. Rev. Lett.*, **56** (1986) 1993.
[70] van Leeuwen R., *Int. J. Mod. Phys. B*, **15** (2001) 1969.
[71] Kim Y. H. and Görling A., *Phys. Rev. Lett.*, **89** (2002) 096402.
[72] Rohlfing M. and Louie S. G., *Phys. Rev. B*, **62** (2000) 4927.
[73] Casida M. E., in *Recent Advances in Density Functional Methods*, edited by D. Chong (World Scientific, Singapore) 1995, p. 155.
[74] See also Petersilka M., Gossmann U. J. and Gross E. K. U., *Phys. Rev. Lett.*, **76** (1996) 1212; Jamorski C., Casida M. E. and Salahub D. R., *J. Chem. Phys.*, **104** (1996) 5134.
[75] Chelikowsky J. R., Kronik L. and Vasiliev I., *J. Phys. Cond. Matt.*, **15** (2003) R1517.

[76] Ambegaokar V. and Kohn W., *Phys. Rev.*, **117** (1960) 423.
[77] Pick R., Cohen M. H. and Martin R., *Phys. Rev. B*, **1** (1970) 910.
[78] Ghosez Ph., Gonze X. and Godby R. W., *Phys. Rev. B*, **56** (1997) 12811.
[79] Sawada K., *Phys. Rev.*, **106** (1957) 372.
[80] Brout R., *Phys. Rev.*, **108** (1957) 515.
[81] Wentzel G., *Helv. Phys. Acta*, **15** (1942) 111.
[82] Ehrenreich H. and Cohen M. H., *Phys. Rev.*, **115** (1959) 286.
[83] Koopmans T., *Physica*, **1** (1934) 104.
[84] Seitz F., *Modern Theory of Solids* (McGraw-Hill, New York) 1940, pp. 313; 314.

I. The hydrophobic effect
II. Interfaces and contact lines in three-phase coexistence

B. Widom

Department of Chemistry, Baker Laboratory, Cornell University
Ithaca, New York 14853-1301, USA

1. – The hydrophobic effect

The hydrophobic effect is seen most clearly in the solubility of a hydrophobic substance in water and in the temperature dependence of that solubility. If Σ is the solubility expressed as the dimensionless ratio of the number density of the hydrophobic molecules in the saturated aqueous solution to that in the equilibrium hydrophobe phase, then an appropriate measure of the solute's hydrophobicity is [1, 2]

$$\Delta F^* = -kT \ln \Sigma, \tag{1.1}$$

with k Boltzmann's constant and T the temperature. This ΔF^* is a part of the free-energy change that accompanies the transfer at a fixed temperature T of a solute molecule from the hydrophobe phase into water; it is the part that results only from the structural and energetic changes that occur in the two phases as a result of the transfer. If the transfer occurs with the volumes of both phases fixed, ΔF^* is a Helmholtz free-energy change; if at fixed pressure, it is a change in Gibbs free energy.

In most cases of interest Σ is virtually a function of the temperature alone. The corresponding changes ΔS^*, ΔE, and ΔC_v in the entropy, energy, and constant-volume heat capacity that accompany the transfer of the hydrophobic molecule with the volumes

TABLE I. – *The values $\Delta F^*/kT$, $\Delta S^*/k$, $\Delta E/kT$, $\Delta C_v/k$ for the transfer of the listed molecule from the gas phase into water at* 300 K.

Molecule	$\Delta F^*/kT$	$\Delta S^*/k$	$\Delta E/kT$	$\Delta C_v/k$
CH_4	3.4	−7.5	−4.1	29
C_2H_6	3.1	−9.7	−6.6	36
C_3H_8	3.4	−11.0	−7.6	41
n-C_4H_{10}	3.6	−12.5	−8.9	40
n-C_5H_{12}	3.9	−14.0	−10.1	

of the two phases fixed are then [1, 2]

$$\Delta S^* = k\,\frac{\mathrm{d}(T\ln\Sigma)}{\mathrm{d}T}\,, \tag{1.2}$$

$$\Delta E = kT^2\,\frac{\mathrm{d}\ln\Sigma}{\mathrm{d}T} \quad (= \Delta F^* + T\Delta S^*), \tag{1.3}$$

$$\Delta C_v = kT\,\frac{\mathrm{d}^2(T\ln\Sigma)}{\mathrm{d}T^2}\,. \tag{1.4}$$

Some values of the dimensionless $\Delta F^*/kT$, $\Delta S^*/k$, $\Delta E/kT$, and $\Delta C_v/k$ at $T = 300\,\mathrm{K}$ calculated with (1.1)-(1.4) from the experimental solubilities [3] are in table I [2]. Each entry is for the transfer of the listed molecule from the gas phase into water (taking account, in the case of pentane, of the known vapor pressure of the liquid and its dependence on temperature).

That ΔF^* is positive and equal to a few multiples of kT is a reflection of the low solubility of the substance in water. That ΔS^* and ΔE are both negative means that the main reason for the low solubility is an unfavorable entropy change accompanying the transfer of the molecule from the gas phase into water in spite of a favorable energy change. This and the large positive heat-capacity change are the defining characteristics of the hydrophobic effect [1, 4]. Except for CH_4, which, as is usual for the first member of a homologous series, is anomalous, $\Delta F^*/kT$ for the hydrocarbons listed increases by about 0.3, $\Delta S^*/k$ becomes more negative by about 1.4, and $\Delta E/kT$ becomes more negative by about 1.2, for each additional CH_2 group. That the incremental changes in $\Delta S^*/k$ and $\Delta E/kT$ are each about four times as great as that in $\Delta F^*/kT$ is an example of an approximate energy-entropy compensation [5-7], with $\Delta F^*/kT = \Delta E/kT - \Delta S^*/k$.

Table II makes some interesting contrasts with table I. The entries are again calculated with (1.1)-(1.3) from the experimental solubilities [3] and vapor pressures. We note that for the transfer of an H_2O molecule from the gas phase into liquid n-pentane, each of ΔF^*, ΔS^*, and ΔE is of sign opposite to what it is for the transfer of an n-C_5H_{12} molecule from the gas phase into liquid water (table I). The transfer is now entropically favorable but energetically unfavorable. There are not the same profound structural rearrangements attendant on transferring H_2O into pentane as on transferring C_5H_{12} into

TABLE II. – *The values* $\Delta F^*/kT$, $\Delta S^*/k$, $\Delta E/kT$ *for molecule transfers at* 288 K.

Molecule transferred	from	into	$\Delta F^*/kT$	$\Delta S^*/k$	$\Delta E/kT$
H_2O	gas phase	liquid n-pentane	−1.1	2.3	1.2
H_2O	liquid water	liquid n-pentane	10.2	8.1	18.3
n-C_5H_{12}	liquid n-pentane	liquid water	9.6	−9.6	0.0

water; in particular, no disturbance of a hydrogen-bonded network. Further, that ΔF^* is now negative means, according to (1.1), that in liquid pentane saturated with water at 288 K the concentration of H_2O molecules is greater than in the equilibrium vapor. Equivalently, for equal concentrations of H_2O in the two phases (now not in equilibrium with each other), the chemical potential of H_2O would be higher in the vapor (in vacuum, essentially) than in the liquid: a water molecule would "rather" be in liquid pentane than in vacuum whereas a pentane molecule would "rather" be in vacuum than in liquid water.

There is a large positive ΔF^* for the transfer of an H_2O molecule from liquid water into liquid pentane because of the very large positive ΔE (in spite of a favorable ΔS^*), due in part to the energy cost of breaking hydrogen bonds to free the H_2O molecule from its liquid. The large positive ΔF^* for transferring a pentane molecule from liquid pentane into water is due entirely to the unfavorable entropy change; there is almost no net change in the energy.

A consequence of the unfavorable free energy of accommodation of a hydrophobic molecule in water is that the free energy of accommodating a pair of such molecules is less when they are close together than when they are far apart, for when they are close together the total volume of structurally affected water is smaller. There is thus an effective, solvent-mediated attraction between them, which is much stronger even than their direct van der Waals attraction [8,9]. Clearly the strength of this solvent-mediated attraction must in some way be quantitatively related to the unfavorable hydration free energy of the isolated hydrophobe, which is the ΔF^* in table I. A lattice model that lends itself to determining this connection has been analyzed [2]. The parameters in the model were chosen to reproduce the ΔF^*, ΔS^*, and ΔE for CH_4 listed in table I, and then the solvent-mediated attraction between a closely spaced pair of CH_4 molecules in water was calculated. It is observed that the strength of the attraction is only about kT, which may have important consequences for the operation of the hydrophobic effect in biological contexts. It is also found that the hydrophobic attraction becomes stronger with increasing temperature, just as ΔF^* becomes more positive (which, via (1.1) and (1.2), follows from ΔS^* being negative); and that the strength of this attraction, expressed as a multiple of the thermal energy kT, is given roughly by $0.7\,\Delta F^*/kT - 1.2$.

2. – Interfaces and contact lines in three-phase coexistence

Three phases may meet in a line of common contact, at which there is in general an excess (positive or negative) adsorption analogous to adsorption at the interface between two phases. Just as the Gibbs adsorption equation relates the interfacial adsorptions Γ_i to the rate at which the interfacial tension σ varies with the chemical potentials μ_i, so also is there a Gibbs adsorption equation relating the linear adsorptions Λ_i to the rate at which the line tension τ —the excess free energy per unit length of the three-phase contact line [10, 11]— varies with the chemical potentials.

As a reminder, the Gibbs adsorption equation for the planar interface between two phases α and β in a system of c components at temperature T is

$$\mathrm{d}\sigma = -\Gamma_1\,\mathrm{d}\mu_1 - \cdots - \Gamma_c\,\mathrm{d}\mu_c - \Gamma_s\,\mathrm{d}T, \tag{2.1}$$

where Γ_s is the surface excess entropy per unit area. The $c+1$ thermodynamic field variables $\mu_1, \ldots, \mu_c, T$ are not independent but are related by the Clapeyron equation

$$(\rho_1^\alpha - \rho_1^\beta)\mathrm{d}\mu_1 + \cdots + (\rho_c^\alpha - \rho_c^\beta)\mathrm{d}\mu_c + (s^\alpha - s^\beta)\mathrm{d}T = 0, \tag{2.2}$$

where ρ_i^α and ρ_i^β are the densities of component i in the bulk α and β phases, respectively, while s^α and s^β are the entropy densities in those phases. In particular, in a 2-component system at fixed temperature, then, there is only one independently variable thermodynamic field, which we may choose to be μ_1, say; and then

$$\frac{\mathrm{d}\sigma}{\mathrm{d}\mu_1} = -\Gamma_1 + \Gamma_2\,\frac{\rho_1^\alpha - \rho_1^\beta}{\rho_2^\alpha - \rho_2^\beta} \quad (T = \text{const}). \tag{2.3}$$

The adsorptions Γ_i in (2.1) and (2.3) (as well as the linear adsorptions Λ_i to be introduced presently) are defined with respect to, and depend on, the arbitrary choice of a Gibbs dividing surface (or surfaces, in the case of the Λ_i), but the particular combination of adsorptions on the right-hand side of (2.3) is invariant to that choice: $\mathrm{d}\sigma/\mathrm{d}\mu_1$ on the left-hand side is an experimentally measurable quantity. If one chooses the dividing surface so as to make $\Gamma_2 = 0$, as one almost always may, then (2.3) becomes

$$\frac{\mathrm{d}\sigma}{\mathrm{d}\mu_1} = -\Gamma_{1(2)}, \tag{2.4}$$

where $\Gamma_{1(2)}$ means the adsorption of component 1 in a frame of reference in which there is no net adsorption of component 2.

The analog of (2.1) for the line of mutual contact of three coexisting phases α, β, γ is [10]

$$\mathrm{d}\tau = -\Lambda_1\,\mathrm{d}\mu_1 - \cdots - \Lambda_c\,\mathrm{d}\mu_c - \Lambda_s\,\mathrm{d}T. \tag{2.5}$$

Now with three phases present only $c-1$ of the $c+1$ field variables $\mu_1, \ldots, \mu_c, T$ are independent because there are two independent Clapeyron equations, (2.2) and

$$(\rho_1^\beta - \rho_1^\gamma)\mathrm{d}\mu_1 + \cdots + (\rho_c^\beta - \rho_c^\gamma)\mathrm{d}\mu_c + (s^\beta - s^\gamma)\mathrm{d}T = 0. \tag{2.6}$$

Then in particular in a 3-component system at fixed temperature there is again only one independently variable thermodynamic field, which we may again choose to be μ_1, so that, analogously to (2.3),

$$\frac{d\tau}{d\mu_1} = -\Lambda_1 - \Lambda_2 \frac{d\mu_2}{d\mu_1} - \Lambda_3 \frac{d\mu_3}{d\mu_1} \quad (T = \text{const}), \tag{2.7}$$

where $d\mu_2/d\mu_1$ and $d\mu_3/d\mu_1$ are expressible in terms of the ρ_i^α, ρ_i^β, ρ_i^γ $(i = 1, 2, 3)$ from (2.2) and (2.6) with $dT = 0$. We may almost always, if we wish, choose dividing surfaces such that $\Lambda_2 = \Lambda_3 = 0$, in which case, as in (2.4),

$$\frac{d\tau}{d\mu_1} = -\Lambda_{1(2,3)}, \tag{2.8}$$

where $\Lambda_{1(2,3)}$ means the linear adsorption of component 1 in a frame of reference in which the linear adsorptions of components 2 and 3 vanish.

When, as we are supposing, there is a line of mutual contact of the three phases, the three interfacial tensions $\sigma^{\alpha\beta}$ etc. satisfy the inequality $\sigma^{\alpha\gamma} < \sigma^{\alpha\beta} + \sigma^{\beta\gamma}$ and the phases meet at the contact line with contact angles all of which are greater than 0 and less than 180°. In the limit in which $\sigma^{\alpha\beta} + \sigma^{\beta\gamma} - \sigma^{\alpha\gamma} \to 0$ the β phase spreads at ("wets") the $\alpha\gamma$ interface, and the contact angle β measured through the β phase, which is related to the tensions by

$$1 - \cos\beta = \frac{\left(\sigma^{\alpha\beta} + \sigma^{\beta\gamma} - \sigma^{\alpha\gamma}\right)\left(\sigma^{\alpha\beta} + \sigma^{\beta\gamma} + \sigma^{\alpha\gamma}\right)}{2\sigma^{\alpha\beta}\sigma^{\beta\gamma}}, \tag{2.9}$$

vanishes. This is the wetting transition. We wish to know what happens to the adsorption $\Lambda_{1(2,3)}$ as such a wetting transition is approached [12].

It is known from theory and numerical studies [13-24] that the line tension τ increases steeply as the controlling thermodynamic field, here μ_1, approaches its value, here $\mu_{1,w}$, at the wetting transition. By (2.8) this implies that the adsorption $\Lambda_{1(2,3)}$ diverges at the wetting transition. From a simple geometric construction [12] we may conclude that the volume of the region around the contact line in which the structure and composition of the system differ significantly from what they are in the bulk phases and interfaces far from the contact line increases proportionally to $1/\beta$ as $\beta \to 0$ at the wetting transition. This region being the seat of the linear adsorption, it suggests that $\Lambda_{1(2,3)}$ diverges proportionally to $1/\beta$. Since $\sigma^{\alpha\beta} + \sigma^{\beta\gamma} - \sigma^{\alpha\gamma} \sim |\mu_1 - \mu_{1,w}|$ [15], and since $\beta \sim (\sigma^{\alpha\beta} + \sigma^{\beta\gamma} - \sigma^{\alpha\gamma})^{1/2}$, from (2.9), this would imply in turn that $\Lambda_{1(2,3)}$ may diverge proportionally to $|\mu_1 - \mu_{1,w}|^{-1/2}$. A more refined argument [21] suggests that with short-range forces there might be an additional logarithmic factor attached to $|\mu_1 - \mu_{1,w}|^{-1/2}$, but such logarithmic factors are notoriously difficult to discern in experiment. If long-range forces are important they, too, will affect the law by which $\Lambda_{1(2,3)}$ diverges at wetting [20]. It would in any case be of both academic and practical interest to detect experimentally this predicted divergence of the linear adsorption at the wetting transition, and there may be systems that lend themselves to such a study [12].

* * *

I wish to acknowledge the contributions of my co-workers PRABHAKAR BHIMALAPURAM and KENICHIRO KOGA to the work reported in ref. [2] and the helpful comments and suggestions of J. O. INDEKEU on an earlier draft of ref. [12]. References [2] and [12] were the main sources for these lectures. This work was supported by the U.S. National Science Foundation and the Cornell Center for Materials Research.

REFERENCES

[1] BEN-NAIM A., *Hydrophobic Interactions* (Plenum Press, New York) 1980.
[2] WIDOM B., BHIMALAPURAM P. and KOGA K., *Phys. Chem. Chem. Phys. (PCCP)*, **5** (2003) 3085.
[3] *IUPAC Solubility Data Series* (Pergamon Press), vols. **9** (1982), **24** (1986), **37** (1989).
[4] FRANK H. S. and EVANS M. W., *J. Chem. Phys.*, **13** (1945) 507.
[5] LUMRY R. and RAJENDER S., *Biopolymers*, **9** (1970) 1125.
[6] GRUNWALD E. and STEEL C., *J. Am. Chem. Soc.*, **117** (1995) 5687.
[7] QIAN H. and HOPFIELD J. J., *J. Chem. Phys.*, **105** (1996) 9292.
[8] ISRAELACHVILI J. N., *Intermolecular and Surface Forces: With Applications to Colloidal and Biological Systems* (Academic Press) 1985, p. 105.
[9] SMITH D. E. and HAYMET A. D. J., *J. Chem. Phys.*, **98** (1993) 6445.
[10] ROWLINSON J. S. and WIDOM B., *Molecular Theory of Capillarity* (Oxford University Press) 1982. It is now known (Y. Djikaev and B. Widom, 2004, submitted) that there are additional terms in (2.5), (2.7), and (2.8) required to preserve the invariance of the right-hand sides to the arbitrary choice of location of the contact line. These do not affect the conclusions about how the linear adsorptions diverge at wetting.
[11] RUSANOV A. I., *Colloid Surf. A*, **156** (1999) 315.
[12] WIDOM B., to be published in *Colloid Surf. A*.
[13] DE GENNES P. G., *Rev. Mod. Phys.*, **57** (1985) 827.
[14] INDEKEU J. O., *Physica A*, **183** (1992) 439.
[15] SZLEIFER I. and WIDOM B., *Mol. Phys.*, **75** (1992) 925.
[16] VAREA C. and ROBLEDO A., *Phys. Rev. A*, **45** (1992) 2645.
[17] VAREA C. and ROBLEDO A., *Phys. Rev. E*, **47** (1993) 3772.
[18] PERKOVIĆ S., SZLEIFER I. and WIDOM B., *Mol. Phys.*, **80** (1993) 729.
[19] DOBBS H. T. and INDEKEU J. O., *Physica A*, **201** (1993) 457.
[20] ROBLEDO A. and INDEKEU J. O., *Europhys. Lett.*, **25** (1994) 17.
[21] INDEKEU J. O., *Int. J. Mod. Phys. B*, **8** (1994) 309.
[22] BLOKHUIS E. M., *Physica A*, **202** (1994) 402.
[23] PERKOVIĆ S., BLOKHUIS E. M. and HAN G., *J. Chem. Phys.*, **102** (1995) 400.
[24] BAUER C. and DIETRICH S., *Eur. Phys. J. B*, **10** (1999) 767.

Fluids under tight confinement

J. K. Percus

Courant Institute and Physics Department, New York University
251 Mercer Street, New York, NY 10012, USA

A preliminary study is made of the new phenomenology encountered when classical fluids are confined to enclosures that are of the order of particle size in all but one spatial dimension. Self-diffusion is taken as the indicator of this phenomenology. The anomalous diffusion occurring in strictly one-dimensional flow is first reviewed, and then its extension to the single-file regime in which particles cannot pass each other. When the system enters the parametric regime in which particle exchange is first possible, a rapid transition to the characteristics of normal diffusion occurs, which is organized by the concept of "hopping time".

Introduction

Both pure and applied physics are rapidly entering the nanoscale regime, in which available space is measured in units of atomic or molecular size. Familiar examples range from transport through pores in biological membranes [1] to the highly effective zeolite catalysts [2]. All are characterized by a very high surface/volume ratio and require a substantial extension of traditional bulk statistical mechanics.

In entering a relatively new field, a proven strategy is to start by examining caricatures of the systems of interest in order to focus more sharply on the novel aspects that characterize the field. We will carry this strategy to an extreme and devote our initial effort to a study of hard disks (2D) or spheres (3D) in a hard cylindrical enclosure of infinite length (fig. 1). The dynamics of a single particle will be taken as *strictly classical*, but will be allowed to range from *inertial* to *diffusive*. The latter is literally meaningful for motion in a dense medium of small particles, but applies as well for an

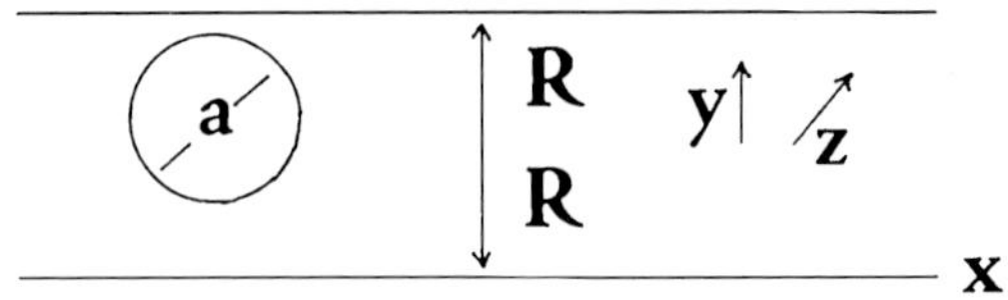

Fig. 1. – Parameters of the 2- (and 3-) dimensional confined hard-sphere fluid.

extensive set of random forces, from radiative throughout the volume to that imposed by thermal wall motion or even by wall roughness [3]. We will attend principally to analysis of *self-diffusion* of a sphere in the fluid of spheres as a signature of many-body system dynamics. Recall, for example, elementary kinetic theory where the basic linear transport coefficients, viscosity, conductivity, etc. are all expressible in terms of the coefficient of self-diffusion [4].

We are dealing with hard spheres (or disks) of diameter a in a hard pore of radius R—so that the center of a core is confined to a radius $\widehat{R} = R - (1/2)\,a$. Parameter space is conveniently decomposed into 3 subspaces

1.	$R = a/2$,	strictly one-dimensional motion of the core centers;
2.	$a/2 < R < a$,	single-file motion, the cores cannot pass each other;
3.	$R > a$,	the passing region,

and we will decompose our discussion in the same fashion.

1. – One dimension (fig. 2)

Here, since we are imagining an infinitely long tube (with an imposed mean particle density per unit length), a given sphere can move *if* its neighbors get out of the way by also moving, and so forth. So one expects sort of a *collective motion* of *clusters*. In fact, in a typical dynamical run, if you look at the number of particles N in a short section of the tube as a function of time, it looks as if they are passing in bunches (fig. 3).

What does this say about the motion of a given particle in the environment of all the others? And what should we be measuring? Presumably, as indicated above, something

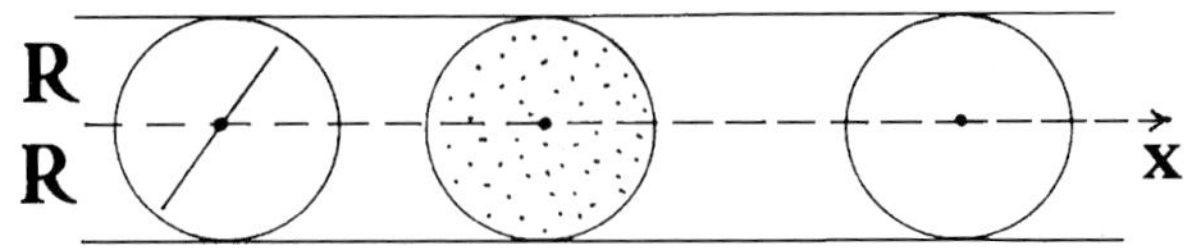

Fig. 2. – One dimensionally confined fluid: $R = a/2$, $(\widehat{R} = 0)$.

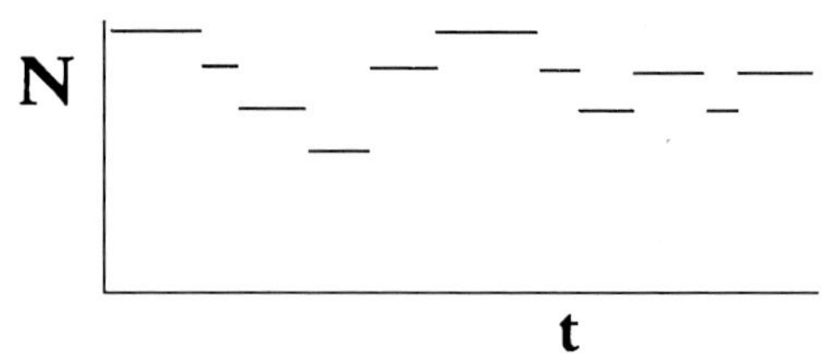

Fig. 3. – Particle clustering in time.

associated with self-diffusion. Traditionally, one examines the average (over all particles you happen to look at) wandering of a particle in the form

$$\left\langle (\Delta x(t))^2 \right\rangle, \tag{1.1}$$

where

$$\Delta x(t) \equiv x(t) - x(0),$$

in comparison with what it would be without the surrounding particles to hinder it. The latter is basic. If there were no other particles around, then one extreme would be free inertial motion,

$$\textit{inertial} \;:\; \left\langle (\Delta x(t))^2 \right\rangle_0 = \left\langle v^2 \right\rangle t^2 \,, \tag{1.2}$$

where $\langle v^2 \rangle$ is the mean square velocity of the individual particles under consideration. The other extreme is that of a particle moving by random kicks; in a "Brownian" medium, a particle behaves diffusively at large enough time,

$$\textit{diffusive} \;:\; \left\langle (\Delta x(t))^2 \right\rangle_0 = Dt \,. \tag{1.3}$$

The subscript 0 in (1.2), (1.3) denotes a particle not interacting with the others.

Collisions with neighbors will clearly slow things down. How much? To find out, even at a heuristic level, a preliminary transformation is enormously helpful [5]: if we look down the line of centers (remember that here $\widehat{R} = 0$), the only thing that the girth of the particles is doing is using up space. So we can *collapse* this space (fig. 4), reducing the system to *hard point particles*, and afterwards put the space back in.

Let us then tag a point particle and follow its motion. This motion will be pretty much that of the center of mass of a cluster of, say M particles (M *will* depend on time), that either have to move ahead of "our" particle, or catch up when they have the space. In this cluster, two particles, when they meet (fig. 5), act as if they just pick up each other's tags—so if you do not need the tag, they act as if they freely pass through each other. Since the motion of the M-particle *set* has therefore been as if they are

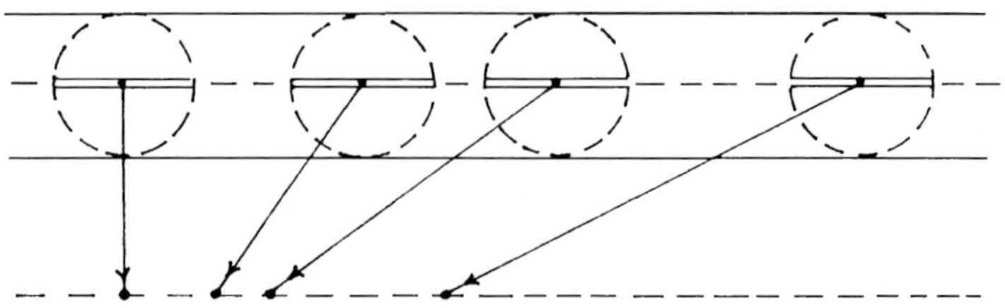

Fig. 4. – Diameter-shrinking map.

independent, we have at once the usual r.m.s. addition of errors

$$\left\langle \left(\Delta x_{\rm cm}(t)\right)^2 \right\rangle = \frac{1}{M^2} \left\langle \sum_{i=1}^{M} \left(\Delta x_i(t)\right)^2 \right\rangle_0 = \frac{1}{M} \left\langle \left(\Delta x(t)\right)^2 \right\rangle_0 . \tag{1.4}$$

On the other hand, characterizing our system by its mean particle *density* n/unit length, or specific length $\ell = 1/n$, and neglecting the fluctuations in ℓ over the M-particle cluster, we clearly have —see diagram (fig. 6)—

$$\begin{aligned} \Delta x_{\rm cm} &= \frac{1}{2}(M_+ \ell - M_- \ell) \\ &= \left(\frac{\ell}{2}\right) \left(\#(x_i) > x(0) - \#(x_i) < x(0)\right) \\ &= \left(\frac{\ell}{2}\right) \sum_{i=1}^{M} \mathrm{sgn}\ (x_i - x(0)) \end{aligned} \tag{1.5}$$

(sgn $\equiv$ the sign of). Hence, by independence again,

$$\langle \Delta\, x_{\rm cm}(t)^2 \rangle = \left(\frac{\ell^2}{4}\right) M\,. \tag{1.6}$$

Since $\Delta x_{\rm cm} \sim \Delta x$, we have [6] on eliminating M from (1.4) and (1.6)

$$\langle \Delta x(t)^2 \rangle = \left(\frac{\ell}{2}\right) \left[\langle \Delta x(t)^2 \rangle_0\right]^{1/2} . \tag{1.7}$$

The result (1.7) is sort of amazing: for inertial one-body motion, with $\langle \Delta x(t)^2 \rangle_0 \propto t^2$, we now have

$$\textit{inertial}, \quad \langle \Delta x(t)^2 \rangle \propto t\,, \tag{1.8}$$

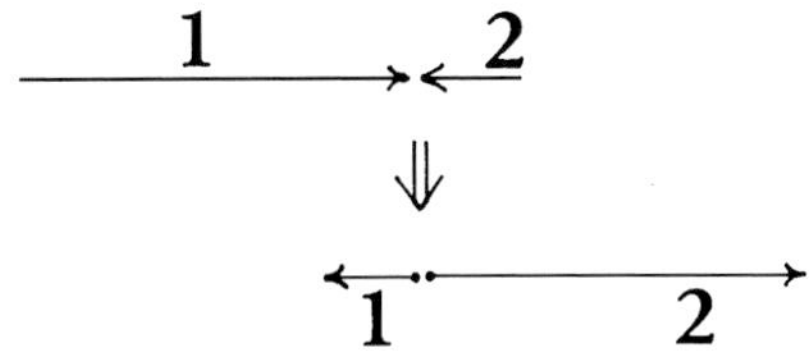

Fig. 5. – Switch of tags on collision.

so self-diffusion looks like ordinary diffusion. But, for basic Brownian diffusion dynamics, with $\langle \Delta x(t)^2 \rangle_0 \propto t$, we find

$$\text{(1.9)} \qquad \textit{diffusive,} \quad \left\langle \Delta x(t)^2 \right\rangle \propto t^{1/2},$$

anomalous or *sub*-diffusion. Since here $\langle \Delta x(t)^2 \rangle / t \xrightarrow[t\to\infty]{} 0$, the diffusion constant has dropped to zero, due to the continued hindrance by neighbors.

Two important comments. First, there was a lot of hand-waving. But it can be done right, and then one finds instead [7] that, at long time,

$$\text{(1.10)} \qquad \left\langle \Delta x(t)^2 \right\rangle = \ell \langle |\Delta x(t)| \rangle_0 ,$$

a small quantitative but no qualitative change.

Second, we can put the cores back in by observing that the available "volume" per particle after the core-shrinking transformation is only $\ell - a$, so that our final expression in this strictly one-dimensional situation becomes

$$\text{(1.11)} \qquad \left\langle \Delta x(t)^2 \right\rangle = (\ell - a) \langle |\Delta x(t)| \rangle_0 ,$$

again in the long-time limit.

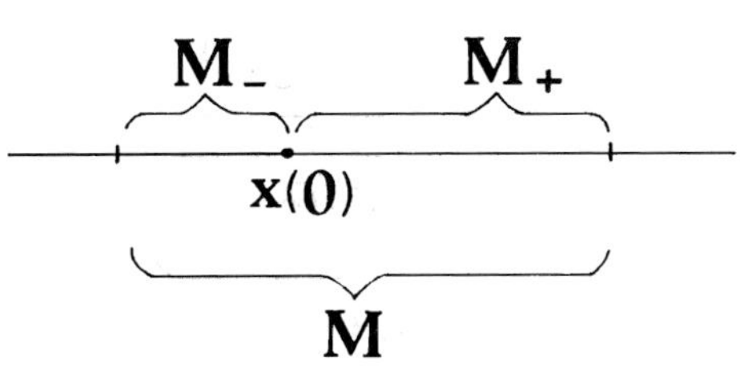

Fig. 6. – Displacement of M-particle cluster.

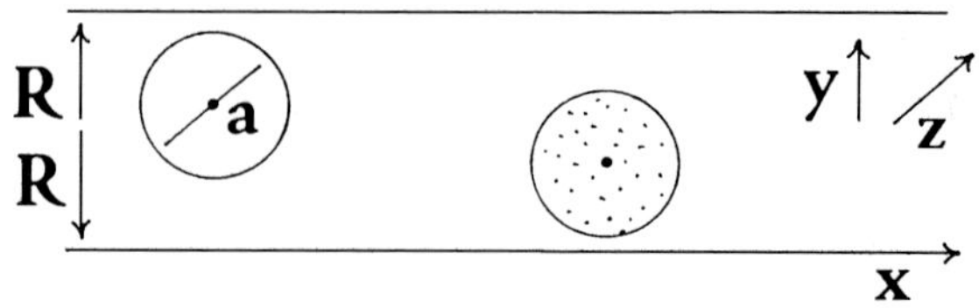

Fig. 7. – Single-file configuration: $a/2 < R < a$ $(0 < \widehat{R} < a/2)$.

2. – Single file, non-passing [7]

Now, we increase the diameter of the confining tube, so that the particles can wiggle *transversely* (fig. 7). But does it matter? Yes, because although we will always be focussing on the unbounded longitudinal motion, two cores, when they meet, can have a c.m. longitudinal separation less than a, *e.g.* (in 2D)

$$x_2 - x_1 = \left[a^2 - (y_2 - y_1)^2\right]^{1/2} \tag{2.1}$$

at collision.

This change in longitudinal turning point separation is as if one had *soft* one-dimensional rods. How soft?

The softness should show up in equilibrium too. For hard cores in one-D, we recall the equation of state $P(\ell - a) = KT$, or

$$\ell = a + \frac{1}{\beta P} \quad \left(\beta = \frac{1}{KT}\right). \tag{2.2}$$

As $\widehat{R}$ increases from 0 and the system really becomes two- or three-dimensional, we can still write the equation of state in the form

$$\ell = a_{\text{eff}}\big(\beta P, \widehat{R}\big) + \frac{1}{\beta P}, \tag{2.3}$$

where a appears to decrease (soften) with increasing $\widehat{R}$. But watch the notation: ℓ is the *linear* or longitudinal mean separation and P the total force on a cross-sectional area.

Perhaps the effective diameter a_{eff} is universal—unaffected by the situation it refers to. In that case, we would preferable rewrite (2.3) as

$$a_{\text{eff}}\big(\ell, \widehat{R}\big) = \ell - \frac{1}{\beta P(\ell, \widehat{R})} \tag{2.4}$$

and the long-time mean-square displacement of (1.11) for hard discs or spheres would become the extremely simple

$$\left\langle \Delta x(t)^2 \right\rangle = \frac{1}{\beta P(\ell, \widehat{R})} \langle |\Delta x(t)| \rangle. \tag{2.5}$$

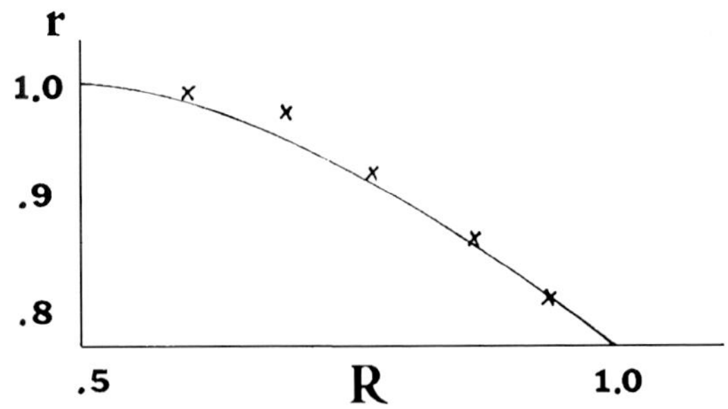

Fig. 8. – $r = a_{\text{eff}}/a$ *vs.* R for Brownian dynamics (X) and thermal equilibrium (–).

Equation (2.5) of course has been obtained by an argument using words as weapons, and it cannot possibly hold (as we will see explicitly) when R passes the threshold value of a. But empirically, it is a different story, and a comparison of a_{eff} obtained (in 3D) from stochastic dynamics and equilibrium data shows (fig. 8) a quite close but not perfect fit. In fact, one can carry out an analytic expansion in which (2.5) serves as leading order.

3. – Passing region

All of this is fine, but a little bit dull, since qualitatively things are still one-dimensional. It all changes dramatically when the container is wide enough that one core can *hop* past another (fig. 9). The drama is most evident for motion in a Brownian medium, as dynamical simulations show [8]: the motion changes—quite abruptly from subdiffusion to ordinary diffusion as R/a passes 1. It is easy to see why [8].

Suppose that τ_{hop} is the mean time that it takes for a core in between two cores to hop past one of them to the middle of the next intercore interval (on either side). Except at the moment of hopping, no exchange can occur, and so "our" core moves by anomalous diffusion:

$$t < \tau_{\text{hop}} : \left\langle \Delta x(t)^2 \right\rangle = C(\ell - a)(Dt)^{1/2}\,; \qquad (3.1)$$

D is the basic diffusion constant and C is a fixed parameter.

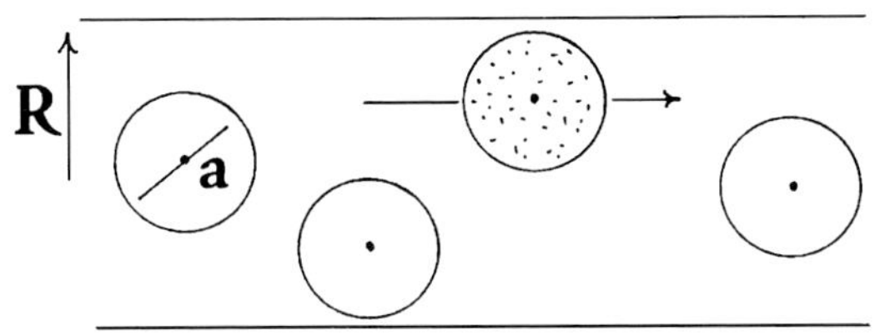

Fig. 9. – Threshold region for hopping: $R > a$ ($\widehat{R} > a/2$).

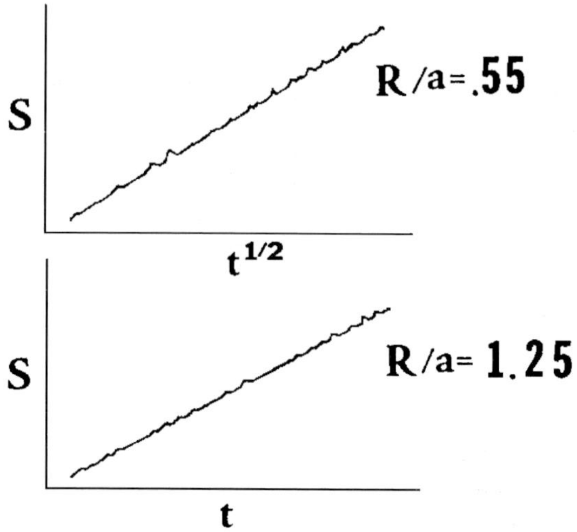

Fig. 10. – $S = \langle(\Delta x)^2\rangle$ below and above hopping threshold.

Now over a time interval T, our core will carry out this maneuver T/τ_{hop} times, randomly to right or left. Hence at T, we will have

$$\langle \Delta x(t)^2 \rangle = \frac{T}{\tau_{\text{hop}}} C(\ell - a)(D\,\tau_{\text{hop}})^{1/2} = KT\left(\frac{D}{\tau_{\text{hop}}}\right)^{1/2}, \tag{3.2}$$

i.e. ordinary diffusion with diffusion constant

$$D_{\text{eff}} = K\left(\frac{D}{\tau_{\text{hop}}}\right)^{1/2}, \tag{3.3}$$

a relation that is very well verified by simulations (fig. 10).

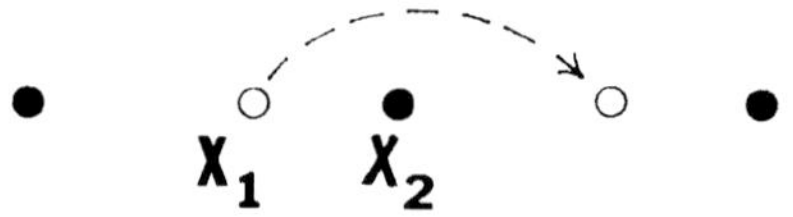

Fig. 11. – Next interval hopping.

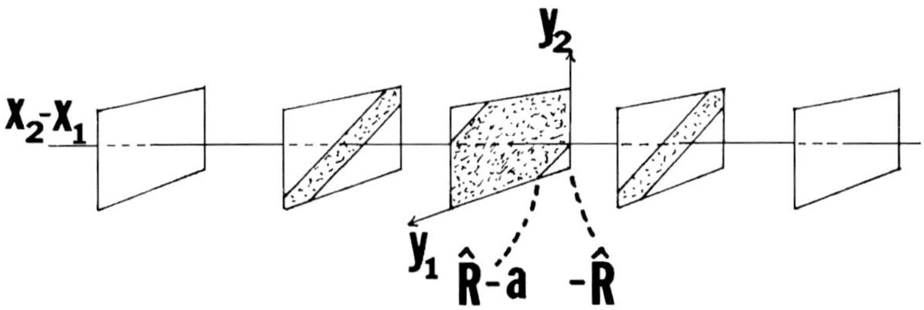

Fig. 12. – Channeling of 2-particle flow in 2 dimensions.

There remains the evaluation of τ_{hop}. Now empirically, in 3D, there is a quite convincing power law dependence:

$$\tau_{\text{hop}} \propto \left(\frac{R}{a} - 1\right)^{-3}. \tag{3.4}$$

Here is the sort of thing that is involved [9]. It is clearly the mean time that it takes from one mid-interval to another (technically, mean first passage time) that we want (fig. 11). Consider the 2D case, for easier visualizaton, with the two disks involved specified by (x_1, y_1) and (x_2, y_2), and suppose that $\widehat{R} - a/2 = \epsilon > 0$. Then the point in $(x_2 - x_1, y_1, y_2)$ space is being channeled through the narrow pair of triangular windows (fig. 12) (windows are unshaded) at $x_2 - x_1 = 0$, with a total area of $(2\widehat{R} - a)^2 = 4a^2((R/a) - 1)^2$. Thus a quick guess would be $\tau_{\text{hop}} \sim ((R/a) - 1)^{-2}$. The full dynamics of the pair is solvable in various dynamical settings, and the power law behavior (fig. 13) is robust, but the exponent does depend upon the dynamical category.

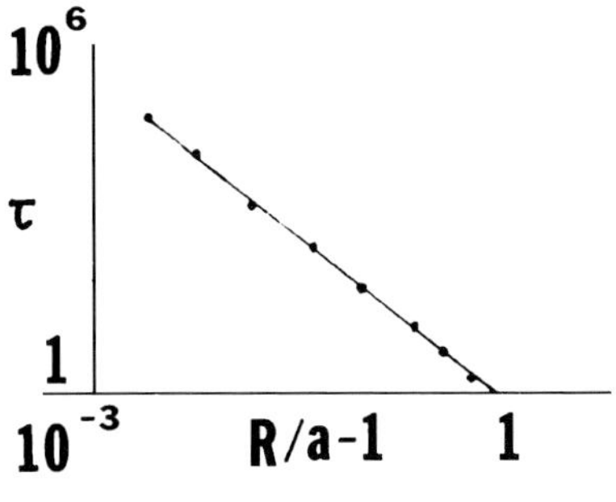

Fig. 13. – Hopping time τ *vs.* excess diameter (3 dimensions).

* * *

The work reported here is the result of joint efforts with R. BOWLES, I. KAMENETSKY, K. K. MON, Q. WANG and J. YAN. Supported in part by DOE Chemical Sciences Division.

REFERENCES

[1] HILLE B., *Ion Channels of Excitable Membranes* (Sinauer, Sunderland, Mass.) 2001.
[2] KÄRGER K. and RUTHVEN D. M., *Diffusion in Zeolites and Other Microporous Solids* (Wiley, New York) 1992.
[3] MON K. K. and PERCUS J. K., *J. Chem. Phys.*, **119** (2003) 3343.
[4] See, *e.g.*, REICHL L. E., *A Modern Course in Statistical Physics* (Wiley, New York) 1998.
[5] See, *e.g.*, LEBOWITZ J. L. and PERCUS J. K., *Phys. Rev.*, **155** (1967) 122.
[6] MON K. K., PERCUS J. K. and YAN J., *Molecular Simulation*, **29** (2003) 721.
[7] PERCUS J. K., *Phys. Rev. A*, **9** (1974) 557.
[8] MON K. K. and PERCUS J. K., *J. Chem. Phys.*, **117** (2002) 2289.
[9] BOWLES R. K., MON K. K. and PERCUS J. K., submitted for publication.

Giant vesicles: Transient pores and tube extraction

N. Borghi and F. Brochard-Wyart

Laboratoire PCC Institut Curie / CNRS UMR 168
11, rue P. & M. Curie, 75231 Paris Cedex 05, France

E. Karatekin

Laboratoire de Biologie Cellulaire et Moléculaire de la Sécrétion IBPC / CNRS UPR 1929
13, rue P. & M. Curie, 75005 Paris, France

I. Derényi

Department of Biological Physics, Eötvös University
Pázmány P. stny. 1A, H-1117 Budapest, Hungary

Introduction

Transport of ions and molecules into and inside the cell is central to many biological processes. Pores allow transport across membranes, and lipidic tubes from an intercellular compartment to another. In order to understand the physics of these two phenomena, we have reproduced transient pores and lipidic tubes on giant unilamellar vesicles (GUVs), whose phospholipidic membrane models the cell membrane. Understanding transmembrane transport is a key ingredient in the development of new technologies such as gene therapy. Moreover, apart from mimicking biological phenomena, another application of tube extraction is the fabrication of microchannel networks for nanofluidic technologies. We discuss here both theoretically and experimentally in sect. **1** the life of transient pores and in sect. **2** the extraction of tubes from giant vesicles.

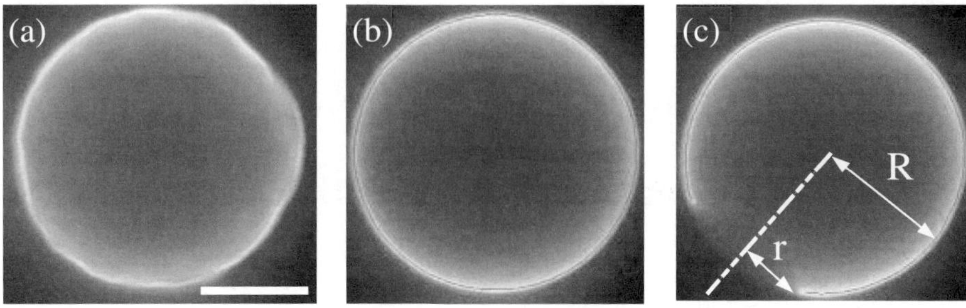

Fig. 1. – Appearance of a pore in a GUV observed with fluorescence microscopy (a lipophilic fluorescent probe is inserted in the membrane). (a) A fluctuating vesicle is illuminated with the microscope's lamp (bar = $10\,\mu$m). (b) Within ≈ 30 min the vesicle in (a) becomes tense and perfectly spherical. (c) Further build-up of membrane tension under illumination causes a pore to open suddenly (extracted from [1]).

1. – Transient pores

1˙1. *Bubbles and vesicles.* – Under ordinary circumstances, the membrane tension of a GUV is essentially nil compared to the surface tension of a soap or viscous bubble. Indeed, the lipids are highly insoluble in water. As they associate into a closed membrane, they minimize their area per polar head so that the membrane tension, σ, is zero. This leads to two consequences:

- The membrane fluctuations are huge and are limited only by the weak curvature energies. The vesicle looks like an empty balloon (fig. 1(a)). It also explains the flicker phenomenon of red blood cells. Typically, the amplitude of fluctuations of a GUV (or a red cell) is of order of $0.1\,\mu$m while for a soap bubble it is a few Angströms.

- A GUV is robust. It does not burst because its membrane tension is zero. A hole in the membrane, instead of opening up further, will cicatrise to decrease its line energy. But what happens when a vesicle is swollen? Does it burst like a soap bubble? Under a mechanical tension, the fluctuations of the membrane are killed and a GUV adopts the spherical shape of a bubble. But for a bubble, the surface tension is constant during the growth of a pore. It leads to its bursting, while for vesicles, we observe transient pores. They open, but close back! We were amazed to observe the closure of pores. To understand the growth and the closures of pores, we shall first remind the model of bursting of viscous bubbles, and extend it to vesicles.

1˙2. *Life of transient pores.* – A typical sequence leading to pore opening is shown in fig. 1. Optical illumination tensed this vesicle within ≈ 30 min (fig. 1(a)-(b)). When GUVs are suspended in a medium of appropriate viscosity, further illumination made a

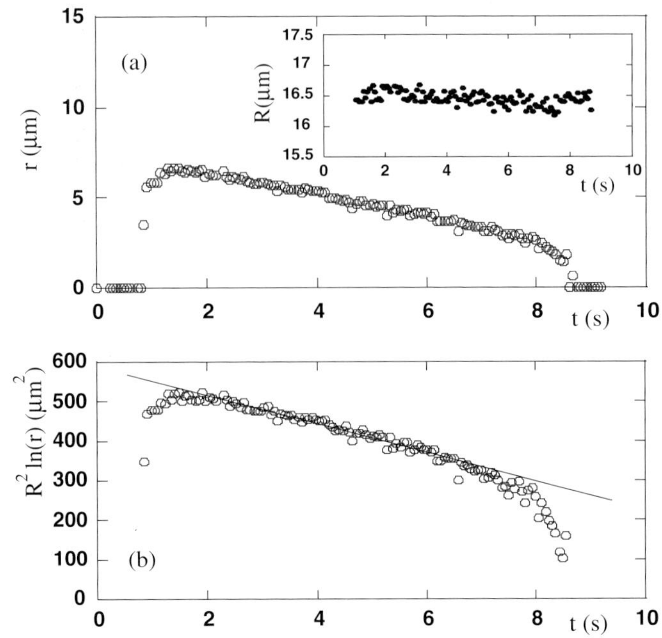

Fig. 2. – (a) Pore (r, ○) and vesicle (R, •, inset) radii as a function of time for a DOPC vesicle in the presence of 4.9×10^{-2} mM Tween 20. (b) $R^2 \ln r$ plotted against time for the same pore. The least-squares fit to the linear part has slope $\simeq -44.7\,\mu\text{m}^2/\text{s}$, which implies $\mathcal{T} \simeq 6.7\,\text{pN}$ (extracted from [1]).

pore appear suddenly (fig. 1(c)). The radii of the pore, r, and the vesicle, R, measured as a function of time from a sequence of images, are plotted in fig. 2(a).

Actually, after the appearance of the first pore, numerous pores occur in succession at regular intervals of a few minutes, on the same vesicle, yet two pores never open simultaneously. Pores can appear anywhere in a vesicle; however, those appearing at the equator, such as the one shown in fig. 1(c), are more easily amenable to image analysis.

The dynamics of a pore of radius r are described by [2, 3]

$$4\pi\eta_s\dot{r} = 2\pi(r\sigma - \mathcal{T}). \tag{1}$$

The driving force for pore opening is the membrane tension, σ, while that for closure is the line tension, $\mathcal{T}$. Energy is dissipated in the membrane, which has a surface viscosity η_s. This is analogous to the opening of holes in viscous bare films [4], wherein σ remains constant and $\mathcal{T}$ is neglected due to its very small value. However, pores in vesicles are more complex since σ relaxes due to two mechanisms: i) the opening of the pore and ii) the leak-out of the internal liquid through it (caused by the excess Laplace pressure); both allow the initial membrane amount of lipid to be distributed over a smaller projected area. The membrane tension is thus a function of the pore (r) and the vesicle (R) radii:

$$\frac{\sigma(r,R)}{\sigma_0} = 1 - \frac{r^2}{r_c^2} - \frac{R_i^2 - R^2}{R_i^2 - R_0^2}. \tag{2}$$

In eq. (2) R_i is the vesicle's initial radius, R_0 its radius under zero tension, and σ_0 the initial membrane tension. The critical pore radius, $r_c = 2\sqrt{R_i^2 - R_0^2}$, corresponds to the complete relaxation of σ in the absence of leak-out. Mechanisms i) and ii) are accounted for by the second and the third terms in eq. (2), respectively. Since the product σr may become very small in eq. (1), $\mathcal{T}$ cannot be neglected here, unlike in the bursting of viscous bare films. In eq. (2), R drops with the opening of a pore. Its dynamics is obtained by considering the flux, Q, of liquid (viscosity η_0) passing through the pore:

$$Q = -4\pi R^2 \dot{R} = \frac{2\sigma r^3}{3\eta_0 R}. \tag{3}$$

Equation (3) is valid as long as r varies slowly compared to the leak-out velocity of the liquid. Equations (1), (2) and (3) describe the entire life of a pore [2, 3].

With the viscous mixtures employed in our experiments, the leak-out of the inner liquid is very slow and so is the pore closure, as seen in fig. 2(a). We focus here on this regime only, in which a measurement of the closure velocity yields $\mathcal{T}$, as follows. Since r varies very slowly, we make the quasi-static approximation $\sigma r - \mathcal{T} \approx 0$, from eq. (1). Equally slowly varying is σ; hence we approximate $\dot{\sigma} \approx 0$ and obtain $4R\dot{R} - r\dot{r} \approx 0$, from eq. (2). With these approximations in eq. (3), we obtain

$$\dot{r} = \frac{-2\mathcal{T} r}{3\pi\eta_0 R^2}. \tag{4}$$

What is remarkable about eq. (4) is that the closure velocity depends only on $\mathcal{T}$ and measurable or known parameters. After integration of eq. (4) with the experimentally justified approximation $R \simeq \text{const}$ (see inset in fig. 2(a)), we see that a plot of $R^2 \ln r$ *vs.* t should be linear with slope $= -2\mathcal{T}/3\pi\eta_0$. Knowing $\eta_0 \approx 32\,\text{cP}$ from an independent measurement (our samples contain 66% v/v glycerol), $\mathcal{T}$ is readily obtained. As an example, the data in fig. 2(a) is replotted in this form in fig. 2(b). A least-squares fit to the straight line portion of this plot yields a line tension of 6.7 pN in the presence of 4.9×10^{-2} mM Tween 20 (a detergent, see later).

1·3. *Edgeactants.* – Like surfactants modify the surface (or interfacial) tension of liquids, edgeactants molecules act on the line tension. In the first case, the amphiphilic property of surfactants makes them form a layer at the interface of water and air (or any hydrophobic medium), lowering the surface tension [5]. In the second case, amhiphilic molecules incorporate naturally in the lipidic membrane but their spontaneous curvature, c_0, affects the line tension as shown in fig. 3. Molecules with positive spontaneous curvature $c_0 > 0$ preferentially incorporate at the pore's border, lowering the stress lipids

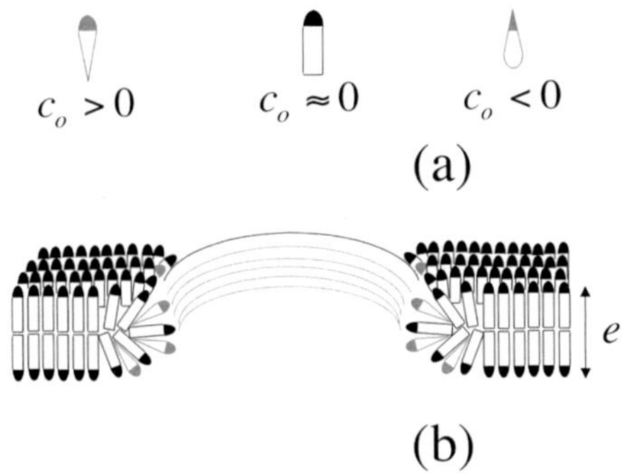

Fig. 3. – (a) Spontaneous curvature. Detergents typically have $c_0 > 0$, while phospholipids have $c_0 \approx 0$. Some molecules, such as cholesterol, can be modelled as inverted cones and possess negative spontaneous curvature $c_0 < 0$. (b) Molecules with a positive curvature adsorb at the pore's edge and decrease the line tension, while molecules with negative spontaneous curvature are depleted at the border, and increase the line tension (taken from [6]).

undergo in such a folded geometry. In a like manner, detergents reduce the energy cost of a pore's border: the line tension. At the contrary, molecules with negative spontaneous curvature $c_0 < 0$ increase the energy cost of a pore's border. Thanks to these properties, amphiphilic molecules with non-zero spontaneous curvature will be called "edgeactants".

i) *Line tension of bare lipid membranes.* – We first consider the case of "bare" lipid vesicles, in which the membrane is devoid of any additives, except for the presence of the fluorescent probes (whose spontaneous curvatures are ≈ 0, due to their structural similarity with double-chained lipid molecules).

As with detergents, measurements of line tension are quite sensitive to the presence of small impurities which may be inserted in the pore's edge (as observed previously elsewhere). In our case, we have consistently measured similar values of $\mathcal{T}$, provided the lipid was purchased from the same supplier. From the first one, we measure $\mathcal{T} \simeq 7\,\text{pN}$, from the second one, $\mathcal{T} \simeq 15\,\text{pN}$. We speculate that the difference obtained may be due to the presence of small amounts of impurities. Thus, in our studies of the effect of inclusion molecules, we were careful to use the same supplier for a given set of (comparative) measurements, *i.e.* the first one with cholesterol addition and the second one for Tween 20 addition.

ii) *Cholesterol increases the line tension.* – Cholesterol, a water insoluble amphiphile shown in fig. 4, is a central component in the regulation of the properties of eukaryotic cell membranes [7, 8]. It can be modelled as an inverted cone [9, 10] (spontaneous curvature $c_0 < 0$), and is thus expected to increase the line tension (fig. 3).

It was already observed that addition of cholesterol into lecithin vesicles reduced permeabilities to ions and small molecules [9], a phenomenon closely linked with membrane pores.

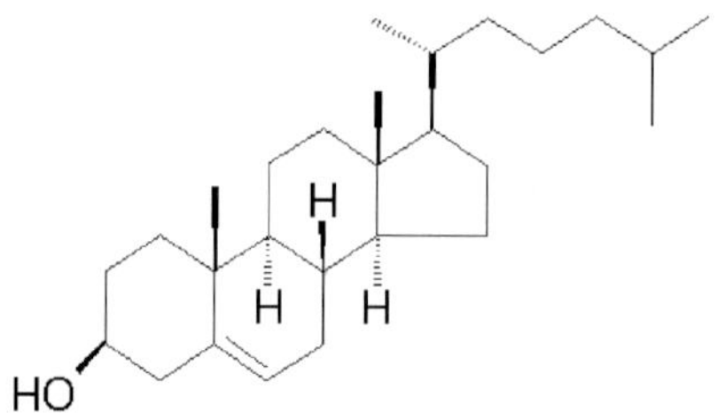

Fig. 4. – Molecular structure of cholesterol. The polar head is the small hydroxyl group. It can be modelled as an inverted cone with negative spontaneous curvature $c_0 < 0$.

Measuring $\mathcal{T}$ for a number of different vesicles and pores in the presence of varying amounts of cholesterol added into the membrane, we have obtained the results shown in fig. 5.

As expected, the line tension increases as the cholesterol mole fraction in the membrane increases.

iii) *Detergents reduce the line tension.* – Detergents are water soluble amphiphiles. They can typically be modelled as cone-shaped molecules possessing positive spontaneous curvature $c_0 > 0$ (fig. 3). At air/water interface, detergents lower the surface tension, γ, by forming a layer [5]. Their surface density, hence the value of γ, depends on the bulk concentration, C, of detergent in water. At the lowest concentrations, detergents are solubilized and a fraction of them adsorbs at the interface. Increasing C, γ decreases until one reaches the "critical micellar concentration" (CMC). Any further increase in C does not affect γ; excess surfactant goes into forming micellar aggregates in the bulk. Classical surface thermodynamics relates the "surface excess", *i.e.* the surface concentration of adsorbants, Γ_S, to γ and C, thanks to the Gibbs' adsorption equation which,

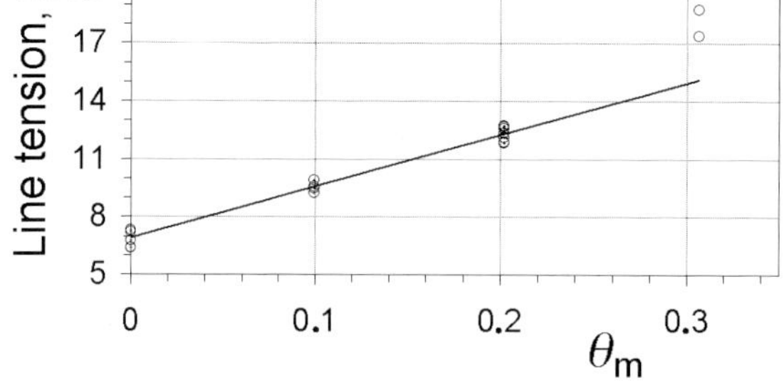

Fig. 5. – Line tension, $\mathcal{T}$, as a function of cholesterol mole fraction, θ_m, for DOPC vesicles. The straight line is a least-squares fit for 0–0.2 mole fraction cholesterol ($\mathcal{T}_{fit} = 26.7\,\theta_m + 6.90$, $R^2 = 0.98$). (Mass fraction cholesterol $\approx \theta_m/2$.) (Extracted from [6].)

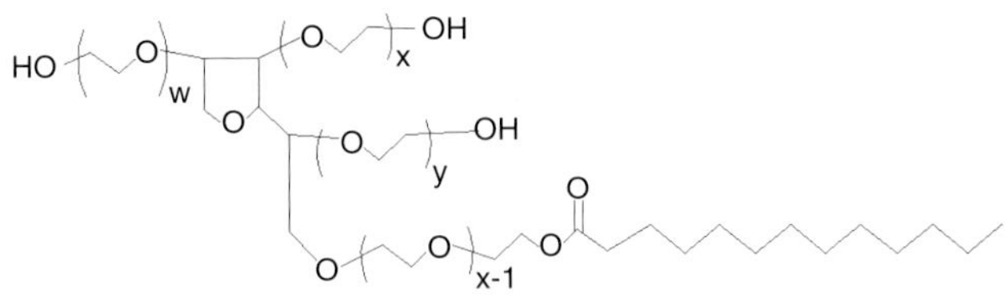

Fig. 6. – Molecular structure of Tween 20: POE 20 sorbitant monolaurate, $w + x + y = 20$. It can be modelled as a cone-shaped molecule with positive spontaneous curvature $c_0 > 0$.

in the dilute regime, is given by

$$\Gamma_{\mathrm{S}} = -\frac{\mathrm{d}\gamma}{\mathrm{d}\mu} = -\frac{\mathrm{d}\gamma}{kT\,\mathrm{d}\ln C}, \tag{5}$$

where $\mu \simeq kT \ln C$ is the solute chemical potential. The 1D counterpart of this adsorption phenomenon, namely the thermodynamics of lines, has only received some theoretical attention [11-13]. In 1D systems, the equivalent of the surface tension is the line tension, $\mathcal{T}$, and Gibbs' relation becomes

$$\Gamma_{\mathrm{L}} = -\frac{\mathrm{d}\mathcal{T}}{\mathrm{d}\mu} = -\frac{\mathrm{d}\mathcal{T}}{kT\,\mathrm{d}\ln C}, \tag{6}$$

where Γ_{L} is the "line excess" density of adsorbants. In our experimental system, measurement of $\mathcal{T}$ as a function of the bulk detergent concentration C should allow the evaluation of the line excess.

The detergent micelles have the ability to dissolve insoluble or sparingly soluble materials in the aqueous solution by incorporating them into the micellar interior to form mixed micelles. Detergents can also solubilize phospholipids and other constituents of biomembranes. Due to this property, detergents are widely used in membranology. Thanks to these important applications, the interaction of detergents with vesicles has received great attention (for recent reviews see [14, 15]).

We have studied the effect of the detergent Tween 20 (fig. 6) on transient pores and the line tension in GUVs. Tween 20 is a common detergent used in the solubilization of membrane proteins [16], having a hydrophilic lipophilic balance (HLB) of 16.7 which places it among good emulsifiers for oil-in-water emulsions [17]. Thanks to its bulky polar head relative to its hydrophobic tail, it can be modelled as a cone-shaped molecule with a positive spontaneous curvature, $c_0 > 0$. Thus, we expect that it should insert itself into the highly curved edge of a pore, as shown schematically in fig. 3(b). This should reduce the line tension.

The propensity of amphiphiles with positive spontaneous curvature for facilitating transient pore formation was already noted [9] in the presence of lysolecithin (which is similar to lecithin, but has only one rather than two hydrophobic tails and hence

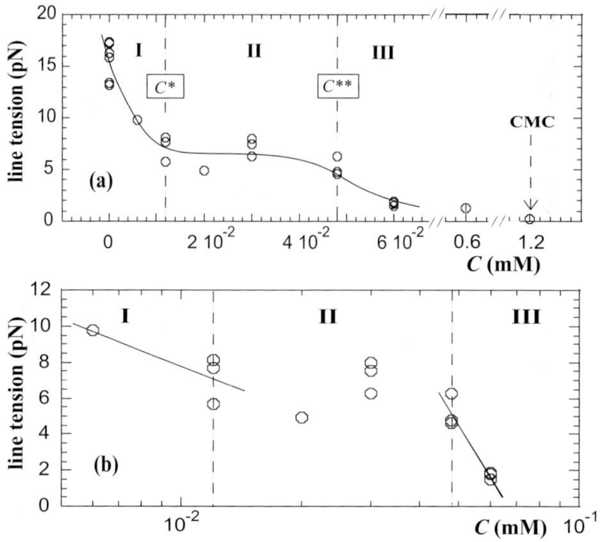

Fig. 7. – (a) Line tension of DOPC membrane edges as a function of bulk Tween 20 concentration, C. (b) The data in (a) (except for $C = 0$ and the two highest concentrations) plotted on semi-logarithmic coordinates. The linear fit in regime I has slope $\simeq -9\,\text{pN}$, while that in regime III has slope $\simeq -40\,\text{pN}$ (extracted from [1]).

possesses positive spontaneous curvature), permeability of lecithin vesicles to ions and small molecules increased.

iv) *Line tension isotherm.* – Measurements of $\mathcal{T}$ as a function of C are shown in fig. 7, up to $C \approx \text{CMC}$.

Looking at the isotherm in fig. 7, we identify three distinct regimes. For the lowest concentrations of Tween 20, the line tension decreases as more detergent is added (regime I). This continues up to a concentration $C = C^* \approx 10^{-2}\,\text{mM}$, beyond which $\mathcal{T}$ reaches a plateau (regime II, $C^* < C < C^{**} \approx 5\times 10^{-2}\,\text{mM}$). There exists a third regime for $C > C^{**}$, in which $\mathcal{T}$ decreases again with increasing C.

The first two regimes in the isotherm of fig. 7(a) are analogous to a classical surface adsorption isotherm for a soluble surfactant. In regime I the detergent is partitioned between bulk solution and the edges of a pore, as drawn schematically in fig. 8(a). The preference of the detergent for the pore's line is due to its conical shape with a large hydrophilic head. It can thus significantly reduce the edge energy of a membrane by adsorbing at the borders. As the bulk detergent concentration is increased, the border becomes saturated with detergent at $C \approx C^*$. Any further detergent added into the system must go into a reservoir. Micelles are excluded as a reservoir, since we are well below the CMC. Instead, we expect that excess detergent goes into the lipid membrane in regime II (fig. 8(b)). For these two regimes, a plot of $\mathcal{T}$ *vs.* $\ln C$ (fig. 7(b)) allows us to

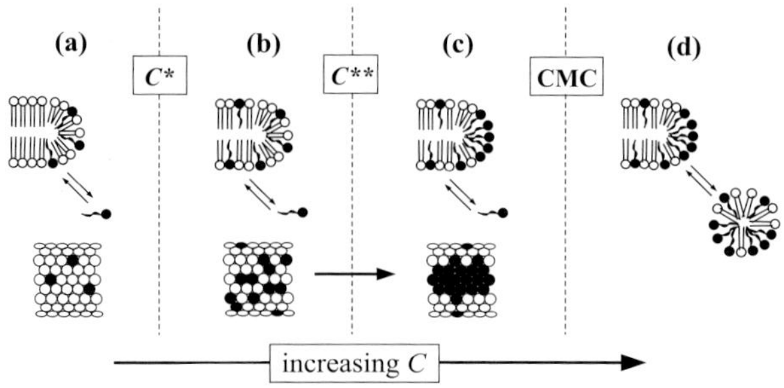

Fig. 8. – Proposed interactions between the pore's edge and detergents in different regimes. Top row: transversal view. Bottom row: front view of the membrane edge, with dark (white) dots representing detergent (lipid) headgroups (extracted from [1]).

estimate the line excess, Γ_L, at $C \approx C^*$. The straight line fitted to the data in regime I in fig. 7(b) has slope $\simeq -9\,\text{pN}$, implying $\Gamma_L \approx 2\,\text{nm}^{-1}$, from eq. (6). This is a reasonable value, assuming that the detergent head group occupies an area similar to that of the lipid ($\approx 60\,\text{Å}^2$), and considering that the line can in fact accommodate a few molecules along its thickness, as drawn schematically in fig. 8.

Increasing the bulk detergent concentration above C^{**} we reach regime III, in which the line tension starts decreasing again. This is an indication of a new organization in the system. In this regime, fitting a straight line to the data near C^{**} in fig. 7(b), we find a slope $\simeq -40\,\text{pN}$, which implies a line density of $\approx 9\,\text{nm}^{-1}$. The line can still accommodate such a high density, provided the detergent molecules are organized in a condensed manner, as shown schematically in fig. 8(c). The sharp change of slope at C^{**} is reminiscent of a transition from a dilute to a condensed phase when a non-soluble amphiphile is compressed at the water/air interface.

For C approaching the CMC (the two highest detergent concentrations in fig. 7(a)), the integrity of the lipid membrane is progressively lost. At such high concentrations we typically observe a single pore which lasts until the whole vesicle is slowly dissolved, possibly via formation of lipid-detergent mixed micelles (fig. 8(d)).

We estimate the energy of adsorption per site, ϵ, as follows. For n molecules, each of size a, adsorbed on a line of length L, the free energy per site is

$$\frac{a}{L}\frac{F}{kT} = -\frac{\phi\epsilon}{kT} + \phi \ln \phi + (1-\phi)\ln(1-\phi), \tag{7}$$

where $\phi = na/L$ is the fraction of occupied sites. The "line osmotic pressure" is then

$$\Pi_L = \left(\frac{\partial F}{\partial L}\right)_n = -\frac{kT}{a}\ln(1-\phi). \tag{8}$$

Calculating $\mu = \partial F/\partial n$ and equating it to $kT\ln(Ca^3)$ gives us $\phi/(1-\phi) =$

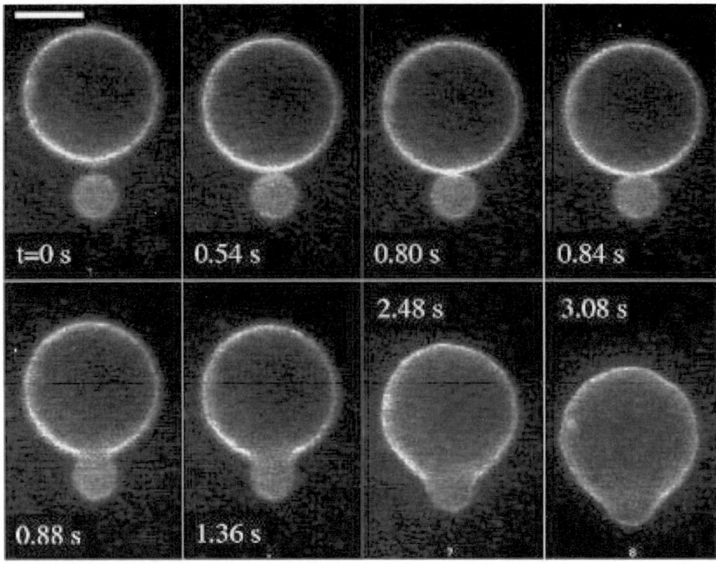

Fig. 9. – Fusion event between two vesicles in the presence of Tween 20. Bar = 10 μm (taken from [20]).

$Ca^3 \exp[\epsilon/kT]$. Using this in the expression for Π_L, we write

$$\mathcal{T} = \mathcal{T}_0 - \Pi_L = \mathcal{T}_0 - \frac{kT}{a} \ln\left(1 + Ca^3 e^{\epsilon/kT}\right). \tag{9}$$

The reader can verify that eq. (9) is consistent with the Gibbs relation in eq. (6). A fit of eq. (9) to our data in regimes I and II with $a \simeq 1$ nm yields $\epsilon \simeq 12kT$, a value comparable to measured heats of adsorption of typical detergent molecules to lipid membranes [18].

v) *Fusion and exocytosis.* – A remarkable phenomenon related to the interaction of amphiphiles with lipid bilayers is membrane fusion, which remains an enigma and a challenge in molecular biology [19]. In fact, the energy barrier to nucleate a pore in a membrane is very sensitive to $\mathcal{T}$: we observe spontaneous fusion events between GUVs when $\mathcal{T}$ is sufficiently lowered in the presence of Tween 20 [20] (fig. 9).

1.4. *Conclusion.* – Lipid membranes do not like having edges. They possess a self-healing capacity: if punctured, a membrane reseals itself. This is essential for maintaining the integrity of all membraneous cellular compartments and hence for maintaining life itself. This amazing capacity of lipid membranes is due to the line tension. Line tension is modulated by the presence of edgeactants in the membrane, whose effect depends on the spontaneous curvature of the molecule. We have shown that a molecule such as cholesterol (present in biomembranes) increases line tension so that it is more difficult

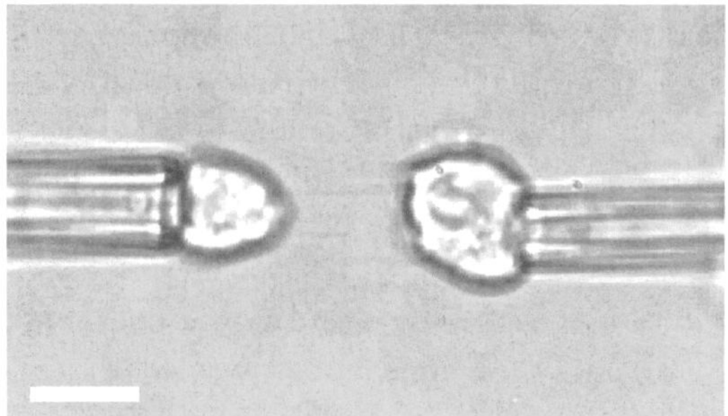

Fig. 10. – Tubes of membrane connecting two cells pulled apart by micromanipulation. Courtesy of Sylvie Dufour and Chu Yeh-Shiu (Institut Curie). Bar = 5 μm (extracted from [21]).

to make and maintain a pore open in the membrane. On the other hand, Tween 20, a detergent molecule, decreases the energy cost of a pore and increases its lifetime. At high concentration, it can even facilitate membrane fusion. Natural edgeactants exist and have crucial biological functions (protein talin [22] and lysolipids [23]).

We have been able to visualize transient pores on vesicles only in viscous solutions. If the inner liquid is a low-viscosity liquid, such as water, leak-out through a pore becomes fast and limits the size and lifetime of pores, making it difficult to visualize them [2, 3]. However, the presence of pores in such systems has been suggested by indirect measurements: anomalously high permeation rates, explained by the presence of small, transient pores, have been observed for vesicles tensed by adsorption in water [3, 24, 25]. Other researchers have been able to visualize long lived, large pores in low-viscosity solutions by creating the pores by electroporation, and maintaining large membrane tensions using micropipette aspiration [26, 27]. If the driving force for pore opening, *i.e.* the membrane tension, is allowed to relax, as is the case with freely floating vesicles, then pores would be aborted before reaching observable sizes in water.

2. – Lipidic tubes

2·1. *Introduction.* – Several cell organelles (such as the Golgi apparatus, the mitochondrial membrane or the endoplasmic reticulum) exhibit tubular structures [28, 29] that grow and retract, probably allowing transport function from an intracellular compartment to another. Tube formation also occurs during cell development and motility when crawling on a substrate. Figure 10 shows two cells pulled apart from each other using micromanipulation. Initially stuck together, they exhibit thin tubes going from one cell membrane to the other where they are bound via adhesion proteins. Pulling tubes from cells provides new insight into membrane-cytoskeleton coupling [30].

To understand the mechanism of tube formation, fluid lipidic tethers have been successfully reproduced in model systems such as vesicles, using a wide range of techniques: micropipette manipulation [31], optical tweezers [32], electrical microelectrode manipulation [33], growth of rigid biopolymers inside vesicles [34], traction by molecular motors [35], hydrodynamical flows [36]. Apart from mimicking biological phenomena, another application of tube extraction is the fabrication of tube networks for nanofluidic and artificial cell design [33]. Tubes offer the possibility to connect vesicles, or even living cells, acting as microreactors when containing reactive molecules.

Theoretical approaches have been carried out to describe the vesicle-tube shape transition [37,38] and the statics of this phenomenon is now well understood (subsect. **2**˙2). Here we present our studies on tubes pulled away by a micromanipulation from an immobile GUV (subsects. **2**˙3 and **2**˙4), and tubes formed from GUVs stretched under hydrodynamical strong flows (subsects. **2**˙5 and **2**˙6).

2˙2. *Threshold force of tube extrusion.* – The formation of tubes when a vesicle is stretched can be understood as a first-order transition at a certain threshold force f_c [37,38]. The following, simple, argument shows this. Let us assume that the tension σ of the vesicle is constant, and that the volume Ω is fixed. Stretched by two opposite distributed forces in a "Gedanken" experiment (fig. 11), we assume that the vesicle is elongated. The shape is ellipsoidal at small forces. If L is the length along the long axis and R_0 the initial radius, the force is proportional to $L - R_0$ ($f \approx \sigma(L - R_0)$). Under strong forces, the shape becomes a "cigar" of length L, radius r. Neglecting end caps effects ($r/L \ll 1$), the energy of the cylindrical vesicle is the sum of the curvature and surface energies:

$$\frac{F}{2\pi} = \frac{\kappa}{2r^2} rL + \sigma rL\,, \tag{10}$$

where κ is the bending rigidity ($\kappa \approx 10kT$). With the conservation of the volume $\pi r^2 L = \Omega$, we can eliminate r and calculate the force $f = \partial F/\partial L$ from eq. (10):

$$\frac{f}{2\pi} = \frac{3}{4}\kappa\sqrt{\frac{L\pi}{\Omega}} + \frac{1}{2}\sigma\sqrt{\frac{\Omega}{L\pi}} = \frac{3\kappa}{4r} + \frac{1}{2}\sigma r\,. \tag{11}$$

We show in fig. 11 a schematic plot of the force *vs.* elongation L: in the case of liquid droplets, the decrease of f *vs.* L is related to the Rayleigh instability of a liquid cylinder. Here, by contrast, for $f = f_c$, we have coexistence between a quasi-spherical vesicle and a thin tube. The pressure in the two compartments is equal. The pressure inside the sphere is $P = 2\sigma/R_0 \approx 0$. Inside the tube, the pressure deduced from eq. (10) is

$$p_t = \frac{\sigma}{r} - \frac{\kappa}{2r^3} \approx 0\,. \tag{12}$$

The radius of the tube at coexistence, called the critical radius r_c, is then

$$r_c = \sqrt{\frac{\kappa}{2\sigma}}\,. \tag{13}$$

The critical force f_c is deduced from eq. (11) with $r = r_c$:

$$f_c = 4\pi\sigma r_c = 2\pi\sqrt{2\kappa\sigma}\,. \tag{14}$$

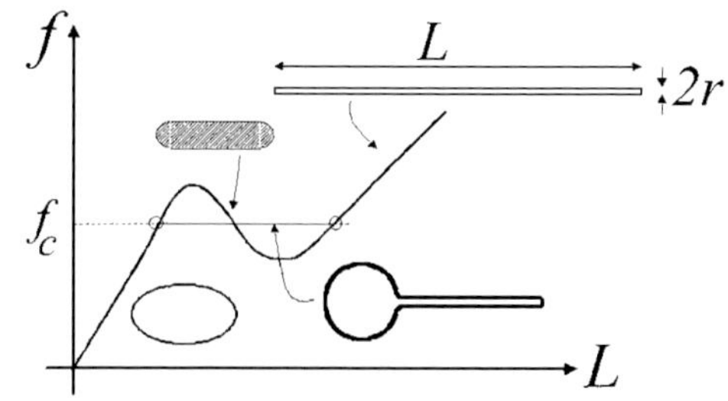

Fig. 11. – Schematic plot of force f *vs.* elongation L of a vesicle of initial radius R_0 and constant tension σ. At small forces, the vesicle has an ellipsoidal shape and the force is proportional to $L - R_0$. Under strong forces, the shape becomes a "cigar" of length L and radius r. The force f decreases with elongation L: this region is unstable and corresponds to the Rayleigh instability of a cylindrical vesicle. At intermediate force $f = f_c$, we have a coexistence between a quasi-spherical vesicle and a tether.

At this point, it is well to make a few remarks before continuing further:

- The capillary force f on the tube is not $2\pi\sigma r$, but $\pi\sigma r$ because one must differentiate the energy (eq. (10)) at constant volume:

$$f = \frac{\partial}{\partial L}(2\pi\sigma r L)_\Omega = 2\pi\sigma r + 2\pi L \frac{\partial}{\partial L}(\sigma r)_\Omega = \pi\sigma r\,. \tag{15}$$

- The shape of the vesicle coexisting with the filament is nearly the shape of a liquid droplet deposited on a fiber of radius $2r_c$, as pointed out by several groups [39, 38, 40]. In the region where surface tension dominates, the shape of the vesicle (given by the profile $z(x)$) can be derived from the conservation of the force $f = 2\pi\sigma z\cos\theta$, where θ is related to the slope ($\tan\theta = \mathrm{d}z/\mathrm{d}x$). The resulting profile $z = z_0\cosh(x/z_0)$ corresponds to the profile of a meniscus of a liquid rising on a fiber of radius z_0. For $f = f_c$, the comparison of eq. (14) with the one on the conservation of the force shows that $z_0 = 2r_c$. A crossover region of extension r_c relates the vesicle to the tube. In this region, both the surface energy and the membrane bending play a role.

- There is a strong analogy between the stretching of vesicles, and the stretching of a polymer chain in a bad solvent. The polymer globule is characterized by a surface tension. Stretched by a force f, Zhulina and Halperin [41] have predicted a first-order coil-stretch transition: the globule coexists with an elongated polymer chain.

2·3. *Structure of the stretching tube.* – Here, the vesicle is immobilized by a micropipette and the tube is pulled at constant velocity U by a bead. The length of the tube is $L = Ut$ at time t of extrusion. The friction on the tube generates a gradient of surface tension $\sigma(x)$. We derive $\sigma(x)$ and the profile of the tube $r(x)$ from the three following equations:

– We assume a quasistatic equilibrium. The pressure inside the tube is uniform and equal to zero. Equation (12) leads to

$$2\sigma r = \frac{\kappa}{r}\,. \tag{16}$$

– The flow of lipids is conserved:

$$U(x)r(x) = Ur_{\mathrm{m}}\,, \tag{17}$$

where r_{m} is the tip radius and U the (fixed) stretching velocity.

– The balance of mechanical and friction forces can be written as

$$f(x + \mathrm{d}x) = f(x) = \zeta_{\mathrm{P}} U(x)\mathrm{d}x\,, \tag{18}$$

where ζ_{P} is the friction coefficient per unit length for a cylinder (length L, radius r) in a flow parallel to the symmetry axis: $\zeta_{\mathrm{P}} = 4\pi\eta/(\ln(L/r) - 1/2)$.

With eqs. (11) and (16), eqs. (17) and (18) lead to

$$-2\pi\frac{\kappa}{r^2}\frac{\mathrm{d}r}{\mathrm{d}x} = \zeta_{\mathrm{P}} U(x) = \zeta_{\mathrm{P}} U\frac{r_{\mathrm{m}}}{r}\,. \tag{19}$$

The structure of the stretching tube is specified by those three equations, and the two boundary conditions $U(L) = U$ at the extremity and $\sigma(0) = \sigma_0$, that is $r(0) = r_{\mathrm{c}} = \sqrt{\kappa/2\sigma}$, at the birth of the tube in connection with the "mother" vesicle. The integral of eq. (19) gives

$$\ln\frac{r}{r_{\mathrm{c}}} = -\frac{\tilde{2}\eta U}{\kappa} r_{\mathrm{m}} x\,, \tag{20}$$

where $\tilde{2}$ is a numerical coefficient including the hydrodynamic interactions: $\tilde{2} \approx 2/\ln(L/r_{\mathrm{c}})$ if the size of the container (or experiment chamber) denoted as d is larger than the length of the tube. In confined geometries, the hydrodynamic interactions are screened and L is replaced by d. Equation (20) involves a characteristic length defined by $l^2 = \kappa/\tilde{2}\eta U$. The radius of the tube r_{m} at the tip is given then by

$$\ln\frac{r_{\mathrm{c}}}{r_{\mathrm{m}}} = \frac{r_{\mathrm{m}}}{r_{\mathrm{c}}}\frac{L}{L_{\mathrm{c}}}\,, \tag{21}$$

where $L_{\mathrm{c}} = l^2/r_{\mathrm{c}}$ is the typical length of variation of radius. With $f = 20\,\mathrm{pN}$, $U = 100\,\mu\mathrm{m/s}$, we find $L_{\mathrm{c}} \approx 200\,\mu\mathrm{m}$. When $L \ll L_{\mathrm{c}}$, eq. (21) gives $r_{\mathrm{m}} = r_{\mathrm{c}}$, meaning that the size of the tube is constant. When $L \gg L_{\mathrm{c}}$, eq. (21) leads to

$$\frac{r_{\mathrm{m}}}{r_{\mathrm{c}}} \approx \frac{L_{\mathrm{c}}}{L}\ln\frac{L}{L_{\mathrm{c}}}\,. \tag{22}$$

Including all logarithmic corrections in the prefactor, eq. (22) leads to $r_{\rm m} \approx \kappa/\eta UL$. In the limit $L \gg L_{\rm c}$, the profile and the membrane tension are given by

$$r(x) = r_{\rm c} \exp\left[\frac{-xr_{\rm m}}{l^2}\right] \qquad \text{and} \qquad \sigma(x) = \sigma_0 \exp\left[\frac{2xr_{\rm m}}{l^2}\right]. \tag{23}$$

The radius of the tube decreases from $r_{\rm c}$ to $r_{\rm m}$, while $\sigma(x)$ increases from σ_0 (the vesicle tension) to $\sigma_{\rm m} = \kappa/2r_{\rm m}^2$ at the tip. The force at this tip $f_{\rm t}$, directly deduced from eq. (22),

$$\frac{f_{\rm t}}{f_{\rm c}} \approx \frac{L}{L_{\rm c}} \approx \frac{\eta U r_{\rm c} L}{\kappa} \tag{24}$$

increases linearly with ηUL, neglecting logarithmic corrections. Along the tube, the increase of $f(x)$ is deduced from eq. (23):

$$f(x) = 2\pi\frac{\kappa}{r} = f_{\rm c} \exp\left[\frac{xr_{\rm m}}{L_{\rm c}r_{\rm c}}\right]. \tag{25}$$

Remark: If the force $f_{\rm t}$ is maintained constant, eq. (24) shows that $L(t)$ increases with $t^{1/2}$ ($f_{\rm t} \approx \tilde{2}\eta LU$, that is $L \approx ((f_{\rm t}/\eta)t)^{1/2}$).

2$\cdot$4. *Tube retraction: "semi-frozen" approximation.* – At $t = 0$, we stop the traction of the tube by the bead (radius $R_{\rm b}$). The tube has a length L_0. To calculate the retraction $L(t)$, we assume that the tube can relax only from its free end. This is called the "Albatros theorem" by polymer physicists: "Ses ailes de géant l'empêchent de marcher", Baudelaire poem [42]. A relaxation in the middle would imply a motion of the whole tube, and would be extremely slow. On the other hand, the free end (here the bead) can collect the tube very quickly [43, 44]. This explains why we state here that the tube is frozen in its initial conformation. The change in the tube radius is extremely low, because it implies motion in ultra-thin capillaries. We first assume in [36] that the forces are frozen.

Here, we improve this model and show that the shape of the tube is frozen, but the tension along the tube is not and relaxes faster. That is why we name our description the "semi-frozen" model.

– Relaxation of $\sigma(x)$.

The fluctuation of the surface tension along the tube is described by Lucassen modes [45] extended to tubes [36]: for a modulation of wave vector q:

$$\frac{1}{\tau_q} = D_\sigma q^2\,, \qquad \text{with} \qquad D_\sigma = \frac{Er}{\eta}. \tag{26}$$

E is the elastic modulus associated to the stretching of the fluctuating membrane:

$$E \approx 3\sigma\frac{8\pi\kappa}{kT}(\sim 10^3\sigma). \tag{27}$$

We look at the relaxation of the gradient along the tube. Equation (26) with $q \sim 1/L$ leads to

$$\tau_{\mathrm{rel}\sigma} = \frac{\eta L^2}{Er} . \tag{28}$$

Because E is much larger than the vesicle tension σ, we will see later that this time is always much smaller than the retraction time τ_{ret}. We shall set $\sigma = \sigma_0$, the vesicle tension imposed by the micropipette.

– Relaxation of $r(x)$.

A periodic modulation $r(x)$ of the tube radius leads to a modulation $\mathrm{d}p/\mathrm{d}x$ of the pressure inside the tube:

$$\frac{\mathrm{d}p}{\mathrm{d}x} = \frac{2\sigma}{r^2}\frac{\mathrm{d}p}{\mathrm{d}x} \tag{29}$$

and a flow $J(x) = -(\pi r^4/8\eta)(\mathrm{d}p/\mathrm{d}x)$. With the volume conservation equation $2\pi r(\mathrm{d}r/\mathrm{d}t) + (\mathrm{d}J/\mathrm{d}x) = 0$, we also find a diffusive relaxation:

$$\frac{1}{\tau_q} = D_r q^2 , \qquad \text{with} \qquad D_r \simeq \frac{\sigma r}{8\eta} . \tag{30}$$

Taking $q = 1/L$, the tube diameter relaxation time $\tau_{\mathrm{rel}r}$ is

$$\tau_{\mathrm{rel}r} = \frac{\eta L^2}{\sigma r} . \tag{31}$$

Because σ is much smaller than E, $\tau_{\mathrm{rel}r}$ is thousand time larger than $\tau_{\mathrm{rel}\sigma}$.

– Retraction of the tube.

The retraction of the tube is calculated by equating the Stokes friction force on the bead moving at velocity $\dot{L}$ to the force $f(x = L(t)) = 2\pi((3\kappa/4r) + (\sigma_0 r/2))$:

$$f(L(t)) = -6\pi\eta R_{\mathrm{b}}\dot{L} . \tag{32}$$

This equation predicts a slow-down of the retraction, in agreement with experimental observations [36], when $L(t = 0)$ is much larger than L_{c}. On the other hand, if $L \ll L_{\mathrm{c}}$, the retraction is at constant velocity ($f = f_{\mathrm{c}}$ in eq. (32)).

2·5. *Vesicles under strong flow.* – We study here the deformation of tethered vesicles under strong flows. The vesicles are attached to micro-beads deposited on the ground of a microfluidic channel or to the tip of a micro-rod. The use of a micro-rod is an efficient but delicate micromanipulation technique, which allows to avoid any wall effect.

When a flow velocity U is applied, the friction force on the vesicle is $f_v = 6\pi\eta UR$. $f_v = f_c$ defines the threshold velocity U_c to extrude a tube:

$$3\eta U_c R = \sqrt{2\kappa\sigma} = \frac{\kappa}{r}\,. \tag{33}$$

In our experiments, the initial tension σ_0 is not imposed and ranges between 10^{-6} and 10^{-5} N/m. From eq. (33), σ_0 defines the threshold velocity U_{c0}. If $U > U_{c0}$, a tube is extruded and grows at a velocity $U' = \mathrm{d}L/\mathrm{d}t$. The velocity seen by the vesicle is $U - \dot{L}$. Assuming local equilibrium at the point of birth of the tube, the friction force $6\pi\eta(U - U')R$ balances the threshold force, leading to

$$U - \dot{L} = U_c\,. \tag{34}$$

At time zero, $\sigma = \sigma_0$ and $U_c = U_{c0}$. As a tube is extruded, the excess surface area of the vesicle decreases, and the vesicle membrane tension σ increases. If ΔA is the surface extruded by the tube, the membrane tension σ is related to the initial tension by [46]

$$\frac{\Delta A}{4\pi R^2} = \frac{kT}{8\pi\kappa}\ln\left(\frac{1 + R^2\sigma/6\kappa}{1 + R^2\sigma_0/6\kappa}\right) + \frac{\sigma - \sigma_0}{\kappa_s}\,, \tag{35}$$

where κ_s is the stretching rigidity. The experimental results show that we always have $R^2\sigma \gg 6\kappa$ and $\sigma \ll \kappa_s \sim 0.1$ N/m, so that the second (Hookean) term can be neglected and the argument in the logarithm simplified. For a tube of length L and radius r, the area extension is $\Delta A = 2\pi Lr$, which leads from eq. (35) to

$$\frac{Lr}{2R^2} = \frac{kT}{8\pi\kappa}\ln\frac{\sigma}{\sigma_0}\,. \tag{36}$$

As L grows, σ increases, and so does U_c according to eq. (33). On the other hand, eq. (34) shows that $\dot{L}$ decreases such that we ultimately reach a stationary state, characterized by $\dot{L} = 0$, $L = L_\infty$ and $\sigma = \sigma_\infty$.

From eq. (34), $\dot{L} = 0$ leads to $U_c(\sigma_\infty) = U$, which fixes σ_∞ and r_∞ according to eq. (33). Inserting σ_∞ and r_∞ into eq. (36), we can derive the stationary tube length $L_\infty(U)$:

$$L_\infty = \tau_0 U \ln\frac{U}{U_{c0}}\,, \qquad \text{with} \qquad \tau_0 = \frac{3\eta R^3 kT}{2\pi\kappa^2}\,. \tag{37}$$

Figure 12 shows a plot of $L_\infty(U)$ from measurements on two vesicles of different sizes. Data are fitted with $L_\infty = \tau_0 U \ln U/U_{c0}$.

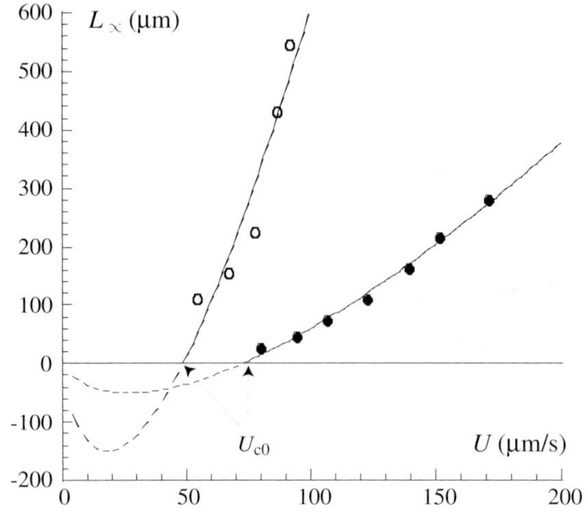

Fig. 12. – Stationary length L_∞ of the tube as a function of flow velocity U for vesicles of radii $R \simeq 7\,\mu$m (black dots) and $R \simeq 11\,\mu$m (white dots). Data are fitted with $L_\infty = \tau_0 U \ln U/U_{c0}$ (taken from [21]).

For the small vesicle ($R = 7\,\mu$m), the fit gives $U_{c0} \simeq 70\,\mu$m/s, which corresponds to $\sigma_0 \simeq 2.7 \times 10^{-5}$ N/m at room temperature. For the big vesicle ($R = 11\,\mu$m), it gives $U_{c0} \simeq 50\,\mu$m/s and $\sigma_0 \simeq 1.9 \times 10^{-5}$ N/m. These are typical values for vesicles under mild tension. The fit gives $\tau_0 \simeq 1.8$ s for the small vesicle, and $\tau_0 \simeq 8.4$ s for the big vesicle. Using typical values of $\eta = 10^{-3}$ Pa s and $\kappa \sim 10kT$, the calculation gives $\tau_0 \simeq 0.9$ s for the small vesicle, and $\tau_0 \simeq 2.9$ s for the big vesicle. The comparison of calculated and measured values of τ_0 can be used as an estimation of κ and gives, respectively, $7 \pm 1kT$ and $6 \pm 1kT$. These values are in reasonable agreement with typical values.

The range of flow velocities used here (60 μm/s to 170 μm/s) corresponds to $f_v \simeq 11$ to 33 pN. Using eq. (33), we find the maximum tensions σ_∞ ranging from 2.6×10^{-5} to 1.6×10^{-4} N/m and the tube radii r_∞ ranging from 25 to 12.5 nm. In [47], a model including the asymmetry between the two membrane leaflets induced by the extrusion of the tube leads to similar values of tube radii, ranging from 50 to 10 nm within the range of forces 10 to 50 pN. Taking into account the uncertainty on κ, our estimations are in reasonable agreement with this previous work.

2·6. *Stop-flows: tube relaxation.* – The vesicle under a flow U reaches a stationary state ($L = L_\infty(U)$, $\sigma = \sigma_\infty(U)$). We stop suddenly the flow and we follow the retraction of the tube. The plot of the relaxation $L(t)$ for three velocities U is shown in fig. 13. We observe that the velocity of retraction decreases from $|\dot{L}| = U$ at time $t = 0$ to $|\dot{L}| = U_{c0}$ at the end, where $\sigma = \sigma_0$, the initial tension. As the vesicle eats the tube, its tension relaxes from σ_∞ at time $t = 0$ to σ_0. The net result is a drop of the retraction force.

More quantitatively, the retraction equation results from a balance of the driving

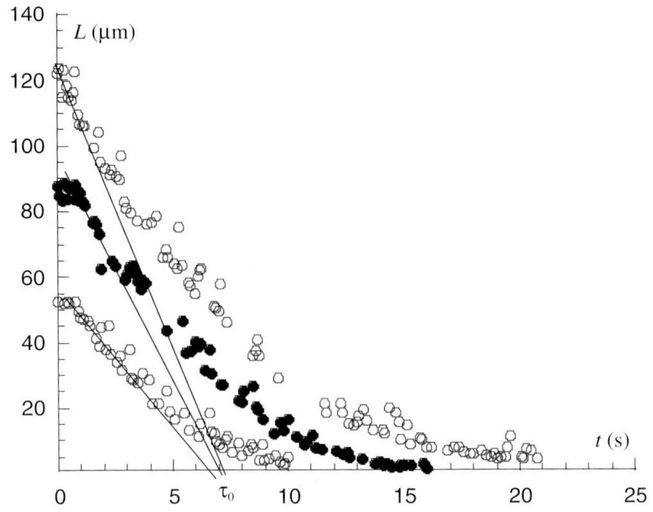

Fig. 13. – Length of the tube during retraction for three initial L_∞. The initial slope gives an estimation of $\tau_0 \simeq 7\,\mathrm{s}$ in agreement with the calculated value from eq. (37).

(retraction) force and the Stokes friction:

$$-3\eta R\dot{L} = \sqrt{2\kappa\sigma} = \frac{\kappa}{r}, \qquad \text{with} \qquad \sigma = \sigma_0 \exp\left[\frac{4\pi Lr\kappa}{R^2kT}\right]. \tag{38}$$

This equation can be rewritten as

$$\frac{-\tau_0\dot{L}}{L} \ln \frac{-\dot{L}}{U_{c0}} = 1 . \tag{39}$$

Equation (39) shows that the tube retracts with a characteristic time $\tau_{\mathrm{ret}} \sim \tau_0$, depending logarithmically upon U. Experimentally, we find $\tau_0 \sim 7\,\mathrm{s}$ from fig. 13, in agreement with τ_0 calculated from eq. (37) with $\kappa \simeq 7kT$ and $R \simeq 10\,\mu\mathrm{m}$.

2·7. *Concluding remarks.* – We have described the structure and retraction dynamics of tubes extruded by pulling beads or hydrodynamical flows from giant unilamellar vesicles.

The tube shape depends on the extraction velocity. The tube retraction dynamics depend only on membrane tension, which relaxes quickly compared to the retraction time. Thus, the retraction dynamics are controlled by the balance between Stokes friction and vesicle extrusion threshold force f_c, which decreases as the tube retracts, in the case where the tube is stuck to a micro-rod and the vesicle is free. Our experiments allow us to know the critical velocity U_{c0} and subsequently the initial tension of the vesicle. A possible application of this work would be the estimation of σ_0 through the measurement of L_∞ for any extrudible liquid membrane.

We now plan to perform similar experiments on cells. Our aim is to check the mechanical properties of cells, and to see whether we can detect tumoral cells, which are known to have different mechanical properties than healthy cells.

* * *

We would like to thank P.-H. Puech, A. Buguin, P. Nassoy, D. Cuvelier and O. Rossier for fruitful discussions and advice in experiments.

REFERENCES

[1] Puech P. H., Borghi N., Karatekin E. and Brochard-Wyart F., *Phys. Rev. Lett.*, **90** (2003) 128304.
[2] Brochard-Wyart F., de Gennes P.-G. and Sandre O., *Physica A*, **278** (2000) 32.
[3] Sandre O., Moreaux L. and Brochard-Wyart F., *Proc. Natl. Acad. Sci. USA*, **96** (1999) 10591.
[4] Debrégeas G., Martin P. and Brochard-Wyart F., *Phys. Rev. Lett.*, **75** (1995) 3886.
[5] Israelachvili J. N., *Intermolecular and Surface Forces*, 2nd edition (Academic Press, London) 1992.
[6] Karatekin E., Sandre O., Guitouni H., Borghi N., Puech P. H. and Brochard-Wyart F., *Biophys. J.*, **84** (2003) 1734.
[7] Simons K. and Ikonen E., *Science*, **290** (2000) 1721.
[8] Rukmini R., Rawat S. S., Biswas S. C. and Chattopadhyay A., *Biophys. J.*, **81** (2001) 2122.
[9] Israelachvili J. N., Marčelja S. and Horn R. G., *Q. Rev. Biophys.*, **13** (1980) 121.
[10] Carnie S., Israelachvili J. N. and Pailthorpe B. A., *Biochim. Biophys. Acta*, **554** (1979) 340.
[11] Rusanov A. I., *Colloids Surf. A*, **156** (1999) 315.
[12] Chen P., *Colloids Surf. A*, **161** (2000) 23.
[13] Babak V. G., *Colloids Surf. A*, **156** (1999) 423.
[14] Inoue T., *Interaction of Surfactants with Phospholipid Vesicles*, in *Vesicles. Surfactant Science Series*, Vol. **62** (Marcel Dekker, New York) 1996.
[15] Lasch J., *Biochim. Biophys. Acta*, **1241** (1995) 269.
[16] Karasawa K., Qiu X. and Lee T.-C., *J. Biol. Chem.*, **274** (1999) 8655.
[17] de Gennes P.-G., Brochard-Wyart F. and Quéré D., *Gouttes, Bulles, Perles et Ondes* (Belin, Paris) 2002.
[18] Heerklotz H. and Seelig J., *Biochim. Biophys. Acta*, **1508** (2000) 69.
[19] Zimmerberg J., *Trends Cell Biol.*, **11** (2001) 233.
[20] Karatekin E., Sandre O. and Brochard-Wyart F., *Polym. Int.*, **52** (2003) 486.
[21] Borghi N., Rossier O. and Brochard-Wyart F., *Europhys. Lett.*, **64** (2003) 837.
[22] Saitoh A., Takiguchi K., Tanaka Y. and Hotani H., *Phys. Rev. Lett.*, **78** (1997) 386.
[23] Chernomordik L. V., Kozlov M. M., Melikyan G. B., Abidor I. G., Markin V. S. and Chizmadzhev Y., *Biochim. Biophys. Acta*, **812** (1985) 643.
[24] Guedeau-Boudeville M.-A., Jullien L. and di Meglio J.-M., *Proc. Natl. Acad. Sci. USA*, **92** (1995) 9590.
[25] Bernard A.-L., Guedeau-Boudeville M.-A., Sandre O., Palacin S., di Meglio J.-M. and Jullien L., *Langmuir*, **16** (2000) 6801.
[26] Zhelev D. V. and Needham D., *Biochim. Biophys. Acta*, **1147** (1993) 89.
[27] Moroz J. D. and Nelson P., *Proc. Natl. Acad. Sci. USA*, **95** (1998) 1032.

[28] Lee C. and Chen L. B., *Cell*, **54** (1988) 37.
[29] Mollenhauer H. H. and Morre D. J., *Histochem. Cell. Biol.*, **109** (1998) 533.
[30] Hochmuth R. M. and Marcus W. D., *Biophys. J.*, **82** (2002) 2964.
[31] Evans E. and Yeung A., *Chem. Phys. Lipids*, **73** (1994) 39.
[32] Dai J. and Sheetz M. P., *Biophys. J.*, **77** (1999) 3363.
[33] Karlsson M., Sott K., Cans A.-S., Karlsson A., Karlsson R. and Orwar O., *Langmuir*, **17** (2001) 6754.
[34] Fygenson D. K., Elbaum M., Shraiman B. and Libchaber A., *Phys. Rev. E*, **55** (1997) 855.
[35] Roux A., Cappello G., Cartaud J., Prost J., Goud B. and Bassereau P., *Proc. Natl. Acad. Sci. USA*, **99** (2002) 5394.
[36] Rossier O., Cuvelier D., Borghi N., Puech P. H., Derényi I., Buguin A., Nassoy P. and Brochard-Wyart F., *Langmuir*, **19** (2003) 575.
[37] Derényi I., Jülicher F. and Prost J., *Phys. Rev. Lett.*, **88** (2002) 238101.
[38] Powers T. R., Huber G. and Goldstein R. E., *Phys. Rev. E*, **65** (2002) 041901.
[39] Heinrich V., Božič B., Svetina S. and Žekš B., *Biophys. J.*, **76** (1999) 2056.
[40] Bruinsma R., *Physica A*, **234** (1996) 249.
[41] Halperin A. and Zhulina E. B., *Europhys. Lett.*, **15** (1991) 417.
[42] Baudelaire C., *Spleen et idéal in Les Fleurs du Mal* (Paris) 1857.
[43] de Gennes P.-G., *J. Chim. Phys.*, **87** (1967) 962.
[44] Buguin A. and Brochard-Wyart F., *Macromolecules*, **29** (1996) 4937.
[45] Lucassen J. and Van den Tempel M., *Chem. Eng. Sci.*, **27** (1972) 1283.
[46] Servuss R. M., Harbich W. and Helfrich W., *Biochim. Biophys. Acta*, **436** (1976) 900.
[47] Waugh R. E., Song J., Svetina S. and Žekš B., *Biophys. J.*, **61** (1992) 974.

Analytical theory of polyelectrolytes

L. BLUM

Department of Physics, University of Puerto Rico
P.O. Box 23343, Río Piedras, Puerto Rico 00931

O. BERNARD

Laboratoire des Propriétés Physico-Chimiques des Electrolytes
URA 430, Université Pierre et Marie Curie
4 Place Jussieu, 75252 Paris Cedex 05, France

1. – Introduction

Polymers in general and polyelectrolytes in particular are of practical interest in a number of applications in industry and biology. In many biological systems the role of water and electrolytes is known to be of primordial relevance. However, with even the most advanced methods [1], it is presently impossible to simulate systems that have both the correct water and the electrolyte. For that reason physically motivated but accurate simple theories are of practical interest [2-10].

A simple and reliable theory that is able to handle both the ions, the water and the polyelectrolyte is the Mean Spherical Approximation (MSA) and its extensions, which are derived from the Mean Spherical Model [11-14]. They provide a systematic way to study complex ionic systems that is coherent and simple. The end result are thermodynamic and structural properties given in terms of a very small set of scaling parameters Γ_α. The general scaling solution of the MSA for polyelectrolytes discussed by Blum *et al.* [15] and by von Solms *et al.* [16, 17]. These solutions do not obey the large n-DH asymptotic limit. The theory of Bernard and Blum [18] obeys this exact limiting law for arbitrary n.

Charged systems obey exact asymptotic relations. The fact that the excluded volumes are included in the MSA makes this theory asymptotically exact at large densities. It has been shown that the MSA satisfies the Onsager high-density limits [19-21], and therefore can be regarded as an interpolation scheme between low density and the high-density situations. It is then possible to construct interpolation schemes that are amazingly simple, and that can be quite accurate. The simplest one is the BIMSA [22, 23], which incorporates another exact asymptotic limit, the highly diluted strong binding limit: If ions dimerize, for example, the charge in the Debye-Hückel limiting relation should be that of the dimer. In the case of flexible polyelectrolytes it should be the total charge of the polymer: Then we get [23, 24] good agreement with computer simulations for a number of properties, including the critical temperature [25, 26].

The MSA is the linearized Poisson-Boltzmann approximation. Like the Debye-Hückel (DH) theory it is solved in terms of a single screening parameter. But unlike the DH, the MSA takes into account the excluded volume of all the ions, those in the center and those in the ionic cloud. The MSA admits simple analytical solutions for the primitive model (continuum dielectric solvent), the ion-dipole model, the ion-YUKAGUA model and more recently for inhomogeneous confined systems. All of these solutions are expressed in terms of a set screening parameters Γ_α that are related to the DH (Debye-Hückel) screening parameter κ

$$\kappa^2 = \frac{4\pi\beta e^2}{\varepsilon} \sum_{j=1}^{m} \rho_j z_j^2 \,, \tag{1}$$

where $\beta = 1/kT$ is the Boltzmann thermal factor, e is the elementary charge, ε is the dielectric constant, ρ_j is the number density and z_j is the electrovalence of ions j. In the single restricted electrolyte of diameter σ case the correct Wiener-Hopf excluded-volume problem of the linear Poisson-Boltzmann equation [2] yields

$$(1 + 2\kappa\sigma) = (1 + 2\Gamma\sigma)^2 \,. \tag{2}$$

The MSA satisfies exact conditions:

1. Low-density high-temperature DH limits. This can be easily verified in the above equation for the primitive case. However, when the temperature goes to zero and there is strong association, this is no longer true. In fact no theory based on the regular Ornstein-Zernike (OZ) equation (which includes the HNC, various flavors of the MHNC, SMSA and Rogers-Young (YR) theory) will go to the right limit.

2. High-density high-temperature Onsager limits: Since the MSA includes all the hard cores in its solution it will satisfy this limit [21].

3. Low-density low-temperature binding limits: The solutions of the Wertheim-Ornstein-Zernike (OZ) will satisfy this condition. The remarkable fact is that the one screening parameter Γ description still holds. However, now the association equilibrium parameters are included in the screening parameter [23].

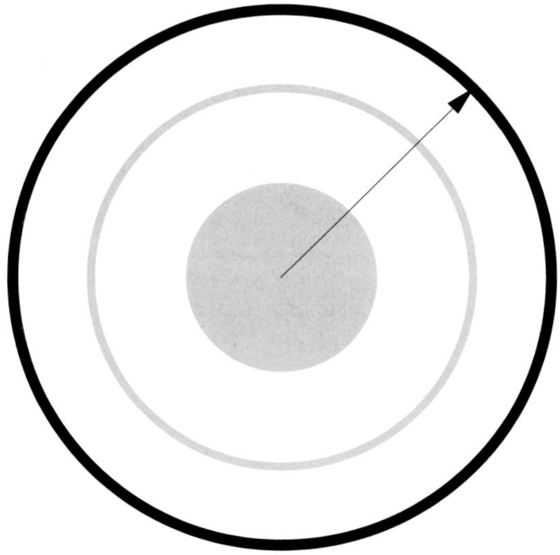

Fig. 1. – The equivalent spherical capacitor: The radius of this capacitor is, from eq. (42), the sum of the hard-sphere radius and the screening length λ_c. There is always an exact value of this length. The MSA circle is the external one, and represents an improvement of the DH estimate, which is the smaller circle.

1˙1. *The perfect screening sum rule* [27]. – Ionic solutions are mixtures of charged particles, the ions, and the neutral solvent particles, most commonly water, which has an asymmetric charge distribution, a large electric dipole and higher electric moments. Because of the special nature of these forces the charge distribution around a given ion and the thermodynamics satisfy a series of conditions or sum rules. One remarkable property of mixtures of classical charged particles is that because of the very long range of the electrostatic forces, they must create a neutralizing atmosphere of counterions, which shields *perfectly* any charge or fixed-charge distribution. Otherwise the partition function, and therefore all the thermodynamic functions, will be divergent [27]. The size of the region where this charge shielding occurs depends not only on the electrostatics, but also on all the other interactions of the system. This means also that for a spherical ion the countercharges have a spherical distribution, and that the internal energy of the ion and its cloud is equal to that of a charged spherical capacitor (see fig. 1).

For spherical ions this means:

1. The internal energy E of the ions is always the sum of the energies of capacitors. For spherical ions the capacitor is a spherical capacitor, and the *exact* form of the energy is

$$\Delta E = -\frac{e^2}{\varepsilon}\sum_i \rho_i z_i \frac{z_i^*}{1/\Gamma_i + \sigma_i}, \tag{3}$$

Fig. 2. – For infinite density the screening sphere collapses on the exclusion sphere of the hard central ion: From eq. (42) we have $\lambda_c = 0$.

where z_i^* is the effective charge, $\beta = 1/kT$ is the usual Boltzmann thermal factor, ε is the dielectric constant, e is the elementary charge, and ions i have charge, diameter and density ez_i, σ_i, ρ_i, respectively. For the continuum dielectric primitive model $\Gamma_i = \Gamma$ for all i.

2. The Onsagerian limits. When the ionic concentration goes to infinity and at the same time the charge diverges, then the limiting energy is bounded by

$$\Delta E = -\frac{e^2}{\varepsilon}\sum_i \rho_i z_i \frac{z_i^*}{\sigma_i}\,, \tag{4}$$

obtained by setting $\Gamma_i \to \infty$ (see fig. 2).

3. A further exact limit is the DH limiting law, which simply requires that for all ions in the system

$$2\Gamma_i \longrightarrow \kappa \qquad \text{with} \qquad \kappa^2 = \frac{4\pi\beta e^2}{\varepsilon}\sum_{j=1}^{m} \rho_j z_j^2\,. \tag{5}$$

4. Finally in systems that are strongly associating in the limit of *total* association the above equation still holds. This means that if component 1 forms a n-mer the DH

limiting law must satisfy

(6) $$\kappa^2 = \frac{4\pi\beta e^2}{\varepsilon}\left[\sum_{j=2}^{m}\rho_j z_j^2 + \rho_1(nz_1)^2\right].$$

This limiting law is not satisfied by any closure of the regular Ornstein-Zernike equation, but only for closures of the Werteim-Ornstein-Zernike equation [28, 29].

The known analytical MSA solutions for dimers and polymers satisfy a "universality" principle for the excess entropy

(7) $$k_B\Delta S^{(\mathrm{MSA})} = -\frac{\Gamma^3}{3\pi}.$$

Then Γ is determined by

(8) $$\frac{\partial[\beta\Delta E(\Gamma) + \Gamma^3/(3\pi)]}{\partial\Gamma} = 0.$$

For the simple restricted case of an equal-size ionic mixture we get eq. (2). For more complex systems, like polyelectrolytes, this equation constitutes a new relation. For flexible polyelectrolytes it can be derived from the BIMSA [15].

It can be further extended in several ways:

- To systems of arbitrary geometry [15,21,23]. In this case the solution is constructed to satisfy the perfect screening theorems [27].

- To approximations that go beyond the MSA. In fact the scaling solutions have been obtained for a quite general closure of the Ornstein-Zernike equation as a sum of Yukawa potentials [9]. The solution of the Ornstein-Zernike equation is obtained explicitly in terms of a new set of scaling parameters $\mathcal{G}_\alpha$, and in some cases, explicit equations of state can be obtained [10].

Furthermore, the direct correlation function is constructed from the screening parameters [19-21] and satisfy the required asymptotic limiting laws.

2. – Theory of electrolytes

We review the classical Debye-Hückel theory, which is the linearized Poisson-Boltzmann theory. There are at least two ways of stating this equation, the differential form and the integral form. Both Fourier transforms are identical.

The electrostatic potential satisfies the Poisson equation

(9) $$\nabla^2\phi_i(r) = -\frac{4\pi}{\epsilon}q_i(r).$$

Then, the Poisson-Boltzmann equation

$$\nabla^2 \phi_i(r) = -e\frac{4\pi}{\epsilon}\sum_{j=1}^{m} z_j \rho_j e^{-\beta e z_j \phi_i(r)}\,, \tag{10}$$

which is linearized as

$$\lim_{r\to\infty} g_{ij}(r) \simeq 1 - \beta e z_j \phi_i(r). \tag{11}$$

From here the DH equation

$$\nabla^2 \phi_i(r) = \kappa^2 \phi_i(r), \tag{12}$$

where

$$\kappa^2 = \frac{4\pi\beta e^2}{\epsilon}\sum_{j=1}^{m} \rho_j z_j^2\,. \tag{13}$$

Transforming to spherical coordinates, one gets

$$\left(\frac{1}{r}\right)\frac{\partial^2[r\phi_i(r)]}{\partial r^2} = \kappa^2 \phi_i(r). \tag{14}$$

The general solution of this equation is

$$r\phi_i(r) = A_i e^{-\kappa(r-\sigma/2)} + B_i e^{\kappa(r-\sigma/2)}\,. \tag{15}$$

To satisfy boundary condition at $r \to \infty$ we require that $B_i = 0$. A_i is obtained requiring continuity at $r = \sigma/2$:

$$\phi_i(r) = \frac{z_i e}{\epsilon(1+\kappa\sigma/2)}\frac{e^{-\kappa(r-\sigma/2)}}{r}\,. \tag{16}$$

Using eq. (11)

$$\frac{\rho_j^{\mathrm{DH}}(r)}{\rho_j} = 1 - \frac{z_j z_i \beta e^2}{\epsilon(1+\kappa\sigma/2)}\frac{e^{-\kappa(r-\sigma/2)}}{r}\,. \tag{17}$$

In a more general form we write

$$\rho_j(r) = \rho_j\left[1 - \frac{z_j z_i \beta e^2 \kappa}{\epsilon}\frac{e^{-r/\lambda_c}}{r}\right], \tag{18}$$

where λ_c is the unknown screening parameter.

Using the electroneutrality (perfect screening, spherical term) we get

$$(19)\qquad q_i^{\text{cloud}} = \sum_j z_j 4\pi \int_0^\infty \mathrm{d}r r^2 \rho_j(r) = -e z_i \kappa^2 \lambda_c^2 = -e z_i\,.$$

Then in the DH

$$\kappa = \frac{1}{\lambda_c}\,.$$

The DH electrostatic energy is the sum of a collection of spherical capacitors

$$(20)\qquad \Delta E^{\text{ch}} = \left(\frac{2\pi e^2}{\epsilon}\right) \sum_{i,j} \rho_i z_i z_j \int_0^\infty \mathrm{d}r r^2 \frac{\rho_j(r)}{r} = -\sum_i \frac{\rho_i (e z_i)^2}{2\epsilon} \left[\frac{1}{\sigma/2 + 1/\kappa}\right] = -\frac{\kappa^3}{8\pi\beta(1+\kappa\sigma/2)}\,.$$

The energy of such a capacitor is its charge squared and divided by its radius (fig. 1):

$$(21)\qquad \Delta E^{\text{cap}} = \frac{q^2}{\epsilon r}$$

$$(22)\qquad \Delta E^{\text{ch}} = -\sum_i \frac{\rho_i (e z_i)^2}{2\epsilon} \left[\frac{1}{\sigma/2 + 1/\kappa}\right].$$

In the limit of zero concentration, from eq. (20)

$$(23)\qquad \Delta E^{\text{ch}} = -\frac{\kappa^3}{8\pi\beta(1+\kappa\sigma/2)} \simeq -\frac{\kappa^3}{8\pi\beta}\,.$$

Consider the Helmholtz free energy

$$(24)\qquad \Delta A = \Delta E - T\Delta S\,.$$

From this equation it is verified that

$$(25)\qquad \left(\frac{\partial \Delta A}{\partial T}\right)_V = -\Delta S\,.$$

Then

$$(26)\qquad \Delta A^{\text{DH}} = -\frac{\kappa^3}{12\pi\beta}$$

is the solution of eqs. (24) and (25). We obtain from the latter

$$\frac{\partial \Delta A^{\mathrm{DH}}}{\partial T} = \frac{\kappa^3}{24\pi T} = -\Delta S^{\mathrm{DH}}, \tag{27}$$

which means that the entropy is

$$\Delta S^{\mathrm{DH}} = -\frac{\kappa^3}{24\pi T}. \tag{28}$$

Equation (26) is verified in the $\kappa \to 0$ limit:

$$\begin{aligned} \Delta A = \Delta E - T\Delta S &\longrightarrow -\frac{\kappa^3}{12\pi\beta} \\ &= -\frac{\kappa^3}{8\pi\beta} + k_{\mathrm{B}}T\frac{\kappa^3}{24\pi} \end{aligned} \tag{29}$$

and from eqs. (22)-(28)

$$\Delta A^{\mathrm{DH}} = -\sum_i \frac{\rho_i (e z_i)^2}{2\epsilon}\left[\frac{1}{\sigma_i/2 + 1/\kappa}\right] + \frac{\kappa^3}{24\pi}. \tag{30}$$

The excess osmotic coefficient is obtained using

$$\frac{\partial \Delta A^{\mathrm{DH}}}{\partial V} = -\Delta P^{\mathrm{DH}}. \tag{31}$$

The excess osmotic coefficient is defined by

$$\varphi = 1 + \frac{\beta \Delta P^{\mathrm{DH}}}{\rho_t}, \tag{32}$$

where

$$\rho_t = \sum_i \rho_i. \tag{33}$$

We get

$$\varphi = 1 - \frac{\kappa^3}{24\pi\rho_t}. \tag{34}$$

Using this result we get the Gibbs free energy ΔG^{DH}:

$$\Delta G^{\mathrm{DH}} = \Delta E^{\mathrm{DH}}, \tag{35}$$

from where the mean activity coefficient is

$$\log \gamma_{\pm} = -\frac{\kappa^3}{8\pi\rho_t}\,. \tag{36}$$

3. – Onsager bounds

In 1939 Onsager [2] observed that the internal energy of charged, conducting systems, made of convex molecules with a hard repulsive core, has an exact lower bound.

Every charged object that is immersed in a conducting medium is surrounded by a neutralizing cloud, that cancels its charge and all its electric moments (Blum, Gruber, Lebowitz, Martin (1982)) [27]. The excess energy of the Debye-Hückel model with a core (see eq. (23)) is

$$\Delta E^{\mathrm{ch}} = -\sum_i \frac{\rho_i (ez_i)^2}{2\epsilon}\left[\frac{1}{\sigma/2 + 1/\kappa}\right]. \tag{37}$$

If either the density and/or the charges goes to infinity, then $\kappa \to \infty$ and

$$\Delta E^{\infty} \simeq -\frac{e^2}{\epsilon}\left[\sum_i \frac{\rho_i z_i^2}{\sigma_i}\right]. \tag{38}$$

The Onsager result, however, goes farther since it also states that, in the limit of infinite charge, the excess entropy, ΔS, diverges at a slower rate. Therefore, in the Onsager limit, we have

$$\lim_{\kappa\to\infty} \frac{\Delta S}{\Delta E} = 0\,. \tag{39}$$

This condition is not obeyed by the DH theory without the cores, since the ratio of ΔS to ΔE is

$$\lim_{\kappa\to\infty} \frac{\Delta S^{\mathrm{DH}}}{\Delta E^{\mathrm{DH}}} = \frac{1}{3T}\,, \tag{40}$$

which is incorrect, we get a divergent result when $T \to 0$!

4. – The mean spherical approximation

The MSA interpolates between the Onsager bounds and the DH theory:

$$\Delta A^{\mathrm{MSA}} = \Delta E^{\mathrm{MSA}} - T\Delta S^{\mathrm{MSA}}\,. \tag{41}$$

The energy density equation (37) is written in terms of

$$\lambda_c = \frac{1}{2\Gamma}, \tag{42}$$

$$\Delta E^{\mathrm{MSA}} = -\frac{e^2}{2\epsilon_0}\sum_i \rho_i \frac{z_i^2}{\sigma_i/2 + \lambda_c} = -\sum_i \frac{\rho_i (ez_i)^2}{2\epsilon}\left[\frac{1}{\sigma_i/2 + 1/2\Gamma}\right]. \tag{43}$$

For the excess entropy we simply substitute $\kappa = 2\Gamma$ in the DH expression (28):

$$\Delta S^{\mathrm{MSA}} = -\frac{\Gamma^3}{3\pi T}. \tag{44}$$

The free energy is

$$\Delta A^{\mathrm{MSA}} = -\sum_i \frac{\rho_i (ez_i)^2}{2\epsilon}\left[\frac{1}{\sigma_i/2 + 1/2\Gamma}\right] + \frac{\Gamma^3}{3\pi}. \tag{45}$$

The MSA is obtained optimizing the free energy ΔA:

$$\delta\left[\Delta A^{\mathrm{MSA}}\right] = 0. \tag{46}$$

Then

$$\frac{\partial \Delta A^{\mathrm{MSA}}}{\partial \Gamma} = 0, \tag{47}$$

or

$$\frac{\partial \Delta A^{\mathrm{MSA}}}{\partial \Gamma} = -\sum_i \frac{\rho_i (ez_i)^2}{\epsilon}\left[\frac{1}{(1+\sigma_i\Gamma)^2}\right] + \frac{\Gamma^2}{\pi} = 0, \tag{48}$$

which is an algebraic equation for the new scaling parameter Γ, and is the correct closure equation for the MSA.

The main difference between the MSA and the DH theories is that in the MSA the ions in the ionic cloud surrounding the central ion are also finite-size spheres. This means that the structure function, and also the potential will be oscillating functions, and not simply exponentially decaying.

When all the ions have the same diameter $\sigma_i = \sigma$, then

$$(1 + 2\kappa\sigma) = (1 + 2\Gamma\sigma)^2. \tag{49}$$

The solution of this equation is

$$\Gamma = \left(\frac{1}{2\sigma}\right)\left(\sqrt{(1+2\kappa\sigma)} - 1\right). \tag{50}$$

Now in the Onsagerian limit,

$$\frac{\Delta S^{\mathrm{MSA}}}{\Delta E^{\mathrm{MSA}}} \simeq \frac{\Gamma^3}{\kappa^2} \simeq \kappa^{-1/2} \longrightarrow 0\,. \tag{51}$$

4·1. *Thermodynamic properties.* – From eq. (43) we immediately get the excess internal energy

$$\beta\Delta E^{\mathrm{MSA}} = -\frac{e^2}{\epsilon}\left[\Gamma\sum_i \frac{\rho_i z_i^2}{1+\Gamma\sigma_i}\right]. \tag{52}$$

The excess free energy is obtained directly from (30) and (32):

$$\beta\Delta A^{\mathrm{MSA}} = -\frac{e^2}{\epsilon}\left[\Gamma\sum_i \frac{\rho_i z_i^2}{1+\Gamma\sigma_i}\right] + \frac{\Gamma^3}{3\pi}\,. \tag{53}$$

The excess osmotic coefficient is obtained as before from the thermodynamic relation (21):

$$\frac{\partial\Delta A^{\mathrm{DH}}}{\partial V} = -\Delta P^{\mathrm{DH}}\,, \tag{54}$$

$$\varphi^{\mathrm{MSA}} = 1 - \frac{\Gamma^3}{3\pi\rho_t}\,, \tag{55}$$

where ΔP^{MSA} is the excess pressure. Finally, we get the activity coefficient from the thermodynamic formula

$$\Delta\ln\gamma_{\mp} = \beta\frac{\Delta A^{\mathrm{MSA}}}{\rho_t} + \Delta\phi = \beta\frac{\Delta E^{\mathrm{MSA}}}{\rho_t} = \frac{e^2}{\rho_t\epsilon}\left[\Gamma\sum_i \frac{\rho_i z_i^2}{1+\Gamma\sigma_i}\right]. \tag{56}$$

5. – The BIMSA for dimers

The model of a charged capacitor with a universal entropy form which depends on a single parameter Γ is also valid for the MSA for binding charged spheres, a dimer. Using the WOZ [28, 29] we show [22, 23] that the MSA solution yields a result that is isomorphic with the results for the non-associating systems. We review our previous results for dimers [23]. Consider the case of a dimer formed in a solution containing m species of ions of diameters σ_i and charges z_i. The dimers are formed by species 1 and 2, and the rest of the ions do not associate (fig. 3).

Fig. 3. – The dimer is formed by taking the full association limit of pairs of charged spheres with sticky interactions.

Now we remark that the excess internal energy is a function of the set of the $\alpha_i^N \mathcal{X}_i^N$:

$$\Delta E = \frac{-e^2}{\varepsilon} \sum_k \rho_k z_k \frac{\Gamma^T z_k + \eta^T \sigma_k}{1 + \Gamma^T \sigma_k} + \frac{e^2}{\varepsilon} \sum_k \rho_k \frac{z_k}{\sigma_k} \sum_{N=0}^{3} \alpha_k^N \mathcal{X}_k^N \tag{57}$$

with the implicit equation

$$\eta^T = \frac{\pi}{2\Delta} \sum_k \rho_k \sigma_k \sum_{N=0}^{2} \alpha_k^N \mathcal{X}_k^N, \qquad \mathcal{X}_k^N = \frac{z - \sigma^2 \eta^T}{1 + \Gamma^T \sigma}. \tag{58}$$

Consider first the case of the charged pair with two components of equal diameter $\sigma_1 = \sigma_2 = \sigma$ and opposite charges $z_1 = -z_2$. In this case, from eq. (57) we deduce

$$\Delta E = -\rho \frac{e^2 z^2}{\varepsilon} \left[\left(\frac{\Gamma}{1 + \Gamma\sigma} \pm \frac{1-\alpha}{2\sigma} \frac{1}{(1+\Gamma\sigma)^2} \right) + 2\rho_i \frac{z_i^2 \Gamma}{1 + \Gamma\sigma} \right]. \tag{59}$$

The degree of dissociation is $0 < \alpha < 1$ and is obtained from the dimerization equilibrium relation:

$$K_a = \rho_1 \frac{(1-\alpha)^2}{\alpha}. \tag{60}$$

Let us discuss a few particular cases: The restricted electrolyte case in which the ions are of opposite charge $z_1 = -z_2$. Using the fact that $\eta^T = 0$ in eq. (57), we get, using (8),

$$\frac{\Gamma^2}{\pi} = \frac{\rho z^2 \beta e^2}{\varepsilon} \left[\frac{\Gamma\sigma + \alpha}{(1+\Gamma\sigma)^3} \right]. \tag{61}$$

In the perfect association limit $\alpha = 0$, and

$$\frac{\Gamma}{\pi} = \frac{\rho z^2 \beta e^2}{\varepsilon} \left[\frac{\sigma}{(1+\Gamma\sigma)^3} \right]$$

in which the DH limit of low concentration corresponds to the zero charge of the dimer. In the high-density Onsager limit we get for any nonzero α

$$\frac{\Gamma^2}{\pi} = \frac{\rho z^2 \beta e^2}{\varepsilon}\left[\frac{1}{(1+\Gamma\sigma)^2}\right], \tag{62}$$

which is identical to the free-particle MSA result.

For $z_1 = z_2 = -z_c$ in the limit of perfect association $\alpha = 0$ we get

$$\left(\Gamma^T\right)^2 = \frac{2\pi\rho\beta e^2 z^2 (2+y)}{\varepsilon}\left[\frac{1}{1+2y^2\pi\rho\sigma^3/\Delta}\right]^2, \qquad y = \frac{1}{1+\Gamma^T\sigma_1}, \tag{63}$$

in which the low-density limit corresponds to the limiting term with charge 2.

This equation does satisfy the variational relation (8). It can also be verified that conditions 3 and 4 are satisfied for all values of α. To our knowledge, no other theory does this. However, condition 2 is not satisfied correctly in the limit of perfect association $\alpha = 0$: The straightforward extension of this procedure to trimers and higher polymers will violate condition 2. For example, for $n = 3$ the limiting DH law has an effective charge 8.5, instead of $3^2 = 9$. The error increases with the monomer number n. For this reason we need to take into account triplet exclusion, which fixes this problem for all n.

6. – The polyelectrolyte MSA with the triplet exclusion chain approximation

The extension of the dimer result to arbitrary n is not straightforward. In fact the correct asymptotic form of the DH limiting law is not recovered in this case. We need to include the triplet exclusion in order to get the proper limits. We follow the treatment of Bernard and Blum [18]. We recall that our equation for Γ^T was for the case of a charged chain with particles $1, 2, \ldots n$ with $\sigma_1 = \sigma_2 = \cdots = \sigma_n$ (see fig. 4). In this case, we obtained

$$\left[\Gamma^T\right]^2 = \frac{\pi e^2\beta}{\varepsilon}\left[\sum_k \rho_k \left(\mathcal{X}_k^0\right)^2 + \rho_p \sum_{k=1}^{n-1} \mathcal{X}_k^0 \sum_{i=1}^{n-k} \frac{\mathcal{X}_{k+i}^0}{(1+\Gamma^T\sigma)^i} + \rho_p \sum_{k=2}^{n} \mathcal{X}_k^0 \sum_{i=1}^{k-1} \frac{\mathcal{X}_{k-i}^0}{(1+\Gamma^T\sigma)^i}\right] \tag{64}$$

with

$$\mathcal{X}_1^0 = \frac{z_1 - \sigma_1^2\eta^T}{1+\Gamma^T\sigma_1}, \qquad \mathcal{X}_c^0 = \frac{z_c - \sigma_c^2\eta^T}{1+\Gamma^T\sigma_c}, \tag{65}$$

where the notation is that of Bernard and Blum [23]:

$$\eta^T = \frac{\pi}{2\Delta}\sigma\left[\sum_k \rho_k \mathcal{X}_k^0 + \rho_p \sum_{k=1}^{n-1}\sum_{i=1}^{n-k} \frac{\mathcal{X}_{k+i}^0}{(i+1)(1+\Gamma^T\sigma)^i} + \rho_p \sum_{k=2}^{n}\sum_{i=1}^{k-1} \frac{\mathcal{X}_{k-i}^0}{(i+1)(1+\Gamma^T\sigma)^i}\right] \tag{66}$$

Fig. 4. – The model of a flexible polyelectrolyte which consists of a collection of n monomers with sticky interactions in the full association limit.

and also

$$\Delta E = \frac{-e^2}{\varepsilon}\left[\sum_k \rho_k z_k \frac{\Gamma^T z_k + \eta^T \sigma}{1+\Gamma^T\sigma} - \frac{\rho_p}{\sigma}\sum_{k=1}^{n-1}\sum_{i=1}^{n-k}\frac{z_k \mathcal{X}^0_{k+i}}{(i+1)(1+\Gamma^T\sigma)^i} - \frac{\rho_p}{\sigma}\sum_{k=2}^{n}\sum_{i=1}^{k-1}\frac{z_k \mathcal{X}^0_{k-i}}{(i+1)(1+\Gamma^T\sigma)^i}\right] \tag{67}$$

From Bernard and Blum [18]:

$$\left[\Gamma^T\right]^2 = \frac{\pi e^2\beta}{\varepsilon}\left[\rho_p\left(\mathcal{X}^0_1\right)^2\left(n + 2\sum_{i=1}^{n-1}(n-i)y^i\right) + \sum_{c=n+1}^{m}\rho_c\left(\mathcal{X}^0_c\right)^2\right] \tag{68}$$

with

$$y = \frac{1}{1+\Gamma^T\sigma_1},$$

we get

$$\left[\Gamma^T\right]^2 = \frac{\pi e^2\beta}{\varepsilon}\left[\rho_p\left(\mathcal{X}^0_1\right)^2\left\{n - \frac{2y}{1-y}\left(n - 1 - \frac{y-y^n}{1-y}\right)\right\} + n\rho_p\left(\mathcal{X}^0_c\right)^2\right]. \tag{69}$$

Consider now the restricted case where all diameters are equal. Then, since $\rho_c = n\rho_p$

$$\mathcal{X}^0 = \frac{z - \sigma^2\eta^T}{1+\Gamma^T\sigma} = y\left(z - \sigma^2\eta^T\right), \tag{70}$$

$$\left[\Gamma^T\right]^2 = \frac{\pi e^2\beta}{\varepsilon}\rho y^2\left(z - \sigma^2\eta^T\right)^2\left[n(n+1) + (y-1)\sum_{i=1}^{n-1} y^{i-1}(n-i)(n-i+1)\right], \tag{71}$$

$$\eta^T = \frac{\pi\rho_p\sigma z}{\Delta\Omega^T}\left(\sum_{i=1}^{n-1}\frac{n-i}{i+1}y^{i+1}\right) \tag{72}$$

and

$$\Omega^T = 1 + \frac{\pi}{\Delta}\sigma^3\rho_p\left(ny + \sum_{i=1}^{n-1}\frac{n-i}{i+1}y^{i+1}\right). \tag{73}$$

We get

$$\eta^T = \frac{z}{\sigma^2}\left[\frac{ny\mathcal{B}_n(y)}{1+ny\mathcal{B}_n(y)}\right] \tag{74}$$

with

$$ny\mathcal{B}_n(y) = \frac{\pi}{\Delta}\rho\sigma^3\sum_{i=1}^{n-1}\frac{n-i}{i+1}y^{i+1}\,. \tag{75}$$

We list a few explicit expressions:

$$\begin{aligned}\mathcal{B}_1(y) &= 0,\\ \mathcal{B}_2(y) &= \frac{\pi}{\Delta}\rho\sigma^3 y,\\ \mathcal{B}_3(y) &= \frac{2\pi}{3\Delta}\rho\sigma^3 y(3+y),\\ \mathcal{B}_4(y) &= \frac{\pi}{2\Delta}\rho\sigma^3 y[6+y(3+y)].\end{aligned}$$

Finally, putting it all together we get from eq. (71)

$$\left(1+\Gamma^T\sigma\right)^2\left[\Gamma^T\right]^2 = \frac{\kappa_n^2}{4} \tag{76}$$

with

$$\kappa_n^2 = \rho_p\frac{4\pi\beta e^2z^2}{\varepsilon}\frac{n(n+1)}{(1+ny\mathcal{B}_n(y))^2}\left[1+(y-1)\sum_{i=1}^{n-1}y^{i-1}\left(1-\frac{i}{n}\right)\left(1-\frac{i}{n+1}\right)\right]. \tag{77}$$

For low concentrations $y \simeq 1$:

$$\kappa_n^2 \simeq \rho_p\frac{4\pi\beta e^2z^2}{\varepsilon}n^2\,. \tag{78}$$

For high concentrations $y \simeq 0$:

$$\kappa_n^2 \simeq \rho_p\frac{4\pi\beta e^2z^2}{\varepsilon}n\,. \tag{79}$$

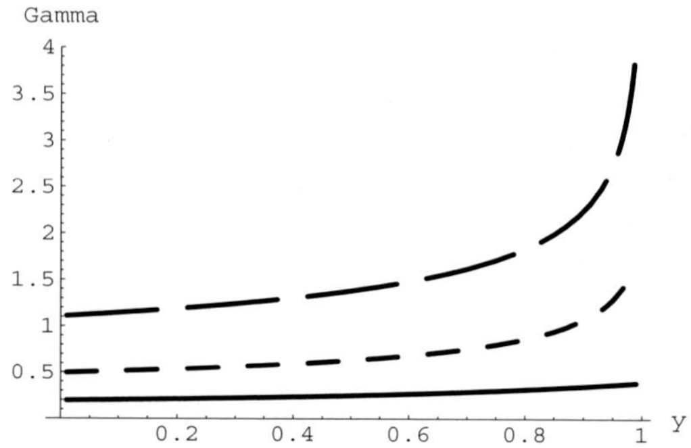

Fig. 5. – The BIMSA screening parameter Γ^T from eq. (75) as a function of y for three different values of the chain length $n = 10, 100$ and 1000.

We remember that the formal solution of eq. (76) is

$$\Gamma_n^T = \frac{1}{2\sigma}\left[\sqrt{1+2\kappa_n\sigma} - 1\right], \tag{80}$$

which now is the starting point of a very rapidly converging iteration scheme. This is so because the correction factor in eq. (77) is bound by the low limiting expression (79) and the high limit (78). The results for different chain lengths are shown in fig. 5.

In the high-density limit the effect of the association is canceled by the crowding effect. This is similar to what happens for the dimers in eq. (62) when $\alpha = 1$. In that case we get again the equation which is that of the fully dissociated system (see eqs. (1), (2)).

7. – Conclusions

The Mean Spherical Approximation (MSA) is the linearized Poisson-Boltzmann theory solved including *all* the excluded-volume constraints for all particles. The theory is simple because all the properties of a large class of systems can be expressed in terms of a single screening parameter Γ, which in the case of a binary mixture of equal-size ions is calculated by solving a simple quadratic equation:

$$\Gamma = \frac{1}{2\sigma}\left[\sqrt{1+2\kappa\sigma} - 1\right], \qquad \kappa^2 = \frac{4\pi\beta e^2}{\varepsilon}\sum_{j=1}^{m} \rho_j z_j^2 .$$

This formalism is extended to arbitrary flexible polyelectrolytes: For a simple chain of n equal hard ions with counterions only one gets

$$\Gamma_n^T = \frac{1}{2\sigma}\left[\sqrt{1+2\kappa_n\sigma} - 1\right], \qquad \kappa_n^2 = \kappa^2\mathcal{F}_n(\kappa_n), \qquad \mathcal{O}(n) < \mathcal{F}_n(\kappa_n) < \mathcal{O}(n^2),$$

where $\mathcal{F}_n$ is a function that interpolates between the high- and low-density exact limits. This result is the solution of the Binding Mean Spherical (BIMSA) approximation for sticky charged hard spheres with triplet exclusion, which satisfies the low-density high-temperature Debye-Hückel (DH) limits, the high-density Onsager limits, the low-density high-coupling association limits, and also includes hard-sphere triplet correlations. The excess thermodynamic functions, the internal energy E, the entropy S have the same simple form for all these cases, and are functions of a single screening parameter Γ_n^T, which however depends on all the sticky interactions.

The agreement with computer simulations is quite good [27].

* * *

This research was supported by DOE grant EDE-FG02-03ER15422.

REFERENCES

[1] CHEUNG M. S., GARCIA A. E. and ONUCHIC J. N., *Proc. Natl. Acad. Sci. USA*, **99** (2002) 685.
[2] BLUM L., *Mol. Phys.*, **30** (1975) 1529.
[3] BLUM L. and HØYE J. S., *J. Phys. Chem.*, **81** (1977) 1311.
[4] BLUM L., *J. Stat. Phys.*, **22** (1980) 661.
[5] BLUM L. and HØYE J. S., *J. Stat. Phys.*, **16** (1977) 399.
[6] GINOZA M., *J. Phys. Soc. Jpn.*, **55** (1986) 95; 1782.
[7] BLUM L., VERICAT F. and HERRERA J. N., *J. Stat. Phys.*, **66** (1992) 249.
[8] BLUM L. and HERRERA J. N., *Mol. Phys.*, **96** (1999) 821.
[9] BLUM L. and UBRIACO M., *Mol. Phys.*, **98** (2000) 829.
[10] BLUM L. and HERNANDO J. A., *J. Condens. Matter Phys.*, **6** (2003) 447.
[11] PERCUS J. K. and YEVICK G. J., *Phys. Rev.*, **110** (1958) 1.
[12] LEBOWITZ J. L. and PERCUS J. K., *Phys. Rev.*, **144** (1966) 251.
[13] WAISMAN E. and LEBOWITZ J. L., *J. Chem. Phys.*, **52** (1970) 4307.
[14] WAISMAN E., *Mol. Phys.*, **25** (1973) 45.
[15] BLUM L., KALYUZHNYI YU. V., BERNARD O. and HERRERA J. N., *J. Phys. Condens. Matter*, **8** (1996) A143; BLUM L., HOLOVKO M. F. and PROTSYKEVYCH I. A., *J. Stat. Phys.*, **84** (1996) 191; PROTSYKEVYCH I. A., KALYUZHNYI YU. V., HOLOVKO M. F. and BLUM L., *J. Mol. Liquids*, **73-74** (1997) 1; KALYUZHNYI YU. V., BLUM L., HOLOVKO M. F. and PROTSYKEVYCH I. A., *Physica A*, **236** (1997) 85.
[16] VON SOLMS N. and CHIEW Y. C., *J. Chem. Phys.*, **111** (1999) 4839.
[17] VON SOLMS N. and CHIEW Y. C., *J. Stat. Phys.*, **100** (2000) 267.
[18] BERNARD O. and BLUM L., *J. Chem. Phys.*, **112** (2000) 7227.
[19] ONSAGER L., *J. Phys. Chem.*, **43** (1939) 189.
[20] ROSENFELD Y. and BLUM L., *J. Chem. Phys.*, **85** (1986) 1556.
[21] BLUM L. and ROSENFELD Y., *J. Stat. Phys.*, **63** (1991) 1177.
[22] BLUM L. and BERNARD O., *J. Stat. Phys.*, **79** (1995) 569.
[23] BERNARD O. and BLUM L., *J. Chem. Phys.*, **104** (1995) 569.
[24] JIANG J. W., BLUM L., BERNARD O. and PRAUSNITZ J. M., *Mol. Phys.*, **99** (2001) 1121.
[25] STEVENS M. J. and KREMER K., *J. Chem. Phys.*, **103** (1995) 1669.
[26] ORKULAS G., KUMAR S. K. and PANAGIOTOPOULOS A. Z., *Phys. Rev. Lett.*, **90** (2003) 048303.
[27] BLUM L., GRUBER CH., LEBOWITZ J. L. and MARTIN PH. A., *Phys. Rev. Lett.*, **48** (1982) 1769.

[28] WERTHEIM M. S., *J. Stat. Phys.*, **35** (1984) 19; **42** (1986) 459; 477.
[29] WERTHEIM M. S., *J. Chem. Phys.*, **85** (1985) 2929; **87** (1987) 7323; **88** (1988) 1214.

Electrostatic applied to biological systems

M. C. Barbosa

Instituto de Física, UFRGS - 91501-970, Porto Alegre, RS, Brazil

1. – Description

Electrostatic plays a relevant role in physics, chemistry and biology. Molecular biology functions in water at room temperature. This is quite a limited temperature range for a physicist if compared with low-temperature physics where a large energy scale can be scanned. However, the presence of long-range interactions in the biological structures gives rise to a rich variety of scales in the problem that ultimately leads to very interesting phenomena. In this lecture, we will review the recent developing in understanding one of these phenomena.

More specifically, let us consider a charged rod, the macroion in the solution of its counterions. Due to the electrostatic interactions, the small ions are bound to the big ion screening the interactions between like-charged rods. When the electrostatic interaction between polyion and the counterion is high enough, the counterions are bound to the polyion even in the infinite dilution limit. This is the case of segments of DNA, modeled as rigid rods, that exhibit condensation at room temperature in water.

These phenomena have many practical applications. Foremost among these is the technology of dressing DNA with cations [1], positively charged starlike polymers called dendrimers [2-4], liposomes in order to produce complexes with DNA. This facilitates the gene delivery through a negative cell surface membrane [5]. Also, one can manufacture nanowires by attaching positively silver or gold colloids to DNA [6].

Besides being strongly bound to DNA, multivalent ions repeal each other forming a correlated liquid. As a result, condensation is enhanced and in the presence of salt the charge of the polyion is reversed. Moreover, in dense polyion solutions, the counterions

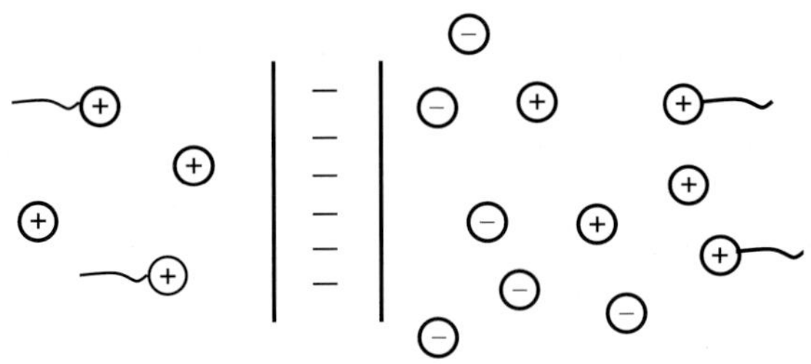

Fig. 1. – The model.

belonging to different macroions correlate so as to create an effective attractive force between the like-charged polyions.

This like-charge attraction manifests itself in *in vitro* formation of toroidal aggregates of concentrated DNA [7, 8], similar to the one observed in bacteriophasage heads [9], and in bundle formation of F-actin and tobacco mosaic virus [10]. The enhancement of condensation and consequent charge reversal [11-13] can be seen in the formation of the chromatin in eukariotic cells. Nature uses proteins having large positive charge, the histones. The histones are released from the protein and associate to the DNA in a necklace shape.

Although the phenomena described above are quite complex, we can go along describing them by employing surprisingly simple models and theories. Particularly, in this lecture we will show that a lot of physics and correlations are within the Debye-Hückel (DH) theory, an old theory introduced primarily to study electrolytes 80 years ago [14].

2. – DNA in the presence of monovalent counterions and surfactants

Let us consider the following system, illustrated in fig. 1, a solution of DNA strands of length L and diameter a_{p}, monovalent surfactant and monovalent salt. In aqueous solution, the polyions become ionized resulting in a negative charge $-Zq$ uniformly distributed with separation $b = L/Z$. The solvent, water, is modeled as a continuous medium of dielectric constant D. The ions of the salt are completely dissociated, forming an equal number of positive and negative ions. Similarly, the surfactants are assumed to be fully dissociated producing negative monovalent coions and polymeric chains with a monovalent cationic head group. For simplicity, all the counterions and coions are treated as identical, independent of the molecules from which they were derived. The electrolytes are depicted as hard spheres of diameter a_{c} and charge $\pm q$ and the surfactant is modeled as a polymer of s monomers, each one considered as a rigid sphere of diameter a_{c} with the head monomer carrying charge $+q$. The interaction between the hydrophobic tails is short ranged and characterized by the parameter χ. The density of DNA strands is $\rho_{\mathrm{p}} = N_{\mathrm{p}}/V$, the density of the monovalent salt is $\rho_{\mathrm{s}} = N_{\mathrm{s}}/V$ and the density of amphiphiles is $\rho_{\mathrm{a}} = N_{\mathrm{a}}/V$.

In a dilute solution, the polyions will arrange themselves uniformly in the solution, while the microions (counterions, salt and surfactant) will stay in the vicinity of the polyion. This picture suggests that there is strong positional correlation between the polyions and strong positional correlation among the microions. In this sense, the strong electrostatic interactions between the polyions, the counterions, salt cations and surfactant heads leads to the formation of complexes, which in thermodynamic equilibrium will be made up of one polyion, n_c monovalent counterions and n_a monovalent surfactants. We do not consider the effects of polydispersity in the size of the complexes, since it does not affect the final result. Due to the association and to the charge conservation, there are only two free quantities and so

$$(1)\quad \rho_\text{c} = (Z - n_\text{c})\rho_\text{p} + \rho_\text{s}\,, \quad \rho_{\text{a}+} = \rho_\text{a} - n_\text{a}\rho_\text{p}\,, \quad \rho_- = \rho_\text{s} + \rho_\text{a}\,, \quad \rho_f = \rho_\text{c} + \rho_{\text{a}+} + \rho_-\,,$$

where ρ_c is the density of free monovalent counterions, $\rho_{\text{a}+}$ is the density of free amphiphiles and ρ_- is the density of negative ions.

The objective of this theory is to determine the number of counterions n_c and surfactants n_a associated to each DNA strand. For this, we construct the Helmholtz free energy of the system and minimize it. The relevant contributions for the Helmholtz free energy density are two, the electrostatic and the entropic, namely:

$$(2)\qquad \frac{F}{V} = f_\text{el} + f_\text{ent}\,.$$

f_el is the electrostatic free energy between the polyion and the counterions and can be derived in the spirit of the DH theory for electrolytes as follows. Suppose we choose a coordinate system so that it is centered on one of the polyions with the z-axis along the polyion's major axis. The electrostatic potential for $r > a$ satisfies the Laplace equation

$$(3)\qquad \nabla\phi(r) = 0\,,$$

while the potential for $r < a$ satisfies the Poisson equation, namely:

$$(4)\qquad \nabla^2\phi(r) = \frac{4\pi}{\epsilon}\rho_q(r)$$

with a charge distribution given by

$$(5)\qquad \rho_q = -Zq\rho_\text{p} + q\rho_q e^{-\beta q\phi(r)}\,.$$

In order to solve this equation, we would like to linearize it. Since the electrostatic interactions are quite strong in the vicinity of the polyion, the polyion-counterions correlation in the vicinity of the polyion must be taken into account. This can be done by introducing the idea of polyion-counterion association. In this context the polyelectrolyte solution can be thought of as composed of free unassociated counterions, salt and surfactant, ρ_q and clusters made of one polyion, n_c counterions and ρ_a surfactant molecules, as illustrated in fig. 2. Once the cluster concept is introduced in the theory, one can expand the

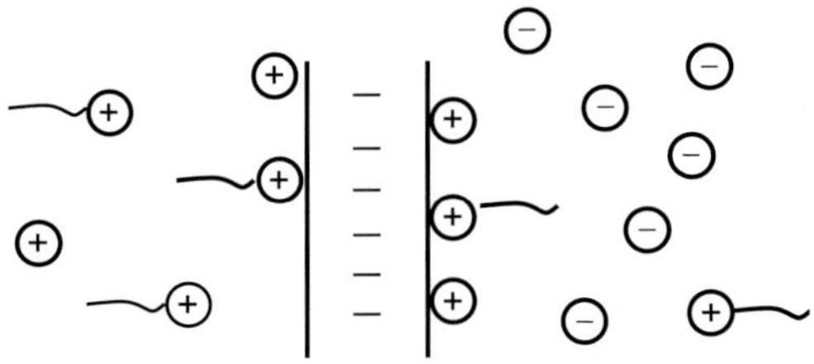

Fig. 2. – Association.

exponential in eq. (5) and use the linearized form in the Poisson equation which gives rise to the Poisson-Boltzmann equation, namely:

$$\nabla^2 \phi(r) = \kappa^2 \phi(r), \tag{6}$$

where

$$\kappa = \sqrt{\frac{4\pi q^2 \rho_f}{k_B T \epsilon}}\,. \tag{7}$$

The solution of the linearized equation outside the cluster at $r > a$ is given by

$$\phi(r) = \frac{2\lambda}{\epsilon} \frac{K_0(\kappa r)}{\kappa a K_\nu(\kappa a)}\,, \tag{8}$$

where K_ν is the modified Bessel function of order ν and $\lambda = (Z - n_c - n_a)/L$. For $r < a$ the potential is given by

$$\phi(r) = \ln \frac{r}{a} + \frac{2\lambda}{\epsilon} \frac{K_0(\kappa a)}{\kappa a K_\nu(\kappa a)}\,. \tag{9}$$

The electrostatic free energy due to the free-ions–cluster interaction can be obtained from the electrostatic energy

$$u_p = \frac{1}{2} \int \left[\rho_q(r) + q_p(r)\right] \phi(r) \mathrm{d}^3 r \tag{10}$$

together with the Debye charging process, namely

$$\begin{aligned} \beta f_{el} &= \int_0^1 \frac{u_q}{2\lambda} \mathrm{d}\lambda \\ &= -\frac{(Z-n)^2 \xi}{Z} \left[\frac{2\gamma_E - 1}{2} + \ln(\kappa a)\right] \rho_p^*\,, \end{aligned} \tag{11}$$

where $\xi \equiv \beta q^2 / Da$ and $\rho^* = \rho a^3$.

The calculation of the entropic contribution can be obtained with the aid of Flory [15] theory of mixing. The free energy is a sum of ideal free energies of various species, namely

$$\beta f_{\text{ent}} = \sum \left[\rho_{\text{s}} - \rho_{\text{s}} \ln \left(\frac{\phi_{\text{s}}}{\zeta_{\text{s}}} \right) \right], \tag{12}$$

where s represents the different species and ζ_{s} is the internal partition of the species s. In the case of the particles without structure, the internal partition function $\zeta_- = \zeta_{\text{c}} = \zeta_{\text{a}+} = 1$. The volume fraction ϕ_{s} of the different species are:

$$\begin{aligned} \phi_{\text{p}} &= \frac{\pi \rho_{\text{p}}^*}{4(a/L)} \left(\frac{a_{\text{p}}}{a} \right)^2 + \frac{Z \pi \rho_{\text{p}}^*}{6} (s_{\text{a}} m_{\text{a}} + m_{\text{c}}) \left(\frac{a_{\text{c}}}{a} \right)^3, \\ \phi_{\text{c}} &= \rho_{\text{c}}^* \frac{\pi}{6} \left(\frac{a_{\text{c}}}{a} \right)^3, \\ \phi_{\text{a}+} &= \frac{s_{\text{a}} \pi \rho_{\text{a}+}^*}{6} \left(\frac{a_{\text{c}}}{a} \right)^3, \\ \phi_- &= \frac{\pi \rho_-^*}{6} \left(\frac{a_{\text{c}}}{a} \right)^3. \end{aligned} \tag{13}$$

Here we introduce the fractions of counterions, $m_{\text{c}} = n_{\text{c}}/Z$, and surfactant, $m_{\text{a}} = n_{\text{a}}/Z$, associated to each DNA strand. Accounting for all these terms, the entropic contribution becomes

$$\begin{aligned} \beta f_{\text{ent}} = \rho_{\text{p}} \ln \left(\frac{\phi_{\text{p}}(1 + m_{\text{c}} + m_{\text{a}})}{\zeta_{\text{cl}}(1 + m_{\text{c}} + m_{\text{a}} s_{\text{a}})} \right) - \rho_{\text{p}} + \\ + \rho_{\text{c}} \ln \phi_{\text{c}} - \rho_{\text{c}} + \rho_{\text{a}_+} \ln \frac{\phi_{\text{a}_+}}{s_{\text{a}}} - \rho_{\text{a}_+} + \\ + \rho_- \ln \phi_- - \rho_-, \end{aligned} \tag{14}$$

where the entropy for the free surfactants arises from the Flory theory for polymers [15] which is also the basis of the entropic contribution for the complex [16]. The internal partition function of the complex, ζ_{cl}, can be calculated by modeling the DNA by a one-dimensional lattice with Z sites. If the number of ions associated to each site can be only zero or one, this problem becomes equivalent to finding the free energy of a one-dimensional array with three different states: empty, an associated surfactant, an associated counterion.

The solution of this model is not trivial, so we employ the mean-field Gibbs-Bogoliubov-Feynman inequality. The resulting partition function is given by

$$\begin{aligned} -\ln \zeta_{\text{c}}[m_{\text{c}}, m_{\text{a}}] = \xi K \left[\frac{{Z_{\text{c}}}^2}{Z^2} - 1 \right] + \beta \chi (Z-1) m_{\text{a}}^2 \\ + Z m_{\text{c}} \ln m_{\text{c}} + Z m_{\text{a}} \ln m_{\text{a}} \\ + Z(1 - m_{\text{c}} - m_{\text{a}}) \ln(1 - m_{\text{c}} - m_{\text{a}}), \end{aligned} \tag{15}$$

where $\xi \equiv \beta q^2/Da$ is the Manning parameter, $K = Z[\psi(Z) - \psi(1)] - Z + 1$, and $\psi(n)$ is the digamma function.

The equilibrium configuration of the system is found by the minimization of the Helmholtz free energy, leading to two equations, namely

$$\frac{\partial f}{\partial m_{\rm c}} \delta m_{\rm c} = 0\,, \qquad \frac{\partial f}{\partial m_{\rm a}} \delta m_{\rm a} = 0\,. \tag{16}$$

Solving this system of two equations, it is possible to obtain the values of $m_{\rm c}$ and $m_{\rm a}$.

We define a "surfoplex" as a complex in which almost all of the DNA's phosphate groups are neutralized by the associated surfactant molecules. As mentioned earlier, we are interested in the minimum amount of cationic surfactant needed to transform naked DNA into surfoplexes. To this effect, we study the dependence of the number of condensed surfactant molecules on the bulk concentration of surfactant $\rho_{\rm a}$. In order to evaluate the relevance of hydrophobic interactions between the amphiphiles, the hydrophobic parameter was varied from 0 to $\beta\chi = -3.5$. The effect of addition of high concentrations of salt to the system was analyzed by varying the amount of salt added to the system $\rho_{\rm s}$.

Figures 3 and 4 illustrate the surfactant binding isotherms, $\beta_{\rm a} = m_{\rm a}$ and $\beta_{\rm c} = m_{\rm c}$, as a function of total amphiphilic concentration for hydrophobicity parameter $\beta\chi = -3.5$, salt concentrations $\rho_{\rm s} = 5\,{\rm mM}$, $18\,{\rm mM}$, DNA concentration $\rho_{\rm p} = 2\,\mu{\rm M}$ for a DNA segment with 220 base pairs. At very low densities of surfactant, the condensation is dominated by the monovalent ions, since the amphiphilic molecules gain more entropy, and thus lower the total free energy, by staying free. However, as the concentration of surfactant increases, the gain in electrostatic energy due to condensation and to the hydrophobicity of the amphiphilic molecules wins over the entropy. Even though the use of monovalent ions guarantees the decrease of the effective charge of the DNA, it does not revert its charge. In order to get charge reversal multibinding and multivalent ions are required. In the next section we will include these two ingredients.

3. – DNA in the presence of multivalent ions

Let us now consider another system made of $\rho_{\rm p}$ DNA molecules, $\rho_{\rm M}$ monovalent molecules, $\rho_{\rm s}$ divalent-salt molecules, $\rho_{\rm c} = Z\rho_{\rm p}$ monovalent counterions and $\rho_{\rm a}$ monovalent surfactant. The strong electrostatic interactions between the polyions, the counterions, salt cations and surfactant heads lead to the formation of complexes, which in thermodynamic equilibrium will be made up of one polyion, $n_{\rm c}$ monovalent counterions, $n_{\rm s}$ divalent salt ions and $n_{\rm a}$ monovalent surfactants. We do not consider the effects of polydispersity in the size of the complexes, since it does not affect the final result. Due to the association and to the charge conservation, there are only two free quantities and so one can use eq. (1).

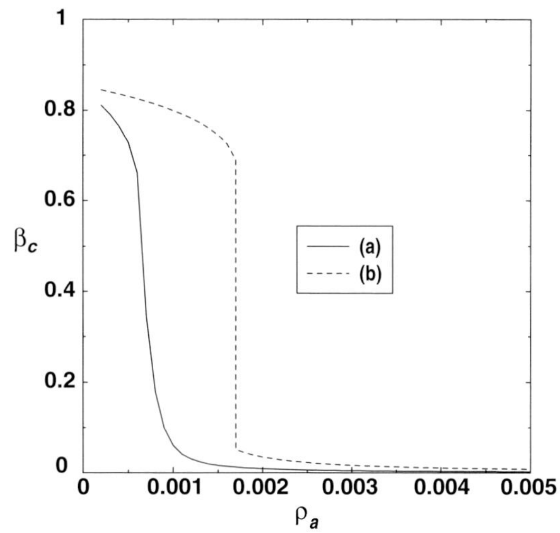

Fig. 3. – The DNA-counterion binding isotherm for various concentrations of added salt: (a) 5 mM, (b) 18 mM.

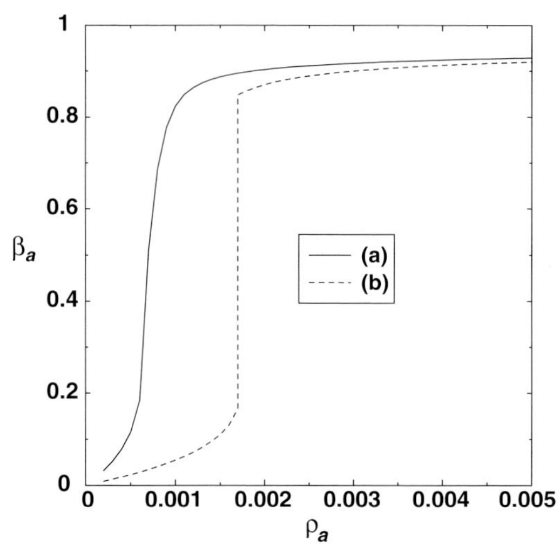

Fig. 4. – The DNA-surfactant binding counterions isotherm for various concentrations of added salt: (a) 5 mM, (b) 18 mM.

The theory developed in the previous section did not account for the correlations between the ions. For dilute systems, the correlation between the free ions is irrelevant, however, the correlations between the associated ions is indeed important, particularly in the case of multivalent ions. So, it becomes important to compute the correlation between the divalent ions of salt. To begin with, let us assume that the n_s associated divalent salt

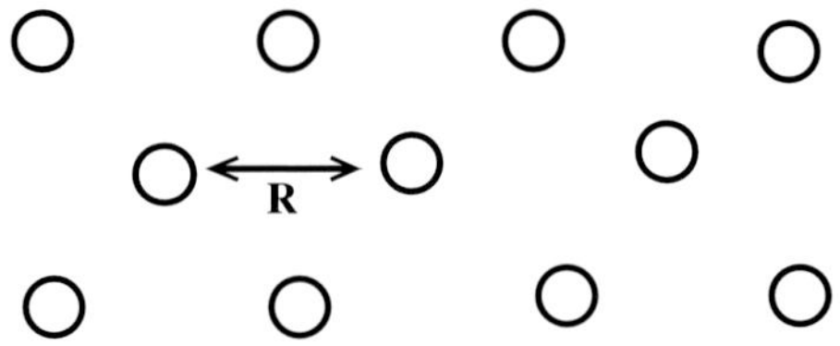

Fig. 5. – Two-dimensional Wigner crystal.

ions are in a planar surface and not in the cylindrical geometry. The dependence on the shape of the polyions for this contribution of the free energy is not dramatic, particularly if the polyion radius is bigger than the distance between the charged groups along the polymer. In this case, the two-dimensional concentration of divalent ions is $(n_{\mathrm{s}})/(2\pi a_{\mathrm{p}}L)$ and the surface area per ion can be characterized by R such that $\pi R^2 = 2\pi a_{\mathrm{p}}L/(n_{\mathrm{s}})$ as illustrated in fig. 5. Thus $R = \sqrt{2a_{\mathrm{p}}L/(n_{\mathrm{s}})}$ and $\Gamma = 4q^2/(k_{\mathrm{B}}T\epsilon R)$, where Γ is the plasma parameter, a Coulomb coupling constant that scales quadratically with the valence. For $\Gamma \approx 1$, correlations are not so relevant. This is the case of monovalent ions. For $\Gamma \gg 1$ the screening atmosphere is narrowly confined as a two-dimensional liquid of classical charged particles in a uniform neutralizing background located at the surface of the polyion. This liquid, also called one-component plasma, at zero temperature acquires the minimal energy state of a Wigner crystal showed in fig. 5. The correlational energy associated to this plasma made of n_{s} ions of valence v is given by

$$e(n_{\mathrm{s}}) = \frac{-3.92q^2}{\epsilon R}\,. \tag{17}$$

Applying the charging process to this energy one gets the free energy associated with the correlational energy of divalent salt ions given by

$$\beta f(n_{\mathrm{s}}) = -0.98\left(\frac{(n_{\mathrm{s}})L/a_{\mathrm{p}}}{2\pi}\right)^{1/2}(n_{\mathrm{s}})\rho_{\mathrm{p}}^{*}\,. \tag{18}$$

Once we have the proper correlational energy given by the above equation, we have to address another important point. In the case of surfactants with the high hydrophobicity parameter, the energy gain when two amphiphilic molecules are close is so high that instead of associating one surfactant per site it might occur that more than one amphiphilic molecule is close to a charge group [17]. In this case the strong electrostatic attraction between the polyions, counterions, and amphiphiles leads to the formation of complexes consisting of one polyion, n_{c} counterions, and n_{a} amphiphilic molecules. We shall assume that to each phosphate group of the DNA molecule can be associated at most one counterion or $l_{\max}$ surfactants. This assumption seems to be quite reasonable in view of the fact that the electrostatic repulsion between the counterions will prevent more than

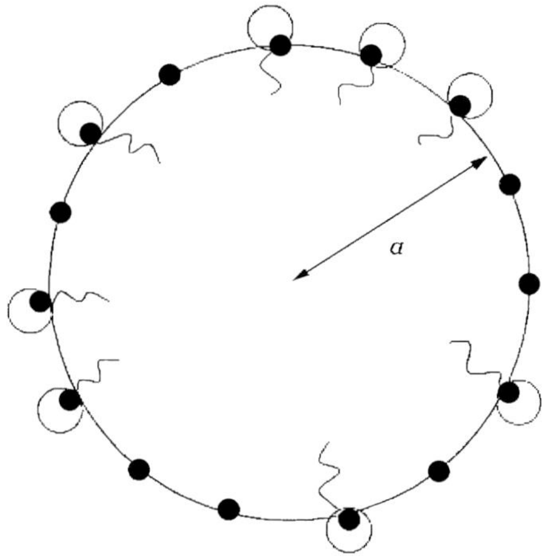

Fig. 6. – l surfactant molecules associated to one charged group forming an l-ring.

one counterion from condensing onto a given monomer. On the other hand, the gain in hydrophobic energy resulting from the close packing of the surfactant molecules might be able to overcome the repulsive electrostatic interaction between the surfactant head groups, favoring condensation of more than one surfactant on a given monomer. The l amphiphilic molecules form a "ring" of radius a around the central negative monomer of the DNA (see fig. 6). If we assume that most of the hydrocarbon chain of the associated surfactants is hidden inside the DNA molecule, the maximum number of surfactants in a ring can be estimated from the excluded-volume considerations, $l_{\max} = 2a/a_{\rm c}$, where $(a_{\rm p} + a_{\rm c})/2$ is the radius of the exclusion cylinder around a polyion.

At equilibrium, each site (monomer) of a polyion can be free or have one counterion or a ring of $l = 1, \ldots, l_{\max}$ surfactants associated to it. We define the surface coverage of counterions as $p_{\rm c} = n_{\rm c}/Z$, and the surface coverage of surfactant rings as $p_l = n_l/Z$, where $n_{\rm c}$ is the number of condensed counterions and n_l is the number of rings containing l surfactants. Each polyion has a distribution of rings containing from one to $l_{\max}$ surfactants. We shall neglect the polydispersity in the size of the complexes, assuming that all the complexes have $n_{\rm c}$ counterions and $n_{\rm s}$ amphiphilic molecules —in rings of $\{p_l\}$. The aim of the theory is to determine the characteristic values of $m_{\rm a}$, $m_{\rm c}$, and the surface coverage by rings. To accomplish this, the free energy of the DNA-surfactant system given by eq. (2),

$$\frac{F}{V} = f_{\rm el} + f_{\rm ent}^{\rm new} + f_{\rm WC}\,, \tag{19}$$

has to be minimized.

In this case, the electrostatic contribution is given by eq. (11) and the entropic contribution is given by

$$\beta f_{\mathrm{ent}}^{\mathrm{new}} = \rho_{\mathrm{p}} \ln \left(\frac{\phi_{\mathrm{p}}(1 + m_{\mathrm{c}} + m_{\mathrm{a}} + m_{\mathrm{s}})}{\zeta_{\mathrm{cl}}^{\mathrm{new}}(1 + m_{\mathrm{c}} + m_{\mathrm{a}} s_{\mathrm{a}} + m_{\mathrm{s}})} \right) - \rho_{\mathrm{p}} + + \rho_{\mathrm{c}} \ln \phi_{\mathrm{c}} - \rho_{\mathrm{c}} + \rho_{\mathrm{s}+} \ln \phi_{\mathrm{s}+} - \rho_{\mathrm{s}+} + \rho_{\mathrm{a}_+} \ln \frac{\phi_{\mathrm{a}_+}}{s_{\mathrm{a}}} - \rho_{\mathrm{a}_+} + + \rho_- \ln \phi_- - \rho_- . \tag{20}$$

The internal partition function of the complex, ζ_{cl}, can be calculated by modeling the DNA by a one-dimensional lattice with Z sites. If the number of ions associated to each site can be only zero or one, this problem becomes equivalent to finding the free energy of a one-dimensional array with four different states: empty, an associated surfactant, an associated counterion and a divalent associated salt.

The solution of this model is not trivial, so we employ the mean-field Gibbs-Bogoliubov-Feynman inequality. The resulting partition function can be found in ref. [17].

The equilibrium configuration of the system is found by the minimization of the Helmholtz free energy, leading to four equations, namely

$$\frac{\partial f}{\partial m_{\mathrm{c}}} \delta m_{\mathrm{c}} = 0$$

$$\frac{\partial f}{\partial m_{\mathrm{s}}} \delta m_{\mathrm{s}} = 0$$

$$\frac{\partial f}{\partial m_{\mathrm{a}}} \delta m_{\mathrm{a}} = 0 \tag{21}$$

$$\frac{\partial f}{\partial p_l} \delta p_l = 0 . \tag{22}$$

Solving this system of four equations, it is possible to obtain the values of m_{c} and m_{a} and the distribution $\{p_l\}$. We define a "surfoplex" as a complex in which almost all of the DNA's phosphate groups are neutralized by the associated surfactant molecules. As mentioned earlier, we are interested in the minimum amount of cationic surfactant needed to transform naked DNA into surfoplexes. To this effect, we study the dependence of the number of condensed surfactant molecules on the bulk concentration of surfactant ρ_{a}. In order to evaluate the relevance of hydrophobic interactions between the amphiphiles, the hydrophobic parameter was varied from 0 to $\beta\chi = -6$. The effect of addition of high concentrations of salt to the system was analyzed by varying the amount of salt added to the system ρ_{s}.

Figure 7 illustrates the case of monovalent surfactant, monovalent and divalent salt. The first graph contains the binding isotherms, $\beta_{\mathrm{a}} = m_{\mathrm{a}}$, $\beta_{\mathrm{c}} = m_{\mathrm{c}}$ and $\beta_{\mathrm{s}} = m_{\mathrm{s}}$, respectively, as a function of total amphiphilic concentration ρ_{a} for $\chi = -6k_{\mathrm{B}}T$. The density of DNA is $\rho_{\mathrm{p}} = 2\,\mu\mathrm{M}$ and the number of base pairs is 220, giving $z = 440$. The concentrations of added salt are: $\rho_{\mathrm{M}} = 5 \times 10^{-4}\,\mathrm{M}$, $\rho_{\mathrm{s}} = 1 \times 10^{-4}\,\mathrm{M}$. The distribution of m-divalent ions on the ring in a complex with n_{s} divalent ions associated is given by the

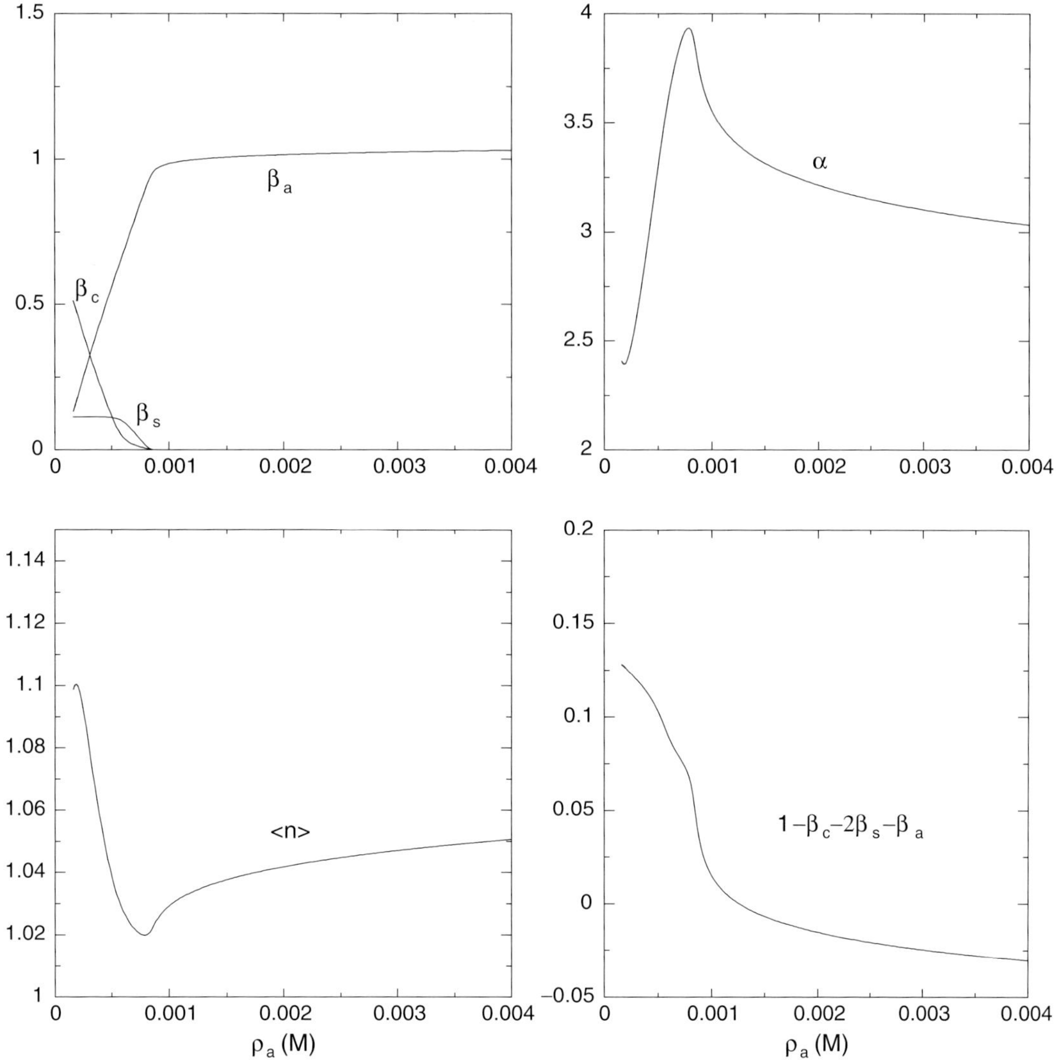

Fig. 7. – The adsorption isotherms for DNA solution from ref. [17].

form

$$p_n = \frac{e^{-\alpha n}}{n_{\rm s} \sum_{l=0}^{l_{\max}} l e^{-\alpha l}} \, . \tag{23}$$

The values of α for different surfactant concentrations are illustrated in fig. 7. The average number of divalent ions, $\langle n \rangle$, in each ring is also shown in fig. 7. Differently from the pure monovalent case where both the correlational term and the ring formation were not taken into account, one can observe from the charge of complex given by $1 - \beta_c - 2\beta_s - \beta_a$ that there is charge reversal. This means that for a certain density of surfactant the DNA is completely neutralized. Further increase of surfactant concentration reverts the charge.

4. – Conclusions

In conclusion, we have presented a simple theory of DNA, for monovalent and multivalent salt and surfactant solutions. Our results should be of direct interest to researchers working on the design of improved gene delivery systems. In particular, we find that addition of cationic surfactants leads to a cooperative binding. This binding happens far below the critical micell concentration. The effect is stronger for divalent ions where overcharging appears.

* * *

This work was supported by the Brazilian Science Agencies CNPq, Capes and CT-PETRO.

REFERENCES

[1] KABANOV A. V. and KABANOV V. A., *Bioconjugate Chem.*, **6** (1995) 7.
[2] TANG M. X., REDERMAN C. T. and SZOKA F. C., *Bioconjugate Chem.*, **7** (1996) 703.
[3] KABANOV V. A., SEREYEV V. G., PYSHKINA O. A., ZINCHENKO A. A., ZEZIN A. B., JOOSTEN J. G. H., BRACKMAN J. and YOSHIKAWA K., *Macromolecules*, **33** (2000) 9587.
[4] EVANS H. M., AHMAD A., PFOHL T., MARTIN A. and SAFINYA C. R., *Bull. Am. Phys. Soc.*, **46** (2001) 391.
[5] FELGNER P. L., *Sci. Am.*, **276** (1997) 102.
[6] BRAUN E., EICHEN Y., SIVAN U. and BEN-YOSEPH G., *Nature*, **391** (1998) 775.
[7] BLOOMFIELD V. A., *Biopolymers*, **31** (1991) 1471.
[8] BLOOMFIELD V. A., *Biopolymers*, **44** (1997) 269.
[9] SLIMENKO S. M., *J. Mol. Biol.*, **23** (1967) 523.
[10] TANG J. X., *Ber. Bunsenges. Phys. Chem.*, **100** (1996) 796.
[11] HA B.-Y. and LIU A., *Phys. Rev. Lett.*, **79** (1997) 1289.
[12] ARENZON J., STILCK J. and LEVIN Y., *Eur. Phys. J. B*, **12** (1999) 79.
[13] SHKLOVSKII B. I., *Phys. Rev. Lett.*, **82** (1999) 3628.
[14] DEBYE P. and HÜCKEL E., *Physik*, **24** (1923) 185.
[15] FLORY P., *Principles of Polymer Chemistry* (Cornell University Press, Ithaca, NY) 1971.
[16] KUHN P. S., LEVIN Y. and BARBOSA M. C., *Chem. Phys. Lett.*, **298** (1998) 51; KUHN P. S., BARBOSA M. C. and LEVIN Y., *Physica A*, **269** (1999) 278; KUHN P. S., LEVIN Y. and BARBOSA M. C., *Physica A* , **274** (1999) 8.
[17] MARCELO B. A. SILVA, PAULO S. KUHN and LIACIR S. LUCENA, *Physica A*, **296** (2001) 31.

Spherical crystallography: Virus buckling and grain boundary scars

D. R. NELSON

Lyman Laboratory of Physics, Harvard University - Cambridge, MA 02138, USA

1. – Introduction

Understanding ordered states on curved surfaces like the sphere requires dealing with "geometrical frustration". Order parameters which describe crystalline or hexatic order in the centers of mass of the microscopic degrees of freedom, or the projections of tilted molecules onto a local tangent plane, often minimize their energy by aligning in neighboring regions. It is natural to assume that this alignment can be achieved by parallel transport of the vector or tensor order parameter which represents the local order [1,2]. In a continuum description, the required parallel transport can be described by connection coefficients which couple to the displacement or phase angle much like a vector potential couples to the phase of a Type-II superconductor [3]. However, such alignment cannot be achieved everywhere when a nonzero Gaussian curvature is present. In particular, an attempt to produce a vector field aligned by parallel transport fails on a loop which contains a net integrated Gaussian curvature [4]. The Gaussian curvature embodied in the line integral of the connection coefficients around the loop resembles the frustration in a superconducting order parameter produced by a nonzero magnetic field [5].

Consider N particles with a repulsive hard-core interaction closed-packed on the surface of a sphere. If R is the radius of the sphere, $N \sim (R/a)^2$, where a is the hard-core diameter. Now imagine "triangulating" this sphere by connecting nearest-neighbor particles using the spherical analogue of the familiar Wigner-Seitz construction of solid state physics. The geometric frustration discussed above appears on a more microscopic level in Euler's theorem [6] on a sphere, which states the number of triangles T, edges E and vertices N of such a tesselation are related,

$$T - E + N = 2. \tag{1}$$

Each edge is shared with two triangles, but a single triangle has three edges, so $T = (1/3)(2E)$ and eq. (1) can be rewritten as $N - (1/3)E = 2$. If the coordination numbers at each node in this triangulation are denoted Z_j, then $E = (1/2)\sum_j Z_j$ and Euler's theorem constrains the deviations of the $\{Z_j\}$ from 6 on any surface with the topology of a sphere,

$$\sum_{j=1}^{N} (6 - Z_j) = 12. \tag{2}$$

Although identical particles with a hard-core repulsion in a plane typically pack into a triangular lattice with six-fold coordination, on a sphere eq. (2) shows that there must be at least 12 five-fold disclinations as well as an arbitrary number of six-coordinated sites. The 12 extra disclination defects (each representing a rotational discontinuity of $2\pi/6$) can be viewed as a consequence of the geometrical frustration associated with the Gaussian curvature of a sphere, which integrates to $4\pi = 12 \times (2\pi/6)$. One plausible candidate for the ground state of an array of particles on the sphere is a "superlattice" of these disclinations (rather like the Abrikosov flux lattice of a Type-II superconductor) arrayed at the vertices of a regular icosahedron.

Although this hypothesis is likely to be approximately correct for the ground state of N interacting particles on a sphere for moderate values of R/a, the very large energy associated with the twelve isolated disclinations makes this guess questionable when R/a becomes large. In fact, the extra energy due to the disclinations (relative to a perfect triangular lattice in flat space) grows like YR^2 (see sect. **2**), where Y is a two-dimensional Young's modulus. In this review, we briefly sketch two ways of reducing this energy. The first, relevant to hollow spherical shells which can be used to model viruses, invokes a buckling transition of the 12 disclinations and leads to a *faceted* icosahedral shape [7]. If κ is the bending rigidity of the spherical shell, the YR^2 behavior of the disclination energies is replaced by $\kappa \ln(R/a)$ for large R. Another mechanism takes over for particles packed on a large spherical droplet, where surface tension limits significant buckling of disclinations out the local tangent plane. Even in this case, the energy can still be reduced by introducing grain boundaries which radiate from the 12 disclinations [8]. The energy now grows like $E_c(R/a)$, where E_c is a dislocation core energy. This energy is larger than if buckling of the disclinations were allowed, but still smaller than the amount $\sim YR^2$ characteristic of isolated disclinations when R is large.

In sect. **2**, we review virus buckling, while sect. **3** presents a brief discussion of grain boundary "scars" which have now been observed experimentally [9].

2. – Theory of virus shapes

Understanding virus structures is a rich and challenging problem [10], with a wealth of new information now becoming available. Although traditional X-ray crystallography still allows the most detailed analysis [11], three-dimensional reconstructions of icosahedral viruses from cryo-electron micrographs are now becoming routine [12]. Electron microscope images of many identical viruses in a variety of orientations are used to reconstruct a three-dimensional image on a computer, similar to CT (computed tomography) scans in medical imaging. There are now, in addition, beautiful single-molecule experiments which measure the work needed to load a virus (bacteriophage $\phi 29$) with its DNA package [13]. It is of some interest to explore the elastic parameters and physical ideas which determine the shapes of viruses with an icosahedral symmetry.

The analysis of approximately spherical viruses dates back to pioneering work by Crick and Watson in 1956 [14], who argued that the small size of the viral genome requires identical structural units packed together with an icosahedral symmetry. These principles were put on a firm basis by Don Caspar and Aaron Klug in 1962 [15], who showed how the proteins in a viral shell (the "capsid") could be viewed as icosadeltahedral triangulations of the sphere by a set of pentavalent and hexavalent morphological units ("capsomers"). The viral shells (there can also be an outer envelope composed of membrane elements from the host cell) are characterized by a pair of integers (P, Q) such that the number of morphological units is $N = 10(P^2 + PQ + Q^2) + 2$. To get from one pentavalent capsomer to another, one moves P capsomers along a row of near-neighbor bonds on the sphere, turns 60 degrees and moves another Q step. The Euler theorem discussed in the introduction insures that the number of capsomers in five-fold environments is exactly 12. A simple icosahedron of 12 morphological units corresponds to $(1, 0)$ while soccer balls and C_60 fullerene molecules are $(1, 1)$ structures with 32 polygons. An icosadeltahedron with symmetry indices $(3, 1)$ is shown in fig. 1. The Polyoma virus (SV40) is a $(2, 1)$ structure with 72 capsomeres, while the much larger Adenovirus and *Herpes simplex* virus are $(5, 0)$ and $(4, 0)$ structures with 252 and 162 morphological units, respectively. Structures like that in fig. 1 with P and Q nonzero and $P \neq Q$ are chiral.

Note that the relatively small Polyoma virus (diameter 440 Å) is round (see fig. 2a), while the much larger *Herpes simplex* virus (diameter 1450 Å) has a more angular or faceted shape [16] (see fig. 2b). Faceting of large viruses is in fact a common phenomenon; the protein subunits of different viruses, moreover, have approximately the same molecular weight [17]. If these protein assemblies are characterized by elastic constants and a bending rigidity [18], it is interesting to ask how deviations from a spherical shape develop with increasing virus size.

The faceting of large viruses may in fact be caused by a buckling transition associated with the 12 isolated points of 5-fold symmetry. These singularities can be viewed as disclinations in an otherwise 6-coordinated medium. It is well known that the large strains associated with an isolated disclination in a *flat* disk of triangular lattice leads to

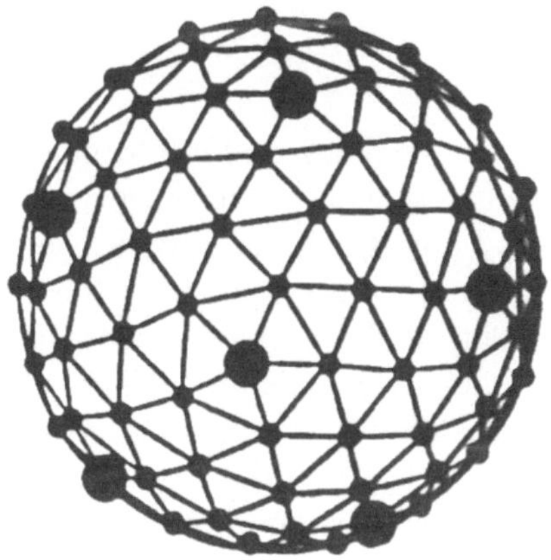

Fig. 1. – A right-handed (3, 1) triangulated net (icosadeltahedron) used to describe virus structure. The (1, 3) structure is left-handed.

buckling into a conical shape for [1, 19]

$$\frac{YR^2}{\kappa} \geq 154, \tag{3}$$

where Y is the two-dimensional Young's modulus, κ is the bending rigidity and R is the disk radius. The energy of a single 5-fold disclination with "charge" $s = 2\pi/6$ centered

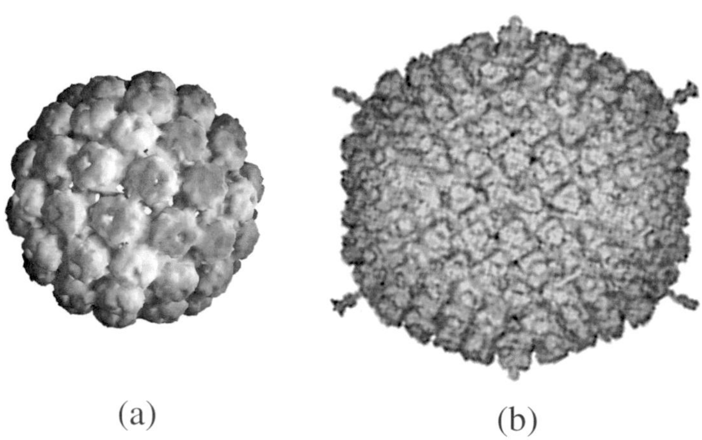

Fig. 2. – The Polyoma virus (a) is approximately spherical while the larger Adenovirus (b) is more faceted (not to scale) [16].

in a flat array of proteins of size R is approximately

$$E_5 \approx \frac{1}{32\pi} s^2 Y R^2. \tag{4}$$

However, above a critical buckling radius $R_b \approx \sqrt{154\kappa/Y}$, is there a conical deformation with [19]

$$E_5 \approx \left(\frac{\pi}{3}\right) \kappa \ln\left(\frac{R}{R_b}\right) + \frac{1}{32\pi} s^2 Y R_b^2 . \tag{5}$$

One might expect a similar phenomenon for 12 disclinations confined to a surface with a spherical topology. Indeed, the elastic energy for 12 disclinations on an undeformed sphere of radius R is has a form similar to eq. (4), namely [8]

$$E \approx 0.604\left(\frac{\pi Y R^2}{36}\right), \tag{6}$$

where the sphere radius R now plays the role of the system size. Although it seems highly likely that these 12 disclinations can lower their energy by buckling for large R, the nonlinear nature of the underlying elastic theory [18] leads to complex interactions between the resulting conical deformations. A boundary layer analysis of the von Karman equations for bent plates predicts anomalous scaling for the mean curvature in the vicinity of the ridges connecting conical singularities [20]. Interesting scaling behavior also arises in the vicinity of the apexes of the cones themselves [21]. Another intriguing physical realization of the buckling problem (with $R/a \gg 1$ so that boundary layers near ridges are important) lies in the faceting of lecithin vesicles at temperatures sufficiently low so that the lipid constituents have crystallized [22].

To search for a buckling transition in viral capsids or crystalline vesicles, one must study the strains and ground states of crystalline particle arrays with 12 disclinations in a spherical geometry [7]. The nonlinear Foppl-von Karman equations for thin shells with elasticity and a bending energy can be treated using a floating mesh discretization developed and studied extensively in the context of "tethered surface" models of polymerized membranes [23]. By taking the nodes of the mesh to coincide with the capsomers, even small viruses can be handled in this way, although any buckling transition will surely be smeared out unless R_b/a is large, where a is the spacing between morphological units. Ideas from continuum elastic theory will, of course, be most applicable for vesicles composed of many lipids and for large viruses —viruses with as many as 1692 morphological units have been reported [24].

There may be inherent limitations on the size of viral capsids that follow from the elastic properties of thin shells. Because larger viruses can accommodate more genetic material, larger sizes could confer an evolutionary advantage. If, however, large viruses buckle away from a spherical shape, the resistance of the capsid to mechanical deformation will degrade. A theory of buckled crystalline order on spheres allows estimates

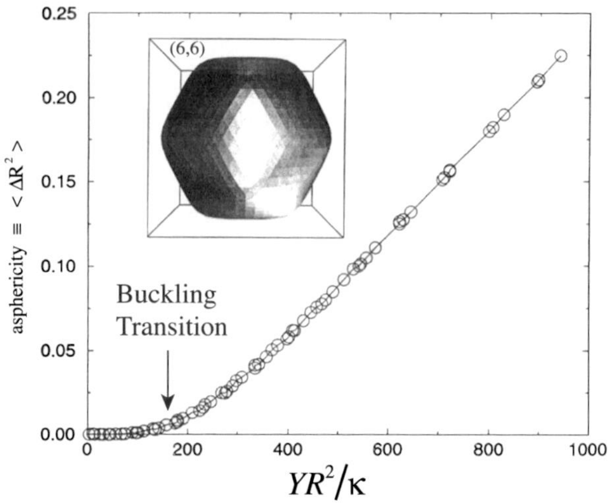

Fig. 3. – Mean square asphericity as a function of YR^2/κ for many different icosadeltahedra. The inset shows a $(6,6)$ structure with $YR^2/\kappa \approx 1000$.

of important elastic parameters of the capsid shell from structural data on the shape anisotropy [7]. Although some aspects of virus structure can be accounted for by the physics of shell theory, other features could be more relevant to cell recognition, avoidance of immune response, etc. Estimates of quantities such as the bending rigidity and Young's modulus of an empty viral shell might allow an understanding of deformations due to loading with DNA or RNA [13].

The results of recent investigations of buckling transitions in a spherical geometry using the methods of refs. [19] and [23] are illustrated in fig. 3 [7]. Spherical shells do indeed become aspherical as the "von Karman number" YR^2/κ increases from values of order unity to $YR^2/\kappa \gg 1$. The mean square "asphericity" (deviation from a perfect spherical shape) departs significantly from zero when YR^2/κ exceeds 154, the location of the buckling transition in flat space. Note that fits of buckled viruses or crystalline vesicles to this universal curve would allow an experimental estimate of the ratio Y/κ. More quantitative information on the buckled shape can be extracted by expanding the radius $R(\theta, \phi)$ in spherical harmonics,

$$R(\theta, \phi) = \sum_{\ell=0}^{\infty} \sum_{m=-\ell}^{\ell} Q_{\ell m} Y_{\ell m}(\theta, \phi), \tag{7}$$

and studying the rotationally invariant quadratic invariants allowed for icosahedral viruses or vesicles, namely

$$\hat{Q}_\ell = \sqrt{\frac{1}{2\ell+1} \sum_{m=-\ell}^{\ell} |Q_{\ell m}|^2} \Big/ Q_{00} \tag{8}$$

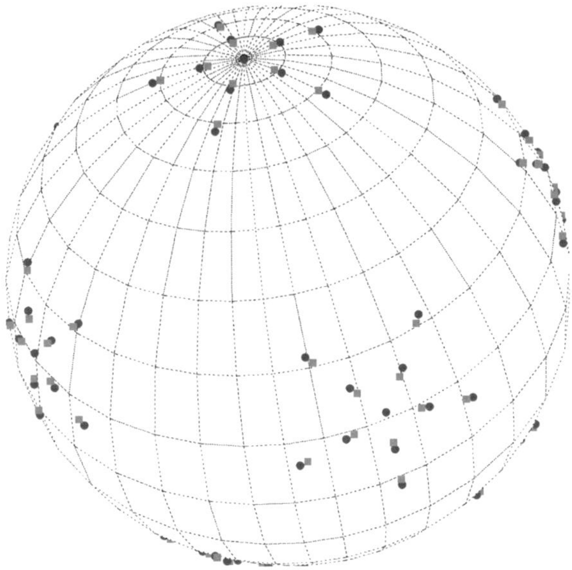

Fig. 4. – Dislocations emerging from 5-fold disclinations (circles) for a crystal of approximately 26000 particles on a sphere [4]. Each dislocation is a 5–7 (circle-square) disclination *pair* in this representation.

with $\ell = 0, 6, 10, 12, 16, 18, \dots$ [25]. Although *any* parameter set of the form $\{\hat{Q}_6, \hat{Q}_{10}, \hat{Q}_{12} \dots\}$ could be consistent with an icosahedral symmetry, all buckled objects describable by the theory of elastic shells should in fact lie on a universal curve parametrized by the value of YR^2/κ. Deviations from this curve are presumably be of some biological interest. In addition to exploring shape changes induced by an internal pressure (a virus loaded with $6\,\mu$m of DNA can have an internal pressure as large as 6 MPa! [13]), it would interesting to study the spontaneous membrane curvature produced by an asymmetric conical shape of the constituent proteins [26]. See ref. [7] for more details and quantitative fits to the shapes of the HK97 virus and a yeast virus.

3. – Grain boundary scars

When particles are packed on a surface where the restoring force is a surface tension, instead of a bending rigidity, buckling is suppressed. For example, water droplets in oil coated with colloidal particles remain spherical even in the presence of disclinations in the colloidal array. Remarkably, the classical ground state of interacting particles confined to a frozen surface like a sphere is still a matter of debate [27], especially in the limit of large sphere radius R compared to the particle spacing a. This problem was first posed for interacting electrons by J. J. Thomson in 1904 [28]. Physical realizations include multi-electron bubbles in superfluid helium [29], and "colloidosomes" [30], which are lipid bilayers or droplets coated with colloidal particles. Colloidosomes are potential delivery vehicles for drugs, flavors and fragrances. To study the ground states of such systems, one can analyze an effective free energy describing the physics of disclination arrays constrained to lie on an arbitrary two-dimensional surface. For spherical surfaces, the ground state for moderate R/a can be approximated by a triangular lattice with twelve 5-fold disclinations at the vertices of an icosahedron. Building on works by Alar Toomre [31] and by Michael Moore and collaborators [32], it is possible to construct a detailed theory of interacting defects on a frozen, spherical topography [8]. Although the buckling found in viruses is not allowed, the ground state for large R/a includes unusual finite-length grain boundaries. The methods of ref. [8] allow the direct minimization of the energy of, say, 26000 interacting particles to be replaced by the much easier problem of finding the ground state of 132 interacting disclination defects, (see fig. 4) with interactions determined by continuum elastic theory on the sphere. Excellent results can be obtained using Young's modulus and the dislocation core energy of particles in flat space as input parameters. Work by Bausch *et al.* [9] on colloidal particles absorbed onto water droplets in oil indicates a conventional ground state with just 12 disclination defects for small R/a, but with extra dislocations associated with grain boundaries for $R/a \geq 5$, corresponding to particle numbers $N \gtrsim N_c \approx 360$. Both this critical value of N and the number of excess dislocations in the grain boundary arms for $N > N_c$ can be estimated using the theory of ref. [8], with good agreement with experiment.

See ref. [33] for a discussion of the energetics of the crystalline state for large particle numbers and power law potentials when *both* buckling and grain boundary scars are forbidden.

* * *

The work described here was carried out in collaboration with M. BOWICK, J. LIDMAR, L. MIRNY and A. TRAVESSET. It is a pleasure to acknowledge these fruitful collaborations, as well as discussions with S. HARRISON and A. TOOMRE. This work was supported by the National Science Foundation, through Grant DMR-0231631 and in part through the Harvard Materials Research Laboratory, via Grant DMR-0213805.

REFERENCES

[1] NELSON D. R. and PELITI L., *J. Phys. (Paris)*, **48** (1987) 1085.
[2] DAVID F., in *Statistical Mechanics of Membranes and Surfaces*, edited by D. R. NELSON, T. PIRAN and S. WEINBERG (World Scientific, Singapore) 1989.
[3] TINKHAM T., *Introduction to Superconductivity*, 2nd edition (McGraw-Hill, New York) 1996.
[4] LANDAU L. D. and LIFSHITZ E. M., *The Classical Theory of Fields* (Pergamon, New York) 1971, Chapt. 11.
[5] See, *e.g.*, NELSON D. R., *Phys. Rev. B*, **28** (1983) 5515.
[6] COXETER H. S. M., *Introduction to Geometry* (Wiley, New York) 1969.
[7] LIDMAR J., MIRNY L. and NELSON D. R., *Phys. Rev. E*, **68** (2003) 051910.
[8] BOWICK M. J., NELSON D. R. and TRAVESSET A., *Phys. Rev. B*, **62** (2000) 8738 and references therein.

[9] BAUSCH A. R., BOWICK M. J., CACCIUTO A., DINSMORE A. D., HSU M. F., NELSON D. R., NIKOLAIDES M. G., TRAVESSET A. and WEITZ D. A., *Science*, **299** (2003) 1716.

[10] CHIU W., BURNETT R. M. and GARCEA R. L. (Editors), *Structural Biology of Viruses* (Oxford University Press, New York) 1997.

[11] HARRISON S. C., *Harvey Lectures*, **85** (1991) 127.

[12] BAKER T. S., OLSON N. H. and FULLER S. D., *Microbiol. Molec. Biol. Rev.*, **63** (1999) 862.

[13] SMITH D. E., TANS S. J., SMITH S. B., GRIMES S., ANDERSON D. L. and BUSTAMANTE C., *Nature*, **413** (2001) 748.

[14] CRICK F. and WATSON J. D., *Nature*, **177** (1956) 473.

[15] CASPAR D. and A. KLUG A., *Cold Spring Harbor Symposium on Quantitative Biology*, **27** (1962) 1.

[16] REDDY V., NATARAJAN P., OKERBERG B., LI K., DAMODARAN K., BROOKS M. R. III and JOHNSON J., *Virus Particle Explorer (VIPER), a Website for Virus Capsid Structures and their Computational Analyses*, *J. Virol.*, **75** (2001) 11943.

[17] MURZIN A. G., BRENNER S. E., HUBBARD T. and CHOTHIA C., *J. Mol. Biol.*, **247** (1995) 536.

[18] LANDAU L. D. and LIFSHITZ I. M., *Theory of Elasticity* (Pergamon, New York) 1975.

[19] SEUNG S. and NELSON D. R., *Phys. Rev. A*, **38** (1988) 1005.

[20] LOBKOVSKY A. E. and WITTEN T. A., *Phys. Rev. E*, **55** (1997) 1577.

[21] CERDA E. and MAHADEVAN L., *Phys. Rev. Lett.*, **80** (1998) 2358.

[22] BLAUROCK A. E. and GAMBLE R. C., *J. Membrane Biol.*, **50** (1979) 187; see also DUBOIS M., DEMÉS B., GULIK-KRZYWICKI T., DEDIEU J.-C., VAUTRIN C., DESERT S., PERZ E. and ZEMB T., *Nature*, **411** (2001) 672.

[23] KANTOR Y. and NELSON D. R., *Phys. Rev. A*, **36** (1987) 4020.

[24] YAN X., OLSON N. H., VAN ETTEN J. L., BERGOIN M., ROSSMANN M. G. and BAKER T. S., *Nature Struct. Biol.*, **7** (2000) 101.

[25] STEINHARDT P., NELSON D. R. and RONCHETTI M., *Phys. Rev. Lett.*, **47** (1981) 1297.

[26] For an analogous exploration of the shapes of *liquid* membranes with a spherical topology, see SEIFERT U., *Advances in Physics*, **46** (1997) 13; for a discussion of viral shells, see BRUINSMA R. F., GELBERT W. M., REGUERA D., RUDNICK J. and ZANDI R., *Phys. Rev. Lett.*, **90** (2003) 248101.

[27] SAFF E. B. and KUIJLAARS A. B. J., *The Math. Intelli.*, **19** (1997) 5.

[28] THOMSON J. J., *Philos. Mag.*, **7** (1904) 237.

[29] LEIDERER P., *Z. Phys. B*, **98** (1995) 303.

[30] RAMOS L., LUBENSKY T. C., DAN N., NELSON P. and WEITZ D. A., "Surfactant-Mediated Two-Dimensional Crystallization of Colloidal Crystals", in *Science*, **86** (1999) 2325.

[31] TOOMRE ALAR, unpublished work.

[32] DODGSON M. J. W. and MOORE M. A., *Phys. Rev. B*, **55** (1997) 3816; PEREZ-GARRIDO A. and MOORE M. A., *Phys. Rev. B*, **60** (1999) 15; 628.

[33] BOWICK M., CACCIUTO A., NELSON D. R. and TRAVESSET A., *Phys. Rev. Lett.*, **89** (2002) 185; 502.

Protein folding yesterday, today and tomorrow

E. Shakhnovich

Department of Chemistry and Chemical Biology, Harvard University
12 Oxford Street Cambridge MA 02138, USA

1. – Introduction

As was noted in recent review [1] protein folding field undergoes cyclic development. Initially protein folding was viewed as a strictly experimental field belonging to the realm of biochemistry where each protein is viewed as a unique system that requires its own detailed characterization —akin to any mechanism in biology. Introduction, in the early nineties, of simplified models to the field and their success in explaining several key aspects of protein folding such as thermodynamically and apparent kinetic two-state folding of many proteins, nucleation mechanism and its relation to native state topology, has pretty much shifted thinking towards views motivated by physics that treat the folding problem in its statistical mechanical aspect emphasizing universality of folding scenarios over uniqueness of folding pathways for each protein. This approach dominated the theoretical thinking in the last decade (reviewed in [1-4]). It was realized that the protein folding problem is where physics chemistry and biology meet. At the very basic level of microscopic models statistical mechanics provided tools that facilitated our understanding of many fundamental properties of proteins. A very basic model of protein, that of a heteropolymeric molecule [5] has provided many insights into basic thermodynamic and kinetic properties of proteins [5-8]. A careful analysis of the basic model —that of a random heteropolymer— showed that random heteropolymers are not likely to exhibit many protein-like properties such as, most importantly, cooperativity of their folding transition [9,6]. This is perhaps not surprising as proteins are biological macromolecules whose sequences underwent evolutionary selection. In particular, it was found that protein-like sequences should have specific properties, such as, most importantly a gap in its density of states [6, 9-11], *i.e.* significant (extensive in protein length) difference in energy between native conformation and closest to its in energy structurally unrelated states. Further, it was shown explicitly that sequences that satisfy this condition of foldability can be found via a simple evolution-like selection procedure and the number of such sequences was estimated for typical proteins [9, 12, 13]. These theoretical developments provided conceptual bridge between two key aspects of proteins: their physical nature as macromolecules and their biological nature as heteropolymers with evolutionary selected sequences. This provides deep connection between physical studies of protein folding and genomic studies that are carried out on numerous organisms leading to a flood of information about sequences and structures of natural proteins.

Revolution in molecular biology that changed paradigm for most researchers in the field had its impact on protein folding studies. We witnessed in recent years a significant explosion in the amount of information about protein sequences, structures and function which caused significant change in paradigm from "one gene, one protein one student" to "one organism, one genome one proteome one student". This aspect is very important as it allowed, for the first time in the history of biochemical sciences to view protein folding and morphogenesis in a truly evolutionary perspective. While numerous earlier studies focused on folding pathways of individual proteins, the present experimental approaches allow us to address a global issues of what are the possible folding scenarios and how have they evolved in response to various evolutionary pressures that were exerted on proteins in the process of their morphogenesis.

2. – Simple models provide valuable insights into protein folding

Studies of simple models indeed contributed considerably to our understanding of protein folding by focusing experimental and subsequent theoretical studies on key common milestones along protein folding pathways such as transition states (and their common structural features such as the folding nucleus [7, 14-17]), on- and off-pathway intermediates [18-21]. Importantly, many of the experimental studies were directly motivated by specific predictions and questions raised in theoretical studies. For example the analysis of nucleation in lattice models and in bioinformatics studies of protein families [14, 22] as well as studies of Go-models of proteins [23-26] suggested that the location of the nucleus may be robust within families of homologous and analogous proteins as well with respect to changes in environment. A number of experimental studies were undertaken to test and further explore these insights from theory [27, 28]. Their results appeared to be in good agreement with theoretical predictions.

These and several other successes of theoretical studies of simple universal models brought harmony to the protein folding field drawing theory closer to experiment and opening a direct channel of communication between theorists and experimentalists in the field.

Subsequent developments followed two routes. First, simple models continued to be instrumental in the explanation of various statistical, generic observations in protein folding. One of the most interesting of such more recent observations was the discovery that

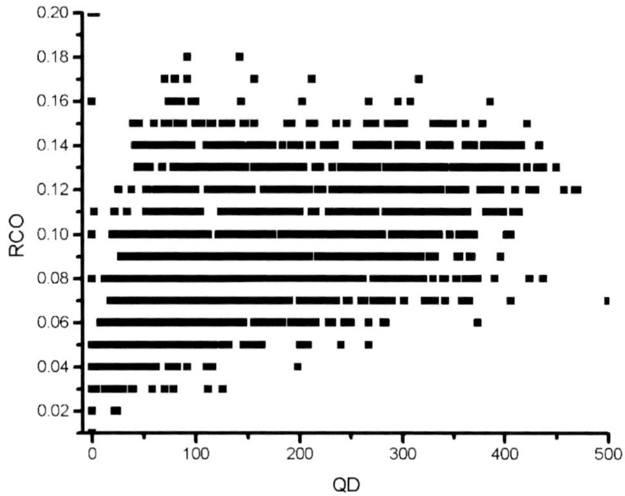

Fig. 1. – The scatter plot between the number of long-range contacts (Q_{D}) and Relative Contact order taken over 3400 DALI domains. Correlation coefficient $R = 0.34$.

the Relative Contact Order (that characterizes the relative prevalence of long-range and short-range contacts) may serve as a good predictor of folding rates in some proteins [29]. This and other observations motivated another class of highly simplified models that allowed detailed analysis under an extremely limiting set of assumptions such as that model proteins should consist of two contiguous "native" parts separated by no more than one disordered fragment [30-32] and similar highly restrictive assumptions. Nevertheless the analysis of the putative transition states in such models revealed some consistency with reality as perceived through the comparison of "predicted" ϕ-values with experimental ones. Overall it is difficult to judge the measure of success of these analyses because in some cases the actual residue-by-residue predictions were not reported. Rather, an overall correlation coefficient between experimental and theoretical ϕ-values was presented along with color-coded pictures of the protein backbone [31]. In some cases the number of adjustable parameters exceeded the number of datapoints that the model sought to predict [30]. This line of inquiry was continued in recent work [33,34], where the so-called "topomer model" was proposed. The main conclusion from the analysis of this model is that folding time grows exponentially with the number of long-range contacts (denoted as Q_{D} in [33]). However this conclusion is puzzling as it implies that the model predicts exponential growth of folding time with chain length (because the number of long-range contacts should be proportional to chain length to maintain protein compactness). Such exponential dependence is a signature of the famous Levinthal paradox. Furthermore it was claimed in [33] that Q_{D} and the Relative Contact Order (RCO) are "strongly correlated". However statistical analysis of DALI domains (fig. 1) does not show strong correlations. Moreover, fig. 1 shows that two structural measures —RCO and Q_{D}— cannot both serve as predictors of folding rate.

3. – News from experiment

On the other hand, a remarkable progress has been achieved over the last several years in experimental studies of protein folding. More advanced experimental techniques were developed that allowed researchers to significantly extend the time resolution of their kinetic experiments to low microseconds, using such approaches as laser T-jump and continuous-flow [35, 36]. Single-molecule techniques are used to probe folding kinetics [37, 38]. These and many other experimental studies provided a much more detailed experimental view on protein folding landscapes that either overcame or have a potential to overcome such traditional limitations as loss of information due to ensemble averaging or lack of time resolution to detect intermediates or properly evaluate burst phases. To this end, the discussion between Roder's group and Baker group concerning the intermediates in protein-G folding is noteworthy: while Baker's experiments using traditional stopped-flow equipment and W43 fluorescence as a single probe revealed no intermediates [39]. In Roder's experiment, the use of a more time-sensitive continuous-flow apparatus made it possible to discern a major on-pathway folding intermediate [21]. Furthermore, a careful analysis of chevron plots for several proteins carried out recently by Kiefhaber and coworkers revealed a slight yet noticeable curvature in the unfolding branch which can serve as evidence of transient intermediates or multiple transition states as well as possible effect of mutations on the unfolded state [40, 41]. Similarly, Clarke and coworkers analyzed non-linearity of chevron plots in several Ig-fold proteins and concluded that its most likely origin is in the existence of parallel folding pathways passing through distinct transition states and that the denaturant may shift the dominant pathway [42]. Radford's group work on helical bacterial immunity proteins also revealed complex pathways, including intermediates stabilized by non-native interactions in some of them and the possibility to change the complexity of folding pathways via mutations [43]. Further deep insights into a detailed picture of protein folding landscape can be obtained from AFM pulling experiments [44, 45] that probe free energy profile along complementary reaction coordinates.

4. – Detailed all-atom simulations —where theory meets experiment

This brief and by far incomplete review of experimental work in protein folding nevertheless illustrates impressive advances that provide more detailed structural information about the folding mechanism. However, here experiment probably reaches the limit of its ability to provide structural insights without simulations. This calls for very accurate computational models that match precision of experimental information and allow unambiguous structural interpretation of experimental data. This call was heeded by several groups: Daggett group employed unfolding simulations analysis based on the premises that unfolding is the microscopic reverse of folding, Pande distributed the computing approach and our lab all-atom simulations of various models of intermediate complexity. There was also an important development by Caflisch [46] and Roitberg [47, 48] groups in folding of peptides of about 20 aminoacid in length. While many of these approaches show promises, they have not yet ripened to become a fully satisfactory computational tool to study protein folding. The premise on which the work of Daggett is based has been questioned by several authors [49,50] who argued that unfolding is simulated at different conditions than those at which normal folding experiments are carried out. Such difference in simulation conditions may results in significant differences in observed pathways. Furthermore, the method to select the transition state in a smooth-looking trajectory has never been understood by many researchers and it is not entirely clear how reliable and "blind" is it. In any case crucial p_{fold} simulations [51,52] (see also below) have never been carried out on any presumed transition states. In that case the proposed transition states structures can be considered putative transition states. These limitations notwithstanding the high-temperature unfolding approach nevertheless provided many useful insights and testable predictions [53].

The distributed computing approach developed by Pande and coworkers aims to substitute long simulations in microsecond or even millisecond time scales by multiple short runs on thousands of computers. This approach is based on the view that the statistics of folding events is exponential so that in many short-time attempts few events can be observed and analyzed [54]. A few successes were claimed in predicting folding times of short proteins —villin headpiece [55] and Trp cage [56] (both proteins were predicted to fold in about $5\,\mu$s time scale). This approach got considerable attention recently and it was scrutinized by several authors [57,58]. In particular Caflisch and coworkers [58] carefully analyzed this method by comparing it with traditional long-time "equilibrium" simulation of short beta-peptide (a control that Pande group should have done in the first place). These authors showed that a distributed computing approach selects very atypical high-energy pathways that are usually not observed in longer "equilibrium" runs. However, the key requirement to any computational method is its ability to provide microscopic structural insights into the folding mechanism that can be then tested experimentally. To this end, Kubelka and coworkers carried out an experimental test of one of the predictions from distributed computing simulations of the villin headpiece with negative outcome, despite the agreement between predicted and observed folding rates [59].

While successes of some of the all-atom simulations are encouraging, they are still limited to very short peptides and many aspects of them are problematic. To this end, a more pragmatic approach was suggested by Vendruscolo and coworkers who used experimental ϕ-values as constraints akin to NMR NOE's [60,61] to build structural model of TSE. However these authors did not test, in direct folding simulations, whether their proposed transition state is a true kinetic transition state, *i.e.* its transmission coefficient $p_{\text{fold}} = 0.5$. This is a serious shortcoming of this very useful approach: as we will see below, many conformations belonging to such constructed *putative* transition state ensemble (TSE) are not true transition state. The work from our group [51] addresses this concern in detail.

The precision level of experimental studies both in traditional ensemble-based approaches such as protein engineering and single-molecule experiments, needs to be matched or exceeded by simulation techniques. Following this route, our group developed a novel and powerful tool for protein folding studies —all-atom Monte-Carlo dynamic simulations [62]. The method takes into account all heavy atoms of the protein and the move set consists of combinations of local and non-local moves. Calibration of the move set appeared to be a major undertaking that included careful comparison with the dynamics of short peptides undergoing helix-coil transition and comparison of rates observed in simulations (in terms of number of Monte-Carlo steps) and experiments where such data are available. This included second beta-hairpin from protein G whose folding rate is known from experiments by Eaton and coworkers [63] as well as α-helices [64] and several small proteins. In all cases the observed folding rates appeared to be highly linearly correlated with experimental ones and the results on the dynamics of helix-coil transitions appeared to be highly consistent with MD simulations data and experiment [64-66]. These results provided sufficient evidence that the developed technique is accurate enough to be useful for modeling folding mechanisms of small proteins, and we embarked on the studies of the protein folding atomic level of detail. As a first step, Go potential was used to model energetics of protein interactions. At an intermediate level of complexity Go models proved useful. In Go model only interactions between groups [24] or atoms [26, 62, 67] that are neighbors in the native state are treated as attractive in any conformation. The benefit of such models is that they "solve" the folding potential problem by guaranteeing that the correct native state is the global energy minimum. Their obvious shortcoming is that knowledge of native structure is needed in order to build such potentials and also they may underestimate non-native interactions in some cases [68]. However, in many cases they are the only potentials that allow full folding simulations from random coil to native state and as such provide extremely detailed insights into folding mechanisms for model proteins, stimulating further experiments and simulations.

The first protein that we studied, crambin, was mostly a proof-of-principle study that showed that, using atomic potentials that includes realistic steric interactions and contact Go atom-atom potentials, we can obtain numerous successful folding trajectories, for real proteins, at the atomic level of detail, using available computational resources (48-node Pentium Athlon-based cluster). Nevertheless even this first study provided strong evidence about complexity of folding pathways and relative role of energetic factors and topological frustration in defining folding pathways.

Our next major undertaking in this direction was to simulate complete folding of a protein that has been well-characterized in experiment. Ig-binding domain of staphylococcal protein G appeared to be an ideal model as its relatively short size and relatively fast-folding (3–5 ms) made it feasible for our method to simulate and, on the other hand, there was a plethora of experimental data to compare with [21, 39, 69]. In this case it was necessary to carefully calibrate short-range potentials (mostly H-bond) relative to long-range Go-energetics. This was accomplished by setting the strength of *non-specific* backbone hydrogen bonds to comply with thermodynamic data on the stability of *isolated* elements of protein-G secondary structure.

The simulation revealed a complex picture of protein-G folding that entails parallel pathways converging to a common Transition State Ensemble (fig. 2). The transition

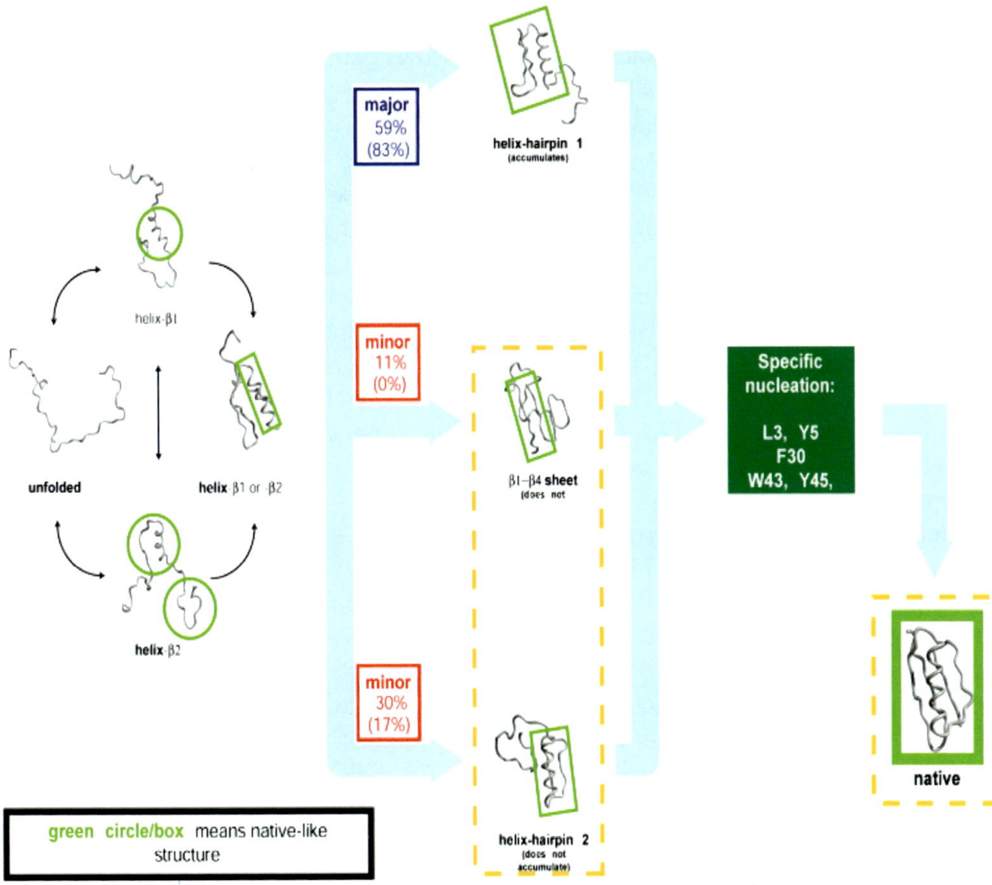

Fig. 2. – Schematic representation of the folding pathway of protein G as revealed from all-atom simulations. Three parallel pathways passing each through its own intermediate converge at the common Transition State Ensemble that features a specific nucleus composed of six hydrophobic residues.

state ensemble contains a specific nucleus of six hydrophobic residues, consistent with the general picture of nucleation mechanism and consistent with available ϕ-values (see below).

This study taught us several lessons, most important of which is that ensemble averaging (as is done in most experiments) and selection of the experimental probe/reaction coordinate (as it is W43 fluorescence) may significantly affect the apparent picture towards sometimes misleading conclusions. It emphasizes the crucial role that simulations must play in interpreting experiment —"Only theory decides what we manage to observe"(A. Einstein).

The developed all-atom simulations tool made it possible to address several problems that belong to the "holy grail" of the protein folding field that all previous coarse-grained models were not able to approach due to their (over) simplified character. While the issue of statistics and dynamics of side-chain packing was recognized by many as being at the cornerstone of the protein-folding problem [70, 71]. The all-atom MC simulations proved instrumental in addressing this problem. First, a direct sampling of side-chain packing states was performed to resolve a long-standing issue [70]: how many side-chain packing arrangements are sterically compatible with a given backbone conformation? The analysis was performed for several models of sterics —from hard-shell to van-der-Waals soft-shell steric interactions— with unexpected conclusion that many (exponential in the number of side-chain degrees of freedom) conformations are compatible with a given backbone conformation [72]. Naturally this degeneracy is broken in real proteins by interactions so that the native conformation of side-chains is energetically favored over alternatives (decoys). The side-chain packing decoys generated by this algorithm are used to develop atom-atom potentials for protein folding using potential optimization techniques [73-76].

The large conformation space of packed side-chains suggests that there may be a peculiar dynamics of their packing during folding. Again all-atom folding simulations proved an invaluable tool to address this difficult question —the analysis of many individual trajectories for protein-G folding allows to develop a very detailed picture of how side-chains get organized in the folding process and the results are quite interesting. It appears that there is a broad distribution in time-scales for side-chain packing times even in an apparently two-state kinetics, but side-chains that constitute the nucleus are the fastest to acquire their native conformation [77]! This result was obtained in [77] in simulations of a new lattice model with side-chains as well as in analysis of trajectories of all-atom simulations of protein G.

Further analysis of protein-G folding trajectories revealed a complex folding scenario whereby major features of protein topology and packing of nucleus side-chains get established first concurrently with nucleation, while side-chain packing of the rest of the structure occurs over longer time scale and is accompanied with backbone fluctuations (see fig. 3).

These longer time-scale fluctuations appear to be of peculiar character resembling glass transition dynamics with its signature power law relaxation of many characteristics such as total energy. The detailed analysis of them requires a new theoretical approach based on mode-coupling theory. Consistent with the specific aims of the project we started to develop such approach by formulating the main theoretical formalism [78] applicable to homo- and hetero-polymer dynamics. We were able to show that indeed in the low-temperature regime a glass transition may occur that would feature a long-time non-exponential relaxation of energy. However, this is only the initial step —a comprehensive theory that would treat directly side-chain relaxation in proteins is a matter of future development.

The all-atom Monte-Carlo algorithm developed in our lab appeared to be a valuable tool to study folding dynamics and thermodynamics. However, any folding study has two major components: a) search strategy/dynamic algorithm and b) energy function that

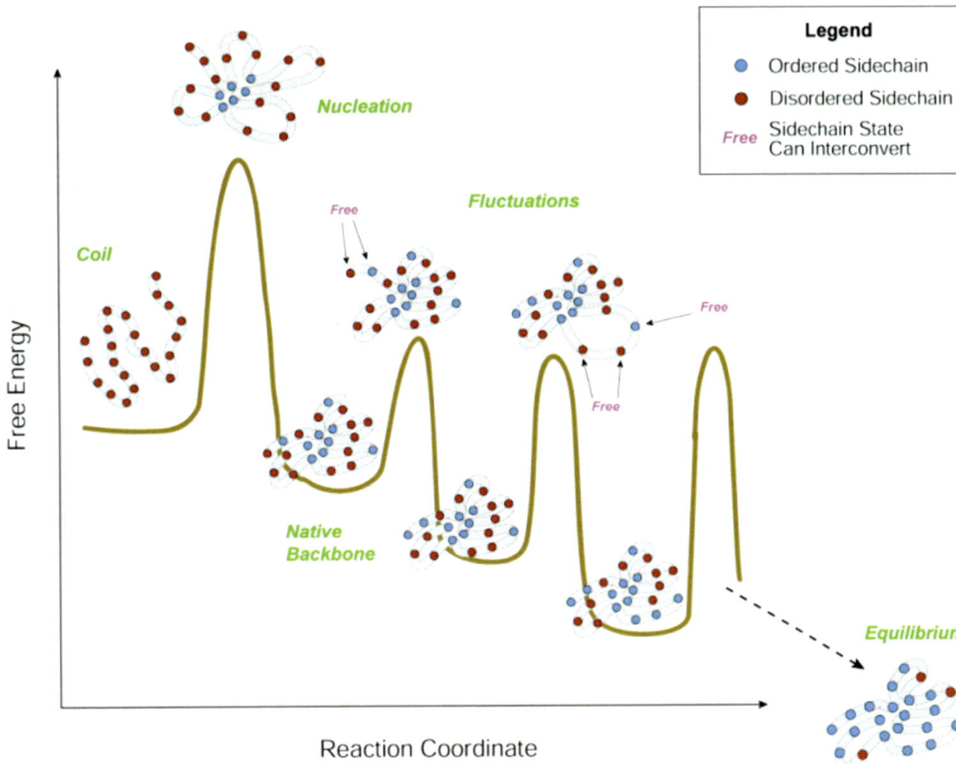

Fig. 3. – Schematic representation of the free energy landscape of a folding protein. After the main (nucleation) transition state has been overcome subsequent conformations remain committed to folding but they have to overcome several barriers related to side-chain packing of various parts of the molecule. That is achieved through local unfolding fluctuations. Such landscape complexity gives rise to a glassy relaxation behavior.

should select native structure as global minimum. The energy function used in all-atom studies described above is based on Go prescription. This may be a good choice for an initial study as it indeed guarantees that the native state is the global energy minimum. However, it requires knowledge of the native structure (or at least NOESY constraints from NMR experiments) and may underestimate energetic contribution and persistence of some non-native contacts. The latter were shown to play a possible role in nucleus formation, as predicted in simulations and bioinformatics analysis [79] and confirmed in experiment [80].

The task to develop atom-base potentials capable to fold a protein in a sequence-based model is extremely challenging as many who work in protein structure prediction and simulations may appreciate. For example simple knowledge-based approaches of the type suggested by Miyazawa and Jernigan [81] do not provide sufficient energy gaps to make it possible to fold proteins at atomic resolution [75]. Z-score and related optimization methods also did not provide adequate energetics. However, a new approach developed in our lab seems to be far more efficient. It is based on selection of atomic potentials to make realistic protein energetics resemble Go-based energetics as much as possible. To this end, in the spirit of knowledge-based potentials the interactions often observed in protein structures are deemed more attractive, while non-existent interactions are more repulsive:

$$E_{\mathrm{AB}} = \frac{-\mu N_{\mathrm{AB}} + (1-\mu)\tilde{N}_{\mathrm{AB}}}{\mu N_{\mathrm{AB}} + (1-\mu)\tilde{N}_{\mathrm{AB}}}, \tag{1}$$

where E_{AB} is the contact interaction energy between atom types A and B, N_{AB} is the number of AB pairs found in contact and $\tilde{N}_{\mathrm{AB}}$ is the number of AB pairs in the database that are not in contact. μ is a parameter that determines the average interaction (repulsion or attraction); it can be chosen to provide uniform and high (10–20%) acceptance rate in Monte Carlo simulation by preventing overly rapid collapse or excessively slow compactization. The preliminary results for folding of a small three-helix bundle protein with these potentials are very encouraging providing repetitive and systematic folding within 2A RMS from crystal structure [82]. Furthermore the potentials derived from different databases seem to correlate well [82] which is an encouraging sign that it may be transferable.

5. – Analysis of the folding nucleus at atomic resolution

Development of the all-atom MC simulations made it possible to overcome a major shortcoming of the present experimental studies of protein folding —their relative disconnect from detailed structural interpretation. Indeed even the most structurally detailed protein engineering method pioneered by Fersht and coworkers [83] used to be qualitatively interpreted as "high ϕ-value residues belong to the nucleus, while low ϕ-value ones do not". Such reasoning being qualitatively acceptable in some cases, often misleads, *e.g.* I76 in CI2 shows low ϕ-value in many mutations [15], however a careful double-mutant study attributes it to the folding nucleus [84]. The qualitative character of the interpretation of the protein engineering method was noted by Fersht and Daggett [85] who insightfully pointed out that ϕ-values should be treated as experimental constraints akin to NOESY in NMR determination of protein structure. This idea was further developed by Vendruscolo [60,61] and coworkers who used ϕ-values to reconstruct *putative* transition state of acylphosphatase —one of the proteins studied by Dobson and coworkers using protein engineering methods. These authors reconstructed putative TSE for the protein using ϕ-values as constraints but they did not test whether the proposed conformations represent true TSE, *i.e.* set of conformations for which the transmission coefficient to the folded state $p_{\mathrm{fold}} = 0.5$ [86]. Our method of all-atom simulations provided a unique opportunity to address this issue. First we carried out the analysis of TSE for CI2 [51] — perhaps the best characterized protein in terms of ϕ-value analysis [15]. We showed there that ϕ-values correctly specify, in general the TSE as $\langle p_{\mathrm{fold}}\rangle$ over putative TSE ensemble appeared to be close to 0.5. The work presented in [51] was like a "proof of principle" both for p_{fold} calculations and ϕ-value analysis. Our subsequent study presented a much more detailed picture of TSE for protein-G folding. In particular it clarified a number of key issues related to the ϕ-value analysis: a) What is the minimal number of ϕ-value constraints to enable reliable reconstruction of TSE? B) What is the relation between ϕ-values of residues reported in various mutations and its role in forming the TSE? The analysis of protein G provides answers to these questions for that protein. It points out that p_{fold} analysis is a crucial part of any structural interpretation of the TSE and that just simplistic assignment of high ϕ-values to nucleus residues and/or low ϕ-values to non-nucleus ones may be in several cases misleading. This analysis along with other all-atom simulation studies of protein G [26, 77] provide a detailed picture of its energy landscape (fig. 3). The landscape shown in fig. 3 also explains the frequently observed bimodal distributions of p_{fold} values by sampling of the surroundings of the TSE. These results show that the all-atom p_{fold} analysis serves as a very powerful tool for detailed structural characterization of TSE.

In addition to careful calibration of the move set as described above, we tested the dynamic realism of the Monte-Carlo by simulating folding using an alternative method — Discontinuous Molecular Dynamics [23,87-92]. This method is based on direct evaluation of dynamics by solving energy and momentum conservation equations each time protein atoms interact between themselves or with "ghost" solvent particles. Several models were studied within the Go model energetic prescription —from generic compact structure [23] to SH3 domain [91] to amyloid-like aggregates [92]. The analysis of these simulations shows that the developed picture of specific nucleation is very robust between models and simulation techniques. Moreover, we developed a very promising model to study protein aggregation and amyloidosis [92]. Energetics of this model is based on specific side-chain–like interactions combined with nonspecific backbone hydrogen bonding. It provided a very plausible experimentally testable generic model of amyloid particle formation.

6. – From ensemble to single molecules —pulling and stretching

Our analysis of protein-G folding [26] suggested (perhaps not surprisingly) that ensemble averaging in experiments may conceal important features of folding pathways. To this end, single-molecule studies appear to be a very important complementary approach to elucidate folding kinetics and landscapes. A number of groups undertook such studies [37,44,93-95]; in most cases protein stability/unfolding was probed in mechanical AFM experiments when the molecule was mechanically stretched.

We started to build theoretical foundation for understanding mechanical response of proteins in single-molecule experiments in the spirit of our usual approach to scientific problems: bottom-up trying first to reach some understanding of most fundamental aspects of the problem from analytical study to proceed further to address more specific aspects of the problem in simulations and experiments. To this end, first, we developed an analytical theory of the mechanical properties of random heteropolymers [86, 96]. This theory predicted the regime of gradual stretching of a heteropolymer when force comes close to a critical value f_c with intermediate structures resembling bead on a string. Furthermore a phase diagram of a stretching heteropolymer was presented in variables temperature —stretching force that outlined the regimes where such intermediates can be observed. The resulting behaviour is quite different from that of mechanical proteins most notable titin where domains unfold in a two-state manner at or around critical force [93]. However, titin is a protein selected by evolution to perform mechanical functions. When a non-mechanical protein underwent stretching it exhibited much more gradual unfolding [95] —in complete agreement with theoretical predictions because from the point of view of the mechanical properties barnase studied in [95] is essentially a random heteropolymer.

7. – Bringing evolution into picture —conservation and designability

So far our discussion concerned single-molecule properties of a protein. While acknowledging that proteins are not random sequences, it did not put much emphasis on the evolutionary origin of proteins that results in highly nonrandom selection of their sequences and structures. However, as indicated in many studies, [6, 97-101] deep understanding of protein folding cannot be achieved without taking into consideration the constraints imposed by evolution. To this end, we undertook a comprehensive analytical and simulation study of the evolution of protein sequences and structures. To this end, we developed a Z-score model of protein design and evolution that seeks sequences that optimize the Z-score:

$$Z = \frac{E_{\mathrm{N}} - E_{\mathrm{av}}}{\sigma}, \tag{2}$$

where E_{N} is the energy of a sequence in native conformation, E_{av} is the average energy of a sequence over a set of alternative conformations (decoys), σ is the standard variation of decoys energies. This approach is based on understanding of the basic principles of protein stability developed in protein-folding studies over the last decade [6, 10, 11, 100, 102]. The dynamic simulation of sequence selection in this model provided a basic understanding of time-scale separation on which protein superfamilies and families are born (fig. 4) and showed a good correlation between the predicted pattern of conservation and that observed in sequence alignments. Moreover several positions where discrepancy between simulated and real conservatism profile is observed can be identified as functional sites giving rise to the approach to predict active sites with significant accuracy [103]. Besides the simulation analysis we also developed an analytical theory based on the "profile approach" akin to the mean-field approximation that substitutes the influence of neighboring aminoacids by effective field which itself is determined self-consistently. This approach provides a rational way to read the messages hidden in the multiple alignment of protein sequences on the residue-by-residue basis.

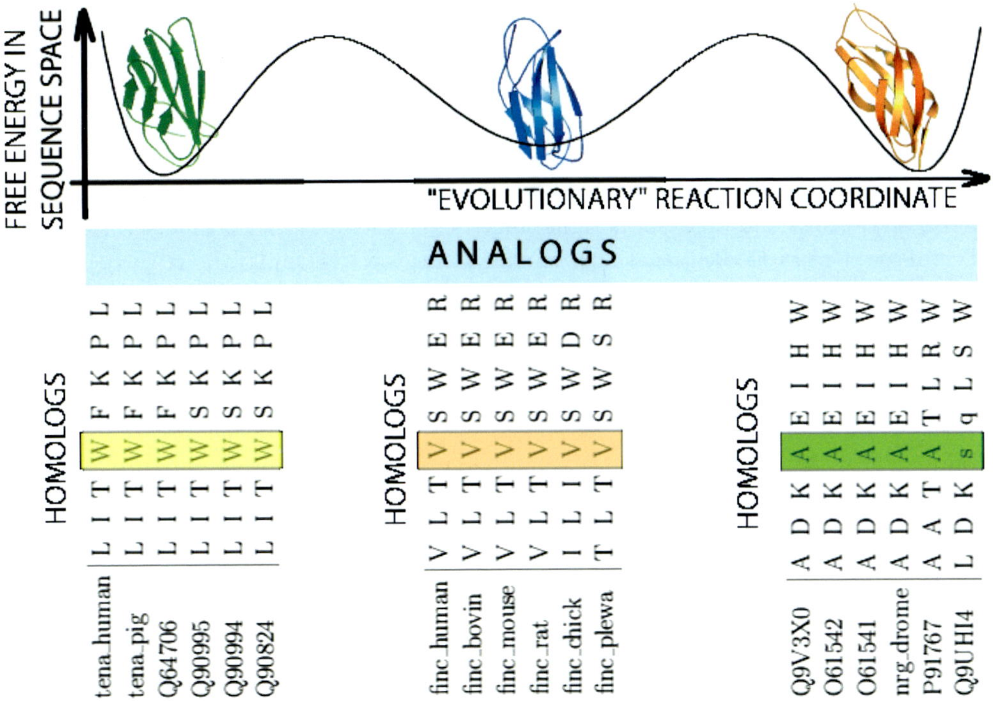

Fig. 4. – A schematic representation of the evolutionary processes that result in conservation patterns of amino acids. For a given family of folds, *e.g.* immunoglobulin (Ig) fold in this diagram, there are several alternative minima (3) in the hypothetical free energy landscape in the sequence space as a function of the "evolutionary" reaction coordinate (*e.g.*, time). Each of these minima are formed by mutations in protein sequences at some typical time scales, τ_0, that do not alter the protein's thermodynamically and/or kinetically important sites, forming families of homologous proteins. Transitions from one minimum to another occur at time scales $\tau = \tau_0 \exp[\Delta G/T]$, where ΔG is the free energy barrier separating one family of homologous proteins from another. At time scale τ mutations occur that would alter several amino acids at the important sites of the proteins in such a way that the protein properties are not compromised. At time scale τ the family of analogs is formed. In three minima we present three families of homologs (1TEN, 1FNF, and 1CFB) each composed of six homologous proteins. We show 8 positions in the aligned proteins: from 18 to 28. It can be observed that at position 4 (marked by blocks) in each of the families presented in the diagram amino acids are conserved within each family of homologs, but vary between these families. This position corresponds to position 21 in Ig fold alignment (to 1TEN) and is conserved. (Courtesy of N. Dokholyan.)

8. – Bringing structure into the picture: designability

The studies of evolution were often focused on selection of sequences. However an equally important aspect is selection of structures to become proteins. Real proteins are known to be unusually symmetric and structurally regular. It has been speculated that real proteins are highly designable, *i.e.* they can accommodate many sequences that would fold into their structure with sufficiently low energy [12, 104-108]. However, up to now the progress in research of designability was hindered by the lack of understanding of which structural features of protein render them more designable. This situation changed with the work from our lab [109] where we showed that a good structural determinant of designability of a structure is the maximal eigenvalue of its contact matrix (CM). A related set of quantities which are traces of higher orders of contact matrices can also serve as a good proxy of designability. This discovery, made in analytical theory and confirmed in lattice calculations, changed the way we think about protein design and designability by transforming this field from mostly speculative or based on highly unrealistic two-dimensional lattice models into a quantitative field where folds can be evaluated based on their designability and predictions can be compared with real data provided by bioinformatics analysis. We are at the beginning of this path but the first results are quite noteworthy. First, we compared designabilities of the proteomes of thermophilic and mesophilic bacteria and found that thermophilic proteomes are composed of statistically more designable proteins [110]. Further, we undertook a direct analysis of correlation between designability (as predicted from theory [109]) and population protein families at various levels: sequence homology, structural or functional relatedness. Strikingly, a strong statistical correlation between designability and family population is observed in the form of linear correlation between designability parameter and (log) of the number of sequences in complete agreement with theoretical predictions.

Summarizing this discussion we note that a remarkable progress in protein-folding studies has been achieved over the past few years. Several concepts developed in studies of simple models that were controversial few years ago are now well established through careful experimental and detailed theoretical study. A good example of such synergetic work between theory and experiment is the establishment of nucleation mechanism as a principal one that governs folding kinetics of many small proteins. What started as a theoretical suggestion derived from simulations in a coarse-grained model [14] became one of the major paradigms in the protein folding field that helps to understand existing experiments and plan new ones. At present we are witnessing the beginning of new era in protein folding studies where detailed characterization of folding pathways at the level of atomic and single-molecule resolution becomes available. This has enormous potential to change the way we think about biology. However, this potential cannot be realized without bringing the folding field in harmony with another revolution that is taking place now —in genomics and proteomics where an avalanche of data about protein sequences structures and functions becomes available. Detailed understanding of protein folding provides rationalization of this data raising the level of proteomics field from purely observational to fundamental physico-chemical basis of life sciences.

Up to now such "marriage" made in heaven has not yet realized. Many researchers in the folding field continued to focus closely their studies on single or few proteins and/or single methods of study. The present situation in life sciences, however calls for new approaches to meet new challenges and take advantage of the wonderful opportunities that modern-day structural genomics and proteomics provide. By attacking over wide front —from detailed atom-based modeling of folding pathways to whole proteomes of various organisms— the researchers will gain unique insights into how physics, chemistry and biology (evolution) work together in an amazing phenomenon of life.

REFERENCES

[1] MIRNY L. and SHAKHNOVICH E., *Protein folding theory: from lattice to all-atom models*, *Annu. Rev. Biophys. Biomol. Struct.*, **30** (2001) 361-96.

[2] ONUCHIC J. N., LUTHEY-SCHULTEN Z. and WOLYNES P. G., *Theory of protein folding: the energy landscape perspective*, *Annu. Rev. Phys. Chem.*, **48** (1997) 545-600.

[3] PLOTKIN S. S. and ONUCHIC J. N., *Understanding protein folding with energy landscape theory. Part II: Quantitative aspects*, *Q. Rev. Biophys.*, **35** (2002) 205-86.

[4] SHEA J. E. and BROOKS C. L., 3rd. *From folding theories to folding proteins: a review and assessment of simulation studies of protein folding and unfolding*, *Annu. Rev. Phys. Chem.*, **52** (2001) 499-535.

[5] SHAKHNOVICH E. I. and GUTIN A. M., *Formation of unique structure in polypeptide chains. Theoretical investigation with the aid of a replica approach*, *Biophys. Chem.*, **34** (1989) 187-99.

[6] BRYNGELSON J. D. and WOLYNES P. G., *Spin glasses and the statistical mechanics of protein folding*, *Proc. Natl. Acad. Sci. USA*, **84** (1987) 7524-8.

[7] SHAKHNOVICH E. I., *Theoretical studies of protein-folding thermodynamics and kinetics*, *Curr. Opin. Struct. Biol.*, **7** (1997) 29-40.

[8] PANDE V. S., GROSBERG A. Y. and TANAKA T., *Statistical mechanics of simple models of protein folding and design*, *Biophys. J.*, **73** (1997) 3192-210.

[9] SHAKHNOVICH E. I. and GUTIN A. M., *Engineering of stable and fast-folding sequences of model proteins*, *Proc. Natl. Acad. Sci. USA*, **90** (1993) 7195-9.

[10] SALI A., SHAKHNOVICH E. and KARPLUS M., *Kinetics of protein folding. A lattice model study of the requirements for folding to the native state*, *J. Mol. Biol.*, **235** (1994) 1614-36.

[11] SALI A., SHAKHNOVICH E. and KARPLUS M., *How does a protein fold?*, *Nature*, **369** (1994) 248-51.

[12] SHAKHNOVICH E. I., *Protein design: a perspective from simple tractable models*, *Fold. Des.*, **3** (1998) R45-58.

[13] SHAKHNOVICH E. I., *Proteins with selected sequences fold into unique native conformation*, *Phys. Rev. Lett.*, **72** (1994) 3907-3910.

[14] ABKEVICH V. I., GUTIN A. M. and SHAKHNOVICH E. I., *Specific nucleus as the transition state for protein folding: evidence from the lattice model*, *Biochem.*, **33** (1994) 10026-36.

[15] ITZHAKI L. S., OTZEN D. E. and FERSHT A. R., *The structure of the transition state for folding of chymotrypsin inhibitor 2 analysed by protein engineering methods: evidence for a nucleation-condensation mechanism for protein folding*, *J. Mol. Biol.*, **254** (1995) 260-88.

[16] FERSHT A. R., *Optimization of rates of protein folding: the nucleation-condensation mechanism and its implications*, *Proc. Natl. Acad. Sci. USA*, **92** (1995) 10869-73.

[17] FERSHT A. R. and DAGGETT V., *Protein folding and unfolding at atomic resolution*, *Cell*, **108** (2002) 573-82.

[18] GUTIN A. M., ABKEVICH V. I. and SHAKHNOVICH E. I., *Is burst hydrophobic collapse necessary for protein folding?*, *Biochem.*, **34** (1995) 3066-76.

[19] SOSNICK T. R., MAYNE L., HILLER R. and ENGLANDER S. W., *The barriers in protein folding*, *Nat. Struct. Biol.*, **1** (1994) 149-56.

[20] QI P. X., SOSNICK T. R. and ENGLANDER S. W., *The burst phase in ribonuclease A folding and solvent dependence of the unfolded state*, *Nat. Struct. Biol.*, **5** (1998) 882-4.

[21] Park S. H., Shastry M. C. and Roder H., *Folding dynamics of the B1 domain of protein G explored by ultrarapid mixing*, *Nat. Struct. Biol.*, **6** (1999) 943-7.

[22] Mirny L. A. and Shakhnovich E. I., *Universally conserved positions in protein folds: reading evolutionary signals about stability, folding kinetics and function*, *J. Mol. Biol.*, **291** (1999) 177-96.

[23] Dokholyan N. V., Buldyrev S. V., Stanley H. E. and Shakhnovich E. I., *Identifying the protein folding nucleus using molecular dynamics*, *J. Mol. Biol.*, **296** (2000) 1183-8.

[24] Shea J. E., Onuchic J. N. and Brooks C. L., 3rd. *Exploring the origins of topological frustration: design of a minimally frustrated model of fragment B of protein A*, *Proc. Natl. Acad. Sci. USA*, **96** (1999) 12512-7.

[25] Shea J. E., Onuchic J. N. and Brooks C. L. III, *Probing the folding free energy landscape of the Src-SH3 protein domain*, *Proc. Natl. Acad. Sci. USA*, **99** (2002) 16064-8.

[26] Shimada J. and Shakhnovich E. I., *The ensemble folding kinetics of protein G from an all-atom Monte Carlo simulation*, *Proc. Natl. Acad. Sci. USA*, **99** (2002) 11175-80.

[27] Chiti F., Taddei N., White P. M., Bucciantini M., Magherini F., Stefani M. and Dobson C. M., *Mutational analysis of acylphosphatase suggests the importance of topology and contact order in protein folding*, *Nat. Struct. Biol.*, **6** (1999) 1005-9.

[28] Martinez J. C., Pisabarro M. T. and Serrano L., *Obligatory steps in protein folding and the conformational diversity of the transition state*, *Nat. Struct. Biol.*, **5** (1998) 721-9.

[29] Plaxco K. W., Simons K. T. and Baker D., *Contact order, transition state placement and the refolding rates of single domain proteins*, *J. Mol. Biol.*, **277** (1998) 985-94.

[30] Munoz V. and Eaton W. A., *A simple model for calculating the kinetics of protein folding from three-dimensional structures*, *Proc. Natl. Acad. Sci. USA*, **96** (1999) 11311-6.

[31] Alm E. and Baker D., *Prediction of protein-folding mechanisms from free-energy landscapes derived from native structures*, *Proc. Natl. Acad. Sci. USA*, **96** (1999) 11305-10.

[32] Galzitskaya O. V. and Finkelstein A. V., *A theoretical search for folding/unfolding nuclei in three-dimensional protein structures*, *Proc. Natl. Acad. Sci. USA*, **96** (1999) 11299-304.

[33] Makarov D. E., Keller C. A., Plaxco K. W. and Metiu H., *How the folding rate constant of simple, single-domain proteins depends on the number of native contacts*, *Proc. Natl. Acad. Sci. USA*, **99** (2002) 3535-9.

[34] Makarov D. E. and Plaxco K. W., *The topomer search model: A simple, quantitative theory of two-state protein folding kinetics*, *Protein. Sci.*, **12** (2003) 17-26.

[35] Eaton W. A., Munoz V., Hagen S. J., Jas G. S., Lapidus L. J., Henry E. R. and Hofrichter J., *Fast kinetics and mechanisms in protein folding*, *Annu. Rev. Biophys. Biomol. Struct.*, **29** (2000) 327-59.

[36] Capaldi A. P., Shastry M. C., Kleanthous C., Roder H. and Radford S. E., *Ultrarapid mixing experiments reveal that Im7 folds via an on-pathway intermediate*, *Nat. Struct. Biol.*, **8** (2001) 68-72.

[37] Schuler B., Lipman E. A. and Eaton W. A., *Probing the free-energy surface for protein folding with single-molecule fluorescence spectroscopy*, *Nature*, **419** (2002) 743-7.

[38] Lipman E. A., Schuler B., Bakajin O. and Eaton W. A., *Single-molecule measurement of protein folding kinetics*, *Science*, **301** (2003) 1233-5.

[39] McCallister E. L., Alm E. and Baker D., *Critical role of beta-hairpin formation in protein G folding*, *Nat. Struct. Biol.*, **7** (2000) 669-73.

[40] Sanchez I. E. and Kiefhaber T., *Hammond behavior versus ground state effects in protein folding: evidence for narrow free energy barriers and residual structure in unfolded states*, *J. Mol. Biol.*, **327** (2003) 867-84.

[41] SANCHEZ I. E. and KIEFHABER T., *Evidence for sequential barriers and obligatory intermediates in apparent two-state protein folding*, *J. Mol. Biol.*, **325** (2003) 367-76.

[42] WRIGHT C. F., LINDORFF-LARSEN K., RANDLES L. G. and CLARKE J., *Parallel protein-unfolding pathways revealed and mapped*, *Nat. Struct. Biol.*, **10** (2003) 658-62.

[43] CAPALDI A. P., KLEANTHOUS C. and RADFORD S. E., *Im7 folding mechanism: misfolding on a path to the native state*, *Nat. Struct. Biol.*, **9** (2002) 209-16.

[44] CARRION-VAZQUEZ M., LI H., LU H., MARSZALEK P. E., OBERHAUSER A. F. and FERNANDEZ J. M., *The mechanical stability of ubiquitin is linkage dependent*, *Nat. Struct. Biol.*, **10** (2003) 738-43.

[45] BROCKWELL D. J., PACI E., ZINOBER R. C., BEDDARD G. S., OLMSTED P. D., SMITH D. A., PERHAM R. N. and RADFORD S. E., *Pulling geometry defines the mechanical resistance of a beta-sheet protein*, *Nat. Struct. Biol.*, **10** (2003) 731-7.

[46] CAVALLI A., HABERTHUR U., PACI E. and CAFLISCH A., *Fast protein folding on downhill energy landscape*, *Protein. Sci.*, **12** (2003) 1801-3.

[47] SIMMERLING C., STROCKBINE B. and ROITBERG A. E., *All-atom structure prediction and folding simulations of a stable protein*, *J. Am. Chem. Soc.*, **124** (2002) 11258-9.

[48] QIU L., PABIT S. A., ROITBERG A. E. and HAGEN S. J., *Smaller and faster: the 20-residue Trp-cage protein folds in 4 micros*, *J. Am. Chem. Soc.*, **124** (2002) 12952-3.

[49] FINKELSTEIN A. V., *Can protein unfolding simulate protein folding?*, *Protein Eng.*, **10** (1997) 843-5.

[50] DINNER A. R. and KARPLUS M., *Is protein unfolding the reverse of protein folding? A lattice simulation analysis*, *J. Mol. Biol.*, **292** (1999) 403-19.

[51] LI L. and SHAKHNOVICH E. I., *Constructing, verifying, and dissecting the folding transition state of chymotrypsin inhibitor 2 with all-atom simulations*, *Proc. Natl. Acad. Sci. USA*, **98** (2001) 13014-8.

[52] GSPONER J. and CAFLISCH A., *Molecular dynamics simulations of protein folding from the transition state*, *Proc. Natl. Acad. Sci. USA*, **99** (2002) 6719-24.

[53] DAGGETT V. and FERSHT A., *The present view of the mechanism of protein folding*, *Nat. Rev. Mol. Cell. Biol.*, **4** (2003) 497-502.

[54] SHIRTS M. R. and PANDE V. S., *Mathematical analysis of coupled parallel simulations*, *Phys. Rev. Lett.*, **86** (2001) 4983-7.

[55] ZAGROVIC B., SNOW C. D., SHIRTS M. R. and PANDE V. S., *Simulation of folding of a small alpha-helical protein in atomistic detail using worldwide-distributed computing*, *J. Mol. Biol.*, **323** (2002) 927-37.

[56] SNOW C. D., NGUYEN H., PANDE V. S. and GRUEBELE M., *Absolute comparison of simulated and experimental protein-folding dynamics*, *Nature*, **420** (2002) 102-6.

[57] FERSHT A. R., *On the simulation of protein folding by short time scale molecular dynamics and distributed computing*, *Proc. Natl. Acad. Sci. USA*, **99** (2002) 14122-5.

[58] PACI E., CAVALLI A., VENDRUSCOLO M. and CAFLISCH A., *Analysis of the distributed computing approach applied to the folding of a small beta peptide*, *Proc. Natl. Acad. Sci. USA*, **100** (2003) 8217-22.

[59] KUBELKA J., EATON W. A. and HOFRICHTER J., *Experimental tests of villin subdomain folding simulations*, *J. Mol. Biol.*, **329** (2003) 625-30.

[60] VENDRUSCOLO M., PACI E., DOBSON C. M. and KARPLUS M., *Three key residues form a critical contact network in a protein folding transition state*, *Nature*, **409** (2001) 641-5.

[61] PACI E., VENDRUSCOLO M., DOBSON C. M. and KARPLUS M., *Determination of a transition state at atomic resolution from protein engineering data*, *J. Mol. Biol.*, **324** (2002) 151-63.

[62] SHIMADA J., KUSSELL E. L. and SHAKHNOVICH E. I., *The folding thermodynamics and kinetics of crambin using an all-atom Monte Carlo simulation*, *J. Mol. Biol.*, **308** (2001) 79-95.

[63] MUNOZ V., THOMPSON P. A., HOFRICHTER J. and EATON W. A., *Folding dynamics and mechanism of beta-hairpin formation*, *Nature*, **390** (1997) 196-9.

[64] THOMPSON P. A., EATON W. A. and HOFRICHTER J., *Laser temperature jump study of the helix↔coil kinetics of an alanine peptide interpreted with a "kinetic zipper" model*, *Biochem.*, **36** (1997) 9200-10.

[65] HUMMER G., GARCIA A. E. and GARDE S., *Helix nucleation kinetics from molecular simulations in explicit solvent*, *Proteins*, **42** (2001) 77-84.

[66] HUMMER G., GARCIA A. E. and GARDE S., *Conformational diffusion and helix formation kinetics*, *Phys. Rev. Lett.*, **85** (2000) 2637-40.

[67] CLEMENTI C., GARCIA A. E. and ONUCHIC J. N., *Interplay among tertiary contacts, secondary structure formation and side-chain packing in the protein folding mechanism: all-atom representation study of protein L*, *J. Mol. Biol.*, **326** (2003) 933-54.

[68] PACI E., VENDRUSCOLO M. and KARPLUS M., *Native and non-native interactions along protein folding and unfolding pathways*, *Proteins*, **47** (2002) 379-92.

[69] KUSZEWSKI J., CLORE G. M. and GRONENBORN A. M., *Fast folding of a prototypic polypeptide: the immunoglobulin binding domain of streptococcal protein G*, *Protein. Sci.*, **3** (1994) 1945-52.

[70] RICHARDS F. M. and LIM W. A., *An analysis of packing in the protein folding problem*, *Q. Rev. Biophys.*, **26** (1993) 423-98.

[71] BROMBERG S. and DILL K. A., *Side-chain entropy and packing in proteins*, *Protein. Sci.*, **3** (1994) 997-1009.

[72] KUSSELL E., SHIMADA J. and SHAKHNOVICH E. I., *Excluded volume in protein side-chain packing*, *J. Mol. Biol.*, **311** (2001) 183-93.

[73] GOLDSTEIN R. A., LUTHEY-SCHULTEN Z. A. and WOLYNES P. G., *Optimal protein-folding codes from spin-glass theory*, *Proc. Natl. Acad. Sci. USA*, **89** (1992) 4918-22.

[74] MAIOROV V. N. and CRIPPEN G. M., *Contact potential that recognizes the correct folding of globular proteins*, *J. Mol. Biol.*, **227** (1992) 876-88.

[75] MIRNY L. A. and SHAKHNOVICH E. I., *How to derive a protein folding potential? A new approach to an old problem*, *J. Mol. Biol.*, **264** (1996) 1164-79.

[76] VENDRUSCOLO M., MIRNY L. A., SHAKHNOVICH E. I. and DOMANY E., *Comparison of two optimization methods to derive energy parameters for protein folding: perceptron and Z score*, *Proteins*, **41** (2000) 192-201.

[77] KUSSELL E., SHIMADA J. and SHAKHNOVICH E. I., *Side-chain dynamics and protein folding*, *Proteins*, **52** (2003) 303-21.

[78] PITARD E. and SHAKHNOVICH E. I., *Mode-coupling theory for heteropolymers*, *Phys. Rev. E S.*, **63** (2001) 041501.

[79] LI L., MIRNY L. A. and SHAKHNOVICH E. I., *Kinetics, thermodynamics and evolution of non-native interactions in a protein folding nucleus*, *Nat. Struct. Biol.*, **7** (2000) 336-42.

[80] VIGUERA A. R., VEGA C. and SERRANO L., *Unspecific hydrophobic stabilization of folding transition states*, *Proc. Natl. Acad. Sci. USA*, **99** (2002) 5349-54.

[81] MIYAZAWA S. and JERNIGAN R. L., *Residue-residue potentials with a favorable contact pair term and an unfavorable high packing density term, for simulation and threading*, *J. Mol. Biol.*, **256** (1996) 623-44.

[82] KUSSELL E., SHIMADA J. and SHAKHNOVICH E. I., *A structure-based method for derivation of all-atom potentials for protein folding*, *Proc. Natl. Acad. Sci. USA*, **99** (2002) 5343-8.

[83] MATOUSCHEK A., KELLIS J. T. jr., SERRANO L. and FERSHT A. R., *Mapping the transition state and pathway of protein folding by protein engineering*, *Nature*, **340** (1989) 122-6.

[84] LADURNER A. G., ITZHAKI L. S. and FERSHT A. R., *Strain in the folding nucleus of chymotrypsin inhibitor 2*, *Fold. Des.*, **2** (1997) 363-8.

[85] DAGGETT V., LI A., ITZHAKI L. S., OTZEN D. E. and FERSHT A. R., *Structure of the transition state for folding of a protein derived from experiment and simulation*, *J. Mol. Biol.*, **257** (1996) 430-40.

[86] DU R., PANDE V., GROSBERG A., TANAKA T. and SHAKHNOVICH E. I., *On the transition coordinate for protein folding*, *J. Chem. Phys.*, **108** (1998) 334-50.

[87] ZHOU Y., HALL C. K. and KARPLUS M., *First-Order Disorder-to-Order Transition in an Isolated Homopolymer Model*, *Phys. Rev. Lett.*, **77** (1996) 2822-2825.

[88] JANG H., HALL C. K. and ZHOU Y., *Protein folding pathways and kinetics: molecular dynamics simulations of beta-strand motifs*, *Biophys. J.*, **83** (2002) 819-35.

[89] ZHOU Y. and KARPLUS M., *Interpreting the folding kinetics of helical proteins*, *Nature*, **401** (1999) 400-3.

[90] DOKHOLYAN N. V., BULDYREV S. V., STANLEY H. E. and SHAKHNOVICH E. I., *Discrete molecular dynamics studies of the folding of a protein-like model*, *Fold. Des.*, **3** (1998) 577-87.

[91] DING F., DOKHOLYAN N. V., BULDYREV S. V., STANLEY H. E. and SHAKHNOVICH E. I., *Direct molecular dynamics observation of protein folding transition state ensemble*, *Biophys. J.*, **83** (2002) 3525-32.

[92] DING F., DOKHOLYAN N. V., BULDYREV S. V., STANLEY H. E. and SHAKHNOVICH E. I., *Molecular dynamics simulation of the SH3 domain aggregation suggests a generic amyloidogenesis mechanism*, *J. Mol. Biol.*, **324** (2002) 851-7.

[93] FOWLER S. B., BEST R. B., TOCA HERRERA J. L., RUTHERFORD T. J., STEWARD A., PACI E., KARPLUS M. and CLARKE J., *Mechanical unfolding of a titin Ig domain: structure of unfolding intermediate revealed by combining AFM, molecular dynamics simulations, NMR and protein engineering*, *J. Mol. Biol.*, **322** (2002) 841-9.

[94] LI H., OBERHAUSER A. F., FOWLER S. B., CLARKE J. and FERNANDEZ J. M., *Atomic force microscopy reveals the mechanical design of a modular protein*, *Proc. Natl. Acad. Sci. USA*, **97** (2000) 6527-31.

[95] BEST R. B., LI B., STEWARD A., DAGGETT V. and CLARKE J., *Can non-mechanical proteins withstand force? Stretching barnase by atomic force microscopy and molecular dynamics simulation*, *Biophys. J.*, **81** (2001) 2344-56.

[96] GEISSLER P. L. and SHAKHNOVICH E. I., *Reversible stretching of random heteropolymers*, *Phys. Rev. E*, **65** (2002) 056110.

[97] TAVERNA D. M. and GOLDSTEIN R. M., *The evolution of duplicated genes considering protein stability constraints*, *Pac. Symp. Biocomput.*, (2000) 69-80.

[98] TAVERNA D. M. and GOLDSTEIN R. A., *The distribution of structures in evolving protein populations*, *Biopolym.*, **53** (2000) 1-8.

[99] PANDE V. S., GROSBERG A. Y. and TANAKA T., *Nonrandomness in protein sequences: evidence for a physically driven stage of evolution?*, *Proc. Natl. Acad. Sci. USA*, **91** (1994) 12972-5.

[100] RAMANATHAN S. and SHAKHNOVICH E., *Statistical mechanics of proteins with "evolutionary selected" sequences*, *Phys. Rev. E*, **50** (1994) 1303-1312.

[101] WILDER J. and SHAKHNOVICH E. I., *Proteins with selected sequences: a heteropolymeric study*, *Phys. Rev. E*, **62** (2000) 7100-10.

[102] SHAKHNOVICH E. I. and GUTIN A. M., *A new approach to the design of stable proteins*, *Protein Eng.*, **6** (1993) 793-800.

[103] PEI J., DOKHOLYAN N. V., SHAKHNOVICH E. I. and GRISHIN N. V., *Using protein design for homology detection and active site searches*, *Proc. Natl. Acad. Sci. USA*, **100** (2003) 11361-6.
[104] LI H., HELLING R., TANG C. and WINGREEN N., *Emergence of preferred structures in a simple model of protein folding*, *Science*, **273** (1996) 666-9.
[105] LI H., TANG C. and WINGREEN N. S., *Are protein folds atypical?*, *Proc. Natl. Acad. Sci. USA*, **95** (1998) 4987-90.
[106] GOVINDARAJAN S. and GOLDSTEIN R. A., *Why are some proteins structures so common?*, *Proc. Natl. Acad. Sci. USA*, **93** (1996) 3341-5.
[107] WOLYNES P. G., *Symmetry and the energy landscapes of biomolecules*, *Proc. Natl. Acad. Sci. USA*, **93** (1996) 14249-55.
[108] FINKELSTEIN A. V., BADRETDINOV A. and GUTIN A. M., *Why do protein architectures have Boltzmann-like statistics?*, *Proteins*, **23** (1995) 142-50.
[109] ENGLAND J. L. and SHAKHNOVICH E. I., *Structural determinant of protein designability*, *Phys. Rev. Lett.*, **90** (2003) 218101.
[110] ENGLAND J. L., SHAKHNOVICH B. E. and SHAKHNOVICH E. I., *Natural selection of more designable folds: a mechanism for thermophilic adaptation*, *Proc. Natl. Acad. Sci. USA*, **100** (2003) 8727-31.

Essential heterogeneities in hole-doped cuprate superconductors

K. A. MÜLLER

Physics Institute, University of Zurich, CH-8057 Zurich, Switzerland

1. – Introduction

An expanded review of recent experiments in hole-doped cuprates is presented [1]. These include photoemission, inelastic neutron scattering, EXAFS, PDFs, electron paramagnetic resonance and susceptibility data. For doping concentrations below optimum all are compatible with Jahn-Teller bipolaron formation at the pseudogap temperature T^*, with a simultaneous presence of fermionic quasiparticles. The theoretically derived superconductivity onset and maximum T_c at optimal doping agree quantitatively with observation. Very large isotope effects at T^* for lower doping supports the vibronic character of the ground state.

2. – Local distortions

The essential structure of high-T_c, superconductors is well-represented by Tokura's [2] picture of donor/acceptor/inert layers bracketing a CuO_2 sheet conductor. The many-body wave function in the material is vibronic, *i.e.* includes both nuclear and electronic parts, for a single ion with the degenerate ground state

$$\Psi = \sum_i \psi_n^i \psi_e^i \tag{1}$$

without being in the adiabatic slaving limit and $i = 2, 3$. The many-body nuclear part ψ_n^i is probed in neutron scattering experiments, while the electronic part ψ_e^i is probed by photoemission. There are in general, two types of excitation energies, that are doping-dependent. This "vibronic" picture, in the sense of an interplay between the electronic charge and the ionic lattice structure, is a major theme of this talk.

Photoemission (ARPES) data by Lanzara *et al.* [3] clearly shows a common feature in different high-T_c superconductors, that is a signature of the lattice playing a role. The quasiparticle energy *versus* (rescaled) wave vector plots for Bi2212, Bi2201 and LSCO show a kink, while NCCO does not show any such behaviour (fig. 1). The kink occurs near 70 meV at a characteristic wave vector half in the Brillouin Zone, and indicates two different group velocities. They are due to two different quasiparticles, one of fermionic character near the Fermi energy and the other of bosonic character at more negative energy.

Probing the other part of the wave function in eq. (1), the inelastic neutron scattering by Egami and collaborators [4] shows that LO phonon spectra in YBCO and LSCO change significantly with oxygen concentration x varying from 0.93 to 0.20 (fig. 2). There is a distinct feature in the dispersion at 60–80 meV that also occurs in the Brillouin zone along a particular wave vector. The intensity of the excitation above the anomaly changes significantly in favour of the one below it. This reflects a change of the ratio of the two types of quasiparticles present.

While structure clearly is expected to play a role, there was already early evidence by the EXAFS measurements of Bianconi *et al.* [5] that this lattice structure is locally varying. They suggested in $x = 0.15$ LSCO that nano-domains occurred, with alternating bands or "stripes", separating charge-rich and charge-poor regions. The stripes consist of distorted unit-cell bands (D) of width ~ 8 Å and undistorted unit-cell bands (U) of width ~ 16 Å (fig. 3). The U regions are locally LTO-like, while the D regions have LTT-like CuO_6 octahedra, relatively tilted by about 16 degrees, and with bonds relatively closer in length. We note that the distorted octahedra are similar to a "Q_2" local mode, familiar in the Jahn-Teller effect, a point we will return to, later.

Charge inhomogeneities are associated with the local patterns of octahedral tilts, or more generally, with a pair distribution (111) of such tilts as found by the Billinge group [6] (PDFs). They find from neutron diffuse scattering, that for LSCO, the $x = 0.1$ data corresponding to "3 degree tilts" can, in fact, be reproduced by combinations of heavily tilted (5 degree) and untilted (0 degree) octahedra.

EPR data by Kochelaev *et al.* [7] show that in LSCO distortions exist (fig. 4). They have been successfully modelled as a 3-spin polaron, again associated with dynamic "Q_2"-like Jahn-Teller distortions.

In short, a variety of probes, including neutron scattering, PDF, ESR and EXAFS show that structural modes are important at various *finite wave vector* values, involving combinations of JT distortions.

3. – Pairing

The intersite JT correlations can in fact, lead to a finite wave vector $q_c = 2\pi/d$ pairing interaction, with a small pair size and short coherence length, $d \sim \xi$.

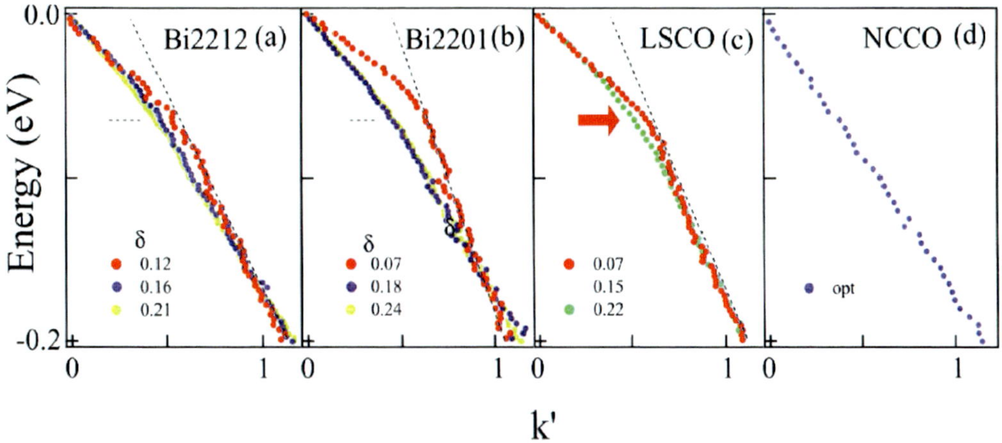

Fig. 1. – The quasiparticle dispersions *versus* the rescaled momentum k' for three p-type material systems (nodal direction): (a) Bi2212; (b) Bi2201; (c) LSCO. The arrow indicates the frequency value obtained by inelastic neutron diffraction data. The dispersions are compared with n-type superconductor NCCO (panel d) along ΓY. The dotted lines are guides to the eye obtained fitting the high-energy part with a linear function (see first ref. [3]).

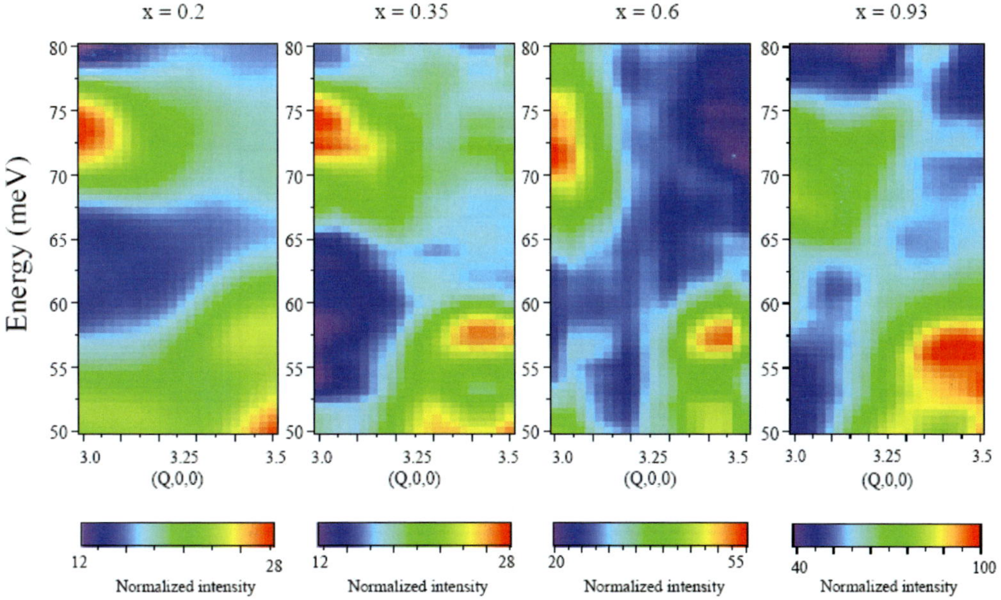

Fig. 2. – Composition dependence of the inelastic neutron scattering intensity from YBCO single crystals with $x = 0.2, 0.35, 0.6$ and 0.93, at $T = 10\,\mathrm{K}$ (see first ref. [4]).

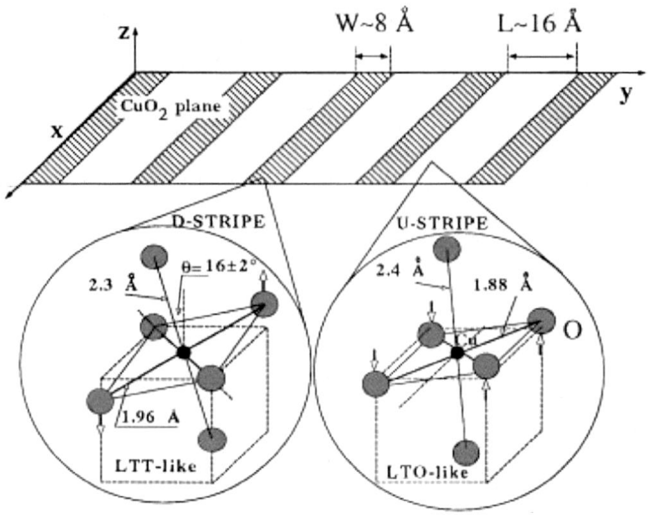

Fig. 3. – Stripe formation at T^* $La_{2-x}Sr_xCuO_4$, $x = 0.150$. Pictorial view of the distorted CuO_6 octahedra, left side, of the "LTT type" assigned to the distorted stripes (D stripes) of width $W \sim 8$ Å and of the undistorted octahedra, right side, of the "LTO type" assigned to the undistorted stripes (U stripes) of width $L \sim 16$ Å. The superlattice of quantum stripes of wavelength $\lambda = L + W$ is shown in the upper part (see ref. [5]).

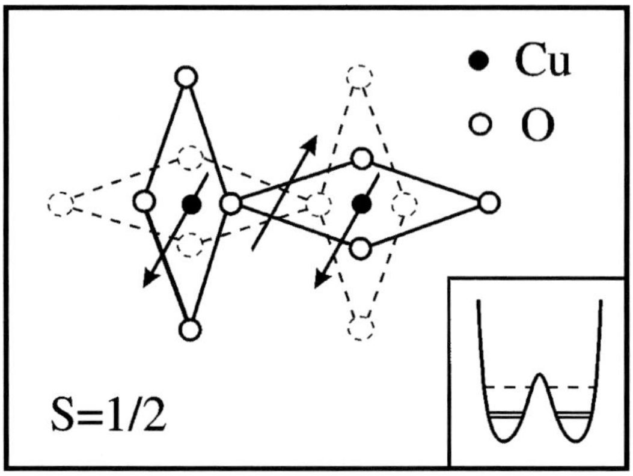

Fig. 4. – 3-spin magnetic polaron which is regarded as the EPR active center in the CuO_2 plane. The Jahn-Teller distorted polaron has two degenerated configurations as indicated by the dashed lines. The inset shows the corresponding double-well potential with the excited vibronic state (dashed line) and the ground state split by tunneling (solid lines) (see ref. [7]).

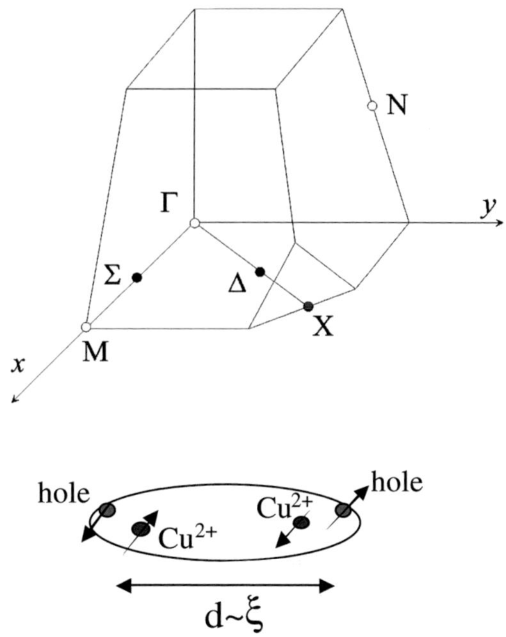

Fig. 5. – The intersite JT-pairing interaction. The Brillouin zone (BZ) of $La_{2-x}Sr_xCuO_4$ corresponding to the tetragonal phase with point group D_{4h} (see ref. [8]).

Kabanov and Mihailovic [8] have introduced a model compatible with these experiments, and to an interaction with a coupling constant of the form

$$g(q) = g_0\left[(q - q_c)^2 + \Gamma^2\right]^{-1}, \tag{2}$$

that is resonant in the wave vector. A group-theoretical analysis showed that, one can have coupling between $q \neq 0$ phonons and two-fold degenerate electron states including spins, (fig. 5) all with the resonant coupling structure of eq. (2).

We note here that Weisskopf had shown for classical superconductors, that

$$\Delta \sim \frac{E_F \lambda}{\xi}, \tag{3}$$

where λ is the screening length. This implies that the superconducting gap $\Delta \sim T_c$ is high when the coherence length ξ is small. Thus a small-pair model can be a high-T_c model.

In fact, if pairs are small, then superconductivity can be established through some kind of percolation, with pair size l_p smaller than the coherence length, and bigger than the lattice scale a:

$$a < \xi < l_p\,. \tag{4}$$

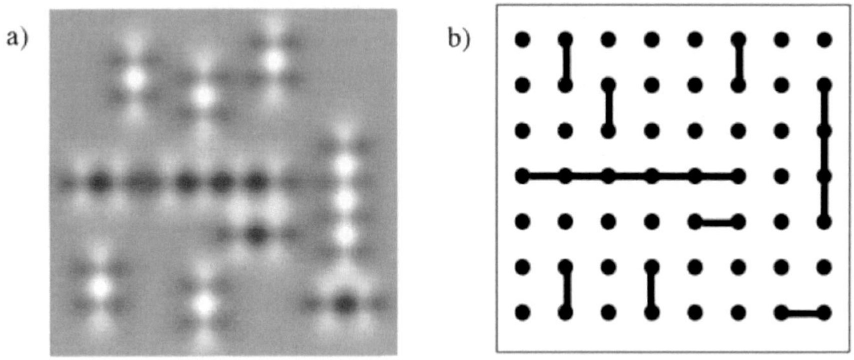

Fig. 6. – a) The amplitude of the lattice deformation caused by pairs described by the mesoscopic Jahn-Teller model. The picture corresponds to a "snapshot" at 6% doping at $T = 0$. b) The bond percolating model describing the situation in a) (see ref. [9]).

The picture that emerges is of Jahn-Teller induced mesoscopic pairs, that fluctuate and percolate [9]. A further development of these ideas yields (fig. 6) quantitatively:
i) an understanding of the minimum coherence length observed;
ii) the right percentage of holes for the onset of cuprate superconductivity ($\sim 6\%$);
iii) the right ${T_c}^{max}$ value.

4. – Pseudogap

Returning to structure, we consider the pseudogap temperature T^* where the local distortions begin to occur. Work by Bussmann-Holder *et al.* [10] has shown that bands of different symmetries $x^2 - y^2$, $3z^2 - r^2$ can be coupled dynamically by tilting octahedra. The model yields both a T_c, and a pseudogap temperature T^*. The pseudogap formation drastically reduces the energy separation between charge and spin levels, from $\sim 2\,\text{eV}$ to $\sim$ meV.

There are also other related analyses [11] based on the JT polarons combining to form bipolarons, whose binding energy is found to fall as $\sim 1/x$ with doping. So the temperature-doping phase diagram would show a bipolaron formation temperature $T^*(x)$, that decreases with doping x; and is above another negative-slope characteristic temperature, where bipolarons cluster to form stripes. The decreasing T^* line would nearly meet the increasing superconducting T_c phase boundary at the optimum doping, *i.e.* maximum temperature ${T_c}^{max}$ for superconductivity.

The local structure, and its formation temperature T^* can be probed by XANES methods, when an X-ray photon ejects an electron from Cu^{2+} and the electron waves interact with the O^{2-} neighbours. Plots in an early paper by Lanzara *et al.* [12] show fluorescence counts *versus* phonon energy, with features that are linked to the neighbouring La/Sr, and in-plane oxygen. NQR is another important probe of local structure, as used by Imai and collaborators [13].

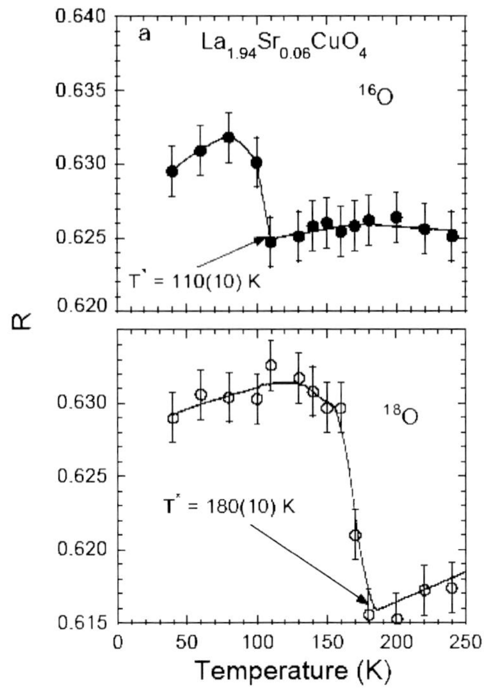

Fig. 7. – The oxygen-isotope effect on the charge-stripe ordering temperature T^* in $La_{1.94}Sr_{0.06}CuO_4$ and the doping dependence of T^*. The temperature dependence of the XANES peak intensity ratio R for the ^{16}O and ^{18}O samples of $La_{1.94}Sr_{0.06}CuO_4$ (see ref. [12]).

The temperature-dependence of the XANES peak intensity ratio shows a dip at $T^* \sim 110\,K$, that is associated with stripe formation. Since this technique probes only oxygen neighbours, it is a site-specific way to investigate effects of isotopic substitutions. In fact there is a large isotope effect, with the $O^{(16)} \to O^{(18)}$ substitution causing a rise to $T^* \sim 180\,K$ (fig. 7).

Several techniques can be used to probe the T^*-*versus*-doping curve for La and the techniques like XANES, NQR and EPR probe widely different time scales (10^{-13} s, 10^{-7} s, 10^{-9} s). Nonetheless, the data fall on the same curve below T_c^{max}.

Other investigations include inelastic time-of-flight neutron scattering measurements by Rubio Temprano *et al.* [14] that substitute for yttrium. This work, on the isotopic series $HoBa_2Cu_4^{(n)}O_8^{(p)}$ where $n = 63, 65$ and $p = 16, 18$, reveals a large T^* isotope effect for oxygen ($\alpha_O^* = -2.2$), and an even larger value for copper ($\alpha^*{}_{Cu} = -4.9$). Thus *both* copper and oxygen play a role in stripe formation of YBCO, on a scale $\sim 10^{-12}$ s, that cannot be accounted for by magnetic interaction.

5. – Concluding comments

Most theories of cuprate superconductors use the machinery of highly correlated conducting matter, and in these the lattice is taken as rigid at the outset. The many experiments of quite different character outlined here tell a quite different story, especially if the time scale of a particular experiment is sufficiently short. Then the heterogeneity becomes manifest in a clear manner. For all of them, the vibronic character of the ground state is present in a consistent way. This however was the concept which led to the discovery of the HTS cuprates [15]. Finally, it should be noted that the JT bipolaron quasiparticles include high correlations as well, which is often overlooked.

* * *

The author is indebted to S. R. SHENOY, A. SAXENA and T. LOOKMAN for compiling and JYOTSANA LAL for editing the manuscript, based on his oral presentation in Trieste and Varenna.

REFERENCES

[1] A shorter version, lacking figures, has appeared in A. R. BISHOP, S. R. SHENOY and S. SRIDHAR (Editors), *Intrinsic Multiscale Structure and Dynamics in Complex Electronic Oxides* (World Scientific, Singapore) 2003, p. 1.

[2] TOKURA Y. and ARIMA T., *Jpn. J. Appl. Phys.*, **29** (1999) 2388.

[3] SHEN Z. X., LANZARA A., ISHIHARA S. and NAGAOSA N., *Philos. Mag. B*, **82**, 13 (2002) 1349; LANZARA A., BOGDANOV P. V., ZHOU X., KELLAR S. A., FENG D. L., LIU E. D., YOSHIDA T., EISAKI H., FUJIMORI A., KISHIO K., SHIMOYAMA J. I., NODA T., UCHIDA S., HUSSAIN Z. and SHEN Z. X., *Nature*, **412** (2001) 510.

[4] PETROV Y., EGAMI T., MCQUEENEY R. J., YETHIRAJ M., MOOK H. A. and DOGAN F., *cond-mat*, **v1** (2000) 0003414; MCQUEENEY R. J., PETROV Y., EGAMI T., YETHIRAJ M., SHIRANE G. and ENDOH Y., *Phys. Rev. Lett.*, **82** (1999) 628; CHUNG J.-H., EGAMI T., MCQUEENEY R. J., YETHIRAJ M., ARAI M., YOKOO T., PETROV Y., MOOK H. A., ENDOH Y., TAJIMA S., FROST C. and DOGAN F., *Phys. Rev. B*, **67** (2003) 014517.

[5] BIANCONI A., SAINI N. L., LANZARA A., MISSORI M., ROSETTI T., OYANAGI H., YAMAGUCHI H., OKA K. and ITO T., *Phys. Rev. Lett.*, **76** (1996) 3412.

[6] BOZIN E. S., BILLINGE S. J. L., KWEI G. H. and TAKAGI H., *Phys. Rev. B*, **59** (1999) 4445.

[7] KOCHELAEV B. I., SICHELSCHMIDT J., ELSCHNER B., LEMOR W. and LOIDL A., *Phys. Rev. Lett.*, **79** (1997) 4274.

[8] MIHAILOVIC D. and KABANOV V. V., *Phys. Rev. B*, **63** (2001) 054505.

[9] MIHAILOVIC D., KABANOV V. V. and MUELLER K. A., *Europhys. Lett.*, **57** (2002) 254.

[10] BUSSMAN-HOLDER A., MUELLER K. A., MICNAS R., BUETTNER H., SIMON A., BISHOP A. R. and EGAMI T., *J. Phys. Condens. Matter*, **13** (2001) L169.

[11] MUELLER K. A., ZHAO G. M., CONDER K. and KELLER H., *J. Phys. Condens. Matter*, **10** (1998) 291.

[12] LANZARA A., ZHAO G. M., SAINI N. L., BIANCONI A., CONDER K., KELLER H. and MUELLER K. A., *J. Phys. Condens. Matter*, **11** (1999) L541.

[13] HUNT A. W., SINGER P. M., THURBER K. R. and IMAI T., *Phys. Rev. Lett.*, **82** (1999) 4300.

[14] RUBIO TEMPRANO D., MESOT J., JANSSEN S., CONDER K., FURRER A., MUTKA H. and MUELLER K. A., *Phys. Rev. Lett.*, **84** (2000) 1990.

[15] BEDNORZ J. G. and MUELLER K. A., *Adv. Chem.*, **100** (1988) 757, Nobel Lecture.

The mystery of superconductivity in the cuprates evinced by London penetration depths measurements

T. SCHNEIDER

Physik-Institut der Universität Zürich - Winterthurerstrasse 190, CH-8057 Zürich, Switzerland

The London penetration depth is a fundamental quantity of a superconductor. It plays a key role in determining and uncovering many properties of a superconductor, including homogeneity, anisotropy, isotope effects, importance of quantum and thermal fluctuations, and facets of the nature of superconductivity in a particular material. In recent years experimental data on the temperature, dopant concentration, magnetic field and oxygen isotope mass dependence of the penetration depth became available for a variety of cuprate superconductors. Here we analyze and discuss the experimental data, guided by the generic phase diagram of the cuprates, depicted in fig. 1. After passing the so-called underdoped limit ($p_{\mathrm{u}} \approx 0.05$), where p is the hole concentration, T_{c} reaches its maximum value $T_{\mathrm{c}}^{\mathrm{m}}$ at $p_{\mathrm{m}} \approx 0.16$. With further increase of p, T_{c} decreases and finally vanishes in the overdoped limit $p_{\mathrm{o}} \approx 0.27$ [1, 2]. There is the line $T_{\mathrm{c}}(p)$ of finite temperature phase transitions, separating the superconducting and non-superconducting states, with critical endpoints at p_{u} and p_{o}. Here T_{c} vanishes and the cuprates undergo at zero temperature doping (p) tuned quantum phase transitions. As far as their nature is concerned, resistivity measurements reveal a quantum-superconductor–to–insulator (QSI) transition in the underdoped limit (p_{u}) and in the overdoped limit (p_{o}) a quantum-superconductor–to–normal state (QSN) transition. Another essential experimental fact is the doping dependence of the anisotropy. In tetragonal cuprates it is defined as the ratio $\gamma = \lambda_c/\lambda_{ab}$ of the London penetration depths due to supercurrents flowing perpendicular (λ_c) and parallel (λ_{ab}) to the ab-planes. Approaching the QSN transition γ remains finite, while at the QSI transition it tends to infinity [3, 4]. When γ remains finite the system exhibits anisotropic but genuine three-dimensional (3D) behaviour, while $\gamma \to \infty$ implies 2D behavior. The resulting competition between anisotropy and superconductivity raises

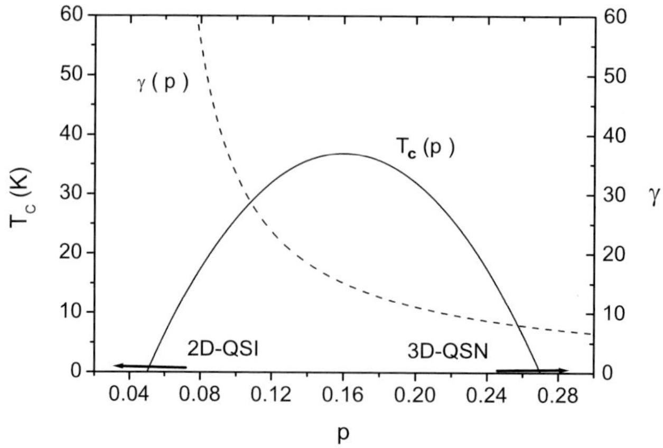

Fig. 1. – Schematic phase diagram of cuprate superconductors. Variation of T_c and $\gamma(T=0)=\lambda_c(T=0)/\lambda_{ab}(T=0)$ with hole concentration p. Quantum phase transitions occur at the endpoints $p=p_u\simeq 0.05$ and $p=p_o\simeq 0.27$ of the critical line $T_c(p)$. At p_u a two-dimensional quantum-superconductor–to–insulator (2D-QSI)- and at $p_o\simeq 0.27$ a 3D quantum-superconductor–to–normal state (3D-QSN) transition occurs.

serious doubts whether 2D mechanisms and models, corresponding to the limit $\gamma=\infty$, can explain the essential observations of superconductivity in the cuprates [4]. There is mounting evidence that close to the phase transition line $T_c(p)$ thermal fluctuations dominate, while quantum fluctuations dominate both the QSI and QSN transitions. Furthermore, due to the 3D to 2D crossover, tuned by the rise of γ with reduced dopant concentration, these fluctuations are enhanced. For these reasons, mean-field treatments, including the BCS theory, are expected to apply far from the critical line only. This singles out the low-temperature region around optimum doping p_m.

In the mean-field approximation the London penetration depth of an anisotropic superconductor in the Meissner state is given by [5,6]

$$(1)\qquad \frac{1}{\lambda_i^2(T)}\cong\frac{e^2}{\pi^2\hbar c^2}\oint \mathrm{d}S_\mathrm{F}\frac{v_{\mathrm{F}_i}v_{\mathrm{F}_i}}{|\mathbf{v}_\mathrm{F}|}\left(1+2\int_{\Delta_k}^{\infty}\mathrm{d}E_k\frac{\partial f(E_k)}{\partial E_k}\frac{E_k}{\sqrt{E_k-\Delta_k^2}}\right).$$

$\mathrm{d}S_\mathrm{F}$, v_{F_i} and $|\mathbf{v}_\mathrm{F}|$ are, respectively, the surface element of the Fermi surface, Fermi velocity in direction i and the magnitude of the Fermi velocity. $f(E_k)$ is the Fermi function and Δ_k the energy gap in direction k. The i index refers to the principal crystallographic directions a, b and c. The second term, which is negative, describes the decrease of $1/\lambda_i^2(T)$ caused by the thermal population of Bogoliubov quasiparticle levels with energy E_k, and it is this quantity where the anisotropy and magnitude of the energy gap enters. In this approximation $1/\lambda_i^2(T)$ vanishes close to T_c as $1/\lambda_i^2(T)=1/\lambda_{i0}^2(1-T/T_c)^\nu$ with $\nu=1/2$, while there is mounting evidence for $\nu\simeq 2/3$ in the experimentally accessible regime. Thus close to T_c where thermal fluctuations dominate the mean-field treatment fails. Otherwise, *e.g.* at sufficiently low temperatures and far away from the

QSI and QSN transitions, the neglect of fluctuations appears to be justified. Here eq. (1) reduces for noninteracting quasiparticle excitations around the four d-wave nodes, which dominate the leading low-temperature behavior, to [7]

$$(2)\quad \frac{1}{\lambda_{ab}^2(T)} = \frac{1}{\lambda_{ab}^2(0)}(1 - AT),\ A = \lambda_{ab}^2(0)\ln 2\frac{k_{\mathrm{B}}e^2 v_{\mathrm{F}}}{\hbar^2 c^2 d v_2},\ \frac{1}{\lambda_i^2(0)} \cong \frac{e^2}{\pi^2\hbar c^2}\oint \mathrm{d}S_{\mathrm{F}}\frac{v_{\mathrm{F}_i} v_{\mathrm{F}_i}}{|\mathbf{v}_{\mathrm{F}}|}.$$

The Fermi velocities v_{F} and v_2 enter the quasiparticles excitation energies $E_k = \hbar\sqrt{v_{\mathrm{F}}^2 k_1^2 + v_2^2 k_2^2}$ and refer to velocities along directions normal and tangential to the Fermi surface at each node. d is the mean interlayer spacing along the c-axis. The velocity ratio v_{F}/v_2 is a fundamental material parameter which measures the anisotropy of the quasiparticle excitation spectrum. This scenario is not restricted to the penetration depth. It predicts simple power-law temperature dependencies in the thermodynamic and transport properties at sufficiently low temperatures. For example, the penetration depth measurements find that $1/\lambda_{ab}^2$ exhibits in the clean limit and at low temperature a linear temperature dependence [8,9], in agreement with eq. (2). The NMR relaxation rate exhibits the expected T^3 temperature dependence [10]. The predicted effect of impurities in giving rise to a universal thermal conductivity [11,12] has been confirmed [13]. The clean-limit specific heat varying as T^2 appears to have been observed [14-16].

For a spherical Fermi surface, one recovers for the zero-temperature penetration depth the standard result, $1/\lambda^2(T = 0) = 4\pi ne^2/(mc^2)$, where n is the number density of the electrons in the normal state [17]. In recent years the penetration depth has been the subject of intense experimental investigation in high-T_{c} compounds [18-24]. It has become common practice to associate the anisotropy with anisotropic effective masses m_i^* and to interpret the experimental data in terms of the London formula

$$(3)\quad \frac{1}{\lambda_i^2(0)} = \frac{4\pi n_{\mathrm{s}} e^2}{m_i^* c^2}$$

by introducing the number density n_{s} of the superfluid. In a real superconductor, the ionic potential modifies the spherical Fermi surface of free electrons drastically. In this case it is not evident what should be taken for the effective mass m_i^* and the number density n_{s}. Although, using eqs. (2) and (3), we can define the ratio n_{s}/m_i^*, which often has been used to interpret experimental results, in terms of

$$(4)\quad \frac{1}{\lambda_i^2(0)} \cong \frac{e^2}{\pi^2\hbar c^2}\oint \mathrm{d}S_{\mathrm{F}}\frac{v_{\mathrm{F}_i} v_{\mathrm{F}_i}}{|\mathbf{v}_{\mathrm{F}}|} = \frac{4\pi n_{\mathrm{s}} e^2}{m_i^* c^2}\ \frac{n_{\mathrm{s}}}{m_i^*}.$$

This relation shows that n_{s}/m_i is just a way of parameterizing experimental results, with no discernible connection to the band mass or carrier concentration. Indeed in the mean-field approximation $1/\lambda_i^2(0)$ is determined by normal-state properties, namely the integral of the Fermi velocity over the Fermi surface. Noting that in high-T_{c} superconductors band structure calculations [25] and ARPES [26, 27] studies uncovered drastic deviations from the free-electron picture, it is evident that interpretations based on the ratio n_s/m_i^* obscure the origin of the anisotropy $\gamma_{ij} = \lambda i/\lambda_j$ and the doping dependence of the zero-temperature penetration depth, as well as the isotope effect on this quantity. Indeed, an inspection of eq. (2) leads to the conclusion that the anisotropy stems from flat portions of the Fermi surface, while the doping dependence reflects that of the Fermi surface. On the other hand, there is little doubt about the importance of residual electron-electron and electron-phonon interactions, not accounted for in eq. (2). Quantifying these interactions is difficult in the normal state of the cuprates, given the lack of well-defined single-particle excitations as revealed by various experiments. Contrariwise, well-defined quasiparticle excitations do exist in the superconducting state, and a description of the low-temperature state in terms of superfluid-Fermi-liquid theory is believed to apply. Fermi-liquid corrections account for the Fermi-liquid interactions between electrons. In the superconductor their effect is the renormalization of the Fermi velocity ratio in terms of $v_F/v_2 \to \alpha^2 v_F/v_2$ [7], where v_F/v_2 is the bare value entering eq. (2). The comparison of $d\lambda^{-2}/dT$ at $T = 0$ in Bi2212, evaluated with eq. (2) and the ARPES estimate for v_F/v_2 [26] with the value deduced from penetration depth measurements [28], points to a substantial Fermi-liquid renormalization, namely $\alpha^2 \approx 0.3$, due to interactions between the nodal quasiparticles in the superconducting state. To explore the doping dependence of the renormalization, we invoke the empirical relation $T_c\, d/dT(\lambda_{ab}^2(0)/\lambda_{ab}^2(T))|_{T=0} \approx -0.6$ [9], which applies to a variety of cuprates with T_c ranging from 30 to 130 K and dopant concentrations extending from the underdoped to the optimally doped regime. It implies with the empirical relation, $T_c \propto 1//\lambda_{ab}^2(0)$ [23], established for underdoped cuprates and eq. (2) that $\alpha^2 v_F/v_2$ is nearly doping independent.

Much less attention has been devoted to the renormalization of the Fermi velocity due to electron-phonon interaction. Isotope substitution, *i.e.* the exchange of ^{18}O by ^{16}O, is a suitable probe, whereby the lattice parameters remain essentially unaffected [29,30], while the phonon frequencies associated with the mass of the oxygen ions, or more generally, the oxygen lattice degrees of freedom are modified [31]. In recent years the isotope effect on the zero-temperature penetration depth has been investigated, using a variety of techniques. In table I we listed the experimental estimates for the relative change

$$\frac{\Delta\lambda_{ab}^2(\widetilde{T})}{\lambda_{ab}^2(\widetilde{T})} = \frac{{}^{n}\lambda_{ab}^2(\widetilde{T}) - {}^{m}\lambda_{ab}^2(\widetilde{T})}{{}^{m}\lambda_{ab}^2(\widetilde{T})}, \tag{5}$$

upon isotope exchange for various cuprates and MgB_2, where $n = {}^{18}O$, $m = {}^{16}O$ in the cuprates and $n = {}^{11}B$, $m = {}^{10}B$ in MgB_2. The data also reveals that within experimental accuracy $\Delta\lambda_{ab}^2(\widetilde{T})/\lambda_{ab}^2(\widetilde{T}) = \Delta\lambda_{ab}^2(0)/\lambda_{ab}^2(0)$. At zero temperature and taking the anisotropy of the quoted materials into account ($\lambda_c \gg \lambda_a \approx \lambda_b$) the mean-field expression (2) reduces to

$$\frac{1}{\lambda_{ab}^2(T=0)} \cong \frac{e^2}{\pi^2 \hbar c^2 \sqrt{2}} \oint dS\, v_{F_{ab}}. \tag{6}$$

TABLE I. – *Experimental estimates for* $\Delta\lambda_{ab}^2(\widetilde{T})/\lambda_{ab}^2(\widetilde{T}) = ({}^n\lambda_{ab}^2(\widetilde{T}) - {}^m\lambda_{ab}^2(\widetilde{T}))/{}^m\lambda_{ab}^2(\widetilde{T})$ *with* $n = {}^{18}O$, $m = {}^{16}O$ *in the cuprates and* $n = {}^{11}B$, $m = {}^{10}B$ *in* MgB_2.

	T_c (K)	$\widetilde{T}$	$\Delta\lambda_{ab}^2(\widetilde{T})/\lambda_{ab}^2(\widetilde{T})$	Ref.
$YBa_2Cu_3O_{7-\delta}$	89.0	10	0.05	[20]
$Y_{0.7}Pr_{0.3}Ba_2Cu_3O_{7-\delta}$	60.6	5	0.13	
$Y_{0.6}Pr_{0.4}Ba_2Cu_3O_{7-\delta}$	45.3	5	0.11	
$YBa_2Cu_3O_{7-\delta}$ (film)	89.3	4	0.05	[21]
$La_{2-x}Sr_xCuO_{4+\delta}$, $x = 0.08$	19.5	0	0.10(2)	[19]
$La_{2-x}Sr_xCuO_{4+\delta}$, $x = 0.086$	22.4	0	0.08(1)	
$Bi_{1.6}Pb_{0.4}Sr_2Ca_2Cu_3O_{10+\delta}$	107	0	0.05	[18]
MgB_2	38.5	0	0.02(2)	[24]

Since the lattice parameters remain essentially unaffected [29, 30] by isotope exchange, while the dynamics, associated with the mass of the respective ions, are modified, the substantial isotope effect on the zero-temperature penetration depth requires a renormalization of the normal-state Fermi velocity $\mathbf{v}_F \to \widetilde{\mathbf{v}}_F = \mathbf{v}_F/(1+f)$, where $\mathbf{v}_F$ is the bare velocity and f the electron-phonon coupling constant which changes upon oxygen isotope exchange in the cuprates but remains nearly unaffected by boron isotope substitution in MgB_2. However, in the Migdal-Eliashberg [32] (ME) treatment of the electron-phonon interaction the coupling constant f is independent of the ionic masses and assumed to be small [33, 34]. This is true if the parameter $\omega_0 f/E_F$ is small, where ω_0 is the relevant phonon frequency and E_F the Fermi energy. Thus the isotope effect on the penetration depth is expected to be small, of the order of the adiabatic parameter $\widetilde{\gamma} = \omega_0/E_F \ll 1$. The ME theory retains terms only of order 0. Cuprates, however, have Fermi energies much smaller than those of conventional metals [35] so that $\widetilde{\gamma}$ is no longer negligibly small. Thus the large oxygen isotope effect on the zero-temperature in-plane penetration depth in the cuprates, listed in table I, poses a fundamental challenge to this understanding and calls for a theory that goes beyond ME [36, 37].

Indeed, the relative shifts $\Delta\lambda_{ab}^2(\widetilde{T})/\lambda_{ab}^2(\widetilde{T})$ are substantial and surprisingly close to $({}^{18}M_o - {}^{16}M_o)/{}^{16}M_o = 0.125$. This differs fundamentally from the behavior of optic (O) phonon frequencies. The expected behavior, $\omega \propto M^{-1/2}$, was confirmed in $YBa_2Cu_3O_{7-\delta}$ by measuring the frequency shift of the transverse optic phonons (copper-oxygen stretching modes), yielding $\Delta\omega/\omega \approx -0.06$, in agreement with $\Delta\omega/\omega \approx ({}^{16}M_o/{}^{18}M_o)^{1/2} - 1 = -0.057$ [31]. In any case the observed oxygen isotope effect on the zero-temperature penetration depth uncovers together with eq. (6), a substantial renormalization of the normal-state Fermi velocity due to oxygen lattice degrees of freedom, while in MgB_2 this renormalization due to boron isotope exchange turns out to be marginal. This renormalization is also expected to affect the superconducting properties. Taking the empirical relation $T_c\, \mathrm{d}/\mathrm{d}T(\lambda_{ab}^2(0)/\lambda_{ab}^2(T))|_{T=0} \approx -0.6$ [9] for granted, eq. (2)

implies that

$$\frac{\Delta T_c}{T_c} = -\frac{\Delta A}{A} = -\frac{\Delta\lambda_{ab}^2(0)}{\lambda_{ab}^2(0)} - \frac{\Delta\left(\widetilde{v_F/v_2}\right)}{\left(\widetilde{v_F/v_2}\right)}, \tag{7}$$

where $\widetilde{v_F/v_2}$ is the bare ratio, renormalized with respect to electron-phonon coupling. Noting that close to optimum doping $\Delta T_c/T_c$ is negligible small, the oxygen isotope exchange uncovers a substantial electron-phonon renormalization of v_F/v_2, characterizing the quasiparticles in the superconducting state. Indeed, close to optimum doping $\Delta(\widetilde{v_F/v_2})/(\widetilde{v_F/v_2}) \cong -\Delta\lambda_{ab}^2(0)/\lambda_{ab}^2(0)$ holds and this quantity varies from 0.05 to 0.11 (see table I). This effect, providing direct evidence of the relevance of electron-phonon coupling in the superconducting state, should be observable with ARPES. Noting that thermally excited quasiparticles destroy superconductivity by driving $1/\lambda_{ab}^2(T)$ to zero, we can estimate T_c by extrapolating eq. (2) to $1/\lambda_{ab}^2(T) = 0$. This yields $T_c A \approx 1$ and confirms eq. (7). In this context it is interesting to note that details of the Fermi-surface topology of deuterated κ-(BEDTTTF)$_2$Cu(NCS)$_2$ have been measured as a function of pressure and compared with equivalent measurements of the undeuterated salt. The data suggest that the negative-isotope effect observed on deuteration is due to small differences in Fermi-surface topology caused by the isotopic substitution [38].

Close to the phase transition line $T_c(p)$ thermal critical fluctuations, neglected in mean-field treatments, dominate the thermodynamic properties. Approaching the phase transition line around $p = p_m$ from below (see fig. 1), there is mounting evidence that the critical behavior of homogeneous cuprates falls in the experimentally accessible temperature regime into the 3D-XY universality class [4,39-52]. Here critical 3D-XY fluctuations dominate because the fluctuations of the vector potential are strongly suppressed due to the small value of the effective charge of the pairs [39]. In the 3D-XY universality class the transition temperature T_c, the critical amplitude of the specific heat A^- and the critical amplitudes of the penetration depths λ_{i0} are universally related by [50]

$$(k_B T_c)^3 = \left(\frac{\Phi_0^2 R^-}{16\pi^3}\right)^3 \frac{1}{\lambda_{a0}^2\lambda_{b0}^2\lambda_{c0}^2 A^-}. \tag{8}$$

$R^- \simeq 0.815$ is a universal number. The critical amplitudes are defined as $c = (A^-/\alpha)t^{-\alpha}$, $t = 1 - T/T_c$ and $\lambda_i^2 = \lambda_{i0}^2 t^{-\nu}$, where α and ν are the critical exponents. Although T_c, A^- and λ_{i0}^2 depend on the dopant concentration, isotope exchange etc., universality implies that this combination does not. Hence this relation puts a crucial constraint on the microscopic theory of superconductivity in cuprates. To illustrate this point it is instructive to consider the doping dependence of T_c, $\lambda_{ab}^2 \simeq \lambda_{a0}\lambda_{b0}$, $\gamma_0 = \lambda_{c0}/\lambda_{ab0}$ and A^- close to the 2D-QSI transition (see fig. 1). Here T_c, $1/\lambda_{ab0}^2$ and γ_0 are known to scale as $T_c \propto 1/\lambda_{ab0}^2 \propto 1/\gamma_0 \propto p - p_u$ so that the critical amplitude of the specific-heat singularity vanishes according to eq. (8) as $A^- \propto (p - p_u)^2 \propto T_c^2$. This is consistent with the specific heat data for underdoped $La_{2-x}Sr_xCuO_4$ and underdoped $Tl_2Ba_2CuO_{6+\delta}$ [53]. Furthermore, the universal relation implies that the relative changes upon isotope substitution satisfy the relation [54]

$$\frac{\Delta T_c}{T_c} = -\frac{\Delta A^-}{3A^-} - \frac{1}{3} \sum_{i=a,b,c} \frac{\Delta \lambda_{i0}^2}{\lambda_{i0}^2} . \tag{9}$$

It explains why close to the transition temperature and optimum doping ($p = p_m$) a substantial isotope effect on the penetration depths is compatible with a negligible effect on T_c.

Approaching the 2D-XY-QSI transition in the underdoped regime (see fig. 1) a crossover to the universal relation [50]

$$T_c = \frac{\Phi_0^2 \overline{R}_2}{16\pi^3 k_B} \frac{d_s}{\lambda_{ab}^2(0)} , \tag{10}$$

takes place and the cuprates correspond to a stack of independent sheets of thickness d_s. $\overline{R}_2$ is a universal number. The flow to 2D–XY-QSI behavior is experimentally well confirmed in terms of Uemura's plot [23]. It is a characteristic 2D property and also applies to the onset of superfluidity in ^{4}He films adsorbed on disordered substrates where it is well confirmed [55]. Although T_c, d_s and $\lambda_{ab}^2(0)$ depend on dopant concentration, isotope substitution, etc., universality implies that this relation does not. This puts yet another constraint on the microscopic theory of superconductivity in the cuprates. Furthermore it yields for the relative changes upon isotope exchange the relation

$$\frac{\Delta T_c}{T_c} = \frac{\Delta d_s}{d_s} - \frac{\Delta \lambda_{0ab}^2(0)}{\lambda_{ab}^2(0)} , \tag{11}$$

applicable close to the underdoped limit ($p = p_u$). Although the experimental data are rather sparse in the underdoped regime [54, 56], suggestive evidence for an isotope effect on the effective thickness d_s of the superconducting sheet emerges, namely $\Delta d_s/d_s = ({}^{18}d_s - {}^{16}d_s)/{}^{16}d_s \approx 0.03$ [54]. Estimates for d_s can be derived from the crossing point phenomenon in the temperature dependence of the magnetization for various applied magnetic fields, applied parallel to the c-axis. In a 3D anisotropic superconductor, falling into the 3D-XY universality class, the magnetization data plotted in terms $m_z/H_z^{1/2}$ *vs.* T will cross at T_c, where z is along the c-axis. Consistency with this behavior was found in a variety of cuprates [50]. On the contrary, in a 2D superconductor, corresponding to a slab of thickness d_s, the crossing point occurs in the plot m_z *vs.* T at the Kosterlitz-Thouless transition temperature T_{KT}, where $m_z \propto -k_B T_{KT}/(\Phi_0 d_s)$. Even though bulk cuprates are strictly 2D only close to the 2D-QSI transition at $p = p_u$ (see fig. 1), in the highly anisotropic materials, such as Bi-2212 and Tl-1223, 2D crossing point features have been observed and used to estimate d_s, yielding values close to the c-axis lattice constant [50]. Since the lattice parameters remain essentially unaffected [29, 30] by isotope exchange, while d_{s} does, the substantial relative change $\Delta d_{\mathrm{s}}/d_{\mathrm{s}} = ({}^{18}d_{\mathrm{s}} - {}^{16}d_{\mathrm{s}})/{}^{16}d_{\mathrm{s}} \approx 0.03$ [54] uncovers again the relevance of electron-lattice coupling. Going further, by combining eqs. (7) and (11), extrapolated to the underdoped regime, we obtain the approximate relation

(12) $$\frac{\Delta d_{\mathrm{s}}}{d_{\mathrm{s}}} = -\frac{\Delta\left(\widetilde{v_{\mathrm{F}}/v_2}\right)}{\left(\widetilde{v_{\mathrm{F}}/v_2}\right)},$$

which provides additional evidence for the coupling between superconducting properties and lattice degrees of freedom. The approximate nature of this relation stems from the fact that the d-wave quasiparticle scenario does not hold down to the 2D-QSI transition due to the neglect of quantum fluctuations, associated with the phase of the order parameter. Indeed, the linear-in-T temperature dependence of $1/\lambda_{ab}^2$ simply follows from the existence of a 2D-QSI transition. The result, $(\lambda_{ab}(0)/\lambda_{ab}(T))^2 - 1 \propto T/T_{\mathrm{c}}$, holds for d- and s-wave pairing [9].

Additional evidence for electron-lattice coupling emerges from the combined oxygen isotope and finite-size effects. Due to inhomogeneities of extrinsic or intrinsic origin, cuprates are homogeneous over a finite extent only. Thus a finite-size effect [57] is expected to occur, whereby the correlation volume cannot grow beyond the volume of the homogeneous domains. When 3D-XY critical fluctuations dominate there is the universal relationship [50]

(13) $$\frac{1}{\lambda_i(T)\lambda_j(T)} = \frac{16\pi^3 k_{\mathrm{B}} T}{\Phi_0^2 \sqrt{\xi_i^t(T)\xi_j^t(T)}},$$

between the London penetration depths λ_i and transverse correlation lengths ξ_i^t in directions i and j. In the presence of inhomogeneities with length scales L_i the $\xi_i^t = \xi_{i0}^t |t|^{-\nu}$, where $t = T/T_{\mathrm{c}} - 1$, cannot diverge but are bounded by

(14) $$\xi_i^t \xi_j^t \leq L_k^2, \quad i \neq j \neq k.$$

A characteristic feature of the resulting finite-size effect is the occurrence of an inflection point at T_{p_k} in $1/\lambda_i(T)\lambda_j(T)$ below T_{c}, the transition temperature of the homogeneous system. Here

(15) $$\xi_i^t(T_{p_k})\xi_j^t(T_{p_k}) = L_k^2, \quad i \neq j \neq k,$$

and eq. (13) reduces to

(16) $$\left.\frac{1}{\lambda_i(T)\lambda_j(T)}\right|_{T=T_{p_k}} = \frac{16\pi^3 k_{\mathrm{B}} T_{p_k}}{\Phi_0^2}\frac{1}{L_k}.$$

In the homogeneous case $1/(\lambda_i(T)\lambda_j(T))$ decreases continuously with increasing temperature and vanishes at T_c, while in the presence of inhomogeneities it exhibits an inflection point at $T_{p_k} < T_c$, so that

$$\mathrm{d}\left(\frac{1}{\lambda_i(T)\lambda_j(T)}\right)\Big/\mathrm{d}T\bigg|_{T=T_{p_k}} = 0. \tag{17}$$

Since the experimental data for the temperature dependence of the penetration depths is available in the form λ_{ab} and λ_c only, we rewrite eq. (16) as

$$L_c = \frac{16\pi^3 k_B T_{p_c} \lambda_{ab}^2(T_{p_c})}{\Phi_0^2}, \qquad L_{ab} = \frac{16\pi^3 k_B T_{p_b} (\lambda_{ab}(T)\lambda_c(T))_{T=T_{p_{ab}}}}{\Phi_0^2}. \tag{18}$$

Apart from the inflection point, an essential characteristic of a finite-size effect is the finite-size scaling function [58]. In the present case and for λ_{ab} it is defined in terms of

$$\left(\frac{\lambda_{0ab}}{\lambda_{ab}(T)}\right)^2 |t|^{-\nu} = g_c(y), \quad y = \mathrm{sign}(t)|t|\left(\frac{L_c}{\xi_{0ab}^t}\right)^{1/\nu} = \mathrm{sign}(t)\left|\frac{t}{t_{p_c}}\right|. \tag{19}$$

For $t = T/T_c - 1$ small and $L_c \to \infty$, so that $\pm y \to \infty$, it should tend to $g_c(y \to -\infty) = 1$ and $g_c(y \to \infty) = 0$, respectively, while for $t = 0$ and $L_c \neq 0$ it diverges as

$$g_c(y \to 0) = g_{0c}|y|^{-\nu} = g_{0c}\left|\frac{t}{t_{p_c}}\right|^{-\nu}, \tag{20}$$

whereby $(\lambda_{0ab}/\lambda_{ab}(T_c, L))^2 = g_{0c}|t_{p_c}|^{\nu} = g_{0c}\xi_{0ab}^t/L_c$. As expected, a sharp superconductor–to–normal-state transition requires domains of infinite extent. Moreover at t_{p_c}, $y = 1$ and there is an inflection point because $\mathrm{d}(\lambda_{0ab}/\lambda_{ab}(T,L))^2/\mathrm{d}t = 0$. Since the scaling function $g_c(y)$ depends on the type of confining geometry and on the conditions imposed (or not, in the case of free boundaries) at the boundaries of the domains, this applies to the amplitude g_{0c} as well. In fig. 2a we displayed the microwave surface impedance data for $\lambda_{ab}^2(T=0)/\lambda_{ab}^2(T)$ and $\mathrm{d}(\lambda_{ab}^2(T=0)/\lambda_{ab}^2(T))/\mathrm{d}T$ *vs.* T of a high-quality $Bi_2Sr_2CaCu_2O_{8+\delta}$ single crystal taken from Jacobs *et al.* [59]. The solid curve indicates the leading 3D-XY critical behavior of the homogeneous system, while the data uncovers a rounded transition which occurs smoothly. This behavior, together with the occurrence of an inflection point around $T_{p_c} \approx 87\,\mathrm{K}$, where $\mathrm{d}(\lambda_{ab}^2(T=0)/\lambda_{ab}^2(T))/\mathrm{d}T$ exhibits an extremum, points to a finite-size effect. With $\lambda_{ab}(T=0) = 1800\,\text{Å}$ obtained from μSR measurements [60], $T_{p_c} \approx 87\,\mathrm{K}$ and $\lambda_{ab}^2(T=0)/\lambda_{ab}^2(T_{P_c}) = 0.066$, we obtain with the aid of eq. (18) the estimate $L_c \approx 68\,\text{Å}$. Although the spatial extent of the homogeneous domains along the c-axis appears to be of nanoscale only, the small critical amplitude of the transverse correlation length, $\xi_{0ab}^t = L_c(1 - T_{p_c}/T_c)^{2/3} \approx 2.2\,\text{Å}$, makes the 3D-XY critical regime $(\lambda_{ab}^2(0)/\lambda_{ab}^2(T) \propto (1-T/T_c)^{2/3})$ attainable. An additional and essential characteristic of a finite-size effect appearing in the temperature dependence of

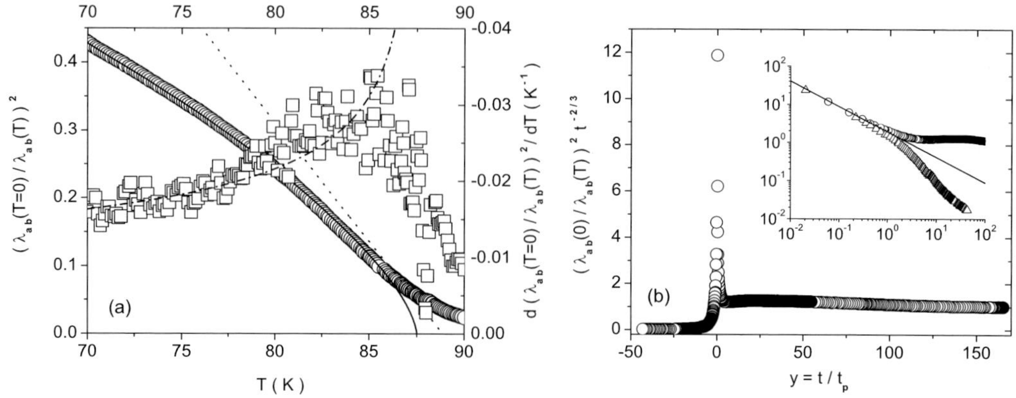

Fig. 2. – (a) Microwave surface impedance data for $\lambda_{ab}^2(T=0)/\lambda_{ab}^2(T)$ ($\bigcirc$) and $\mathrm{d}(\lambda_{ab}^2(T=0)/\lambda_{ab}^2(T))/\mathrm{d}T$ ($\square$) *vs.* T of a high-quality $Bi_2Sr_2CaCu_2O_{8+\delta}$ single crystal taken from Jacobs *et al.* [59]. The solid line is $\lambda_{ab}^2(0)/\lambda_{ab}^2(T) = 1.2(1-T/T_c)^{2/3}$ and the dash-dot line its derivative with $T_c = 87.5\,\mathrm{K}$, indicating the leading critical behavior of the homogeneous system. The dotted line is the tangent to the inflection point at $T_p \approx 87\,\mathrm{K}$, where $\mathrm{d}(\lambda_{ab}^2(0)/\lambda_{ab}^2(T))/\mathrm{d}T$ is maximum. (b) Finite-size scaling function $g(y) = (\lambda_{0ab}/\lambda_{ab}(T))^2|t|^{-\nu}$ *vs.* $y = t/|t_p|$ for the data shown in (a). The solid line in the inset is eq. (19) with $g_{0c} = 1.6$.

the in-plane penetration depth is the consistency of $(\lambda_{0ab}/\lambda_{ab}(T))^2|t|^{-\nu}$ *vs.* $t/|t_{p_c}|$ with the shape and limiting behavior of the finite-size scaling function (see eq. (19)). In fig. 2b we displayed $(\lambda_{0ab}/\lambda_{ab}(T))^2|t|^{-\nu}$ *vs.* $t/|t_{p_c}|$. The apparent agreement with the aforementioned characteristic behavior of this function provides strong evidence for a finite-size effect, due to the limited extent L_c of homogeneous superconducting domains along the c-axis. Clearly, such a finite-size scaling analysis, performed on one set of data for one particular sample and material only, cannot distinguish between an intrinsic or extrinsic origin of the inhomogeneity. Noting that this behavior was found in a variety of cuprates and for data obtained with different techniques [61], one is led to the conclusion that inhomogeneities, giving rise to a finite-size effect, are yet another facet of the mystery of superconductivity in the cuprates. Clearly this finite-size effect is not restricted to the penetration depth but should be visible in other thermodynamic properties. In the specific heat it leads to a rounding of the peak and its consistency with a finite-size effect was established for the data taken on $YBa_2Cu_3O_{7-\delta}$ high-quality single crystals [50]. In these samples the domain size was found to range from 300 to 400 Å. Furthermore nanoscale spatial variations in the electronic characteristics have also been observed in underdoped $Bi_2Sr_2CaCu_2O_{8+\delta}$ with scanning tunnelling microscopy (STM) [62-65]. They reveal a spatial segregation of the electronic structure into 3 nm diameter superconducting domains in an electronically distinct background.

Supposing that the limiting length scales change upon isotope exchange, we obtain from eq. (18) the relation

$$\frac{\Delta T_{p_c}}{T_{p_c}} = \frac{\Delta L_c}{L_c} - \frac{\Delta\lambda_{ab}^2(T_{p_c})}{\lambda_{ab}^2(T_{p_c})}, \tag{21}$$

TABLE II. – *Finite-size estimates for* $\Delta T_{p_c}/T_{p_c}$, $\Delta L_c/L_c$, $\Delta\lambda^2_{ab}(T_{p_c})/\lambda^2_{ab}(T_{p_c})$, ${}^{16}T_{p_c}$, ${}^{18}T_{p_c}$, ${}^{16}L_c$ *and* ${}^{18}L_c$ *for an* ^{18}O *content of* 89% *taken from* [66].

x	0	0.2	0.3
$\Delta T_{p_c}/T_{p_c}$	−0.000(2)	−0.015(3)	−0.021(5)
$\Delta L_c/L_c$	0.12(5)	0.13(6)	0.16(5)
$\Delta\lambda^2_{ab}(T_{p_c})/\lambda^2_{ab}(T_{p_c})$	0.11(5)	0.15(6)	0.15(5)
${}^{16}T_{p_c}$ (K)	89.0(1)	67.0(1)	52.1(2)
${}^{18}T_{p_c}$ (K)	89.0(1)	66.0(2)	51.0(2)
${}^{16}L_c$ (Å)	9.7(4)	14.2(7)	19.5(8)
${}^{18}L_c$ (Å)	10.9(4)	16.0(7)	22.6(9)

which matches eq. (11), applicable in the 2D limit where the limiting length is set by d_s, the thickness of the independent sheets. Furthermore, this relation opens the possibility to probe the coupling between superfluidity and lattice degrees of freedom close to criticality, where mean-field treatments fail. Indeed, given the fact that the lattice parameters remain essentially unaffected [29, 30] by isotope exchange, a purely electronic mechanism requires $\Delta L_c = 0$. The effect of oxygen isotope substitution on the inhomogeneity-induced finite-size effect has been explored in $Y_{1-x}Pr_xBa_2Cu_3O_{7-\delta}$ [66]. From the resulting estimates, listed in table II, several observations emerge. First, L_c increases systematically with reduced T_{p_c}. Second, L_c grows with increasing x and upon isotope exchange (^{16}O, ^{18}O). Third, the relative shift of T_{p_c} is very small. This reflects the fact that the change of L_c is essentially due to the superfluid, probed in terms of $1/\lambda^2_{ab}$. Accordingly, $\Delta L_c/L_c \approx \Delta\lambda^2_{ab}(T_{p_c})/\lambda^2_{ab}(T_{p_c})$ for $x = 0$, 0.2 and 0.3.

To appreciate the implications of these estimates, we note again that for fixed Pr concentration the lattice parameters remain essentially unaffected [29, 30]. Accordingly, an electronic mechanism without coupling to lattice degrees of freedom implies $\Delta L_c = 0$. On the contrary, a significant change of L_c upon oxygen exchange uncovers the coupling between the superfluid, probed by λ^2_{ab}, and the oxygen lattice degrees of freedom. A glance to table II shows that the relative change of the superconducting domains along the c-axis upon oxygen isotope exchange is significant, ranging from 12 to 16%, while the relative change of the inflection point T_{p_c} is an order of magnitude smaller. For this reason the significant relative change of L_c at fixed Pr concentration is accompanied by essentially the same relative change of λ^2_{ab}, which probes the superfluid. This uncovers unambiguously the existence and relevance of the coupling between the superfluid and oxygen lattice degrees of freedom. Furthermore, this behavior agrees with the isotope effect on d_s [54], the limiting length scale close to the 2D-QSI transition. Potential candidates for the relevant lattice degrees of freedom are the Cu-O bond-stretching-type phonons showing temperature dependence, which parallels that of the superconductive order parameter [67].

An additional probe to unravel the mystery of superconductivity in the cuprates is the response to a magnetic field. In the early discussion of the symmetry of the order parameter, Yip and Sauls [68] proposed that the angular position of the gap nodes could be probed by a measurement of the magnetic-field dependence of the penetration depth. In the local limit and for $T \to 0$, they predicted the linear relationship, $(\lambda(H=0,T=0)/\lambda(H,0))^2 - 1 \propto -H$, where the factor of proportionality is independent of temperature. Several experimental groups tried to verify this prediction, but failed to identify a linear H term which scaled with temperature according to the theory [69-72]. On the other hand, calculations based on a d-wave model, treated in the quasiclassical approximation, suggest that $(\lambda(H=0,T=0)/\lambda(H,0))^2 - 1 \propto -\sqrt{H}$ [73]. However, in these treatments fluctuations have been neglected. Close to the phase transition line $T_c(p)$, where thermal fluctuations dominate, the in-plane penetration depth scales as $\lambda_{ab}^2 \propto \xi_c = \xi_{ab}/\gamma$, while the magnetic field applied parallel to the c-axis scales as $H_c \propto \Phi_0/\xi_{ab}^2$. Thus the in-plane penetration satisfies at T_c the scaling form

$$\frac{1}{\lambda_{ab}^2(T_c, H_c)} \propto \sqrt{H_c}\,, \tag{22}$$

revealing that a superconductor is dramatically influenced by an applied magnetic field. This behavior can be understood by noting that in an applied magnetic field the correlation length cannot grow indefinitely. For nonzero magnetic field H_c there is the limiting length scale $L_{H_c} \simeq \sqrt{\Phi_0/(aH_c)}$ with $a \simeq 3.12$ [61], related to the average distance between vortex lines. Indeed, as the magnetic field increases the density of vortex lines becomes greater, but this cannot continue indefinitely. The limit is roughly set on the proximity of vortex lines by the overlapping of their cores. Due to this limiting length scale the phase transition of a homogeneous superconductor is rounded and occurs smoothly. At $T=0$ and close to the 2D-QSI transition the in-plane penetration depth scales as $\lambda_{ab}^2 \propto \xi_{ab}^z$, where z is the dynamic critical exponent of this transition. With $H_c \propto \Phi_0/\xi_{ab}^2$ this yields the scaling form

$$\left(\frac{\lambda(0,0)}{\lambda(H_c, T=0)}\right)^2 - 1 \propto -H_c^{z/2}. \tag{23}$$

Contrariwise at $T=0$ and close to the 3D-QSN transition the in-plane penetration depth scales as $\lambda_{ab}^2 \propto \xi_{ab}^{z+1}$, so that

$$\left(\frac{\lambda(0,0)}{\lambda(H_c, T=0)}\right)^2 - 1 \propto -H_c^{(z+1)/2}. \tag{24}$$

Since the order parameter is assumed to be a complex scalar, these scaling forms hold for both s-wave and d-wave pairing. Taking the evidence for a 2D-QSI transition with $z=1$ and a 3D-QSN transition with $z=2$ into account [4], the scaling form $(\lambda(H=0,T=0)/\lambda(H,0))^2 - 1 \propto -\sqrt{H_c}$ is expected to hold in the underdoped regime, while in the overdoped limit $(\lambda(H=0,T=0)/\lambda(H,0))^2 - 1 \propto -H_c^{3/2}$ should apply. In any case more extended experimental investigations are required, including samples covering the full doping range, to overcome the present impasse. Another property suited to shed light

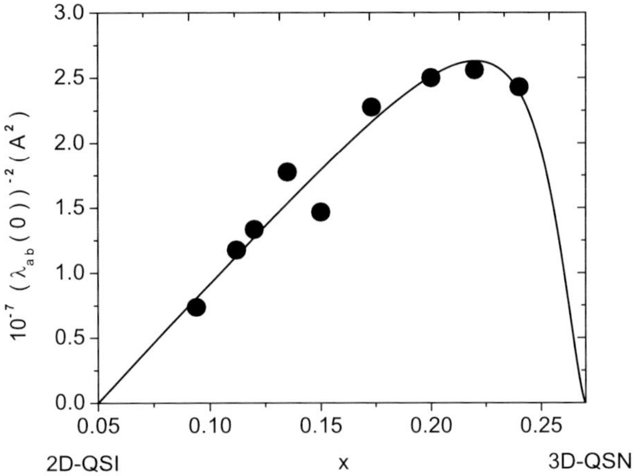

Fig. 3. – $1/\lambda_{ab}^2(0)$ *vs.* $p \simeq x$ for $La_{2-x}Sr_xCuO_4$. •: experimental data taken from Panagopoulos *et al.* [75]. The solid line indicate the crossover from a 2D-QSI transition with $z = \overline{\nu} = 1$ to a 3D-QSN transition with $z = 2$ and $\overline{\nu} = 1/2$.

on the critical properties of the quantum transitions is the magnetic-field dependence of the zero-temperature specific-heat coefficient. At $T = 0$ and close to the 2D-QSI or 3D-QSN transitions it scales as $(c/T)_{T=0} \propto \xi_{ab}^{D-z} \propto H_c^{(D-z)/2}$. The data taken on $La_{2-x}Sr_xCuO_4$ [74] points to $(D-z)/2 \simeq 1/2$, irrespective of the dopant concentration. This suggests, respectively, $z = 1$ for the 2D-QSI and $z = 2$ for the 3D-QSN transition.

The doping dependence of the zero-temperature penetration depths provides another link between the quantum critical behavior in the underdoped and overdoped regimes. In fig. 3 we displayed $1/\lambda_{ab}^2(0)$ *vs.* $p \simeq x$ for $La_{2-x}Sr_xCuO_4$ taken from Panagopoulos *et al.* [75]. Close to the quantum phase transitions $1/\lambda_{ab}^2(0)$ scales as $1/\lambda_{ab}^2(0) \propto \delta^{\overline{\nu}(D+z-2)}$ [50], where $\delta = p - p_u$ at the 2D-QSI and $\delta = p_o - p$ at the 3D-QSN transition. The solid line indicates the crossover from a 2D-QSI transition with $z = \overline{\nu} = 1$ to a 3D-QSN transition with $z = 2$ and $\overline{\nu} = 1/2$. While the flow to the 2D-QSI transition is apparent, the data does not extend sufficiently close to 3D-QSN criticality to confirm this crossover unambiguously. In any case, it emerges that the properties of the ground state are controlled by the crossover from the 2D-QSI to the 3D-QSN critical point. For this reason it can be understood that away from these quantum critical points, around optimum doping ($p = p_m$, see fig. 1), quantum fluctuations are suppressed to the extent that Bogoliubov quasiparticle features are observable and mean-field treatments represent a reasonable starting point. A 2D-QSI transition with $z = 1$ and $\overline{\nu} = 1$ coincides with the theoretical prediction for a 2D disordered bosonic system with long-range Coulomb interactions [76-78]. A potential candidate for 3D-QSN criticality is the disordered d-wave superconductor to disordered metal transition at weak coupling considered by Herbut [79,37], with $z = 2$ and $\overline{\nu} = 1/2$ energy. Here the disorder destroys superconductivity, while at the 2D-QSI transition it localizes the pairs.

To summarize, in the regime where mean-field treatments are expected to apply, the substantial isotope effect on the zero-temperature penetration depth, established by a variety of experimental techniques, implies a renormalization of the normal-state Fermi velocity due to an electron-lattice coupling, beyond the band velocity and the ME theory. This coupling also affects the superconducting properties and should be observable with ARPES. Close to the critical line $T_c(p)$ and the critical endpoints, either thermal or quantum fluctuations, not included in mean-field treatments, dominate. In these regions of the phase diagram the theory of quantum and thermal critical phenomena applies. Given the universality class of the respective transition, there are universal relations between critical properties, putting stringent constraints on the microscopic theory of superconductivity in the cuprates. Along the phase transition line there is in the experimentally accessible temperature regime mounting evidence for 3D-XY universality and for a 3D-2D crossover as the underdoped limit is approached. Here a 2D-QSI transition occurs. This crossover, measured in terms of the ratio $\gamma = \lambda_c/\lambda_{ab}$ (see fig. 1), is well documented [3,4] and implies that superconductivity in the cuprates is a genuine 3D phenomenon. The resulting competition between anisotropy and superconductivity raises serious doubts whether 2D mechanisms and models, corresponding to the limit $\gamma = \infty$, can explain the essential observations of superconductivity in the cuprates. Indeed, as the dopant concentration is increased, the cuprates undergo in the ground state a crossover from 2D-QSI to 3D-QSN criticality. For this reason the observation of Bogoliubov quasiparticle feature far away from these quantum critical points, around optimum doping ($p = p_{\mathrm{m}}$, see fig. 1), can be understood. Here quantum fluctuations do not dominate. Yet another facet of the mystery emerges from the evidence for a finite-size effect in the temperature dependence of the in-plane penetration depth. Although the limiting length scales may depend on the history of the sample, their dependence on oxygen isotope substitution reveals and confirms the coupling between superconducting properties and lattice degrees of freedom. Although the majority opinion on the mechanism of superconductivity in the cuprates is that it occurs via a purely electronic mechanism, and lattice degrees of freedom are supposed to be irrelevant, we have shown that the oxygen isotope effect on the in-plane penetration depth uncovers yet another facet, the hitherto ignored coupling between lattice degrees of freedom and both normal-state and superconducting-state properties, the existence of inhomogeneities giving rise to a finite-size effect. Finally we note that a variety of other properties also display pronounced phonon and electron-lattice effects: superconductivity-induced lattice changes [45, 51, 80-82], superconductivity-induced phonon renormalization [67, 83-87], tunnelling phonon structures [88, 89], etc., give additional evidence of significant electron-lattice coupling. Furthermore we have seen that the occurrence of power law terms in the low-temperature and low-magnetic-field dependence of the in-plane penetration depth do not necessarily single out d-wave pairing, but stem from fluctuations at work, associated with a complex scalar order parameter which is compatible with both s- and d-wave pairing.

* * *

The author is grateful to M. Cohen, R. Khasanov, H. Keller, K. A. Müller and D. R. Nelson for very useful comments and suggestions on the subject matter, and to C. Ambrosch-Draxl and R. Zeyher for clarifying correspondence.

REFERENCES

[1] TALLON J. L. *et al.*, *Phys. Rev. B*, **51** (1995) 12911.
[2] PRESLAND M. R. *et al.*, *Physica C*, **176** (1991) 95.
[3] SCHNEIDER T., *Europhys. Lett.*, **60** (2002) 141.
[4] SCHNEIDER T., *Physica B*, **326** (2003) 289.
[5] CHANDRASEKHAR B. S. and EINZEL D., *Ann. Phys. (Leipzig)*, **2** (1993) 535.
[6] HIRSCHFELD P. J. and GOLDENFELD N., *Phys. Rev. B*, **48** (1993) 4219.
[7] DURST A. C. and LEE P. A., *Phys. Rev. B*, **62** (2000) 1270.
[8] HARDY W. N. , BONN D. A., MORGAN D. C., LIANG R. and ZHANG K., *Phys. Rev. Lett.*, **70** (1993) 3999.
[9] SCHNEIDER T. and SINGER J. M., *J. Superconductivity*, **13** (2000) 789.
[10] MARTINDALE J. A., BARRETT S. E., O'HARA K. E., SLICHTER C. P., LEE W. C. and GINSBERG D. M., *Phys. Rev. B*, **47** (1993) 9155.
[11] LEE P. A., *Phys. Rev. Lett.*, **71** (1993) 1887.
[12] GRAF M. J., YIP S.-K., SAULS J. A. and RAINER D., *Phys. Rev. B*, **53** (1996) 15147.
[13] TAILLEFER L., LUSSIER B., GAGNON R., BEHNIA K. and AUBIN H., *Phys. Rev. Lett.*, **79** (1997) 483.
[14] MOLER K. A., SISSON D. L., URBACH J. S., BEASLEY M. R., A. KAPITULNIK A., BAAR D. J., LIANG R. and HARDY W. N., *Phys. Rev. B*, **55** (1997) 3954.
[15] WRIGHT D. A., EMERSON J. P., WOODFIELD B. F., GORDON J. E., FISHER R. A. and PHILLIPS N. E., *Phys. Rev. Lett.*, **82** (1999) 1550.
[16] WANG Y. *et al.*, *Phys. Rev. B*, **63** (2001) 094508.
[17] TINKHAM M., *Introduction to Superconductivity* (McGraw Hill, New York) 1975.
[18] ZHAO G., KIRTIKAR V. and MORRIS D. E., *Phys. Rev. B*, **63** (2001) 024503.
[19] HOFER J. *et al.*, *Phys. Rev. Lett.*, **84** (2000) 4192.
[20] KHASANOV R. *et al.*, *J. Phys. Condens. Matter*, **15** (2003) L17.
[21] KHASANOV R. *et al.*, cond-mat/0305477.
[22] TALLON J. L. *et al.*, cond-mat/0211071.
[23] UEMURA Y. J., *Solid State Commun.*, **126** (2003) 2338.
[24] DI CASTRO D. *et al.*, cond-mat/0307330.
[25] PICKETT W. E., *Rev. Mod. Phys.*, **61** (1989) 433.
[26] CAMPUZANO J. C., NORMAN M. R. and RANDERIA M., cond-mat/0209476.
[27] LANZARA A. *et al.*, *Nature*, **420** (2001) 511.
[28] LEE S. F. *et al.*, *Phys. Rev. B*, **53** (1996) 11825.
[29] CONDER K. *et al.*, in *Phase Separation in Cuprate Superconductors*, edited by SIGMUND E. and MULLER K. A. (Springer, Berlin) 1994, p. 210.
[30] RAFFA F. *et al.*, *Phys. Rev. Lett.*, **81** (1998) 5912.
[31] WANG N. L. *et al.*, *Phys. Rev. Lett.*, **89** (2002) 087003.
[32] MIGDAL A. B., *Sov. Phys. JETP*, **7** (1958) 996; ELIASHBERG G. M., *Sov. Phys. JETP*, **11** (1960) 696.
[33] MAKSIMOV E. I., *Zh. Eksp. Teor. Fiz.*, **37** (1969) 1562.
[34] CARBOTTE J. P., *Rev. Mod. Phys.*, **62** (1990) 1027.
[35] PARAMEKANTI A., RANDERIA M. and TRIVEDI N., cond-mat/0305611.
[36] DEPPELER A. and MILLIS A. J., *Phys. Rev. B*, **65** (2002) 100301.
[37] DEPPELER A. and MILLIS A. J., *Phys. Rev. B*, **65** (2002) 224301.
[38] BIGGS T. *et al.*, cond-mat/0203392.
[39] FISHER D. S., FISHER M. P. A. and HUSE D. A., *Phys. Rev. B*, **43** (1991) 130.
[40] SCHNEIDER T. and KELLER H., *Int. J. Mod. Phys. B*, **8** (1993) 487.
[41] OVEREND N., HOWSON M. A. and LAWRIE I. D., *Phys. Rev. Lett.*, **72** (1994) 3238.

[42] HUBBARD M. A. *et al.*, *Physica C*, **259** (1996) 309.
[43] KAMAL S. *et al.*, *Phys. Rev. Lett.*, **73** (1994) 1845.
[44] KAMAL S. *et al.*, *Phys. Rev. B*, **58** (1998) 8933.
[45] PASLER V. *et al.*, *Phys. Rev. Lett.*, **81** (1998) 1094.
[46] SCHNEIDER T., HOFER J., WILLEMIN M., SINGER J. M. and KELLER H., *Eur. Phys. J. B*, **3** (1998) 413.
[47] SCHNEIDER T. and SINGER J. M., *Physica C*, **313** (1999) 188.
[48] HOFER J. *et al.*, *Phys. Rev. B*, **60** (1999) 1332; **62** (2000) 631.
[49] SCHNEIDER T. and SINGER J. M., *Physica C*, **341-348** (2000) 87.
[50] SCHNEIDER T. and SINGER J. M., *Phase Transition Approach to High Temperature Superconductivity* (Imperial College Press, London) 2000.
[51] MEINGAST C. *et al.*, *Phys. Rev. Lett.*, **86** (2001) 1606.
[52] OSBORN K. D., VAN HARLINGEN D. J., AJI VIVEK, GOLDENFELD N., OH S. and ECKSTEIN J. N., cond-mat/0204417.
[53] LORAM J. W. *et al.*, *Physica C*, **235-240** (1994) 134.
[54] SCHNEIDER T., *Phys. Rev. B*, **67** (2003) 134514.
[55] CROWELL P. A., VAN KEULS F. W. and REPPY J. R., *Phys. Rev. B*, **55** (1997) 12620.
[56] SCHNEIDER T. and KELLER H., *Phys. Rev. Lett.*, **82** (2001) 4899.
[57] FISHER M. E. and BARBER M. N., *Phys. Rev. Lett.*, **28** (1972) 1516; FISHER M. E., *Rev. Mod. Phys.*, **46** (1974) 597; PRIVMAN V., *Finite Size Scaling and Numerical Simulation of Statistical Systems* (World Scientific, Singapore) 1990.
[58] SCHULTKA N. and MANOUSAKIS E., *Phys. Rev. Lett.*, **75** (1995) 2710.
[59] JACOBS T., SRIDHAR S., LI Q., GU G. D. and KOSHIZUKA N., *Phys. Rev. Lett.*, **75** (1995) 4516.
[60] LEE S. F. *et al.*, *Phys. Rev. Lett.*, **71** (1993) 3862.
[61] SCHNEIDER T., cond-mat/0306668.
[62] LIU J., WAN J., GOLDMAN A., CHANG Y. and JIANG P., *Phys. Rev. Lett.*, **67** (1991) 2195.
[63] CHANG A., RONG Z., IVANCHENKO Y., LU F. and WOLF E., *Phys. Rev. B*, **46** (1992) 5692.
[64] CREN T., RODITCHEV D., SACKS W., KLEIN J., MOUSSY J.-P., DEVILLE-CAVELLIN C. and LAGUËS M., *Phys. Rev. Lett.*, **84** (2000) 147.
[65] LANG K. M., MADHAVAN V., HOFFMAN J. E., HUDSON E. W., EISAKI H., UCHIDA S. and DAVIS J. C., *Nature*, **415** (2002) 413.
[66] SCHNEIDER T., KHASANOV R., CONDER K. and KELLER H., cond-mat/0302403.
[67] CHUNG J.-H. *et al.*, *Phys. Rev. B*, **67** (2003) 014517.
[68] YIP S. K. and SAULS J. A., *Phys. Rev. Lett.*, **69** (1992) 2264.
[69] MAEDA A., IINO Y., HANAGURI T., MOTOHIRA N., KISHIO K. and FUKASE T., *Phys. Rev. Lett.*, **74** (1995) 1202; MAEDA A., HANAGURI T., IINO Y., MASUOKA S., KAKATA Y., SHIMOYAMA J., KISHIO K., ASAOKA H., MATSUSHITA Y., HASEGAWA M. and TAKEI H., *J. Phys. Soc. Jpn.*, **65** (1996) 3638.
[70] CARRINGTON A., GIANNETTA R. W., KIM J. T. and GIAPINTZAKIS J., *Phys. Rev. B*, **59** (1999) 14 173.
[71] BIDINOSTI C. P., HARDY W. N., BONN D. A. and LIANG R., *Phys. Rev. Lett.*, **83** (1999) 3277.
[72] SONIER J. E., BREWER J. H., KIEFL R. F., MORRIS G. D., MILLER R. I., BONN D. A., CHAKHALIAN J., HEFFNER R. H., HARDY W. N. and LIANG R., *Phys. Rev. Lett.*, **83** (1999) 4156.
[73] SHARAPOV S. G., GUSYNIN V. P. and BECK H., *Phys. Rev. B*, **66** (2002) 012515.
[74] CHEN S. J. *et al.*, *Phys. Rev. B*, **58** (1998) 14573.

[75] PANAGOPOULOS C., XIANG T., ANUKOOL W., COOPER J. R., WANG Y. S. and CHU C. W., *Phys. Rev. B*, **67** (2003) 220502.
[76] FISHER M. P. A. *et al.*, *Phys. Rev. Lett.*, **64** (1990) 587.
[77] CHA M. C., FISHER M. P. A., GIRVIN S. M., M. WALLIN M. and YOUNG P. A., *Phys. Rev. B*, **44** (1991) 6883.
[78] HERBUT I. F., *Phys. Rev. B*, **61** (2000) 14723.
[79] HERBUT I. F., *Phys. Rev. Lett.*, **85** (2000) 1532.
[80] YOU H., WELP U. and FANG Y., *Phys. Rev. B*, **43** (1991) 3660.
[81] CHMAISSEM O., JORGENSEN J. D., SHORT S., KNIZHNIK A., ECKSTEIN Y. and SHAKED H., *Nature*, **397** (1999) 45.
[82] SANI N. L., BIANCONI A. and OYANAGI A., *J. Phys. Soc. Jpn.*, **70** (2001) 2092.
[83] PINTSCHOVIUS L. and REICHARDT W., in *Physical Properties of High Temperature Superconductors IV*, edited by GINSBERG D. (World Scientific, Singapore) 1994, p. 295.
[84] EGAMI T. and BILLINGE S. J. L., in *Physical Properties of High Temperature Superconductors V*, edited by GINSBERG D. (World Scientific, Singapore) 1996, p. 265.
[85] HADJIEV V. G., ZHOU X. J., STROHM T., CARDONA M., LIN Q. M. and CHU C. W., *Phys. Rev. B*, **58** (1998) 1043.
[86] MOOK H. A. and DOGAN F., *Nature (London)*, **401** (1999) 145.
[87] LANZARA A., BOGDANOV P. V., ZHOU X. J., KELLER S. A., FENG D. L., LU E. D., YOSHIDA T., EISAKI H., FUJIMORI A., KISHIO K., SHIMOYAMA J., NODA T., UCHIDA S., HUSSAIN Z. and SHEN Z.-X., *Nature (London)*, **412** (2001) 510.
[88] SHIMADA D., SHIINA Y., MOTTATE A., OHYAGI Y. and TSUDA N., *Phys. Rev. B*, **51** (1995) R16495.
[89] GONNELLI R. S., UMMARINO G. A. and STEPANOV V. A., *Physica C*, **275** (1997) 162.

The effect of long-term correlations on the statistics of the return intervals between rare events

A. BUNDE, J. F. EICHNER and J. W. KANTELHARDT(*)

Institut für Theoretische Physik III, Justus-Liebig-Universität - 35392 Giessen, Germany

S. HAVLIN

Minerva Center and Department of Physics, Bar-Ilan University - Ramat-Gan 52900, Israel

1. – Introduction

Extreme catastrophic events such as floods, very high temperatures or earthquakes usually occur very rarely in nature. However, in some cases it happens that not long after an extreme event another one happens, an effect which is called clustering of rare events. The origin of this clustering is not clear, and in the last decades it has been sometimes attributed to the global warming of the atmosphere. Here, we suggest that the existence of long-term correlations may partially explain such a tendency.

When investigating the occurrence of rare events (for reviews, see, *e.g.*, [1-4]) one usually considers a record of N elements x_i, $i = 1, 2, \ldots, N$, that can be, *e.g.*, annual river flows or annual temperatures at a given hydrological or meteorological station. The main quantity of interest is the time interval between events x_i that exceed a certain threshold q. When the threshold is small, the return intervals are short, when the threshold is large, the return intervals are long (see fig. 1(a)).

(*) Present address: Computational Nanoscience, Fachgruppe Theorie, Fachbereich Physik, Martin-Luther-University Halle-Wittenberg, 06099 Halle (Saale), Germany.

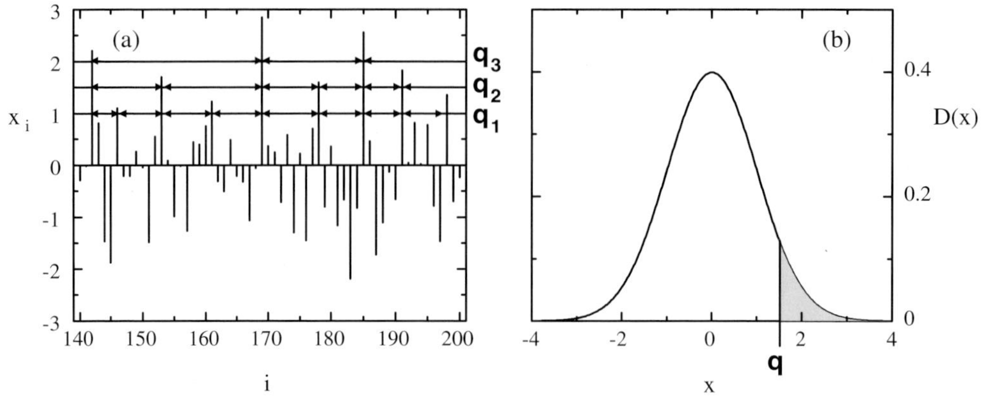

Fig. 1. – (a) Illustration of the definition of the return intervals $r_q(l)$, $l = 1, \ldots, N_q$, from a record x_i, $i = 1, \ldots, N$. The return intervals for three threshold values q_1, q_2, q_3 are indicated by arrows. (b) Illustration of the distribution function $D(x)$ of the values x_i in a record. Here, a Gaussian with zero mean and variance one is shown. The ratio between the grey area in the figure and the total area under the distribution function is the probability that an event of size greater than or equal to q occurs.

We are mainly interested in three quantities: the first one is the mean return time of an event of size q or greater, which we denote by R_q. The second quantity is the distribution of the return intervals, which we denote by $P_q(r)$. (Note that the mean return interval is just the first moment of $P_q(r)$.) In addition, we are interested in the correlation behavior of the return intervals. If they are correlated, the corresponding autocorrelation function $C_r(s)$ between return intervals separated by $s-1$ return intervals is non-zero in a certain range of s. If they are uncorrelated, the correlation function is zero except for $s = 0$. As an important consequence of long-term correlations, both R_q and $P_q(r)$ can become dependent upon the return interval r_0 immediately preceding r. Thus, we will also study conditional mean return times and their corresponding distributions.

In conventional extreme value statistics, the basic assumption is that the events are uncorrelated, at least when they are far away from each other. In this case, one can obtain the mean return time of large events of size greater than q just by considering the distribution function $D(x)$ of the x_i record. In the example in fig. 1(b), $D(x)$ is a Gaussian with zero mean and variance one. The ratio between the grey area in the figure and the total area under the distribution function is the probability that an event of size greater than or equal to q occurs, and the mean return time is just the inverse of this probability. When the events are uncorrelated, also the return intervals are uncorrelated, and the distribution function $P_q(r)$ is the well-known Poisson distribution,

$$P_q(r) = \left(\frac{1}{R_q}\right) \exp\left[-\frac{r}{R_q}\right]. \tag{1}$$

In this case, the only quantity that remains to be calculated is the mean return time, which directly follows from the tail of the distribution D. However, when q is large, an event of size $\geq q$ is a rare event. In this case, the tail of the distribution function is not well known. Accordingly, in conventional extreme value statistics a great amount of work has been devoted for obtaining reasonable extrapolations for the tail of the distribution. However, instead of analyzing $D(x)$ directly, mostly the distributions of maxima within fixed periods of time have been considered, for example the distribution $D_R(x)$ of the maxima of x_i within periods of R years. In most practical cases $D_R(x)$ converges to a Fisher-Tippet-Gumbel distribution for large R values [1-6].

In this paper we are interested in the effect of long-term correlations on the statistics of the return intervals. Accordingly, we start with a discussion of long-term correlated records in nature.

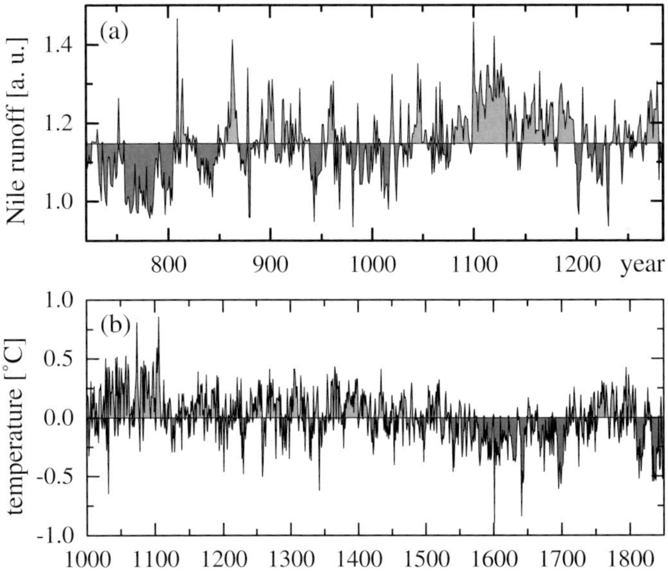

Fig. 2. – (a) Annual water levels of the Nile river (from 722 to 1284) [7] and (b) reconstructed annual temperature deviations of the northern hemisphere (1000-1850) [8]. The light grey areas indicate positive deviations of the data from their average, and the dark grey areas indicate negative deviations. Since the "patchy" look of the plots, that indicates the correlations, is more pronounced in (a), we can already guess that river runoff exhibits stronger correlations than temperatures.

2. – Long-term correlated signals

Long-term correlated sequences x_i, with zero mean, are characterized by an autocorrelation function

$$C_x(s) = \langle x_i \, x_{i+s} \rangle = \frac{1}{N-s} \sum_{i=1}^{N-s} x_i \, x_{i+s} \,, \tag{2}$$

that decays very slowly, by a power law, with the time span s separating the pairs of elements,

$$C_x(s) \sim s^{-\gamma}, \quad 0 < \gamma < 1. \tag{3}$$

The mean correlation time is the integral over $C_x(s)$. It is easy to see that this time diverges when γ is between 0 and 1, and that is why we call these records *long-term correlated*. Long-term correlated data are characterized by the tendency to stick to their momentary value for a more or less extended period of time. This may lead to pronounced positive or negative deviations from the average value for quite long periods of time. Such behavior can directly be observed, *e.g.*, in long records of river runoff data (fig. 2(a)) and in long-term temperature reconstructions (fig. 2(b)). In the figure, the red areas indicate positive deviations from the mean and the blue areas indicate negative deviations. Since the "patchy" look of the plots, that indicates the correlations, is more pronounced in fig. 2(a), we can already guess that river runoff exhibits stronger correlations than temperatures.

In the last years it has become clear that many physiological and hydrometeorological records are long-term correlated [9]. For example, the sequence of heartbeat intervals of a healthy person, during wake [10,11] and during REM sleep [12], is long-term correlated with an exponent γ close to 0.3 for REM sleep. We know, actually already since the pioneering work of Hurst [13], that daily or annual river flows measured at a hydrological station are usually long-term correlated with γ between 0.3–0.5 (see also [14, 15]). The data shown in fig. 2(a) is characterized by $\gamma \approx 0.3$. Similar values have also been obtained for many other geophysical records [16,17]. Other examples include the volatility of stock prizes with $\gamma \approx 0.15$ [18], traffic in the internet [19, 20], and atmospheric as well as sea surface temperatures [21-24]. For temperature records at continental stations, γ is always very close to 0.7 [21, 22]. For island stations as well as for sea surface temperatures, γ is around 0.4 [23,24]. This behavior of the sea surface temperatures leads to $\gamma \approx 0.4$ for the reconstructed annual temperature data of the northern hemisphere shown in fig. 2(b). The fact that for all continental stations, irrespective of their climate zone and their distance from the oceans, the range of γ values is very narrow [25], has made this power law an ideal (and uncomfortable) test bed for global climate models [26].

Next we consider how the mean return interval R_q, the distribution function $P_q(r)$ of the return intervals and their correlation behavior are changed in the presence of long-term correlations. We will always assume that the x_i values are chosen from a

Gaussian distribution with zero mean and unit variance. We have generated the long-term correlated series using the Fourier transform technique (see, *e.g.*, [27] and references therein). First we consider the mean return interval R_q.

3. – Mean return interval R_q

For simplicity, we assume periodic boundary conditions. In this case, for a given threshold q, there exist N_q return intervals $r_q(l)$, $l = 1, 2, \ldots, N_q$, which then satisfy the sum rule

$$\sum_{l=1}^{N_q} r_q(l) = N. \tag{4}$$

When we shuffle the data randomly, thereby destroying the long-term correlations, the sum rule also applies (with the same N_q value). Accordingly, for both long-term correlated and uncorrelated records, R_q is given by

$$R_q = \frac{1}{N_q} \sum_{l=1}^{N_q} r_q(l) = \frac{N}{N_q}, \tag{5}$$

i.e. the mean return interval is not changed in the presence of long-term correlations, and the techniques developed for uncorrelated data to calculate R_q can be used also for records with long-term correlations. Next we consider the distribution $P_q(r)$ of the return intervals.

4. – Distribution function $P_q(r)$ of the return intervals

To investigate $P_q(r)$ as a function of γ, we have generated long records ($N = 10^6$) for $\gamma = 0.1, 0.4, 0.7$, and without correlations ($\gamma = \infty$). For each γ value we calculated $P_q(r)$ for several threshold values q. The result for $\gamma = 0.4$ and $q = 1.5$ is shown in fig. 3(a) (in grey), where also $P_q(r)$ for uncorrelated data is shown (in black) for comparison. This semi-logarithmic plot $P_q(r)$ for $\gamma = 0.4$ differs considerably from the Poisson distribution for uncorrelated data. The probability of having return intervals well below R_q and well above R_q is strongly enhanced compared with the uncorrelated case. To determine the functional form of $P_q(r)$, we have plotted, in a double-logarithmic fashion, $-\ln(P_q(r)/P_q(1))$ as a function of r/R_q. The results for $\gamma = 0.1, 0.4, 0.7, \infty$ and $q = 1.5, 2.0, 2.5$ are shown in fig. 3(b). For each value of γ, the curves with different q-values collapse to a single line. Each line, in the double-logarithmic plot, has a slope very close to γ. Therefore, for long-term correlated records with exponent γ between 0 and 1, the distribution function of the return intervals becomes simply a stretched exponential,

$$\ln P_q(r) \sim -\left(\frac{r}{R_q}\right)^{\gamma}, \tag{6}$$

where the exponent is identical with the correlation exponent.

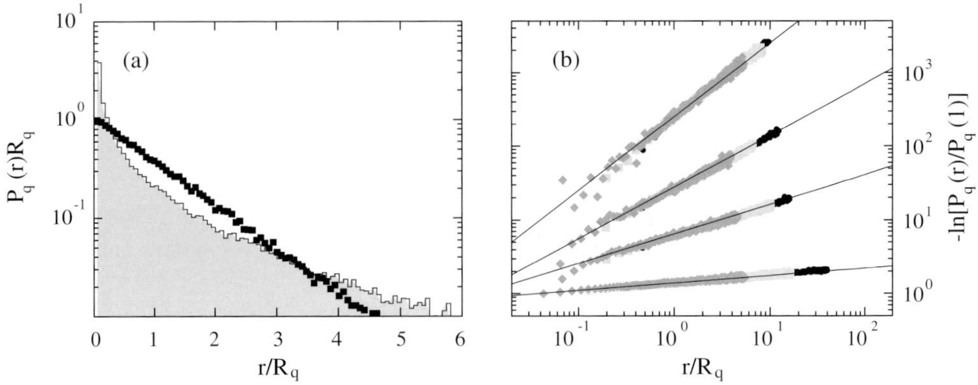

Fig. 3. – (a) Distribution $P_q(r)$ of the return intervals r for the quantile $q = 1.5$ for uncorrelated data (filled squares) and for long-term correlated data ($\gamma = 0.4$, grey histogram). (b) Double-logarithmic plot of $-\ln(P_q(r)/P_q(1))$ as a function of r/R_q for $\gamma = 0.1, 0.4, 0.7, \infty$ (from bottom to top) and $q = 1.5$ (black), 2.0 (light grey), 2.5 (dark grey). In this presentation, the slope of the curves corresponds to the exponent γ in the stretched exponential (6). The straight lines are shown for comparison and they have the slope γ for the long-term correlated data and one for the uncorrelated data ($\gamma = \infty$). (After [28].)

5. – Clustering of the return intervals

Equation (6) indicates that return intervals both well below and well above R_q are considerably more frequent in long-term correlated data than in uncorrelated data. It does not quantify, however, if the return intervals themselves are arranged in a correlated or in an uncorrelated fashion. To study the correlations, we evaluated the autocorrelation function $C_r(s) = \langle r_q(l)r_q(l+s)\rangle - R_q^2$. The results for $\gamma = 0.4$ and 0.7 and three q-values each (q = 1, 1.5, and 2) are shown in fig. 4. In the double-logarithmic presentation, for each γ-value, the three curves are parallel straight lines with the slopes $-\gamma$. This suggests that also the return intervals are long-term correlated, with the same exponent γ as in the original record. Accordingly, large and small return intervals are not arranged in a random fashion. Instead, we expect them to form clusters.

The calculation of the autocorrelation functions requires records of more than 10^6 data points. Real or reconstructed records, like the annual water levels of the river Nile [7] or the northern hemisphere temperature reconstruction [8], consist typically of 10^3 data points. Thus, for quantifying the clustering of extreme events in real data, we need to consider quantities that require considerably less statistics. Such a quantity is the conditional mean return interval R_{q,r_0}, which is defined as the mean value of those intervals in the record that immediately follow an interval of length r_0. For uncorrelated systems, subsequent return intervals are independent of each other and $R_{q,r_0} = R_q$. For long-term correlated records, we expect $R_{q,r_0}/R_q < 1$ for r_0 well below R_q and $R_{q,r_0}/R_q > 1$ for r_0 well above R_q. Figure 5 shows the numerical results for $R_{q,r_0}/R_q$ as a function of r_0/R_q, for $\gamma = 0.4$ and q = 1, 1.5, and 2. Within the numerical accuracy, the

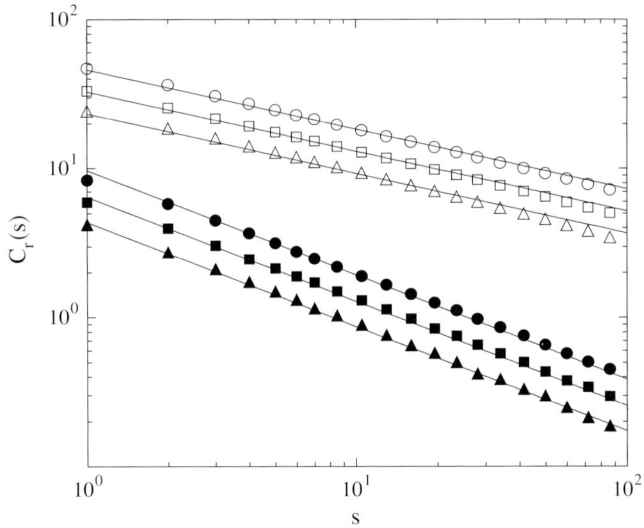

Fig. 4. – Autocorrelation function $C_r(s)$ of the return intervals for $q = 1.0$ (circles), $q = 1.5$ (squares), and $q = 2.0$ (triangles) for long-term correlated data with $\gamma = 0.4$ (open symbols) and $\gamma = 0.7$ (filled symbols). In the double-logarithmic plot, the slopes of the straight lines are equal to γ. The plot suggests that also the return intervals are long-term correlated, with the exponent γ of the original records. The length of the series was $N = 2^{21}$, and averages over 1000 configurations have been performed. (After [28].)

data for different q-values scale. The figure quantifies the way clumping and stretching of large events occur when the data are long-term correlated. For $\gamma = 0.4$, for example, R_{q,r_0} can be as low as $0.7R_q$ for $r_0 \simeq R_q/5$ and as large as $1.8R_q$ for r_0 close to $5R_q$. Next, we study how the condition r_0 affects the distribution of the following return intervals r.

6. – Conditional distribution function $P_q(r|r_0)$

In sect. **4** we have shown that the distribution of all return intervals r, $P_q(r)$, is a stretched exponential characterized by the mean return interval R_q and the correlation exponent γ. Now, since R_{q,r_0} significantly depends upon the value of the immediately preceding interval r_0, it is likely that also the corresponding conditional distribution function $P_q(r|r_0)$ depends on r_0. $P_q(r|r_0)$ is the probability of finding a return interval r provided that the preceding interval is r_0. Figure 6 shows the histograms corresponding to $P_q(r|r_0)$ for long-term correlated data and for three ranges of r_0. One can see easily that for all three cases the functional forms are identical, if we plot $P_q(r|r_0)$ *vs.* $r/R_{q,r_0}$. Hence we have, in generalization of eq. (6),

$$\ln P_q(r|r_0) \sim -\left(\frac{r}{R_{q,r_0}}\right)^{\gamma} \tag{7}$$

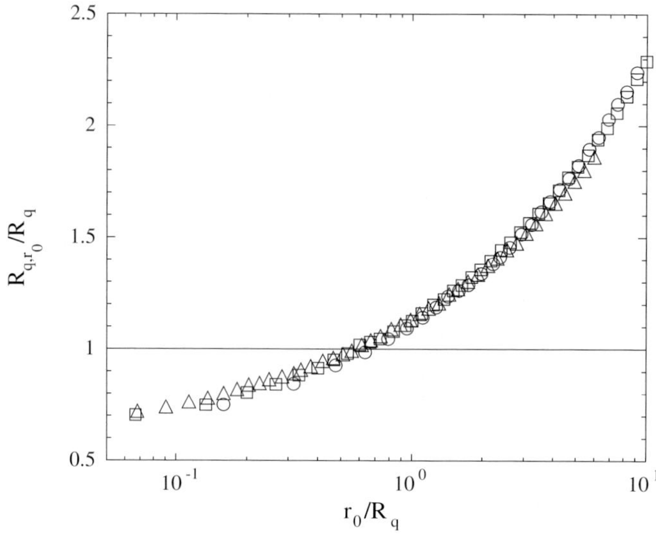

Fig. 5. – Mean conditional return interval R_{q,r_0} *vs.* preceding return interval r_0 (normalized by R_q) in a double-logarithmic plot for $\gamma = 0.4$, and for $q = 1.0$ (circles), $q = 1.5$ (squares), and $q = 2.0$ (triangles). The straight line is the result for the shuffled uncorrelated data. The length of the series was $N = 2^{21}$, and averages over 1000 configurations have been performed. (After [28].)

for large $r/R_{q,r_0}$. To understand the r_0-dependence of the prefactor, we note that to a first-order approximation (which is rigorous in the uncorrelated case) $P_q(r|r_0)$ is given by

$$P_q(r|r_0) \simeq P_q(r)P_q(r_0). \tag{8}$$

Correlations and clustering will lead to deviations from this behavior. Figure 7 shows $P_q(r|r_0)/[P_q(r)P_q(r_0)]$ for long-term correlated data. Deviations of this quantity from 1 indicate the clustering. The plot is clearly symmetric regarding r and r_0, reflecting the time reversal symmetry. The probability of finding two successive return intervals of similar size, *i.e.* $r \approx r_0$, is significantly increased, although the non-universal strength of the increase hinders a full data collapse. On the other hand, the probability of finding very different r and r_0 immediately following each other is drastically reduced. The plot can be used to estimate the change in the probability for a certain return interval r after the occurrence of a given interval r_0. It shows that this change in the probability mainly depends upon the ratio r/r_0, at least for rather small or rather large ratios.

7. – Evaluation of observational and reconstructed datasets

For testing the relevance of our results for real records, we have analyzed the historical record of water levels of the Nile river [7] and an 851 year long reconstructed record of annual northern hemisphere temperatures [8]. Both data sets are long-term correlated,

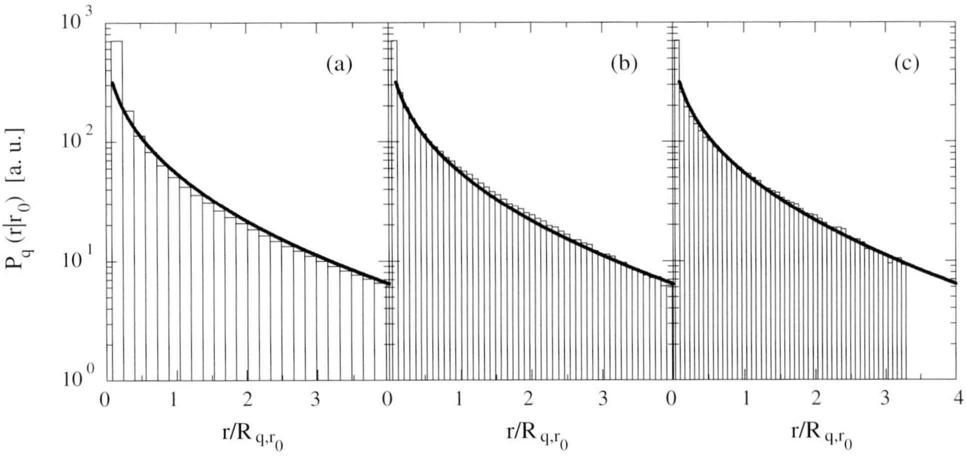

Fig. 6. – Histograms for the conditional probability $P_q(r|r_0)$ for the occurrence of return intervals r immediately after return intervals r_0 in the ranges (a) $r_0 \in (0.0, 0.1R_q)$, (b) $r_0 \in (2.0R_q, 2.1R_q)$, (c) $r_0 \in (4.0R_q, 4.1R_q)$, showing 50 bins. After the rescaling procedure, where r was divided by the conditional mean return interval R_{q,r_0}, all distributions have the same stretched exponential form. The data was long-term correlated with $\gamma = 0.4$, and 1500 configurations of length $N = 2^{21}$ were considered. The quantile was $q = 2.0$.

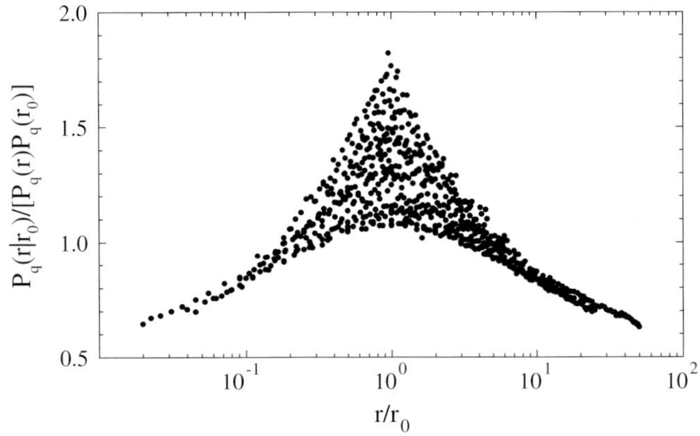

Fig. 7. – Collapse of the distributions $P_q(r|r_0)$ for many different values of r and r_0. By dividing by $P_q(r)$ and $P_q(r_0)$, we obtain a data collapse asymptotically, although there is no collapse for $r \approx r_0$. The data was long-term correlated with $\gamma = 0.4$, and 400 configuration were considered. The quantile was $q = 1.5$.

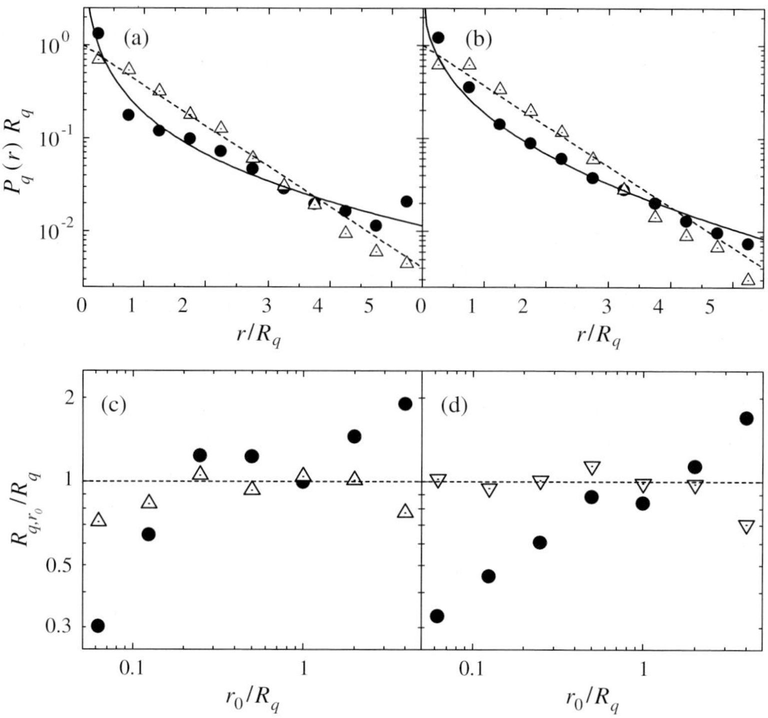

Fig. 8. – Distribution function $P_q(r)$ and mean conditional return interval R_{q,r_0} for the data shown in fig. 2, (a, c) annual water levels of the Nile river (from 722 to 1284) [7] and (b, d) reconstructed annual temperature data of the northern hemisphere (1000-1850) [8]. The full circles represent the observed data sets, while the open triangles represent data sets that are obtained by randomly shuffling the observed original records. The original records exhibit long-term correlations with $\gamma \approx 0.3$ for the Nile and $\gamma \approx 0.4$ for the temperature record. Since $P_q(r)R_q$ as well as $R_{q,r_0}/R_q$ depend only on r/R_q (see figs. 3(b) and 5), we averaged $P_q(r)R_q$ and $R_{q,r_0}/R_q$ over several q-values in order to improve the statistics. For the temperature record we limited ourselves to data up to 1850 to exclude possible clustering of rare events due to global warming. (After [28].)

with $\gamma \approx 0.3$ for the Nile and $\gamma \approx 0.4$ for the temperature records. For comparison with the theoretical results, we have standardized the records to zero mean and unit variance. The results are presented in fig. 8. Figures 8(a, b) show the distribution $P_q(r)$ of the return intervals r for the original data (filled circles) and for the shuffled data (open triangles). While the shuffled data clearly follow the Poisson distribution (dashed line), the original data are close to a stretched exponential decay (continuous line) with $\gamma = 0.3$ (Nile) and $\gamma = 0.4$ (temperatures) in agreement with the prediction (6). The fluctuations of the results are due to the short length of the records.

Figures 8(c, d) show the mean conditional return interval R_{q,r_0}, for both the water levels of the Nile river and the temperature data. The figures indicate clustering of small and large return intervals, in agreement with the predicted behavior of fig. 5. For the shuffled data, $R_{q,r_0}/R_q$ is close to one and the clustering disappears, as expected. The scattering on both the original and the shuffled data is due to the short length of the records.

8. – Conclusion

In this paper, we have shown that for long-term correlated records with a Gaussian distribution, the distribution of the return intervals follows a stretched exponential with an exponent γ identical to the correlation exponent. The distribution of conditional return times follows the same functional dependence, and only the mean conditional return interval varies with the value of the preceding interval. We also found that this distribution cannot fully characterize the appearance of rare events since the return intervals appear in a long-term correlated fashion. The decay of the correlations in the return interval is again described by the same exponent γ. It is important to emphasize that both the distribution function and the long-term correlations between the return intervals scale the same for different q-values (figs. 3(b) and 5). Due to the scaling, we can evaluate the behavior of return intervals of extremely rare events (very large q-values) from the behavior of intermediate q-values that are accessible in the observational data. We also presented evidences which indicate that both the stretched exponential behavior and the clustering of rare events can be seen in long observational and reconstructed records.

Clustering of rare events (*e.g.* floods) has been reported for the Middle Ages (see, *e.g.*, [29,30]), where anthropogenic forcings were clearly absent, as well as in the past 100 years. It has been argued that the clustering of catastrophic events in the last century is due to the global warming of the atmosphere. Our results may explain the clustering of catastrophic events in the Middle Ages and indicate that at least part of the clustering of rare events may be due to long-term persistence.

* * *

This work has been supported by the Bundesministerium für Bildung und Forschung and the Minerva Foundation.

REFERENCES

[1] GUMBEL E. J., *Statistics of Extremes* (Columbia University Press, New York) 1958.
[2] GALAMBOS J., LECHNER J. and SIMIN E. (Editors), *Extreme Value Theory and Applications* (Kluwer, Dordrecht) 1994.
[3] EMBRECHTS P., KLÜPPELBERG C. and MIKOSCH T., *Modelling Extremal Events*, edited by KARATZAS I. and YOR M. (Springer, Berlin) 1997.
[4] V. STORCH H. and ZWIERS F. W., *Statistical Analysis in Climate Research* (Cambridge University Press, Cambridge) 2002.
[5] FISHER R. A. and TIPPETT L. H. C., *Proc. Camb. Philos. Soc.*, **24** (1928) 180.
[6] ANTAL T., DROZ M., GYÖRGYI G. and RACZ Z., *Phys. Rev. Lett.*, **87** (2001) 240601.

[7] Tousson O., *Memoire sur l'histoire du Nil*, in *Memoiries a l'Institut d'Egypte*, **18** (1925) 366; Popper W., *The Cairo Nilometer*, in *Publications in Semitic Philology 12* (University of California Press, Berkeley) 1951; data obtained from: `http://sunsite.univie.ac.at/statlib/S/beran`; see also: Whitcher B., Byers S. D., Guttorp P. and Percival D. B., *Wat. Resourc. Res.*, **38** (2002) 12.

[8] Jones P. D., Briffa K. R., Barnett T. P. and Tett S. F. B., *The Holocene*, **8** (1998) 455; data obtained from: IGBP PAGES/World Data Center-A for Paleoclimatology Data Contribution Series #1998-039; NOAA/NGDC Paleoclimatology Program, Boulder CO, USA.

[9] Bunde A., Kropp J. and Schellnhuber H.-J. (Editors), *The Science of Disasters — Theories of Disasters in Weather, Biology and Finance* (Springer, Berlin) 2002.

[10] Peng C.-K., Mietus J., Hausdorff J. M., Havlin S., Stanley H. E. and Goldberger A. L., *Phys. Rev. Lett.*, **70** (1993) 1343.

[11] Ivanov P. Ch., Bunde A., Amaral L. A. N., Havlin S., Fritsch-Yelle J., Baevsky R. M., Stanley H. E. and Goldberger A. L., *Europhys. Lett.*, **48** (1999) 594.

[12] Bunde A., Havlin S., Kantelhardt J. W., Penzel T., Peter J. H. and Voigt K., *Phys. Rev. Lett.*, **85** (2000) 3736.

[13] Hurst H. E., *Trans. Am. Soc. Civil Eng.*, **116** (1951) 770.

[14] Tessier Y., Lovejoy S., Hubert P., Schertzer D. and Pecknold S., *J. Geophys. Res. Atmosph.*, **101(D21)** (1996) 26427.

[15] Koscielny-Bunde E., Kantelhardt J. W., Braun P., Bunde A. and Havlin S., submitted to *J. Hydrology*.

[16] Hurst H. E., Black R. P. and Simaika Y. M., *Long-Term Storage: an Experimental Study* (Constable, London) 1965.

[17] Mandelbrot B. B. and Wallis J. R., *Wat. Resour. Res.*, **5** (1969) 321.

[18] Liu Y., Cizeau P., Meyer M., Peng C.-K. and Stanley H. E., *Physica A*, **245** (1997) 437.

[19] Leland W. E., Taqqu M. S., Willinger W. and Wilson D. V., *IEEE Trans. Network.*, **2** (1994) 1.

[20] Paxson V. and Floyd S., *IEEE Trans. Network.*, **3** (1995) 226.

[21] Koscielny-Bunde E., Bunde A., Havlin S. and Goldreich Y., *Physica A*, **231** (1996) 393.

[22] Koscielny-Bunde E., Bunde A., Havlin S., Roman H. E., Goldreich Y. and Schellnhuber H.-J., *Phys. Rev. Lett.*, **81** (1998) 729.

[23] Pelletier J. D. and Turcotte D. L., *J. Hydrology*, **203** (1997) 198.

[24] Monetti R. A., Havlin S. and Bunde A., *Physica A*, **320** (2003) 581.

[25] Eichner J. F., Koscielny-Bunde E., Bunde A., Havlin S. and Schellnhuber H.-J., *Phys. Rev. E*, **68** (2003) 046133.

[26] Govindan R. B., Vyushin D., Bunde A., Brenner S., Havlin S. and Schellnhuber H.-J., *Phys. Rev. Lett.*, **89** (2002) 028501.

[27] Makse H. A., Havlin S., Schwartz M. and Stanley H. E., *Phys. Rev. E*, **53** (1996) 5445.

[28] Bunde A., Eichner J. F., Havlin S. and Kantelhardt J. W., *Physica A*, **330** (2003) 1.

[29] Pfister C., *Wetternachhersage. 500 Jahre Klimavariationen und Naturkatastrophen 1496-1995* (Verlag Paul Haupt, Bern) 1998, p. 244.

[30] Glaser R., *Klimageschichte Mitteleuropas* (Wissenschaftliche Buchgesellschaft, Darmstadt) 2001.

POSTERS

Exchanges in complex networks: Income and wealth distributions

T. Di Matteo

Applied Mathematics, Research School of Physical Sciences
Australian National University - 0200 Canberra, Australia
INFM-Dipartimento di Fisica "E. R. Caianiello"
via S. Allende, 84081 Baronissi (SA) Italy

T. Aste and S. T. Hyde

Applied Mathematics, Research School of Physical Sciences
Australian National University - 0200 Canberra, Australia

1. – Introduction

Empirically the literature reports several behaviors for the income and wealth distributions in different countries. A century ago, the Italian social economist Pareto suggested a power law [1] distribution in the high-income range, namely, in terms of cumulative distribution: $P_>(w) \propto w^{-\alpha}$, with α being the Pareto index [2]. On the other hand, Montroll [3] suggested a lognormal distribution with power law tail for the USA personal income. More recently, wealth and income distributions in the USA and in the United Kingdom have been described by an exponential distribution with power law high-end tails [4]. Whereas, the Japanese personal income distribution appears to follow lognormal distributions also with power law tails [5, 6]. In some recent papers Zipfs law has also been proposed [7]. In this paper we add to the above empirical investigations an analysis of the income distribution in Australia (fig. 1).

From the theoretical side, it has been shown that pure multiplicative stochastic (MSP) processes can explain the lognormal income distribution but they fail to explain the power law tails [8]. Power law tails can be obtained extending MSP processes by including —for instance— additive noise and boundary constraints [9-12]. These models explain well the emergence of power law distributions, but they are incomplete, neglecting interactions between agents. Hence, MSPs with interacting agents connected through a network have

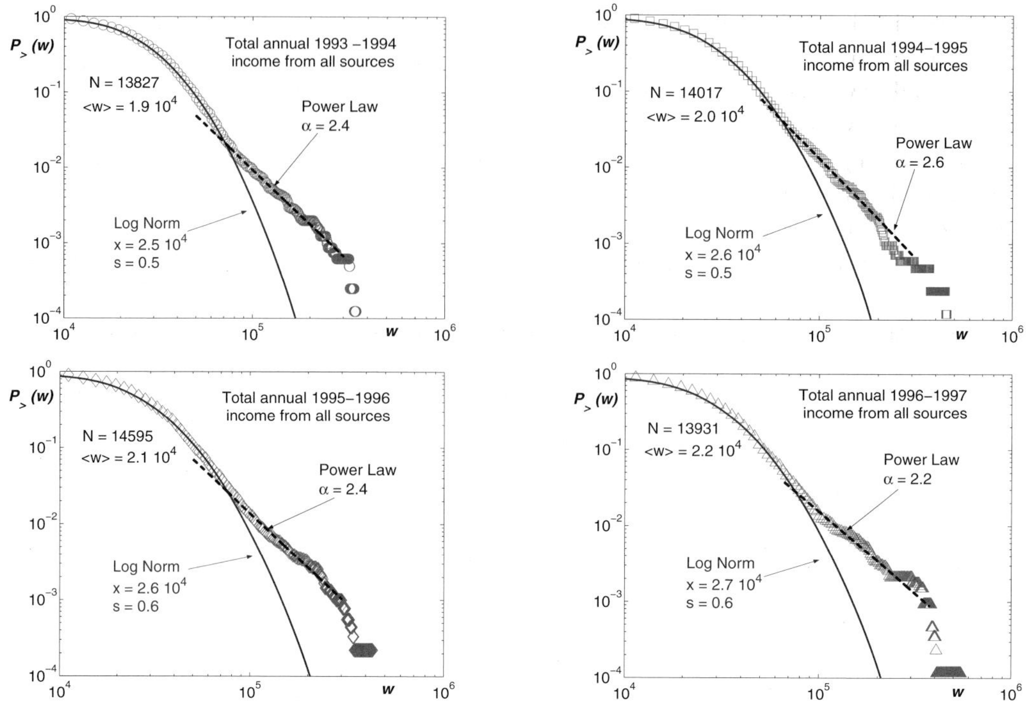

Fig. 1. – Complementary cumulative distributions for the total annual income from all sources in Australia in the years 1993-1997.

been developed [13-16]. These models retrieve power law tails with exponents α which are related to the network properties.

In this paper, we show that distributions with power law tails can emerge also from additive stochastic processes with interacting agents. In this case, we show that the network of connections among agents plays a crucial role. Indeed, the resulting wealth distribution is shaped directly by the degree distribution of the network. The original purpose of the present work was not to construct any realistic model for the wealth distribution. Our aim was simply to demonstrate the possibility to obtain "fat" tails also without the use of multiplicative stochastic processes. Rather surprising, we find out that the results from such an additive process are in good qualitative agreement with the empirical data for the income distribution in Australia.

2. – Income distribution in Australia

Let us briefly start with the empirical analysis of the data for the incomes in Australia. We analyze data from the Australian Bureau of Statistics: "Survey of Income and Housing Costs Confidentialised Unit Record Files". In fig. 1 we report the complementary cumulative distributions ($P_>(w) = 1 - \int_{-\infty}^{w} p(\xi)\mathrm{d}\xi$) for the total annual income from all sources in the years 1993-94, 1994-95, 1995-96, 1996-97. These data are compared with two possible trends in two different regions: lognormal at medium incomes (above 10^4 A \$) and power law at high incomes. As one can see the large income region is rather well described with power-law–like tails: $P_>(w) \propto w^{-\alpha}$ with exponents α, respectively, equal to 2.4, 2.6, 2.4, 2.2. Whereas the small-incomes region is in better agreement with a lognormal distribution: $P(w) = 1/(ws\sqrt{2\pi}) \exp[-\log^2(w/x)/(2s^2)]$ (with the values for s and x reported in the figures). Let us now introduce the theoretical framework and show how these behaviors can be accounted by using an additive interacting stochastic process.

3. – Wealth distribution from interacting additive stochastic processes

Consider N agents which interact through a social network and suppose that at the time t a given agent l has a wealth $w_l(t)$. Within the same framework of other models proposed in the literature [14-16], let us first introduce a rather general expression for the wealth evolution:

$$w_l(t+1) - w_l(t) = A_l(t) + B_l(t)w_l(t) + \sum_{j(\neq l)} w_j(t)Q_{j\to l}(t) - \sum_{j(\neq l)} w_l(t)Q_{l\to j}(t), \tag{1}$$

where the coefficient $A_l(t)$ is an additive noise and the factor $B_l(t)$ is a multiplicative noise. These are stochastic variables which reflect market and social fluctuations. In addiction with these stochastic terms eq. (1) describes the exchanges between agents through a network: agent l receives a faction $Q_{j\to l}(t)$ of the wealth of agent j and gives a fraction $Q_{l\to j'}(t)$ of its wealth to agent j'. The MSP model mentioned above takes into account only the multiplicative term $B_l(t)$; their extensions introduce also the additive noise $A_l(t)$ and the interactions $Q_{j\to l}(t)$. Differently, in this paper we neglect the multiplicative term and take into account only the additive noise and the interactions. In particular we assume that: i) there are no stochastic multiplicative terms ($B_l(t) = 0$); ii) the additive term $A_l(t)$ is a Gaussian noise with average zero and variance σ_0^2; iii) each agent distributes a portion q_0 of its wealth equally among the other agents which are in contact with it through the social network. This last assumption implies

$$Q_{j\to l}(t) = \begin{cases} \dfrac{q_0}{z_j}, & \text{if } l \in \mathcal{I}_j\,, \\ 0, & \text{elsewhere,} \end{cases} \tag{2}$$

where z_j is the number of agents in contact with agent j and $\mathcal{I}_j$ represents the set of the agents which exchange with agent j. Equation (1) becomes

$$w_l(t+1) = A_l(t) + (1-q_0)w_l(t) + \sum_{j\in\mathcal{I}_l} \frac{q_0}{z_j} w_j(t). \tag{3}$$

Note that in our case $\langle A_l(t)\rangle_t = 0$ and eq. (3) describes a system which conserves in average the total wealth.

The probability $P_{t+1}(x,l)\mathrm{d}x$ that the agent l at the time $t+1$ has a wealth between x and $x+\mathrm{d}x$ is related to the probabilities to have a set $\{Q_{j\to l}(t)\}$ of exchange coefficients and a set of additive coefficient $\{A_l(t)\}$ such that a given distribution of wealth $\{w_j(t)\}$ at the time t yields, through eq. (3), to the wealth x for the agent l at time $t+1$. That is

$$P_{t+1}(x,l) = \int_{-\infty}^{\infty} \mathrm{d}a\, \Lambda_t(a,l) \int_{-\infty}^{\infty} \mathrm{d}w_1 \cdots \int_{-\infty}^{\infty} \mathrm{d}w_N P_t(w_j,j) \times \times \delta\left(x - a - (1-q_0)w_l - \sum_{j'} \frac{q_0}{z_{j'}} w_{j'}\right), \tag{4}$$

where $\delta(x)$ is the Dirac delta-function and $\Lambda_t(a,l)$ is the probability density to have at time t on site l an additive coefficient $A_l(t) = a$.

The Fourier transform of eq. (4) reads

$$\hat{P}_{t+1}(\varphi,l) = \frac{e^{-\sigma_0^2\varphi^2/2}}{\sqrt{2\pi}} \hat{P}_t((1-q_0)\varphi,l) \prod_{j\in\mathcal{I}_l} \hat{P}_t\left(\frac{q_0}{z_j}\varphi, j\right). \tag{5}$$

By definition the cumulants of the wealth probability distribution are given by the expression

$$k_l^{(\nu)}(t) = (-i)^\nu \frac{\mathrm{d}^\nu}{\mathrm{d}\varphi^\nu} \ln \hat{P}_t(\varphi,l)\bigg|_{\varphi=0}, \tag{6}$$

where the first cumulant $k_l^{(1)}(t)$ is the expectation value of the stochastic variable w_l at the time t ($\langle w_l(t)\rangle$) and the second moment $k_l^{(2)}(t)$ is its variance ($\sigma_l^2(t)$).

By taking the logarithm of eq. (5) and applying eq. (6), we get

$$k_l^{(\nu)}(t+1) = c^{(\nu)} + (1-q_0)^\nu k_l^{(\nu)}(t) + \sum_{j\in\mathcal{I}_l} \left(\frac{q_0}{z_j}\right)^\nu k_j^{(\nu)}(t), \tag{7}$$

with $c^{(2)} = \sigma_0^2$ and $c^{(\nu)} = 0$ for any $\nu \neq 2$. This equation describes the propagation of the cumulants of the wealth distribution. A consequence of this equation is that the only moments which asymptotically can be different from zero are the first (the mean) and the second (the variance). With this last being directly proportional to σ_0^2.

We now seek for stationary solutions of eq. (7), *i.e.* situations in which at infinitely large times, the cumulants do not change in time: $k_l^{(\nu)}(t) = \bar{k}_l^{(\nu)}$.

3·1. *Self-consistent solution: crystal.* – Let us first consider an "ideal" social network where every agent is connected with an equal number of other agents ($z_j = \bar{z}$) in a perfectly ordered "crystalline" structure. In this case, each agent is equivalent to each other and the asymptotic wealth distribution must be the same for every one (*i.e.*, $k_j^{(\nu)}$

independent of j). From eq. (7) it follows that the expectation value for the wealth on each site is a constant and it is equal to the average wealth at $t = 0$:

$$\langle w_l \rangle = \bar{k}_l^{(1)} = \frac{1}{N} \sum_j w_j(0). \tag{8}$$

Its variance is

$$\sigma_l^2 = \bar{k}_l^{(2)} = \frac{\sigma_0^2}{1 - q_0^2/\bar{z} - (1 - q_0)^2}. \tag{9}$$

Whereas all the other moments $\bar{k}^{(\nu)}$ are equal to zero for $\nu \geq 3$.

3·2. *Self-consistent solution: general case.* – We now consider the more general case of a non-regular network. From eq. (7) it follows that a self-consistent solution for the average stationary wealth on each node of the network is

$$\langle w_l \rangle = \bar{k}_l^{(1)} = \frac{z_l}{\bar{z}} m, \tag{10}$$

with m the average wealth on the ensemble of agents ($m = (1/N) \sum_j w_j(t)$) and $\bar{z}$ the average network connectivity ($\bar{z} = (1/N) \sum_j z_j$). Therefore, the expectation value for the wealth of a given agent results proportional to its number of connections in the social network. On the other hand, we mentioned above that the only other moment which can be different from zero is the second. Therefore, we expect that the probability to find a given wealth on a given agent is a Gaussian distribution with average $(z_l/\bar{z})m$ and finite variance.

3·3. *Wealth distribution —analytical.* – The wealth probability distribution in the ensemble of agents is given by the sum of the distributions for each agent divided by the total number of agents. We have seen above that each agent has a wealth within a Gaussian distribution with average directly proportional to its connectivity (eq. (10)) and finite variance. The resulting wealth distribution for the ensemble of agent is therefore a weighted sum of Gaussian distributions with averages proportional to the network connectivity and weights given by the degree distribution. This overall distribution is shaped by the underlying distribution of the connectivity between agents (the degree distribution). It has been observed that in many social systems the degree distribution typically follows a power law behavior in the region of large number of connections [17]. This power law behavior in the network connectivity will be therefore reflected in the wealth distribution which will assume a power law tail terminated by an exponential cut-off (for a finite system). This behavior is qualitatively in agreement with what observed empirically. A comparison between the empirical data and the distribution resulting by summing a set of 3000 Gaussian distributions with averages proportional to the connectivity $x = (z/\bar{z})x_0$ (with $x_0 = 30000$, $\bar{z} = 1.28$), equal variances $s = 18000$ and power law degree distribution $p(z) = p_0 z^{-a}$ (with $a = 3.2$, $z_{\min} = 1$, $z_{\max} = 3000$) is shown in fig. 2. As one can see the qualitative agreement is quite satisfactory.

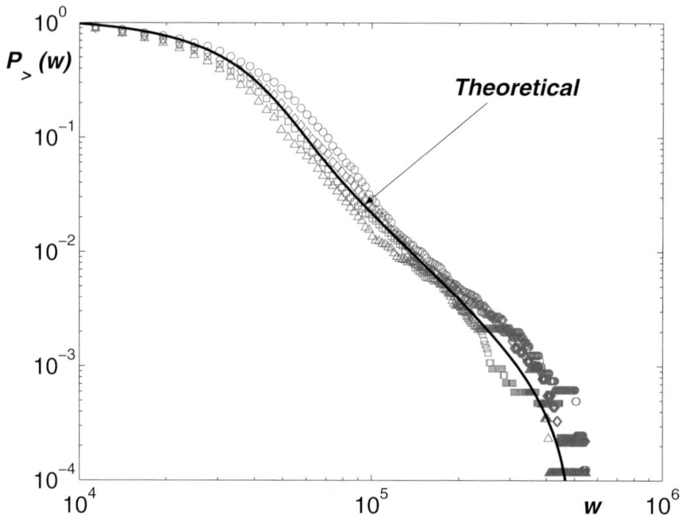

Fig. 2. – Comparison between the empirical data and the theoretical (complementary cumulative) distribution associated with a scale-free network.

3˙4. *Wealth distribution —numerical simulation.* – We generated large networks, with $N = 30000$ agents, by iteratively performing switching of neighbors in a triangulation embedded in a manifold with genus $g = 10000$. We introduce an "energy" $E = \sum_j (z_j - \bar{z})^2$ and we perform a Glauber-Kawasaki type of dynamics. This procedure is an extension to $g \neq 1$ and negative "temperatures" of the method presented in [18]. Two different networks were generated by performing 600000 switches from a disordered start respectively at inverse "temperatures" $\beta = -0.5$ (network 1) and $\beta = +0.5$ (network 2). At positive temperature a rather homogeneous network emerges with degree distribu-

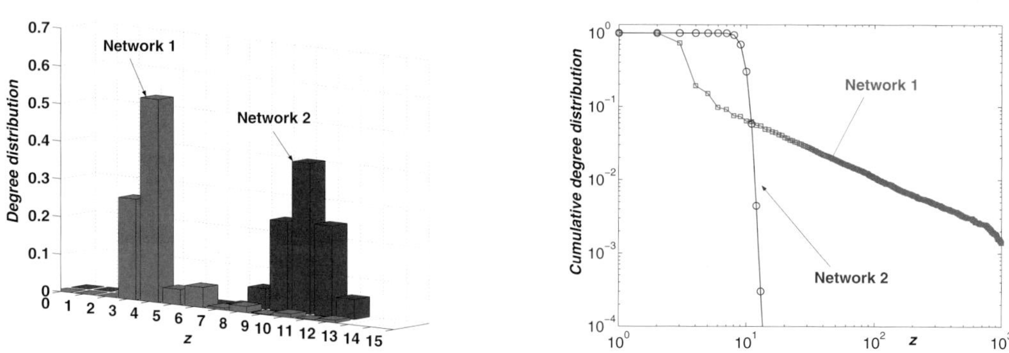

Fig. 3. – The two degree distributions for the two networks used in the simulations. (Left: degree distributions. Right: complementary cumulative degree distributions.)

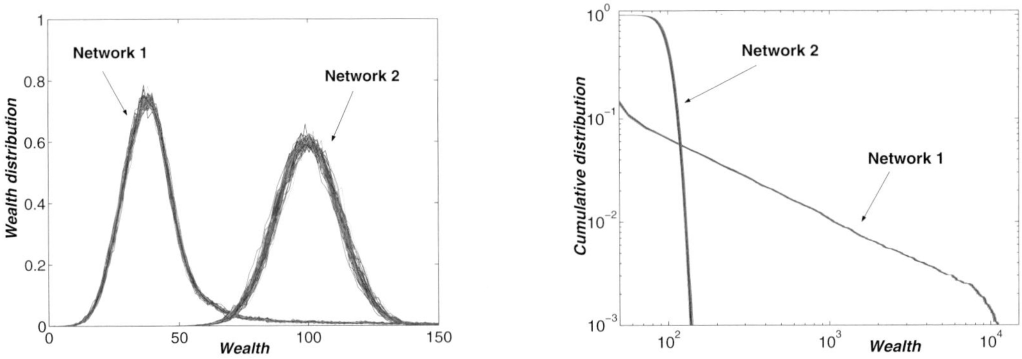

Fig. 4. – The wealth distributions (left) and their complementary cumulative distributions (right) resulting from 50 simulations performed on two the different networks (1 and 2).

tion centered around the average ($\bar{z} = 6 + 12(g-1)/N$) and with exponentially fast decreasing tails (fig. 3, network 2). On the other hand, negative temperatures favour the formation of inhomogeneous-scale free networks, with power law tails in the degree distribution (fig. 3, network 1).

Once the networks are generated, we associate to each agent an equal initial wealth of m (arbitrary) units. We set $m = 100$, $q_0 = 0.1$, $\sigma = 0.05/m$ and we run the simulation by updating at each time step all the agent's wealth by using eq. (3) up to a maximum time T. We verify that a steady state distribution is achieved after about 100 time steps and therefore we set $T = 1000$. The resulting wealth distributions for 50 simulations over fixed underlying networks are reported in fig. 4. We see that the network properties have a dramatic effect on the overall behavior of the wealth distribution. We observe an exponentially fast decay in the homogeneous network, whereas we obtain a power

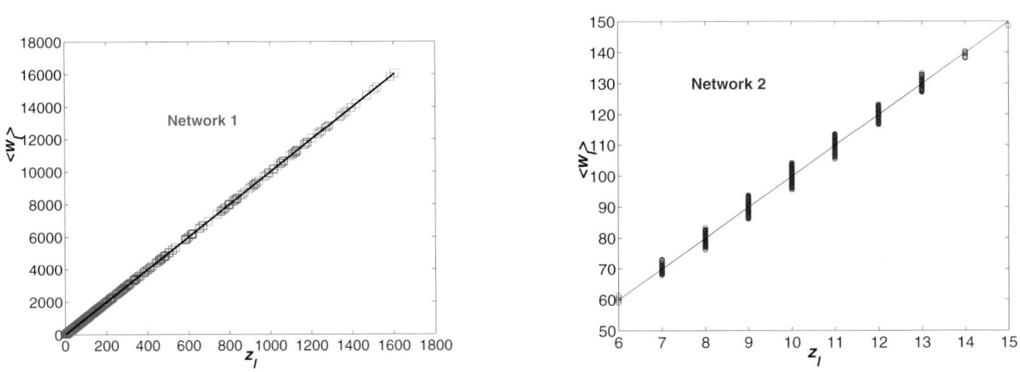

Fig. 5. – Comparison between the average values for the wealth of a given agent l (symbols) calculated from 50 simulations on the two networks (left: network 1; right: network 2) with the theoretical predictions from eq. (10) (lines).

law "fat" tail in the scale-free one. We verify that, in agreement with eq. (10), the expectation value for the wealth on each agent is proportional to its connectivity in the social network. Figure 5 reports the average (over the 50 simulation) wealth on each agent *vs.* its connectivty. The theoretical prediction (eq. (10)) is also reported showing a remarkable agreement for both the networks. The small spreading of the data (more evident for network 2) indicates that non-local effects might also have some relevance.

4. – Conclusion

We have shown that a mechanism of wealth exchange with additive Gaussian noise can produce distributions with power law tails when the network which connects the agents is of a scale-free type. Although the original purpose of this work was not to produce a realistic model for the wealth evolution, we find a good qualitative agreement between the empirical data and the theoretical prediction. More realistic models will be proposed in future works by introducing also multiplicative stochastic terms and a dynamical evolution in the network connectivity.

* * *

T. Di Matteo wishes to thank the Research School of Social Sciences, ANU, for providing the ABS Data. This work was partially founded by the ARC Discovery Project: DP0344004. We acknowledge the Australian Partnership for Advanced Computing (APAC) National Facility.

REFERENCES

[1] Gabaix X., Gopikrishnan P., Plerou V. and Stanley H. E., *Nature*, **423** (2003) 267.
[2] Pareto V., *Cours d' Èconomique Politique* (Macmillan, London) 1897.
[3] Montroll E. W. and Badger W. W., *Introduction to Quantitative Aspects of Social Phenomena* (Gordon and Breach, New York) 1974.
[4] Dragulescu A. and Yakovenko V. M., *Eur. Phys. J. B*, **20** (2001) 585; *Physica A*, **299** (2001) 213.
[5] Souma W., *Fractals*, **9** (2001) 463.
[6] Fujiwara Y., Souma W., Aoyama H., Kaizoji T. and Aoki M., *Physica A*, **321** (2003) 598.
[7] Aoyma H., Souma W., Nagahara Y., Okazaki M. P., Takayasu H. and Takayasu N., *Fractals*, **8** (2000) 293.
[8] Gibrat R., *Les Inégalités Économiques* (Sirey, Paris) 1932.
[9] Levy M. and Solomon S., *Int. J. Mod. Phys. C*, **7** (1996) 595.
[10] Sornette D., *Phys. Rev. E*, **57** (1998) 4811.
[11] Zhi-Feng Huang and Solomon S., *Physica A*, **294** (2001) 503.
[12] Solomon S. and Richmond P., *Physica A*, **299** (2001) 188.
[13] Amaral L. A. N., Buldyrev S. V., Havlin S., Salinger M. A. and Stanley H. E., *Phys. Rev. Lett.*, **80** (1998) 1385.
[14] Biham O., Malcai O., Levy M. and Solomon S., *Phys. Rev. E*, **58** (1998) 1352.
[15] Bouchaud J.-P. and Mézard M., *Physica A*, **282** (2000) 536.
[16] Solomon S. and Richmond P., *Eur. Phys. J. B*, **27** (2002) 257.
[17] Newman M. E. J., *SIAM Rev.*, **45** (2003) 167.
[18] Aste T. and Sherrington D., *J. Phys. A: Math. Gen.*, **32** (1999) 7049.

Multi-scaling of 2D impact fragmentation

H. KATSURAGI, D. SUGINO and H. HONJO

Department of Applied Science for Electronics and Materials
Interdisciplinary Graduate School of Engineering Sciences, Kyushu University
6-1 Kasugakoen, Kasuga, Fukuoka 816-8580, Japan

1. – Introduction

While the impact fragmentation phenomenon is very ubiquitous, the underlying physical mechanism has not been understood yet. Because the impact fragmentation occurs in a moment, it is difficult to study this problem. Much scientific and engineering efforts have been paid for the investigation of this phenomenon [1]. However, many unsolved problems still remain now. One of the most interesting problems for impact fragmentation is its criticality. When the imparted energy is not enough, the sample does not break. And a sufficiently large imparted energy yields many fragments. A focus of the work is on the low imparted energy region, where the energy transferred during the collision is near the point at which the samples just start to break.

Recently, the critical property of impact fragmentation has been studied by Kun and Herrmann [2], and Åström *et al.* [3]. Kun and Herrmann concluded that the criticality of 2D impact fragmentation belongs to the percolation universality class [2]. On the other hand, Åström *et al.* revealed that there is another universality for 2D impact fragmentation [3]. In addition, Hayakawa examined the dependency on dimensionality of the universality class [4]. He predicted that the universality of impact fragmentation depends not only on the dimensionality of fractured samples, but also on the dimensionality of the propagating failure wave. All those works are numerical simulations. Therefore, we investigated the criticality of impact fragmentation by experiments in order to check those numerical results. Multi-scaling was found near the fragmentation critical point, and we introduced a binomial multiplicative model to explain this multi-scaling.

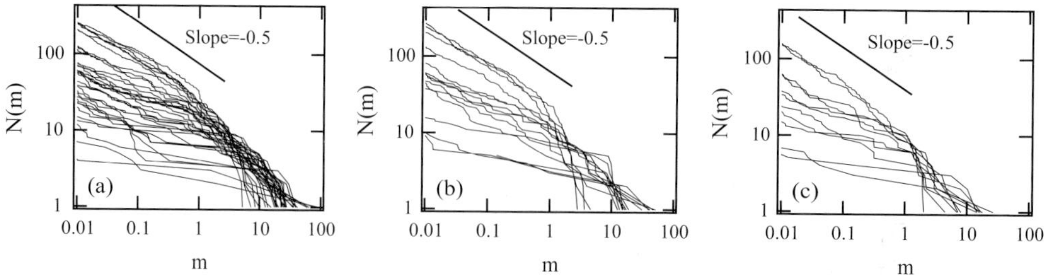

Fig. 1. – Cumulative distributions of fragment mass. The plots (a), (b), and (c) correspond to the results of 150 mm, 100 mm, and 50 mm length samples, respectively. They contain (a) 34, (b) 15, and (c) 11 samples data. For highly fragmented data, clear scaling parts, in which the exponent is 0.5, can be seen. However, the data near the fragmentation critical point do not show a clear trend.

2. – Experiment

We used a simple experimental apparatus. The samples we used are glass tubes (50 mm outside diameter and 46 mm inside diameter; 150 mm, 100 mm, or 50 mm long). Tubes have 2D geometry and stand by themselves. Therefore, we can easily apply a flat impact to one end of 2D objects using the tube geometry. We performed experiments of 2D impact fragmentation with glass tubes. The tube sample was sandwiched between a stainless-steel stage and a stainless-steel plate. The heavy brass weight was dropped onto the stainless-steel plate. The control parameter of this experiment was the height of the weight dropped. More details about the experimental setup can be found in ref. [5]. We controlled the dropping height h within the vicinity of the critical point at which the sample did not break up. In other words, we focused on the low-energy regime.

3. – Results and discussion

Cumulative distributions of the fragment mass of all fragmentation events are shown in fig. 1. The cumulative distribution $N(m)$ is defined as

$$N(m) = \int_m^\infty n(m')\mathrm{d}m', \tag{1}$$

where $n(m)$ is the fragments number of mass m. As can be seen in fig. 1, there are universal scaling parts $N(m) \sim m^{-0.5}$, independent of sample size. This scaling behavior is consistent with results of numerical simulations [3, 4]. On the other hand, this value is in conflict with the percolation scaling hypothesis [6]. Thus, we suppose that the universality class of the 2D fragmentation with flat impact differs from that of percolation. While clear scaling can be observed for highly fragmented data, the data near the critical point do not show a clear power law. If we fit those data with power law form, the resulting scaling exponents include much fluctuations, and do not show a clear tendency.

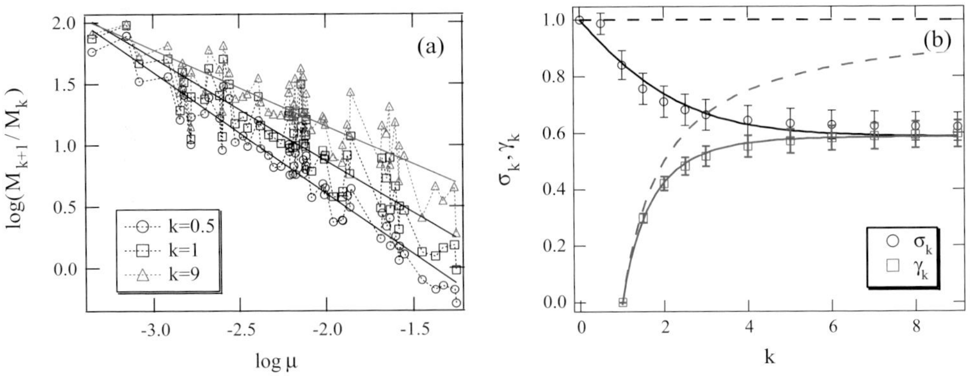

Fig. 2. – Multi-scaling exponents σ_k and γ_k. (a) Some plots of $\log(M_{k+1}/M_k)$ as functions of $\log\mu$. (b) The values of σ_k and γ_k *vs.* k. Solid lines correspond to the result of the binomial multiplicative model ($a = 2/3$ of eqs. (5a) and (5b)). Broken lines indicate the trivial mono-scaling case corresponding to $a = 1/2$.

Certainly, the larger the imparted energy is, the smaller the fragments size becomes. Kun and Herrmann used the largest fragment mass as an order parameter. However, our experimental data show much fluctuations again by this analysis method. Therefore, we employed the pseudo control parameter *multiplicity* μ. The μ was introduced by Campi [7], and should be written as,

$$\mu = m_{\min}\frac{M_0}{M_1}, \tag{2}$$

where $m_{\min}$ is the smallest limit of mass measurement, and we fix it as 0.01 g for all data. And the k-th order moment of fragment mass distribution M_k is written as

$$M_k = \sum_m m^k n(m). \tag{3}$$

Therefore, M_0 and M_1 correspond to the total mass and the total number of fragments, respectively. Here, we assume two kinds of simple scaling between M_k and μ as,

$$\frac{M_{k+1}}{M_k} = \mu^{-\sigma_k}, \tag{4a}$$

$$\frac{M_k}{M_1} = \mu^{-k\gamma_k}. \tag{4b}$$

In fig. 2(a), we show some plots of $\log(M_{k+1}/M_k)$ *vs.* $\log\mu$. Each slope indicates each corresponding σ_k. Since it contains a little large uncertainty, one should estimate the σ_k with error. In a similar way (measuring slopes of $\log(M_k/M_1)$ *vs.* $\log\mu$), we can obtain the values of γ_k. Figure 2(b) shows σ_k and γ_k obtained from the fitting of the form eqs. (4a) and (4b). These data show nontrivial multi-scaling nature as shown in fig. 2.

We consider a simple binomial multiplicative model to interpret this multi-scaling property. Let us consider a unit mass fragment at the initial state. At the first step, the initial fragment cleaves into two pieces whose masses are a and $1-a$ ($0 \le a \le 1$). The same process occurs on each fragment at the next step. We assume that this cleaving continues until the imparted energy dissipates. Then, we can easily calculate the exponents σ_k and γ_k as follows:

$$\sigma_k = -\frac{\ln\left[a^{k+1} + (1-a)^{k+1}\right] - \ln\left[a^k + (1-a)^k\right]}{\ln 2}, \tag{5a}$$

$$\gamma_k = -\frac{\ln\left[a^k + (1-a)^k\right]}{k \ln 2}. \tag{5b}$$

In fig. 2(b), we show the curves obtained from the case of $a = 2/3$ of eqs. (5a) and (5b), as solid lines. These curves agree with the experimental results very well. If we substitute $a = 1/2$ into eqs. (5a) and (5b), we obtain the trivial mono-scaling property. We show the curves for $a = 1/2$ as broken lines in fig. 2(b).

4. – Conclusions

We examined experimentally the possible critical behaviour of fragmentation induced by dropping a weight on a glass tube close to the limit at which the tube did not break any more. It was the 2D critical impact fragmentation experiment. We obtained the scaling exponent of cumulative distribution of fragment mass as 0.5 that conflicts with the percolation universality. Therefore, we concluded that the criticality of brittle fragmentation differs from that of percolation universality. We obtained critical multi-scaling and proposed a binomial multiplicative model for this phenomenon. Two kinds of exponents σ_k and γ_k were calculated to investigate the validity of the model. The model can yield proper results to interpret the experimental data, however, the parameter a needs to be tuned at the particular value 2/3. The origin of the value $a = 2/3$ is an open problem.

* * *

The authors would like to thank Prof. S. OHTA and Prof. H. SAKAGUCHI for helpful comments.

REFERENCES

[1] BEYSENS D., CAMPI X. and PEFFERKORN E. (Editors), *Fragmentation Phenomena* (World Scientific, Singapore) 1995.
[2] KUN F. and HERRMANN H. J., *Phys. Rev. E*, **59** (1999) 2623.
[3] ÅSTRÖM J. A., HOLIAN B. L. and TIMONEN J., *Phys. Rev. Lett.*, **84** (2000) 3061.
[4] HAYAKAWA Y., *Phys. Rev. B*, **53** (1996) 14828.
[5] KATSURAGI H., SUGINO D. and HONJO H., *Phys. Rev. E*, **68** (2003) 046105.
[6] STAUFFER D. and AHARONY A., *Introduction to Percolation Theory* (Taylor & Francis, London) 1992.
[7] CAMPI X., *Phys. Lett. B*, **208** (1988) 351; CAMPI X. and KRIVINE H., *Observables in Fragmentation*, pp. 312-321 in ref. [1].

Numerical and mean-field study of a lattice glass model

M. Tarzia, M. Pica Ciamarra, A. de Candia and A. Coniglio

Dipartimento di Scienze Fisiche, Università di Napoli "Federico II"
Istituto Nazionale di Fisica della Materia, Unità di Napoli
Monte S. Angelo, via Cintia, 80126 Napoli, Italy

The glass transition is a phenomenon that is observed when a liquid is cooled well below its melting temperature, avoiding the crystallization of the system [1-4]. The motion of the molecules is slowed down more and more by lowering the temperature, until at some point it becomes so slow that the liquid appears as a disordered solid. There are many valuable theories based on free volume [2], cooperative rearrangments of particles [3], inherent structures approach [4, 5], mode coupling theory [6] and replica method [7]. A further insight in the understanding of the glass transition was achieved through the study of the p-spin glass [8], a mean-field model with p-body interactions and quenched disorder, which reproduces many features of glass-forming liquids. More recently, Biroli and Mézard have introduced an interesting lattice glass model [9], where occupation variables interact via p-body potentials. The mean-field solution on the Bethe lattice shows a scenario identical to p-spin models, and in numerical simulations show a behavior typical of glass-forming liquids. In order to reduce the tendency to crystallize, the authors considered a mixture of two types of particles [10]. More recently we have introduced a simple lattice gas model with only two-body interactions [11]. The model reproduces in mean field again the scenario typical of p-spin models, and numerical simulations in three dimensions show the same behavior typical of glass-forming liquids, including cage effect and two-step relaxation. Moreover, while having a crystalline phase at low temperature, the model does not crystallize when cooled, even at extremely low cooling rates. This makes the model more suitable to study the glass transition, and allows to obtain very clear results.

The model is defined as follows. We partition the space in regular cells, such that not more than one particle can have its center of mass inside the cell. The position and orientation of the particle inside the cell is represented by a coarse-grained discrete

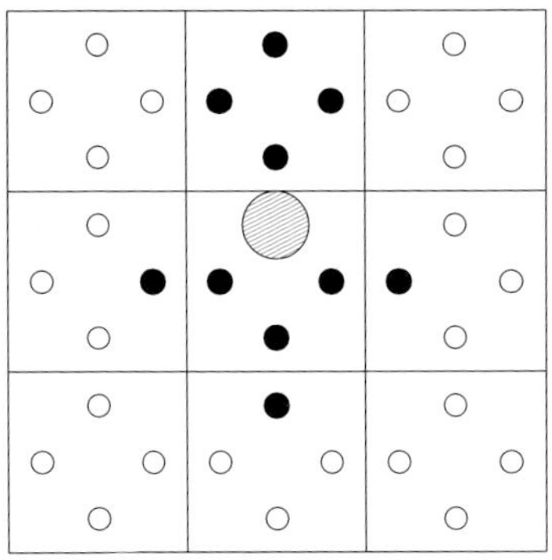

Fig. 1

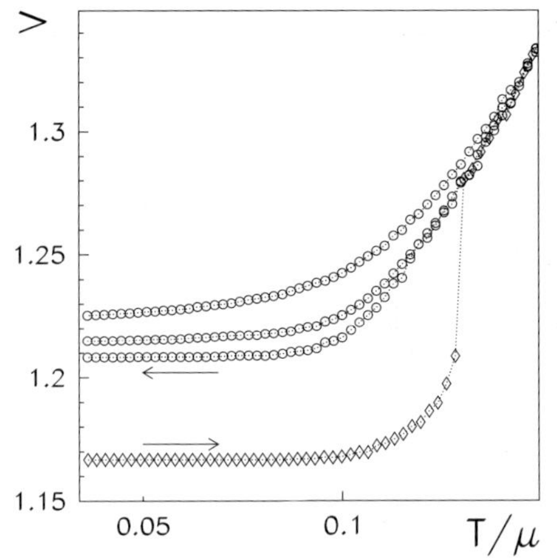

Fig. 2

Fig. 1. – The model in two dimensions: the space is partitioned in square cells, and each cells can be occupied by at most one particle in one of four positions (little circles). A particle in a given position (big shaded circle) forbids the presence of another particle in the positions colored in black.

Fig. 2. – Specific volume ($1/\rho$) as a function of temperature, for a three-dimensional system of size 28^3. Circles: the system is cooled starting from high temperature, for cooling rates $\dot{T}/T = -10^{-4}, -10^{-5}, -10^{-7}$. Diamonds: the system is heated starting from the crystalline ground state, with heating rate $\dot{T}/T = 10^7$.

internal degree of freedom, that can assume a finite number q of states, and the interaction between nearest-neighbor particles depends explicitly on their position inside the cell. The model is therefore described by the following Hamiltonian:

$$\mathcal{H} = \sum_{\langle ij \rangle} n_i n_j \phi_{ij}(\sigma_i, \sigma_j) - \mu \sum_i n_i \,, \tag{1}$$

where $n_i = 0, 1$ whether the i-th cell is occupied by a particle or not, $\sigma_i = 1, \ldots, q$ represents the internal position of the particle, and $\phi_{ij}(\sigma_i, \sigma_j)$ is the interaction energy between two particles in cells i and j with internal positions σ_i and σ_j. It is clear that, by choosing a sufficiently large number q of internal positions, and an opportune interaction matrix $\phi_{ij}(\sigma_i, \sigma_j)$, one can approximate as closely as desired any model defined in the continuum. On the other hand, it is plausible that a few number of internal states may be enough to catch the fundamental characteristics of glass-forming liquids.

We considered a particularly simple realization of the model (1). In two dimensions, we partition the space in square cells, and subdivide each cell into four internal positions. When a cell is occupied by a particle in a given position, a hard-core repulsion forbids the presence of another particle in some of the internal positions of the neighboring cells

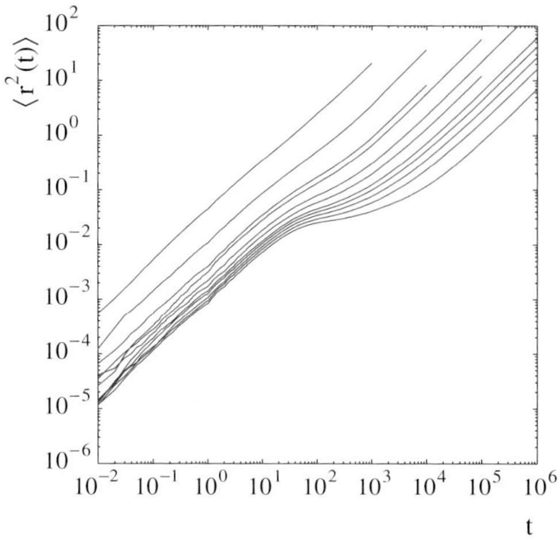

Fig. 3

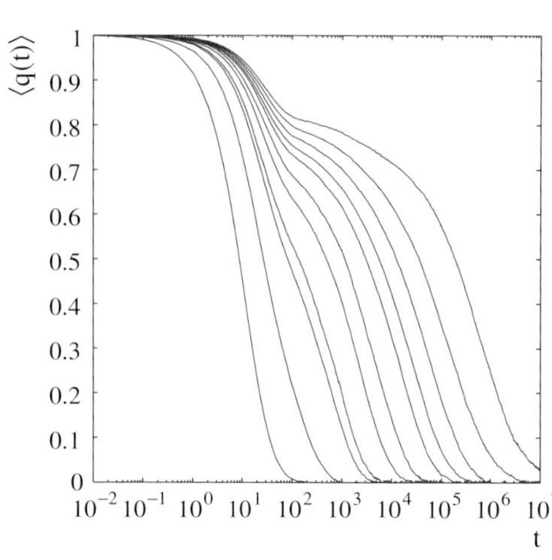

Fig. 4

Fig. 3. – Mean square displacement $\langle r^2(t)\rangle$ in a three-dimensional system of size 15^3, for densities $\rho = 0.6, 0.7, 0.75, 0.76, 0.78, 0.79, 0.8, 0.805, 0.81, 0.815, 0.82$.

Fig. 4. – Self-overlap $\langle q(t)\rangle$ as a function of time for the same densities of fig. 3.

(see fig. 1). The extension to three dimensions is straightforward [12]. For every spatial dimension d, the model has a crystalline ground state with density $2d/(2d+1)$, on lattices with sides multiples of $2d+1$ lattice spacings [11].

We have simulated the model in three dimensions, by means of grand-canonical and canonical dynamics. A compression experiment on a system of size 28^3 is shown in fig. 2. A simple Monte Carlo grand-canonical (variable density) dynamics is performed, and temperature is slowly decreased, starting from some high value. The results are shown as open circles in fig. 2, for various cooling rates. The final high-density state is strongly dependent on the cooling rate, as observed in glass-forming liquids. On the other hand, no tendency to crystallization is observed: even at the slowest cooling rate ($\dot{T}/T = -10^{-7}$) no transition to the crystalline state was observed [13]. After having equilibrated the system at a certain density in the liquid phase, we have switched to a canonical (fixed density) dynamics. These simulations are performed on lattices of size 15^3. The mean square displacement $\langle r^2(t)\rangle$ of the particles is show in fig. 3. In fig. 4 instead we plot the self-overlap, defined as $\langle q(t)\rangle = (1/N)\sum_i \langle n_i(t)n_i(0)\,\sigma_i(t)\cdot\sigma_i(0)\rangle$, where $\sigma_i(t)$ are unit length vectors, pointing in one of the six coordinate directions, representing the position of the particle inside the cell, and N is the number of particles. As happens in glass-forming liquids, at high density a *plateau* develops both in the mean square displacement and in the self-overlap. This is the signature of the so-called "cage effect". Particles are trapped in cages formed by neighboring particles, and vibrate rapidly around some equilibrium position, giving rise to the first decay in correlation functions. Only after a

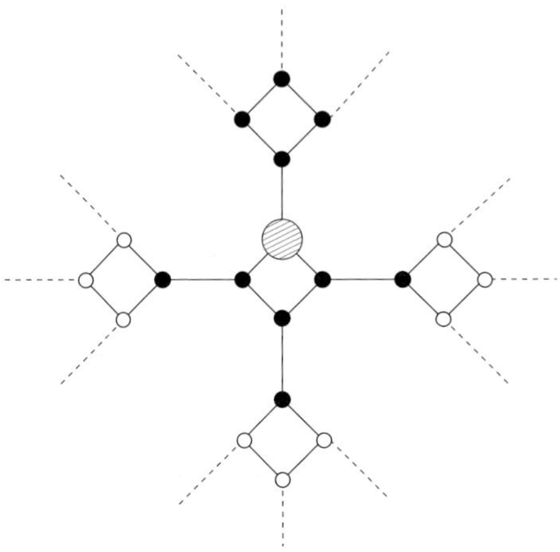

Fig. 5

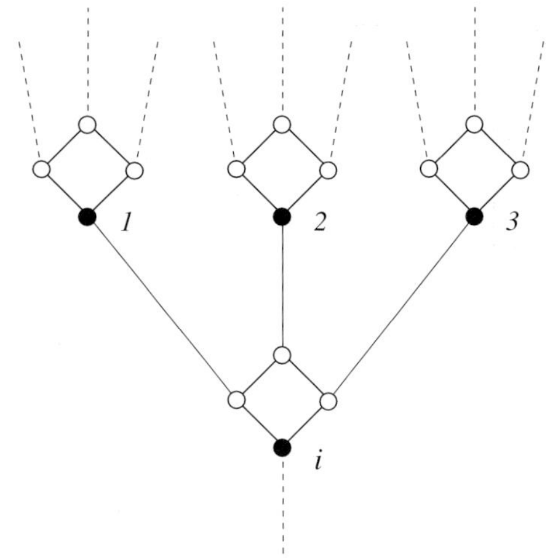

Fig. 6

Fig. 5. – The model on a tree-like graph (here with $k = 3$): each site is subdivided in $k + 1$ positions (little circles), and connected to $k+1$ randomly chosen neighbors. A particle in a given position (big shaded circle) forbids the presence of another particle in the positions colored in black.

Fig. 6. – Merging of the branches $j = 1, \ldots, k$ (here with $k = 3$) onto the site i: the "external" positions are those colored in black, the others are the "internal" ones.

long time they manage to escape from their cages, giving rise to the second final decay to equilibrium. The time of the slow relaxation diverges when the glass transition is approached [14].

We have then studied the model on the Bethe lattice, namely a random lattice with fixed connectivity $k + 1$ [9, 15] (see fig. 5). Consider a "branch" ending on site i, that is a graph in which site i has only k neighbors, and let $Z_0^{(i)}$, $Z_{\rm ext}^{(i)}$ and $Z_{\rm int}^{(i)}$ be the partition functions of the branch, restricted respectively to configurations where site i is empty, occupied on the "external" position, and occupied on one of the k "internal" positions (see fig. 6). We define the "local fields" a_i and h_i acting on the site i by the relations $e^{\beta h_i} = Z_{\rm int}^{(i)}/Z_0^{(i)}$, $e^{\beta a_i} = (Z_{\rm ext}^{(i)} + Z_{\rm int}^{(i)})/Z_0^{(i)}$. We have the following recursion relations for the local fields:

$$e^{\beta a_i} = e^{\beta\mu}\left(\prod_{j=1}^{k}\frac{1+e^{\beta h_j}}{1+e^{\beta a_j}}\right)\left(1+\sum_{j=1}^{k}\frac{1}{1+e^{\beta h_j}}\right), \tag{2a}$$

$$e^{\beta h_i} = e^{\beta\mu}\left(\prod_{j=1}^{k}\frac{1+e^{\beta h_j}}{1+e^{\beta a_j}}\right)\left(\sum_{j=1}^{k}\frac{1}{1+e^{\beta h_j}}\right). \tag{2b}$$

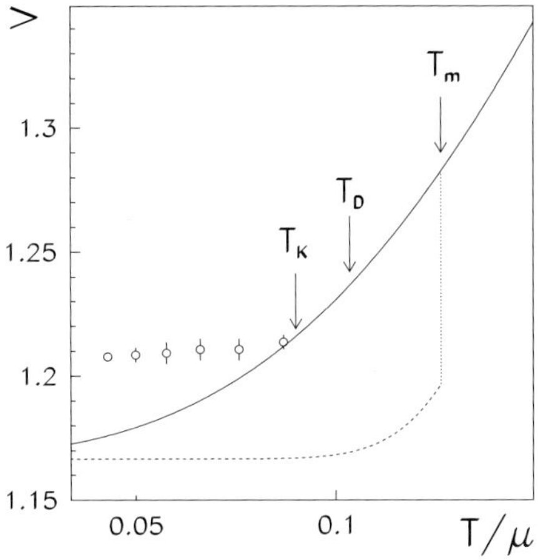

Fig. 7

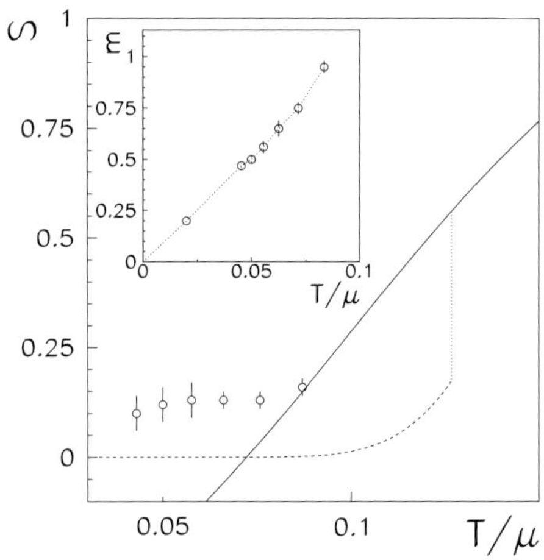

Fig. 8

Fig. 7. – Specific volume (inverse density) as a function of the temperature, on the random regular graph with $k = 5$. Solid line: replica symmetric liquid phase; broken line: crystalline phase; open circles: glassy phase in the 1-step RSB approximation.

Fig. 8. – Entropy per site as a function of the temperature, on the random regular graph with $k = 5$. The notation is the same as in fig. 7. Inset: parameter m as a function of the temperature. The line is a guide for the eye.

The free-energy shift of the process is given by $e^{-\beta\Delta F} = Z_0^{(i)} / \prod_{j=1}^k Z_0^{(j)} = \prod_{j=1}^k (1 + e^{\beta a_j})$. To evaluate the total free energy, one has to consider also the free-energy shifts when merging $k+1$ branches on a new site $\Delta F^{(1)}$, and when merging two branches $\Delta F^{(2)}$ [15]. The replica symmetric solution of the problem corresponds to the case in which there is only one pure state, and the fields do not fluctuate from site to site. In this case, the total free energy is given by $F = \Delta F^{(1)} - (k+1)\Delta F^{(2)}/2$, from which one can find the specific volume and the entropy per site, that are shown as solid lines in figs. 7 and 8 (in the case $k = 5$). As happens in the model studied by Biroli and Mézard [9], the entropy becomes negative for temperatures below some threshold [16].

We have then looked for a solution at the level of 1-step replica symmetry breaking (RSB) [17, 18, 15]. In this case, many pure states exist, and the local fields fluctuate also on the single site. The self-consistency equation, when merging k branches on a site i, now reads

$$(3) \qquad P(a,h) \propto \int \prod_{j=1}^{k} \mathrm{d}a_j \, \mathrm{d}h_j \, P(a_j, h_j) \times \delta(a - a_i)\delta(h - h_i) \exp[-\beta m \Delta F],$$

where a_i, h_i and ΔF are functions of a_j and h_j via eqs. (2), m is a real parameter, and the total free energy is

$$(4)\qquad F = -\frac{1}{\beta m}\log\int\prod_{j=1}^{k+1}\mathrm{d}P(a_j,h_j)e^{-\beta m\Delta F^{(1)}} + \frac{k+1}{2\beta m}\log\int\prod_{j=1}^{2}\mathrm{d}P(a_j,h_j)e^{-\beta m\Delta F^{(2)}}.$$

One then has to maximize the free energy with respect to the parameter $m \in [0,1]$.

At temperature $T = 0$ the solution of (3) corresponds to the close-packing limit and it is easily found. Indeed, in that case the fields a_i and h_i can only have the finite set of values: $a_i = (1-k)\mu, \ldots, 0, \mu$ and $h_i = a_i$ if $a_i < \mu$ or $0, \mu$ if $a_i = \mu$. The distribution $P(a,h)$ is then a sum of $k+2$ delta-functions, and the integral equation (3) gives rise to $k+2$ algebraic equations for the weights of the delta-functions. One has to give a finite value to the parameter $\beta m\mu$, and maximize the free energy with respect to it. For $k = 5$, the maximum is $F = -0.8265\,\mu$, for $\beta m\mu = 8.078$. At temperature $T > 0$ the solution of (3) can be found iteratively until the procedure converged. The results for the volume and entropy, and for the parameter m, are shown as open circles in figs. 7 and 8, in the case $k = 5$. The 1-step RSB solution appears below $T_{\mathrm{D}} \simeq 0.105\,\mu$, and becomes stable below $T_{\mathrm{K}} \simeq 0.087\,\mu$, where $m = 1$ and the configurational entropy vanishes.

In conclusion, we have studied a lattice glass model with two-body interactions. On the Bethe lattice, it shows the same scenario of discontinuous spin glasses, and of other recently studied lattice models of glass. Moreover, it shows no tendency to crystallize when simulated in three dimension with a simple Monte Carlo dynamics, which avoids the need to introduce a mixture, as in many models of glass.

REFERENCES

[1] ANGELL C. A., NGAI K. L., MCKENNA G. B., MCMILLAN P. F. and MARTIN S. W., *J. Appl. Phys.*, **88** (2000) 3113.
[2] CHOEN M. H. and TURNBULL D. J., *J. Chem. Phys.*, **31** (1959) 1164.
[3] GIBBS J. H. and DI MARZIO E. A., *J. Chem. Phys.*, **28** (1958) 373; ADAMS G., *J. Chem. Phys.*, **43** (1965) 139.
[4] DEBENEDETTI P. G. and STILLINGER F. H., *Nature*, **410** (2001) 259.
[5] STILLINGER F. H. and WEBER T. A., *Phys. Rev. A*, **25** (1982) 978; SASTRY S., DEBENEDETTI P. G. and STILLINGER F. H., *Nature*, **393** (1998) 554; SCIORTINO F., KOB W. and TARTAGLIA P., *Phys. Rev. Lett.*, **83** (1999) 3214.
[6] GÖTZE Y. W., in *Liquids, Freezing, and Glass Transition*, edited by J. P. HANSEN, D. LEVESQUE and J. ZINN-JUSTIN (Elsevier, Amsterdam) 1991.
[7] MÉZARD M. and PARISI G., *J. Chem. Phys.*, **111** (1999) 1076; MÉZARD M., in *More is Different*, edited by M. P. ONG and R. N. BHATT (Princeton University Press, Princeton, NJ) 2001.
[8] KIRKPATRICK T. R. and THIRUMALAI D., *Phys. Rev. Lett.*, **58** (1987) 2091.
[9] BIROLI G. and MÉZARD M., *Phys. Rev. Lett.*, **88** (2002) 025501.
[10] Further study of the model was done in LAWLOR A., REAGAN D., MCCULLAGH G. D., DE GREGORIO P., TARTAGLIA P. and DAWSON K. A., *Phys. Rev. Lett.*, **89** (2003) 245503. Very similar behaviors are found in HARTMANN L. and WEIGT M., *Europhys. Lett.*, **62** (2003) 533.

[11] PICA CIAMARRA M., TARZIA M., DE CANDIA A. and CONIGLIO A., *Phys. Rev. E*, **67** (2003) 057105.

[12] In that case the space is partitioned in cubic cells with six internal positions.

[13] We have also computed the density in the crystalline state, starting from the ground state at very low temperature, and heating up slowly (diamonds in fig. 2). This ordered state becomes unstable at $T \simeq 0.150\,\mu$, where the system falls into the liquid phase.

[14] The diffusivity goes to zero as a pawer law: $D \simeq (\rho - \rho_c)^\gamma$, ($\rho_c = 0.844$, and $\gamma = 3.45$). The dynamical susceptibility $\chi(t) = N(\langle q(t)^2 \rangle - \langle q(t) \rangle^2)$, shows a behavior typical of p-spin models and Lennard Jones binary mixture, with a peak at finite time that grows when approaching the glass transition, due to the presence of dynamical heterogeneities.

[15] MÉZARD M. and PARISI G., *Eur. Phys. J. B*, **20** (2001) 217; *J. Stat. Phys.*, **111** (2003) 1.

[16] One can also find a (replica symmetric) crystalline solution, for $k = 5$, this solution appears below $T_{\rm ms} = 0.1633\,\mu$, and becomes stable below the melting temperature $T_{\rm m} = 0.1444\,\mu$. The corresponding volume and entropy are shown as dashed lines in figs. 7 and 8.

[17] MÉZARD M., PARISI G. and VIRASORO M. A., *Spin Glass Theory and Beyond* (World Scientific, Singapore) 1987.

[18] MONASSON R., *J. Phys. A*, **31** (1998) 513.

Localization of waves and particles on the comb lattice

G. BALDI

Dipartimento di Fisica, Università di Trento, Via Sommarive 14, 38050 Povo, Trento
Istituto Nazionale di Fisica della Materia, Sezione di Trento, Italy

D. CASSI and R. BURIONI

Dipartimento di Fisica, Università di Parma, Parco Area delle Scienze 7A, 43100 Parma
Istituto Nazionale di Fisica della Materia, Sezione di Parma, Italy

1. – Introduction

Recent developments in microelectronics have allowed the construction of quantum devices characterized by a complex geometrical arrangement of some discrete elements, such as Josephson junctions or quantum dots. This possibility has stimulated the study of physical models defined on general discrete structures, with the aim of better understanding the physical properties of these systems, such as their vibrational dynamics or their quantum behavior.

The topology of a complex structure can be described in terms of an abstract model called a graph, which is made of sites and links. The sites correspond to the discrete elements of the system and the links describe their mutual relations. A graph is algebraically described by its adjacency matrix, which is defined as follows:

$$A_{i,j} = \text{number of links between site } i \text{ and site } j.$$

One of the most relevant operators in physics is the Laplacian which, on a graph, is defined [1] as $L = Z - A$, where Z is the diagonal matrix of the coordination numbers z_i, the number of nearest neighbors of site i.

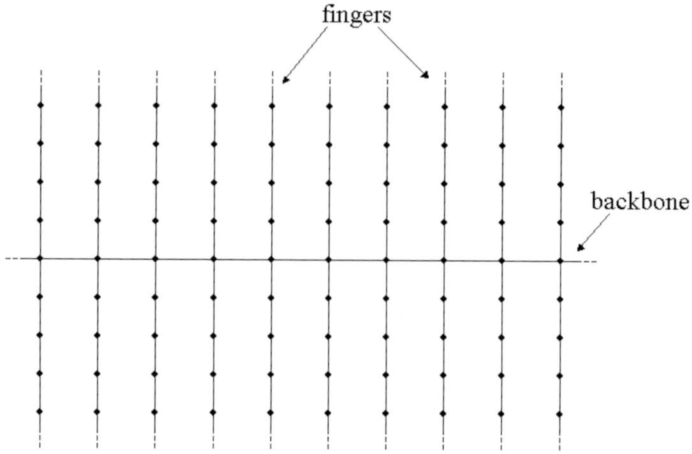

Fig. 1. – Two-dimensional comb lattice.

A very important complex network is the comb lattice [2, 3], an infinite graph which can be obtained connecting a one-dimensional chain, called the finger, to each site of a linear chain, the backbone (see fig. 1). The increasing interest in the study of this particular network is due to the existence in nature of a great variety of systems, such as polymers or biological molecules, that share the structure of the comb lattice and to the possibility of producing quantum devices, such as arrays of Josephson junctions [4], micro bridges or quantum wires characterized by the geometry of a comb graph.

The description of the large-scale topology of complex networks is based on the definition of a parameter [5, 6], called the spectral dimension $\bar{d}$, that generalizes the Euclidean dimension of regular lattices to a generic graph. However, in addition to the large-scale description, the complex structure of a graph can induce a great variety of relevant effects due to its local geometrical details. Recent works [7] on Bose-Einstein condensation on inhomogeneous networks have put into evidence an example of peculiar behavior due to the geometrical inhomogeneity, the presence of condensation at finite temperature for a system of non-interacting bosons in the absence of an external confining potential on the comb lattice, where $\bar{d} = 1$.

The simplest significant physical models on a graph are described by means of a particular matrix, that we will call the Hamiltonian, composed of a kinetic and a potential term. The former is directly linked to the topology of the graph while the latter describes the existence of defects or impurities in the material or the presence of an external field.

It is well known that on a regular lattice the eigenvectors of the Hamiltonian are either localized or extended over the entire graph. Moreover, the study of the spectral

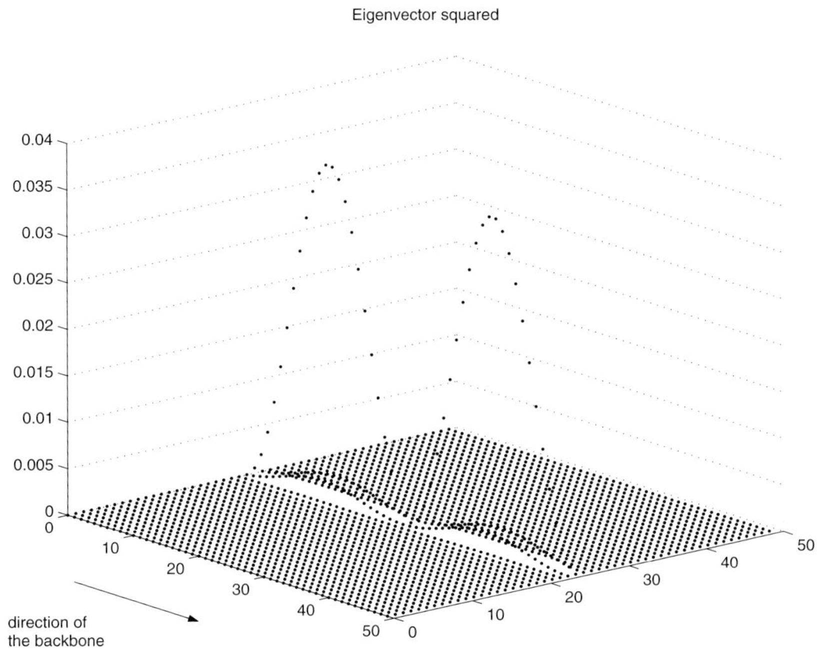

Fig. 2. – One of the eigenvector of a finite comb lattice made of 50 × 50 sites, corresponding to the high-energy region of the spectrum.

problem for the Hamiltonian is simplified by the translational invariance of the lattice, that allows us to introduce concepts like the reciprocal lattice and the Block's theorem.

In this paper we will briefly describe the main results that we have recently obtained for the spectral problem for the Hamiltonian operator over a generic graph, where the absence of translational invariance gives rise to a great variety of eigenstates, that can be localized on particular domains, even of infinite size (see fig. 2). The analytical details of our approach are reported in ref. [8].

The existence of a relationship between the localization problem and the properties of a random walker [9] moving on the graph is used as a method to solve the spectral problem on an arbitrary network. The effects of the potential matrix on the density of states of the Hamiltonian are investigated and in particular the relation between the existence of localized states and the form of the potential is analyzed.

The random walk method is then applied to study the effects of the potential on the comb lattice, which has a peculiar continuous spectrum divided into two section, one that corresponds to generalized eigenstates extended on the whole graph and another one that presents states localized on the backbone and exponentially decreasing on the fingers of the comb (see fig. 2). It is shown that the existence of these peculiar kind of states in the high-energy region of the continuous spectrum can influence the presence of localized states corresponding to a positive potential. In particular the spectrum of the Hamiltonian is analytically determined in the presence of a generic potential which differs from zero only on a finite subgraph of the comb lattice.

2. – Analytical methods

The Laplacian operator L is directly related to the vibrational dynamics of the graph and to the quantum-mechanical description of the motion of a particle on the graph.
In the first case we look at the graph as a system of masses, represented by the sites of the graph, tied together by springs situated along the links of the graph. In the harmonic approximation the dynamics of this system is equivalent to the eigenvalue problem for the Laplacian operator.
The motion of a quantum particle on a graph is described by a discrete Schrödinger equation where the kinetic term of the Hamiltonian is proportional to the Laplacian and the potential is proportional to a diagonal real matrix U.

These two physical examples show the importance of the study of the spectral problem for the operator $H = L + U$. In the following we will focus our attention on the study of this problem over a generic graph. In this situation it is not possible to apply powerful mathematical tools typical of Euclidean lattices, such as the Fourier transform and the reciprocal lattice.

We will introduce a method that will enable us to face the problem over a generic graph and with every potential, exploiting the relationship between this problem and the one of a random walker moving on the same graph.

2·1. *Random walks.* – The walker is a classical particle moving from one site to another of a graph G. Every instant of discrete time is associated with the site where the walker is in that moment. The walk is said to be random because the site where the particle moves is chosen in a random way on the base of a given probability $p_{i,j}$ that the walker jumps from one site to another.
We are interested in the random walk with trap and sources, that is the one generated by a jump matrix probability of the following kind:

$$p_{i,j} = \frac{A_{i,j}}{z_i + d_i}, \quad d_i \in \mathbb{R}. \tag{1}$$

If $d_i > 0$ the walker can die when it passes through the site i and we say that there is a trap on that site, while if $-z_i < d_i < 0$ it is possible that some other particles will born when the walker passes through the site, which is then called a source. The situation with $d_i < -z_i$ is meaningless because in that case $p_{i,j}$ can be negative. One of the quantity of interest in the study of the random walk problem is the probability $P_{i,j}(t) = (p^t)_{i,j}$ that the walker starting from the site i will reach the site j at time t. It is more useful for our purposes to determine another quantity, the so-called generating function of random walks $\tilde{P}(\lambda)$, defined as the discrete Laplace transform of $P(t)$.

2·2. *Resolvent operator and generating function.* – One way of finding the spectrum of $H = L + U$, on a given graph G, is to determine its resolvent:

$$R(w) = (H - w)^{-1}.$$

Our main result states that it is possible to obtain this operator by studying the properties of a suitable random walk with traps and sources on the graph G. This random walk is the one defined by the following probability jump matrix:

$$p(w) = (Z + U - w\,I)^{-1} A, \tag{2}$$

that is now a function of the variable w of the resolvent operator, in the same way as the generating function is a function of λ and w. With some manipulation one can demonstrate the following relation between $R(w)$ and $\tilde{P}(\lambda, w)$:

$$R(w) = \tilde{P}(1, w)\,(Z + U - w\,I)^{-1}. \tag{3}$$

From the knowledge of the resolvent operator one can derive all the spectrum of H. In particular the pure points spectrum coincides with the values of w, where $R(w)$ is not defined, while the continuous part of the spectrum corresponds to the points of discontinuity of $R(w)$.

The relationship between the random walk problem and the eigenvalue problem for H is very useful because it allows us to compute the resolvent operator of H on every graph where we can analytically determine the generating function $\tilde{P}(\lambda, w)$ by means of some geometrical and combinatorial techniques.

This is possible, for example, on the comb graph. However, the result (3) is very useful also in that cases where the analytical approach is of difficult application, because one can always simulate the motion of a random walker as a Monte Carlo method.

3. – Conclusions

In the example of the comb lattice we have shown [8] how it is possible to analytically determine the spectrum of the Hamiltonian operator on a complex network, when the potential differs from zero only on a finite subgraph of the considered graph, using the random walks' method. This relationship between the motion of the random walker and the spectral problem for the Hamiltonian can also be exploited for numerical computations. The random walker is naturally implemented as a Monte Carlo method, thus allowing the analysis of complex potentials on every finite graph.

The Monte Carlo method consists in evaluating the generating function $\tilde{P}_{i,i}(1, w)$ of a random walker simulated on a computer. We can obtain an approximation of the function $P_{i,i}(t)$ by repeating the simulation a great number of times and counting the number of transitions of the walker from site i at time t; $\tilde{P}_{i,i}(1, w)$ is then evaluated by summing over t.

In the case of the comb lattice the number of localized states is always less than or equal to the number of non-zero elements in the potential matrix. It will be of great interest to see if this situation also occurs on other graphs and if it is possible to devise some regularity in the dependence of the number of localized states as a function of number, position and strength of holes and walls.

This is strictly related to one of the main open questions in the study of the localization problem on inhomogeneous complex networks, that is whether or not it is possible to extend the Anderson's theorem [10] to disordered structures. This problem can be fruitfully investigated using the Monte Carlo method, based on the motion of a random walker, to study the localization over a generic graph with a random potential.

A further promising development of the present work will be the study of the relationship between the random walks' properties and the shape of the eigenvectors of the Hamiltonian.

REFERENCES

[1] WOESS W., *Bull. London Math. Soc.*, **26** (1994) 1.
[2] CASSI D. and REGINA S., *Mod. Phys. Lett. B*, **6** (1992) 1397.
[3] CASSI D. and REGINA S., *Phys. Rev. Lett.*, **76** (1996) 2914.
[4] SIMANEK E., *Inhomogeneous Superconductors: Granular and Quantum Effects* (Oxford University Press) 1994.
[5] ALEXANDER S. and ORBACH R., *J. Phys. (Paris) Lett.*, **43** (1982) 625.
[6] HATTORI K., HATTORI T. and WATANABE H., *Prog. Theor. Phys. Suppl.*, **66** (1987) 108.
[7] BURIONI R., CASSI D., RASETTI M., SODANO P. and VEZZANI A., *J. Phys. B: At. Mol. Opt. Phys.*, **34** (2001) 4697.
[8] BALDI G., CASSI D. and BURIONI R., *Localized states on comb lattices*, to be published in *Phys. Rev. E*.
[9] MONTROLL E. W. and WEISS G. H., *J. Math. Phys.*, **6** (1965) 167.
[10] ANDERSON P. W., *Phys. Rev.*, **109** (1958) 1492.

Synchronization properties of coupled FitzHugh-Nagumo systems

C. J. TESSONE and R. TORAL

Instituto Mediterráneo de Estudios Avanzados (IMEDEA) CSIC-UIB, Ed. Mateu Orfila
Campus UIB, 07122-Palma de Mallorca, Spain
Departament de Física, Universitat de les Illes Balears
Campus UIB, 07122-Palma de Mallorca, Spain

C. R. MIRASSO

Departament de Física, Universitat de les Illes Balears
Campus UIB, 07122-Palma de Mallorca, Spain

J. D. GUNTON

Department of Physics, Lehigh University - Bethlehem, PA 18015, USA

1. – Model studied

In the last years, the phenomenon of synchronization in coupled limit cycle oscillators has been extensively investigated [1, 2]. However, not such a thorough study has been carried out for excitable systems, despite the fact that many features of the dynamics of biologically relevant systems, for example, neurons, can be described by simple excitable models. For example, it is found that epileptic crises are characterized by a particularly large amount of neurons firing simultaneously [3].

It is the goal of this paper to delve into some aspects of the synchronization properties of coupled excitable systems under the presence of noise. Although we do not have any specific applications in mind, we believe that our results are quite general, since we use a prototypical model of excitable dynamics. It is known that in those systems noise can induce phenomena such as stochastic resonance [4] (under the presence of an external forcing) or coherence resonance [5-7]. The latter is a mechanism by which an unforced excitable system shows a maximum degree of regularity in the period between emitted pulses in the presence of the right amount of noise. We focus here on the stationary synchronization properties of the common firings [8], and a more detailed study including the coherence resonance aspects is left for future work. We find that there is a non-equilibrium phase transition between synchronized and desynchronized states. We discuss the proper order parameter to characterize this transition and obtain numerically the phase diagram.

We consider the FitzHugh-Nagumo model which provides the simplest representation of firing dynamics and has been widely used as a model for spiking neurons as well as for cardiac cells [9, 10]. The model is defined in terms of activation x and inhibition y variables, as follows:

$$\epsilon \dot{x} = x - \frac{1}{3}x^3 - y, \tag{1}$$

$$\dot{y} = x + a + D\xi(t), \tag{2}$$

where, following ref. [5], a Gaussian white noise $\xi(t)$ of zero mean and correlations $\langle \xi(t)\xi(t')\rangle = \delta(t-t')$ has been added to the slow variable y. D will be called the noise intensity. The difference in the time scales of x and y is measured by ϵ, a small number. We work exclusively in the so-called excitable regime, characterized by $|a| > 1$.

There is a single stable fixed point (x_0, y_0) which, in the absence of any external perturbation, $D = 0$, is reached independently of the initial condition. When random perturbations are present, the trajectories eventually exit the basin of attraction of the stable fixed point and return to it after making an excursion in phase space, *i.e.* a pulse.

The next step is to consider an ensemble of N globally coupled systems:

$$\epsilon \dot{x}_i = x_i - \frac{1}{3}x_i^3 - y_i + \frac{k}{N}\sum_{j=1}^{N}(x_j - x_i), \tag{3}$$

$$\dot{y}_i = x_i + a + D\xi_i(t), \quad i = 1, \ldots, N, \tag{4}$$

with independent noises, $\langle \xi_i(t)\xi_j(t')\rangle = \delta_{ij}\delta(t-t')$. The systems are globally coupled by a gap-junctional form, as indicated by the last term of eq. (3), where k is the coupling strength.

Numerical simulations of this coupled system of equations (1) show that, for some range of parameter values, the different units fire pulses at the same times. Notice that, although some amount of noise is needed in order to induce firings and hence observe synchronized behavior, a too large noise finally degrades the quality of the synchronized

(1) The numerical integration of eqs. (3), (4) use a stochastic Runge-Kutta method (known as the Heun method [11]) with a time step $h = 10^{-4}$.

state. A general framework to study such synchronization phenomena is given by the work by Kuramoto [1]. He considers coupled phase variables $\phi_i(t)$ following a stochastic dynamics and discusses the existence of a synchronized regime in terms of the coupling strength and the noise intensity. It turns out that the Kuramoto model displays a genuine phase transition in which synchronization disappears if the noise surpasses a given critical value. We will show that the same behavior can be observed in our model.

The first step consists in defining phase-like variables ϕ_i for our model. They should satisfy the condition that their variation between 0 and 2π represents the pulse movement starting from the fixed point, traveling through all the cycle, and ending again at the fixed point. Several different approaches have been taken in order to evaluate the phases ϕ_i. The most naïve, definition is based upon the fact that the limit cycles in which the variables (x_i, y_i) evolve are approximately centered around the origin. Then, the easiest choice is

$$\phi_i = \arctan\left(\frac{y_i}{x_i}\right). \tag{5}$$

However, this choice is only valid for particular cases of the parameters of the FitzHugh-Nagumo model. For large noise intensities, for example, the pulses are not so clearly centered around the origin. A definition of more general validity uses the so-called Hilbert transform [2]. Let us consider the variable $x_i(t)$. From it we can construct the so-called "analytic signal", $s_i(t) = x_i(t) + \mathrm{i}\hat{x}_i(t)$, where $\hat{x}_i(t)$ denotes the Hilbert transform of the function $x_i(t)$. For a general function, $g(t)$, such a transform is defined as

$$\hat{g}(t) = -\frac{1}{\pi}\mathsf{PV}\int_{-\infty}^{\infty}\frac{g(\tau)}{t-\tau}\mathrm{d}\tau, \tag{6}$$

where PV denotes the principal value of the integral. The phase is defined as the argument of $s_i(t)$, *i.e.*

$$\phi_i(t) = \arctan\left(\frac{\hat{x}_i}{x_i}\right). \tag{7}$$

From a computational point of view, it is very costly to perform the convolution involved in the Hilbert transform. We will show now that the same phase can be obtained by a much more efficient procedure. This is based upon the equality

$$g(t) + \mathrm{i}\hat{g}(t) = 2\mathcal{F}^{-1}\left[\mathcal{F}[g(t)]\cdot\Theta(\omega)\right], \tag{8}$$

involving the Fourier transform operator $\mathcal{F}$. Here $\Theta(\omega)$ is the Heaviside function: $\Theta(\omega) = 0$ for $\theta < 0$, $\Theta(\omega) = 1$ for $\theta \geq 0$ defined in the Fourier space ω.

This relation can be proved by replacing

$$g(t) = \int_{-\infty}^{\infty} g(t_0)\delta(t-t_0)\mathrm{d}t_0\,, \tag{9}$$

on the right-hand side of (8):

$$2\mathcal{F}^{-1}\left[\mathcal{F}\left[\int_{-\infty}^{\infty} g(t_0)\delta(t-t_0)\mathrm{d}t_0\right]\cdot\Theta(\omega)\right] = 2\int_{-\infty}^{\infty}\mathcal{F}^{-1}\big[\mathcal{F}[g(t_0)\delta(t-t_0)]\cdot\Theta(\omega)\big]\mathrm{d}t_0$$
$$= 2\int_{-\infty}^{\infty}\mathcal{F}^{-1}\left[g(t_0)\delta(t-t_0)\frac{\mathrm{e}^{\mathrm{i}t_0\omega}}{\sqrt{2\pi}}\cdot\Theta(\omega)\right]\mathrm{d}t_0$$
$$= 2\int_{-\infty}^{\infty}\left(\frac{1}{2}g(t_0)\delta(t-t_0)+\frac{\mathrm{i}}{2\pi}\frac{g(t_0)}{t_0-t}\right)\mathrm{d}t_0$$
$$= g(t)+\mathrm{i}\hat{g}(t).$$

Thus one can achieve the calculation of the Hilbert transform by using two Fourier transforms. This leads to a very efficient numerical algorithm since the use of the fast Fourier transform involves a computer time of order $\mathcal{O}(T\log T)$ instead of $\mathcal{O}(T^2)$ which would be the case if one evaluates directly the convolution that defines the Hilbert transform (T is the length of the time series considered).

2. – Synchronization properties

We define an order parameter that allows us to measure the degree of synchronization in the coupled system. In order to follow the Kuramoto scheme, we use the phases ϕ_i introduced before in terms of the Hilbert transform, eq. (7). Collective amplitude, $\rho(t)$,

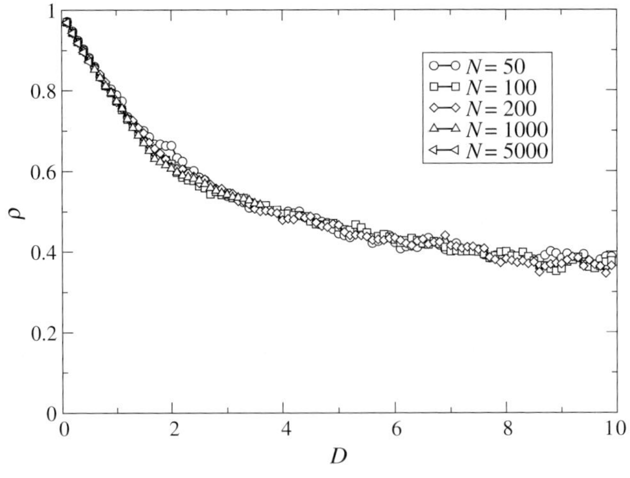

Fig. 1. – Order parameter ρ as a function of noise intensity for a system of globally coupled FitzHugh-Nagumo systems. Values of the parameters: $a = 1.001$, $\epsilon = 0.01$, $k = 1$.

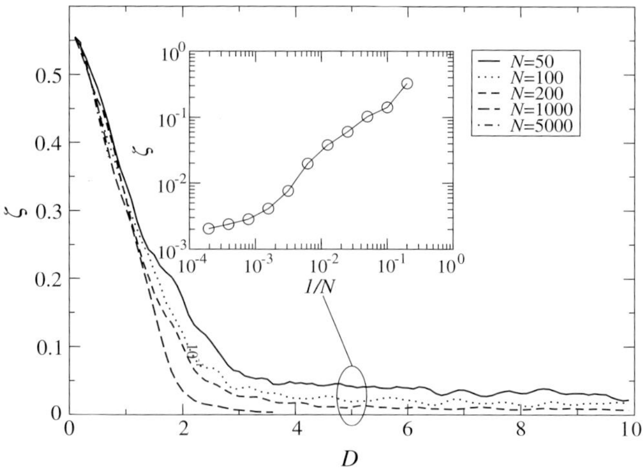

Fig. 2. – In this figure we show the order parameter ζ for various system sizes, from $N = 50$ to $N = 10000$. It is also shown the dependence of the order parameter on system size. The parameters of the simulation were $a = 1.001$ and $k = 1$. The inset shows that in the unsynchronized state, the finite-size effects scale as $\zeta = \zeta_\infty + \mathcal{O}[N^{-1/2}]$ for sufficiently large N.

and phase, $\psi(t)$, variables are defined as

(10) $$\rho(t)\mathrm{e}^{\mathrm{i}\psi(t)} = \frac{1}{N}\sum_{i=1}^{N} \mathrm{e}^{\mathrm{i}\phi_i(t)}.$$

Initially, the order parameter ρ introduced by Kuramoto is defined as the time average:

(11) $$\rho \equiv \langle \rho(t) \rangle_t .$$

In fig. 1 we plot ρ as a function of the noise intensity D for different number of coupled systems. It turns out that the order parameter continuously decreases with increasing D, thus showing that the quality of the synchronization worsens for large noise intensity. The dependence of the order parameter ρ for relatively large system size ($N > 100$) disappears, showing that finite-size effects are very small for these systems sizes. Figure 1 shows that the Kuramoto order parameter for this coupled FitzHugh-Nagumo model does not decay to zero with increasing noise intensity. This is due to the fact that for an excitable system, most of the time all the units oscillate near the fixed point. Hence, the order parameter ρ is different from zero, even in the case in which all the units are uncoupled and fire unsynchronizedly. Since we are interested in measuring the deviations from this unsynchronized state, we use a different order parameter, ϱ, first introduced in ref. [12]:

(12) $$\zeta = \left\langle \left| \rho(t)\mathrm{e}^{\mathrm{i}\,\psi(t)} - \left\langle \rho(t)\,\mathrm{e}^{\mathrm{i}\,\psi(t)} \right\rangle_t \right| \right\rangle_t .$$

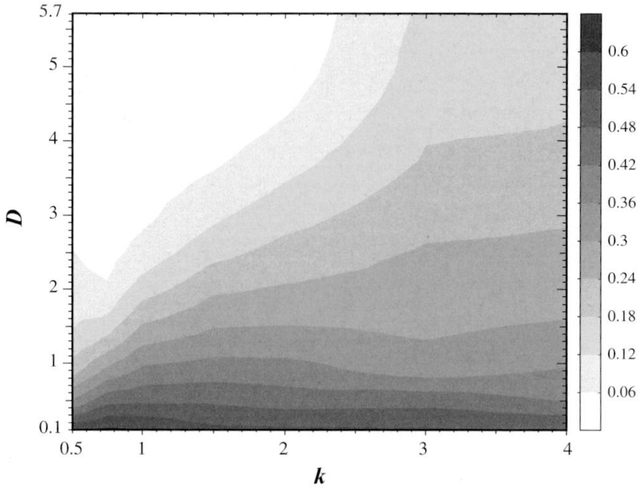

Fig. 3. – The order parameter ϱ is plotted against the control parameters k, and D; it is apparent a phase transition to desynchronization. In the simulations $a = 1.001$.

In the case of complete desynchronization, $\rho(t)\mathrm{e}^{\psi(t)}$ is almost constant except for finite system-size fluctuations (see the inset in fig. 2):

$$\rho(t)\,\mathrm{e}^{\mathrm{i}\,\psi(t)} = \rho \mathrm{e}^{\mathrm{i}\,\langle\psi(t)\rangle_t} + \mathcal{O}\big[N^{-1/2}\big]. \tag{13}$$

Replacing into eq. (12), we get that $\zeta \cong \mathcal{O}[N^{-1/2}]$, and $\zeta = 0$ in the thermodynamic limit. Therefore, any value of $\zeta > 0$ indicates that the units have synchronized, and the larger the value of ζ the higher the degree of synchronization.

In fig. 2, we plot the new order parameter ζ, for the same values of the parameters as in the previous figure. In this case, we notice the vanishing of the order parameter, indicating clearly the existence of a phase transition at a critical value $D_\mathrm{c} \approx 2.1$ separating the regime of synchronization/desynchronization. Note that the location of this transition could not be easily derived from the data in fig. 1. The complete phase diagram for a wide range of values of noise intensity D and coupling constant k is plotted in fig. 3.

3. – Conclusions

In summary, we have shown that an ensemble of globally coupled FitzHugh-Nagumo excitable systems subjected to independent noises experience a loss of synchronization for increasing noise intensity. Paradoxically, it is noise what initially induces the firings and sets the possibility of observing synchronized pulses.

The synchronization/desynchronization transition requires a proper definition of the order parameter for its characterization, since the usual measures used in coupled oscillators do not properly identify the transition point. We have found that a modified definition of the usual Kuramoto order parameter clearly displays such a transition. This order parameter is obtained from phase-like variables defined through the use of the Hilbert transform and we have given details of a numerically efficient method to compute the phase variables. Further work will aim to characterize this non-equilibrium transition and its universality class. Preliminary results show that the transition is present in locally coupled systems in $d = 2$ dimensions, but not in $d = 1$.

As stated before, we do not have any specific applications in mind, but since the FitzHugh-Nagumo equations have been widely used to model some biological systems, we believe that our results can be relevant when analyzing the collective response of such systems in a noisy environment.

* * *

This work is supported by the Ministerio de Ciencia y Tecnología (Spain) and FEDER, projects BFM2001-0341-C02-01 and BFM2000-1108, and NSF grant DMR9813409.

REFERENCES

[1] KURAMOTO Y., in *Chemical Oscillations, Waves and Turbulence* (Springer-Verlag, New York) 1984.
[2] PIKOVSKY A. S., ROSEMBLUM M. and KURTHS J., in *Synchronization: A Universal Concept in Nonlinear Sciences* (Cambridge University Press, Cambridge, UK) 2001.
[3] NIEDEMEYER N. and LOPES DA SILVA F. H., in *Electroencephalography, Basic Principles, Clinical Applications and Related Fields* (Urban & Schwarzenberg, Baltimore) 2001.
[4] WIESENFELD K., PIERSON D., PANTAZELOU E., DAMES C. and MOSS F., *Phys. Rev. Lett.*, **72** (1994) 2125.
[5] PIKOVSKY A. S. and KURTHS J., *Phys. Rev. Lett.*, **78** (1997) 775.
[6] TORAL R., MIRASSO C. R. and GUNTON J. D., *Europhys. Lett.*, **61** (2003) 162.
[7] LINDNER B. and SCHIMANSKY-GEIER L., *Phys. Rev. E*, **60** (1999) 7270.
[8] ZHOW C. and KURTHS J., *Chaos*, **13** (2003) 401.
[9] KOCH C., in *Biophysics of Computation* (Oxford University Press, New York) 1999.
[10] GLASS L., HUNTER P. and MCCULLOCH A. (Editors), *Theory of Heart* (Springer-Verlag, New York) 1991.
[11] SAN MIGUEL M. and TORAL R., *Stochastic effects in physical systems*, in *Instabilities and Nonequilibrium Structures VI*, edited by E. TIRAPEGUI, J. MARTÍNEZ and R. TIEMANN (Kluwer Academic Publishers, Dordrecht, The Netherlands) 2000.
[12] SHINOMOTO S. and KURAMOTO Y., *Prog. Theor. Phys.*, **75** (1986) 1105.

Preliminary results on the granular drag force

M. Pica Ciamarra

Center for Nonlinear Dynamics, The University of Texas at Austin - Austin, TX 78712, USA
Dipartimento di Scienze Fisiche, Università di Napoli "Federico II" and
INFM, Unità di Napoli - 80126 Napoli, Italy

A. H. Lara, A. T. Lee, D. I. Goldman and H. L. Swinney

Center for Nonlinear Dynamics, The University of Texas at Austin - Austin, TX 78712, USA

1. – Introduction

Several experiments on the impact of projectiles in granular media have been conducted in recent years [1-6]. This work has been motivated by interest in understanding of craters on planets (*e.g.*, Venus) and moons (*e.g.*, Jupiter's Callisto), as well as craters formed by explosions and the impact of projectiles on the earth's surface [7]. During impact, a granular medium has properties of both fluids and solids. While the initial impact of a projectile in a liquid also creates a transient crater, the crater in a liquid collapses and the surface of the medium becomes flat [8]. In contrast, because of dissipation and friction, a granular medium freezes before the surface becomes flat, and a static crater remains visible on the surface.

To understand granular materials near the transition between fluid and solid, it is of interest to determine the *granular drag force* (GDF) that stops a projectile. Some conclusions about the GDF are drawn in [4,6], where the final penetration depth h of the projectile is related to its impact energy E_i and to the mean value of the GDF $\langle F\rangle$ by

$$E_i + Mgh = \langle F\rangle h, \tag{1}$$

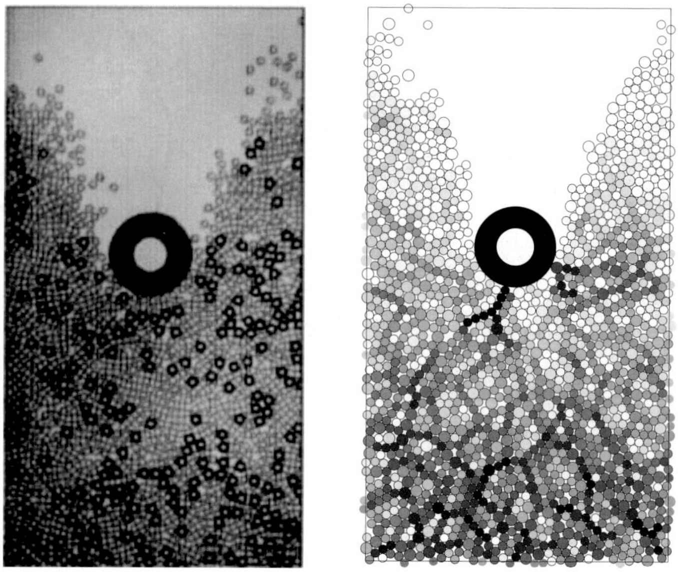

Fig. 1. – Snapshot of a projectile penetrating a bidisperse mixture of small disks with size ratio 1.4(16%) : 1(84%). Left image: experiment, where the larger disks of the bidisperse mixture have darker shading. Right image: simulation with each disk i darkened according to the sum of the magnitudes of all the forces acting on the particle, $\sum_j |\vec{f}_{ij}|$, where j is a contacting particle.

where M is the mass of projectile and g the gravitational acceleration. Surprisingly, there have been no systematic studies of the time evolution of the GDF, which is the subject of our study.

We have obtained preliminary results on the GDF from laboratory experiments and simulations on two-dimensional granular systems. The two-dimensionality enables us to track the projectile in the experiment, and to numerically simulate the system with a smaller number of grains than in a three-dimensional system. Our experiments and simulations reveal three distinct regimes of motion. The first regime is the initial impact, during which time the projectile rapidly decelerates and the granular bed in the neighboring region fluidizes. The second and longest regime corresponds to the penetration of the bed by the projectile, during which time the projectile creates a transient crater, as fig. 1 illustrates. As the projectile penetrates the medium, it slows down and comes almost to a stop. Then the projectile motion enters the final regime, during which time the transient crater collapses, leaving a permanent shallow crater visible on the surface. At high impact velocities, the collapsing crater can result in the formation of a jet, as observed in [2,3]. Similar jets have also been obtained in our simulations.

2. – Experiment

The experimental particle bed and the impact projectile (a steel disk of diameter $D = 4.46\,\text{cm}$ and mass $M = 30.2\,\text{g}$) are confined between two glass plates with a plate separation 1.1 times the length of the cylindrical particles. This setup allows us to image the impact and track the instantaneous position of the projectile as it penetrates the granular bed. We can control the energy (and velocity $v_0 < 400\,\text{cm/s}$) of the impact projectile by adjusting the drop height ($h < 80\,\text{cm}$). The bed of grains is composed of a bidisperse mixture of nylon disks ($N_1 = 84\%$, diameter $d_1 = 0.456\,\text{cm}$ and mass $m_1 = 0.049\,\text{g}$, and $N_2 = 16\%$, diameter $d_2 = 0.635\,\text{cm}$ and mass $m_2 = 0.097\,\text{g}$). A bidisperse rather than monodisperse granular medium is used so that the bed does not crystallize and retains the disorder that is typical of granular materials in three dimensions. To create an initial random configuration of the grains prior to each measurement, the bed is fluidized with a uniform flow of air that is slowly reduced until the bed is motionless. This protocol allows us to begin each observation with a reproducible area fraction.

3. – Simulation

The impact is studied with a soft-core molecular dynamics simulation [9-11], where two disks (or a disk and the projectile) can overlap. The interaction force between disks or between a disk and the projectile has components both normal ($\hat{n}$) and tangential ($\hat{s}$) to a particle's surface. We use a normal interaction similar to those used in previous work [9-11], except that we use a Heaviside step function θ to create a continuum force that mimics the elastic-plastic model (see, for example, fig. 8 of [12]). The two force components are

$$\vec{F}_n = -\left[k\delta + m_\mathrm{r}\gamma_n|\vec{v}_n|\theta(\vec{v}_n)\hat{n}\right]\hat{n}, \tag{2}$$

$$\vec{F}_s = \min\left[m_\mathrm{r}\gamma_s|\vec{v}_s|, \mu|\vec{F}_n|\right]\hat{s}, \tag{3}$$

where $\vec{v}_n$ and $\vec{v}_s$ are the normal and tangential components of the surface velocity. The hardness of the particles is set by $k = 3.2 \cdot 10^3\,\text{kg s}^{-2}$. The other parameters are the viscoelastic constants and the static friction coefficient, which are taken to be, respectively, $\gamma_n = 10^4\,\text{s}^{-1}$, $\gamma_s = 8 \cdot 10^3\,\text{s}^{-1}$, and $\mu = 0.28$. Finally, m_r is the reduced mass of the colliding particles.

4. – Results

Experimental and numerical results for the penetration distance $y(t)$ and the penetration velocity $v_y(t)$ for an impact velocity $v_0 = 238\,\text{cm/s}$ are shown in fig. 2. The vertical lines divide the three regimes: impact, penetration, and collapse. As in liquid [13] and solid [8] impacts, the projectile decelerates quickly immediately after contact. In the subsequent penetration regime, we find that the velocity decreases linearly in time, and

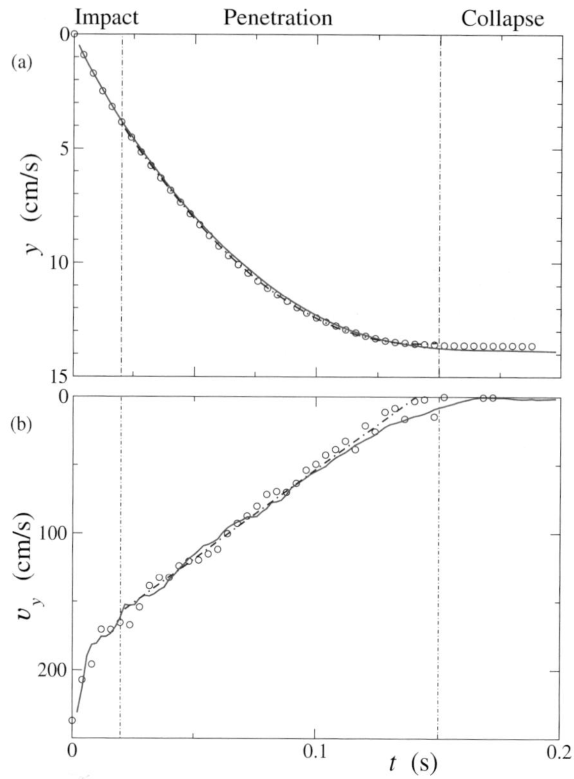

Fig. 2. – (a) Penetration distance and (b) penetration velocity from experiment (○) and simulation (lines) for an impact velocity of $v_0 = 238\,\text{cm/s}$. The dashed line is a parabolic fit for $y(t)$; correspondingly, $v_y(t)$ is fit to a straight line.

consequently $y(t)$ is fit to a parabola

$$y(t) = v_0 t + \frac{1}{2} a_y(v_0) t^2, \tag{4}$$

where $a_y(v_0) = g + \mathcal{F}(v_0)/M$ is the total acceleration, g the gravity, and $\mathcal{F}/M$ the acceleration due to the GDF. Equation (4) implies that the granular drag force is constant in the penetration regime. Finally, during the collapse regime, the velocity approaches zero very slowly, over a time span that is very long compared to the penetration time.

An absence of velocity dependence of the force acting on a projectile has also been found in [14], where the force acting on a cylinder dragged in a granular medium with constant slow velocity ($v \simeq 1\,\text{mm/s}$) was studied. It is remarkable that we also find the same result in a non-stationary situation characterized by velocities 10^3 times larger.

The granular drag force fluctuates strongly, as fig. 3 illustrates. The force rises to a high value immediately after contact, and then drops and fluctuates about a constant

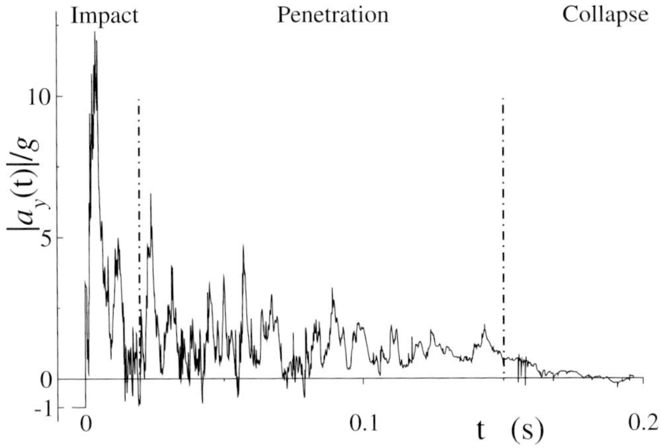

Fig. 3. – Time dependence of the normalized acceleration of the projectile.

value during the penetration. The strong fluctuations have two origins. One is the small ratio between the linear size of the projectile and that of the grains ($D/d \simeq 8$). The second is the inhomogeneous response of granular media, which are known to form force chains in response to an applied stress [15]. The fluctuations are characterized by the power spectrum of the normalized acceleration, which is described by a $1/f^2$ decay at high frequencies in the penetration regime. Granular studies of media with an imposed shear have also yielded a $1/f^2$ decay of the acceleration power spectrum [16]. More recently [17] reported the same observation for the power spectrum of the angular position of a probe partially dipped in a granular medium subject to continuous shaking or tapping. The $1/f^2$ behavior of the power spectrum (and the exponential decay of the acceleration autocorrelation function in the penetration regime, not shown) implies a Brownian motion of the velocity, *i.e.*, a series of random jumps in the detected force [16].

5. – Discussion and conclusions

The existence of a constant mean deceleration in the penetration regime can be plausibly explained in terms of the creation and destruction of force chains. When the projectile falls a distance Δ, a number $n(\Delta)$ of force chains are created and broken. Assuming that the average energy required to break a force chain $\bar{e}$ does not depend on the time taken to break it, the work required to move the projectile by Δ, $E(\Delta) = n(\Delta)\bar{e}$, does not depend on the velocity of the projectile. This plausibility argument explains why the drag force acting on the projectile is independent of the instantaneous velocity of the projectile.

In conclusion, we have conducted preliminary experimental and molecular dynamics studies of the dynamics of a projectile impacting a granular medium. Previous work on impacts in granular media [1-5] have studied the properties of craters left on the surface after an impact but did not study the time evolution of the position of the projectile.

We find that after impact, the force acting on the projectile fluctuates around a constant value. The fluctuations arise from the formation and breaking of force chains in front of the projectile.

Our preliminary study of the properties of the granular drag force raises many questions. Particularly, one could study the dependence of the properties of the granular drag force on various parameters, such as the impact energy or the impact velocity. A natural extension of this work would investigate the forces acting on the individual grains of the medium during the impact [18].

* * *

This work was supported by the U.S. Department of Energy Office of Basic Energy Sciences.

REFERENCES

[1] MIZUTANI H., SHIN KAWAKAMI, YASUHIKO TAKAGI, MANABU KATO and MINEO KUMAZAV, *J. Geophys. Res.*, **88** (1983) A835.
[2] THORODDSEN S. T. and SHEN A. Q., *Phys. Fluids*, **12** (2001) 4.
[3] MIKKELSEN R., VERSLUIS M., KOENE E., BRUGGERT G. W., VAM DER MEER D. and LOHSE D., *Phys. Fluids*, **14** (2002) S14.
[4] UEHARA J. S., AMBROSO M. A., OJHA R. P. and DURIAN D. J., *Phys. Rev. Lett.*, **90** (2003) 194301.
[5] WALSH A. M., HOLLOWAY K. E., HABDAS P. and DE BRUYN J. R., *Phys. Rev. Lett.*, **91** (2003) 104301.
[6] NEWHALL K. A. and DURIAN D. J., preprint cond-mat/0309161.
[7] WORTHINGTON A. M., *Proc. R. Soc.*, **34** (1882) 217.
[8] MELOSH H. J., *Impact Cratering: A Geologic Process* (Oxford University Press, New York) 1989.
[9] BUCHHOLTZ V. and PÖSCHEL T., *Gran. Matt.*, **22** (1998) 1.
[10] LANDRY J. W., GREST G. S., SILBERT L. E. and PLIMPTON S. J., *Phys. Rev. E*, **67** (2003) 041303.
[11] CUNDALL P. A. and STRACK O. D. L., *Geotechnique*, **29** (1979) 47.
[12] LABOUS L., ROSATO A. D. and DAVE R. N., *Phys. Rev. E*, **56** (1997) 5717.
[13] GLASHEEN J. W. and MCMAHON T. A., *Phys. Fluids*, **8** (1996) 2078.
[14] ALBERT R., PFEIFER M. A., BARBÁSI A. L. and SCHIFFER P., *Phys. Rev. Lett.*, **82** (1999) 205.
[15] MUETH D. M., JAEGER H. M. and NAGEL S. R., *Phys. Rev. E*, **57** (1998) 3164.
[16] MILLER B., O'HERN C. O. and BEHRINGER R. P., *Phys. Rev. Lett.*, **77** (1996) 3110.
[17] D'ANNA G. and GREMAUD G., *Nature*, **413** (2001) 407.
[18] PICA CIAMARRA M., LARA A. H., LEE A. T., GOLDMAN D. I., VISHIK I. and SWINNEY H. L., *Phys. Rev. Lett.*, **92** (2004) 194301.

A granular fluid modeled as a driven system of elastic hard spheres

A. Astillero and A. Santos

Departamento de Física, Universidad de Extremadura - E-06071 Badajoz, Spain

1. – Introduction

The simplest model of a granular fluid in the rapid-flow regime consists of a gas of (smooth) *inelastic* hard spheres (IHS) characterized by a constant coefficient of normal restitution α [1]. The inelasticity of collisions produces a decrease of the mean kinetic energy (or granular temperature) with a *cooling rate* $\zeta \propto 1-\alpha^2$. Interestingly, a cooling effect can also be generated in a gas of *elastic* hard spheres (EHS) by the application of an effective drag force with a *friction coefficient* $(1/2)\zeta$. At a macroscopic level of description, the hydrodynamic balance equations of mass, momentum, and energy for the IHS gas are (formally) identical to those for the frictional EHS gas. However, the microscopic dynamics is physically quite different in both systems: in the IHS gas i) the particles move freely between two successive collisions but ii) each colliding pair loses energy upon collision; in the EHS case i) the particles lose energy between collisions due to the action of the drag force but ii) energy is conserved by collisions. This implies that during a certain small time step, only the small fraction of colliding particles are responsible for the cooling of the system in the IHS case, whereas all the particles contribute to the cooling in the EHS case. Therefore, there is no reason in principle to expect that the relevant physical properties (*e.g.*, the velocity distribution function) are similar for IHS and frictional EHS under the same conditions. For instance, in the so-called homogeneous cooling state the solutions to the respective Boltzmann equations for IHS and EHS differ: while the distribution function is a (time-dependent) Gaussian for EHS [2], deviations from a Gaussian (as exemplified by a nonzero kurtosis and by an overpopulated high-energy tail) are present in the case of IHS [3]. Notwithstanding this, the differences between the homogeneous solutions for IHS and EHS are not quantitatively important in the domain of thermal velocities and so it is still possible that both systems exhibit comparable departures from equilibrium in inhomogeneous states where transport of momentum and/or energy is the relevant phenomenon. The investigation of this possibility is the main aim of this work.

2. – Model of driven elastic hard spheres

The Boltzmann equation for a gas of inelastic hard spheres (IHS) is [4, 5]

$$(\partial_t + \mathbf{v}\cdot\nabla)f(\mathbf{v}) = \sigma^{d-1}\int \mathrm{d}\mathbf{v}_1 \int \mathrm{d}\widehat{\boldsymbol{\sigma}}\,\Theta(\mathbf{g}\cdot\widehat{\boldsymbol{\sigma}})(\mathbf{g}\cdot\widehat{\boldsymbol{\sigma}})\left[\alpha^{-2}f(\mathbf{v}')f(\mathbf{v}_1') - f(\mathbf{v})f(\mathbf{v}_1)\right] \equiv J^{(\alpha)}[f,f]. \tag{1}$$

In this expression $f(\mathbf{r},\mathbf{v};t)$ is the one-particle distribution function, σ is the diameter of a sphere, d is the dimensionality of the system, Θ is the Heaviside step function, $\widehat{\boldsymbol{\sigma}}$ is a unit vector directed along the centers of the two colliding spheres at contact, $\mathbf{g}=\mathbf{v}-\mathbf{v}_1$ is the relative velocity, and α is the coefficient of normal restitution. The precollisional or restituting velocities $\mathbf{v}'$ and $\mathbf{v}_1'$ are given by

$$\mathbf{v}' = \mathbf{v} - \frac{1+\alpha}{2\alpha}(\mathbf{g}\cdot\widehat{\boldsymbol{\sigma}})\widehat{\boldsymbol{\sigma}}, \qquad \mathbf{v}_1' = \mathbf{v}_1 + \frac{1+\alpha}{2\alpha}(\mathbf{g}\cdot\widehat{\boldsymbol{\sigma}})\widehat{\boldsymbol{\sigma}}. \tag{2}$$

The collision operator for elastic hard spheres (EHS), $J^{(1)}[f,f]$, is obtained from eqs. (1) and (2) by setting $\alpha = 1$.

The first $d+2$ moments of the distribution function define the number density n, the nonequilibrium flow velocity $\mathbf{u}$ and the *granular* temperature T. The most important properties of $J^{(\alpha)}[f,f]$ are those that determine the form of the macroscopic balance equations for mass, momentum and energy, namely

$$\int \mathrm{d}\mathbf{v}\{1,\mathbf{v},mV^2\}J^{(\alpha)}[f,f] = \{0,\mathbf{0},-dnT\zeta\}, \tag{3}$$

where m is the mass of a particle, $\mathbf{V}(\mathbf{r},t)\equiv\mathbf{v}-\mathbf{u}(\mathbf{r},t)$ is the peculiar velocity and $\zeta(\mathbf{r},t)$ is the *cooling rate* due to the inelasticity of the collisions. Although it is a nonlinear functional of the distribution function f [5, 6], a simple *estimate* $\zeta\approx\zeta_0$ is obtained by replacing the actual distribution function f by the *local equilibrium* distribution $f_0 = n(m/2\pi T)^{d/2}\exp[-mV^2/2T]$. The result is [6, 7]

$$\zeta_0(\mathbf{r},t) = \nu_0(\mathbf{r},t)\frac{d+2}{4d}(1-\alpha^2), \quad \nu_0 \equiv \frac{8\pi^{(d-1)/2}\sigma^{d-1}}{(d+2)\Gamma(d/2)}n\left(\frac{T}{m}\right)^{1/2}. \tag{4}$$

According to the arguments of sect. **1**, our model consists of the replacement

$$J^{(\alpha)}[f,f] \longrightarrow \beta(\alpha)J^{(1)}[f,f] + \frac{1}{2}\zeta_0(\alpha)\frac{\partial}{\partial\mathbf{v}}\cdot(\mathbf{V}f), \tag{5}$$

where $\beta(\alpha)$ is a positive constant to be determined. The model (5) complies with the properties (3), except that the true cooling rate is replaced by the local equilibrium estimate (4). According to eq. (5), the gas of *inelastic* hard spheres is replaced by an "equivalent" gas of *elastic* hard spheres subjected to the action of a drag force $\mathbf{F}_{\text{drag}} = -(m\zeta_0/2)\mathbf{V}$ proportional to the peculiar velocity with a friction constant that depends on the local density and temperature. This drag force mimics the cooling effect due to dissipative collisions in the underlying granular system. The parameter β accounts for the fact that, in principle, the gas of EHS that more efficiently succeeds in capturing the main properties of the granular gas is made of particles with a diameter $\sigma' = \beta^{1/(d-1)}\sigma$ that does not necessarily coincide with the diameter σ of the inelastic spheres. Alternatively, we can view β as a correction factor to modify the collision rate of the equivalent system of EHS. A comparison between the transport coefficients of IHS [7, 8] and those of the "equivalent" EHS [9] suggests the choice

$$\beta(\alpha) = \frac{1}{2}(1+\alpha). \tag{6}$$

3. – Simulations

In order to test the model (5) in inhomogeneous states far from equilibrium, we have carried out computer simulations of the Boltzmann equation by means of the DSMC method [10] in both systems (IHS and EHS) for the simple shear flow problem. In the simple shear flow [1, 11], the gas is enclosed between two infinite parallel plates located at $y = \pm L/2$ and moving with velocities $\pm U/2$ along the x-axis. When a particle crosses one of the plates it is re-entered through the opposite plate by applying the standard Lees-Edwards boundary conditions [12]. This produces a viscous heating effect that tends to increase the temperature of the system, whereas the inelastic cooling (in the IHS system) or the drag force (in the EHS system) tend to decrease the temperature. Eventually, a nonequilibrium steady state (NESS) is reached when both effects cancel each other. If the size of the system is large enough as to avoid clustering effects, the NESS is characterized by uniform density and temperature, and a linear velocity profile $u_x(y) = ay$, where $a = U/L$ is the constant shear rate.

As a test case, we have considered three-dimensional systems with $L = 2.5\lambda$, where $\lambda = (\sqrt{2}\pi\bar{n}\sigma^2)^{-1}$ is the average mean free path of the IHS gas ($\bar{n}$ being the average density), and $U = 10v_0$, where $v_0 = \sqrt{2T_0/m}$ is the initial thermal velocity (T_0 being the initial temperature). The shear rate is then $a = 4\tau_0^{-1}$, where $\tau_0 = \lambda/v_0$ is the initial mean free time of the IHS gas. The coefficient of restitution for the IHS gas has been taken as $\alpha = 0.9$. As for the EHS gas, its collision rate has been reduced by a factor $\beta = 0.95$, in agreement with eq. (6). In both cases the viscous heating effect dominates during the transient regime until the NESS is reached. In the simulations we have considered a number $N = 10^4$ of simulated particles, a width layer $\delta L = 0.05\lambda$ and a time step $\delta t = 10^{-3}\tau_0\sqrt{T_0/T}$.

Figure 1(a) shows the velocity and temperature profiles at times $t/\tau_0 = 0.13, 0.5, 1, 1.5, 2$ for both systems, starting from an initial condition of *total* equilib-

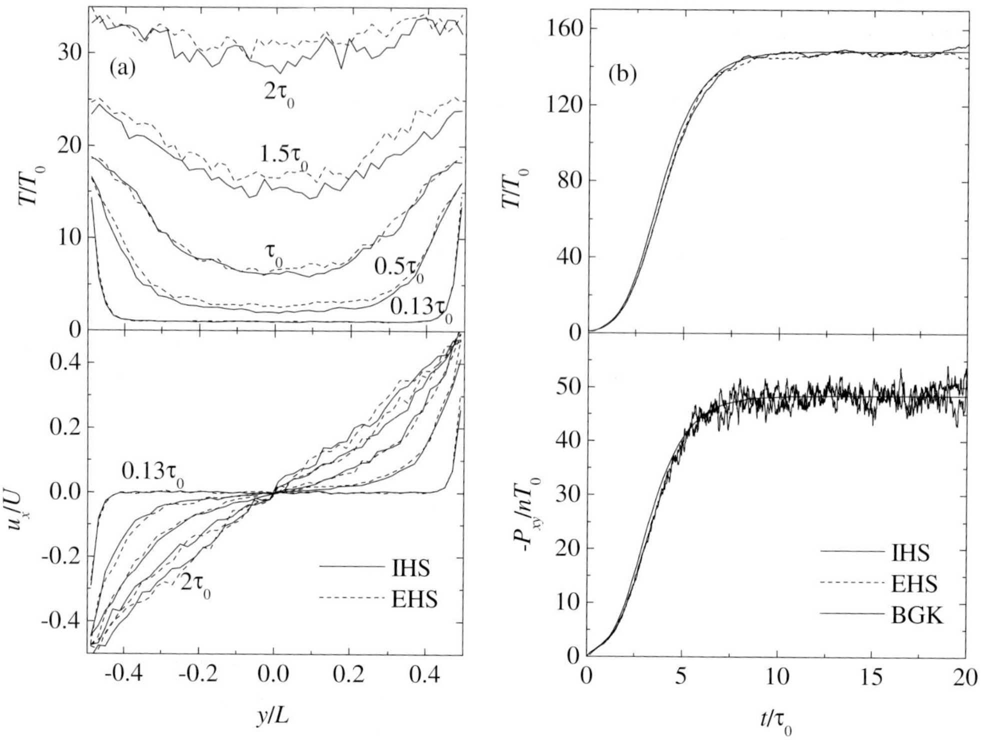

Fig. 1. – (a) Velocity and temperature profiles. (b) Time evolution of the temperature and the shear stress.

rium. By time $t = 2\tau_0$ (what corresponds to about 2.4 collisions per inelastic particle) the velocity profile is practically linear and the temperature profile is almost uniform. However, the global temperature keeps growing in time until it reaches a stationary value $T \simeq 146T_0$ for $t \gtrsim 10\tau_0$ (*i.e.* after about 48 collisions per inelastic particle). The time evolution of T/T_0 and of $-P_{xy}/nT_0$ (where P_{xy} is the shear stress) is shown in fig. 1(b) for an initial condition of *local* equilibrium, so that the system is initially prepared with a linear velocity profile. The NESS value of the shear stress represents an effective shear viscosity about 14% smaller than the Navier-Stokes value corresponding to $\alpha = 0.9$. Figure 1, along with a more comprehensive comparison that will be reported elsewhere [9], shows that the equivalent EHS system succeeds in capturing the main nonequilibrium transport properties of the underlying IHS system.

4. – Kinetic modeling

4·1. *BGK and ES models.* – The mapping IHS → EHS allows one to take advantage of the existence of simple kinetic models for EHS to extend them straightforwardly to

IHS. For instance, consider the so-called ellipsoidal statistical (ES) model [11,13]

$$J^{(1)}[f,f] \longrightarrow -\nu_0(1-\epsilon)(f-f_\epsilon), \tag{7}$$

where

$$f_\epsilon(\mathbf{v}) = n\left(\frac{mn}{2\pi}\right)^{d/2}(\det \mathsf{R})^{-1/2}\exp\left[-\frac{mn}{2}\mathsf{R}^{-1}:\mathbf{VV}\right], \quad \mathsf{R}=\frac{1}{1-\epsilon}(p\mathsf{I}-\epsilon\mathsf{P}), \tag{8}$$

P being the pressure tensor. The parameter $\epsilon \le d^{-1}$ can be freely chosen. The choice $\epsilon = d^{-1}$ makes the ES model reproduce the correct value $\mathrm{Pr}=(d-1)/d$ of the Prandtl number. The simplest choice, however, is $\epsilon = 0$, in which case $\mathsf{R}=p\mathsf{I}$, the reference function $f_\epsilon = f_0$ becomes the local equilibrium distribution and the ES model reduces to the well-known BGK model.

In the spirit of (5), the extension of the ES model to IHS is

$$J^{(\alpha)}[f,f] \longrightarrow -\beta(\alpha)\nu_0(1-\epsilon)(f-f_\epsilon)+\frac{1}{2}\zeta_0(\alpha)\frac{\partial}{\partial \mathbf{v}}\cdot(\mathbf{V}f). \tag{9}$$

In particular, setting $\epsilon = 0$ we get a simplified version of the BGK-like model for IHS that was proposed in ref. [6]. This BGK model is easy to solve for the simple shear flow problem considered in sect. **3**. The thin solid lines in fig. 1(b) represent the evolution of the temperature and the shear stress according to such a solution. As can be observed, the BGK solution exhibits an excellent agreement with the DSMC results for both IHS and EHS.

4'2. *Mixtures*. – The same idea behind (5) can be extended to a multi-component granular gas [9]. In the special case where all the species have the same flow velocity ($\mathbf{u}_i = \mathbf{u}$), our model becomes

$$J_{ij}^{(\alpha_{ij})}[f_i,f_j] \longrightarrow \beta_{ij}(\alpha_{ij})J_{ij}^{(1)}[f_i,f_j]+\frac{1}{2}\zeta_{ij}(\alpha_{ij})\frac{\partial}{\partial \mathbf{v}}\cdot[(\mathbf{v}-\mathbf{u}_i)f_i] \tag{10}$$

with

$$\beta_{ij}=\frac{1+\alpha_{ij}}{2}, \qquad \zeta_{ij}=\frac{\sqrt{2}\pi^{(d-1)/2}}{\Gamma(1+d/2)}n_j\mu_{ji}^2\sigma_{ij}^{d-1}\left(\frac{T_i}{m_i}\right)^{1/2}\left(1+\frac{m_iT_j}{m_jT_i}\right)^{3/2}(1-\alpha_{ij}^2), \tag{11}$$

where T_i is the granular temperature of species i and $\mu_{ji} \equiv (1+m_i/m_j)^{-1}$. The model (10) preserves the first $d+2$ collision integrals of IHS in the leading Sonine approximation with $\mathbf{u}_i = \mathbf{u}$. The important point is that the approximation (10) allows one to transfer any given kinetic model

$$J_{ij}^{(1)}[f_i,f_j] \longrightarrow K_{ij}^{(1)} \tag{12}$$

for *elastic* mixtures [11] into an equivalent model for *inelastic* mixtures:

$$J_{ij}^{(\alpha_{ij})}[f_i, f_j] \longrightarrow K_{ij}^{(\alpha_{ij})} = \frac{1+\alpha_{ij}}{2} K_{ij}^{(1)} + \frac{\zeta_{ij}}{2} \frac{\partial}{\partial \mathbf{v}} \cdot [(\mathbf{v}-\mathbf{u}_i) f_i]. \tag{13}$$

5. – Conclusion

In summary, we have shown that the nonequilibrium transport properties of a Boltzmann gas of inelastic hard spheres can be satisfactorily captured by an equivalent gas of elastic hard spheres driven by a dissipative drag force $\mathbf{F}_{\text{drag}} = -(m\zeta_0/2)\mathbf{V}$, where $\zeta_0(\alpha)$ is the (local equilibrium) cooling rate. Besides, the elastic particles must reduce their collision rate by a factor $\beta(\alpha) \approx (1/2)(1+\alpha)$ in order to "disguise" as a granular gas. While the "equivalent" system of EHS does not retain finer details of the true IHS gas (*e.g.*, high-energy tails, velocity correlations, ...), it is able to account for those phenomena (*e.g.*, inelastic clustering) that can be described at a hydrodynamic level. Finally, we have exploited the possibility of reverting the mapping IHS → EHS to construct kinetic models for granular gases as natural extensions of known kinetic models originally proposed for elastic particles.

* * *

This work has been supported by the Ministerio de Ciencia y Tecnología (Spain) through grant No. BFM2001-0718. A.A. is grateful to the Fundación Ramón Areces for a predoctoral fellowship.

REFERENCES

[1] CAMPBELL C. S., *Annu. Rev. Fluid Mech.*, **22** (1990) 12; GOLDHIRSCH I., **35** (2003) 267.
[2] GARZÓ V., SANTOS A. and BREY J. J., *Physica*, **163** (1990) 651.
[3] VAN NOIJE T. P. C. and ERNST M. H., *Gran. Matt.*, **1** (1998) 57.
[4] GOLDSHTEIN A. and SHAPIRO M., *J. Fluid Mech.*, **54** (1987) 47.
[5] BREY J. J., DUFTY J. W. and SANTOS A., *J. Stat. Phys.*, **87** (1997) 1051.
[6] BREY J. J., DUFTY J. W. and SANTOS A., *J. Stat. Phys.*, **97** (1999) 281.
[7] BREY J. J., DUFTY J. W., KIM C. S. and SANTOS A., *Phys. Rev. E*, **58** (1998) 4638.
[8] BREY J. J. and CUBERO D., in *Granular Gases*, edited by T. PÖSCHEL and S. LUDING, *Lect. Notes Phys.*, Vol. **564** (Springer, Berlin) 2001, pp. 59-78.
[9] ASTILLERO A. and SANTOS A., unpublished.
[10] BIRD G., *Molecular Gas Dynamics and the Direct Simulation of Gas Flows* (Clarendon, Oxford) 1994.
[11] GARZÓ V. and SANTOS A., *Kinetic Theory of Gases in Shear Flows. Nonlinear Transport* (Kluwer Academic Publishers, Dordrecht) 2003.
[12] LEES A. W. and EDWARDS S. F., *J. Phys. C: Solid State*, **5** (1972) 1921.
[13] CERCIGNANI C., *Mathematical Methods in Kinetic Theory* (Plenum Press, New York) 1990.

A simple model for a transverse dune field

E. J. R. Parteli

Institut für Computerphysik ICP, Universitaet Stuttgart
Pfaffenwaldring 27, 70569, Stuttgart, Germany

H. J. Herrmann

Institut für Computerphysik ICP, Universitaet Stuttgart
Pfaffenwaldring 27, 70569, Stuttgart, Germany
Departamento de Física, Universidade Federal de Fortaleza
60455-970, Fortaleza, CE, Brazil

1. – Introduction

There are over one hundred different types of dunes that have been classified by geomorphologists according to their shape and size [1]. Dunes can be found in deserts, in coastal areas, on the sea-bottom [2], and even on Mars [3]. Although no general theoretical model has been developed up to now to explain this wide variety of dune shapes, it is well known that the dune morphology is mainly determined by the amount of available sand and by the behavior of the wind over the year. The simplest type of dunes occurs when there is not very much sand, and when the direction of the blowing wind does not change with time. These dunes have a crescent-like shape and are called "barchans". As the sand availability becomes larger, parallel dunes perpendicular to the wind direction, which are called "transverse dunes", arise for a wind blowing steadily from the same direction. Ensembles of transverse dunes, with crest to crest distances ranging from a few meters to over 3 km, are found to cover about 40% of all terrestrial sand seas, being common in the Northern Hemisphere [4, 5], and also dominating the sand landscape on Mars [3]. The last few years have seen an increasing interest in the investigation of the geomorphological aspects and the evolution of transverse dunes [6-9].

This work reports on a simple one-dimensional model [10] for the evolution of transverse dunes in a field, which reproduces the observation that these dunes have generally the same height and velocity of propagation, independent of the initial configuration of the field. By investigating coalescence of transverse dunes with different velocities, we extend here the idea recently proposed by H. Besler [11] that barchan dunes behave like big solitons in the desert.

2. – The model

We consider a set of N dunes distributed along the x-axis, each one with a variable height h_i, but zero width. The dunes are allowed to move in the x-direction with a velocity $v_i \equiv \mathrm{d}x_i/\mathrm{d}t$ given by

$$v_i = \frac{a}{h_i}\,, \tag{1}$$

where x_i is the position of the i-th dune, $i = 1, 2, \ldots, N$, and a is a constant phenomenological parameter that contains information about the wind speed, the grain size, etc. The behavior of the sand flux on a dune field is crucial to the evolution of the field dynamics, as interactions between dunes strongly depend on the sand transport among them. For a dune that migrates shape invariantly through sand avalanches on its slip face, the sand flux ϕ over the dune behaves as $\mathrm{d}\phi/\mathrm{d}h$ = constant, where $h(x)$ is the height profile of the dune. For the model studied in this paper, we consider the height h_i of the dune at the position of the crest, and set for the i-th dune:

$$\phi_i = b\,h_i\,, \tag{2}$$

where b is our second phenomenological parameter, again depending on wind speed, grain size, etc. Note that eq. (2) means that the rate of sand transport over the crest is larger for the bigger dunes than for the smaller ones. As a consequence, exchange of sand is expected between neighbouring dunes with different heights, as a bigger dune tends to loose sand more rapidly than a smaller one.

To model the inter-dune interaction, we first observe that for a two-dimensional slice of a dune, the mass of sand in the dune, or equivalently, its area A, increases with the height of the dune as $A \sim h^2$. On the other hand, the amount of sand transported depends on the difference between the sand fluxes on the crests of the neighbouring dunes, *i.e.* if $\mathrm{d}A_i$ is the volume change for the i-th dune during a time interval $\mathrm{d}t$, then $\mathrm{d}A_i \sim (\mathrm{d}h_i)^2 \sim (\phi_i - \phi_{i-1})$. Thus, we write

$$\frac{\mathrm{d}h_i}{\mathrm{d}t} = -c \times \sqrt{|\phi_i - \phi_{i-1}|} \times \mathrm{sign}(\phi_i - \phi_{i-1}), \tag{3}$$

where c is the third parameter of our model that depends on the shape of the dune, and the last factor on the right side ensures that more sand is transported from the larger dune to the smaller one. As in a real transverse-dune field, we set the sand influx to

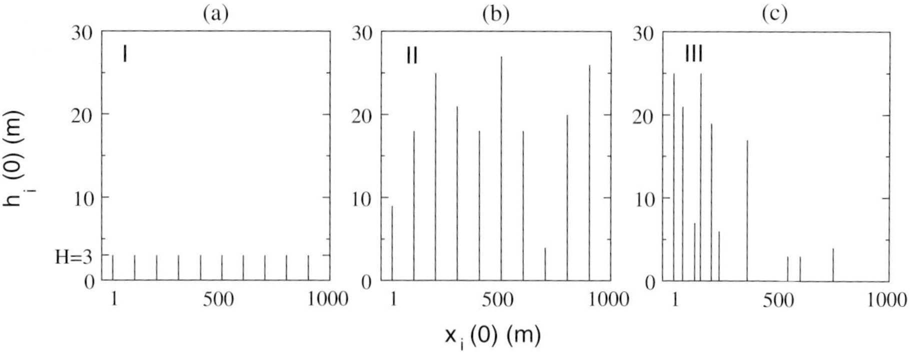

Fig. 1. – Different realizations of a set of N dunes with positions $x_i(0)$ and heights $h_i(0)$ at $t = 0$. In (a) the dunes are equally spaced, and have the same height H (group I). Dunes in group II still have regular spacing but random heights (b), while in group III, both the positions and heights are random (c).

point in the positive wind (x) direction and assume it to be constant. We call the sand influx ϕ_0, which effectively acts as a dune fixed at the position $x_0 = 0$, with a constant height $h_0 = \phi_0/b$. Equations (1), (2) and (3) are repeatedly iterated, which gives the time evolution of the dune field.

3. – Results and discussion

We studied the evolution of a set of N dunes with initial positions $x_i(0)$ and heights $h_i(0)$, for different configurations with $N = 10$, as shown in fig. 1. In the first group of simulations (I), the dunes are equally spaced along the x-axis at $t = 0$, and have the same heights $h_i(0) = H$, $i = 1, 2, \ldots, N$ (fig. 1(a)). In the second group (II), we take random values for the heights $h_i(0)$ chosen from a homogeneous distribution between 0 and 1, but still the dune spacing is regular (fig. 1(b)). And in the third group (III), both the values for the heights and the positions of the dunes along the x-axis are chosen randomly (fig. 1(c)). The trajectory of the dunes along the x-axis and the changes in their heights are shown as a function of time in figs. 2(a) and 2(b), respectively, for particular realizations corresponding to each one of the initial states I, II and III. The values of the phenomenological parameters are $a = 300\,\mathrm{m}^2/\mathrm{year}$, $b = 1500\,\mathrm{year}^{-1}$ and $c = 0.05$, and the sand influx is chosen to be $\phi_0 = 2.25 \times 10^4\,\mathrm{m/year} \approx 4.3\,\mathrm{cm/min}$. In the simulations, each iteration corresponds to 0.01 year. As can be seen from these figures, after a certain number of iterations $t \sim T_\infty$, all dunes reach the same height $h_i(T_\infty) = \phi_0/b = 15\,\mathrm{m}$, and consequently the same velocity $v_i(T_\infty) = a/h(T_\infty) = (a \times b)/\phi_0 = 20\,\mathrm{m/year}$, independently of the initial configuration.

In the discussion above, the distances between the dunes were large enough that their relative order was the same during the whole evolution of the field. However, if a smaller, faster migrating dune is very closely behind a larger, slower one, they can reach the same

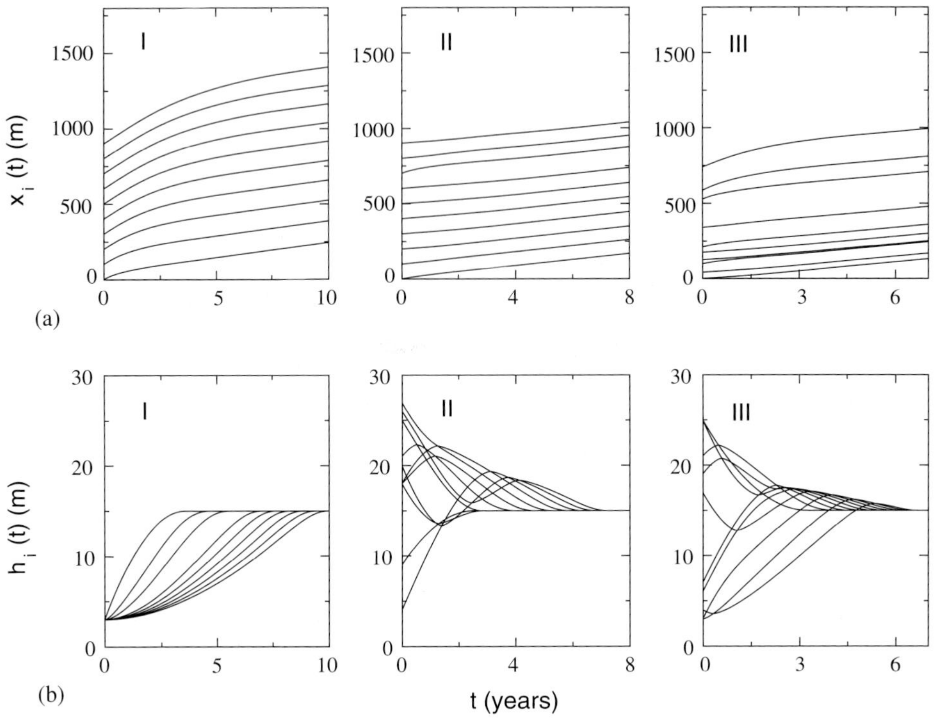

Fig. 2. – (a) Trajectories of the $N = 10$ dunes along the x-axis as function of time, for each of the realizations shown in fig. 1. The evolution of the heights of the dunes for each case is shown in (b). The values of the parameters used are $a = 300\,\mathrm{m}^2/\mathrm{year}$, $b = 1500\,\mathrm{year}^{-1}$, $c = 0.05$, and $\phi_0 = 2.25 \times 10^4\,\mathrm{m/year}$. See text for details.

position after some finite time. For simplicity, dune coalescence is implemented here through the emergence of a single dune, with height $h_{\mathrm{c}} = \sqrt{h_i^2 + h_{i+1}^2}$. Consequently, the number of dunes N in the field decreases by unity, whenever two dunes coalesce. The main plot of fig. 3 shows the variable $x_i(t)$ for a set of $N(0) = 20$ dunes with random values of $x_i(0)$ and $h_i(0)$ for $t = 0$. The values of ϕ_0 and of the parameters a, b and c are the same as in figs. 2(a) and 2(b), but now their distances in the initial state are much smaller. The number of dunes as a function of time for the realization in this figure is shown in the upper-left-hand-corner inset. Note that this number decreases monotonically until the last coalescence between dunes has occured, and the final number of dunes, $N(\infty) = 3$ has been reached. As we can see from the bottom-right-hand-corner inset, coalescence of dunes does not modify the result that the dunes in the field do all have the same height in the final state.

For a given distribution of dunes in the field at $t = 0$, the number $N(\infty)$ of dunes in the final state tends to be larger for smaller values of the parameter a, the other parameters b, c and ϕ_0 being fixed. This is because a controls the velocity of the dunes,

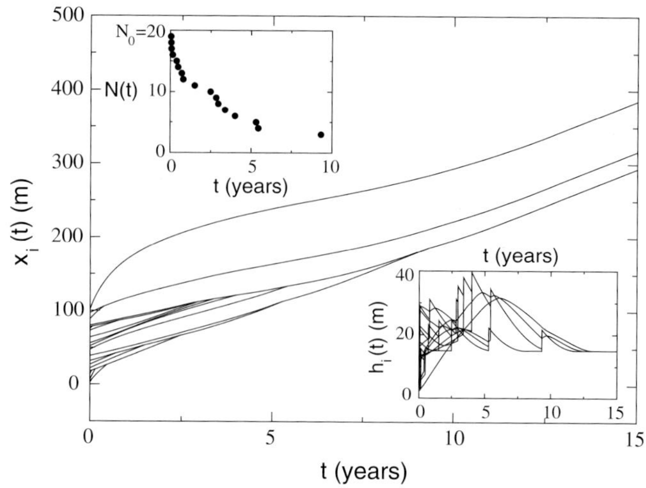

Fig. 3. – The main plot shows the time evolution of the position $x_i(t)$ for a set of $N(0) = 20$ dunes with coalescence for the values of the parameters used in fig. 2. The number of dunes as a function of time, $N(t)$, is shown in the upper-left-hand-corner inset of this figure. From the bottom-right-hand-corner inset, we can see that dune coalescence does not change the fact that all dunes have the same height for $t \to T_\infty$.

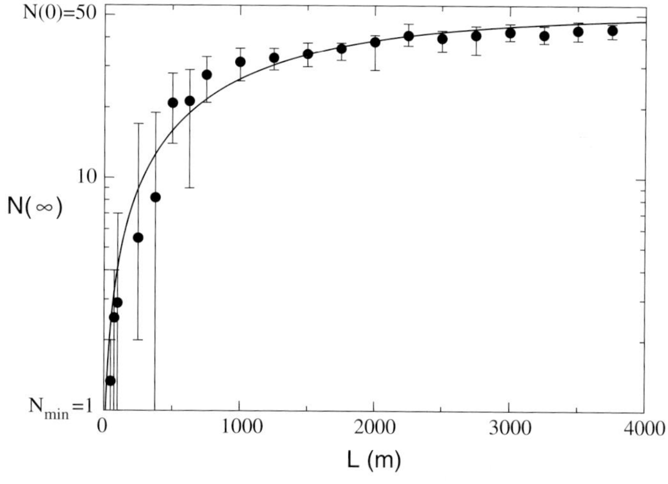

Fig. 4. – Average number of dunes $N(\infty)$ for $t = T_\infty$, as function of the size L of the system, *i.e.* the distance between the first dune and the last one in the initial state. The statistical error bars are obtained from an average over different realizations $(x_i(0), h_i(0))$ of $N(0) = 50$ dunes at $t = 0$. The full line corresponds to the fit using the equation $N(\infty) = N(0)\,[1 - q \times \exp[-L/L^*]]$, with $q \simeq 1$ and $L^* \simeq 1.34 \times 10^3$ m.

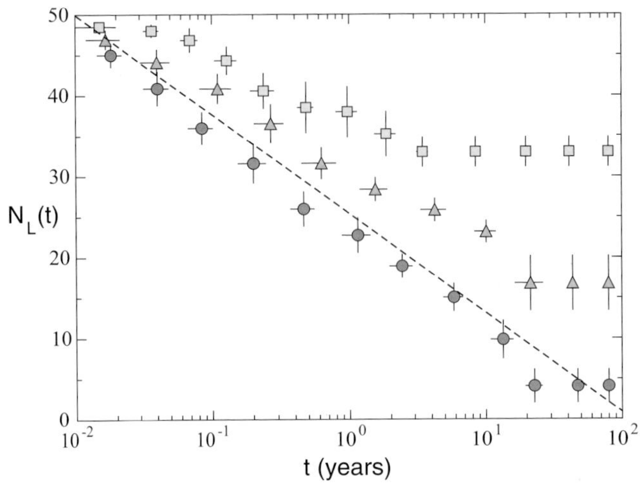

Fig. 5. – Three $N_L(t) \times t$ curves corresponding to averages over different initial realizations of the type III $(x_i(0), h_i(0))$ of a set of $N(0) = 50$ dunes, which are let to evolve with coalescence. The curves correspond to $L = 100\,\mathrm{m}$ (circles) $L = 500\,\mathrm{m}$ (triangles) and $L = 1000\,\mathrm{m}$ (squares), where L is the distance between the first dune and the last one in the initial state of the field. Time is plotted on a logarithmic scale. The dashed straight line is a guide to the eye meant to fit the curve for $L = 100\,\mathrm{m}$ (circles).

and coalescence is more frequent in a field with faster dunes. In the same way, the closer the dunes are in the initial state, the more they are going to coalesce on average. As an example, we show in fig. 4 the behavior of $N(\infty)$ as a function of the length L of the system, *i.e.* the initial distance between the first dune and the last one in the field, for a set having a fixed number of $N(0) = 50$ dunes, using the same values of the phenomenological parameters as in fig. 1. As we can see in this figure, the smaller (larger) the value of L, the more (less) the dunes tend to coalesce. The statistical error bars are obtained from different realizations of dunes with random heights and positions at time $t = 0$. The full line shown in fig. 4 corresponds to the fit

$$N(\infty) = N(0)\left[1 - q \times \exp\left[\frac{-L}{L^*}\right]\right], \tag{4}$$

where $N(0) = 50$, $q \simeq 1$, and $L^* \simeq 1.34 \times 10^3\,\mathrm{m}$. In fig. 5, we show for three different values of L the corresponding curves $N_L(t)$, averaged over five realizations of $(x_i(0), h_i(0))$, as a function of time t. The values of L shown are $L = 100\,\mathrm{m}$ (circles), $L = 500\,\mathrm{m}$ (triangles) and $L = 1000\,\mathrm{m}$ (squares). The dashed line corresponds to a straight line in a linear-log fit to the curve corresponding to $L = 100\,\mathrm{m}$. We can observe that $N_L(t)$ decays approximately logaritmically with time, independently of L. After a variable time interval $T_c \leq T_\infty$, coalescence stops and an asymptotic value $N(\infty)$ is reached. We have found that the number of iterations to reach the final state decreases as $1/c$, which in fact is independent of the initial height H of the dunes.

* * *

EJRP acknowledges a fellowship from CAPES - Brasília/Brazil.

REFERENCES

[1] LANCASTER N., *Geomorphology of Desert Dunes* (Routledge, London) 1995.
[2] BERNE S., BOURILLET J. F., DURANT J. and LERICOLAIS G., *Bull. Centres Rech. Explor. Prod. Elft Aquitaine*, **13** (1989) 395.
[3] THOMAS P. C., MALIN M. C., CARR M. H., DANIELSON G. E., DAVIES M. E., HARTMANN W. K., INGERSOLL A. P., JAMES P. B., MCEWEN A. S., SODERBLOM L. A. and VEVERKA J., *Nature*, **397** (1999) 592.
[4] TSOAR H. and BLUMBERG D. G., *Earth Surf. Process. Landforms*, **27** (2002) 1147.
[5] BURKINSHAW J. R., ILLENBERGER W. K. and RUST I. C., *Geol. Soc. Spec. Publ.*, **72** (1993) 23.
[6] WERNER B. T., *Geology*, **23** (1995) 1107.
[7] MOMIJI H., CARRETERO-GONZALEZ R., BISHOP S. R. and WARREN A., *Earth Surf. Process. Landforms*, **25** (2000) 905.
[8] MOMIJI H. and WARREN A., *Earth Surf. Process. Landforms*, **25** (2000) 1069.
[9] SCHWAEMMLE V. and HERRMANN H. J., *Earth Surf. Process. Landforms*, **29** (2004) 769, cond-mat 0301589.
[10] PARTELI E. J. R. and HERRMANN H. J., *Physica A*, **327** (2003) 554, cond-mat 0305036.
[11] BESLER H., *Phys. Bl.*, **53** (1997) 10.

Simple model for mechanical responses of collapsed DNA chains

H. WADA, Y. MURAYAMA and M. SANO

Department of Physics, University of Tokyo - Hongo, Tokyo 113-0033, Japan

1. – Introduction

The elastic property of single polymer chains is a fundamental and well-established subject of polymer physics [1]. However, as recent developments of micromanipulation techniques of single molecules, detailed experimental studies of single biopolymers such as double-stranded DNA (dsDNA) [2] and proteins [3] have opened up a rich and fascinating physics. In particular, the elastically driven unfolding (or coil-globule) transition found within a single polymer subjected to a stretching force in poor-solvent condition is one of the most salient phenomena [4].

In our recent works [5, 6], experimental results on the elastic properties of dsDNA collapsed by trivalent cations spermidine^{3+} (SPD) have been reported. We have found that the force *versus* extension curves showed two striking phases, *i.e.*, force plateaus and stick-release patterns, depending on the concentration of SPD. When the concentration of SPD is relatively low and the DNA is just below the coil-globule transition point, force plateaus of 1–2 pN are observed during both the stretching and the relaxing processes. On the other hand, in higher range of concentration of SPD, stick-release patterns appear in the stretching processes, while force plateaus of 0.5–2 pN are observed only during the relaxing processes. The latter finding implies the formation and subsequent unfolding of globular objects.

In this paper, we present a simple phenomenological model to describe these elastic responses of collapsed DNA. The model is represented by the elastic theory for the semiflexible polymer (the wormlike chain model) [7] combined with the order parameter equation, which accounts for the intramolecular transition kinetics [8]. It is shown that our continuum and discrete model successfully reproduce the observed force plateau and stick-release pattern in the force-extension curves, respectively.

2. – Model

When an intramolecular phase segregation has occurred and an elongated coil state and a collapsed state coexist within a single polymer, an apparent contour length L_{a} differs from the original one L_0 [9]. Presumably, the collapsed globule is hard to be stretched compared with the coil-like chain. It is thus considered that the fraction of the coil state virtually determines the flexibility of the polymer. If we identify this polymer with an ideal WLC having contour length L_{a}, the force F and the extension x of the chain can be connected by the formula [7]

$$F(x) = \frac{k_{\mathrm{B}}T}{l_{\mathrm{p}}}\left(\frac{x}{L_{\mathrm{a}}} + \frac{1}{4(1-x/L_{\mathrm{a}})^2} - \frac{1}{4}\right), \tag{1}$$

where l_{p} is the persistence length, k_{B} the Boltzmann constant, and T the temperature. In order to calculate the effective contour length L_{a}, we introduce an order parameter ρ along the coordinate s, where s is the original contour distance from one chain end. We define that $\rho(s)\mathrm{d}s$ represents the local contour length of the coil-like flexible part within a small section $s \sim s + \mathrm{d}s$. Hence, $\rho = 1$ corresponds to a fully random coil state, while $\rho = 0$ indicates an entirely collapsed state. In a bimodal state, ρ varies from 0 to 1. We also assume that a collapsed part $(1-\rho)\mathrm{d}s$ acts like a coil part with its length $\epsilon(1-\rho)\mathrm{d}s$, where $\epsilon \ll 1$. Then the effective contour length is now expressed as

$$L_{\mathrm{a}} = \int_0^{L_0} \{\epsilon + (1-\epsilon)\rho(s)\}\mathrm{d}s. \tag{2}$$

Since ρ is a non-conservative quantity by definition, we can assume the following relaxational dynamics:

$$\frac{\partial \rho}{\partial t} = -\frac{\lambda}{k_{\mathrm{B}}T}\frac{\delta \tilde{F}}{\delta \rho} + \theta(s,t), \tag{3}$$

where the thermal noise source θ is related to the kinetic coefficient λ via the usual fluctuation-dissipation relation. The free energy functional $\tilde{F}$ is assumed to be of the form

$$\frac{\tilde{F}}{k_{\mathrm{B}}T} = \int_0^{L_0}\left\{\frac{K}{2}\left(\frac{\partial \rho}{\partial s}\right)^2 + W(\rho, f)\right\}\mathrm{d}s. \tag{4}$$

The effective thermodynamic potential W includes the dimensionless force $f \equiv l_{\mathrm{p}}F/k_{\mathrm{B}}T$ as a control parameter, and has the local minima at $\rho = 0$ and $\rho = 1$ for each f. The profile of W and its dependency on f are shown in fig. 1. Choosing the polynomial form for W, we arrive at $\partial W/\partial \rho = -\mu\rho(1-\rho)(\rho - b(f))$. For the numerical stability,

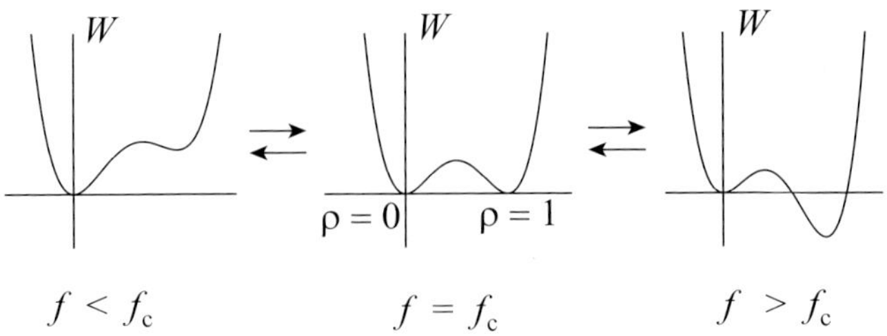

Fig. 1. – Schematic representation of a double-well potential W controlled by the parameter f.

we adopt here the specific form of $b(f)$ as $b(f) = (1/2)\{1 - \tanh(f - f_c/v)\}$, where the positive constants K, μ, f_c and v are material dependent. After rescaling the variables as $x \to L_0 x$, $t \to t/(\lambda\mu)$ and $\theta \to (\lambda\mu)\theta$, we obtain the equation of motion for ρ of the form

$$(5) \qquad \frac{\partial \rho}{\partial t} = \nu^2 \frac{\partial^2 \rho}{\partial s^2} - \rho(\rho - b)(\rho - 1) + \theta,$$

where $\nu^2 = K/(\mu L_0^2)$ is the dimensionless diffusion constant, and the amplitude of the noise correlation is now given by $M = (\mu L_0)^{-1}$. Throughout this paper, we impose the zero-flux boundary conditions given by $\partial_s \rho(s = 0, 1, \tau) = 0$.

3. – Numerical simulations

By controlling the extension x and solving numerically eqs. (1)-(5), we can obtain a F-x curve as a set of consecutive data points (x_i, F_i), $(i = 1, 2, \ldots, N_D)$. In each measurement of (x_i, F_i), the extension is fixed at $x = x_i$ during a certain waiting time T after its instantaneous change. Since the force fluctuates, we average the force for the waiting time in each measurement:

$$(6) \qquad F_i = \frac{k_B T}{l_p}\left[\frac{1}{T}\int_0^T \mathrm{d}t f(x_i, t)\right].$$

The parameters are fixed at $L_0 = 5\,\mu$m, $l_p = 25$ nm, $T = 296$ K corresponding to the setup of the experiment [6]. Because the thickness of the interface $\xi = 2^{3/2}\nu$ between $\rho = 0$ and $\rho = 1$ is expected to be comparable to the scaled persistence length $l_p/L_0 = 5 \times 10^{-3}$, we set $\nu \sim 10^{-3}$. The extension x is varied from $x_{\min} = 1.0\,\mu$m to $x_{\max} = 4.8\,\mu$m with $\Delta x = |x_{i+1} - x_i|$ set to 19 nm. The noise intensity M, ϵ, f_c and v are chosen to be 0.0005, 0.15, 12.5 and 3.7, respectively. The simulation is performed for $T = 3.0$, which is long enough for a polymer to reach quasi-stationary state for the above parameter set.

We use the standard implicit method (Crank-Nicolson scheme) for the numerical integration by approximating the nonlinear term as $\rho(\rho - b)(\rho - 1) \approx (1/2)[\rho(t + \Delta t) +$

$\rho(t)][\rho(t) - b][\rho(t) - 1]$ with the time increment Δt and grid size Δs set to 0.005 and 0.0025, respectively. Throughout the following calculations, the initial condition of the chain is set to the full random-coil state $\rho = 1$ with the extension $x_{max} = 4.8\,\mu$m. We first relax the polymer extension to $x_{min} = 1\,\mu$m, and again stretch it back to $x_{max} = 4.8\,\mu$m.

3'1. *Force plateau.* – The calculated F-x curve is shown in fig. 2 (a). The elastic response is obviously different from that of the ideal WLC. The F-x curve clearly shows force plateaus of 1-2 pN during both the stretching and the releasing processes. In the releasing process, a depression in the force curve can be found associated with the onset of the force plateau. The energy barrier for generating a critical globule nucleus may be responsible for this characteristic behavior. In addition, a marked hysteresis appears between the stretching and the releasing process. It should be emphasized that these features are also generically observed in the experiment [5, 6].

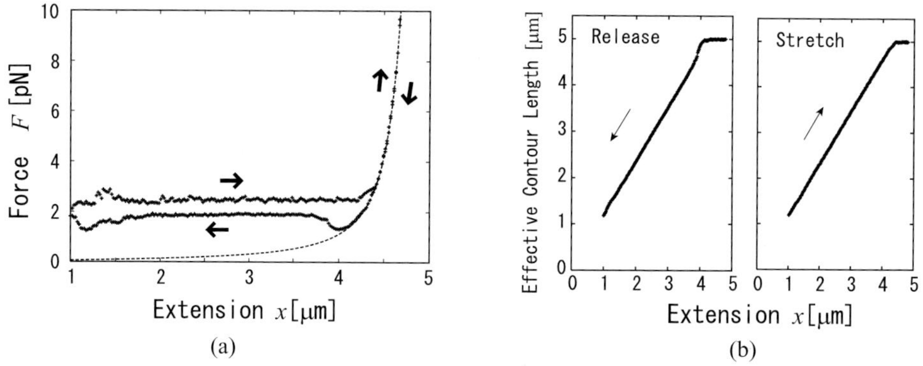

Fig. 2. – (a) The force *vs.* extension curve obtained from the numerical simulation. The dashed line is the ideal WLC curve for the parameters $l_p = 25$ nm, $L = 5\,\mu$m, and $T = 296$ K. (b) The effective contour length L_a as a function of x in the stretching and releasing process. The parameters for these plots are $M = 5 \times 10^{-4}$, $\nu = 0.001$, $f_c = 12.5$, $v = 3.7$ and $T = 3.0$.

In fig. 2 (b), the effective contour length L_a is plotted as a function of the extension x. One can clearly see that L_a is proportional to x while the force plateau is observed. This is a natural consequence because, in our model calculation, eq. (1) takes constant value when x/L_a is constant. Although data are not shown here, the temporal evolution of the spatial structure of $\rho(s)$ clearly demonstrates the force-driven phase transition within the single polymer. The physical mechanism underlying the observed hysteresis and its relation with the waiting time is also discussed in detail in ref. [8] in terms of the motion of the domain boundaries.

3'2. *Stick-release phenomena.* – When the concentration of SPD is large enough to induce deep condensation of a single polymer, a force-extension curve shows a stick-release pattern. Although the electrostatic effect may play an essential role in this stage, the problem is beyond the scope of this paper. Instead, we solely present a rather simple

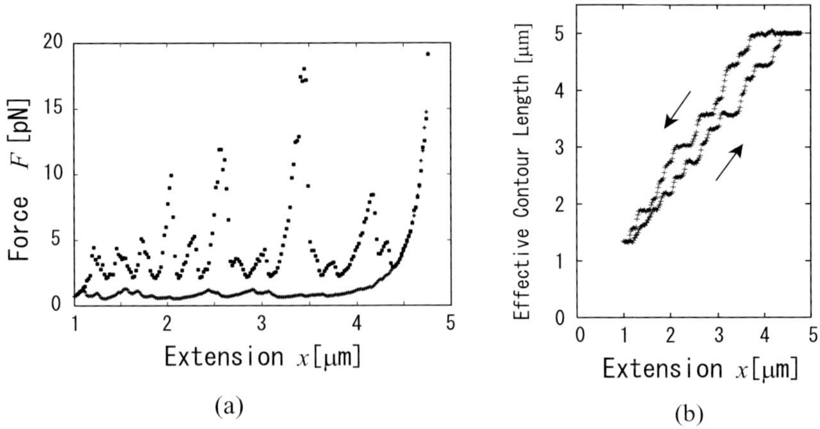

Fig. 3. – (a) The force *vs.* extension curve obtained from the numerical simulation. (b) Plots of effective contour length L_a as a function of x in the stretching and the releasing processes. The parameters used are $M = 6 \times 10^{-5}$, $\nu = 0.001$, $f_c = 12.0$, $v = 6.0$ and $T = 3.0$.

description of this phenomenon here.

In the experiment [6], there is a tendency for the stick-release response to occur periodically with a certain characteristic length $x_u \sim 300\,\mathrm{nm}$. As mentioned in the introduction, this finding suggests that the stick-release patterns correspond to the subsequent abrupt releases of condensed globular objects connected with short coils. Furthermore, since x_u is comparable to the experimentally observed length of one turn of toroid [10], it is expected that a globular bead forms a toroid-like structure with its typical size $d \sim 100\,\mathrm{nm}$.

Based on these considerations, we shall employ a further coarse-graining of the internal structure of the molecule. We assume that the chain is now composed of $Q = L_0/x_u$ pieces of "modules" that can undergo a transition between two energetically different states. We assume, however, that there still exists relevant short-range interactions between polymer segments to recover its structural uniformity. Then the procedure to modify the original model is simply given by $\rho(s,\tau) \to \rho_i(\tau)$, $\int_0^{L_0} \mathrm{d}s \to x_u \sum_{i=1}^{Q}$. This corresponds to the discrete version of the plateau model.

The numerical simulation is performed with the parameters $Q = 16$, $\epsilon = 0.10$ and $M = 6 \times 10^{-5}$. Since the effective potential W should have the deeper minima, it is appropriate here to choose a value of M smaller than that in the force plateau case. Other conditions are the same as those specified in the previous subsection.

A selected plot of the F-x curve is shown in fig. 3 (a). In the stretching process, the stick-release pattern appears, while the force plateau of 0.5-2 pN in magnitude is observed during the relaxing process. Force magnitudes at the peak positions are also generally within the range of 4-12 pN. Therefore, the present calculation captures the main features of our previous measurements [6] well.

In fig. 3 (b), we plot the effective contour length L_a as a function of the extension x. The curve looks rather irregular compared to that of the force plateau, fig. 2 (b). Sharp changes of the contour length may correspond to the abrupt releases of the globular objects or the rapid condensations of the short coils.

4. – Conclusion

Our theoretical calculations suggest that two types of elastic responses, *i.e.*, force plateau and stick-release pattern can be understood as WLC behaviors of an apparent contour length which can vary with the force-dependent transition rate. It is proved that combining the elastic theory of a single polymer with the kinetics of a metastable system is efficient to describe the main features of the measured force-extension curves of collapsed DNA.

* * *

We are grateful to T. OHTA for variable discussions. This work was partially supported by Grant-in-Aid for JSPS Fellows, from Ministry of Education, Culture, Sports, Science, and Technology, Japan.

REFERENCES

[1] DOI M. and EDWARDS S. F., *The Theory of Polymer Dynamics* (University Press, Oxford) 1986.
[2] BUSTAMANTE C., MARKO J. F., SIGGIA E. D. and SMITH S., *Science*, **265** (1994) 2599.
[3] RIEF M., GAUTEL M., OESTERHELT F., FERNANDEZ J. F. and GAUB H. E., *Science*, **276** (1997) 1109.
[4] BAUMANN C. G., BLOOMFIELD V. A., SMITH S. B., BUSTAMANTE C., WANG M. D. and BLOCK S. M., *Biophys. J*, **78** (2000) 1965.
[5] MURAYAMA Y. and SANO M., *J. Phys. Soc. Jpn.*, **70** (2001) 345.
[6] MURAYAMA Y., SAKAMAKI Y. and SANO M., *Phys. Rev. Lett.*, **90** (2003) 018102.
[7] MARKO J. F. and SIGGIA E. D., *Macromolecules*, **28** (1994) 8759.
[8] WADA H., MURAYAMA Y. and SANO M., *Phys. Rev. E*, **66** (2002) 061912.
[9] YOSHIKAWA K. and MATSUZAWA Y., *J. Am. Chem. Soc.*, **118** (1996) 929.
[10] ARSCOTT P. G., LI A.-Z. and BLOOMFIELD V. A., *Biopolymer*, **30** (1990) 619.

Actin-based motility of droplets

O. CAMPÀS

Institut Curie, UMR 168, Laboratoire Physico-Chimie Curie
11, rue Pierre et Marie Curie, F-75248 Paris cédex 05, France
Departament d'Estructura i Constituents de la Matèria, Universitat de Barcelona
Av. Diagonal 647, E-08028 Barcelona, Spain

J. F. JOANNY and J. PROST

Institut Curie, UMR 168, Laboratoire Physico-Chimie Curie
11, rue Pierre et Marie Curie, F-75248 Paris cédex 05, France

1. – Introduction

Force generation by polymerization of actin filaments plays a central role in biology. At molecular scales, the bending of single filaments driven by thermal fluctuations allows for polymerization against applied forces. Each single actin filament is able to generate a force of several piconewtons. Cooperative phenomena between a large number of actin filaments are needed to induce movement at cellular scales. Cellular motility involves in most cases the growth of an actin gel made of a filamentous actin network. Several bacterial pathogens (*Listeria monocytogenes, Shigella flexneri, Rickettsia conorii*) use the growth of an actin gel to generate forces that allow them to propel inside a host cell [1, 2]. At the level of the cell, the movement of endocytic vesicles is also driven by the growth of an actin gel [3-5]. Although actin polymerization is very important for cell crawling, in this case the cell movement is due to a subtle interplay between the dynamic actin network and molecular motors. On the contrary, force generation in bacteria does not involve molecular motors. Clearly, actin-based motility is a robust force generation mechanism. *Listeria* has extensively been studied because it provides the simplest system to understand actin-based motility. Different models have been proposed to understand

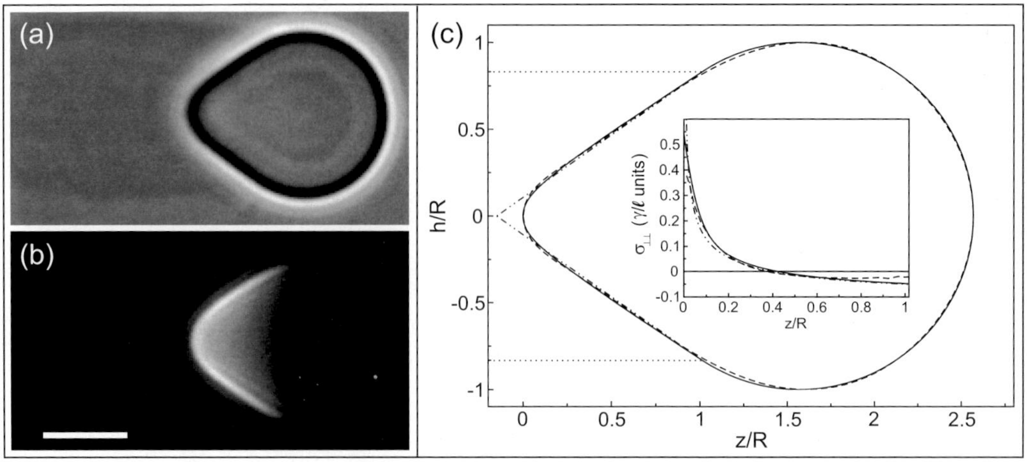

Fig. 1. – Oil droplet covered with fluorescent labeled actin polymerization promoters (VCA) and placed in HeLa cell extracts observed (a) by phase contrast microscopy and (b) by fluorescence microscopy. The VCA appears bright. Bar $3\,\mu$m. (c) Droplet shape and stress distribution (inset). Numerical integration for $\epsilon = 0.049$ and $\sin\theta_0 = 0.58$ (continuous line); experimental obtained profile (dashed line); zeroth-order asymptotic expansion (dash-dotted line). Experiments done by Hakim Boukellal and Cécile Sykes (Institut Curie).

the underlying physical mechanisms leading to the observed motility process. Molecular models account for the Brownian flexibility of single actin filaments (but depend strongly on biochemical details) [6]. On the other hand, mesoscopic approaches lead to robust theories as they are based on general conservation laws [7, 8].

Our bio-mimetic system is an oil droplet covered on the surface by proteins that trigger actin polymerization. This system undergoes a spontaneous symmetry breaking from the original spherical symmetry. Once the droplet escapes from the original actin cloud, an actin comet tail progressively develops and is deformed near the droplet surface due to its own growth (fig. 1a). The energy produced by actin polymerization is stored in the form of elastic energy of the gel. Actin polymerization does not directly propel the droplet but maintains the steady shape and stress distribution. The propulsive force comes from a subtle mechanism which couples the peculiar stress distribution to the surface density of nucleating proteins. Eventually, the system reaches a steady state and moves at a constant velocity V. By studying these systems, we aim to understand the force generation mechanism due to the elastic stresses. In this sense, the stress distribution on the droplet surface carries important information about the motility process.

The movement of the droplet under the action of elastic stresses is, physically speaking, a free-boundary problem. We have introduced a theoretical model based on mechanical equilibrium and on an exponential dependence of the growth velocity of the gel with normal elastic stresses, which accounts well for the experimental observations. We have solved analytically and numerically this free boundary problem and, by comparing

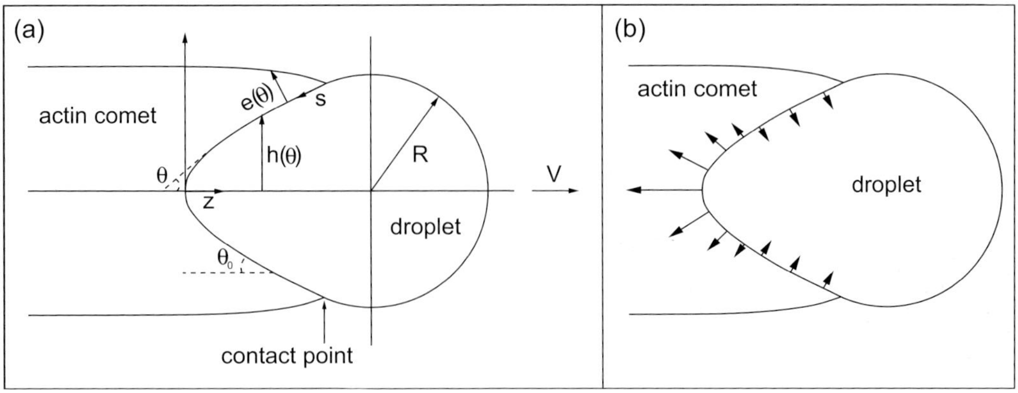

Fig. 2. – (a) Sketch of the droplet. The shape of the droplet is characterized by $h(z)$. (b) Qualitative stress distribution for a droplet in the steady state. The gel pulls the droplet backwards at the back, and forward at the sides.

theory to experiments, we have been able to estimate various essential parameters of actin-based motility. The experiments on this bio-mimetic system have been performed by Hakim Boukellal and Cécile Sykes (Institut Curie). Previous bio-mimetic studies on *Listeria*-like systems can be found in ref. [9].

2. – Elastic model

Unlike that of bacteria, the surface of the liquid droplet is characterized by a surface tension γ; it can be deformed under the action of elastic stresses. In a steady state, the shape of the droplet is specified by the balance between normal elastic stresses and the restoring force due to surface tension. As there exists a clear colocalization of nucleating proteins and actin gel (fig. 1a, b), the front part of the droplet, which is not coated by the actin gel, is a sphere of radius R (specified by volume conservation). In this region, mechanical equilibrium is expressed by the Laplace equation $P_{\text{i}} - P_{\text{o}} = 2\gamma/R$ (where P_{i}, P_{o} are the pressures of the fluids inside and outside the droplet, respectively). The difference in pressures $P_{\text{i}} - P_{\text{o}}$ is the same at every point of the surface. Hence, the local normal force balance is written as

$$\frac{2\gamma}{R} = \gamma\left(\frac{\cos\theta}{h} + \frac{\mathrm{d}(\cos\theta)}{\mathrm{d}h}\right) - \sigma_{\perp\perp}, \tag{1}$$

where $-\sigma_{\perp\perp}$ is the normal force per unit area applied by the actin gel on the droplet, h is the fluid thickness and θ is the angle between the local tangent and the axis defined by the velocity V (sketched on fig. 2).

The actin gel grows in the outward normal direction to the surface, and its growth velocity changes under the action of elastic stresses, but the precise functional dependence is not known. We assume a Boltzmann dependence according to classical rate theories [10]

$$v_p(h) = v_p^0 \exp\left[\frac{\sigma_{\perp\perp}(h)}{\sigma_0}\right], \quad \sigma_0 \equiv \frac{k_B T}{a^2 \delta}, \tag{2}$$

where a is the mean distance between nucleating proteins on the surface and δ is a length of the order of the size of an actin monomer. We have checked that the results of the model are robust under changes of the functional form of eq. (2). As a first approximation, we assume that the gel is incompressible [11] and that the actin comet tail is a perfect cylinder. Under these conditions, local mass conservation takes the following form:

$$v_p(h) = V \sin\theta(h). \tag{3}$$

The equations written so far are able to specify a family of shapes for the droplet, but in order to fully characterize the shape, we need to fix the only dynamical parameter in the model, the velocity V. The physically meaningful solutions are those for which the system as a whole is not accelerated. Therefore, global force balance specifies the value of the velocity V. The global force acting on the droplet has different sources. A propulsive force coming from the integrated contribution of the normal elastic stresses F_p, and a friction force between the droplet and the actin gel (hydrodynamic friction is smaller by several orders of magnitude) [8]. This friction is essential if one expects the system to reach a steady state. It is important to note that, with the equations written so far, the propulsive force exactly vanishes for any shape of the deformed droplet (keeping the rotational symmetry around the axis defined by the velocity). The propulsive force comes from small variations in surface tension induced by the local stresses. Even if these variations in surface tension are essential to understand the origin of the propulsive force, their effect on the steady shape of the droplet is negligible. Thus, the family of shapes for the droplet in the steady state can be calculated at zero propulsive force (constant surface tension). In ref. [5] Giardini *et al.* estimate the propulsive force done by the actin gel by using eq. (1). From our discussion above on the force generation mechanism, it is clear that this method cannot give a finite value of the propulsive force because it vanishes in this approximation if calculated properly.

The shape of the region where the gel interacts with the droplet only deviates from a perfect cone when the stresses applied by the gel are large compared to the intrinsic stress scale σ_0. This defines an intrinsic length scale for the system $\ell \equiv \gamma a^2 \delta / k_B T$, and the stress scale σ_0 may be rewritten as $\sigma_0 = \gamma/\ell$. Typical values for actin suggest $\ell \sim 0.1\,R$. Thus, we define $\epsilon \equiv \ell/R$ and we solve the equations by using a singular perturbation scheme. At lowest order, the stress is zero and the shape is a perfect cone with an opening angle θ_0 given by $\sin\theta_0 = v_p^0/V$. The lowest non-zero order ($\mathcal{O}(\epsilon)$) for the stress has the following form:

$$\sigma_{\perp\perp}(h) = \frac{\gamma}{R}\left(\frac{R\cos\theta_0}{h} - 2\right). \tag{4}$$

The contact point between the gel and the droplet is given by $h_c = R\cos\theta_0$ (see fig. 2). The stress is positive at the back of the droplet ($h < h_c/2$), indicating that the gel is pulling the droplet in the direction opposite to motion. On the other hand, at the sides of the cone-like shape ($h_c/2 < h < h_c$) the stress is negative, meaning that the gel is pushing the droplet forward, in the direction of motion. The typical stress distribution is depicted in fig. 2b.

The model contains one material parameter ϵ and one dynamical parameter V (or $\sin\theta_0$), as we have already pointed out. We have also numerically integrated the equations in order to find the shape of the droplet, and the stress distribution, for a given value of both parameters. Hakim Boukellal and Cécile Sykes were able to obtain experimentally the shape of the deformed oil droplet, its velocity, and its surface tension among other quantities [12]. The velocity of the advancing droplet for a given cell extract was $V = 0.15 \pm 0.03\,\mu\text{m}\,\text{min}^{-1}$, and the surface tension $\gamma = 4.0 \pm 0.6\,\text{mN}\,\text{m}^{-1}$. Thereby, by fitting the experimental observed shape with the theoretical predictions (fig. 1c), we are able to obtain the actin polymerization velocity at zero stress on surfaces $v_p^0 = 1.5\,\text{nm}\,\text{s}^{-1}$ and the magnitude of the stresses developed by the actin gel $\sigma_0 = 32\,\text{nN}\,\mu\text{m}^{-2}$. In other words, our theoretical framework provides a quantitative way of analyzing actin-based motility.

In fig. 1c, we compare the experimental shape and stress distribution to the ones obtained from the model described so far. Very close to the contact point experimental observations and theory differ slightly. This is due to the fact that we assumed the actin comet tail to be a perfect cylinder. Near the contact point this is no longer true. In fact, it is easy to check from eq. (4) that the model predicts a finite stress at the contact point, which is not possible. We then developed a more refined analysis of the system in this region. We only assume that the actin gel is incompressible and we do not constrain the shape of the actin tail which is described by the function $e(s)$. Under these conditions, local mass conservation is written as

$$v_p(s) = V\cos\theta \frac{\mathrm{d}e(s)}{\mathrm{d}s}, \tag{5}$$

where s is the arclength from the contact point. In the case where the actin tail is a perfect cylinder, $\mathrm{d}e/\mathrm{d}s = \tan\theta$ and eq. (5) reduces to eq. (3). At a scaling level, the variation of the tensile stresses along the surface is given by $\mathrm{d}\sigma_{ii}/\mathrm{d}s = (E/R_i V\cos\theta)(V\sin\theta - v_p)$, where i can either be the azimuthal direction ϕ or the tangent direction t, E is an elastic modulus of the gel and R_i is the local radius of curvature along the i direction. Therefore, at lowest order, both tensile stresses are constant. Furthermore, close enough to the contact point, the actin gel can be described as a thin shell, allowing us to write the normal stress as $\sigma_{\perp\perp} = -e(\sigma_{tt}/R_t + \sigma_{\phi\phi}/R_\phi)$. Asymptotic matching with the solution found for the shape far away from the contact point imposes $\sigma_{\phi\phi} = 0$. Defining $\tilde{\sigma}_{tt} \equiv \sigma_{tt}/\sigma_0$ and solving the equations at lowest order, we obtain the following normal stress as a function of the gel thickness e

$$\sigma_{\perp\perp} = -\frac{\gamma}{R}\frac{e\,\tilde{\sigma}_{tt}}{\ell + e\tilde{\sigma}_{tt}}. \tag{6}$$

Thus, when the gel thickness vanishes, normal stresses strictly vanish. Far away from the contact point (distances larger than $\ell/\tilde{\sigma}_{tt}$, which is the size of the boundary layer), we recover the cylindrical shape for the actin tail and we match normal stresses ($\sigma_{\perp\perp} = -\gamma/R$). For a detailed analysis see ref. [12].

Both analytical and numerical results reveal a non-trivial stress distribution where the gel is pulling the droplet backwards at the back, and pushing it forward at the sides. We expect the stress distribution to be similar for *Listeria* or hard beads. Furthermore, by comparing the theoretical predictions to the experimental results, we can estimate various parameters associated to the motility process whose knowledge clarify the underlying mechanisms of actin-based motility. We also propose a new force generation mechanism based on a subtle coupling between the stress distribution and surface tension gradients. Microscopic models [6], where actin polymerization directly propels the object, are inconsistent with the experimental observations and with the results obtained by our mesoscopic approach. A quantitative analysis of the force generation mechanism will be published soon. In future work we aim to quantify the effects of applied external forces on the moving droplet, to find out different dynamical regimes and to characterize non-local effects.

REFERENCES

[1] GOUIN E., GANTELET H., EGILE C., LASA I., OHAYON H., VILLIERS V., GOUNON P., SANSONETTI P. J. and COSSART P., *J. Cell Sci.*, **112** (1999) 1697.
[2] GOLDBERG M. B., *Microbiol. Mol. Biol. Rev.*, **65** (2001) 595.
[3] TAUNTON J., ROWNING B., COUGHLIN M., WU M., MOON R., MITCHISON T. and LARABELL C., *J. Cell Biol.*, **148** (2000) 519.
[4] UPADHYAYA A., CHABOT J. R., ANDREEVA A., SAMADANI A. and VAN OUDENAARDEN A., *Proc. Nat. Acad. Sci.*, **100** (2003) 4521.
[5] GIARDINI P., FLETCHER D. and THERIOT J., *Proc. Nat. Acad. Sci.*, **100** (2003) 6493.
[6] MOGILNER A. and OSTER G., *Biophys. J.*, **84** (2003) 1591.
[7] GERBAL F., CHAIKIN P., RABIN Y. and PROST J., *Biophys. J.*, **79** (2000) 2259.
[8] PROST J., in *Physics of Biomolecules and Cells, Les Houches Lecture Notes LXXV*, edited by H. FLYVBERG, F. JÜLICHER, P. ORMOS and F. DAVID (EDP Sciences, Springer) 2002, pp. 215-236.
[9] BERNHEIM-GROSWASSER A., WIESNER S., GOLSTEYN R. M., CARLIER M. F. and SYKES C., *Nature*, **417** (2002) 308.
[10] EYRING H., *J. Chem. Phys.*, **3** (1935) 107; KRAMERS H. A., *Physica (Utrecht)*, **7** (1940) 284.
[11] GERBAL F., LAURENT V., OTT A., CARLIER M. F., CHAIKIN P. and PROST J., *Eur. Biophys. J.*, **29** (2000) 134.
[12] BOUKELLAL H., CAMPÀS O., JOANNY J. F., PROST J. and SYKES C., *Phys. Rev. E*, **69** (2004) 061906.

Computer simulation of structural properties of dilute aqueous solutions of rare gases from ambient to supercritical conditions

V. De Grandis, P. Gallo and M. Rovere

Dipartimento di Fisica, Università "Roma Tre"
Istituto Nazionale per la Fisica della Materia, Unità di Ricerca Roma Tre
and Democritos National Simulation Center - Via della Vasca Navale 84, 00146 Roma, Italy

1. – Introduction

The properties of hydration of apolar solutes have become of large interest for the applications in many chemical processes in industrial technology and biochemistry [1]. Computer simulation studies on aqueous solutions of rare gases indicate that at ambient conditions the solvation is governed by the formation of hydration cages around the solute [2, 3]. On raising temperature the solute tends to disrupt the cage [2].

Recent neutron diffraction studies performed at supercritical conditions in mixtures of water and rare gases found that the structure of water is compressed by the presence of the solutes [4]. The solubility of rare gases shows a very interesting temperature dependence [5]: it presents a minimum between room temperature and the water boiling point, then it increases with raising temperature and more steeply above the water critical point. It has also been observed that the solubility of rare gases is larger for larger size of the solute [2, 6].

In this paper we study with computer simulation the structure of mixtures of Ar and water from ambient to supercritical conditions and we present a comparison of the pair correlation functions of these mixtures at supercritical conditions with recent experimental data [4]. Results about the solubility of Ar and Ne in water are also shown and discussed. In the numerical calculations water is simulated with the SPC/E potential [7] and the polarizable model developed by Brodholt, Sampoli and Vallauri (BSV) [8].

2. – Computer simulation of aqueous solutions

We performed the computer simulations of a mixture of 256 water molecules and Ar at the concentration of one atom of solute for 40 molecules of solvent [4]. The water has been described by the polarizable BSV model (see model 4 in ref. [8]) and, for comparison, by the SPC/E model [7]. The simulations have been realized in the NVE ensemble with the minimum image convention. The interaction between the oxygen and the solute has been modeled with a Lennard-Jones(LJ) potential. The parameters of the LJ potential of Ar are assumed as $\epsilon_{\mathrm{Ar}}/k_{\mathrm{B}} = 125\,\mathrm{K}$, $\sigma_{\mathrm{Ar}} = 3.415\,\text{Å}$ [2]. We explored a range of temperature and density from ambient conditions to the supercritical state ($T = 673\,\mathrm{K}$ $\rho = 0.499\,\mathrm{g/cm^3}$), where experimental data are available [4]. Below $T = 673\,\mathrm{K}$ the calculations were carried out at the points which correspond to the liquid branch of the coexistence curve [9] of the SPC/E model. More details about the simulation are reported in ref. [10].

3. – Solute-solvent structure

In fig. 1(a) we report the pair correlation functions of argon-oxygen $g_{\mathrm{OAr}}(r)$ obtained with the BSV (on the left) and the SPC/E (on the right). At ambient conditions in both the models the $g_{\mathrm{OAr}}(r)$ shows a well defined first peak. The $g_{\mathrm{HAr}}(r)$ (fig. 1(b)) shows a first peak at the same position as $g_{\mathrm{OAr}}(r)$ indicating that the solute is located equidistant on the average from the oxygens and hydrogens. The shoulder around 3.8 Å in the $g_{\mathrm{HAr}}(r)$ indicates however that some of the O-H bonds point radially towards the second shell. Comparing the position of the second peak of the $g_{\mathrm{OAr}}(r)$ with the one of the $g_{\mathrm{HAr}}(r)$ it is evident that in the second shell the water molecules are preferentially oriented with the oxygen atoms towards the solute and the hydrogen atoms towards the bulk [2]. On increasing temperature the second hydration shell tends to disappear. In the $g_{\mathrm{HAr}}(r)$ only the shoulder remains at the supercritical conditions indicating that the solute tends to expel the water molecules which are in its solvation shell. In the supercritical state the BSV potential, which is more attractive with respect to the SPC/E model, predicts a more structured $g_{\mathrm{OAr}}(r)$ [10]. We can infer that in the process of expelling the solvent the depth of the solute-solvent potential well plays an important role [2, 11]. In fig. 2(a)-(b) the argon-water radial distribution functions obtained from our simulation are compared with the experimental results [4]. In both $g_{\mathrm{OAr}}(r)$ and $g_{\mathrm{HAr}}(r)$ the BSV model results agree very well with the experiment [10].

4. – Solubility of non polar solutes in water

The solubility of a solute in a solvent is measured by the parameter γ, given by $\gamma = \exp[-\mu_{\mathrm{ex}}/k_{\mathrm{B}}T]$, where μ_{ex} is the excess chemical potential of the solute in the liquid phase [12]. We have evalued μ_{ex} of the solute in the solvent with the Test Particle Method proposed by Widom [13]. In this approach the excess chemical potential of a solute dissolved in an N-particle system is given by $\mu_{\mathrm{ex}} = -k_{\mathrm{B}}T \ln\langle \exp[-\beta\Phi]\rangle_N = -k_{\mathrm{B}}T \ln\gamma$, where $\langle \ldots \rangle_N$ represents a canonical average of the Boltzmann factor of the solute-solvent

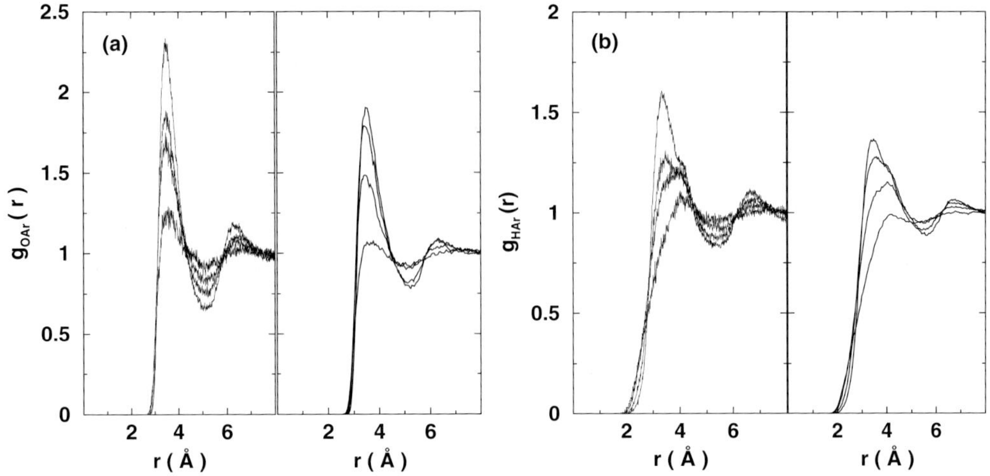

Fig. 1. – Oxygen-argon (a) and hydrogen-argon (b) pair correlation functions on increasing temperatures ($T = 300\,\mathrm{K}$, $\rho = 0.998\,\mathrm{g/cm^3}$; $T = 373\,\mathrm{K}$, $\rho = 0.949\,\mathrm{g/cm^3}$; $T = 473\,\mathrm{K}$, $\rho = 0.841\,\mathrm{g/cm^3}$; $T = 673\,\mathrm{K}$, $\rho = 0.499\,\mathrm{g/cm^3}$): comparison between the BSV model (panel on the left) and the SPC/E model (panel on the right).

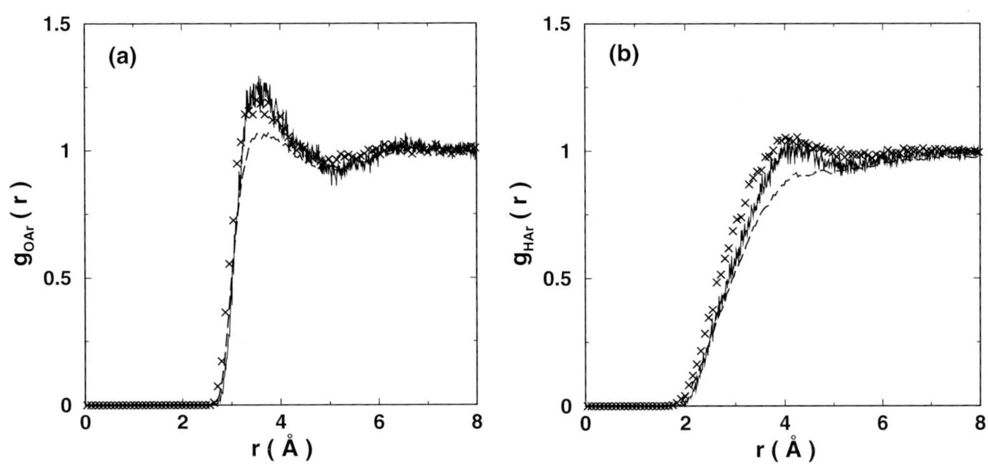

Fig. 2. – Oxygen-argon (a) and hydrogen-argon (b) pair correlation function comparison between the BSV model (solid line), the SPC/E model (long dashed line) and the experimental results [4] (crosses) at $T = 673\,\mathrm{K}$.

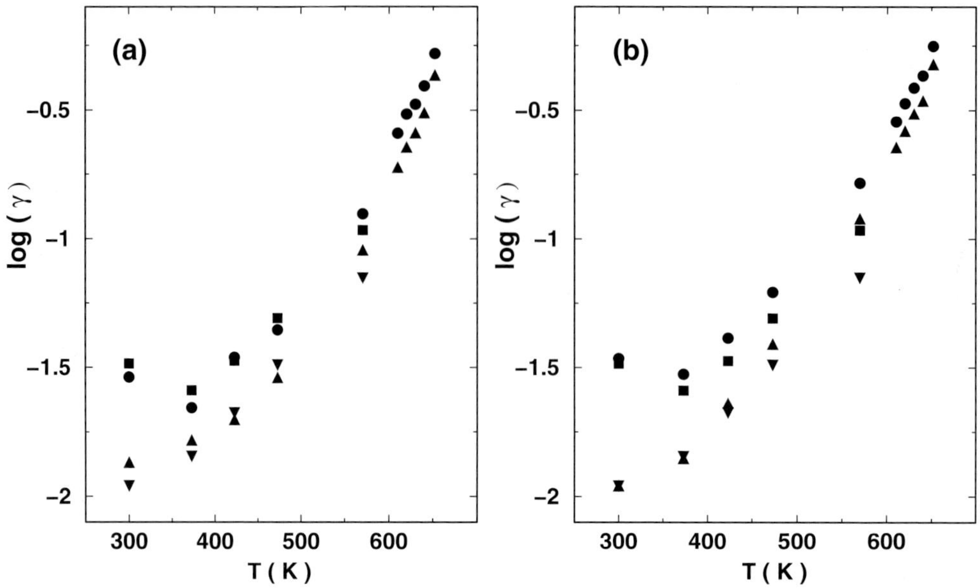

Fig. 3. – Computed values of the solubility parameter obtained by BSV (a) and SPC/E (b) for Ne (up triangles) and Ar (circles). Experimental data [5] are also reported for comparison (experimental solubility of Ne is represented by triangles down and squares for Ar).

interaction energy Φ evaluated over the configurations of the pure solvent (see [2, 14] for more details). The chemical potential can be expressed as the sum of an energetic and an entropic term $\mu_{\rm ex} = E_\mu - T\Delta S_\mu$: the term E_μ is the average solute-solvent interaction energy and the term ΔS_μ represents the entropic cost to accomodate the solute in the system. The interaction between the solute and the solvent has been modeled with a Lennard Jones potential, as described in sect. **2**, plus a polarization term which describes the polarization of the solute by the electric field generated by water: $\Phi^{\rm pol} = -(1/2)\alpha|\mathbf{E}|^2$, where α is the polarizability, equal to $\alpha_{\rm Ne} = 0.396\,\text{Å}^3$ for Ne and $\alpha_{\rm Ar} = 1.642\,\text{Å}^3$ for Ar [2]. We have performed our calculations at thermodynamical points along the saturation line of the SPC/E water [9], and a simulation run was carried out near the critical point ($T = 652\,{\rm K}$ $\rho = 0.326\,{\rm g/cm^3}$). In fig. 3 we report the temperature dependence of the solubility of Ne and Ar obtained by the BSV (a) and the SPC/E (b) models together with the experimental data of Fernandez-Prini and Crovetto [5]. The solubilities calculated with both models compare well with the experimental data. Both the potentials reproduce very well the data near ambient conditions. On raising the temperature, we observe that the BSV keeps on following closely the experimental curve while the SPC/E results are quantitatively less similar.

In fig. 4 we report the quantities $\mu_{\rm ex}$, E_μ and $-T\Delta S_\mu$ as function of temperature for Ne and Ar mixtures. We observe, in agreement with experimental results, that the

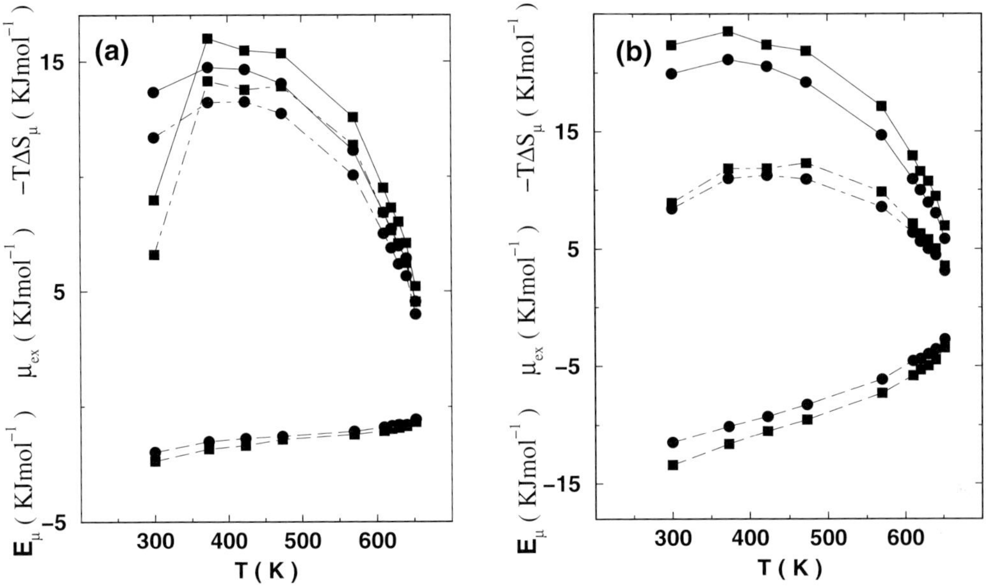

Fig. 4. – Temperature dependence of μ_{ex} (dot-dashed line), E_μ (dashed line) and $-T\Delta S_\mu$ (solid line) for Ne (a) and Ar (b) obtained by BSV (squares) and SPC/E (circles).

solubility increases with the size of the solute. To rationalize this behaviour, we note that the energetic term E_μ and the entropic one $-T\Delta S_\mu$ contribute to μ_{ex} in opposite ways [2], as shown in fig. 4, for Ne (a) and Ar (b). Therefore it is the interplay between E_μ and $-T\Delta S_\mu$ which determines how the solubility depends on the temperature and the solute sizes. The balance between the two terms is more evident for the bigger solute.

The trend of the energetic term E_μ with the solute size is such that it favors the solubility of larger solutes, because their polarizability is bigger and the interaction energy solute-solvent is more attractive. On the contrary the entropic term favors the solubility of smaller solutes, because the cost to insert the solute decreases with decreasing size. The net result is that the term E_μ determines the hierarchy between rare gases solubilities in water, while the contribution $-T\Delta S_\mu$ governs the temperature dependence of the parameter γ [2]. The position of the minimum point of the solubility is the same of the maximum in the term $-T\Delta S_\mu$ and nearly corresponds with the water boiling point. Below and above the temperature corresponding to this maximum, the entropic cost tends to favor the hydration of non polar solutes, but the microscopic mechanisms wich underline this behaviour are very different in these two temperature regions [2]. Below the boling point, water's structure is characterized by an extended network of hydrogen bonds. The presence of this open structure decreases the entropic cost to insert the solute, because it can be accomodated in the interstitial positions of the network [3]. Above this temperature however water's structure experiences strong modifications associated with a drastic decrease of the degree of H-bonding and a weakening of the orientational correlations in the system [15]. Thus in the temperature range above the water boiling point the decrease of the $-T\Delta S_\mu$ term is due to the decrease of the density, which occurs when we move along the coexistence curve.

5. – Conclusions

We performed a computer simulation study of a solution of water and Ar from ambient to supercritical conditions. For comparison water has been modeled with the SPC/E [7] and the polarizable BSV [8] models. We found that at supercritical conditions the introduction of the polarizability is essential in order to reach an agreement with the experimental argon-water radial distribution functions [4]. They are in fact well reproduced by the BSV model. We have also calculated the solubility of Ne and Ar in water and we have compared our results with the experimental data available in the literature [5]. The solubilities calculated with both models compare well with the experimental data. While both potentials agree with the behaviour of the solubility at ambient conditions, on increasing temperatures the BSV model better follows the experimental curve.

Considering the trend of both the structure and the solubility, we can conclude that our findings support the idea that the BSV model works quite well at high temperatures [16].

REFERENCES

[1] Debenedetti P. G., *Metastable Liquids: Concepts and Principles* (Princeton University Press, Princeton) 1997.
[2] Guillot B. and Guissani Y., *J. Chem. Phys.*, **99** (1993) 8075.
[3] Alagona G. and Tani A., *J. Chem. Phys.*, **72** (1980) 580.
[4] Botti A., Bruni F., Isopo A., Modesti G., Oliva C., Ricci M. A., Senesi R. and Soper A. K., *J. Chem. Phys.*, **118** (2003) 235.
[5] Fernandez-Prini R. and Crovetto R., *J. Phys. Chem. Ref. Data*, **18** (1989) 1231.
[6] Lynden-Bell R. M. and Rasaiah J. C., *J. Chem. Phys.*, **107** (1997) 1981.
[7] Berendsen H. J. C., Grigera J. R. and Straatsma T. P., *J. Chem. Phys.*, **91** (1987) 6269.
[8] Brodholt J., Sampoli M. and Vallauri R., *Mol. Phys.*, **86** (1995) 149.
[9] Guissani Y. and Guillot B., *J. Chem. Phys.*, **98** (1993) 8221.
[10] De Grandis V., Gallo P. and Rovere M., *J. Chem. Phys.*, **118** (2003) 3646.
[11] Debenedetti P. G. and Mohamed S. R., *J. Chem. Phys.*, **90** (1989) 4258.
[12] Kennan R. P. and Pollack G. L., *J. Chem. Phys.*, **93** (1990) 2724.
[13] Widom B., *J. Chem. Phys.*, **39** (1963) 2808.
[14] Frenkel D. and Smit B., *Understanding Molecular Simulation* (Academic Press, London) 2002.
[15] Botti A., Bruni F., Ricci M. A. and Soper A. K., *J. Chem. Phys.*, **109** (1998) 3180.
[16] Jedlovszky P., Brodholt J. P., Bruni F., Ricci M. A., Soper A. K. and Vallauri R., *J. Chem. Phys.*, **108** (1998) 8528.

Dielectric spectroscopy of aqueous sodium dodecyl sulfate micellar solutions by differential measurements in transmission lines

LEANDRO LANZI, M. CARLÀ, C. M. C. GAMBI and LEONARDO LANZI

Department of Physics - University of Florence
and I.N.F.M. (Istituto Nazionale per la Fisica della Materia)
Via G. Sansone 1, 50019, Sesto Fiorentino, Firenze, Italy

1. – Introduction

In this paper, the Broad Band Dielectric Spectroscopy technique (BBDS) in the frequency range $f = 100\,\mathrm{kHz}$–$3\,\mathrm{GHz}$ has been used to investigate the dielectric relaxation processes of aqueous micellar solutions. This technique makes it possible to monitor the motion of dipolar species at molecular and supra-molecular level, leading to investigate the dynamics of such species. Whereas the structure of micellar solutions is well known for several types of micelles, on the contrary the dynamic properties of the micelles are not yet fully understood. We have selected sodium dodecyl sulfate (SDS, $C_{12}\,H_{25}\,S\,O_4^-\,Na^+$) micellar solutions because they have been intensively investigated by several techniques and in particular, the microstructure of the micelle is known from small-angle neutron scattering [1-4]. The micelle, dispersed in aqueous phase, is an aggregate of self-assembled surfactant molecules with the hydrophobic tails mainly located in the core and the hydrophilic heads at the surface. For the surfactant of this work the head is composed by the sulfate group with one negative charge and by the positive sodium counterion. Some sodium counterions are bound to the heads, the others are in a diffuse or Gouy-Chapman layer [5-7]. Thus we can distinguish among three concentric spherical regions: a) the internal core of the micelle, b) the interfacial layer in which head groups, some counterions and water molecules are located and c) the diffuse layer with the unbound counterions in the aqueous phase. If we name micelle the sum of core and interfacial layer, the negatively charged micelle is surrounded by a region with equal charge of opposite sign, the diffuse layer. The aqueous phase in which micelles are dispersed is composed of water and surfactant molecules in the monomeric state at the concentration of the critical micellar concentration (*cmc*) (8.3×10^{-3} M at 25 °C [8] for SDS micelles). The dielectric behavior of the bulk water is known from the literature: two relaxation processes characterized by relaxation times of 8.38 ps and 1.1 ps, respectively at 25 °C and their change *vs.* temperature are reported in [9]. The faster process is due to the mobile water molecules and the other is connected with the kinetics of the hydrogen-bond network.

Dielectric spectroscopy measurements on a micellar solution give the complex dielectric permittivity $\varepsilon^* = \varepsilon' - j\,\varepsilon''$ (where $j^2 = -1$) as a function of frequency, as well as the Ohmic conductivity σ. From the experimental spectra it is possible to obtain the dielectric permittivity contribution of each dipolar species and its relaxation time, this last being related to the dynamics of the dipolar species [10-12].

2. – Materials and methods

Micellar solutions of SDS in water were prepared at different surfactant concentrations: 27.4 mM, 134 mM and 259 mM. SDS was from BDH, England (purity 99%). Water was from a Millipore Milli-Q apparatus. The measurements were performed at a temperature of 22.0 ± 0.5 °C.

An Anritsu MS4661A Vector Network Analyzer (VNA) was employed and a new double-differential method was used [13]. The measuring cells were pieces of coaxial transmission line machined from ISO316 steel and filled with the sample as dielectric; the outer and inner diameters of the cavity were 7.70 and 3.35 mm, respectively. Cell lengths of 3, 6 and 8 cm were used. The cells had identical characteristics but for the length. All dimensions had an accuracy of 0.05 mm. The cell was connected to the VNA by two coaxial cables (Sucoflex 104 with type-N connectors). The scattering parameters of the cell were measured by the VNA to obtain the cell propagation constant $\gamma = \sqrt{(r + j\,\omega\,l)\,(g + j\,\omega\,c)}$, where r, l, g and c are the resistance, inductance, conductance and capacitance per unit length of the cell and ω is the angular frequency, $\omega = 2\,\pi\,f$. For the sample of this work, that is not magnetic, the constant r and l depend only on the cell material and do not change between empty and filled cells, while g and c are affected by the sample properties: $g = (\sigma + \omega\,\varepsilon_0\,\varepsilon'')/k$ and $c = \varepsilon_0\,\varepsilon'/k$, where k is a cell geometric constant and ε_0 is the vacuum dielectric permittivity. By repeating the measurements with two cells of different length with both cells filled and empty, it is possible to cancel the contribution of the cell geometry and to extract the complex permittivity and the Ohmic conductivity of the sample. The details of the calculations, the drawing of the cells and a full discussion of the method can be found in [13].

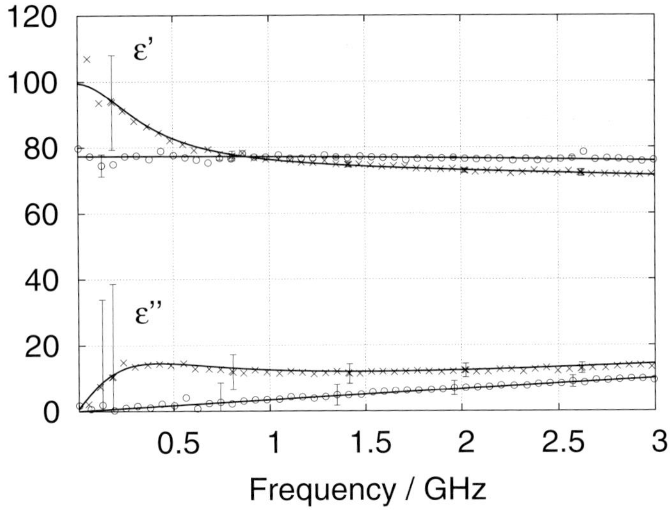

Fig. 1. – Experimental points of ε' and ε'': (○) water, (×) micellar solution with surfactant concentration 134 mM. Solid line represents the fit using eqs. (1) and (2). Only a few experimental points are drawn.

3. – Results

The micellar solutions were investigated at 22 °C with surfactant concentration c of 27.4 mM, 134 mM and 259 mM, all above the critical micellar concentration. Measurements have been performed taking 500 points evenly spaced on the frequency range from 100 kHz to 3 GHz. Figure 1 shows the dielectric spectra of water and of the micellar solution with surfactant concentration 134 mM.

The water permittivity spectrum is well described by a single Debye relaxation process

$$\varepsilon^*_{\mathrm{w}}(\omega) = \varepsilon_{\mathrm{H}} + \frac{\Delta\varepsilon_{\mathrm{w0}}}{1 + j\,\omega\,\tau_{\mathrm{w}}}, \tag{1}$$

where ε_{H} is the water permittivity at high frequency adjusted for the contribution of single water molecule relaxation, $\Delta\varepsilon_{\mathrm{w0}}$ and τ_{w} are, respectively, the step amplitude of ε' and the relaxation time constant due to the relaxation of the hydrogen network. The value for ε_{H} was taken from [9] ($\varepsilon_{\mathrm{H}} = 6.43$) as its determination is well beyond our frequency range. The fit (solid line in fig. 1) gives values in good agreement with data in [9]: $\Delta\varepsilon_{\mathrm{w0}} = 71.2 \pm 0.2$, $\tau_{\mathrm{w}} = 8.0 \pm 0.2$ ps.

The addition of surfactant to water leads to an increase of both real and imaginary parts of the dielectric permittivity in the frequency range 100 kHz–0.6 GHz suggesting the presence of some relaxation processes due to micelles. However, from the observation of the spectra it is not possible to distinguish among the several processes. Thus we have

TABLE I. – *Results of the fit of the dielectric permittivity by a Debye and a Cole-Cole relaxation processes.*

e (mM)	$\Delta\varepsilon_{\mathrm{w}}$	$\Delta\varepsilon$	τ (ns)	σ $(\Omega^{-1}\,\mathrm{m}^{-1})$	h
0	71.2 ± 0.2				
27.4	69.6 ± 0.3	5.5 ± 1.7	0.29 ± 0.15	0.04 ± 0.02	0.000 ± 0.003
134	66.1 ± 0.2	27 ± 3	0.45 ± 0.04	0.38 ± 0.02	0.07 ± 0.02
259	60.9 ± 0.7	50 ± 7	0.53 ± 0.08	0.78 ± 0.02	0.13 ± 0.04

added a Cole-Cole relaxation process to the Debye one, to take into account the micelles contribution. The whole expression of dielectric permittivity is

$$\varepsilon^*(\omega) = \varepsilon_{\mathrm{H}} + \frac{\Delta\varepsilon_{\mathrm{w}}}{1 + j\,\omega\,\tau_{\mathrm{w}}} + \frac{\Delta\varepsilon}{1 + (j\,\omega\,\tau)^{1-h}} + \frac{\sigma}{j\,\omega\,\varepsilon_0}, \tag{2}$$

where $\Delta\varepsilon$, τ and h are, respectively, the step amplitude in ε', the mean relaxation time and the parameter that takes into account the spread of the relaxation times. σ is the whole Ohmic conductivity of the micellar solution.

Experimental data were fitted with eq. (2) and the free fitting parameters σ, $\Delta\varepsilon_{\mathrm{w}}$, $\Delta\varepsilon$, τ and h were obtained. The results are reported in table I.

In figs. 2 and 3, the step amplitudes $\Delta\varepsilon_{\mathrm{w}}$ and $\Delta\varepsilon$ are reported as a function of volume fraction of water and of surfactant, respectively. In fig. 4 the conductivity is reported as a function of SDS concentration.

4. – Discussion

In fig. 2 the data for the water step amplitude $\Delta\varepsilon_{\mathrm{w}}$ as a function of water volume fraction are reported (points). The $\Delta\varepsilon_{\mathrm{w}}$ of the solution is the contribution of bulk water in the sample: it decreases starting from the $\Delta\varepsilon_{\mathrm{w0}}$ value for the pure water as the water volume fraction is reduced. The dashed line represents the expected behavior of the $\Delta\varepsilon_{\mathrm{w}}$ linear decrease to zero. This behavior is clearly not compatible with the experimental data, whose linear fit is represented by the solid line. The discrepancy may be explained if a layer of water with dielectric constant lower than bulk is considered around the micelles.

In fig. 3 the step amplitude $\Delta\varepsilon$ as a function of surfactant addition is reported (points); the linear fit of experimental data (solid line) crosses zero, as expected from theory [10].

In fig. 4 a comparison between the conductivity experimental data (points) and the theoretical behavior [11, 14] (solid line) is shown:

$$\sigma = e^2\, N_{\mathrm{A}} \left[(u_{\mathrm{Na^+}} + u_{\mathrm{DS^-}})\mathit{cmc} + \left(\alpha\, u_{\mathrm{Na^+}} + N\,\alpha^2\, u_{\mathrm{M}}\right)(c - \mathit{cmc}) \right], \tag{3}$$

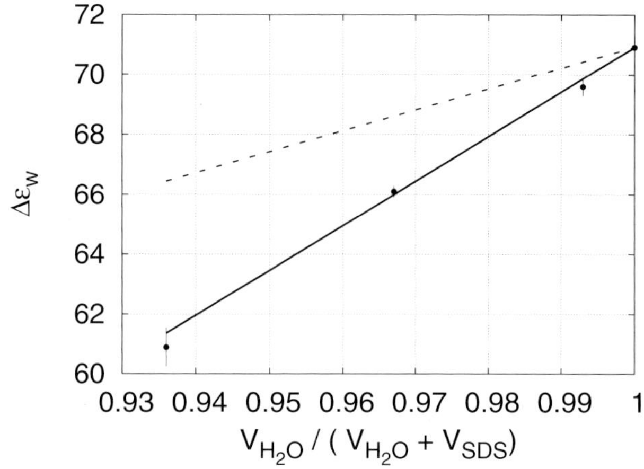

Fig. 2. – Step amplitudes of water contribution to dielectric permittivity as a function of volume fraction of water. Points are the experimental values obtained by the fit of the dielectric permittivity; dashed line is the expected behavior taking into account the water excluded volume from the core of the micelles; solid line is the linear fit of data points.

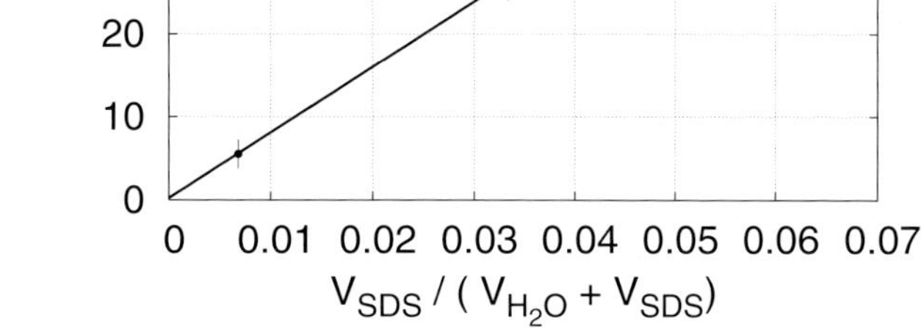

Fig. 3. – Step amplitudes of SDS contribution to dielectric permittivity as a function of volume fraction of SDS. Points are the experimental values obtained by the fit of the dielectric permittivity; solid line is the linear fit of data points.

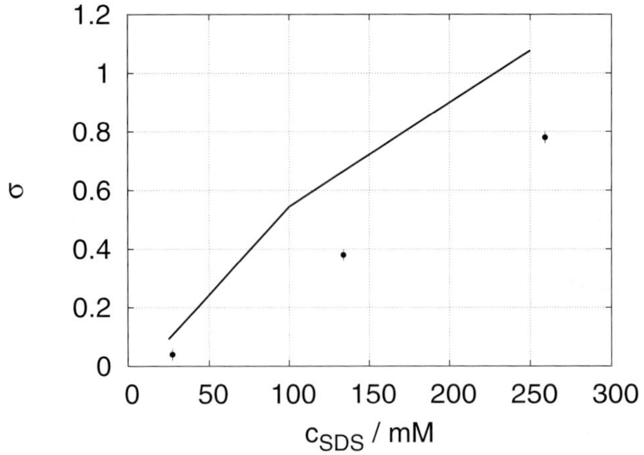

Fig. 4. – Conductivity as a function of SDS concentration. Experimental points measured at 22 °C and the solid line from theory.

where c is the SDS concentration, e is the electron charge, N_A is the Avogadro number, $u_{\mathrm{Na}^+} = 3.24 \times 10^{11}\,\mathrm{s\,kg}^{-1}$ is the Na^+ mobility [15], $u_{\mathrm{DS}^-} = 1.48 \times 10^{11}\,\mathrm{s/kg}$ is the DS^- mobility [16], N is the micelle mean aggregation number, α is the fraction of counterion dissociation in the interfacial region. u_M is the mobility of the micelle core together with the interfacial layer, with total radius a. From Stokes law $u_\mathrm{M} = (6\,\pi\,\eta\,a)^{-1}$, where η is the water viscosity ($\eta = 0.955\,\mathrm{cP}$ at 22 °C [15]) and a is 22.5 Å, 23.3 Å and 24.6 Å at the concentration of 27.4 mM, 134 mM and 259 mM [1], respectively. Both α and N depend on the concentration of SDS and on temperature; the values, available at 25 °C from small-angle neutron scattering experiments [1], were used. There is a discrepancy among the theoretical and the experimental values well beyond experimental errors. Work is in progress to clarify this point.

The values of h are reported in table I; the micellar solution with surfactant concentration 27.4 mM has a zero value of h, suggesting that it may be sufficient to use a Debye relaxation model ($h = 0$) instead of a Cole-Cole one to fit the data. The value of h of the other two solutions is sufficiently different from zero to justify the Cole-Cole model.

The values of τ, reported in table I, are fairly constant within the experimental error. Thus we accept the averaged value $\tau = 0.46\,\mathrm{ns}$ for the micellar relaxation process.

* * *

Thanks are due to A. CATELANI, M. FALORSI, M. MERCIAI and N. PASQUALETTI for the mechanical construction of the cells.

This work was supported by MIUR (Ministero dell'Istruzione, dell'Università e della Ricerca) and INFM (Istituto Nazionale per la Fisica della Materia).

REFERENCES

[1] PAYNE K. A., MAGID L. J. and EVANS D. F., *Progr. Colloid Polymer Sci.*, **73** (1987) 10.
[2] HAYTER J. B. and PENFOLD J., *Colloid Polymer Sci.*, **261** (1983) 1022.
[3] HAYTER J. B. and PENFOLD J., *J. Chem. Soc. Faraday Trans.*, **77** (1981) 1851.
[4] SHEU E. Y. and CHEN S.-H., *J. Phys. Chem.*, **92** (1988) 4466.
[5] HUNTER R. J., in *Foundation of Colloid Science*, Vol. **I** (Clarendon Press, Oxford) 1987.
[6] DEGIORGIO V. and CORTI M. (Editors), *"Physics of Amphiphiles: Micelles, Vescicles and Microemulsions" Proceedings of the International School of Physics "Enrico Fermi", Course XC* (North-Holland Physics Publishing of Elsevier Science, Amsterdam) 1985.
[7] EVANS D. F. and WENNERSTROM H., in *The Colloidal Domain: where Physics, Chemistry, Biology, and Technology Meet* (VCH Publishers. Inc.) 1994.
[8] WILLIAMS R. J., PHILLIPS J. N. and MYSELS K. J., *Trans. Faraday Soc.*, **51** (1955) 728.
[9] BUCHNER R., BARTHEL J. and STAUBER J., *Chem. Phys. Lett.*, **306** (1999) 57.
[10] GROSSE C., *J. Phys. Chem.*, **92** (1988) 3905.
[11] BARCHINI R. and POTTEL R., *J. Phys. Chem.*, **98** (1994) 7800.
[12] BAAR C., BUCHNER R. and KUNZ W., *J. Phys. Chem.*, **105** (2001) 2914.
[13] LEANDRO LANZI, MARCELLO CARLÀ, CECILIA M. C. GAMBI and LEONARDO LANZI, *Rev. Scient. Instrum.*, **73**-8 (2002) 3085.
[14] KALLAY N., TOMISIC V., HRUST V., PIERI R. and CHITTOFRATI A., *Colloid Surf. A*, **222** (2003) 95.
[15] WEAST R. C. (Editor), *Handbook of Chemistry and Physics*, 70th edition (CRC Press, FL) 1989-1990.
[16] PARFITT G. D. and SMITH A. L., *J. Phys. Chem.*, **60** (1962) 942.

Protein folding kinetics: Insights from lattice simulations

P. F. N. Faísca and M. M. Telo da Gama

Centro de Física Teórica e Computacional da Universidade de Lisboa
Complexo Interdisciplinar - Av. Prof. Gama Pinto 2, 1649-003 Lisboa Codex, Portugal

R. C. Ball

Department of Physics, University of Warwick - Coventry CV4 7AL, United Kingdom

1. – Introduction

It is widely accepted that the folding of small proteins (with < 100 amino acids) is well described by a two-state model; this means that, prior to folding, a protein must cross one major energy barrier separating the unique native fold from the ensemble of unfolded conformations and thus the folding time follows a single exponential kinetic law [1].

A major challenge in protein research is to establish correlations between the two-state kinetics and the native state equilibrium properties, *i.e.*, chain length, stability etc. Experimental data shows that the folding rate of two-state folders varies over several orders of magnitude, even for chains of comparable chain length, N [2,3]. Consequently, no correlation between the folding time, t, and N, of two-state folders has been observed so far. Moreover, the same data set shows no correlation between t and the native state stability [3]. What does then control the folding rate?

Recent experimental data for a set of 24 two-state folders indicates the existence a strong correlation between the folding time and a simple, empirical parameter of geometric complexity named relative contact order (CO) [3,4]: The CO measures the average sequence separation of contacting residue pairs in the native structure relative to the protein chain length.

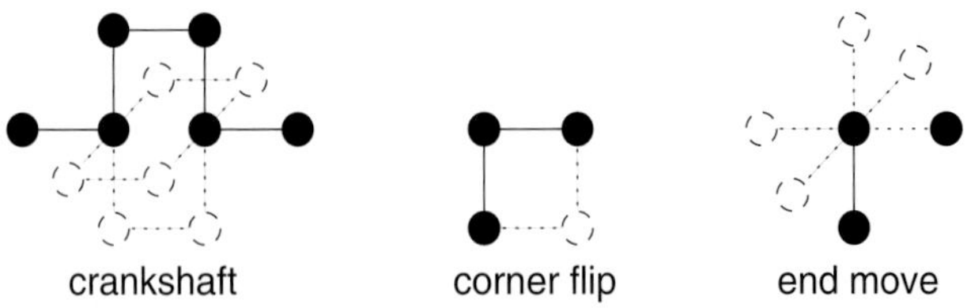

Fig. 1. – The local move set used to generate the folding dynamics.

The majority of protein folding (PF) theory is based, not on studies of the folding of proteins *in vitro*, but on a vast number of findings obtained within the scope of lattice models and Monte Carlo (MC) simulations. Can such simplified models reproduce the experimental features of PF observed so far?

Here we review some of our MC results for a lattice-polymer model that address this issue in what concerns evidence for two-state kinetics and dependence of the folding time on chain length, native state stability and contact order parameter.

2. – The lattice-polymer model

The lattice-polymer model is a coarse-grained representation of the protein backbone. Each amino acid is reduced to a single bead and the peptide bond is represented by a unit length stick. Beads interact in a pairwise manner via the contact Hamiltonian

$$H(\{\sigma_i\},\{\vec{r}_i\}) = \sum_{i>j}^{N} \epsilon(\sigma_i,\sigma_j)\Delta(\vec{r}_i - \vec{r}_j), \quad (1)$$

where $\{\sigma_i\}$ represents an amino acid sequence composed of N beads, σ_i standing for the chemical identity of the i-th bead, while $\{\vec{r}_i\}$ is the set of bead coordinates defining each conformer. The contact function Δ is 1 if beads i and j are in contact but not covalently linked and is 0 otherwise. We follow many previous studies by taking the interaction parameters ϵ from the 20×20 Miyazawa-Jernigan matrix, derived from the distribution of contacts in native proteins [5].

3. – Monte Carlo methods

Beads are allowed to move, via Metropolis MC [6], to sites satisfying excluded volume and fixed bond length constraints on a three-dimensional, infinite lattice. The kink-jump move set (fig. 1) is used to mimic protein movement. Each run starts from a random-coil conformation and the folding dynamics is followed by the fraction of native contacts, Q ($0.0 \leq Q \leq 1.0$, with $Q = 1.0$ corresponding to the native state). The folding time, t, is measured by the first passage time and the folding dynamics is studied at the optimal folding temperature, that is, the temperature that minimises t.

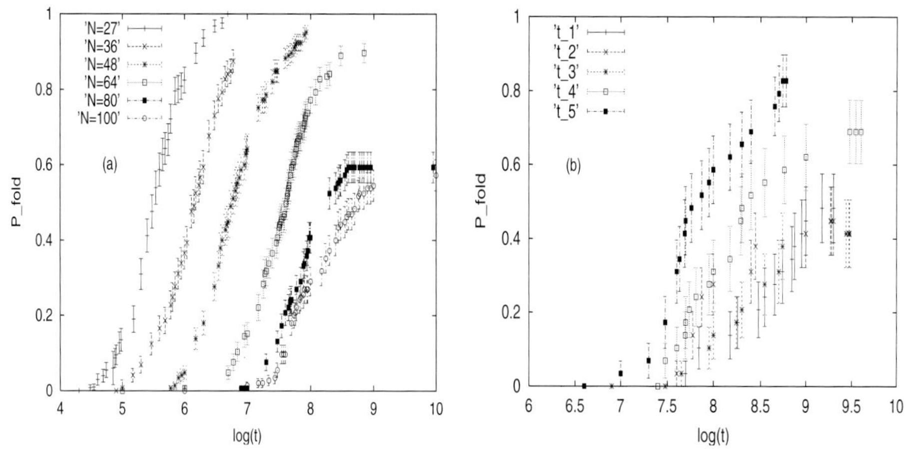

Fig. 2. – Dependence of the folding probability, P_{fold}, on $\log(t)$. (a) For each of the chain lengths $N = 27, 36, 48, 64, 80$ and 100 five target structures were considered and 30 sequences were designed per target. P_{fold}, the probability of the chain having visited its target after time t, was computed as the fraction of simulation runs which ended up in time t. (b) Separate contribution of each of the 100 bead long targets for the dependence of $P_{\text{fold}}(t)$ on $\log(t)$ [8].

The sequences studied were prepared using the method developed by Shakhnovich and Gutin (SG) [7] based on random heteropolymer theory and simulated annealing techniques. The SG method is a thermodinamically oriented design procedure that creates sequences that fit the native conformation with low enough energy to allow for stable folding, that is, for a high target average time occupancy.

4. – Results

4·1. *Dynamical regimes in protein folding.* – Figure 2(a) shows evidence for two distinct folding regimes depending on the protein chain length: In the regime found for $N < 80$ the kinetics is single exponential (fig. 3(a)) and the folding time scales with chain length as $t \approx N^5$ ($r = 0.99$) [8]. In the second dynamical regime ($N \geq 80$) the folding performance is sensitive to target conformation (fig. 2(b)) with some targets being more foldable than others. In this case thermodynamic stability is not the dominant drive of folding and, in particular, it does not solve the kinetic accessibility problem.

For two-state folders with $N = 48$ we investigated the relation between thermodynamic stability, as measured by the sequence energy in the target, and kinetic accessibility, as measured by the first passage time. Results reported in fig. 3(b) show that there is no simple correlation as only a small fraction of the most stable sequences are also relatively quick folders.

4·2. *Evidence for a topology-dependent kinetics.* – We studied the folding kinetics of ≈ 5000 sequences and 100 target conformations distributed over the chain lengths

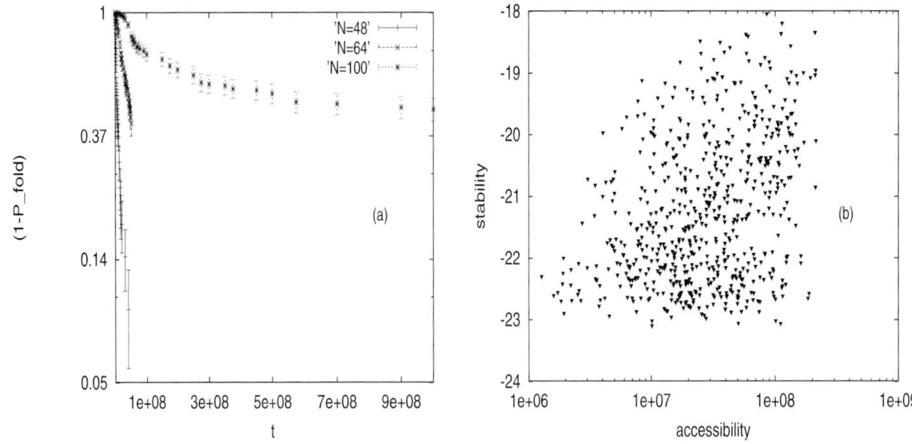

Fig. 3. – (a) Evidence for a single exponential regime for chain length $N = 48$ and $N = 64$ but not for $N = 100$. (b) The accessibility-stability scatter plot for an ensemble of ≈ 1000 sequences with chain length $N = 48$ [8].

$N = 36, 48, 54, 64$ and 80. Targets were found by homopolymer relaxation and selected so as to cover the observed range of CO ($\approx 0.12 < \text{CO} < 0.26$).

We have found evidence for a strong correlation between the contact order parameter, CO, and the folding time. This correlation is particularly strong ($r = 0.80$) for chains with number of beads $N \geq 80$ and higher contact order, CO > 0.17 [9].

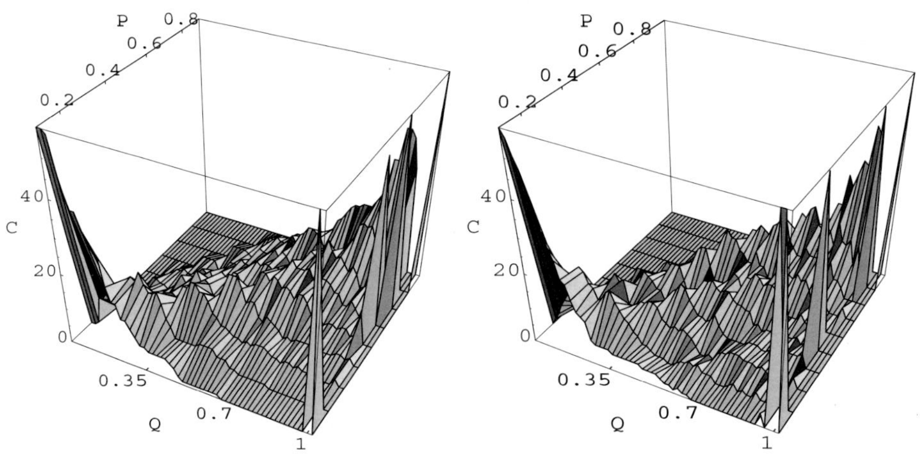

Fig. 4. – Dependence of the number of contacts, C, with a given probability of being formed, P, as a function of Q, the fraction of native contacts for the low-CO target (left) and for the high-CO target (right). These results are averaged over 100 MC folding runs, each MC run corresponding to a different protein sequence.

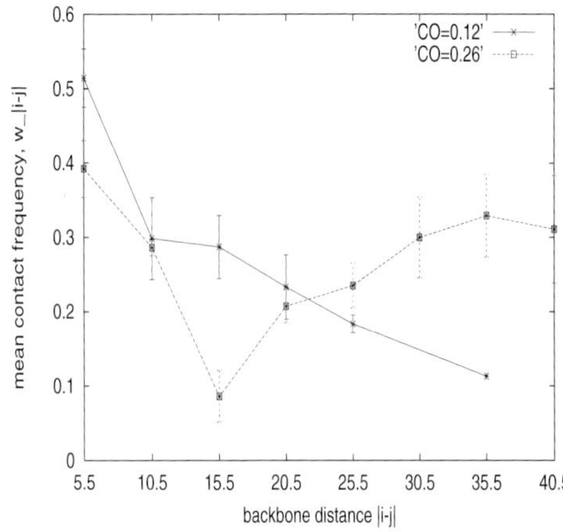

Fig. 5. – Mean native contact frequency, $\bar{\omega}_{|i-j|}$, as a function of the backbone separation, $|i-j|$, for the low- and the high-CO targets.

In order to investigate the influence of CO in the folding kinetics, we have thoroughly analysed the dynamics of $N = 48$ sequences with the lowest and the highest-CO targets [10]. Results show that, independently of target geometry, the folding dynamics is dominated by local contacts in the sense that these are the most frequent during the course of a folding simulation. Similar results have been observed elsewhere [11]. The short folding times ($\ln t = 14.99 \pm 0.13$) observed for the low-CO structure correspond to a folding process where the cumulative gain in native structure during folding (fig. 4, left) is associated with a monotonic decrease of contact frequency with increasing contact range (fig. 5). On the other hand, the longer folding times ($\ln t = 17.59 \pm 0.12$) associated with the high-CO native structure correspond to folding processes where even in late stages the chain explores a large set of unrelated conformations (fig. 4, right).

5. – Conclusions

These results show that, despite its simplicity, lattice models are capable of reproducing some fundamental experimental features observed in the folding of real proteins. Moreover the results also strengthen the idea that, apart from the thermodynamic drive imposed by the free energy of folding, the dynamics is also largely dependent on structural factors related to target geometry, namely the CO parameter. A matter of current interest is to establish the physical mechanism behind such geometry-dependent kinetics. Recent results suggest that this mechanism may be related to some kind of cooperative phenomenon [12]. Our result that, for the low-CO target, the structural organization towards the native fold is driven by the contact range suggests the possibility of developing new, more effective, design methods that take into account structural information.

REFERENCES

[1] Jackson S. E., *Fold. Des.*, **3** (1998) 81.
[2] Wittung-Stafshede P., Lee J. C. and Gray H. B., *Proc. Natl. Acad. Sci USA*, **96** (1999) 6587.
[3] Plaxco K. W., Simmons K. T. and Baker D., *J. Mol. Biol.*, **277** (1998) 985.
[4] Plaxco K. W., Simmons K. T., Baker D. and Ruczinski I., *Biochemistry*, **39** (2000) 11177.
[5] Miyazawa S. and Jerningan R., *Macromolecules*, **18** (1985) 534.
[6] Metropolis N., Rosenbluth A., Rosenbluth M. N., Teller A. H. and Teller E., *J. Chim. Phys.*, **21** (1953) 1087.
[7] Shakhnovich E. I. and Gutin A. M., *Proc. Natl. Acad. Sci. USA*, **90** (1993) 7195.
[8] Faísca P. F. N. and Ball R. C., *J. Chem. Phys.*, **116** (2002) 7231.
[9] Faísca P. F. N. and Ball R. C., *J. Chem. Phys.*, **117** (2002) 8587.
[10] Faísca P. F. N., Telo da Gama M. M. and Ball R. C., *Phys. Rev. E*, **69** (2004) 051917.
[11] Abkevich V. I., Gutin A. M. and Shakhnovich E. I., *J. Mol. Biol.*, **252** (1995) 460.
[12] Jewett A. I., Pande V. S. and Plaxco K. W., *J. Mol. Biol.*, **326** (2003) 247.

Molecular dynamic investigations of liquid nickel's structural properties

A. KIMMEL and V. LAD'YANOV

Physical-Technical Institute of Ural Branch of Russian Academy of Sciences 426001, Izhevsk, Russia

I. GUSENKOV

Udmurt State University, 426000, Izhevsk, Russia

1. – Introduction

In the last few years, extensive studies have been performed on a variety of ionic, covalent, and metallic liquids in order to understand their structural and dynamic behavior from a microscopic perspective. Transition metals, such as nickel, are very interesting in this record. The structure of liquid nickel has been studied in many works by means of direct diffraction methods and investigations of structural sensitive properties [1-3]. Significant modifications of structure parameters were obtained above the nickel melting point. Authors of the work [2] have obtained the fuzzy maximum located close to the line (200) of the hypothetical BCC structure. However, by means of the levitation techniques in conjunction with the neutron scattering experiment the authors of the work [3] unambiguously proved the Franks hypothesis [4], which states that the icosahedral short range order is already revealed above the melting temperature in transition metals, such as Ni, Fe, Co.

Computer simulations give us a new information about systems, such as coordinates of all particles of modelling system, which can not be obtained by means of usual experimental methods. However, using atomic coordinates to calculate the pair correlative function (PCF) or structure factors to compare them with experimental data is not an effective method for investigating the liquid structure. One of the best prospective approach to solving such tasks is the Voronoi polyhedra (VP) and Delaunay simplices (DS) method [5, 6].

Computer simulation methods allow us to advance in understanding the atomic structure of the amorphous and liquid state of matter. But the main problem of methods, such as molecular dynamic (MD) or Monte-Carlo, is a choice of adequate potentials of the interactions between atoms. Usually, the disparities of the computer and experimental results are connected with a non-correspondence of used potentials.

Generally, the task of finding the exact and universal interatomic potentials is quite difficult. Therefore, inverse methods [7, 8], in which the effective pair potentials are derived from the experimental structural data, are one of the best methods for obtaining the effective pair potentials and computing structure, corresponding with real matter. Moreover, the deriving potentials are effective and can be considered as potentials that include many-particle correlation, which is typical for liquid state.

In this work we used the self-consistence inverse method [8] for obtaining the effective pair potentials of liquid nickel at different temperatures above its melting point from the experimental structural data [9]. The potentials allow us to get an adequate structure of liquid nickel, carry out Voronoi-Delaunay statistical-geometrical analysis of local atomic structure and investigate its temperature evolution.

2. – Formalism

2·1. *Self-consistent method for deriving the effective pair potentials.* – In this study the predictor-corrector method, originally proposed by Schommers [8], was used as an accurate method for solving the inverse problem. This self-consistent method is based on the integral equations theory and computer simulations. The procedure of deriving $\varphi(r)$ is iterative. The zero-approximation of the effective potential called predictor, is given by

$$\varphi^0(r) = -k_\mathrm{b} T \ln g^{\exp}(r), \tag{1}$$

where $g^{\exp}(r)$ is the experimental PCF, k_b is Boltzman constant and T is the absolute temperature of the system. Atomic configuration, simulated by means of MD with known potential $\varphi^0(r)$, corresponds to the pair correlation function $g^{\tau-1}(r)$. The next iteration, called corrector, constructs the corrected potential $\varphi^\tau(r)$:

$$\varphi^\tau(r) = \varphi^0(r) - k_\mathrm{b} T \ln \frac{g^{\exp}(r)}{g^{\tau-1}(r)}\,. \tag{2}$$

The agreement between experimental and model PCFs is improved with each next iteration. The iterative process is repeated until the discrepancy χ between the criterion (experimental) and simulated PCFs becomes smaller than the desired accuracy:

$$\chi = \frac{1}{n_1 - n_2} \sum \left[g^{\exp}(r) - g^{\tau}(r)\right]^2, \tag{3}$$

where n_1 and n_2 are the numbers of initial and final points of the histogram, respectively.

2˙2. *Voronoi-Delaunay statistical-geometrical analysis.* – Models of dense disordered systems of spherical atoms have definite statistical motives in the arrangement of these structural elements. Unlike crystals, in which the structural motives are described in terms of the translation order, in this case the methods need to be different from those in crystallography. Thus, Voronoi-Delaunay structural analysis was carried out for describing the local atomic configurations as a function of the atomic coordinates.

The most widely applicable forms of the statistical-geometrical analysis of random systems by VP are topological (distribution of the number of faces and number of edges per face and topological types) and metric (distribution of VP volumes, face areas, the sphericity coefficient etc.) properties [5]. The distribution of the coefficient of the sphericity K_{sph} of VP was chosen in the present work, where the sphericity coefficient is defined by the expression

$$K_{\mathrm{sph}} = \frac{36\pi V^2}{S^2} \,. \tag{4}$$

Here, V is the volume, and S is the surface area of a given polyhedron. Thus, this measure defines a deviation of the shape of VP from a regular sphere and characterizes the homogeneity of the atomic environment.

It is well known, that packing of four spherical particles in the vertices of a regular tetrahedron is the most energetically advantageous atomic configuration. Since each DS is a tetrahedron, it is convenient to choose the deviation of the simplex shape from an ideal tetrahedron as a quantitative measure of the DS. So, the measure of tetrahedricity, Tr, we used was proposed in [6]:

$$Tr = \frac{\sum (l_i - l_j)^2}{15\langle l\rangle^2}\,, \tag{5}$$

where l_i is the length of the i-th edge; $\langle l\rangle$ is the average length for this simplex. This measure was constructed to be zero for an ideal tetrahedron and to increase with distortion. The relative weight of tetrahedricity, $Tr_b = 0.018$, is taken from [6], where it was proposed as boundary value for perfect tetrahedra.

However the most interesting features of disordered systems are seen in the space structural correlation. Since, there is unique correspondence between the Voronoi and Deluanay networks, so using Tr as quantitative measures of DS each vertex of Voronoi network can be assigned a value equal to this measure. To discriminate aggregates of atoms with Tr surrounding them it is convenient to use coloring. In this case only vertices which have values less than the chosen boundary value are colored. Thus, the analysis of the spatial arrangement of atoms with different structural patterns in the structure of liquid reduces to the analysis of typical graphs in the Voronoi network.

2·3. *Time correlative Van Hove function.* – Time correlation Van Hove function describes the temporal relaxation of atoms local order and allows us to determine the local order destruction time. The Van Hove function determines the number of particles $N(r,t)$ at the time t, were situated on the distance interval from r to $r + \mathrm{d}r$, under the condition, that the particle was at the center point at the initial time and Ω_0 is the particle's volume:

$$Gd(r,t) = \Omega_0 \frac{N(r,t)}{4\pi r^2\,\mathrm{d}r}\,. \tag{6}$$

However, it would be interesting to examine the Van Hove function only for particles that remain in the tetrahedral formations during time interval $\triangle t$. So, the function allows us to estimate the life time of such atomic configurations.

3. – Simulation procedure

3·1. *Liquid nickel's effective pair potentials.* – In this work the experimental PCFs, $g^{\exp}(r)$, obtained for liquid nickel at 1773, 1873, 1923, 2023 K [9], are used as criterion functions in the inverse method. The derived effective pair potentials are shown in fig. 1. They have some variations for the four temperatures and display long-ranged oscillations, that can be associated with Friedel's oscillations. In our study performed in the micro-canonical ensemble, we used molecular dynamical models with $N = 1688$ spherically symmetric particles, arranged in a cubical box with the density from 0.07921 to 0.07942 Å^{-3}, corresponding to investigated temperatures [10]. The MD computations, based on the effective potentials $\varphi(r)$, reproduce the PCFs, which are given in insertions on fig. 1. The deviation χ between the calculated and experimental functions reaches 10^{-4}.

The good agreement between the computed and experimental structural properties allows us to use Voronoi-Delaunay analysis of liquid nickel's local structure at different temperatures.

3·2. *Local structure analysis.* – Distributions of the sphericity coefficient for different temperatures are presented in fig. 2. Asymmetrical K_{sph}-distribution, with a most probable value $K_{\mathrm{sph}} = 0.712$ and a shoulder at $K_{\mathrm{sph}} = 0.69$, corresponds to the temperature $T = 1773\,\mathrm{K}$. A contribution to the main maximum forms 12-14 facets VP which have a prevalent content of 5-edged facets. Polyhedron in shape of the dodecahedron (formed from twelve regular 5-edged facets) corresponds to full icosahedral atom packing. So, atoms, which correspond to VP with value $K_{\mathrm{sph}} = 0.712$ and have edge numbers higher than 12, can only correspond to segments of dense icosahedral packing. We can suggest, that two competitive types of VP exist in liquid nickel's structure at $T = 1773\,\mathrm{K}$.

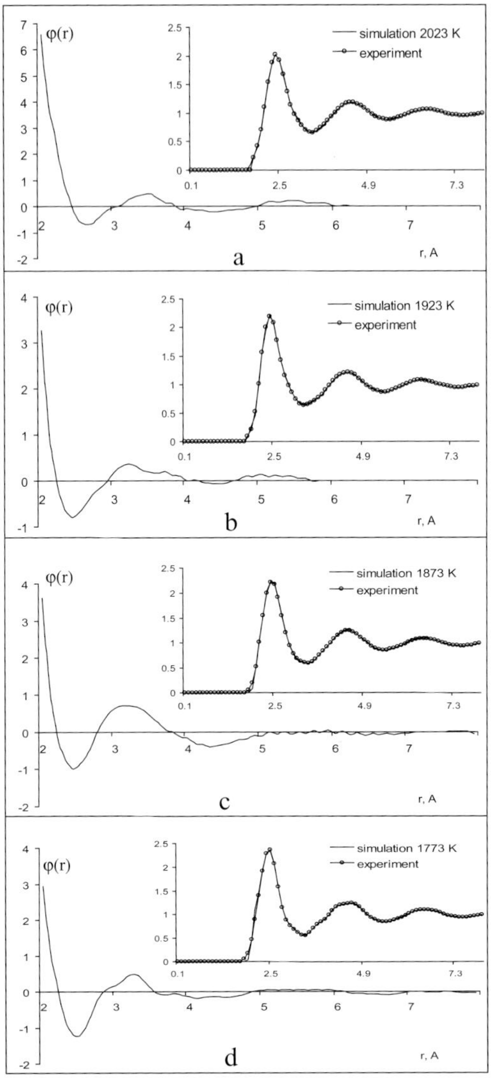

Fig. 1. – The effective pair potentials of liquid nickel at different temperatures a) 1773 K, b) 1823 K, c) 1873 K, d) 1923 K and corresponding pair correlation functions (circles: experimental data, solid lines: simulated results, based on $\varphi(r)$).

K_{sph}-distributions, corresponding to $T > 1773\,\mathrm{K}$, became symmetrical with maxima position at $K_{\mathrm{sph}} = 0.69$. This value coincide with shoulder position for the K_{sph}-distribution at $T = 1773\,\mathrm{K}$. At the same time, in these structures there are no VP having several bordering regular 5-edged facets. This fact confirms that MD-models at $T > 1773\,\mathrm{K}$ do not contain sufficiently large segments of icosahedral atom packing.

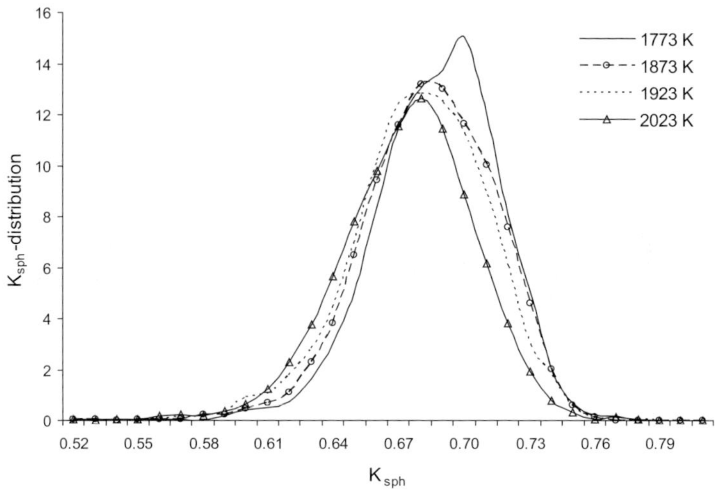

Fig. 2. – K_{sph}-distributions for the liquid nickel configurations.

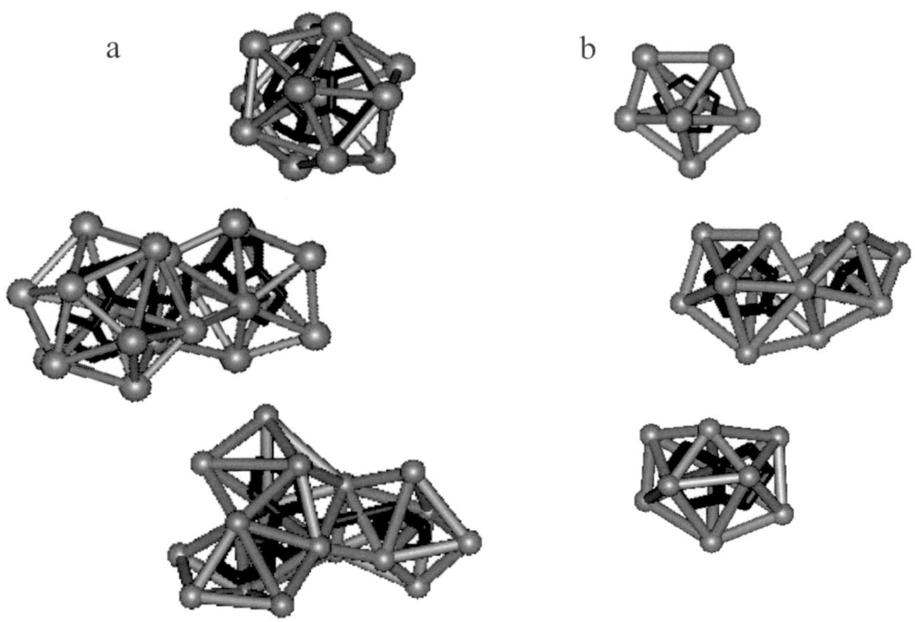

Fig. 3. – a) Tetrahedral formations on the Voronoi network (thick lines) and their corresponding atomic configurations (spheres) in MD-structure at $T = 1773\,\text{K}$. The dodecahedron on the Voronoi network corresponds to full icosahedral atomic configuration. b) Tetrahedral formations at $T > 1773\,\text{K}$. Tr-clusters became isolated and compact. Their typical element is the pentagonal biprism.

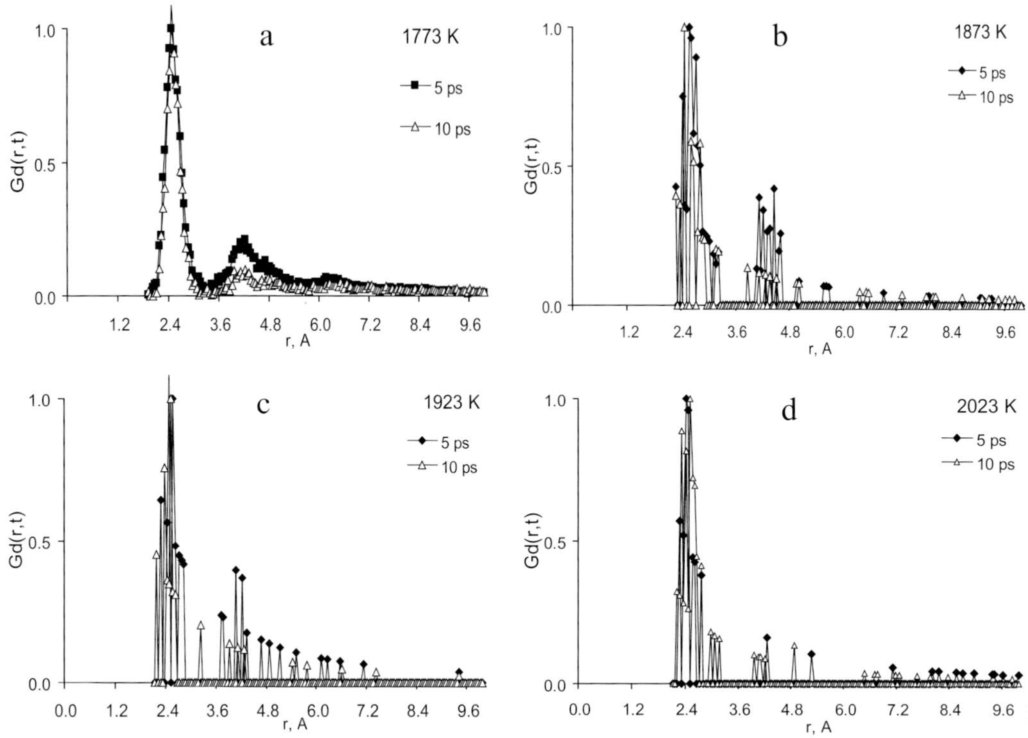

Fig. 4. – Van Hove functions, calculated for atoms that remain in the Tr-clusters during time interval $\triangle t = 5$ and 10 ps. a) 1773 K, the first and the second shells of the neighbors in the Tr-clusters are preserved during $\triangle t = 10^{-11}$ s, b) 1873 K, c) 1923 K, d) 203 K, only the first coordination shell is preserved during $\triangle t = 5 \cdot 10^{-12}$ s.

Investigation of the space structural correlations on the Voronoi network by means of the vertex coloring under condition $Tr < Tr_b$, has shown, that tetrahedral configurations tend to unit into spatial formations (Tr-clusters) with a regular simplicial surrounding. However, properties of these clusters, such as size, local order, lifetime are changed with temperature rise.

Figure 3a demonstrates typical tetrahedral formations in the structure at $T = 1773$ K. The size of Tr-clusters at this temperature reaches 5–10 Å. On the Voronoi network such clusters are mainly composed by 5-membered graphs bounded by the mutual edge. This graph corresponds to the atomic configuration with 5-folder symmetry in shape of the pentagonal biprism. So, in the simulated liquid nickel's structure at $T = 1773$ K the Tr-clusters are represented by various polytetrahedral atomic configurations with 5-folder local symmetry with an icosahedral configuration among them.

At temperatures $T > 1773$ K the structure of tetrahedral formations is simplified.

Tr-clusters become isolated and compact, their size reaches as little as 5–6 Å. A typical element of these formations is presented by 5-membered graphs.

3·3. *Van Hove function.* – Advantageous energetic atomic disposition in the *Tr*-clusters allows us to expect, that such formations should be time stable. We checked this suggestion with help of analysis of time correlative Van Hove functions. The Van Hove functions, calculated at $T = 1773\,\mathrm{K}$, are shown in fig. 4a. The first and the second shells of the neighbors in the *Tr*-clusters are preserved at least during $\triangle t = 10^{-11}\,\mathrm{s}$, that is larger than thermal fluctuation's time.

Figures 4b, c, d displays the Van Hove functions, calculated at temperatures $T > 1773\,\mathrm{K}$. Only the first coordination shell is preserved during $\triangle t = 5 \cdot 10^{-12}\,\mathrm{s}$. We suggest, that it is connected on the one hand, with the increase of the dynamic mobility of particles with temperature rise, on the other hand, with the absence of pronounced icosahedral segments in liquid structure at high temperatures.

4. – Conclusions

In the work the effective pair potentials were derived from the experimental data for liquid nickel at four temperatures. Good agreement between computed and experimental structural properties of the simulated system has made it possible to perform Voronoi-Delaunay analysis of local structure.

The results of these simulations allow us to draw the conclusion that in liquid nickel's local structure in the vicinity of the melting point there are spatial polytetrahedral formations with local 5-folder symmetry (with an icosahedron among them, as the best-known candidate for a basic structural unit of liquid). The results obtained complement a well-known point of view about the existence of the icosahedral short-range order in liquid metals, which does not depend on their pre-melting structural type [3]. With temperature rise tetrahedral formations evolve to isolate and compact ones with preserving of the local five-folder symmetry. At the same time their lifetime is decreased due to the increase of diffusion atom motions.

The appearance of the spatial formations in liquid structure can leads to the essential change of structural viscosity [11]. Existent results of the experimental investigations of the transition metals alloys of the iron group (Fe, Co) [12, 13] demonstrate the uneven decreasing of the alloys viscosity at rising temperature. This does not contradict the above-mentioned arguments. These facts can be the evidence of the structural transitions, which exist in the alloys with rising temperature. Hence, one can expect that the nickel melt will exhibit the ordinary Newton-liquids properties, since the structural clusters are very small. However, the clusters size and their concentration growth with temperature decrease. Because of these processes the melt can have structured-liquid properties, in particular, viscosity dependence on flow velocity gradient can appear.

REFERENCES

[1] BAUM B. A., KHASIN G. A., TAYGUNOV G. V. *et al.*, *Liquid Steel* (Mettalurgiya, Moskow) 1984.
[2] VATOLIN N. A. and PASTUKHOV E. A., *Diffraction Studies of the Structures of High-Temperature Melts* (Nauka, Moscow) 1980.
[3] SCHENK T., HOLLAND-MORITZ D., SIMONET V., BELLISSENT R. and HERLACH D. M., *Phys. Rev. Lett.*, **89** (2002) 075507.
[4] FRANK F. C., *Proc. R. Soc. London A*, **215** (1952) 43.
[5] MEDVEDEV N. N. and NABERUKHIN Y. I., *J. Computer Phys.*, **67** (1986) 223.
[6] MEDVEDEV N. N., *J. Phys. Condens. Matter*, **2** (1990) 9145.
[7] REATTO L., LEVESQUE D. and WEIS J. J., *Phys. Rev. A*, **33** (1986) 3451.
[8] SCHOMMERS W., *Phys. Rev. A*, **28** (1983) 3599.
[9] WASEDA Y., *The Structure of Non-Crystalline Materials* (McGraw-Hill, New York) 1980, an electronic database can be found at `http://www.iamp.tohoku.ac.jp/database/scm/index.html`.
[10] SMITHELL C. J., *Smithells Metals Reference Book* (Butterworths, London) 1983.
[11] FRENKEL YA. I., *Kinetic Theory of Liquids* (Nauka, Leningrad) 1975.
[12] LAD'YANOV V. I., BELTYUKOV A. L. *et al.*, *JETP Lett.*, **72** (2000) 301.
[13] LAD'YANOV V. I. and BELTYUKOV A. L., *JETP Lett.*, **7** (2000) 128.

Monte Carlo simulations of two-dimensional polymer blends

A. CAVALLO, M. MÜLLER and K. BINDER

Institut für Physik, WA331, Johannes Gutenberg-Universität
D-55099 Mainz, Staudinger Weg 7, Germany

The study of the miscibility properties of polymer blends has been of long-standing interest [1, 2], both because of practical reasons (polymeric materials in daily life are generally multi-component systems) and because of the important role played in the general theory of phase transitions [3].

In the last years, due to the enormous developments in nanoscale technology, there has been a growing interest in studying the behavior of polymer mixtures in thin film geometry [4,5]. The consequences of the confinement by parallel walls on multi-component polymeric materials have been investigated in many contexts [6, 7]. Recent theoretical studies on binary mixtures in thin-film geometry [8], supported by Monte Carlo simulations [9] and experiments [10], state that the unmixing behavior and the location of the critical temperature are essentially controlled by the interplay of two parameters: the film thickness Δ and the chain length N. A crossover from the 3D Ising critical behavior to the 2D Ising critical behavior is expected if one considers film thickness Δ of the order of the correlation length ξ of composition fluctuation. On the other hand, in the long chain length limit, the mean-field description due to Flory and Huggins becomes quantitatively correct and a crossover from the Ising to the mean-field universality class occurs [9, 11].

The mean-field picture predicts that the critical temperature T_c of unmixing varies with the chain length N as $T \propto N$ [1]. This behavior is controlled by a balance between two terms: the translational entropy of mixing per monomer, of the order of $1/N$, and the enthalpic contribution, measured in terms of the so-called Flory-Huggins parameter $\chi_{\rm FH}$, which describes the interaction between different molecules. The Flory-Huggins

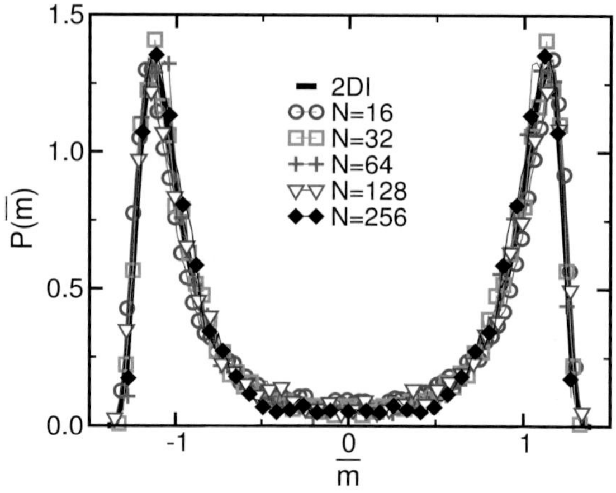

Fig. 1. – Accurate location of the critical temperature via mapping the probability distribution function of the order parameter, normalized to unit variance, onto the universal distribution of the 2D Ising universality class. Here the symbol $\overline{m}$ denotes the deviation of the order parameter $m = n_A/n$ from its critical value $m_c = 1/2$ divided by $\langle (m - m_c)^2 \rangle^{1/2}$. The exact 2D Ising probability distribution is shown as a bold line, while the other symbols denote our results for chain length $N = 16$ to 256.

parameter can easily be related to the effective coordination number z_c, *i.e.* the effective number (on average) of *inter*molecular contacts per monomer. It can give us a quantitative information regarding the degree of interdigitation of the chains in the system.

Monte Carlo simulations [12-16], experiments [10] and analytical theory [17] agree with the linear variation of T_c for polymer mixture in the bulk, even for very short chains. However, due to the fact that also *intra*molecular nearest-neighbor contacts have been taken into account by the simple Flory-Huggins theory, the values of T_c are substantially overestimated. In a thin film geometry the effect of the confinement is drastically reflected in the coordination number which becomes smaller and strongly fluctuating. Of course, this effect will be particularly pronounced for ultrathin films. In these conditions we expect that the Flory-Huggins theory and the linear behavior of T_c with N, valid for the bulk case, completely fail.

This paper is devoted to the study of the miscibility properties of a symmetrical AB-polymer blend ($N_A = N_B = N$) confined in a quasi–two-dimensional thin film. Particularly, here we will show the results obtained by Monte Carlo simulations for the relation T_c *vs.* N and the intermolecular correlation function $g(r)$ which, for the limiting case of ultrathin films, have not been considered before. The polymer mixture is simulated by the bond fluctuation model [13-15, 18] on the simple cubic lattice in the $L \times L \times \Delta$ geometry. In this model, each effective monomer blocks all 8 sites of an elementary cube. A polymer chain is represented on the lattice by N monomers (cubes) linked together by 108 allowed

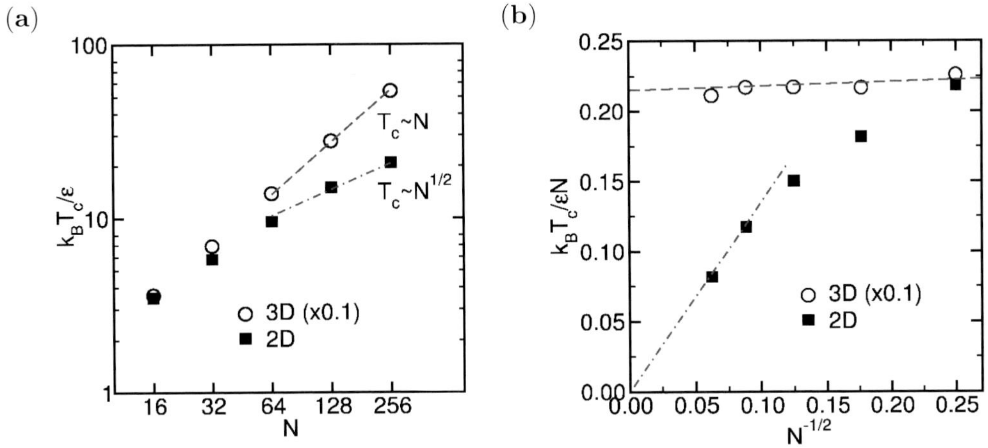

Fig. 2. – a) Critical temperature $k_\mathrm{B}T_\mathrm{c}/\epsilon$ plotted *vs.* chain length N for the two-dimensional bond fluctuation model at volume fractions $\rho = 0.5$. Lattice sizes used range from $L = 256$ ($N = 16$) to $L = 512$ ($N = 256$), respectively. The straight lines indicate the power laws $T_\mathrm{c} \sim N$ and $T_\mathrm{c} \sim N^{1/2}$, respectively. The best fit to all two-dimensional data yields an effective power law $T_\mathrm{c} \sim N^{0.65}$. b) Scaling of T_c/N and $T_\mathrm{c}/N z_\mathrm{c}$ *vs.* $N^{-1/2}$. For comparison the results for the three-dimensional model are included [13].

bond vectors $\vec{b}$ belonging to the set $\{P[200], P[210], P[211], P[300], P[310]\}$. Here the symbol $P[b_x b_y b_z]$ denotes the equivalency class of bonds with length $b = \sqrt{b_x^2 + b_y^2 + b_z^2}$ (measured in units of the lattice spacing) obtained from $\vec{b} \equiv (b_x, b_y, b_z)$ performing all the symmetry operations in the simple cubic lattice. We choose periodic boundary conditions in x and y directions, and hard impenetrable but neutral walls at $z = 0$ and at $z = \Delta + 1 = 3$. The chain length N is varied from $N = 16$ to $N = 256$, choosing volume fractions $\rho = 0.5$ of occupied sites. We use $L = 256$ for $N \leq 128$ and $L = 512$ for $N = 256$. The A and B monomers interact via the most symmetric square well potential $\epsilon_{AA} = \epsilon_{BB} = -\epsilon_{AB} = -\epsilon$ of range $d_\epsilon = \sqrt{5}$. For planar configurations the square-well potential comprises 12 lattice sites instead of 54 in the bulk.

We work in the semi-grandcanonical ensemble [12-14], *i.e.*, we fix the temperature T, the total number of polymers $n = n_A + n_B$ and the chemical-potential difference $\Delta\mu$ between the species A and B. With these choices, the order parameter of the unmixing transition, defined as $m = n_A/n$, is relaxed by randomly attempted interchanges of A and B labels at fixed polymer conformation [12]. Chain configurations are updated by the random hopping algorithm [19] and slithering-snake movements [14]. These Monte Carlo moves do not alter the composition of the mixture. Typically Monte Carlo runs with a length of 2×10^5 Monte Carlo steps per monomer were generated. According to the finite-size scaling theory [20], to locate the critical temperature we match the

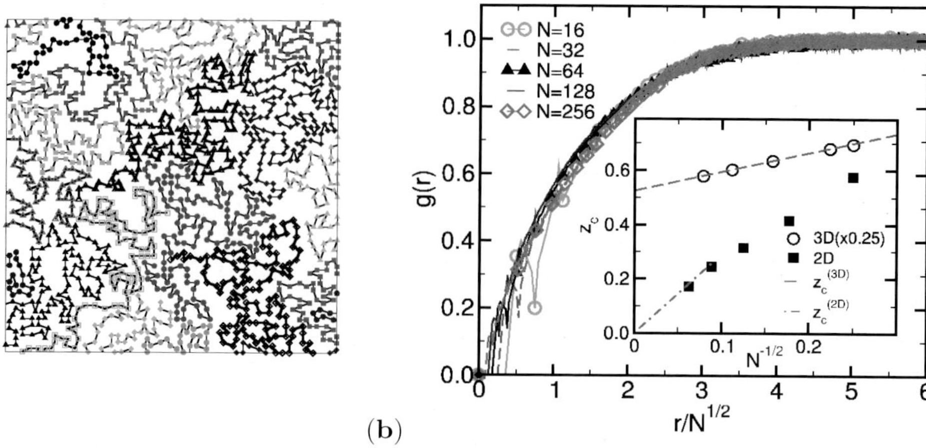

(a) (b)

Fig. 3. – (a) Snapshot of a typical configuration for chain length $N = 128$ at density $\rho = 0.5$. Different chains are drawn in different grey shades. Only a fraction of size 70×70 out of the total simulation cell ($L = 256$) is shown. $\epsilon/k_{\mathrm{B}}T = 0.0665$. (b) Scaling plot of the intermolecular pair correlation function $g(r)$ *versus* $r/N^{1/2} \sim r/R$. Data are always taken close to criticality, but refer to all chains independent of their species. $\rho = 0.5$. Chain lengths $N = 16, 32, 64, 128$, and 256 are included, as indicated in the key. For short chain lengths oscillations on the length scale of a few lattice units are visible and arise from packing effects. The inset shows the effective coordination number *versus* $N^{-1/2}$ for density $\rho = 0.5$. The best fit to the data yields an effective power law $z_{\mathrm{c}} \sim N^{-0.43}$. For long chain lengths the data are compatible with $z_{\mathrm{c}}^{(2\mathrm{D})} = 2\pi\rho d_\epsilon^3/(3\sqrt{6}R_{\mathrm{g}})$ with $d_\epsilon = \sqrt{5}$ and $R_{\mathrm{g}}(N = 128) = 18.7$. In the three-dimensional model one obtains $z_{\mathrm{c}}^{(3\mathrm{D})} = 2.1 + 2.8/\sqrt{N}$ [14].

probability distribution of the order parameter (normalized at unit variance) onto a universal curve which, in the case under study, characterizes the 2D Ising universality class (fig. 1). Quantitative information regarding the interdigitation of the chains and the geometric conformation of the molecules are given plotting the radial intermolecular pair correlation function $g(r)$ and the effective coordination number $z_{\mathrm{c}} \equiv \rho \int_{r \le d_\epsilon} \mathrm{d}^2 r g(\mathbf{r})/4$. Figure 2 shows our results for the critical temperature plotted *vs.* chain length N. The results are compared with the ones obtained for the bulk case [13]. One sees that the increase of the critical temperature with N is much weaker than linear, unlike the result for bulk melts [12, 13]. If one tries a power law fit, we get $T_{\mathrm{c}} \propto N^{\alpha}$, with $\alpha \simeq 0.65$. This result can be explained by some simple considerations on the topology of confined chains. Naively, in two-dimensions we might expect that the coils assume disk-like shapes. This implies that contacts between monomers of different chains are possible only along the perimeter of the disks. Hence, each chain can have only of the order of $\sqrt{N}$ monomer-monomer contacts with its neighbours. On the contrary, in the bulk system, the strong interdigitation between the chains allows intermolecular contacts of the order N. Thus, one can argue that for a two-dimensional system the critical temperature increases with N as $T_{\mathrm{c}} \sim \sqrt{N}$ unlike $T_{\mathrm{c}} \sim N$ as predicted by the mean-field theory. However the situation is not quite so simple as this naive interpretation suggests: a direct visual inspection of the chain conformation shows that the chains typically do not take disk-like shapes. They are much more elongated and irregular (fig. 3). Nevertheless, each chain screens a large part of itself from interactions with other chains, there is a very pronounced correlation hole effect: when we plot the interchain part of the monomer-monomer radial distribution function $g(r)$ *vs.* $r/\sqrt{N} \sim r/R_{\rm g}$, where $R_{\rm g}$ is the radius of gyration, we find that $g(r)$ is rather compatible with a gradual shape of the type $1 - f(r)$ being $f(r) = \exp[2\sqrt{6}r/3R_{\rm g}]$ the correlation function of a two-dimensional Gaussian walk [21]. With this choice, at the leading order in N, we obtain $z_{\rm c} \sim N^{-1/2}$ which is very close to the simulation data (fig. 3).

In summary, we have found that in ultrathin polymer films, where the chain configurations are quasi–two-dimensional, the compatibility of polymer blends is dramatically enhanced in comparison with the bulk, since $T_{\rm c} \sim \sqrt{N}$ rather than $T_{\rm c} \sim N$. This qualitative failure of Flory-Huggins theory is due to the strong screening of interchain interactions, that necessarily occurs in two dimensions: the enthalpy of interaction of one chain with *other* chains scales linear with N in the bulk, but is only proportional to $\sqrt{N}$ in planar geometry. We expect such behavior not only for strictly two-dimensional chains, but also for film thickness which are on the order of the excluded-volume screening length.

REFERENCES

[1] FLORY P. J., *J. Chem. Phys.*, **9** (1941) 660; HUGGINS M. J., *J. Chem. Phys.*, **9** (1941) 440.

[2] BINDER K., *Advances in Polymer Sci.*, **112** (1994) 181; *Computer Phys. Commun.*, **147** (2002) 22.

[3] DE GENNES P. G., *Scaling Concepts in Polymer Physics* (Cornell University Press, Ithaca, N.Y.) 1979.

[4] BINDER K., *Adv. Polym. Sci.*, **138** (1999) 1.

[5] BUDKOWSKI A., *Adv. Polym. Sci.*, **148** (1999) 1.

[6] KARIM A. and KUMAR S. K. (Editors), *Polymer Surfaces, Interfaces, and Thin Films* (World Scientific, Singapore) 2000.

[7] GARBASSI F., MORRA M. and OCCHIELLO E., *Polymer Surface: From Physics to Technology* (Wiley, Chichester) 2000.

[8] TANG H., SZLEIFER I. and KUMAR S. K., *J. Chem. Phys.*, **100** (1994) 5367.

[9] ROUAULT Y., DÜNWEG B., BASCHNAGEL J. and BINDER K., *J. Stat. Phys.*, **80** (1995) 1009.

[10] GEHLSEN M. D., ROSEDALE J. H., BATES F. S., WIGNALL G. D. and ALMDAL K., *Phys. Rev. Lett.*, **68** (1992) 2452.

[11] DEUTSCH H. P. and BINDER K., *J. Phys. II*, **3** (1993) 1049.

[12] SARIBAN A. and BINDER K., *J. Chem. Phys.*, **86** (1987) 5859.

[13] DEUTSCH H. P. and BINDER K., *Macromolecules*, **25** (1992) 6214; *Europhys. Lett.*, **17** (1991) 697.

[14] MÜLLER M. and BINDER K., *Macromolecules*, **28** (1995) 1825.

[15] MÜLLER M., *Macromol. Theory Simul.*, **8** (1999) 343.

[16] ESCOBEDO F. A. and DE PABLO J. J., *Macromolecules*, **32** (1999) 900.

[17] SCHWEIZER K. S. and YETHIRAJ A., *J. Chem. Phys.*, **98** (1993) 9053; YETHIRAJ A. and SCHWEIZER K. S., *J. Chem. Phys.*, **98** (1993) 9080.
[18] CARMESIN I. and KREMER K., *J. Phys. II*, **51** (1990) 915.
[19] CARMESIN I. and KREMER K., *Macromolecules*, **21** (1988) 1988; DEUTSCH H. P. and BINDER K., *J. Chem. Phys.*, **94** (1991) 2294.
[20] BRUCE A. D. and WILDING N. B., *Phys. Rev. Lett.*, **68** (1992) 193.
[21] CAVALLO A., MÜLLER M. and BINDER K., *Europhys. Lett.*, **61** (2003) 214.

Dense binary mixtures of hard spheres: Two mixing scenarios

TH. VOIGTMANN

Fachbereich Physik, Universität Konstanz - Universitätsstraße 10, 78457 Konstanz, Germany

1. – Introduction

Binary hard-sphere mixtures composed of not too differently sized species show two mixing scenarios at high densities. This has been predicted [1] by the mode-coupling theory of the glass transition (MCT) and recently been found in computer simulations [2]. One mixing scenario occurs for size ratios of, say, $\delta = d_{\mathrm{s}}/d_{\mathrm{l}} = 0.6$ and smaller, where d_{s} and d_{l} are the diameters of the small and large spheres, respectively. If one, starting out from the one-component system consisting of either all large or all small spheres, admixes some small concentration of the other species, the glass transition is shifted toward higher packing (volume) fractions. This entropy-driven "plasticization" effect has been observed in colloidal suspensions earlier [3]. Perhaps surprisingly, the scenario changes if one considers mixtures with slightly larger size ratios, $0.8 \leq \delta < 1$, say. Then admixing small concentrations of the second species to a one-component system shifts the glass transition to lower packing fractions, *i.e.* stabilizes the glass instead of destabilizing it.

Experimental confirmation notwithstanding, it seems desirable to shed more light on the physical mechanisms distinguishing between binary mixtures of hard spheres with size ratios as similar as $\delta = 0.6$ and $\delta = 0.8$. The rather involved full MCT equations for binary mixtures aggravate intuitive arguments as to how their results change with a change in input quantities.

One approach, more common for binary mixtures with extreme size ratios ($\delta \ll 1$) is that of integrating out the degrees of freedom of one species. In this contribution, I demonstrate that the above two mixing scenarios for the glass transition are also present in such an effective one-component description (OCD), which proceeds by eliminating the degrees of freedom of the minority species. Note that this is not clear *a priori*, since the OCD only accounts for the minority species on a static level, whereas a crucial ingredient for the glass transition phenomena is the proper treatment of the collective long-time dynamics.

2. – Mode-coupling theory of the glass transition

A summary of the MCT equations suitable for binary mixtures is contained in ref. [1]. Within that theory, the glass transition is a purely dynamical transition at which, without singular changes in the static structure, dynamical relaxation times diverge. The main quantity in this context is the matrix of partial density correlation functions for wave number q and time t, $\Phi_{\alpha\beta}(q,t)$, where α and β are matrix indices labeling the species. The $t \to 0$ limit is just the static structure factor, $\boldsymbol{\Phi}(q,t=0) = \boldsymbol{S}(q)$, adopting matrix notation as indicated by bold symbols. The long-time limit of $\boldsymbol{\Phi}(q,t)$, denoted $\boldsymbol{F}(q)$, is used to distinguish between liquid states, where all density correlations eventually decay, $\boldsymbol{F}(q) = \boldsymbol{0}$, from arrested (glassy) states, where a non-zero portion of the density fluctuation survives for all times, $\boldsymbol{F}(q) \neq \boldsymbol{0}$.

The quantity $\boldsymbol{F}(q)$ can within MCT be calculated as a particular solution of the set of implicit equations [4]

$$\boldsymbol{S}(q) - \boldsymbol{F}(q) = \left[\boldsymbol{S}(q)^{-1} + \mathcal{F}[\boldsymbol{F}](q)\right]^{-1}. \tag{1}$$

Here, $\mathcal{F}$ is the mode-coupling functional, to be defined below. In the liquid, the only admissible solution of eq. (1) is $\boldsymbol{F} \equiv \boldsymbol{0}$. But at certain points in parameter space, a new branch of solutions obeying $\boldsymbol{F} \neq \boldsymbol{0}$ becomes the acceptable one, and they are identified as points where a liquid-to-glass transition occurs. These singular points of eq. (1) can be found by a numerically stable algorithm [1,4]. The $\boldsymbol{F}(q)$ describe the spatial structure of the arrested state and are thus called glass form factors.

The mode-coupling functional $\mathcal{F}$ is given entirely in terms of static correlation functions. After some technical approximation, only $\boldsymbol{c}(q)$, the matrix of partial direct correlation functions (DCF), remains:

$$\mathcal{F}_{\alpha\beta}[\boldsymbol{F}](q) = \frac{\varrho}{2q^4 x_\alpha x_\beta} \sum \int \frac{\mathrm{d}^3k}{(2\pi)^3} V_\alpha^{\alpha'\alpha''}(\vec{q},\vec{k}) F_{\alpha'\beta'}(k) F_{\alpha''\beta''}(p) V_\beta^{\beta'\beta''}(\vec{q},\vec{k}), \tag{2a}$$

$$V_\alpha^{\alpha'\alpha''}(\vec{q},\vec{k}) = \left[(\vec{q}\vec{k}) c_{\alpha\alpha'}(k)\delta_{\alpha\alpha''} + (\vec{q}\vec{p}) c_{\alpha\alpha''}(p)\delta_{\alpha\alpha'}\right]. \tag{2b}$$

In the above expression, the sum runs over all primed indices, and the short-hand $\vec{p} = \vec{q} - \vec{k}$ along with $q = |\vec{q}|$ and similar was used. We let ϱ denote the total number density, and x_α the relative concentration of species α. The packing fraction is given as $\varphi = \sum_\alpha (\pi/6)\varrho x_\alpha d_\alpha^3 = \sum_\alpha \varphi \hat{x}_\alpha$.

The above expressions readily reduce to the one-component limit by dropping all Greek indices and setting $x = 1$. In this case, the static structure enters only through the single DCF $c(q)$. These one-component MCT equations are used for the OCD results shown below.

3. – Effective one-component description

The matrix of partial DCF, $\boldsymbol{c}(q)$, comprising the control-parameter–dependent input for the MCT calculations is defined through the Ornstein-Zernike (OZ) equation [5]:

$$\boldsymbol{h}(q) = \boldsymbol{c}(q) + \varrho \boldsymbol{c}(q)\boldsymbol{X}\boldsymbol{h}(q), \tag{3}$$

in matrix notation, setting $X_{\alpha\beta} = x_\alpha \delta_{\alpha\beta}$. The $\boldsymbol{h}(q)$ are the partial total correlation functions, connected to the static structure factors by $\boldsymbol{S}(q) = \boldsymbol{X} + \varrho \boldsymbol{X}\boldsymbol{h}(q)\boldsymbol{X}$. One can rewrite the OZ equations to include a specified subset of species only through a set of effective DCF [6]. For a binary mixture one thus gets [7]

$$h_{11}(q) = c_{11}^{\text{eff}}(q) + \varrho x_1 h_{11}(q) c_{11}^{\text{eff}}(q), \quad \text{with} \tag{4a}$$

$$c_{11}^{\text{eff}}(q) = c_{11}(q) + \frac{\varrho x_2 c_{12}^2(q)}{1 - \varrho x_2 c_{22}(q)}. \tag{4b}$$

This reformulation is still exact. It specifies the effective potential that is present between the particles of species 1 due to the presence of species 2 in terms of a DCF. As such, it serves as input for an effective OCD MCT calculation.

We use the Percus-Yevick (PY) approximation to the DCF of hard-sphere mixtures [8], giving analytical expressions of $\boldsymbol{c}(q)$, and thus for $c_{11}^{\text{eff}}(q)$. The known deficiencies of the PY approximation do not affect the qualitative trends for mixing close to the glass transition [9]. It is then possible to analytically evaluate the r-dependent effective potential from eq. (4) in mean-spherical approximation, where one gets $-u^{\text{eff}}(r)/(kT) = \varrho x_2 FT^{-1}[c_{12}^2(q)] + \mathcal{O}(x_2^2)$ for $r > d_1$ [7]. One finds an attractive potential with range δ outside the hard-sphere core. It shows only smooth and monotonic dependences on δ. Thus the characteristics of $u^{\text{eff}}(r)$ alone give no clear indication of the existence of two different mixing scenarios.

4. – Mode-coupling results

Figure 1 shows the OCD MCT results for the change in the glass-transition packing fraction due to mixing a small amount of species 2 into a one-component system of species 1, $\delta\varphi^c$, evaluated from $\Delta\varphi^c/\Delta\hat{x}_2$ as $\hat{x}_2 = \varphi_2/\varphi \to 0$. Values $\delta < 1$ correspond to admixing smaller spheres, while values $\delta > 1$ correspond to introducing a small amount of larger spheres. Note that for $\delta = 1$ (*i.e.* the one-component case), $\delta\varphi^c < 0$. The OCD in this case already accounts for a fraction of the volume occupied by the spheres, leaving the remaining particles of species 1 to freeze at a lower effective packing fraction.

One can understand the increase of $\delta\varphi^c$ with increasing δ at $\delta \approx 1$ by assuming that in the "true" binary mixture, the change in excluded volume due to mixing is too small

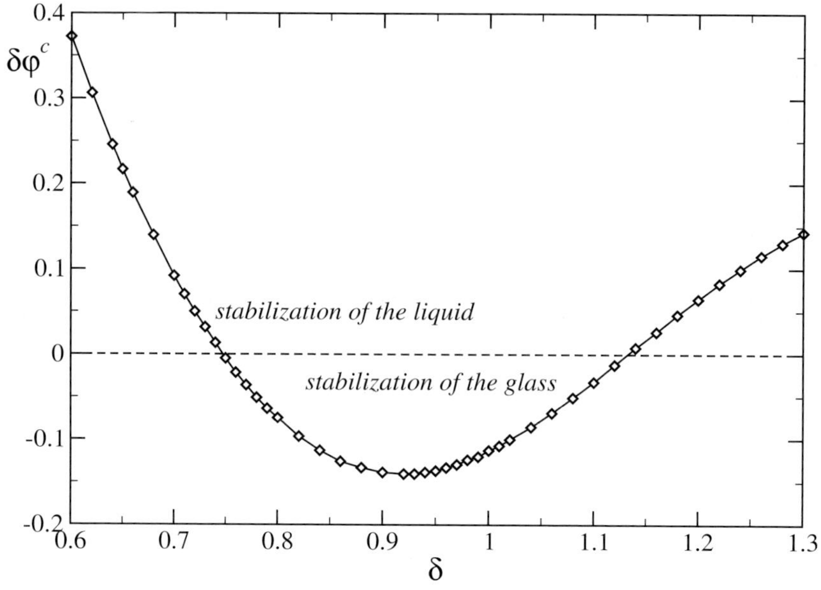

Fig. 1. – Change of the glass transition packing fraction for binary hard-sphere mixtures within the effective one-component description, $\delta\varphi^c = \Delta\varphi^c/\Delta\hat{x}_2$ as $\hat{x}_2 \to 0$, *versus* size ratio δ. $\hat{x}_2 = \varphi_2/\varphi$ is the concentration by volume of the minority species.

to induce significant changes in the microscopic structure of the system. Then the glass transition will happen at about the same number density ϱ_0^c. This in turn implies the packing fraction φ^c to increase (decrease) for $\delta > 1$ ($\delta < 1$).

On the other hand, at small δ, we can make connection to colloidal systems exhibiting short-ranged depletion interaction [10]. Typically, these are realized as some binary mixtures with small size ratio, $\delta \leq 0.1$, say. The short-ranged attraction arises in an effective OCD similar to the one used here. In these systems it has been discovered that introducing a weak, sufficiently short-ranged attraction at high densities destabilizes the glass, *i.e.* shifts φ^c to higher values (cf. ref. [11]). The maximum range of the effective potential at which this occurs, depends on the potential shape [12]. As a rule of thumb, one needs potential-driven bonding to disturb the nearest-neighbour cages of the particles. Then, occasionally the attractive potential will drive two particles closer together than the average, allowing for holes in the cages to be created, and thus for liquid-like flow to occur, more easily. The results shown in fig. 1 suggest that for the OCD potential used here, remnants of this attraction-driven cage breaking can be seen up to $\delta \approx 0.7$. Only then, the excluded-volume–driven effect for $\delta \approx 1$ begins to take over, resulting in a negative minimum of $\delta\varphi^c$ in the OCD at about $\delta \approx 0.9$.

Thus the following picture arises: there is a range of size ratios δ close to unity, where the OCD predicts a decrease of the glass-transition packing fraction with mixing. This arises for $0.88 \leq \delta < 1$, where both introducing small amounts of small spheres into larger

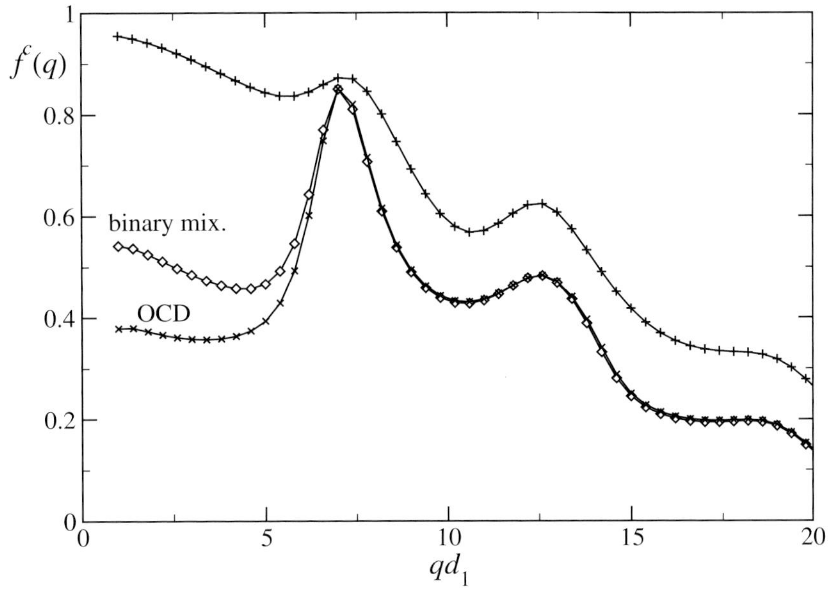

Fig. 2. – Glass form factors $f^c(q)$ at the glass transition for $\delta = 0.6$ and $\hat{x}_2 = 0.01$, calculated from the effective one-component description (crosses), and from the full binary-mixture theory (diamonds). Plus symbols indicate the full binary-mixture result for $\hat{x}_2 = 0.2$. For the binary systems, $f^c_{11}(q)$ is shown.

ones as well as small amounts of large spheres into smaller ones stabilizes the glass. The latter case is represented in fig. 1 as $1 < \delta \leq 1.12 \approx 1/0.88$. For values $d_\mathrm{s}/d_\mathrm{l} \leq 0.74$, the opposite effect holds: here, mixing in either way leads to a "melting" of the glass. For intermediate values, $0.76 \leq \delta \leq 0.87$, the situation depends on whether one considers the admixture of small spheres to large ones or vice versa. Thus the OCD qualitatively recovers the two mixing scenarios of ref. [1].

The limitations of the OCD become more prominent if one looks at the structure of the arrested state at the glass transition points. It is given through the glass form factors $f^c(q)$ in the OCD, and can be compared to the normalized 11 matrix element of $\boldsymbol{F}^c(q)$ in the binary-mixture description, $f^c_{11} = F^c_{11}(q)/S^c_{11}(q)$. Figure 2 exhibits these quantities for a representative size ratio, $\delta = 0.6$, with $\hat{x}_2 = 0.01$. For $\delta < 1$ there is a δ-independent trend for $f^c_{11}(q)$ upon mixing [1]: at almost any wave number q, $f^c_{11}(q)$ increases with increasing concentration of species 2. The reason for this is twofold: there is an increasing contribution stemming from the second species, and a decrease in localization length for particles of species 1 upon mixing. The latter leads to a broadening of the $f^c_{11}(q)$-*versus*-q curve, and altogether an increase of $f^c_{11}(q)$ at fixed q results. This increase is most pronounced for $q \to 0$, where there is a crossover from $f^c_{11}(q = 0) \approx 0.4$ at $x_2 = 0$ to $f^c_{11}(q = 0) = 1$ at $x_2 = 1$. To demonstrate this, fig. 2 includes the binary-mixture result for $f^c_{11}(q)$ for a larger amount of small spheres, $\hat{x}_2 = 0.2$, as plus symbols.

As fig. 2 reveals, this effect is missing in the effective OCD. Particularly at low q, the OCD results stay close to those for $x_2 = 0$, while even at the low x_2 investigated in the figure, the increase in $f^c_{11}(q)$ is up to 30% there. The strong discrepancy at $q \to 0$ shows that the hydrodynamic properties of the mixture are not captured correctly in the OCD. Note that this wave-vector region is accessible to dynamic light-scattering experiments, and so the difference can be expected to be relevant there.

5. – Conclusion

Two mixing scenarios exist for binary hard-sphere systems at their glass transition, depending on the particles' size ratio. One is a stabilization of the glass due to mixing, the other is the melting of the glass due to mixing. The two scenarios already emerge qualitatively correctly from an effective one-component description that contains the second species only through an effective attraction imposed on the first one. Remarkably, the effective potential itself shows no non-monotonic size-ratio dependence that could distinguish the observed two cases. They rather arise as a combination of short-ranged attraction and excluded-volume effects. Limitations of the effective one-component model become evident when one studies the spatial structure of the glassy states, which in both the full theory and simulations is rather different from what results in the one-component description.

* * *

I thank M. FUCHS for his hospitality during my stay in Konstanz, and H. H. VON GRÜNBERG for suggesting this approach. Financial support came from Deutsche Forschungsgemeinschaft, SFB 563, and from SIF for staying at the "Enrico Fermi" school.

REFERENCES

[1] GÖTZE W. and VOIGTMANN TH., *Phys. Rev. E*, **67** (2003) 021502.
[2] FOFFI G. *et al.*, *Phys. Rev. Lett.*, **91** (2003) 085701.
[3] WILLIAMS S. R. and VAN MEGEN W., *Phys. Rev. E*, **64** (2001) 041502.
[4] FRANOSCH T. and VOIGTMANN TH., *J. Stat. Phys.*, **109** (2002) 237.
[5] HANSEN J. P. and MCDONALD I. R., *Theory of Simple Liquids* (Academic Press, London) 1986.
[6] BERESFORD-SMITH B., CHAN D. Y. C. and MITCHELL D. J., *J Colloid Interface Sci.*, **105** (1985) 216; GONZÁLEZ-MOZUELOS P. and CARBAJAL-TINOCO M. D., *J. Chem. Phys.*, **109** (1998) 11074.
[7] MÉNDEZ-ALCARAZ J. M. and KLEIN R., *Phys. Rev. E*, **61** (2000) 4095.
[8] BAXTER R. J., *J. Chem. Phys.*, **52** (1970) 4559.
[9] FOFFI G. *et al.*, *Phys. Rev. E*, **69** (2004) 011505.
[10] POON W. C. K., *J. Phys. Condens. Matter*, **14** (2002) R859.
[11] DAWSON K. *et al.*, *Phys. Rev. E*, **63** (2001) 011401.
[12] GÖTZE W. and SPERL M., *J. Phys. Condens. Matter*, **15** (2003) S869.

On the way to a lattice Boltzmann model for glassy systems?

A. Lamura

Istituto Applicazioni Calcolo, CNR, Sezione di Bari - Via Amendola 122/D, 70126 Bari, Italy

S. Succi

Istituto Applicazioni Calcolo, CNR - V.le del Policlinico 137, 00161 Roma, Italy

1. – Introduction

The behavior of glassy materials and related disordered systems is one of the most debated and still open problems in modern condensed matter physics. Glasses are undercooled liquids which show a spectacular increase in viscosity (14 orders of magnitude) under rapid cooling below a given temperature threshold T_{g}, known as the glass transition temperature. In particular, they exhibit anomalous relaxation, namely long-lived departures from local equilibria, which results from the trapping of atoms around fixed positions in space according to a disordered pattern. The numerical simulation of glassy behavior relies mainly on Molecular Dynamics and Monte Carlo techniques [1, 2]. Both methods incorporate details of microscopic behavior (intermolecular potentials) and, as a result, they fall short of reaching spatio-temporal scales of macroscopic interest. Coarse-grained lattice gas models represent a step ahead in reaching larger scales: On the one hand, there are spin-glasses [3, 4] and, on the other, their modern lattice glass variants [5, 6]. Clearly, these lattice models cannot reach the same degree of physical fidelity as truly atomistic methods.

It is in the spirit of mesoscopic models that the lattice Boltzmann equation (LBE) was originally introduced for simulating hydrodynamic behavior in ordinary fluids [7]. Over the years, LBE has proved quite flexible in describing situations beyond the mere fluid dynamics framework [8]. It appears therefore reasonable to explore whether LBE can be extended in such a way to cover glassy behavior as well. We present in the following a lattice Boltzmann model which allows to catch some features of glassy systems [9].

2. – The model

We use a lattice Boltzmann equation with a single relaxation parameter [10] which reads as follows:

$$f_i(\mathbf{r}+\mathbf{c}_i, t+1) - f_i(\mathbf{r}, t) = -\omega\big(f_i(\mathbf{r}, t) - f_i^{\mathrm{eq}}(\mathbf{r}, t)\big), \tag{1}$$

where $f_i(\mathbf{r}, t) \equiv f(\mathbf{r}, \mathbf{v} = \mathbf{c}_i, t)$ is a discrete distribution function of particles moving along the direction i with discrete speed $\mathbf{c}_i$. We use a nine-velocity model on a square lattice with $c_i = (\pm 1, 0)$, $(0, \pm 1)$, $(\pm 1, \pm 1)$. The right-hand side represents the relaxation to a local equilibrium f_i^{eq} in a time lapse of the order of ω^{-1}. This local equilibrium is taken in the form of a quadratic expansion of a local Maxwellian:

$$f_i^{\mathrm{eq}} = \rho w_i \left[1 + \frac{\mathbf{u}\cdot\mathbf{c}_i}{c_{\mathrm{s}}^2} + \frac{\mathbf{u}\mathbf{u}\cdot(\mathbf{c}_i\mathbf{c}_i - c_{\mathrm{s}}^2 I)}{2c_{\mathrm{s}}^4}\right], \tag{2}$$

where w_i is a set of weights normalized to unity, c_{s} the lattice sound speed (a constant equal to $1/\sqrt{3}$ in our case) and, finally, I stands for the unit tensor. The distribution functions are related to the fluid density ρ and to the fluid momentum $\rho\mathbf{u}$ by

$$\rho = \sum_i f_i\,, \qquad \rho\mathbf{u} = \sum_i f_i \mathbf{c}_i\,. \tag{3}$$

In order to describe light and heavy phases with a single species fluid we follow the approach of Shan and Chen [11] and introduce a non ideal-gas behavior within the LBE formalism by means of a phenomenological pseudo-potential which describes the effects of potential energy interactions.

We introduce a repulsive Lennard-Jones potential interaction, which takes the following expression on the lattice:

$$V(r) = \begin{cases} G\Psi(\rho) r^{-12}, & \text{if } 1 \le r \le \sqrt{2}, \\ 0, & \text{otherwise}, \end{cases} \tag{4}$$

where G is a positive parameter controlling the strength of the interaction. The structure of $\Psi(\rho)$ is in the form of nested polynomials, whose density zeroes are distributed according to a binary tree whose depth depends on the number of generations N_{g}:

$$\Psi(\rho) = (\rho - \rho_0) \prod_{g=2}^{N_{\mathrm{g}}} \frac{-1}{(1 + ((\rho-\rho_0)/(\overline{\rho}_0/2))^{2^{g-1}})^2} \prod_{k=1}^{2^{g-1}} \frac{(\rho - \rho_{gk})}{(\rho_{gk} - \rho_0)}\,, \tag{5}$$

where $\overline{\rho_0}$ controls the density gap. This hierarchical potential implements the presence of $N_\rho = 2^{N_g} - 1$ competing density extrema.

The denominator in eq. (4) is such to let $\Psi \to 0$ outside the hierarchical range of zeroes ρ_{gk}. The result is an effective potential alternating stable repulsive regions ($\Psi' = (\partial\Psi/\partial\rho) > 0$) with unstable attractive ones ($\Psi' < 0$), distributed approximately in the density range $\rho_{\mp} = \rho_0(1 \mp \overline{\rho}_0(1 - 2^{-N_g}))$. The presence of unstable regions causes the appearance of phase transitions which determine the separation of light and heavy phases inside the fluid. In this case we should not observe the typical patterns of phase separation in binary fluids where coherent regions of the two phases grow in time until they span completely the system. In a mode-coupling perspective [12], we expect to be able to generate a disordered short-range coexistence of two competing species: "voids" (regions with $\rho < \rho_0$) and "cages" (regions with $\rho > \rho_0$). In real glasses, the dense phase is responsible for cage effects which ultimately lead to mechanical arrest of the system with an extraordinary increase of the viscosity. The glassy LBE takes the final form

$$f_i(\mathbf{r} + \mathbf{c}_i, t+1) - f_i(\mathbf{r}, t) = -\omega\big(f_i(\mathbf{r}, t) - f_i^{\mathrm{eq}}(\mathbf{r}, t)\big) + F_i(\mathbf{r}, t), \tag{6}$$

where $F_i = -G(\rho/\rho_0)\boldsymbol{\nabla}\Psi(\rho) \cdot \mathbf{c}_i$.

3. – Numerical results

We have simulated the fluid on a 2D lattice of size 256×256 with the following parameters: $\omega = 1.0$, $\rho_0 = 1.0$, $N_g = 3$, $\overline{\rho}_0 = 0.5$, G is varied in the range $[0, 1]$. The initial condition is $\rho(\mathbf{r}) = \rho_0\ (1 + \xi)$, where ξ is a random perturbation uniformly distributed in the range $[-0.01, 0.01]$. The effects due to propagation (left-hand side of eq. (6)) between neighboring sites and to the non-linear potential $\Psi(\rho)$ make the density distribution spread out in time, so that an instability is triggered as soon as G exceeds the critical threshold G_c. Such a value can be found by solving the systems of equations

$$\begin{cases} \dfrac{\partial p'}{\partial \rho} = 0, \\ \dfrac{\partial^2 p'}{\partial \rho^2} = 0, \end{cases} \tag{7}$$

where $p' = c_s^2\rho + V$ is the equation of state when the potential (4) is switched on.

Solving the system (7), we find $G_c \sim 0.089$. Numerically we can verify this prediction by monitoring the value of G for which the quantity $\Delta\rho \equiv \rho_{\max} - \rho_{\min}$ becomes non zero, $\rho_{\max}$ and $\rho_{\min}$ being the maximum and minimum values of ρ, respectively. The density gap $\Delta\rho$ starts up at $G = G_c \sim 0.09$ in good agreement with the analytical result and rapidly fills the available range of states between ρ_- and ρ_+, as the interaction strength is increased. The density ρ distributes in time within the available range of values according to a Gaussian distribution, characterized by decreasing fluctuations around the mean value as time goes on.

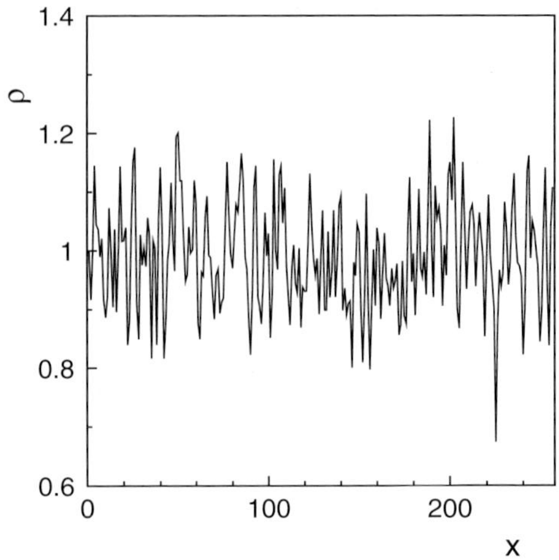

Fig. 1. – The density profile along the x-direction at $y = 128$ at $t = 200$ for the case $G = 0.45$.

In order to check whether we are simulating a phase-separating fluid or a glassy system, we looked at the spatial distribution of the density field $\rho(x, y)$. In fig. 1 we present a cross-section of the density field ρ taken at $y = 128$ at time $t = 200$ (steady state). The pictures clearly indicate a disordered coexistence of voids and cages. This confirms that the glassy potential fulfils the minimal requisite for glassy behavior: sharp density contrasts over a disordered spatial pattern.

As a further observable in fig. 2 we plot the density-density correlation:

$$g(r) = \frac{\langle \delta\rho(x + r_x, y + r_y)\delta\rho(x, y)\rangle}{\langle \rho(x, y)\rho(x, y)\rangle}, \tag{8}$$

where $\delta\rho \equiv \rho - \langle\rho\rangle$ is the density fluctuation around its spatial average and $r^2 = r_x^2 + r_y^2$. This function shows an oscillating-decay behavior similar to the case of real glasses. Being in a lattice, it is clear that data are only available at discrete positions $1, \sqrt{2}, 2, \ldots$. However, a spline interpolation (solid line) indicates that the observed oscillations are not a numerical artifact. In the case of no glassy potential ($G = 0$), $g(r)$ is virtually zero as a consequence of the uniform and constant density distribution. In fig. 3 we present the density-density time-correlation function $h(\tau)$, taken at $x = 128$, $y = 128$. Here, the analogy with glasses is weaker. Instead of anomalous slow relaxation, we find persistent oscillations around zero, possibly reflecting an everlasting competition between the various density minima. This is at odd with the typical signature of glassy behaviour, namely a plateau in the time correlation function. To this purpose, further research work is required.

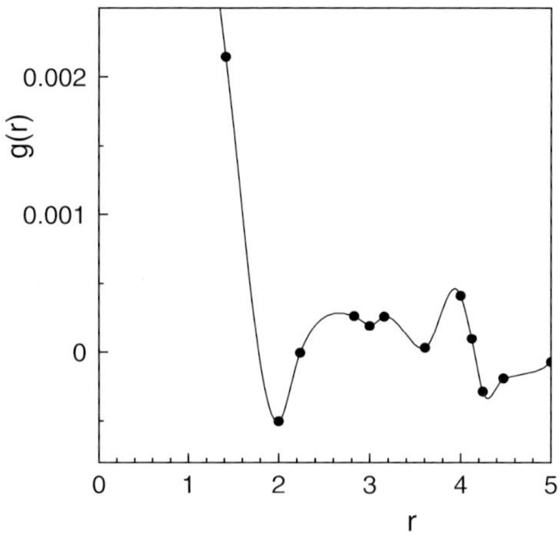

Fig. 2. – Density-density correlation function at $t = 200$.

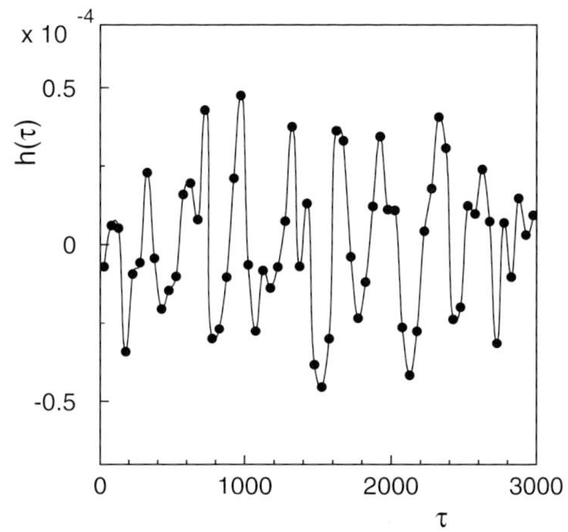

Fig. 3. – Density-density time correlation function.

4. – Conclusion

The proposed glassy LBE produces a number of encouraging results. Geometrical frustration is qualitatively captured by a hierarchical density-dependent potential, alternating repulsive and attractive interactions. This potential gives rise to disordered patterns with short-range, sharp density contrasts. It also yields qualitatively reasonable spatial density-density correlation functions. However, anomalous time relaxation is not captured. We believe that the introduction of a conditional propagation is a route to be explored in future work: possible strategies include a density-dependent relaxation frequency ω and/or density-dependent site-to-site hopping rates [13].

* * *

We thank Profs. K. BINDER, E. MARINARI and G. PARISI for helpful discussions.

REFERENCES

[1] BINDER K., BASHNAGEL J. and PAUL W., *Prog. Polym. Sci.*, **28** (2003) 115.
[2] MÉZARD M. and PARISI G., *J. Phys. Condens. Matter*, **12** (2000) 6655.
[3] BINDER K. and YOUNG A. P., *Rev. Mod. Phys.*, **58** (1986) 801.
[4] MÉZARD M., PARISI G. and VIRASORO M., *Spin Glass Theory and Beyond* (World Scientific, Singapore) 1987.
[5] BOUCHAUD J. P., CUGLIANDOLO L., KURCHAN J. and MÉZARD M., *Spin Glasses and Random Fields*, edited by A. P. YOUNG (World Scientific, Singapore) 1998.
[6] BIROLI G. and MÉZARD M., *Phys. Rev. Lett.*, **88** (2002) 025501.
[7] HIGUERA F., SUCCI S. and BENZI R., *Europhys. Lett.*, **9** (1989) 345; BENZI R., SUCCI S. and VERGASSOLA M., *Phys. Rep.*, **222** (1992) 145; CHEN S. and DOOLEN G., *Ann. Rev. Fluid Mech.*, **30** (1998) 329.
[8] SUCCI S., *The Lattice Boltzmann Equation* (Oxford University Press, Oxford) 2001.
[9] LAMURA A. and SUCCI S., *Physica A*, **325** (2003) 477.
[10] BHATNAGAR P., GROSS E. P. and KROOK M. K., *Phys. Rev.*, **94** (1954) 511.
[11] SHAN X. and CHEN H., *Phys. Rev. E*, **49** (1995) 2941; **47** (1993) 1815.
[12] GOETZE W., *J. Phys. Condens. Matter*, **11** (1999) A1.
[13] IVANOV V. A., JUNG B., SEMENOV A., NYRKOVA I. A. and KHOKHLOV A. R., *J. Chem. Phys.*, **104** (1996) 4214.

Dynamics at a buried polymer interface

XUESONG HU and JYOTSANA LAL

Intense Pulsed Neutron Source, Argonne National Laboratory - Argonne, IL-60439, USA

XUESONG JIAO and L. B. LURIO

Department of Physics, Northern Illinois University - DeKalb, IL-60115, USA

SURESH NARAYANAN

Advanced Photon Source, Argonne National Laboratory - Argonne, IL-60439, USA

1. – Introduction

Polymer/Polymer interfaces play an important role in the mechanical toughness of polymer blends, adhesion properties in coatings and coextrusion of polymers. Interface properties are also critical in many industrial applications. In case of polymer thin films, the presence of surfaces greatly modifies the interplay of entropic and enthalphic forces, leading to properties distinct from those of bulk interfaces. Recently the technique of X-ray Photon Correlation Spectroscopy (XPCS) at grazing incidence was used [1] to probe the dynamics of thermally driven capillary wave fluctuations at the surface of a polystyrene thin film. This work demonstrates that coherent X-ray scattering can be used to measure surface viscosity and surface tension or the ratio of these quantities at very small length scales. Neutron Reflectivity was also used recently [2], to prove the existence of capillary waves at immiscible polymer/polymer interfaces by measurement of the logarithmic dependence of the interface width on film thickness. Surprisingly, however, there is not much other experimental data which has probed the existing theoretical predictions [3] at these length scales.

In this paper, we present a further development of XPCS to study polymer interfaces.

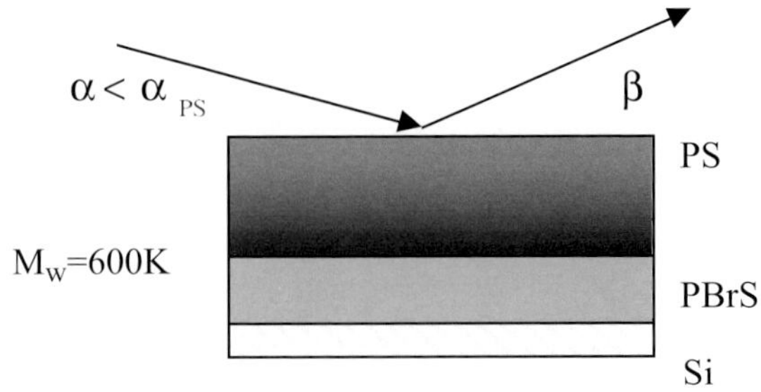

Fig. 1. – Schematic of the polymer bilayer on silicon substrate.

We have probed capillary wave dynamics not just at a free surface, but also at buried polymer/polymer interface within a bilayer film. The bilayer was chosen so that the critical angle for total external reflection for the top layer is smaller than that for the bottom layer. When X-rays are incident below the critical angle of the top layer only the structure and dynamics of the top layer are probed. When X-rays are incident above the critical angle of the top layer but below that of the bottom layer, a standing wave is set up. The phase of this standing wave can be adjusted to have a high intensity at the polymer-polymer interface and simultaneously a node at the polymer-air interface. Consequently, one can isolate for the first time the static scattering and XPCS from a single buried layer. Results on a system consisting of a 120 nm polystyrene film on top of an 80 nm poly(4-bromo styrene) film, supported on a Si substrate will be discussed.

2. – Experimental section

2'1. *Samples.* – Silicon wafers were cleaved into 15 mm×15 mm squares. A sulfuric acid solution was used to remove any ionic/metallic impurities and create a thin oxide layer. Monodisperse Poly(4-bromo styrene) (PBr_xS, $M_w = 670\,k$, $x = 0.972$), Polystyrene (PS, $M_w = 508\,k$) *(Polymer Source)* were used in these experiments. Bilayer substrates were prepared as follows: initially a layer of PBrS was spun cast from toluene solution directly onto HF stripped silicon wafers. The thickness of this layer, typically 80 nm, was measured with ellipsometry. Another layer of PS, around 120 nm, was spun cast onto a glass slide and floated from de-ionized water onto the PBrS layer. A schematic of the sample geometry is shown in fig. 1. The PBrS and PS bilayer samples were annealed at 180 °C, in an oil-trapped vacuum of 10^{-3} Torr for 2 hours. The XPCS experiments were performed at 8-ID beam line at the Advanced Photon Source (APS). The details of the experimental set up are as described previously [4].

2'2. *X-ray reflectivity.* – Specular reflectivity obtained from the PS/PBrS/Si film at 180 °C is shown in fig. 2. Reflectivity was measured *in-situ* in the diffuse scattering setup by using a current integrating PIN diode to measure the total reflected intensity. No slits were used, and the detector size was approximately 5 mm × 5 mm, located at a distance of 3479 mm. The solid line displays the results of a fit to the reflectivity using a

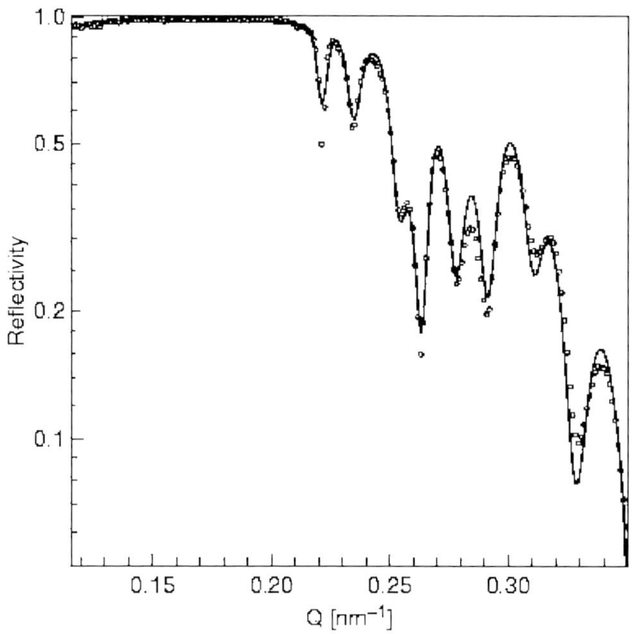

Fig. 2. – Specular reflectivity measured from a bilayer at 180 °C. Points are experimental data. The solid line is the model described in the text.

two-layer model with Gaussian roughness at each interface [5]. The best-fit values for the film thickness were 124 nm and 81 nm for the PS, and PBrS layers, respectively. These values are within 10% of the ellipsometry parameters. The roughness at the PS/PBrS layer was 4.4 nm (Gaussian sigma). The roughness at the Vacuum/PS interface was set to be 0.6 nm without fitting it, since the range of wave vector measured was too small to resolve this length scale. Rather 0.6 nm is the theoretical value expected from capillary wave roughness [4]. As will be discussed below, the ratio of the Vacuum/PS to PS/PBrS interfacial roughnesses can be extracted from the diffuse scattering *vs.* incident angle and this was consistent with the numbers given here. Reflectivity data was collected both before and after the XPCS measurements and showed no significant sample damage.

2˙3. *Diffuse scattering.* – Measurements of dynamics (XPCS) were made by performing X-ray time correlation on the diffuse scattering. Data were collected at two incident angles; $\alpha = 0.16°$ was chosen to yield scattering from only the top interface, and $\alpha = 0.203°$ to yield scattering dominated by the buried interface. Figures 3a and b show the calculated electric field intensities throughout the films for these two incident angles. The calculations were made using the parameters from the fits to the reflectivity. As can be seen in fig. 3a, at $\alpha = 0.16°$ the standing wave does not penetrate down to the PS/PBrS interface (fig. 1), and has a maximum at the position of the vacuum/PS interface of nearly four times the incident intensity. Thus, the lower angle is only sen-

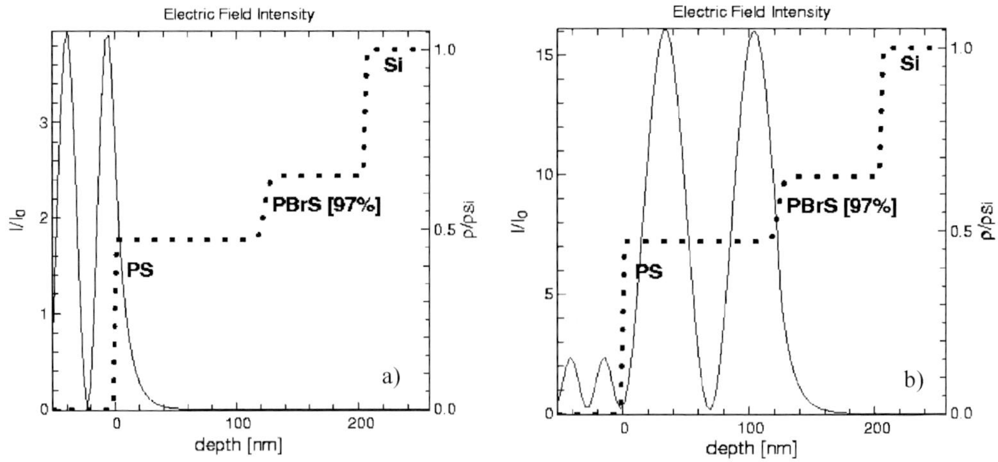

Fig. 3. – Electric-field intensity within bilayer *vs.* depth for incident angle $\alpha = 0.160°$ (a) and $\alpha = 0.203°$ (b). The data are normalized to the intensity of the incident field.

sitive to the top interface. At an angle of 0.203°, the standing wave has a node at the vacuum/PS interface producing almost no scattering and simultaneously the intensity at the PS/PBrS interface is increased by a factor of 8 over the incident beam. Consequently, if resolution could be neglected, the scattering at the larger angle is due nearly entirely to the buried interface.

In order to verify that the distribution of the electric field inside the material was in agreement with this model, a measurement was made of the diffuse scattering intensity at fixed exit angle, β, while varying the incident angle α. Figure 4, displays the measured scattering at an outgoing angle $\beta = 0.45 \pm 0.05$ and $\chi = 0.035 \pm 0.005$. This is compared with the calculated diffuse scattering [6] based on the parameters used in the fit to the specular reflectivity described above. The model was also convolved with the same angular resolution function. The dotted line represents the scattering from just the PS/PBrS interface. This accounts for 80% of the expected scattering at $\alpha = 0.203°$ and would account for nearly 100% of the scattering if resolution effects were not included. As can be seen, the data (squares) show similar peaks as in the model, located at the same positions and of approximately the same magnitude. Both the model and the data have been arbitrarily normalized to 1 at $\alpha = 0.180°$. The fits are therefore only sensitive to the ratio of the surface roughnesses at each interface to each other, not to the absolute value of the surface roughness. Nonetheless, the best agreement was found using the theoretical value for the capillary wave roughness of free PS, and the measured roughness from the specular reflectivity for the buried interface. There is not, however, exact agreement between the model and the data. The deviations are also consistent with the fits to the specular data. The resolution function yields a good match between theory and data over the region near the peak at $\alpha = 0.203°$, and this region also yielded the best fit for the specular reflectivity data. This lends confidence that the scattering at $\alpha = 0.203°$ is well

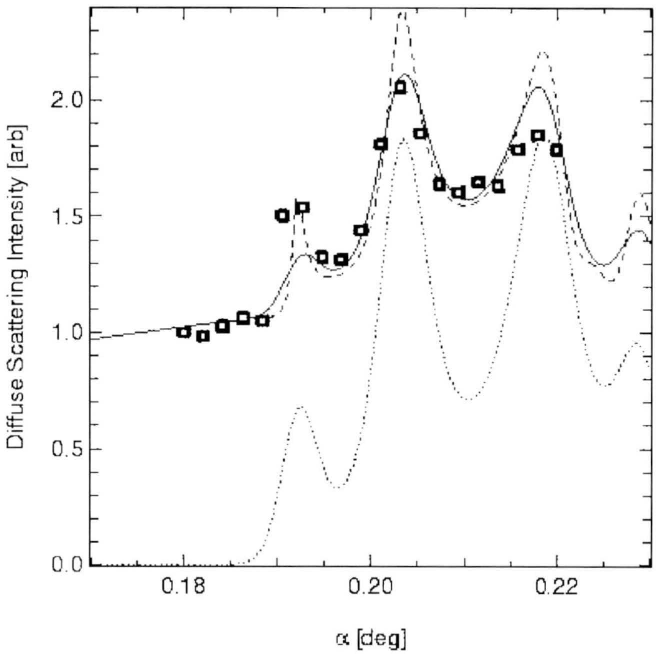

Fig. 4. – Diffuse scattering *vs.* incident angle, α, at fixed output angles $\beta = 0.45 \pm 0.05°$ and $\chi = 0.035 \pm 0.005°$. Squares are measured data integrated over the range of angles given above. The solid line is the theoretical prediction based on the model described in the text, and convoluted with an experimental resolution of 0.0037°. The dashed line is the same model without resolution taken into account. The dotted line is the contribution due only to the PS/PBrS interface, convolved with resolution. The data and model are both normalized to unity at $\alpha = 0.18°$.

predicted by the model, and can thus be assumed to be dominated by the buried interface. At $\alpha = 0.160°$ the penetration depth of the standing wave is relatively insensitive to small deviations in angle, and hence we can be confident in this instance that the scattering at low angle is due entirely to the Vacuum/PS boundary. Note also, that in neither case will the evanescent wave penetrate down to the Si. Scattering from the substrate is always negligible.

2'4. *Surface and interface dynamics.* – The polymer surface and interface are partially coherently illuminated, giving rise to a speckled scattering pattern, which varies in time as the surface and interface modes experience random thermal fluctuations. The normalized intensity-intensity time autocorrelation function, g_2, then yields the sample's surface dynamics. To avoid sample damage, the X-ray exposure of any position on the sample was limited to about 10 minutes, after which time the sample was shifted to a fresh area.

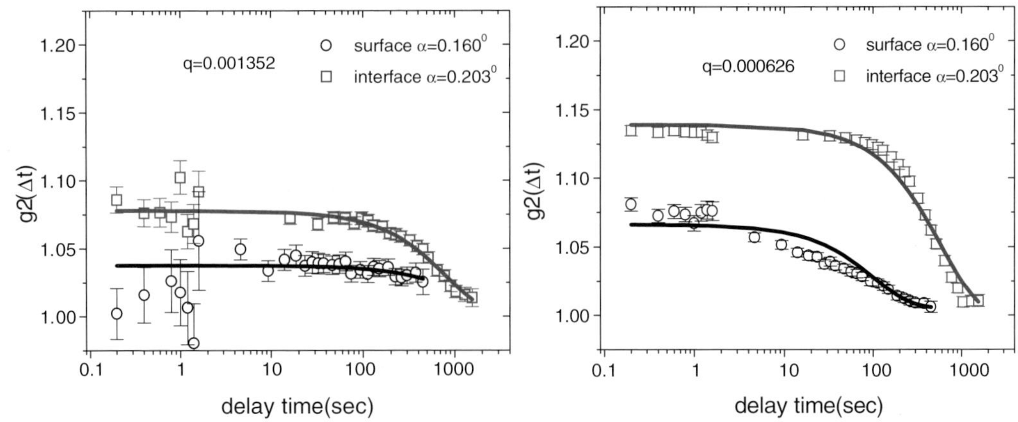

Fig. 5. – Autocorrelation functions for in plane wave vector $q_{||} = 6.26 \times 10^{-3}\,\text{nm}^{-1}$ and $q_{||} = 1.352 \times 10^{-2}\,\text{nm}^{-1}$ at 180 °C. The solid lines are single-exponential fits.

The intensity autocorrelation function is given by

$$g_2(q,t) = \frac{\langle I(q,t')I(q,t'+t)\rangle}{\langle I(q,t')\rangle^2}, \tag{1}$$

where $I(q,t')$ is the scattering intensity at wave vector transfer q at time t'. In eq. (1), the angular brackets in the numerator refer to averages over t'. The value of $I(q)$ in the denominator was obtained not from the time average of the intensity within a single pixel, but rather from the fits to the structure factor of the scattering pattern. This allows for the correlation function to be measured without self-biasing due to possible time-correlations longer than the measurement time. The normalized intensity autocorrelation were calculated pixel-by-pixel from a series of sequential two-dimensional off-specular diffuse-scattering patterns recorded on a charge-coupled device (CCD) camera. This is followed by an appropriate averaging over all pixels corresponding to the same narrow range of in-plane wave vectors $q_{||}$. Figure 5 shows experimental results of incident angle $\alpha = 0.160°$ and $\alpha = 0.203°$, obtained from the PS/PBrS/Si film at 180 °C. The solid lines in fig. 5 represent the best single-exponential fits to the measured autocorrelation at fixed value of $q_{||}$, $g_2(t) = 1 + B\exp[-2t/\tau]$, where B is the speckle contrast and τ is the relaxation time for equilibrium surface height fluctuations. The time constant for buried interface ($\alpha = 0.203°$) is found to be significantly slower than that of the top surface ($\alpha = 0.160°$). The relaxation time τ depends on the ratio of viscosity to surface tension (η/γ) in case of overdamped capillary waves on viscous liquids (eq. (2), ref. [6]). The relaxation times measured here show the correct behavior, since we expect that PBrS has much higher viscosity than PS (higher glass transition temperature) and the interface tension is lower compared to surface tension at the top interface. Thus the factor η/γ is expected to be higher for the PS/PBrS interface than the vacuum/PS interface. A very accurate extraction of τ is not possible with the present data since a clear long time baseline in g_2 (fig. 5) is not reached at low $q_{||}$ ($0.00626\,\text{nm}^{-1}$) for interface measurement. This limitation resulted from the lack of stability of the beam line over periods longer than $\sim 20\,\text{min}$. Stability was measured by the time correlation functions from a purely elastic scattering aerogel sample. At high $q_{||}$ ($0.01352\,\text{nm}^{-1}$) the surface dynamics are too fast for the time window of the experiment as shown by the measured g_2. In future an independent estimate of γ at the PS/PBrS interface will be made and the full $q_{||}$-dependence of the autocorrelation function g_2 at the two interfaces as a function of film thickness and temperature will be measured.

In conclusion, we have demonstrated a new technique to measure the dynamics of a buried interface. By careful adjustment of the incident angle, and consequent standing wave in the film, the static diffuse scattering signal from surface and interface can be clearly discriminated. The measured autocorrelation functions from the surface and interface show a marked difference in the relaxation rates which is consistent with the expected lower surface tension at the PS/PBrS interface and the higher viscosity of PBrS relative to PS. Future measurements should provide quantitative estimates of the surface tension and wave vector dependent viscosity at this buried interface.

* * *

This work was funded by the U.S. Department of Energy, BES-Materials Science, under Contract W-31-109-ENG-38 and by NSF Grant DMR-0209542. The Advanced Photon Source is supported by the U.S. Department of Energy, Office of Basic Science, under W-31-109-ENG-38.

REFERENCES

[1] KIM H., RUHM A., LURIO L. B., BASU J. K., LAL J., SINHA S. K., LUMMA D. and MOCHRIE S. G. J., *Phys. Rev. Lett.*, **90** (2003) 068302.
[2] SFERRAZZA M., XIAO C. and JONES R. A. L., *Phys. Rev. Lett.*, **78** (1997) 3693.
[3] DE GENNES P. G., *Eur. Phys. J. E*, **2** (2000) 201.
[4] LURIO L. B., KIM H., RÜHM A., BASU J. K., LAL J., SINHA S. K. and MOCHRIE S. G. J., *Macromolecules*, **36** (2003) 5704.
[5] TOLAN M., SEECK O. H., WANG J., SINHA S. K., RAFAILOVICH M. H. and SOKOLOV J., *Physica B*, **283** (2000) 22.
[6] HOLY V., KUBENA J., OHLIDAL I., LISCHKA K. and PLOTZ W., *Phys. Rev. B*, **47** (1993) 15896.

Aging properties of critical ferromagnets: A field-theoretical approach

A. Gambassi

Max-Planck-Institut für Metallforschung - Heisenbergstr. 3, D-70569 Stuttgart, Germany
Institut für Theoretische und Angewandte Physik, Universität Stuttgart
Pfaffenwaldring 57, D-70569 Stuttgart, Germany

P. Calabrese

Scuola Normale Superiore and INFN - Piazza dei Cavalieri 7, I-56126 Pisa, Italy

Understanding the non-equilibrium dynamics of physical systems is currently one of the most challenging problems in statistical physics. The absence of a general framework such as that provided by equilibrium statistical mechanics makes it worthwhile to study simple models of systems evolving under non-equilibrium conditions. Unexpected dynamical behaviors have been found in models of disordered and complex systems. One of the most striking of them is *aging*: After a perturbation, the system persists in a non-equilibrium state even if a macroscopic time has elapsed since then. Therefore the thermal history of the system influences, for instance, its response to an external (weak) field, and its behavior in general [1, 2]. Such a phenomenon, however, can also be observed in critical non-disordered models, as pointed out in ref. [3]. In these cases slow-relaxing modes keep the system in a non-equilibrium state even for macroscopically large times, giving rise to an interesting dynamical behavior. In view of this fact, the non-equilibrium critical dynamics of ferromagnetic spin systems has been recently the subject of a renewed interest (see ref. [4] for a brief account). Indeed, as one of such systems consider a model with order parameter $\varphi_{\mathbf{x}}$ and critical temperature T_c, in a generic configuration and, at time $t = 0$, take it in contact with a thermal bath at temperature T. The consequent relaxation process is characterized by two regimes: A transient one with off-equilibrium evolution, for $t < t_R$, and a stationary one with equilibrium evolution for $t > t_R$, where t_R is the relaxation time. In the former the behavior of the system depends on the initial conditions, while in the latter time homogeneity and time-reversal symmetry (in the absence of external fields) are recovered and such a dependence is lost; fluctuations are thus described by the equilibrium dynamics. As long as t_{R} is finite for all the fluctuation modes, equilibrium is always attained. This is the case for $T > T_{\mathrm{c}}$ but generally not for $T < T_{\mathrm{c}}$, when phase ordering takes place [5,6]. In the following we focus on the case of an instantaneous quench from high temperature to T_{c}. A convenient way of describing the dynamics is to study two-time response and correlation functions. The former is defined by $R_{\mathbf{x}}(t,s) = \delta\langle\varphi_{\mathbf{x}}(t)\rangle/\delta h(s)$, where h is a small external field applied at time $s > 0$ in the point $\mathbf{x} = \mathbf{0}$, and $\langle\cdot\rangle$ stands for the mean over the stochastic dynamics. The latter, instead, is defined by $C_{\mathbf{x}}(t,s) = \langle\varphi_{\mathbf{x}}(t)\varphi_{\mathbf{0}}(s)\rangle$. If the system does not reach the equilibrium (*i.e.* $t_{\mathrm{R}} = \infty$), $R_{\mathbf{x}}$ and $C_{\mathbf{x}}$ will depend both on s (the "age" of the system, also called "waiting time") and on the observation time t. This behavior is usually referred to as aging [1,2]. To characterize the distance from equilibrium of an aging system, in contact with a thermal bath at temperature T, the fluctuation-dissipation ratio (FDR) is usually introduced [7,3,8]:

$$X_{\mathbf{x}}(t,s) = \frac{T\,R_{\mathbf{x}}(t,s)}{\partial_s C_{\mathbf{x}}(t,s)}\,. \tag{1}$$

When $t,\, s \gg t_{\mathrm{R}}$ the fluctuation-dissipation theorem holds and thus $X_{\mathbf{x}}(t,s) = 1$. This is no longer true in the aging regime [3,8]. It has been argued that the long-time limit of the FDR at criticality

$$X^{\infty} = \lim_{s\to\infty}\lim_{t\to\infty} X_{\mathbf{x}=0}(t,s), \tag{2}$$

is a novel *universal* quantity of non-equilibrium critical dynamics [9,10,4]. For $T < T_{\mathrm{c}}$ it is possible to show, by means scaling arguments, that $X^{\infty} = 0$ (see, *e.g.*, ref. [4]), whereas for $T > T_{\mathrm{c}}$ the equilibrium value $X^{\infty} = 1$ is found.

Correlation and response functions have been computed for some exactly solvable models, while Monte Carlo simulations (MCs) have been carried out for the two- and three-dimensional Ising Model with Glauber dynamics, always finding $X^{\infty} \leq 1/2$. The corresponding results are summarized in table I.

Field-theoretical methods, and in particular Renormalization Group (RG), are powerful tools for the computation of *universal* quantities (such as critical exponents) and scaling properties of critical phenomena (for an updated review see ref. [14]). In the RG framework the problem of the critical relaxation from a macroscopically prepared initial state has been analyzed since some years, and a new universal exponent associated to an additional time-surface renormalization has been introduced [15].

It is possible to take advantage of these findings to compute the critical FDR (eventually verifying its universality) and the associated *universal* scaling functions for the mesoscopic models of dynamics, overcoming most of the analytical difficulties encountered in the exact solutions of (lattice) models with aging dynamics.

TABLE I. – *Values of* X^∞ *in some models.*

Model	Ref.	$T < T_c$	$T = T_c$	$T > T_c$
Random Walk[a]	[3]	—	1/2	—
Free Gaussian Field[a]	[3]	—	1/2	1
d-dim. Spherical Model[a]	[10]	0	$1 - 2/d^c$	1
1-dim. Ising-Glauber Model[a]	[11]	—	1/2	1
2-dim. Ising-Glauber Model[b]	[10]		0.26(1)	
	[12]		0.340(5)	
	[13]		0.33(2)	
3-dim. Ising-Glauber Model[b]	[10]		0.40	

[a] Exact solution.
[b] Monte Carlo simulations.
[c] $2 < d < 4$.

Here we give a brief account of some recent works we have done [16-19] in order to provide (perturbative) analytical predictions of the critical FDR for models which have not yet been exactly solved (*e.g.*, the d-dim Ising-Glauber Model with $d > 1$, cf. table I). Of course, to give accurate quantitative predictions in physical dimensions $d = 2, 3$, the knowledge of higher-order terms in the perturbative expansion is generally required.

The scaling form of $R_{\mathbf{q}}$ ($C_{\mathbf{q}}$), the Fourier transform in space of $R_{\mathbf{x}}$ ($C_{\mathbf{x}}$), can be obtained from RG analysis [15] in terms of well-known universal exponents. At criticality and for $\mathbf{q} = \mathbf{0}$, one obtains

$$T_c R_{\mathbf{q}=\mathbf{0}}(t,s) = A_R\,(t-s)^a \left(\frac{t}{s}\right)^\theta F_R\left(\frac{s}{t}\right), \tag{3}$$

$$C_{\mathbf{q}=\mathbf{0}}(t,s) = A_C\, s(t-s)^a \left(\frac{t}{s}\right)^\theta F_C\left(\frac{s}{t}\right), \tag{4}$$

where $a = (2 - \eta - z)/z$ and θ is the initial-slip exponent [15]. The functions F_C and F_R are universal once the non-universal normalizations A_R and A_C have been fixed so that $F_i(0) = 1$. Let us mention that local scale invariance (LSI, see ref. [21] for a comprehensive introduction) predicts $F_R(s/t) \equiv 1$ [20, 21]. In ref. [16] the quantity $\hat{X}(t,s) = T\,R_{\mathbf{q}=\mathbf{0}}(t,s)/\partial_s C_{\mathbf{q}=\mathbf{0}}(t,s)$ has been introduced, showing that $\lim_{s\to\infty}\lim_{t\to\infty}\hat{X}(t,s) = X^\infty$. Then, using eqs. (3) and (4), it is easy to see that X^∞ at criticality can be expressed as an *amplitude ratio* [10,16]: $X^\infty = A_R/[A_C(1-\theta)]$. Therefore, to compute it, one has to determine A_R and A_C for the models of interest. (For those considered in refs. [16-19], θ and the standard RG properties had already been obtained in the references cited therein.)

We do not explain here the method used for the computation, being quite standard in the field of dynamic critical phenomena, referring the reader to the literature [22, 15]. Instead we give a summary of the models we analyzed and of the results we obtained.

The universality hypothesis states that the critical phenomena can be described in terms of quantities independent of the microscopic details of the systems, but depending only on global properties such as symmetries, dimensionality, conservation laws, etc. According to this approach the critical properties of a system of N-component spins $S_{\mathbf{x}}$ on a d-dimensional lattice, interacting through a short-range $O(N)$ symmetric potential and with local relaxational dynamics (no conservation laws), can be described in terms of Model A of ref. [23]. This is specified in terms of an N-component field $\varphi(\mathbf{x}, t)$ (the coarse-grained version of $S_{\mathbf{x}}$) evolving according to the stochastic Langevin equation

$$\partial_t \varphi(\mathbf{x}, t) = -\Omega \frac{\delta \mathcal{H}[\varphi]}{\delta \varphi(\mathbf{x}, t)} + \xi(\mathbf{x}, t), \tag{5}$$

where Ω is a kinetic coefficient, $\xi(\mathbf{x}, t)$ a zero-mean stochastic Gaussian noise with $\langle \xi_i(\mathbf{x}, t) \xi_j(\mathbf{x}', t') \rangle = 2\Omega\, \delta(\mathbf{x} - \mathbf{x}') \delta(t - t') \delta_{ij}$, and $\mathcal{H}[\varphi]$ is the static Hamiltonian. It may be assumed, near the critical point, to be of the Landau-Ginzburg form

$$\mathcal{H}[\varphi] = \int \mathrm{d}^d x \left[\frac{1}{2} (\nabla \varphi)^2 + \frac{1}{2} r_0 \varphi^2 + \frac{1}{4!} g_0 \varphi^4 \right], \tag{6}$$

where $r_0 \propto T - T_{\mathrm{c}}$ and g_0 is the coupling constant. Correlation and response functions can be obtained by introducing a field-theoretical action [22] and the (high-temperature) initial condition can be accounted for according to ref. [15]. In terms of the resulting action, all the quantities of interest can be obtained by using the standard ϵ-expansion, with $\epsilon = 4 - d$. In a two-loop expansion we find (see refs. [16, 17] for details)

$$\frac{(X^\infty)^{-1}}{2} = 1 + \frac{N+2}{4(N+8)} \epsilon + \epsilon^2 \frac{N+2}{(N+8)^2} \left[\frac{N+2}{8} + \frac{3(3N+14)}{4(N+8)} + c \right] + O(\epsilon^3), \tag{7}$$

with $c = -0.0415\ldots$. From this two-loop series it is in principle difficult to extract accurate quantitative information. Using [0, 2] and [2, 0] Padé approximants we estimate $X^\infty = 0.429(6)$ for the three-dimensional Ising Model ($\epsilon = 1$, $N = 1$) with Glauber dynamics and $X^\infty = 0.30(5)$ for the two-dimensional one ($\epsilon = 2$, $N = 1$). This latter estimate has been recently confirmed by MCs [13, 12] (see table I([1])).

A question of theoretical and experimental interest is whether and how the critical behavior is altered by introducing in the systems a small amount of uncorrelated non-magnetic quenched impurities (dilution). According to the Harris criterion the phase transition of the three-dimensional Ising model is consequently altered, leading to a new universality class (see ref. [14] for a review) to which the weakly diluted Random Ising Model (RIM) belongs. The RIM is a model of spins s_i on a lattice, with nearest-neighbor interaction Hamiltonian $\mathcal{H}_{\mathrm{RIM}} = -\sum_{\langle ij \rangle} \rho_i \rho_j\, s_i s_j$, where ρ_i are uncorrelated quenched random variables such that $\rho_i = 1$ with probability c ($0 < c \leq 1$ is the spin concentration in the lattice), $\rho_i = 0$ with probability $1 - c$. The coarse-grained version of RIM is given

([1]) The results reported in refs. [13, 12] differ significantly from the early estimate of ref. [10].

by the random temperature Landau-Ginzburg Hamiltonian (see ref. [14] for details). The mesoscopic equation describing its purely relaxational dynamics is given by eq. (5) with $N = 1$ (Ising case) and

$$\mathcal{H} \longmapsto \mathcal{H}_\psi[\varphi] = \mathcal{H}[\varphi] + \frac{1}{2}\int \mathrm{d}^d x\, \psi(\mathbf{x})\varphi^2, \tag{8}$$

where $\psi(\mathbf{x})$ is a spatially uncorrelated random field with Gaussian distribution $P[\psi] \propto \exp[-\psi^2/4w]$. Taking advantage of the known RG analysis of this problem, we find that (see ref. [18] for details and specific references)

$$X^\infty = \frac{1}{2} - \frac{1}{4}\sqrt{\frac{6\epsilon}{53}} + O(\epsilon), \tag{9}$$

that, for $\epsilon = 1$ leads to $X^\infty \sim 0.416$. It would be interesting to check weather this one-loop result agrees so nicely with MCs as in the case of the pure model (cf. refs. [16, 17]), since the numerical convergence of the $\sqrt{\epsilon}$-expansion is known to be very slow and poorly effective also at higher loops [14]. To this order it is not even clear whether the dilution affects X^∞ or not. MCs could clarify this point.

In ref. [19] we extended our analysis to a different model of dynamics, in which the order parameter is coupled to a conserved density (Model C of ref. [23]). The mesoscopic equation of motion is given by eq. (5) with

$$\mathcal{H} \longmapsto \mathcal{H}_{\mathrm{C}}[\varphi, \varepsilon] = \mathcal{H}[\varphi] + \int \mathrm{d}^d x \left[\frac{1}{2}\varepsilon^2 + \frac{1}{2}\gamma_0 \varepsilon \varphi^2\right], \tag{10}$$

where $\varepsilon(\mathbf{x}, t)$ is the coarse-grained version of the (non-critical) conserved density and γ_0 a coupling constant. ε evolves according to a conserved dynamics, given by

$$\partial_t \varepsilon(\mathbf{x}, t) = \Omega\rho\nabla^2 \frac{\delta \mathcal{H}_{\mathrm{C}}[\varphi, \varepsilon]}{\delta \varepsilon(\mathbf{x}, t)} + \zeta(\mathbf{x}, t), \tag{11}$$

where ρ is a kinetic coefficient and $\zeta(\mathbf{x}, t)$ is a zero-mean stochastic Gaussian noise with $\langle \zeta(\mathbf{x}, t)\zeta(\mathbf{x}', t')\rangle = -2\rho\Omega\,\nabla^2\delta(\mathbf{x} - \mathbf{x}')\delta(t - t')$. The field-theoretical formulation of the problem follows in a standard way [22]. The influence of initial conditions and RG properties have been discussed in the literature cited in ref. [19], where we report the computation of X^∞ in details. For the physically relevant fixed point (with $N = 1$) we find

$$X^\infty = \frac{1}{2}\left(1 - \frac{\epsilon}{12}\right) + O(\epsilon^2), \tag{12}$$

up to this order it is exactly the same as in Model A; only higher-loop calculations may clarify whether this is only a coincidence at one-loop or whether it is a deeper property.

In sum we applied field-theoretical methods to various models of aging ferromagnetic systems in order to provide analytic predictions for the corresponding fluctuation-dissipation ratios, which have been confirmed by recent MCs. The universality of X^∞ clearly emerges from our computations. Moreover, the prediction for the scaling function F_R (see eq. (3)) coming from LSI disagrees with our findings (see refs. [17,19] for details and discussion). This casts serious doubts about the applicability of LSI to critical phenomena. A numerical evidence of that could be more easily found for Model C, where the disagreement with LSI already emerges at one-loop order (see ref. [19] for details) and should be large enough to be detected in MCs. Finally we remark that $\hat{X}(t,s)$ is a non-trivial function of t and s. The recent numerical observation in ref. [12], that in the two-dimensional Ising-Glauber model $\hat{X}$ is almost constant, is not surprising because corrections to the asymptotic value are small in the case of $O(N)$ Model A [17]. Such time-dependent corrections should be more easily detectable in the cases of RIM and Model C [18,19].

REFERENCES

[1] STRUIK L. C. E., *Physical Aging in Amorphous Polymers and Other Materials* (Elsevier, Amsterdam) 1978.
[2] VINCENT E., HAMMANN J., OCIO M., BOUCHAUD J. P. and CUGLIANDOLO L. F., *Lect. Notes Phys.*, **492** (1997) 184; BOUCHAUD J. P., CUGLIANDOLO L. F., KURCHAN J. and MÉZARD M., in *Spin Glasses and Random Fields, Directions in Condensed Matter Physics*, edited by A. P. YOUNG, Vol. **12** (World Scientific, Singapore) 1998.
[3] CUGLIANDOLO L. F., KURCHAN J. and PARISI G., *J. Phys. I*, **4** (1994) 1641.
[4] GODRÈCHE C. and LUCK J. M., *J. Phys. Cond. Matt.*, **14** (2002) 1589.
[5] BRAY A. J., *Adv. Phys.*, **43** (1994) 357.
[6] CUGLIANDOLO L. F., *Dynamics of glassy systems*, in *Slow Relaxation and Non Equilibrium Dynamics in Condensed Matter*, Les Houches Session 77 (July 2002), edited by J. L. BARRAT, J. DALIBARD, J. KURCHAN and M. V. FEIGEL'MAN, eprint `cond-mat/0210312`.
[7] CUGLIANDOLO L. F. and KURCHAN J., *Phys. Rev. Lett.*, **71** (1993) 173; *J. Phys. A*, **27** (1994) 5749.
[8] For a review see: CRISANTI A. and RITORT F., *J. Phys. A*, **36** (2003) R181.
[9] GODRÈCHE C. and LUCK J. M., *J. Phys. A*, **33** (2000) 1151.
[10] GODRÈCHE C. and LUCK J. M., *J. Phys. A*, **33** (2000) 9141.
[11] LIPPIELLO E. and ZANNETTI M., *Phys. Rev. E*, **61** (2000) 3369.
[12] MAYER P., BERTHIER L., GARRAHAN J. P. and SOLLICH P., *Phys. Rev. E*, **68** (2003) 016116.
[13] CHATELAIN C., *J. Phys. A*, **36** (2003) 10793.
[14] PELISSETTO A. and VICARI E., *Phys. Rep.*, **368** (2002) 549.
[15] JANSSEN H. K., SCHAUB B. and SCHMITTMANN B., *Z. Phys. B*, **73** (1989) 539.
[16] CALABRESE P. and GAMBASSI A., *Phys. Rev. E*, **65** (2002) 066120; *Acta Phys. Slov.*, **52** (2002) 335.
[17] CALABRESE P. and GAMBASSI A., *Phys. Rev. E*, **66** (2002) 066101.
[18] CALABRESE P. and GAMBASSI A., *Phys. Rev. B*, **66** (2002) 212407.
[19] CALABRESE P. and GAMBASSI A., *Phys. Rev. E*, **67** (2003) 036111.
[20] HENKEL M., PLEIMLING M., GODRÈCHE C. and LUCK J. M., *Phys. Rev. Lett.*, **87** (2001) 265701.
[21] HENKEL M., *Nucl. Phys. B*, **641** (2002) 405.
[22] ZINN-JUSTIN J., *Quantum Field Theory and Critical Phenomena*, 3rd edition (Clarendon Press, Oxford) 1996.
[23] HOHENBERG P. C. and HALPERIN B. I., *Rev. Mod. Phys.*, **49** (1977) 435.

Solvation shell of OH^- ions in water

S. Imberti, A. Botti, F. Bruni and M. A. Ricci

Dipartimento di Fisica "E. Amaldi", Università di Roma Tre, and INFM, Unità di Roma Tre Via della Vasca Navale 84, 00146 Roma, Italy

A. K. Soper

ISIS Facility, Rutherford Appleton Laboratory - Chilton, Didcot, Oxon, OX11 0QX, UK Department of Physics and Astronomy, University College London Gower Street, London WC1E 6BT, UK

1. – Introduction

Under normal conditions in pure water a small fraction (10^{-7}) of molecules dissociates spontaneously to form H^+ and OH^- ions; these so-called *water ions* are strongly hydrated and have unusually high mobilities in water. It is well established [1] that the hydrogen bonds (H bonds) in their solvation shells are continuously breaking and re-forming, and this microscopic dynamics is thought to drive the macroscopic charge transport via a process that is known as *structural diffusion*. This means that it is the pattern of hydrogen bonds characterizing the hydration complex that migrate, in a continuous inter-conversion between covalent and H bond, and not its individual constituents. Consequently, the knowledge of the microscopic structure of water molecules around these solutes is of crucial relevance to the understanding of transport dynamics. As far as the hydrated H^+ is concerned, recent *ab initio* studies [2] addressed this question, finding that the hydrated proton is better described as a defect in the hydrogen bond network, delocalized over several bonds. There are two "limiting" structures occurring in the structural diffusion process: initially the defect is localized as an $H_9O_4^+$ structure, composed by an H_3O^+ core that donates three H bonds to neighbouring water molecules; subsequently, due to thermal fluctuations in the second shell, the proton migrates along an H bond and becomes equally shared between two water molecules, substantially forming an $H_5O_2^+$ complex; an $H_9O_4^+$ complex is eventually reformed as the transfer is complete. Conversely, the detailed structure of solvation complexes of the hydroxide ions received far less attention, and this fact may be ascribed to the old idea [3] that this ion can be regarded as a *proton hole* and that the transport mechanism can be inferred from that of the H^+ ion simply "reversing" donor/acceptor H bonds.

This picture provides the existence in water of two specific limiting hydration complexes, $H_7O_4^-$ and $H_3O_2^-$, revealed by spectroscopic experiments and gas-phase mass spectrometry. These findings are in agreement with simulations and can be rationalized in the proton hole picture frame according to Agmon [4], giving a solvation number of three for the non-transient structure $H_7O_4^-$.

Such view is not supported by Car-Parrinello *ab initio* simulations without [5] or with [6] the presence of a counter-ion, neither by recent dielectric relaxation experiments [7], both giving evidence of a higher solvation number.

Here we present the structure of the solvation shell of OH^- ion, as measured in a neutron diffraction experiment, performed at the SANDALS diffractometer [8] installed at the ISIS spallation source. Since the concentration of OH^- ions in water at normal conditions is below the level required for a neutron diffraction study of the ion hydration shell, we have performed the experiment on an electrolytic solution, namely NaOH in water at concentrations of 1 solute per 12 water molecules, *i.e.* both ions are completely hydrated (see also [9, 10]).

2. – Experiment and data reduction

Neutron diffraction experiments with isotopic substitution (NDIS) [11] aim at the determination of the atom pair correlation functions $g_{\alpha\beta}(r)$, which are related to the probability that an atom of the type α is found at a distance r from an atom of the type β. The differential cross-section $\Sigma(Q)$ (Q is the exchanged wave vector) is measured for three samples, NaOH in H_2O, NaOD in D_2O and a mixture (50%) of the previous two; the different scattering lengths of hydrogen and deuterium allow the extraction of three partial structure factors, traditionally referred to as $S_{XX}(Q)$, $S_{XH}(Q)$ and $S_{HH}(Q)$, where H labels the isotopically substituted atoms, while X indicates all non-substituted atoms.

SANDALS [8] is a small-angle diffractometer installed at the spallation neutron source ISIS, Rutherford Appleton Laboratory (UK); it is especially built for investigating the structure of liquids and amorphous materials over a wide range of momentum transfers. The availability of a large number of detectors at low angles makes SANDALS particularly suited for measuring structure factors of systems containing light atoms such as hydrogen and deuterium, in comparison with standard diffractometers installed at reactor sources. It is worth to remind indeed that although neutrons are the principal and still the best spectroscopic probe for studying hydrogen containing systems, nevertheless low diffraction angles and high-energy neutrons are needed to reduce the effect of systematic errors due to inelastic interactions between neutrons and nuclei. The sample can was a standard SANDALS TiZr flat cell, internal width 1 mm. $Ti_{0.67}Zr_{0.33}$ is a special alloy with almost null coherent scattering cross-section and resistent to chemical attack from the solution. The sample thickness is a compromise between good statistic and reduction of multiple scattering and neutron absorption. The acquisition was split into three blocks lasting about 4 hours, to check the stability of the system.

The data reduction was performed following a tested procedure, reported in the ATLAS manual [12], using the routines available at the ISIS computer system to get three composite partial structure factors $S_{XX}(Q)$, $S_{XH}(Q)$ and $S_{HH}(Q)$. The structure factors are related by Fourier transform to the corresponding pair correlation functions $g(r)$ and give complete information, although spherically averaged, in case of molecules formed by two unlike atoms (*e.g.* H_2O). Conversely in the case of hydroxide solutions each pair correlation function is a linear combination of several site-site contributions, with coefficients depending on the relative abundance and scattering length of individual atomic species. This prevent the possibility to get out from the experimentally determined partial structure factors a complete picture of site-site correlations both in terms of distances and orientations. This can be accomplished by the EPSR analysis.

3. – EPSR analysis

The Empirical Potential Structure Refinement (EPSR) code is a powerful method to extract structural information from partial structure factor, recently developed by A. K. Soper [13]. It performs a standard Monte Carlo computer simulation of the system in question, starting from realistic constraints on the intra-molecular geometry and interaction potentials and uses the diffraction data to generate an additional empirical potential in an iterative manner, so that the simulation reproduces the data as close as possible.

As the simulation converges, what is obtained is a set of "real space" configuration of molecules giving a good fit of the experimental structure factors $S_{XX}(Q)$, $S_{XH}(Q)$ and $S_{HH}(Q)$. This allows to recover smooth site-site correlation functions $g_{XX}(r)$, $g_{XH}(r)$ and $g_{HH}(r)$ simultaneously. In addition, a complete structural information is available, because, as in a standard MC simulation, we can calculate each site-site correlation functions, that means 15 partial $g(r)$ given 5 different atoms, *i.e.* Na, O, H, O_w and H_w, where the subscript w indicates water oxygens and hydrogens. Another important information extracted is orientational correlations: given a molecule as OH^-, with at least one preferential axis (in this case the dipole direction), we can collect positions of water molecules around it avoiding angle averaging. This wealth of structural information allows a better comparison between experiments and radial distribution functions extracted from *ab initio* or classical molecular dynamics simulations.

The EPSR simulation was performed on 490 water molecules, 41 Na^+ and 41 OH^- in a cubic box of side 25.41 Å. We have used the SPC/E model as seed potential for the water-water interactions. The Lennard-Jones parameters for the interactions between the individual solutes sites, as well as the fractional charges on the ions that best fit the data are reported in table I. The corresponding parameters for the interactions between different species have been calculated according to the Lorentz-Berthelot rules. The

TABLE I. – *Parameters of the reference potential used in the EPSR code. The subscript w labels the water atoms as opposite to the hydroxyl ones.*

	ϵ/k_{B} [kJ/mol]	σ [Å]	m [a.m.u.]	q/e
Na	0.251	2.75	23	+0.6791
O	0.184	1.44	16	+0.4238
H	0.125	2.50	2	−1.1029
O_w	0.65	3.166	16	−0.8476
H_w	0	0	2	+0.4238

bond length of the hydroxyl ion has been taken equal to 0.98 Å, the same as for O_wH_w intra-molecular bond. After convergence of the fit were recorded 2500 simulation steps.

4. – Results and conclusions

Here we report the significant site-site radial distribution functions for the hydroxyl-water interaction, divided in three groups. The first evidence is that the water-hydroxyl first neighbour distances are shorter than the corresponding water-water distances in all radial distribution functions shown (compare the straight line with dashed one in all three graphics); this is a signature of the strength of the interactions.

Calculation from the configuration of molecules gives 3.9±0.3 first neighbouring water molecules per hydroxyl oxygen. Hence there are about four water molecules pointing their hydrogens toward the hydroxyl oxygen. This sharp peak signals the formation of hydrogen bonds much shorter than those formed between water molecules and hence very close to linearity. On the contrary hydroxyl hydrogens have their first neighbour water oxygens at 2.5 Å, *i.e.* at about the same distance as the second neighbors of the hydroxyl (compare $g_{\mathrm{HOw}}(r)$ and $g_{\mathrm{OHw}}(r)$ functions in fig. 1). This latter peak is quite broad, with a minimum approach distance of 1.78 Å; since the signature of strong H bonds (namely a well-resolved peak between 1.5 and 1.9 Å) is missing in the $g_{\mathrm{HOw}}(r)$ function, the hydroxyl hydrogen seems to form only very weak H bonds, if any. These findings can be visualized by looking at the three dimensional spatial distribution functions of water around the hydroxyl ion.

In fig. 2 the first two panels from the left show the isosurfaces of the $g_{\mathrm{OOw}}(r,\mathbf{\Omega})$ spatial distribution function, namely the probability of finding a water molecule, whatever oriented, at distance $\vec{r}$ from a hydroxyl ion (that is without averaging over Euler angles), where the hydroxyl dipole moment is oriented along the vertical axis with the oxygen in the origin of reference frame. The first panel shows the region between 2.0 Å and 2.75 Å, where four water molecules are sitting. These are hydrogen bonded with the OH^- ion to form the $H_9O_5^-$ complex, as demonstrated by the presence of a clear sharp peak in the $g_{\mathrm{OHw}}(r)$ radial distribution function. The characteristic cup shape of the spatial distribution function identified here suggests that the about four hydrogen bonds cannot be coplanar. The second panel shows the region between 2.75 Å and 3.2 Å, where

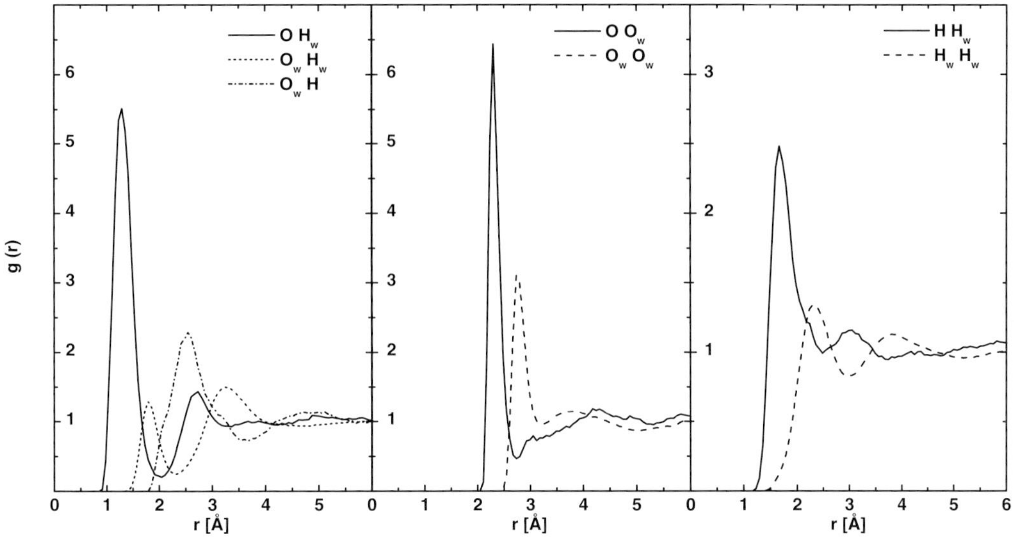

Fig. 1. – Site-site radial distribution function, the subscript w indicates water hydrogens and oxygens with respect to the hydroxyl ones; intra-molecular peaks are not shown.

a fifth water molecule is found in the vicinity of the hydroxyl hydrogen. In the third panel on the right-hand side of this figure we report the spatial distribution function for the direction of dipole moment of the fifth water molecule, with respect to the OH^- dipole direction. The anisotropy of this function suggests that, although this molecule lies too far away from the hydroxyl hydrogen to form hydrogen bonds, nevertheless its orienta-

Fig. 2. – The spatial distribution functions for water molecules whatever oriented around a central hydroxyl ion; reference frame fixed on the OH^- as shown. Left: water molecules exceeding the average value per more that 60% selected in the radial range from 2.0 Å to 2.75 Å; middle: the same in the range from 2.75 Å to 3.2 Å, suggesting the presence of a fifth neighbouring water molecule; right: in the range from 2.75 Å to 3.2 Å the distribution function for the orientation of the dipole moment of water molecules, *i.e.* of the fifth molecule.

tion is correlated to that of the $H_9O_5^-$ complex; as a consequence it can be considered as weakly bound to it.

The picture that emerges from the above analysis is in good agreement with the results of *ab initio* simulations [2,5] concerning the structure of the complex $H_9O_5^-$, formed upon hydration of OH^-. In addition it has to be mentioned that we have clear evidence from the analysis of the orientational correlations of the presence of a weakly bound fifth molecule close to the hydrogen atom: this molecule can play a significant role in the proton transport mechanism as envisaged by *ab initio* simulations [5]. Finally our data do not reveal the presence of other complexes evidenced by the simulations, suggesting that these might be formed only transiently, to trigger the proton transport process.

REFERENCES

[1] see for example EISENBERG D. and KAUZMANN W., *The Structure and Properties of Water* (Oxford University Press) 1969, p. 225.

[2] MARX D., TUCKERMANN M. E., HUTTER J. and PARRINELLO M., *Nature*, **397** (1999) 601 and references therein.

[3] HUCKEL E., *Z. Elektrochem.*, **34** (1928) 546.

[4] AGMON N., *Chem. Phys. Lett.*, **319** (2000) 247.

[5] TUCKERMANN M. E., MARX D. and PARRINELLO M., *Nature*, **417** (2002) 925.

[6] CHEN B., IVANOV I., PARK J. M., PARRINELLO M. and KLEIN M. L., *J. Phys. Chem.*, **106** (2000) 12006.

[7] BUCHNER R., HEFTER G., MAY P. M. and SIPOS P., *J. Phys. Chem. B*, **103** (1999) 11186.

[8] SOPER A. K., *Proceedings of the Conference on Advanced Neutron Sources 1988* in *D. K. Hyer, IOP Conf. Proc. no. 97* (Institute of Physics and Physical Society, London) 1989, p. 353; detailed information on the SANDALS diffractometer can be also found at the web site: `www.isis.rl.ac.uk`.

[9] BOTTI A., BRUNI F., IMBERTI S., RICCI M. A. and SOPER A. K., *J. Chem. Phys.*, **119** (2003) 5001.

[10] BOTTI A., BRUNI F., IMBERTI S., RICCI M. A. and SOPER A. K., *J. Chem. Phys.*, **120** (2004) 10154.

[11] BRUNI F., RICCI M. A. and SOPER A. K., *J. Chem. Phys.*, **114** (2001) 8056.

[12] SOPER A. K., HOWELLS W. S., HANNON A. C., TURNER J. Z. and BOWRON D. T., ATLAS 2.0, Analysis of time-of-flight diffractometer data from liquid and amorphous samples, R.A.L. Report No. RAL-89.046 (1989).

[13] SOPER A. K., *Chem. Phys.*, **202** (1996) 295; *Mol. Phys.*, **99** (2001) 1503.

Predicting protein folding pathways from Mean First-Passage Time analysis

M. A. Micheelsen

Complexity Laboratory, Niels Bohr Institute, University of Copenhagen
Blegdamsvej 17, DK-2100 København Ø, Denmark

1. – Introduction

It was long thought that proteins fold via some uniquely defined pathway, but during the last decade it has become widely accepted that proteins fold via diffusion on complex energy landscapes, ref. [1, 2], where the transition state is thought to consist of many different states. This makes protein folding a far more complex problem, and raises questions about the nature of the surface.

Proteins are involved in a wide range of cell activities like cell structure, signalling, regulatory functions and enzymatic catalysis. They are also responsible for the conversion of chemical energy into mechanical energy in muscles. The crucial factor for proteins to be able to perform these functions is the unique 3-dimensional structure called the native state that is specific to each protein. The failure of the proteins to fold into the correct native state can be fatal and lead to diseases such as alzheimer, the prion disease and cancer. Thus we are very interested in knowing what drives the folding of proteins.

With the increasing computational power, simulations have become an invaluable tool to analyze the folding process. The most accurate approach is provided by Molecular Dynamics (MD) —the direct solution of the equation of motion. However MD simulations are severely limited by the computational time required to probe even small timescales of the dynamics. The main strategy to improve this situation is to increase the elementary time step by tracing out some of the degrees of freedom. Experimental studies of small two-state proteins show that the transition state has a near native topology, ref. [3, 4]. This is used to exclude non-native bonds that are considered less likely, and thereby reduce the possible conformations of the residues or bonds to be either native or unfolded. This reduces the state space to 2^N conformations, where N is the length of the protein. The latter approach is used in this proceeding.

One way of getting information about the kinetics of a biological process is using the reaction rate, but since a complex energy landscape does not present us with a natural way of calculating this parameter that is defined for two-state kinetics, in this proceeding we look at the inverse rate, namely the "Mean-First Passage Time" that can be calculated for each configuration, ref. [5], to investigate the folding pathway of small two-state proteins. We have previously presented this method in ref. [6].

2. – Theory

2·1. *Fold-X.* – Presently there is no known force field that can be applied to a known sequence of amino acids and correctly predict its native state. Instead we work with the Fold-X force field, ref. [3], that has been calibrated against a large set of point mutations, ref. [7].

Fold-X, like previous similar energy models, is based on the experimental evidence that the folding rate of small two-state proteins is highly correlated to the native topology. Topology has been shown to account for up to 3/4 of the folding rate for two-state proteins, ref. [8]. On the other hand, proteins of similar topology have shown differences in the reaction rate of up to 2 orders of magnitude, ref. [9], and Φ-value analysis, ref. [10], demonstrates that some bonds are formed earlier than others in the folding process. These two parts show the necessity of including specific energy terms of the amino acid interactions. In Fold-X these are hydrogen bonds, van der Waals interactions, solvation effects, water bridges, electrostatic and entropy effects for both the side chains, the backbone and loops. The total free energy for a protein state is thus the sum over all interactions and entropy cost regarding the terms mentioned:

$$G = W_{\mathrm{vdw}} G_{\mathrm{vdw}} + W_{\mathrm{solvH}} G_{\mathrm{solvH}} + G_{\mathrm{hbond}} + G_{\mathrm{el}} + T\big(W_{\mathrm{mc}} S_{\mathrm{mc}} + W_{\mathrm{sc}} S_{\mathrm{sc}} + S_{\mathrm{loop}}\big),$$

where W are the weights for the various terms. (See ref. [11] for details.)

2·2. *Mean First Passage Time.* – As previously defined in ref. [6] the equation for MFPT $\tau_i(n)$ of going from a state i to another state n is given in matrix form by

$$[T(n) - 1]\tau(n) = -\beta, \tag{2.1}$$

with $T(n)$ and $\beta(n)$ defined by

$$T_{ij}(n) = \begin{cases} P(i \to j), & i \neq n, \\ 0, & i = n, \end{cases} \qquad \beta_i(n) = \begin{cases} \Delta t, & i \neq n, \\ 0, & i = n. \end{cases} \tag{2.2}$$

$P(i \to j)$ is the probability of moving from state i to state j given by $P(i \to j) = \max(M_i, M_j)^{-1} \max(1, \exp[-(G_j - G_i)/k_{\mathrm{B}}T])$ where M_i and M_j are the number of moves available from state i and j, respectively. This equation obeys detailed balance. The probability for remaining is $P(i \rightarrow i) = 1 - \sum_{j \neq i} P(i \rightarrow j)$. See ref. [6] for a more detailed review.

2˙3. *Simulation.* – The state space of a protein with N residues and two conformations (native and unfolded) counts 2^N states, so to make the MFPT calculations tractable for each residue, we only allow a maximum of 2 continuous folded regions. This still captures the essential thermodynamics of small single-domain proteins as shown by unrestricted Monte-Carlo–based sampling, ref. [12].

The matrix equation was solved using the `linbcg` sparse matrix routine, ref. [13], and the data presented in a condensed form:

$$\Gamma(\nu, \tau) = \sum_j \exp\left[-\frac{G_j}{k_{\mathrm{B}}T}\right] \delta(\tau - \tau_j)\delta(\nu - \nu_j), \tag{2.3}$$

where δ is the Kronecker delta, and ν is the number of ordered residues.

We also present the probability of a residue i being folded along the reaction coordinate ν:

$$P(\nu, i) = \sum_j \exp\left[-\frac{G_j}{k_{\mathrm{B}}T}\right] \delta(i)\delta(\nu - \nu_j), \tag{2.4}$$

where $\delta(i)$ tells weather the i-th residue is folded or unfolded.

Finally we present the free energy in the form

$$G_{\mathrm{tot}}(\nu) = -k_{\mathrm{B}}T \ln\left[\sum_\tau \Gamma(\nu, \tau)\right]. \tag{2.5}$$

3. – Results

Several small two-state proteins have been analyzed, the results for Ci2 are presented in detail in this proceeding. For the analysis of the SH3 domain α-spectrin, see ref. [6].

3˙1. *Ci2.* – Chymotrypsin inhibitor 2 (Ci2) (Protein Data Bank file 2ci2, ref. [14]) is a single module of structure and a true two-state folder. Its major structural elements are: one α-helix, 4 β-strands forming one β-sheet and a large reactive loop. The free energy profile shows the transition state is around 42 ordered residues.

According to Φ-value analysis, ref. [15], the α-helix (residue 31-44) and the β-strand running form residue 47 to 53 are the most stable elements in the transition state. These are also present in the transition state of our model as seen in upper panel, left, fig. 1. The last structural element to be formed is the reactive loop (55-62) according to our model, which is confirmed by its low Φ-values, ref. [15], and expected due to the necessity for a reactive site to be able to change conformation without affecting the structure of the rest of the protein.

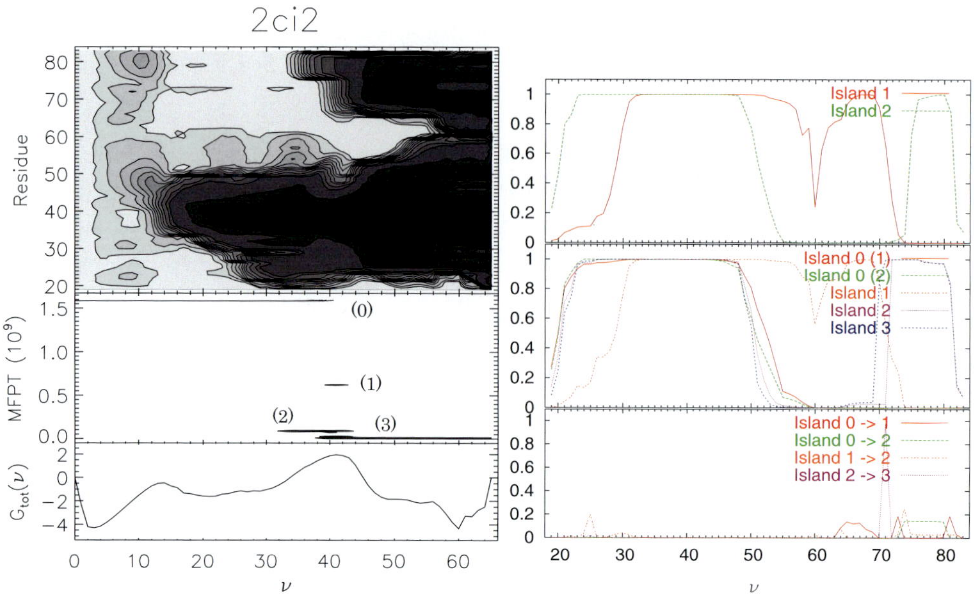

Fig. 1. – Results for the MFPT analysis on Ci2. Left: the upper panel shows the probability of a residue being folded as a function of order parameter ν see eq. (2.4). The middle panel shows the Boltzmann-weighted mean first-passage times see eq. (2.3). The lower panel shows the free energy profile eq. (2.5). Right: the top plot shows the Boltzmann-weighed distribution of folded residues in the 2 islands seen in the middle panel, left. The middle plot shows the Boltzmann-weighed probability that a given residue is folded before a rate limiting step. The lower plot shows in which residues the change happen.

Our analysis clearly shows two parallel folding pathways. Before the first rate limiting step the β-strand running from residue 23-47 as well as the α-helix (shown as dark blue in fig. 2(a)) are structured. Then the folding can proceed in either of two ways with almost equal probability:

1) The structural parts of the protein indicated in red in fig. 2(a) folds to the structure shown in fig. 2(b). This corresponds to the jump to island (1) in middle panel, left, fig. 1. From here it jumps to island (2) by folding the structure shown in purple in fig. 2(b).

2) In the second path the rate limiting step is to jump straight to island (2) or structure (c) in fig. 2. The structural parts that sets of this step is the C-terminus β-strand indicated in green in fig. 2(a).

From island (2) the last rate limiting step is taken folding the orange structural part shown in fig. 2(c).

Both island (1) and island (2) have broad overlaps in MFPT, this happens because structures with a larger ν value are less progressed in the folding process than other structures with smaller ν value. This brings about the question of the validity of ν as the reaction coordinate for protein folding. In our previous work, ref. [6], we tried to improve on this coordinate with some success. The overlaps are caused by orthogonal

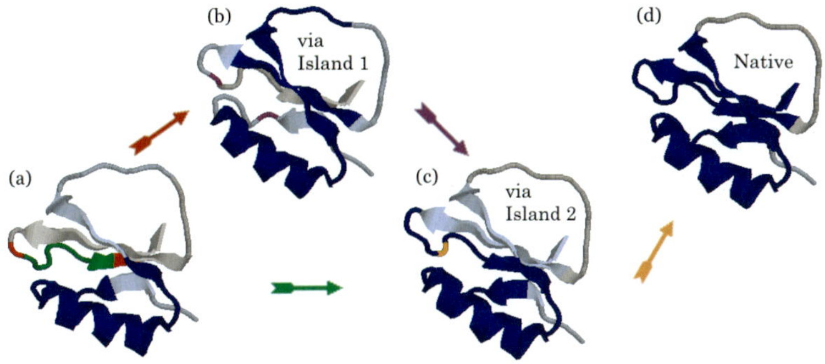

Fig. 2. – The folding pathway of Ci2. The pathway is drawn according to the plots in fig. 1, left. Dark blue indicates the residues that are folded at a given time step corresponding to the residues that are folded with probability 1 in the middle plot of fig. 1, left, while the light blue indicates structures that rapidly fold and unfold (probability of less than 1 in fig. 1). Other colors indicate what residue is changed during a given rate-limiting step, and are color-coordinated in accordance with the lower plot, right in fig. 1.

movements, residues that rapidly fold and unfold at a given time without affecting the overall folding progress.

4. – Discussion

The MFPT model reproduces the experimentally found features of the folding properties for small two-state proteins. It also reveals parallel pathways and kinetic traps (for the latter see ref. [6]).

We analyzed some interesting properties, I shall discuss briefly here:

For all the proteins we used the MFPT approach on we found a strong correlation between the transition state (or rate-limiting steps) and the beginning of formation of non-local bonds. This is again in agreement with the results that topology is a determining factor for the protein folding rate. Although there are some biasing assumptions in this analysis based on this topology dependence on the folding path and rate, the specific energy terms included in Fold-X makes this result nontrivial.

We tried to see if we could find a 1:1 connection between the experimentally found reaction rates and our MFPT from a subset of known two-state folder, ref. [16]. Fitting the data points (MFPT, k) using the function: $\ln(k) = -\ln(\text{MFPT}) + a$ gives $a = 23.73 \pm 1.17$(1). The value of a corresponds to a rate of $10^{9.8}\,\text{s}^{-1}$ for each elementary folding step. This rate should be compared with the attempt rate for bond dissociation

(1) However, a test to see if the 1:1 connection held via the function: $\ln(k) = -b\ln(\text{MFPT}) + c$ gave a much worse result with: $d = 9.45 \pm 6.49$ and $c = 0.28 \pm 0.32$.

which is of the order $\simeq 10^8\,\mathrm{s}^{-1}$, ref. [8] (see below). Previous attempts to calculate reaction rates, ref. [17], have proven quite successful after adjusting the difference in native and nonnative energy to experimental values. This would correspond to a fine tuning of the force field and native structure, and we believe this is the way to improve the model prediction for absolute reaction rates.

We are aware of the restriction the biased state space as well as the residue rather than bond sampling puts on our result. But the agreement with experimental results and the degree of detailed information available though this simple analysis is remarkable. In conclusion, we have demonstrated the use of MFPT calculation for characterizing protein folding pathways. We find it interesting that the simple two-state folders we have analyzed in our present and previous study, ref. [6], both reveals the presence of parallel pathways, and we envisage that this is a generic property of proteins.

* * *

The generous help of JESPER FERKINGHOFF-BORG, with the calculations as well as with this paper and the collaboration with JESPER and CHRISTIAN RISCHEL as my supervisor in developing this method is gratefully acknowledged.

REFERENCES

[1] ONUCHIC J. N., LUTHEY-SCHULTEN Z. and WOLYNES P. G., *Ann. Rev. Phys. Chem.*, **48** (1997) 545.
[2] ONUCHIC J. N., NYMEYER H., GARCIA A. E., CHAHINE J. and SOCCI N. D., *Adv. Prot. Chem.*, **53** (2000) 87.
[3] GUEROIS R. and SERRANO L., *J. Mol. Biol.*, **304** (2000) 967.
[4] ALM E. and BAKER D., *Proc. Natl. Acad. Sci. USA*, **96** (1999) 11305.
[5] PONTRYAGIN L., ANDRONOV A. and VITT A., *Zh. Eksp. Teor. Fiz.*, **3** (1933) 165.
[6] MICHEELSEN M. A., RISCHEL C., FERKINGHOFF-BORG J., GUEROIS R. and SERRANO L., *Europhys. Lett.*, **61** (2003) 561.
[7] GUEROIS R., NIELSEN J. E. and SERRANO L., *J. Mol. Biol.*, **320** (2002) 369.
[8] MAKAROV D. E. and PLAXCO K. W., *Protein Science*, **12** (2003) 17.
[9] FERSHT A. R., *Proc. Natl. Acad. Sci. USA*, **97** (2000) 1525.
[10] TERNSTRÖM T., MAYOR U., AKKE M. and OLIVEBERG M., *Proc. Natl. Acad. Sci. USA*, **96** (1999) 14854.
[11] COLONNA-CESARI F. and SANDER S., *Biophys. J.*, **57** (1990) 1103.
[12] FERKINGHOFF-BORG J., private communication.
[13] PRESS W. H., TEUKOLSKY S. A., VETTERLING W. T. and FLANNERY B. P., *Numerical Recipies in C* (Cambridge University Press) 1992.
[14] MCPHALEN C. A. and JAMES M. N., *Biochemistry*, **26** (1987) 261.
[15] ITZHAKE L. S., OTZEN D. E. and FERSHT A. R., *J. Mol. Biol.*, **254** (1995) 260.
[16] GALZITSKAYA O. V., GARBUZYNSKIY S. O., IVANKOV D. N. and FINKELSTEIN A. V., *Proteins: Struc., Func. & Genetics.*, **51** (2003) 162.
[17] MUÑOZ V. and EATON W. A., *Proc. Natl. Acad. Sci. USA*, **96** (1999) 11311.

Dynamic density functional study of driven colloidal particles: The effect of the system dimension

F. PENNA and P. TARAZONA

Departamento de Física Teórica de la Materia Condensada
Universidad Autónoma de Madrid, E-28049 Madrid, Spain

The dynamical properties of classical fluids at pico-liter scale attract experimentally and theoretically much attention in the soft-matter and biophysics communities, due to the appearance of the microfluidics, ref. [1], also called "lab-on-a-chip", ref. [2], as one of the most relevant 21st century technologies. Theses devices try to put a full chemical laboratory onto your hand through a sophisticated dispositive that includes particles pump, ref. [3], and gates, chemical reaction, ref. [4], etc. In this work we focus our attention on one of the key factors for the design and construction of this type of devices: *the system dimension*. We consider a generic system formed by a dilute solution of colloidal particles dragged (at a constant rate c in the $\hat{\mathbf{z}}$-direction) by a moving potential barrier, modeled by a time-dependent external potential ($V_{\text{ext}}(\mathbf{r},t) = V_{\text{ext}}(\mathbf{r}-ct\hat{\mathbf{z}})$) acting on the colloidal particles but with no effect on the solvent. We base our results on a new technique named Dynamic Density Functional (DDF) theory, ref. [5, 6], which is a generalization of Density Functional (DF) theory to out of equilibrium, for systems with relaxative molecule dynamics (those in which hydrodynamic modes are not relevant), under the hypothesis that the dynamic correlations can be approximated by those in a system with the same density distribution at equilibrium; the DDF is a promising way of describing classical fluids dynamics at a microscopic level including our knowledge of the very well developed DF theory for the equilibrium.

The central deterministic DDF equation is

$$\frac{\partial \rho(\mathbf{r},t)}{\partial t} = \Gamma_0 \nabla \left[\rho(\mathbf{r},t) \nabla \left(\frac{\delta \mathcal{F}[\rho]}{\delta \rho(\mathbf{r},t)} + V_{\text{ext}}(\mathbf{r},t) \right) \right]; \tag{1}$$

where $\mathcal{F}$ is the Helmholtz free energy in DF theory, usually split into the ideal-gas contribution and the excess due to the interactions ($\Delta\mathcal{F}$), and Γ_0 is the mobility of an isolated colloidal particle in the solvent.

In the case of the shifting external potential we are dealing with, $V_{\text{ext}}(\mathbf{r} - ct\hat{\mathbf{z}}) \equiv V_{\text{ext}}(\mathbf{r}')$ (where $\mathbf{r}'$ is the coordinate in the reference framework of the external potential), the most relevant feature comes from the stationary state (achieved after a transient period from any initial state) in which $\rho(\mathbf{r}, t) = \rho(\mathbf{r} - ct\hat{\mathbf{z}}) \equiv \rho(\mathbf{r}')$, *i.e.* when the density distribution is shifted at the same rate c as the potential barrier, so that, eq. (1) reduces to

$$\nabla\left[\rho(\mathbf{r}')\,\nabla\left(\frac{\delta\mathcal{F}[\rho]}{\delta\rho(\mathbf{r}')} + V_{\text{ext}}(\mathbf{r}') + \frac{cz'}{\Gamma_0}\right)\right] = 0. \tag{2}$$

For non-interacting particles, such that $\Delta\mathcal{F} = 0$, eq. (2) becomes a linear Fokker-Planck equation $\nabla^2\rho + \nabla(\rho\cdot\nabla\beta V_k) = 0$, with a "kinetic potential" given by $\beta V_k(\mathbf{r}') = \beta V_{\text{ext}}(\mathbf{r}') + \bar{c}z'$, as a function of the reduced shifting rate $\bar{c} \equiv \beta c/\Gamma_0$, with inverse length units and $\beta = 1/k_{\text{b}}T$. In this case, the effective dimension behaviour is completely determined by the difference between the system dimension (container) and the dimensions over which the external potential is homogeneous, so that an effective 1D system got through a 2D plane barrier dragging particles in a 3D container, is equivalent to a point-like barrier dragging particles in a narrow channel, ref. [7]; similarly, an effective 2D system may be observed for a cylindrical barrier in a 3D container, or with a point-like barrier in a 2D container. For $\Delta\mathcal{F} \neq 0$ (interacting particles), ref. [8-11], this is no longer true, because the excess free energy depends explicitly on the dimension of the colloidal particles container.

Since the system dynamic behaviour is determined mainly through the particles interaction and the effective system dimension, and we are interested here in the characteristics induced by this last factor, we consider in this work only non-interacting particles which would represent diluted solutions of colloidal particles. We set the notation before we start solving eq. (2) for each dimension. In the following, we refer to the *front* (*wake*) region as that away from the external potential, well beyond (behind) the z' positions with $V_{\text{ext}} \neq 0$; if we assume that this external potential barrier is located around the origin of the shifting reference framework, and restricted to values $|z'| \approx \sigma$ with σ the molecular diameter, then the points with $z' \gg \sigma$ are in the *front* while those with $z' \ll -\sigma$ are in the *wake*. In both regions $V_{\text{ext}} = 0$, eq. (2) reduces to $\nabla^2\rho + \bar{c}\,(\partial\rho/\partial z') = 0$, and the analytic form of $\rho(\mathbf{r}')$ may be obtained as a series with the appropriate asymptotic limit in each dimension. Notice that only for effective 1D systems, the *front* and the *wake* are fully separated by the external potential. In higher dimensions, they are connected through those regions with $|z'| \leq \sigma$ but $|\mathbf{r}'| \gg \sigma$, for which $V_{\text{ext}} = 0$. Our results are presented as follow: first we solve analytically eq. (2) (where $V_{\text{ext}} = 0$), then we find numerically the full solution, and finally we compare both.

For a shifting repulsive barrier in effective 1D systems, ref. [8], the solution of eq. (2) is given by $\rho(z') = \rho_0 + A\exp[-\bar{c}z']$ at the *front*, where A/ρ_0 is a constant, fixed by the nature of the barrier and the boundary conditions, and by $\rho(z') = \rho_0$, without any kind of structure, in the *wake* region. It may be proved by a formal reduction of the stationary DDF equation with $\Delta\mathcal{F} = 0$ in any dimension, that this exponential decay behaviour at the *front*, and this absence of excess molecules in the *wake*, are generic characteristics of the total excess density distribution, under constant bulk boundary conditions ($\lim_{|\mathbf{r}'|\to\infty} \rho = \rho_0$), for the integrated excess density distribution across the transverse directions, *i.e.* $(\Delta n(z'))/\rho_0 = (1/\sigma)\int \mathrm{d}x(\rho(\mathbf{r}')/\rho_0 - 1)$ in effective 2D systems, and $(\Delta N(z'))/\rho_0 = (2\pi/\sigma^2)\int \mathrm{D}rR(\rho(\mathbf{r}')/\rho_0 - 1)$ in effective 3D systems.

In effective 2D systems, for a shifting cylindric external potential (*e.g.* the dielectric potential created by a laser beam through a bulk solution), eq. (2), for the region with $V_{\text{ext}}(\mathbf{r}') = 0$, reduces to

$$\frac{\partial^2 \rho}{\partial x^2} + \frac{\partial^2 \rho}{\partial z'^2} + \bar{c}\frac{\partial \rho}{\partial z'} = 0. \tag{3}$$

Through a Fourier transform, the solution is given by

$$\rho(x, z') = \rho_0 + \int_0^\infty \mathrm{d}\alpha \cos[\alpha x]\, e^{-\beta z'} g(\alpha), \tag{4}$$

where

$$\beta_\pm = \frac{\bar{c}}{2} \pm \sqrt{\left(\frac{\bar{c}}{2}\right)^2 + \alpha^2}\,. \tag{5}$$

Far away from the external potential, a fast convergence of the Fourier components $g(\alpha)$ is observed, and the relevant features come from their behavior for $\alpha \ll \bar{c}$. Hence we may use $\beta_+ \approx \bar{c} + \alpha^2/\bar{c}$ and $\beta_- \approx -\alpha^2/\bar{c}$, for the *front* and *wake* regions, respectively. The expansion of $g_\pm(\alpha)$ as an even polynomial function for small α, and the zero *wake* requirement, lead in the *wake* to $g_-(\alpha) \approx A_1\alpha^2 + A_2\alpha^4 + \cdots$, while at the *front* we expect $g_+(\alpha) \approx B_0 + B_1\alpha^2 + B_2\alpha^4 + \cdots$. Thus,

$$\rho(x, z') \approx \rho_0 + \sqrt{\pi} e^{-\bar{c}z'} e^{-a^2/4}\left[B_0\frac{w}{2} + B_1\frac{w^3}{8}(a^2 - 2) + B_2\frac{w^5}{32}(a^4 - 12a^2 + 12) + \cdots\right], \tag{6}$$

for $z' \gg \sigma$, and

$$\rho(x, z') \approx \rho_0 + \sqrt{\pi} e^{-a^2/4}\left[A_1\frac{w^3}{8}(a^2 - 2) + A_2\frac{w^5}{32}(a^4 - 12a^2 + 12) + \cdots\right], \tag{7}$$

for $z' \ll -\sigma$; where $w = \sqrt{\bar{c}/\mid z' \mid}$ and $a = wx$.

In effective 3D system, for a shifting spherical external potential (*e.g.* a driven colloid particle perturbing a bulk solution of others colloids), we use cylindrical coordinates $\mathbf{r}' = (R, \phi, z')$, and eq. (2), for the region with $V_{\text{ext}}(\mathbf{r}') = 0$, reduces to

$$\frac{1}{R}\frac{\partial}{\partial R}\left(R\frac{\partial \rho(\mathbf{r}')}{\partial R}\right) + \frac{\partial^2 \rho(\mathbf{r}')}{\partial z'^2} + \bar{c}\frac{\partial \rho(\mathbf{r}')}{\partial z'} = 0. \tag{8}$$

Through a Hankel transform, the solution is given by

$$\rho(R, z') = \rho_0 + \int_0^\infty d\alpha\, \alpha\, f(\alpha)\, J_0[\alpha\, R]\, e^{-\beta\, z'}, \tag{9}$$

where J_0 is the zeroth-order Bessel function. Proceeding with the Hankel components as in the effective 2D with the Fourier ones, we might take $f_-(\alpha) \approx C_1\alpha^2 + C_2\alpha^4 + \cdots$ in the *wake*, and $f_+(\alpha) \approx D_0 + D_1\alpha^2 + D_2\alpha^4 + \cdots$ at the *front*. Thus,

$$\rho(R, z') \approx \rho_0 + e^{-\bar{c}z'} e^{-h^2/4}\left[D_0\frac{w^2}{2} + D_1\frac{w^4}{8}(h^2-4) + D_2\frac{w^6}{32}(h^4-16h^2+32) + \cdots\right], \tag{10}$$

for $z' \gg \sigma$, and

$$\rho(R, z') \approx \rho_0 + e^{-h^2/4}\left[C_1\frac{w^4}{8}(h^2-4) + C_2\frac{w^6}{32}(h^4-16h^2+32) + \cdots\right]; \tag{11}$$

for $z' \ll -\sigma$; where $h = wR$.

The coefficients $(A, B, C$ and $D)$ are determined by the particular choice of the external potential but, for each dimension, the asymptotic decay forms are generic. In the transverse plane (for fixed z') the excess stationary density distribution $(\rho(\mathbf{r}') - \rho_0)$ is given by a Gaussian times a polynomial function, and along the direction of the movement (z'-axis) for a fixed distance to the origin ($a =$ const in effective 2D systems and $h =$ const in the 3D) is given, at the *front*, by an exponential decay ($\exp[-\bar{c}z']$) and an inverse power in the *wake*. Over the transverse directions, at the *front*, the maximum density is at the z'-axis, while in the *wake*, this axis is a minimum between two symmetric maxima which produce a cap-like structure with parabolic shape. The distance between the maximum and the z'-axis is $2\sqrt{2 \mid z' \mid /\bar{c}\sigma}$ in effective 3D systems and $\sqrt{6 \mid z' \mid /\bar{c}\sigma}$ in the 2D one. In fig. 1 we present the numerical solution for the stationary density distribution of colloidal particles under the influence of an external potential $V_{\text{ext}} = V_0 e^{-|\mathbf{r}'/\sigma|^6}$, with $\beta V_0 = 10$, shifting at the velocity $\bar{c}\sigma = 10$, in a) an effective 3D system and b) a 2D one. The clear cap-like structure, extended further away from the external potential, is clearly observed.

We have mentioned before that only in effective 1D systems the *front* and *wake* are fully separated by the external potential, which implies that as the solution of eq. (2), the *front* and *wake* should share a unique analytic expression in effective dimensions beyond 1D. This unique expression produces, at the *front* (*wake*), an exponential growth of the $\beta_+(\beta_-)$ contribution, which only by the appropriate behaviour of $g(\alpha)(f(\alpha))$ for large α be canceled. We have compared our analytical prediction with the numerical solution of eq. (2), the concordance reveal a local asymptotic convergence which makes those asymptotic expressions useful (eqs. (6), (7), (10), (11)), to represent separately (at the *front* and in the *wake* regions) the stationary state. The use of polar (spherical) coordinates, instead of the rectangular (cylindric) ones, would not solve the problem,

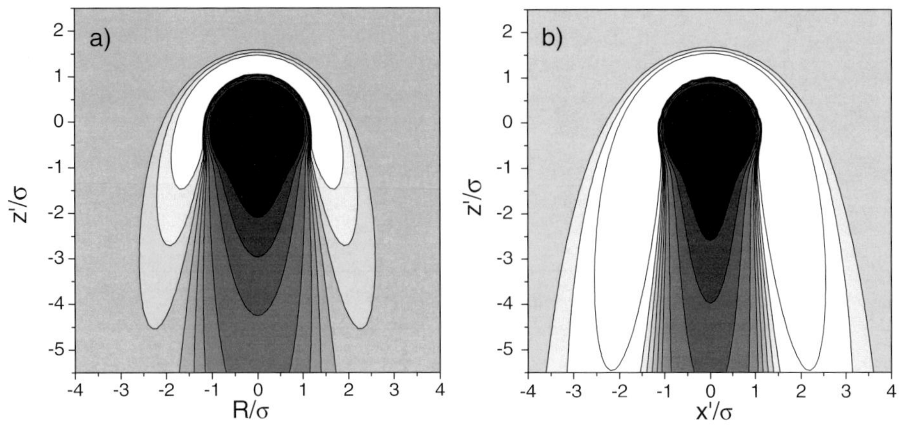

Fig. 1. – Steady-state contour density field of non-interacting colloidal particles a) for an effective 3D system and b) for an effective 2D system, dragging at constant velocity $\sigma\bar{c} = 10$ by an external potential $V_{\text{ext}} = V_0 e^{-|\mathbf{r}'/\sigma|^6}$ with $\beta V_0 = 10$. Dark regions correspond to low densities while the bright one shows high densities.

because the parabolic structure of the *wake* implies the entanglement of the radial and the angular coordinates.

Since this work may be useful in the construction of particles pumps, we want to point out some effects to be considered in the design of an efficient device to drag particles. For the same radial form of the external potential it is clear that an effective 2D system is more efficient dragging particles than a 3D one, because it produces a higher density at the *front* peak. In fig. 2 we present the numerical stationary density distribution of an effective 3D (dashed and dotted lines) and a 2D system (solid and dot-dashed lines), for different fixed values of x and R. The higher *front* density peak in the effective 2D system compared with the 3D one is clearly observed. For the same reason, an effective 1D system is more efficient than the 2D one.

For an effective 1D system we obtained in our previous work, ref. [8], a non-linear dependence of the total current ($j = \bar{c} \int (\rho(\mathbf{r}') - \rho_0)\mathrm{d}\mathbf{r}'$) on the drift velocity. If we denote by $\bar{c}_{\max}$, the velocity at which the total current is maximum for a given external potential, the efficiency of a device as a particles pump increases with $\bar{c}$ until $\bar{c}_{\max}$ is reached, and then diminishes for higher $\bar{c}$. As we have mentioned before, the integrated excess of the stationary density distribution across the transverse directions, Δn in effective 2D systems and ΔN in the 3D one, satisfies the same stationary DDF equation as the 1D density distribution, so we may expect for the total current this non-linear behaviour too. To support numerically our analytical result in fig. 3, we present the integrated excess of the stationary density distribution, a) in an effective 2D system and b) in a 3D one, the exponential decay behaviour at the *front* and the absence of structure in the *wake* as for the 1D stationary density distribution is clearly observed. We have estimated that the

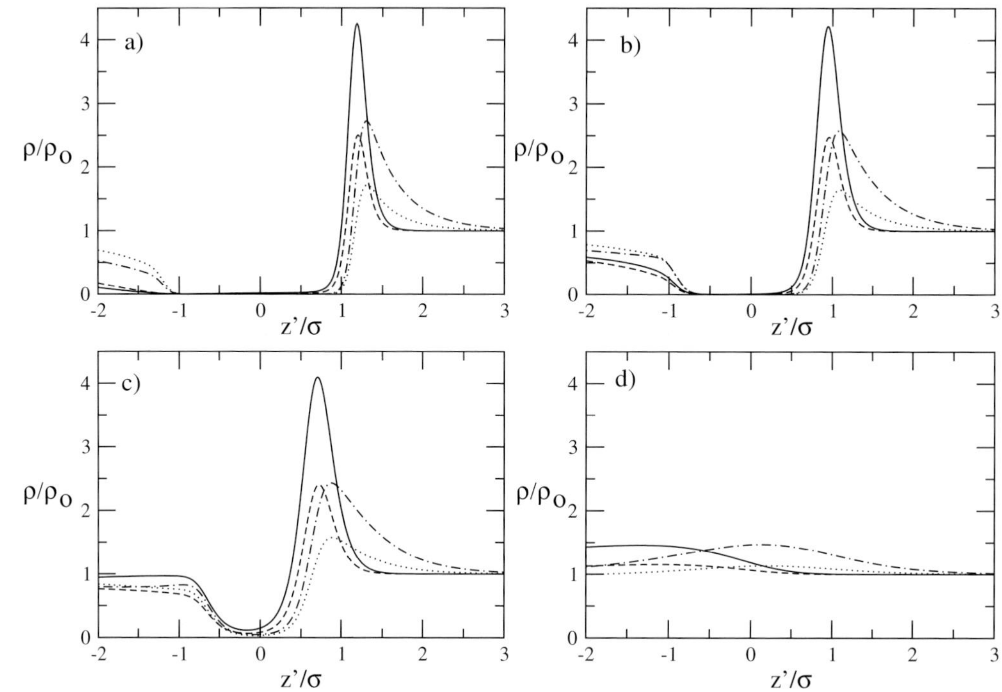

Fig. 2. – Steady-state density profile of colloidal particles under the influence of an external potential $V_{\rm ext} = V_0 e^{-|\mathbf{r}'/\sigma|^6}$, with $\beta V_0 = 10$, for an effective 2D system at drift velocity $\sigma\bar{c} = 10$ (solid line) and $\sigma\bar{c} = 2$ (dotted-dashed line) and for an effective 3D system at drift velocity $\sigma\bar{c} = 10$ (dashed line) and $\sigma\bar{c} = 2$ (dotted line). The fixed radial and perpendicular distance to the z'-axis are a) $R = x = 0$, b) $R = x = 0.75\sigma$, c) $R = x = \sigma$, d) $R = x = 2\sigma$.

maximum total currents are obtain for $\sigma\bar{c}_{\rm max} \approx 0.5$ in 1D, ≈ 12.41 in 2D, and ≈ 14.76 in 3D, which implies that the value of $\bar{c}_{\rm max}$, that determines when the system is saturated by the drift velocity, depends both on the external potential and the effective dimension. Increasing the dimension, we increase the paths for the molecules to avoid the external potential so a higher velocity is needed to make the molecules go thought the external potential.

We have considered the geometrical effects of the effective dimension in systems of non-interacting colloidal particles driven by external barriers with constant drift rates. The main difference between the 2D and 3D cases analyzed here and the previously reported 1D case is the appearance of long wake structures behind the moving barrier. In contrast with the common exponential decay $\rho(\mathbf{r}') - \rho_0 \sim \exp[-\bar{c}z']$ for the structure at barrier *front*, the structure of the *wake* decays only as an inverse power, $\rho(\mathbf{r}') - \rho_0 \sim 1/|z'|^2$ for 3D and $\rho(\mathbf{r}') - \rho_0 \sim 1/|z'|^{3/2}$ for 2D. However, the transverse integral of the wake, over directions other than z', gives no excess of particles, with respect to the unperturbed density ρ_0 far from the potential barrier. This is in agreement with the complete lack

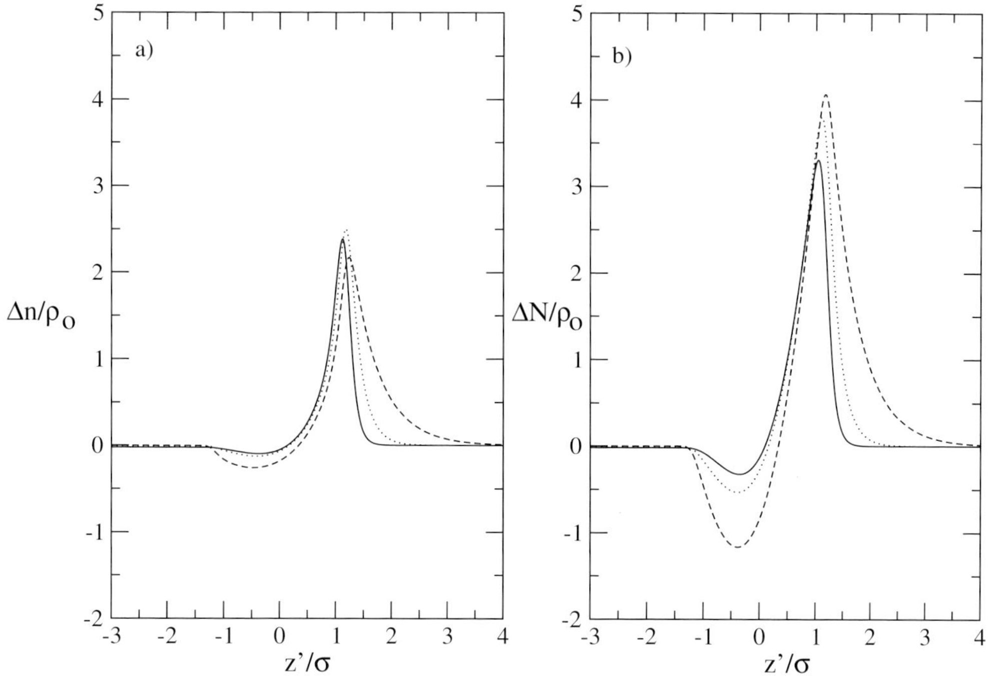

Fig. 3. – Integrated excess of the stationary density distribution over the transverse directions, a) for an effective 2D system and b) for a 3D one. The external potential $V_{\rm ext} = V_0 e^{-|\mathbf{r}'/\sigma|^6}$, with $\beta V_0 = 10$ is shifting at the velocities $\sigma\bar{c} = 10$ (solid lines), $\sigma\bar{c} = 5$ (dotted lines) and $\sigma\bar{c} = 2$ (dashed lines).

of *wake* structure behind a 1D barrier, and it implies that the global current induced by the shifting potentials, calculated as the integral of $\bar{c}(\rho(\mathbf{r}') - \rho_0)$ comes from the balance between the positive contribution from the density peak at the *front*, and the void ($\rho(\mathbf{r}') < \rho_0$) within the barrier. This total current follows a non-monotonic dependence on the shifting rates, as reported in 1D [8], with an optimum shifting rate $\bar{c}_{\rm max}$ which increases with the effective dimensions. The colloid-colloid interactions effects, which have being leaved out of this work, may be included within the general framework of the DDF [8-11] and may help to the design of efficient particle pump devices.

* * *

FP gratefully acknowledges J. DZUBIELLA for discussions. This work has been supported by the Dirección General de Investigación Científica (MCyT) under grant BMF2001-1679-C03-02, and a FPU grant AP2001-0074 from the MECD of Spain.

REFERENCES

[1] GAU H., HERMINGHAUS S., LENZ P. and LIPOWSKY R., *Science*, **283** (1999) 46.
[2] SERVICE ROBERT F., *Science*, **282** (1998) 399.
[3] TERRAY A., OAKEY J. and MARR D. W. M., *Science*, **296** (2002) 1841.
[4] KUKSENOK O., JASNOW D., YEOMANS J. and BALAZS A. C., cond-mat/0305225 (2003).
[5] MARINI BETTOLO MARCONI U. and TARAZONA P., *J. Chem. Phys.*, **110** (1999) 8032.
[6] MARINI BETTOLO MARCONI U. and TARAZONA P., *J. Phys. Condens. Matter*, **12** (2000) A413.
[7] CUI B., DIAMANT H. and LIN B., *Phys. Rev. Lett.*, **89** (2002) 188302.
[8] PENNA F. and TARAZONA P., *J. Chem. Phys.*, **119** (2003) 1766.
[9] DZUBIELLA J. and LIKOS C. N., *J. Phys. Condens. Matter*, **15** (2003) L147.
[10] PENNA F., DZUBIELLA J. and TARAZONA P., *Phys. Rev. E*, **68** (2003) 061407.
[11] DZUBIELLA J., LIKOS C. N. and LÖWEN H., to be published (2003).

Self-poisoning of crystal nuclei in hard-rod liquids

T. SCHILLING and D. FRENKEL

FOM Institute for Atomic and Molecular Physics
Kruislaan 407, 1098 SJ Amsterdam, The Netherlands

Recent experiments on nucleation in suspensions of colloidal rods indicate that it is difficult to grow monocrystals from these particles [1]. Over the last years computer simulation methods have been developed to compute the pathway and rate of crystal nucleation for spherical colloidal particles [2]. In this study we employ similar simulation methods to investigate the nucleation of crystals made from rod-like particles.

When spherical particles crystallize, the crystal nuclei that form tend to be roughly spherical [2,3]. In contrast, non-spherical molecules may form crystals that are strongly anisometric. The reason is twofold: First of all, the solid-liquid interfacial free-energy density may be very different for crystal faces parallel and perpendicular to the molecular axes. As a consequence, the equilibrium shape of small crystal nuclei is non-spherical. In addition, once crystals grow beyond the size of the critical nucleus, the rate at which molecules are incorporated into the crystal may depend strongly on the nature of the crystal face. In some extreme cases, this may lead to the formation of extremely elongated (liquid) crystal fibers [4,5].

Sometimes the growth rate of one or more crystal facets can be inhibited by the addition of an impurity that strongly adsorbs onto the growth front and thereby "poisons" the incorporation of new molecules into that facet. In fact, the molecules that lead to poisoning of a growth surface need not be impurities in the chemical sense. They can simply be misaligned or misfolded specimens of the crystallizing species (see, *e.g.*, [6,7]). In either case, one can argue that the crystal growth is blocked by incorporation of defects in the crystal. In this article, we report a numerical study that reveals a different kind of self-poisoning: the crystal nucleus itself is free of defects, but it is covered by a liquid layer of molecules that do not have the correct orientation to be incorporated in the crystal. This has consequences both for crystal nucleation and for subsequent crystal growth.

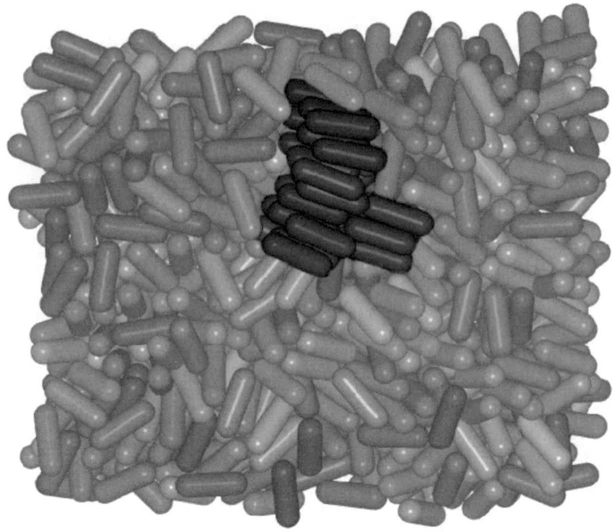

Fig. 1. – A metastable liquid configuration at $p = 6$, containing a (subcritical) crystalline cluster (dark).

The system that we studied is a model for a colloidal suspension of hard, rodlike particles. These particles are modeled as hard spherocylinders with a diameter D and a cylindrical segment of length L. In our simulations, we considered 2400 rods with an aspect ratio $L/D = 2$. Hard spherocylinders have a relatively simple phase diagram [8]: at low pressures, the system forms an isotropic fluid. At a pressure $p = 5.64\ k_{\mathrm{B}}T/D^3$, the isotropic fluid coexists with an orientationally ordered crystal phase. (In the following the unit of energy is $k_{\mathrm{B}}T$ and the unit of length is D.) We are interested in crystal nucleation from the meta-stable fluid. To study this phenomenon, we compressed the fluid beyond coexistence, to a pressure of $p = 6$. At this pressure, the chemical potential difference between the (metastable) fluid and the solid phases is approximately $0.2\ k_{\mathrm{B}}T$ per particle. In spite of the fact that, at this pressure, the solid is more stable than the liquid, spontaneous crystallization never occurred on the time-scales of our (quite long) simulations. However, at higher supersaturations (*e.g.*, $p = 8$, $\Delta\mu = 1.3\ k_{\mathrm{B}}T$/particle) spontaneous crystallization did occur on the timescale of a simulation. By working at $p = 6$, a pressure where spontaneous crystallization is suppressed, we can study the free energy and the structure of small crystal nuclei (see fig. 1), using the biased sampling techniques described in refs. [2, 3].

In order to follow the formation of crystallites from solution, we need an order-parameter that distinguishes particles in a crystalline environment from those in the liquid phase. In the crystal phase, all particles within one crystal tend to be strongly aligned. We employed this fact to define our crystallinity criterion: two particles i and j are said to belong to the same crystallite if: a) their surface-to-surface distance is less than $0.5\ D$ and b) the dot product of the unit vectors along their respective axes satisfies

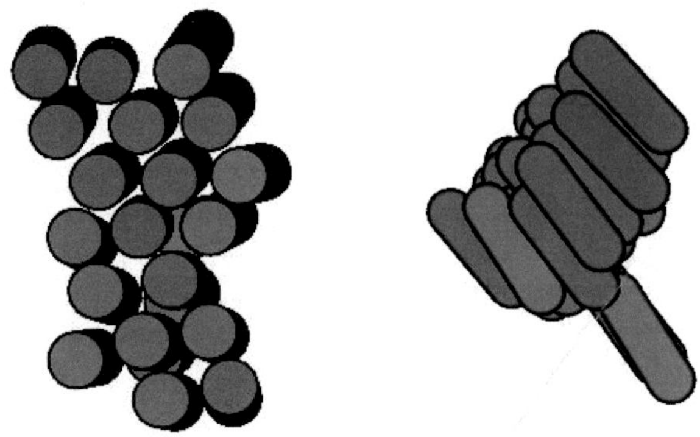

Fig. 2. – Top and side view of a typical crystalline cluster. Most rods are incorporated in a single crystalline layer.

$|\mathbf{u}_i \cdot \mathbf{u}_j| > 0.995$. With this criterion, 99.8% of all particles in an equilibrium crystal are recognized as belonging to that crystal. As the criterion only considers the relative orientation of neighboring particles, it does not distinguish between crystalline, smectic and (dense) nematic structures. This means that we do not impose strong constraints on the structure of the crystal nucleus. A side effect of this rather loose criterion is that many particles in the liquid are identified as belonging to small crystallites. However, we found that the number of rods in these crystallites was rarely larger than 2: about 5% of all particles belong to "crystal" dimers and less than one in a thousand belongs to a larger cluster. We never observed spontaneous fluctuations that resulted in crystalline clusters containing more than 10 particles.

We denote the average number of clusters consisting of n particles by N_n. A good measure for the probability to find a cluster of size n is the ratio $P(n) \equiv N_n/N$, where N is the total number of particles in the system. We define an intensive free energy barrier for the formation of a crystal nucleus of n particles as $\Delta G(n) = -k_{\mathrm{B}}T \ln P(n)$. To probe $P(n)$ for $n > 10$, we use biased, multi-histogram simulations (for details, see ref. [2]).

When we perform such a multi-histogram simulation, we find that we can grow large, ordered clusters. However, somewhat surprisingly, we find that most particles in the ordered cluster are located in a single hexagonally ordered layer (see, *e.g.*, fig. 2). As we bias the cluster-size distribution towards larger crystallites, we observe no tendency to form crystallites that contain multiple layers. To be more precise: a few particles may order on top of the crystallite, but these embryonic new layers quickly dissolve again. All growth of the cluster concentrates on the edges, and the free energy of the cluster rises monotonically with size (see fig. 3). In other words: the system never crossed a nucleation barrier beyond which crystallites would grow spontaneously.

Interestingly, a rough analysis of the data shown in fig. 3 suggests that a flat crystallite is not the most stable one. If we assume that the crystallite is cylindrical, with a thickness

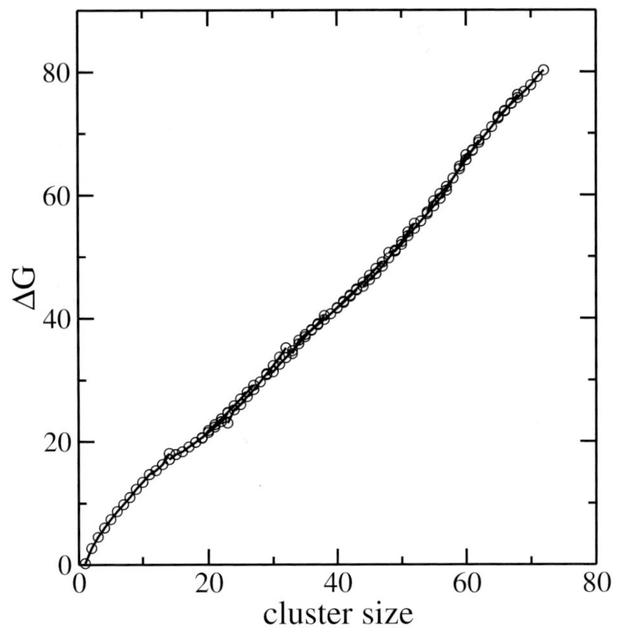

Fig. 3. – Size dependence of the free energy of a small crystal nucleus. For the low supersaturations that we studied, ΔG for a single crystalline layer grows monotonically with cluster size. In order to nucleate a stable crystal, a multilayer nucleus must first form.

$h \equiv L + D$ and a radius r, then we can approximate the free energy of this cluster by

$$\Delta G(r) = 2\pi r h \gamma_{\|} + \pi r^2 \left(2\gamma_{\perp} - h\rho_s|\Delta\mu|\right), \tag{1}$$

where $\gamma_{\|}$ and $\gamma_{\perp}$ are, respectively, the free energy densities of the side and of the top/bottom surfaces. ρ_s is the density of the crystalline phase.

In a lamellar crystal, both the bulk free energy and the (dominant) surface free energy scale with the cross-sectional area. A lamella can only grow spontaneously when $(2\gamma_{\perp} - \rho_s|\Delta\mu|) < 0$. At the low supersaturation that we use ($|\Delta\mu| = 0.2$), this condition is not satisfied. In fact, a fit of eq. (1) to the free-energy barrier in fig. 3 yields: $\gamma_{\perp} = 0.6$. Moreover, we find that $\gamma_{\|} \ll \gamma_{\perp}$. This is interesting, because in this case, we should expect multilayer clusters to be more stable than flat disks. The fact that, in our simulations, we observe the formation of disk-like crystallites suggests that there are *kinetic* reasons that inhibit the formation of multilayer crystallites. The simplest assumption is that multilayer growth is difficult because every new layer has to nucleate on top of an existing layer [5]. To test this, we used the same biased Monte Carlo procedure to compute the free energy of a crystallite that forms on top of an ordered crystalline substrate. However, before we discuss these simulations, it is instructive first to look at the ordering in the liquid near the crystal surface.

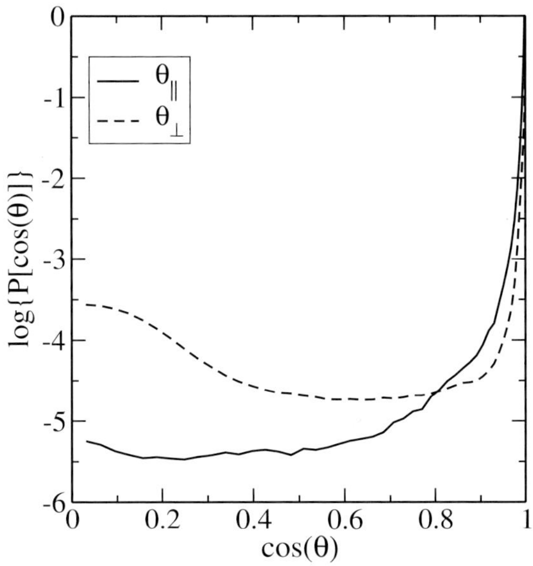

Fig. 4. – Angular distribution of liquid particles close to a surface of a hard-rod crystal. $\theta_{||}$ denotes the surface parallel to the director field ("side surface"), $\theta_{\perp}$ the surface perpendicular to it ("top surface").

Figure 4 shows the angular distribution of particles in the liquid close to the top surface (the surface perpendicular to the director field) and the side surface (the surface parallel to it) of a perfect crystal of hard spherocylinders. The peaks at $|\cos\theta| = 1$ for the side surface and at $\cos\theta = 0$ for the top surface indicate that a large fraction of the particles align parallel to the surface. In addition, for the top surface, the angular distribution also peaks for orientations perpendicular to the surface.

The preference of particles to lie flat on the crystal surface facilitates growth of the side surface, but it makes growth of the top/bottom surface difficult. Particles have to overcome a barrier of roughly 2 kT/particle in order to stand up on the surface and align with the director. In contrast, incorporation of particles on the side face of a crystal is relatively easy. Indeed, we find no inhibition of the lateral growth of spherocylinder crystals, even when the growth of additional layers is effectively suppressed.

The discussion above suggests that, in order to form a second crystal layer, an appreciable fraction of the particles on the top or bottom surface must spontaneously align parallel to the surface normal. The "reaction" path for second-layer formation would then be orientational alignment, followed by crystallization. To explore this possible route, we performed simulations where we introduced a second biasing parameter that facilitates the alignment of the particle orientation $\vec{u_i}$ with the surface normal $\vec{n}$, for particles close to the (top or bottom) surface: $S_{\text{liquid}} = \langle(\vec{n}\cdot\vec{u})^2\rangle$. Unlike the order parameter that we used to measure crystallinity, S increases continuously with alignment. In contrast, particles are only counted as crystalline if they are already well aligned. This implies that

S is sensitive to the pre-alignment that is, presumably, a prerequisite for crystallization.

Indeed, we find that applying an orientational bias greatly facilitates the growth of a second crystal layer.

The experiments by Maeda and Maeda [1] on the isotropic-to-smectic transition in colloidal hard rods of β–FeOOH rods, showed that these particles tend to form disklike crystallites. The present simulations suggest that this experimental observation is a logical consequence of the fact that, at sufficiently high supersaturation, multilayer growth is still kinetically inhibited, due to orientational "self-poisoning" of the top and bottom growth surfaces, whereas lateral growth proceeds relatively unhampered.

* * *

This work is part of the research program of "Stichting Fundamenteel Onderzoek der Materie (FOM)" which is financially supported by "Nederlandse Organisatie voor Wetenschapelijke Onderzoek (NWO)". T. S. acknowledges support through a European Community Marie Curie Fellowship.

REFERENCES

[1] MAEDA H. and MAEDA Y., *Phys. Rev. Lett.*, **90** (2003) 018303.
[2] AUER S. and FRENKEL D., *Nature*, **409** (2001) 1020.
[3] TEN WOLDE P. R., RUIZ-MONTERO M. J. and FRENKEL D., *Faraday Discuss.*, **104** (1996) 93.
[4] DOGIC Z. and FRADEN S., *Philos. Trans. R. Soc. London A*, **359** (2001) 997.
[5] FRENKEL D. and SCHILLING T., *Phys. Rev. E*, **66** (2002) 041606.
[6] ASTHAGIRI D., LENHOFF A. M. and GALLAGHER D. T., *J. Cryst. Growth*, **212** (2000) 543.
[7] UNGAR G., MANDAL P. K., HIGGS P. G., DE SILVA D. S. M., BODA E. and CHEN C. M., *Phys. Rev. Lett.*, **85** (2000) 4397.
[8] BOLHUIS P. and FRENKEL D., *J. Chem. Phys.*, **106** (1997) 666.

Restricted-orientation models of hard-rod fluids

K. SHUNDYAK and R. VAN ROIJ

Institute for Theoretical Physics, Utrecht University - Leuvenlaan 4
3584 CE Utrecht, The Netherlands

Colloidal suspensions of rodlike particles may be considered as mesoscopic analogs of molecular liquid crystals. Even in the simplest case of pure hard-core interparticle interactions they exhibit a rich phase behavior as a function of density [1,2]. Theoretical studies of such systems were pioneered by Onsager who analyzed the transition from a uniform isotropic phase to an orientationally ordered nematic phase in a fluid of hard needles. He realized that the average pairwise (rod-rod) excluded volume is reduced in the nematic phase compared to that in the isotropic phase, and argued that the resulting gain of free volume (and hence translational entropy) compensates the loss of orientation entropy (due to the nematic ordering) at sufficiently high concentrations of rods [3]. Onsager derived a nonlinear integral equation for the orientation distribution function, a key quantity of the theory, which is constant in the isotropic phase and peaked about the director in the nematic phase. He circumvented the problem of explicitly calculating the nematic orientation distribution function by adopting a variational Ansatz, but later it was calculated numerically [1].

Another simplifying approach is to study models with a finite number N of allowed orientations while the positions of the centers of mass of the rods remain continuous. Such a model was first proposed by Zwanzig [4], with orientations of a rod to be restricted to $N = 3$ mutually perpendicular directions $\hat{n}_i$, $i = \{1, \ldots, N\}$. Despite its inability to resolve the orientational structure of the one-particle distribution function in any detail, it has been successfully applied to explore wetting phenomena near a single hard wall and in a slit [5], phase diagrams of polydisperse systems [6], etc.

The possibility of a continuous interpolation between results for the discrete models on the one hand and Onsager-like solutions on the other has first been questioned by Straley [7] in studies of models with dodecahedral ($N = 6$) and icosahedral ($N = 10$)

symmetries. He concluded that they do not trend towards the Onsager solution due to the single allowed orientation within the typical opening angle ($\approx \pi/9$) of the nematic distribution at coexistence. Here we derive a connection between continuous and discrete models in general, and explicitly construct such an interpolation for a specific model.

For a fluid of hard rods of length L and diameter D ($L \gg D$) in a macroscopic volume V at temperature T and chemical potential μ the grand potential functional $\Omega[\rho]$ of the one-particle distribution function $\rho(\hat{\omega})$ can be written, within second virial approximation, as [3]

$$\beta\Omega[\rho(\hat{\omega})] = \int \mathrm{d}\hat{\omega}\rho(\hat{\omega})\big(\ln[\rho(\hat{\omega})\nu] - 1 - \beta\mu\big) + \frac{1}{2}\int \mathrm{d}\hat{\omega}\,\mathrm{d}\hat{\omega}' E(\hat{\omega},\hat{\omega}')\rho(\hat{\omega})\rho(\hat{\omega}'), \tag{1}$$

where $\beta = (kT)^{-1}$ is the inverse temperature, ν is the rod's thermal volume, and $E(\hat{\omega},\hat{\omega}')$ is the excluded volume of rods with orientations $\hat{\omega}$ and $\hat{\omega}'$. The function $\rho(\hat{\omega})$ is normalized as $n = \int \mathrm{d}\hat{\omega}\rho(\hat{\omega})$, with n the bulk number density (which depends on $\beta\mu$). The minimum condition $\delta\Omega[\rho(\hat{\omega})]/\delta\rho(\hat{\omega}) = 0$ on the functional leads to the set of nonlinear integral equations

$$\ln[\rho(\hat{\omega})\nu] + \int \mathrm{d}\hat{\omega}' E(\hat{\omega},\hat{\omega}')\rho(\hat{\omega}') = \beta\mu, \tag{2}$$

to be solved for the equilibrium distribution $\rho(\hat{\omega})$.

Models with a discrete number N of allowed rod orientations can be systematically derived from the continuous model (1) by dividing the unit sphere into solid sectors $\Delta\hat{\omega}_i$, ($i = 1,\ldots,N$) around vectors $\hat{\omega}_i$, and fixing the rod density $\rho(\hat{\omega}) = \rho(\hat{\omega}_i)$ within each sector as well as the excluded volume $E(\hat{\omega},\hat{\omega}') = E(\hat{\omega}_i,\hat{\omega}_j)$ for every pair of sectors. The grand potential functional $\Omega[\rho_i]$ of such "orientationally discretized" fluid with the density $\rho_i = \rho(\hat{\omega}_i)\Delta\hat{\omega}_i$ is

$$\beta\Omega[\rho_i] = \sum_{i=1}^{N}\rho_i\big(\ln[\rho_i\nu] - 1 - \beta\mu\big) + \frac{1}{2}\sum_{i,j=1}^{N} E_{ij}\rho_i\rho_j - \sum_{i=1}^{N}\rho_i \ln \Delta\hat{\omega}_i\,, \tag{3}$$

with normalization $n = \sum_{i=1}^{N}\rho_i$. The last term in (3) represents the contribution due to the discretization procedure into the grand potential Ω, *i.e.* the intrinsic difference between continuous and discrete models. For a homogeneous distribution of vectors $\hat{\omega}_i$ on the unit sphere and $\Delta\hat{\omega}_i = \Delta\hat{\omega}$ (*i.e.* for the models with $N = 3$, 6 and 10), it trivially shifts the chemical potential $\beta\mu_d = \beta\mu + \ln\Delta\hat{\omega}$, which does not have any consequence for the solutions ρ_i at a fixed n, and for the thermodynamics of the isotropic-nematic transition. However, when $\Delta\hat{\omega}_i$ is not the same for all i, it acts as an external orientational field that tends to favor the larger sectors over the smaller ones. This becomes explicit if we consider the Euler-Lagrange equations that corresponds to the discrete functional (or equivalently the analog of eqs. (2))

$$\ln[\rho_i \nu] + \sum_{j=1}^{N} E_{ij}\rho_j = \beta\mu + \ln \Delta\hat{\omega}_i \,, \tag{4}$$

now to be solved for ρ_i. Note, that the equation of state $p = p(n, T)$ does not pick up an additional term from discretization,

$$\beta p = n + \frac{1}{2}\sum_{j=1}^{N} E_{ij}\rho_i\rho_j \,, \tag{5}$$

but the distributions ρ_i to be inserted into it *do* depend on the discretization.

Further discussion requires a specification of the set of allowed orientations $\hat{\omega}_i$, $i = 1, \ldots, N$, and the associated solid angles $\Delta\hat{\omega}_i$. Unfortunately, it is impossible to completely cover a surface of the unit sphere by equal regular spherical M-polygons, where M indicates the number of polygon's sides (only 5 Platonic solids exist). But symmetries of the function $\rho(\hat{\omega})$ can be explicitly included into the set of vectors $\hat{\omega}_i$ in order to simplify the problem. For the present study we fix the $\hat{z}$-axis of the coordinate system to be parallel to the nematic director $\hat{n}$ and assume uniaxial symmetry of the function $\rho(\hat{\omega}) = \rho(\theta)$, with $\theta = \arccos(\hat{\omega}\cdot\hat{n})$ the angle between $\hat{\omega}$ and the nematic director $\hat{n}$. The azimuthal angle is denoted by ϕ, and hence we characterize a vector $\hat{\omega} = (\sin\theta\cos\phi, \sin\theta\sin\phi, \cos\theta)$ by the angles θ and ϕ. The "up-down" symmetry of the nematic phase reduces orientational space to the half of the upper-hemisphere, *i.e.* $\theta \in [0, \pi/2]$ and $\phi \in [0, \pi]$. As we do not expect any azimuthal symmetry breaking we restrict attention to N_ϕ uniformly distributed values for ϕ for every allowed θ. We have considered a uniform distribution of N_θ polar angles $\theta \in [0, \pi/2]$, *i.e.*

$$(\theta_k, \phi_l) = \left(\frac{\pi(k-1)}{2(N_\theta - 1)}, \frac{\pi(l-1)}{N_\phi - 1}\right), \quad k = 1, \ldots, N_\theta, \; l = 1, \ldots, N_\phi, \tag{6}$$

as well as a uniform distribution of N_θ values of $\cos(\theta) \in [0, 1]$, *i.e.*

$$(\theta_k, \phi_l) = \left(\arccos\left[1 - \frac{k-1}{N_\theta - 1}\right], \frac{\pi(l-1)}{N_\phi - 1}\right), \quad k = 1, \ldots, N_\theta, \; l = 1, \ldots, N_\phi, \tag{7}$$

with the conventional definition of $\Delta\hat{\omega}_i = \int_{\Omega_i} \sin\theta \,\mathrm{d}\theta \,\mathrm{d}\phi$.

Figure 1 shows the dimensionless pressure $p^* = \beta p L^2 D$ as a function of the dimensionless bulk density $n^* = nL^2D$ for the grid (6) with different N_θ and $N_\phi = 5$. The plateaux (of the solid lines) correspond to the isotropic-nematic coexistence, obtained by equating pressure and chemical potential in the two phases. For $N_\theta \leq 9$ the transition occurs between the isotropic phase and an (almost) perfectly aligned nematic phase with the pressure essentially equal to the ideal gas pressure. Note that such grids correspond to a single θ within the "Onsager" opening angle, *i.e.* $0 \leq \theta_1 \leq \pi/18 < \theta_2$. As soon as

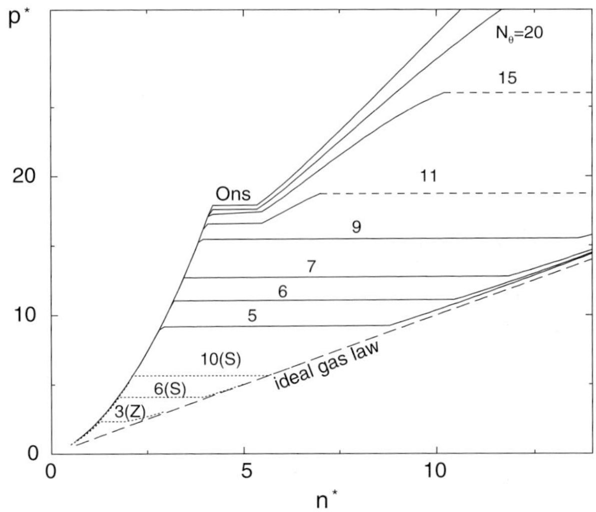

Fig. 1. – Equation of state for models with different number of allowed polar angles N_θ. Positions of the phase transitions are indicated by the horizontal lines. The continuous Onsager solution (Ons) can be reproduced in the present density interval with $N_\theta \geq 50$. Dotted lines correspond to equations of state for the Zwanzig 3(Z), dodecahedral 6(S) and icosahedral 10(S) models [7]. The dashed horizontal lines correspond to a spurious nematic-nematic transition due to poor discretization of the allowed orientations.

$0 \leq \theta_1 < \theta_2 \leq \pi/18$, or $N_\theta > 9$, the distribution function $\rho(\theta)$ at isotropic-nematic coexistence starts to converge to the continuous solution. These results are in full agreement with the previous explanation of Straley [7]. Equations of state for the model (7) are very similar to fig. 1 but start to resemble Onsager-like distribution function for $N_\theta > 80$ due to the poor sampling of $\hat{\omega}_i$ near the nematic director. Equations of state for the Zwanzig ($N = 3$), dodecahedral ($N = 6$) and icosahedral ($N = 10$) models were calculated using the original formulations, and are included for comparison. Our results for $N_\theta < 9$ seem to converge well to these existing results.

For $N_\theta > 20$ the pressure of the high-density nematic phase clearly demonstrates linear dependence on the bulk density, *i.e.* $\beta p(n) \sim n$. With increasing N_θ it gradually approaches a limiting scaling behavior $\beta p(n) = 3n$, established for the continuous Onsager solution by means of a scaling argument [8].

The discretization of the rod's allowed orientations shows the existence of an "artificial transition" from a less-ordered nematic phase to a near-perfectly aligned phase (indicated by the dashed horizontal lines in fig. 1), even for $N_\theta > 9$. This occurs due to the same competition between excluded volume and orientational entropy, and puts another constraint on the description of the nematic bulk state by restricted-orientation models.

In conclusion, we have explored a connection between continuous and restricted-orientation models. A discrete set of orientations starts to resemble Onsager's continuous results when several orientations are present within an opening angle of about $\pi/9$. These results are important in the study of spatially inhomogeneous systems of rods, where a fully continuous approach seems to be particulary difficult due to the coupling between orientational and translational degrees of freedom of the constituent particles. One should then be aware of artifacts due to too coarse a discretization of the allowed angles.

* * *

We would like to thank J. P. STRALEY for useful correspondence. This work is part of the research program of the "Stichting voor Fundamenteel Onderzoek der Materie (FOM)", which is financially supported by the "Nederlandse organisatie voor Wetenschappelijk Onderzoek (NWO)".

REFERENCES

[1] VROEGE G. J. and LEKKERKERKER H. N. W., *Rep. Progr. Phys.*, **55** (1992) 1241.
[2] FRADEN S., in *Observation, Prediction and Simulation of Phase Transitions in Complex Fluids*, edited by M. BAUS *et al.* (Kluwer, Dordrecht) 1995, p. 113.
[3] ONSAGER L., *Ann. N.Y. Acad. Sci.*, **51** (1949) 627.
[4] ZWANZIG R., *J. Chem. Phys.*, **39** (1963) 1714.
[5] VAN ROIJ R., DIJKSTRA M. and EVANS R., *Europhys. Lett.*, **49** (2000) 350.
[6] CLARKE N., CUESTA J. A., SEAR R., SOLLICH P. and SPERANZA A., *J. Chem. Phys.*, **113** (2000) 5817.
[7] STRALEY J. P., *J. Chem. Phys.*, **57** (1972) 3694.
[8] VAN ROIJ R. and MULDER B., *Europhys. Lett.*, **34** (1996) 201.

Wavelet analysis and scaling properties of electropolished surfaces

A. MORA and M. HAASE

Institute for Computer Applications II (ICA 2), University of Stuttgart - Germany

Electropolishing is a technical process to obtain smooth and shiny surfaces by means of electrochemical removal of metal. The complex surface structures are formed through the interplay of two phenomena: a) isotropic structuring due to the formation and growth of oxygen bubbles and b) an anisotropic structuring that occurs when the gas bubbles release from the metal surface and rise in front of it, creating a characteristic pattern of channels or lines of the electropolished surfaces.

Gerlach [1] has obtained electropolished brass samples with two different electrolyte solutions (among others) containing phosphoric acid and alcohol, denoted as METHANOL-elec. and GLYCERINE-elec. (See fig. 1.) The methanol-electrolyte–treated surface is very smooth, shiny with tiny ripples in the vertical direction. On the other hand, the glycerine-electrolyte–treated surface has high peaks, it is rougher and dull and suggests self-affinity. Figure 2 displays the characteristic height profiles $z(x)$ for both electrolytes. In constrast with the large scales, a zoom into the METHANOL-elec. characteristic profile shows that at small scales the surface is rougher than in the GLYCERINE-elec. case (see figs. 2b and d).

Fourier analysis. By means of the power spectral density (PSD) two scaling regions for each surface are identified (see fig. 3). Our interpretation is that the large scales (low wave numbers) provide information about the process of formation of gas lines. The scaling region at small scales (high wave numbers) should correspond to the elementary electrochemical processes that occurs independently of the rising gas bubbles action [2]. We measured the anisotropy introduced by the rising bubbles comparing the power spectral density transversal and parallel to the gas lines (see fig. 4a). Considering the maximum of the ratio of the two spectra, we found about 24 lines per cm for the characteristic width of the gas lines (see fig. 4b). In contrast, the glycerine-electrolyte–treated surface does not display a clear anisotropy.

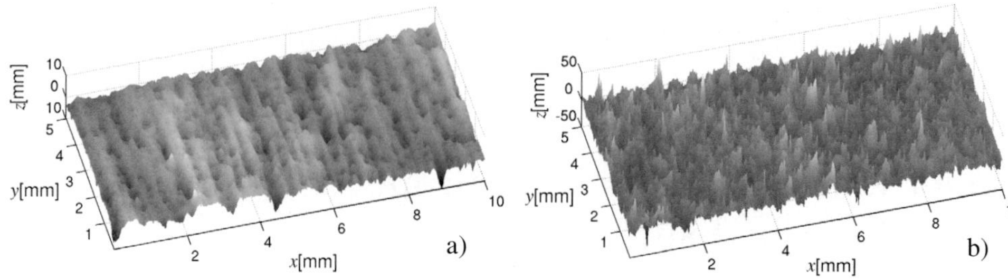

Fig. 1. – Electropolished surfaces: a) METHANOL-elec. b) GLYCERINE-elec.

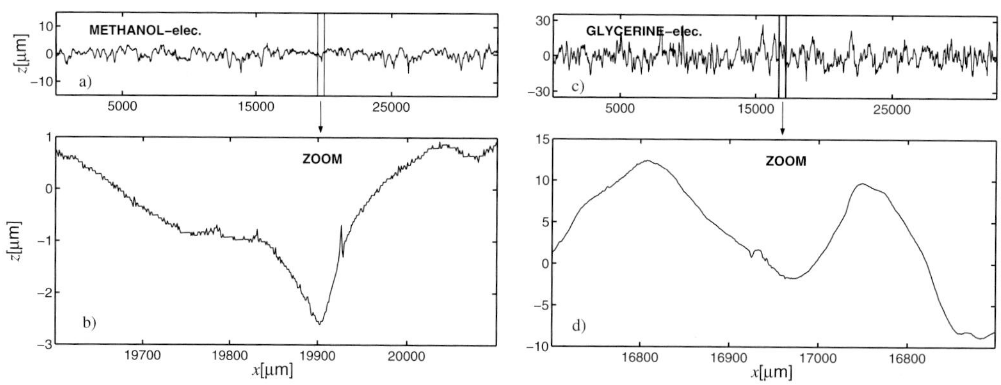

Fig. 2. – For METHANOL-elec.: a) characteristic profile and b) zoomed region. For GLYCERINE-elec.: c) characteristic profile and d) zoomed region.

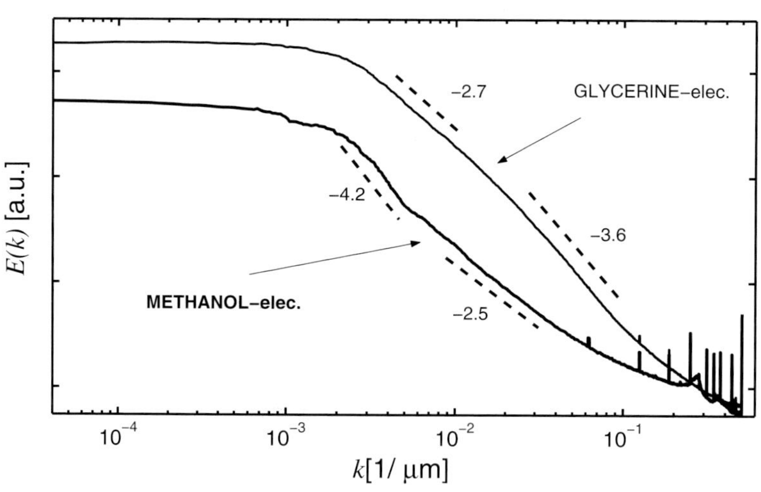

Fig. 3. – Power spectral densities $E(k)$ for METHANOL-elec. and GLYCERINE-elec.

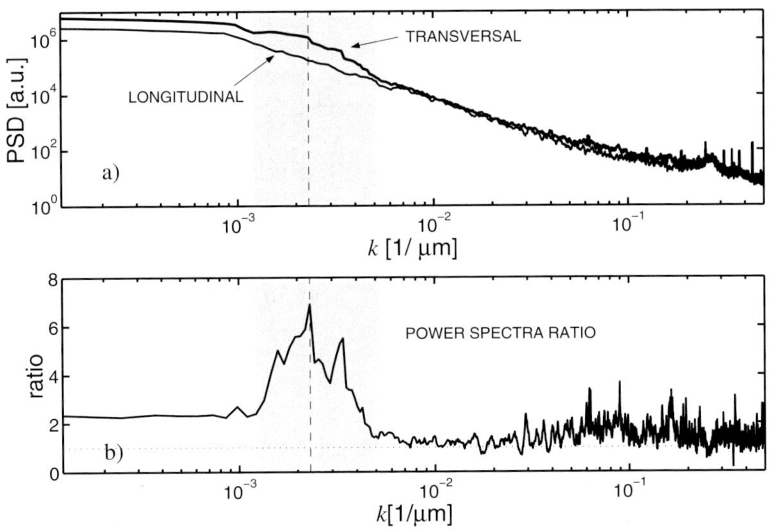

Fig. 4. – a) Power spectra comparison of ensemble averages transversal and parallel to the gas lines for METHANOL-elec. b) The power spectra ratio.

Wavelet analysis. The power spectral density can only provide estimates of the global roughness and give us limited information about the monofractal properties of the profiles. In order to explore the scaling properties of the surface roughness, we have applied recently developed wavelet techniques [3]. Wavelets are defined as two parameter families of functions that are localized in space and scale and can be used to unfold the hierarchical structure of the analysed signal. The continuous wavelet transform (CWT),

$$W_\psi f(a,b) = \frac{1}{a}\int_{-\infty}^{+\infty} f(x)\,\overline{\psi\left(\frac{x-b}{a}\right)}\mathrm{d}x \quad (a, b \in R,\ a > 0), \tag{1}$$

decomposes the function $f(x)$ hierarchically in terms of elementary components $\psi((x-b)/a)$ which are obtained from a single *mother wavelet* $\psi(x)$ by dilations and translations. Here, $\overline{\psi}(x)$ denotes the complex conjugate of $\psi(x)$, a the scale and b the shift (space) parameter.

While the Continuous Wavelet Transform (CWT) provides highly redundant information, the wavelet transform maxima lines contain the essential information about the evolution of scaling properties of irregularities across scales. Thus, they can be considered as a fingerprint of the signal [4,5]. The Wavelet Transform Modulus Maxima (WTMM) method allows to extract the scaling characteristics of irregular functions, which may even follow different power laws in different regions [6]. The method is a generalization of the classical multifractal formalism. Instead of relying on box-counting techniques, wavelets are introduced as oscillating variants of box functions. The singularity of a

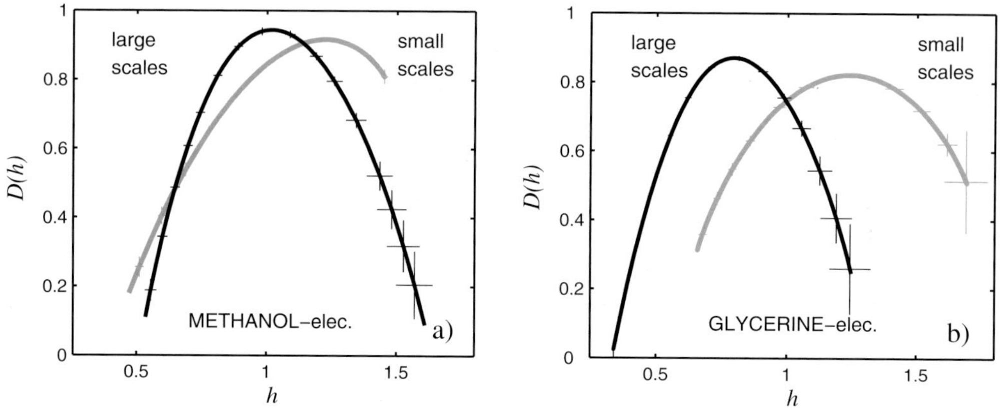

Fig. 5. – Spectra of Hölder exponents $D(h)$ for a) METHANOL-elec. and b) GLYCERINE-elec. for two different scaling regions.

function at a point x_0 can be characterized by the Hölder exponent h, which can be considered as extension of the "order of differentiability" to non-integers. The higher the strength of the singularity the lower the Hölder exponent. For multifractal functions, the singularity spectra $D(h)$ measures the global distribution of singularities having different Hölder exponents and can be calculated by means of the WTMM method [3, 6].

To characterize the multifractal behavior of the two electropolished surfaces, we have calculated for the two identified scaling regions the corresponding distribution of Hölder exponents. The spectra of the Hölder exponents displayed in fig. 5 show similarities at the small scales for the two electrolytes. The most frequently found singularity corresponds to the Hölder exponent h_0 with the highest value of $D(h_0)$. In both cases this value is around 1.25. $D(h_0)$ is smaller for GLYCERINE-elec. indicating that the corresponding singularities are less frequent. For large scales, however, the GLYCERINE-elec. spectrum is shifted to smaller h-values which corroborates the observation, that the GLYCERINE-elec. produces sharper high peaks on large scales than METHANOL-elec. The singularity spectra $D(h)$ can therefore be considered as a quantification of the qualitative conclusions about roughness at different scales drawn from inspecting the profiles.

Stochastic analysis. Next we describe the scale dependence of surface roughness as a stochastic process, because the singularity spectra give us no information about the possible correlations of the roughness measures on different scales. The probability density functions (PDF) of the height increments $z_r(x) = z(x+r/2) - z(x-r/2)$ for various scales r are plotted in fig. 6. The PDF's are normalised to their respective standard deviations σ_r and shifted in vertical direction for clarity. For small scales the shapes of the curves deviate strongly from Gaussian distributions indicating pronounced intermittency effects.

In a series of papers (see for example [7, 8]), a new approach for the stochastic analysis has been proposed which allows to reconstruct equations of the underlying stochastic process from experimental data, provided the process is Markovian. The r-dependency of the height increments z_r can be regarded as a stochastic process evolving in r driven

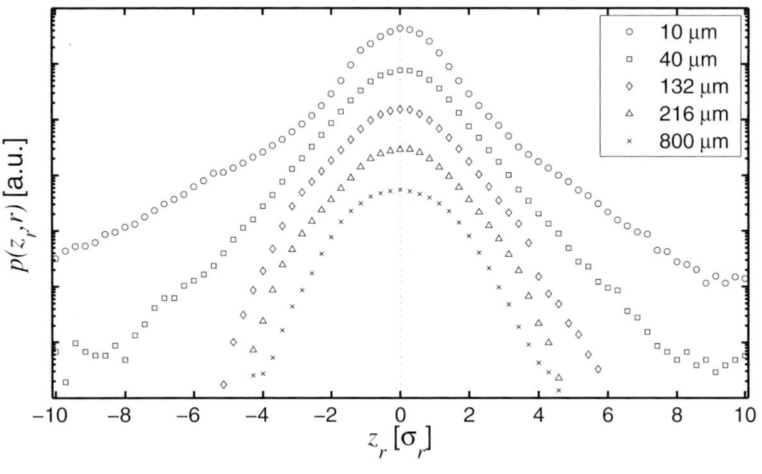

Fig. 6. – PDF(s) of height increments at different scales r for METHANOL-elec.

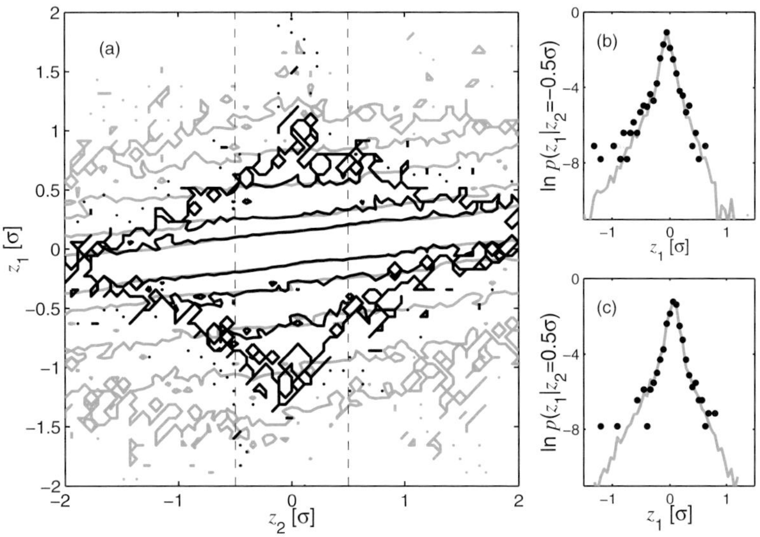

Fig. 7. – a) Contour lines of the single conditional PDFs $p(z_1, r_1|z_2, r_2)$ (gray lines) and double conditional PDFs $p(z_1, r_1|z_2, r_2; z_3 = 0, r_3)$ (black lines). b) and c) Cuts through the conditional PDF for $z_2 = \pm\sigma/2$. Lines: $p(z_1, r_1|z_2, r_2)$, symbols: $p(z_1, r_1|z_2, r_2; z_3 = 0, r_3)$.

by deterministic and random forces and governed by a Fokker-Planck equation. This equation could describe the evolution of the probability density functions of the height increments through different scales.

A preliminary analysis of multiconditional PDF's suggests that the statistics of the height increments for scales larger than a certain threshold has Markovian properties. To fulfil this condition the equation

$$p(z_1, r_1 | z_2, r_2; \ldots; z_n, r_n) = p(z_1, r_1 | z_2, r_2), \quad \text{where} \quad r_1 < r_2 < \cdots < r_n . \tag{2}$$

must hold. However, given the amount of data the verification of this condition is only possible for $n = 3$. In fig. 7a the contour lines of the single conditional PDF $p(z_1, r_1|z_2, r_2)$ (gray lines) are compared with the corresponding lines for the double conditional PDF $p(z_1, r_1|z_2, r_2; z_3 = 0, r_3)$ for the increments $r_1 = 10\,\mu\text{m}$, $r_2 = 108\,\mu\text{m}$, $r_3 = 216\,\mu\text{m}$. The good correspondence of the contour lines gives preliminary evidence of the validity of the Markovian property (with $n = 3$). This correspondence can also be seen in the cuts through the conditional PDF for fixed $z_2 = \pm\sigma/2$ (figs. 7b and c).

Conclusion. For two electropolished metallic surfaces, we have identified scaling regions that can be related with two interacting processes: the basic electropolishing and the anisotropic structuring caused by the rising bubbles. The singularity spectra give quantitative information about the surface roughness. In order to go beyond the multifractal description, we are exploring the possibility to apply the mathematical framework of the Markov processes to determine a Fokker-Planck equation for the evolution PDF's of the height increments over scales.

* * *

This work was supported by the VolkswagenStiftung grant I/77315: *Nichtlineare Dynamik des lasergestützten elektrochemischen Jet-Verfahrens zur Ultrapräzisionsmikrostrukturierung von Metallen für optische Anwendungen.*

REFERENCES

[1] Gerlach C., *Räumliche und zeitliche Instabilitäten in einem technischen Prozess, Elektropolieren von Messing*, Bremen University, Dissertation, 2002. `http://elib.suub.uni-bremen.de/publications/dissertations/E-Diss427_Gerlach_2002.pdf`

[2] Mora A., Gerlach C., Rabbow T., Plath P. J. and Haase M., *Wavelet analysis of electropolished surfaces*, in *Nonlinear Dynamics of Production Systems*, edited by Radons G. and Neugebauer R. (Wyley-VCH, Weinheim) 2003.

[3] Mallat S., *Wavelet Tour of Signal Processing* (Academic Press) 2001.

[4] Haase M., Widjajakusuma J. and Bader R., *Scaling laws and frequency decomposition from wavelet transform maxima lines and ridges*, in *Emergent Nature*, edited by M. M. Novak (World Scientific, Singapore) 2002, pp. 365-374.

[5] Haase M. and Widjajakusuma J., *Int. J. Engin. Sci.*, **41** (2003) 1423.

[6] Muzy J. F., Bacry E. and Arneodo A., *Int. J. Bif. Chaos*, **4** (1994) 245.

[7] Renner C., Peinke J. and Friedrich R., *J. Fluid Mech.*, **433** (2001) 383.

[8] Wächter M., Riess F., Kantz H. and Peinke J., *Europhys. Lett.*, **64** (2003) 579.

International School of Physics "Enrico Fermi"

Villa Monastero, Varenna

Course CLV

1 - 11 July 2003

"The Physics of Complex Systems"
(New Advances and Perspectives)

Directors

Francesco MALLAMACE
Dipartimento di Fisica
Università di Messina
Vill. S. Agata, CP 55
98166 MESSINA
Italy
Tel.: ++39 090 6765016
Fax: ++39 090 395004
Francesco.Mallamace@unime.it

H. Eugene STANLEY
Department of Physics
Center for Polymer Studies
Boston University
590 Commonwealth Avenue
BOSTON, MA 02215
USA
Tel.: ++1 617 3532617
Fax: ++1 617 3533783
hes@meta.bu.edu

Scientific Secretary

Giuseppe MAINO
ENEA, Divisione Fisica Applicata
Via Don Fiammelli 2
40129 BOLOGNA
Italy
Tel.: ++39 051 6098206
Fax: ++39 051 6098359
giuseppe.maino@bologna.enea.it

Lecturers

Kurt BINDER
Institut für Physik
Johannes Gutmnberg Universität Mainz
Staudinger Weg 7
55099 MAINZ
Germany
Tel.: ++49 6131 3923348
Fax: ++49 6131 3925441
Kurt.Binder@uni-mainz.de

Sow-Hsin CHEN
Department of Nuclear Engineering
24-209, MIT
CAMBRIDGE, MA 02139
USA
Tel.: ++1 617 253 3810/3794
Fax: ++1 617 258 8863
sowhsin@mit.edu

Morrel H. COHEN
Department of Physics and Astronomy
Rutgers
The State University of New Jersey
PISTCATAWAY, NJ 08854-8019
USA
Tel.: ++1 908 525 3101
Fax: ++1 908 626 6879
mhcohen@prodigy.net

Antonio CONIGLIO
Dipartimento di Scienze Fisiche
Università di Napoli "Federico II"
Monte S. Angelo
Via Cintia
80126 Napoli
Italy
Tel.: ++39 081 676485
Fax: ++39 081 676346
coniglio@na.infn.it

Pierre-Gilles DE GENNES
Ecole Supérieure de Physique
et de Chimie Industrielle
10 rue Vauquelin
75231 Paris Cedex 05
France
Tel.: ++33 1 4234 6797
Fax: ++33 1 4051 0636
Pierre-Gilles.degennes@espci.fr
pgg@curie.fr

Michael E. FISHER
Institute for Physical Science
and Technology
University of Maryland
College Park, MD 20742
USA
Tel.: ++1 301 405 4819.
Fax: ++1 301 314 9404.
claremon@ipst.umd.edu

K. Alex MÜLLER
Physik Institut
Universität Zürich
Winterthurerstr. 190
8057 Zürich
Switzerland
Tel.: ++41 1 6355749
Fax: ++41 1 6355704
abdalla@physik.unizh.ch

David R. NELSON
Department of Physics
Harvard University
Cambridge, MA 02138
USA
Tel.: ++1 617 495 4331
Fax: ++1 617 496 2545
nelson@cmts.harvard.edu

Giorgio PARISI
Dipartimento di Fisica
Università di Roma "La Sapienza"
P.le A. Moro 2
00185 Roma
Italy
Tel.: ++39 06 49913481/4463158
Fax: ++39 06 4463158
giorgio.parisi@roma1.infn.it

Toni SCHNEIDER
Physikinstitut
Universität Zurich
Winterthurerstr. 190
8057 Zurich
Switzerland
Tel.: ++41 1 7718822
Fax: ++41 635705
toni.schneider@swissonline.ch

Eugene I. SHAKHNOVICH
Department of Chemistry
and Chemical Biology
Harvard University
12 Oxford Street
Cambridge, MA 02138
USA
Tel.: ++1 617 495 4130
Fax: ++1 617 496 5948
eugene@belok.harvard.edu

Harry L. SWINNEY
Physics Department and Center
for Nonlinear Dynamics
University of Texas at Austin
AUSTIN, TX 78712
USA
Tel.: ++1 512 471 4619
Fax: ++1 512 471 1558
swinney@chaos.ph.utexas.edu

Ben WIDOM
Department of Chemistry
Baker Laboratory
Cornell University
ITHACA, NY 14853-1301
USA
Tel.: ++1 607 255 3363
Fax: ++1 607 2554137
widom@vdwaals.chem.cornell.edu

Seminar Speakers

Marcia BARBOSA
Instituto de Fisica
UFRGS
Caixa Postal 15051
91501-970 PORTO ALEGRE, RS
Brazil
Tel.: ++55 51 33166516
Fax: ++55 51 33167286
barbosa@pcstat4.if.ufrgs.br
barbosa@if.ufrgs.br
barbosa@argento.bu.edu

Lesser BLUM
Physics Department
University of Puerto Rico
POB 23343, RIO PIEDRAS
PR 00931-3343
Puerto Rico
Tel.: ++1 787 764-0000 ext. 13576
lesblum@yahoo.com
blum@phys.rush.edu
lblum@upracd.upr.clu.edu
lblum@rrpac.upr.clu.edu
blum@web.up\rr.pr

Françoise BROCHARD
Institut Curie, SPC
Université Pierre et Marie Curie
Rue Pierre et Marie Curie
75231, PARIS Cedex 05
France
Tel.: ++33 1 4051 6778
Fax: ++33 1 4051 0636
francoise.brochard@curie.fr

Armin BUNDE
Institut für Theoretische Physik III
Universität Giessen
35392 GIESSEN
Germany
Tel.: ++49 641 99 33360
Fax: ++49 641 99 33361
Armin.Bunde@theo.physik.uni-giessen.de

Shlomo HAVLIN
Department of Physics
Bar-Ilan University
52900, RAMAT-GAN
Israel
Tel.: ++972 3 5318436
Fax: ++972 3 5357678
havlin@ophir.ph.biu.ac.il

Hans J. HERRMANN
ICA1
University of Stuttgart
Pfaffenwaldring 27
70569 STUTTGART
Germany
Tel.: ++49 711 685 3701
Fax: ++49 711 685 3658
hans@ica1.uni-stuttgart.de

Jerome K. PERCUS
Department of Mathematics
Courant Institute of Mathematical Sciences
251 Mercer St.
NEW YORK, NY 10012
USA
Tel.: ++1 212 998 3130
Fax: ++1 212 995 4121
percus@cims.nyu.edu

Piero TARTAGLIA
Dipartimento di Fisica
Università di Roma "La Sapienza"
P.le A. Moro 2
00185 ROMA
Italy
Tel.: ++39 06 49913499
Fax: ++39 06 4463158
piero.tartaglia@phys.uniroma1.it

Constantino TSALLIS
Centro Brasileiro de Pesquisas Fisicas
Rua Dr. Xavier Sigaud 150
22290-180 RIO DE JANEIRO
Brazil
Tel.: ++55 21 21417190
Fax: ++55 21 21417515
tsallis@cbpf.br

Students

Jean-Noel AQUA
Institute for Physical Science and Technology
University of Maryland
COLLEGE PARK, MD 20742
USA
Tel.: ++1 301 4054821
Fax: ++1 301 3149404
jnaqua@Glue.umd.edu

Antonio ASTILLERO
Departamento de Fisica
Universidsd de Estremadura
Av.da de Alvas s/n
E-06071 BADAJOZ
Spain
Tel.: ++34 924 289300
Fax: ++34 924 289651
aavivas@unex.es

Giacomo BALDI
Dipartimento di Fisica
Università di Trento
Via Sommarive, 14
38050 POVO TN
Italy
Tel.: ++39 0461 882073
baldi@science.unitn.it

Gareth BAXTER
Department of Physics and Astronomy
University of Manchester
Oxford Road
MANCHESTER, M13 9PL
UK
Tel.: ++ 44 161 2754214
Fax: ++ 44 161 2754218
gareth@theory.ph.man.ac.uk

Giacomo BORMETTI
Dipartimento di Fisica Nucleare
e Teorica
Università di Pavia
Via Bassi 6
27100 Pavia
Italy
Tel.: ++39 0382 507703
Fax: ++39 0382 526938
Giacomo.Bormetti@pv.infn.it

Matteo BROCCIO
Dipartimento di Fisica
Università di Messina
Villaggio S. Agata, CP 55
98166 Messina
Italy
Tel.: ++39 090 6765120
Fax: ++39 090 395004
mbroccio@unime.it

Pascal BUENZLI
SB-ITP-GMQ
Ecole Polytechnique Fédérale
de Lausanne
PHB Ecublens
CH-1015 Lausanne
Switzerland
Tel.: ++41 21 6933417
Fax: ++41 21 6935419
pascal.buenzli@epfl.ch

Pasquale CALABRESE
Scuola Normale Superiore
Piazza dei Cavalieri 7
56127 Pisa
Italy
Tel.: ++39 050 2214897
Fax: ++39 050 2214887
calabres@df.unipi.it

Otger CAMPAS
CNRS/Institut Curie
UMR 168, Section Recherche
11, Rue Pierre et Marie Curie
75005 Paris Cedex 05
France
Tel.: ++33 1 42346471
otger@ecm.ub.es
otger.campas@curie.fr

Anna CAVALLO
Institut für Physik
Joannes Gutemberg-Universität Mainz
Staudingerweg 7
D-55099 Mainz
Germany
Tel.: ++49 6131 3923645
Fax: ++49 6131 3925441
acavallo@uni-mainz.de

Roberto CERBINO
Dipartimento di Fisica
Sezione di Ottica
Università di Milano
Via Celoria 16
20133 Milano
Italy
Tel.: ++39 02 50317211
Fax: ++39 02 50317712
roberto.cerbino@fisica.unimi.it

Apratim CHATTERJI
Institut für Physik
Joannes Gutemberg-Universität Mainz
Staudingerweg 7
D-55099 Mainz
Germany
Tel.: ++49 6131 3924104
Fax: ++49 6131 3925441
apratim@uni-mainz.de

Anna CHERNOVA
Centre for Mathematical Biology
Mathematical Institute
University of Oxford
24-29 St Giles'
OXFORD, OX1 3LB
UK
Tel.: ++44 1865 280610
Fax: ++44 1865 270515
chernova@maths.ox.ac.uk

Subir Kumar DAS
Institut für Physik
Joannes Gutemberg-Universität Mainz
Staudingerweg 7
D-55099 MAINZ
Germany
Tel.: ++49 6131 3924104
Fax: ++49 6131 3925441
subir@uni-mainz.de

Valentina DE GRANDIS
Dipartimento di Fisica
Università degli Studi di Roma Tre
Via della Vasca Navale 84
00146 ROMA
Italy
Tel.: ++39 06 55177296
Fax: ++39 06 5579303
degrandis@fis.uniroma3.it

Natasha DENESYUK
Department of Chemistry
University of Cambridge
CAMBRIDGE, CB2 1EW
UK
Tel.: ++44 1223 516696
Fax: ++44 1223 336362
nad28@hermes.cam.ac.uk

Jakob Kisbye DREYER
Niels Bohr Institute
University of Copenhagen
Blegdamsvej 17
DK-2100 COPENHAGEN
Denmark
Tel.: ++45 353 25430
Fax: ++45 353 25425
kisbye@nbi.dk

Joachim DZUBIELLA
Department of Chemistry
University Chemical Laboratory
University of Cambridge
CAMBRIDGE, CB2 1EW
UK
Tel.: ++44 1223 336353
Fax: ++44 1223 336362
jd319@cam.ac.uk

Jan EICHNER
Institut für Theoretische Physik III
Universität Giessen
Heinrich-Buff-Ring 16
D-35392 GIESSEN
Germany
Tel.: ++49 641 99 33376
Fax: ++49 641 99 33369
jan.f.eichner@physik.uni-giessen.de

Patricia FAISCA
Centro de Fisica Teorica e Computacional
Universidade de Lisboa-CFTC-UL
Av. Prof. Gama Pinto 2
1649-003 LISBOA
Portugal
Tel.: ++351 21 7904869
Fax: ++351 21 7954288
patnev@cii.fc.ul.pt

Nobuhisa FUJITA
Arrhenius Laboratory
Structural Chemistry
Stockholm University
10691 STOCKHOLM
Sweden
Tel.: ++46 8 162381
Fax: ++46 8 163118
nobuhisa@struc.su.se

Akira FURUKAWA
Department of Physics
Graduate School of Science
Kyoto University
66-8502 KYOTO
Japan
Tel.: ++81 075 7533770
Fax: ++81 075 7533751
furukawa@scphys.kyoto-u.ac.jp

Nicola FUSCO
Dipartimento di Fisica "E. Caianiello"
Univiversità di Salerno
Via S. Allende
84081 BARONISSI SA
Italy
Tel.: 339 7018359
Tel.: ++39 089 965299
Fax: ++39-089-965275
nicola.fusco@sa.infn.it

Andrea GAMBASSI
Max-Planck-Institut für Metallforschung
Heisenbergstrasse 3
Office 5E13
D-70569 STUTTGART
Germany
Tel.: ++49 711 6891938
Fax: ++49 711 6891922
andrea.gambassi@sns.it
gambassi@mf.mpg.de

Azam GHOLAMI
Hahn-Meitner Institut
Abteilung Theorie (SF5)
Glienicker Str. 100
D-14109 BERLIN
Germany
Tel.: ++49 030 80622678
Fax: ++49 030 80622098
azam@hmi.de

Avi GOZOLCHIANI
Department of Physics
Bar-Ilan University
52900, RAMAT-GAN
Israel
Tel: ++972 3 5313247
Fax: ++972 3 5353298
gozolaa@shoshi.ph.biu.ac.il

Angus GRAY-WEALE
Physical and Theoretical Chemistry Lab.
Oxford University
South Parks Road
OXFORD, OX1 3QZ
UK
Tel.: ++44 7900 681877
angus.gray-weale@jesus.ox.ac.uk

Tom HANNEY
University of Edinburg, JCMB
Kings Building
Mayfield Road
EDINBURG, EH9 3JZ
Scotland
Tel.: ++44 0131 6505289
Fax: ++44 0131 6624712
thanney@ph.ed.ac.uk
tom.hanney@ed.ac.uk

Andreas HILFINGER
Max-Planck Institute for Physics
of Complex Systems
Biophysics Group
Nöthnitzer Str. 38
01187 Dresden
Germany
Tel.: ++49 351 8711216
Fax: ++49 351 8711299
hilfing@mpipks-dresden.mpg.de

Silvia IMBERTI
Dipartimento di Fisica "E. Amaldi"
Università degli Studi di Roma Tre
Via della Vasca Navale 84
00146 Roma
Italy
Tel.: ++39 06 55177255
Fax: ++39 06 5579303
imberti@fis.uniroma3.it

Stelios KARANIKAS
University Chemical Laboratory
Cambridge University
Lensfield Road
Cambridge, CB2 1EW
UK
Tel.: ++44 1223 763872
Fax: ++44 1223 336362
sk361@cam.ac.uk

Hiroaki KATSURAGI
Department of Applied Science
for Electronics and Materials
Interdisciplinary Graduate School
of Engineering Sciences
Kyushu University
6-1 Kasugakoen, Kasuga
Fukuoka, 816-8580
Japan
Tel.: ++81 92 5838387
Fax: ++81 92 5838387
katsurag@asem.kyushu-u.ac.jp

Alfredo Alexander KATZ
Physics Department
University of California
Santa Barbara, CA 93106
USA
Tel.: ++1 805 8937172
Fax: ++1 805 8933307
ajar@physics.ucsb.edu

Anna KIMMEL
Physical-Technical Institute
of the Ural Branch of the
Russian Academy of Sciences
132, Kirova St.
Udmurtia 426000
Russia
Tel.: ++7 3412 524449
Fax: ++7 3512 635672
anny@udm.ru
las@pti.udm.ru

Alexander KUDLAY
Department of Material Science
and Engineering
Northwestern University
Cook Hall
Campus Drive
Evanston, IL 60208-3108
USA
Tel.: ++1 847 4915958
Fax: ++1 847 4917820
kudlay@northwestern.edu

Leandro LANZI
Dipartimento di Fisica
Università di Firenze
Via G. Sansone 1
50019 Sesto Fiorentino FI
Italy
Tel.: ++39 055 4572011/2056
Fax: ++39-055 457-2014
leandro.lanzi@fi.infn.it

Hynek LAVICKA
Department of Physics
Faculty of Nuclear Sciences
and Physical Engineering
Brehova 7
PRAGUE 115 19
Czech Republic
Tel.: ++420 608 613766
h.lavicka@email.cz

Valerie LIVINA
Department of Physics
Bar-Ilan University
52900, RAMAT-GAN
Israel
Tel.: ++972 3 531885
Fax: ++972 3 5317678
livina@.ory.ph.biu.ac.il

Mille A. MICHEELSEN
Niels Bohr Institute
Blegdamsvej 17
DK-2100 COPENHAGEN
Denmark
Tel.: ++45 35 325326
Fax: ++45 35 325016
micmi@nbi.dk

Fawzi MOHAMED
ETH Zurich
Centro Svizzero di Calcolo Scientifico
(CSCS)
Via Cantonale, Galleria 2
CH-6928 MANNO
Switzerland
Tel.: ++41 91 6108243
Fax: ++41 91 6108282
fawzi@cscs.ch

Shay MOSHEL
Department of Physics
Bar-Ilan University
52900, RAMAT-GAN
Israel
Tel.: ++972 3 5313247
Fax: ++972 3 5353298
moshels@shoshi.ph.biu.ac.il

Björn NADROWSKI
Max-Planck Institute
for Physics of Complex Systems
Nöthnitzer Str. 38
01187 DRESDEN
Gremany
Tel.: ++49 351 8711211
Fax: ++49 351 8711299
bn@mpipks-dresden.mpg.de

Alexandru NICOLIN
c/o Prof. M.H. Jensen
Niels Bohr Institute
University of Copenhagen
Blegdamsvej 17
COPENHAGEN
Denmark
Permanent address:
Politehnica University of Bucharest
313 Splaiul Independentei
77206-BUCHAREST
Romania
Tel.: ++40 2122 36713
Fax: ++45 35325016
nicolin@nbi.dk

Anna OCHAB-MARCINEK
Statistical Physics Division
Theory Department
Institute of Physics
Jagellonian University
Ul. Reymonta 4
30-059 Cracow
Poland
Tel.: ++48 012 6324888 ext. 5567
Fax: ++48 012 6334079
ochab@th.if.uj.edu.pl

Pietro PARRUCCINI
Dipartimento di Fisica
Università di Pisa
Via Buonarroti 2
56100 Pisa
Italy
Tel.: ++39 050 2214897
Fax: ++39 050 2214887
parrucci@df.unipi.it

Eric J. Ribeiro PARTELI
Institute of Computer Applications (ICA1)
University of Stuttgart
Pfaffenwaldring 27
70569 Stuttgart
Germany
Tel.: ++49 711 6853594
Fax: ++49 711 6853658
parteli@ica1.uni-stuttgart.de

Florencia PENNA
Dep.to Fisica Teorica
de la Materia Condensada C-5 512
Universidad Autonoma de Madrid
Carretera Colmenar km. 15
28049 Madrid
Spain
Tel.: ++34 91 3978628
Fax: ++34 91 3974950
flor.penna@uam.es

Massimo PICA-CIAMARRA
Dipartimento di Scienze Fisiche
Università di Napoli "Federico II"
Monte S. Angelo
Via Cintia
80126 Napoli
Italy
Tel.: ++39 081 676433
Fax: ++39 081 676346
massimo.picaciamarra@na.infn.it

Paolo PIEROBON
Hahn-Meitner Institut
Abteilung Theorie (SF5)
Glienicker Str. 100
D-14109 Berlin
Germany
Tel.: ++49 030 80622045
Fax: ++49 030 80622098
pierobon@hmi.de

Emanuele PONTECORVO
Dipartimento di Fisica
Università di Roma "La Sapienza"
P.le A. Moro 2
00185 Roma
Italy
Tel.: ++39 06 49913435
Fax: ++39 06 4463158
emanuele.pontecorvo@phys.uniroma1.it

Diego RYBSKI
Institut für Theoretische Physik III
Universität Giessen
Heinrich-Buff-Ring 16
D-35392 Giessen
Germany
Tel.: ++49 641 9933371
Fax: ++49 641 9933369
diego.rybski@physik.uni-giessen.de

Tanja SCHILLING
Instituut Voor Atoom en Molecuulfysica
Computational Physics Group
AMOLF
Kruislaan 407
1098 SJ Amsterdam
The Netherlands
Tel.: ++31 20 6081392
Fax: ++31 20 6684106
t.schilling@amolf.nl

Marcello SEGA
Dipartimento di Fisica
Università di Trento
Via Sommarive 14
38050 Povo TN
Italy
Tel.: ++39 0461 882091
Fax: ++39 0461 881696
sega@science.unitn.it

Kostyantyn SHUNDYAK
Institute for Theoretical Physics
Utrecht University Minnaertgebouw
Leuvenlaan 4
3584 CE Utrecht
The Netherlands
Tel.: ++31 30 253 7558
Fax: ++31 30 253 5937
k.y.shundyak@phys.uu.nl

Yair SREBRO
Technion-Israel Institute of Technology
Physics Department
3200 Haifa
Israel
Tel.: ++972 4 8293836
ysrebro@physics.technion.ac.il

Marco TARZIA
Dipartimento di Scienze Fisiche
Università di Napoli "Federico II"
Monte S. Angelo
Via Cintia
80126 Napoli
Italy
Tel.: ++39 081 676433
Fax: ++39 081 676346
Marco.Tarzia@na.infn.it

Claudio Juan TESSONE
IMEDEA
Physics Department/Edifici Mateu Orfila
Universitat de les Illes Balears
E-07122 Palma de Mallorca
Spain
Tel.: ++34 971 172537
Fax: ++34 971 173426
tessonec@imedea.uib.es

Antonio TROVATO
INFM, Dipartimento di Fisica
Università di Padova
Via Marzolo 8
35131 Padova
Italy
Tel.: ++39 049 8277251
Fax: ++39 049 8277102
trovato@pd.infn.it

Federica TRUDU
ETH Zurich
Centro Svizzero di Calcolo Scientifico
(CSCS)
Via Cantonale, Galleria 2
CH-6928 Manno
Switzerland
Tel.: ++41 91 6108203
Fax: ++41 91 6108282
trudu@cscs.ch

Vincenzo VITELLI
Physics Department
Harvard University
17 Oxford Street
CAMBRIDGE, MA 02138
USA
Tel.: ++1 617 2302231
Fax: ++1 617 4962545
vitelli@fas.harvard.edu

Thomas VOIGTMANN
Physik-Department, T37
TU Muenchen
James-Franck-Str.
85748 GARCHING
Germany
Tel.: ++49 89 289 12388
Fax: ++49 7531 88 3090
thomas.voigtmann@ph.tum.de

Martin WACKENHUT
Institute for Computer Applications (ICA1)
University of Stuttgart
Pfaffenwaldring 27
70569 STUTTGART
Germany
Tel.: ++49 711 6857610
Fax: ++49 711 6853658
mawa@ica1.uni-stuttgart.de

Hirofumi WADA
Department of Mathematical and Life Sciences
Hiroshima University
1-3-1, Kagamiyama
HIGASHI-HIROSHIMA 739-8526
Japan
Tel.: ++81 824 247326
Fax: ++81 824 247327
wada@math.sci.hiroshima-u.ac.jp
wada@daisy.phys.s.u-tokyo.ac.jp

Francesco ZAMPONI
Dipartimento di Fisica
Università di Roma "La Sapienza"
P.le A. Moro 2
I-00185 ROMA
Italy
Tel.: ++39 06 49913435
Fax: ++39 06 4463158
francesco.zamponi@phys.uniroma1.it

Stephan ZSCHIEGNER
Institut für Theoretische Physik III
Universität Giessen
Heinrich-Buff-Rong 16
D-35392 GIESSEN
Germany
Tel.: ++49 641 9933371
Fax: ++49 641 9933369
stephan.zschiegner@physik.uni-giessen.d

Observers

Tomaso ASTE
Department of Applied Mathematics
Research School of Physical Sciences and Engineering
The Australian National University
0200 CANBERRA ACT
Australia
Tel.: ++61 2 61252846
Fax: ++61 2 61250732
tomaso.aste@anu.edu.au

Michele BATTEZZATI
Istituto di Fisica dello Spazio Interplanetario del CNR
Corso Fiume 4
10133 TORINO
Italy
Tel.: ++39 011 6306817
Fax: ++39 011 6630-4056
battezzati@to.infn.it

Ronald COHEN
Geophysical Laboratory
Carnegie Institution of Washington
5251 Broad Branch Rd., NW
WASHINGTON, DC 20015
USA
Tel.: ++1 202 4788937
Fax: ++1 253 3222539
cohen@gl.ciw.edu

Antonio FARAONE
Department of Nuclear Engineering
Massachussetts Institute of Technology
Room 24-212
77 Massachussets Avenue
CAMBRIDGE, MA 02139
USA
Tel.: ++1 617 4523017
Fax: ++1 617 2588863
afaraone@mit.edu

Wenbing HU
Instutuut Voor Atoom en Molecuulfysica
Computational Physics Group
AMOLF
Kruislaan 407
1098 SJ AMSTERDAM
The Netherlands
Tel.: ++31 20 6081276
Fax: ++31 20 6684106
hu@amolf.nl

Jyotsana LAL
IPNS Division
Argonne National Labs.
9700 South Cass Avenue
ARGONNE, IL-60439
USA
Tel.: ++1 630 2526042
Fax: ++1 630 2524163
jlal@anl.gov

Antonio LAMURA
Istituto Applicazioni Calcolo del CNR
Sezione di Bari
Via Amendola 122/D
70126 BARI
Italy
Tel.: ++39 080 5530719
Fax: ++39 080 5588235
a.lamura@area.ba.cnr.it

Roman MARTONAK
Centro Svizzero di Calcolo Scientifico
(CSCS)
ETH Zurich
Via Cantonale, Galleria 2
CH-6928 MANNO
Switzerland
Tel.: ++41 91 6108226
Fax: ++41 91 6108282
martonak@cscs.ch

Alejandro MORA
Institute for Computer Applications
(ICA2)
University of Stuttgart
Pfaffenwaldring 27
70569 STUTTGART
Germany
Tel.: ++49 711 6855927
Fax: ++49 711 6853758
ica2am@csv.ica.uni-stuttgart.de

Alessandro PLUCHINO
Dipartimento di Fisica e Astronomia
Università di Catania
Via S. Sofia 64
95123 CATANIA
Italy
Tel.: ++39 095 3785239
alessandro.pluchino@ct.infn.it